AFRICA, NEAR AND MIDDLE EAST	AMERICAS	ASIA, AUSTRALASIA	EUROPE
1851 British occupy Lagos; slave trade ends	**1850** Compromise of 1850 (USA)	**1850–64** Taiping Rebellion	**1851** Great Exhibition in London
1852 Umar ibn Said Tal begins *jihad*	**1853–4** Gadsden Purchase	**1851** Gold found in Australia	**1852** Fall of French Second Republic; Napoleon III emperor
1852–6 Livingstone journeys across Africa	**1854** Kansas–Nebraska Act; US Republican Party formed	**1852** Second Anglo-Burmese War	**1853–6** Crimean War
1854 Umar ibn Said Tal founds Tukolor empire	**1857** Dred Scott case	**1853** First railway and telegraph in India; Perry visits Japan	**1856** Congress of Paris
1854–61 France occupies interior of Senegal	**1858–61** Mexican War of Reform	**1854** Treaty of Kanagawa; Eureka Rebellion	**1859** Battles of Magenta and Solferino
1858–64 Livingstone explores Central Africa	**1859** John Brown seizes arsenal at Harper's Ferry; first oil well (Pennsylvania)	**1856** Britain annexes Oudh (India)	
		1856–60 Second Opium War	
		1857–8 Indian Mutiny	
1862 Umar conquers Massina	**1860** Lincoln elected US President	**1860** Taranaki Wars begin	**1861** Victor Emanuel II becomes first king of Italy; emancipation of Russian serfs
1866–73 Livingstone's third expedition	**1861–5** American Civil War	**1862** Cochin-China (southern Vietnam) becomes French colony	**1862** Bismarck minister-president of Prussia
1867 Diamonds discovered in Orange Free State	**1863** US emancipation of slaves	**1863** Cambodia (Kampuchea) becomes French protectorate	**1863** Russia suppresses Polish revolt; first underground railway
1868 Lobengula King of Ndebele	**1864–7** French occupy Mexico	**1868** Meiji restoration and end of shogunate in Japan	**1864** First International
1868–73 Fante Confederation	**1864–70** Paraguayan War		**1866** Austro-Prussian War
1869 Suez Canal opens	**1865** Lincoln assassinated; Jamaican rebellion		**1867** North German Confederation and Austro-Hungarian empire founded
	1866 Ku Klux Klan formed		**1868** First Trades Union Congress (Britain)
	1867 Alaska Purchase; British North America Act		
	1867–8 Reconstruction Acts		
	1868 Fourteenth Amendment		
	1869 Women's suffrage (Wyoming)		
1874 British colony of Gold Coast (Ghana) established	**1871** Vancouver and British Columbia join Dominion of Canada	**1873–1903** Acheh War	**1870** French Third Republic founded
1876 Leopold II founds Association of the Congo; British missionaries in Nyasaland; Franco-British dual control in Egypt	**1876** Telephone patented; battle of Little Big Horn; Díaz takes control of Mexico	**1875** Establishment of central Parliament in New Zealand	**1870–1** Franco-Prussian War
1877 Britain annexes Transvaal	**1879–84** War of the Pacific	**1877** Victoria proclaimed Empress of India	**1871** German Second empire founded
1877–9 Last Xhosa War		**1878–80** Second Anglo-Afghan War	**1876** Bulgarian massacre
1879 Zulu War			**1877–8** Russo-Turkish War
			1878 Congress of Berlin
			1879 Dual Alliance (Germany/Austro-Hungary); Irish Land League formed
1880–1 First Boer War	**1883** Pendelton Act	**1884–5** Sino-French War	**1881** Alexander II assassinated; Social Insurance proposed in Prussia
1881 'Scramble for Africa' begins	**1885** Canadian Pacific Railway completed	**1885** Indian National Congress founded; third Anglo-Burmese War	**1882** Triple Alliance (Germany/Italy/Austria)
1882 Britain occupies Egypt	**1886** Slavery abolished in Cuba	**1887** French Indo-China established	**1884** Berlin Conference on Africa
1884 Germany acquires SW Africa, Togoland, Cameroons	**1888** Slavery abolished in Brazil	**1889** Japanese Meiji constitution completed	**1885** First motor cars
1885–90 Britain and Germany partition East Africa	**1889** Oklahoma Territory settled; Brazil a republic		**1889** London Dockers' Strike
1886 Gold discovered in Transvaal			
1891 Young Turk movement formed	**1890** Battle of Wounded Knee	**1890** Maritime Strike	**1890** Bismarck dismissed
1893 Matabele War; Ivory Coast becomes French colony	**1893** US recession	**1891** First Shearers' Strike	**1891** Trans-Siberian Railway begun
1899 Sudan under Anglo-Egyptian protection	**1894** Pullman Strike	**1893** France annexes Laos; women's suffrage (NZ)	**1892** First British Labour MP
1899–1902 Second Boer War	**1895** Cuba rebels against Spain	**1894–5** Sino-Japanese War	**1894** Dreyfus Affair begins; Alexander III forms Franco-Russian Alliance
	1898 Spanish-American War; US acquires Cuba, Puerto Rico, Guam, and Philippines	**1895** Japan occupies Taiwan	**1896** Herzl founds Zionism
		1898 Hundred Days Reform	
		1899–1905 Curzon viceroy in India	

OXFORD
ILLUSTRATED ENCYCLOPEDIA

General Editor   Harry Judge

Volume 4
WORLD HISTORY
FROM 1800 TO THE PRESENT DAY

Volume Editor   Robert Blake

OXFORD
OXFORD UNIVERSITY PRESS
NEW YORK   MELBOURNE
1988

Oxford University Press, Walton Street, Oxford, OX2 6DP

Oxford New York Toronto
Delhi Bombay Calcutta Madras Karachi
Petaling Jaya Singapore Hong Kong Tokyo
Nairobi Dar es Salaam Cape Town
Melbourne Auckland

and associated companies in
Berlin Ibadan

Oxford is a trademark of Oxford University Press

British Library Cataloguing in Publication Data
Oxford illustrated encyclopedia.
Vol. 4: World history from 1800 to
the present day.
1. Encyclopedias and dictionaries
I. Judge, Harry   II. Blake, Robert
Blake, Baron
032   AE5
ISBN 0-19-869136-X

Library of Congress Cataloging in Publication Data
World History—from 1800 to the present day.
(Oxford illustrated encyclopedia ; v. 4)
1. History, Modern—19th century. 2. History,
Modern—20th century. I. Blake, Robert, 1916-
II. Series.
D358.W74   1988   909.8   87-20408
ISBN 0-19-869136-X

Text processed by the Oxford Text System
Printed in Hong Kong

General Preface

The *Oxford Illustrated Encyclopedia* sets out, in its eight independent but related volumes, to provide a clear and authoritative account of all the major fields of human knowledge and endeavour. Each volume deals with one such major field and is devoted to a series of short entries, arranged alphabetically, and furnished with illustrations. One of the delights of the Encyclopedia is that it serves equally those who wish to dip into a particular topic (or settle a doubt or argument) and those who wish to acquire a rounded understanding of a major field of human knowledge.

The first decision taken about the Encyclopedia was that each volume should be dedicated to a clearly defined theme—whether that be the natural world, or human society, or the arts. This means that each volume can be enjoyed on its own, but that when all the volumes are taken together they will provide a clear map of contemporary knowledge. It is not, of course, easy to say where one theme begins and another ends and so each volume foreword (like the one that follows) must therefore attempt to define these boundaries. The work will be completed with a full index to guide the reader to the correct entry in the right volume for the information being sought.

The organization within each volume is kept as clear and simple as possible. The arrangement is alphabetical, and there is no division into separate sections. Such an arrangement rescues the reader from having to hunt through the pages, and makes it unnecessary to provide a separate index for each volume. We wanted, from the beginning, to provide as large a number as possible of short and clearly written entries, since research has shown that readers welcome immediate access to the information they seek rather than being obliged to search through longer entries. The decision, which has been consistently applied, must obviously determine both the number and nature of the entries, and (in particular) the ways in which certain subjects have been divided for ease of reference. Care has been taken to guide the reader through the entries, by providing cross-references to relevant articles elsewhere in the same volume.

This is an illustrated encyclopedia, and the illustrations have been carefully chosen to expand and supplement the written text. They add to the pleasure of those who will wish to browse through these varied volumes as well as to the profit of those who will need to consult them more systematically.

Volumes 1 and 2 of the series deal respectively with the physical world and the natural world. Volumes 3 and 4 are concerned with the massive subject of the history of the world from the evolution of *Homo sapiens* to the present day. Volume 5 introduces the reader to the visual arts, music, and literature, and Volume 6, on technology, both historical and contemporary, provides a guide to human inventiveness. The last two volumes help to place all that has previously been described and illustrated in wider contexts. Volume 7 offers a comprehensive guide to the various ways in which people live throughout the world and so includes many entries on their social, political, and economic organization. The series is completed with Volume 8, which places the life of our shared world in the framework of the universe, and includes entries on the origins of the universe as well as our rapidly advancing understanding and exploration of it.

I am grateful alike to the editors, consultants, contributors, and illustrators whose common aim has been to provide such a stimulating summary of our knowledge of the world.

HARRY JUDGE

CONTRIBUTORS

Dr Ron Adams

Dr M. M. Ally

John Bailie

Jan Bassett

N. H. Brasher

R. A. Burchell

Dr Peter Carey

Dr S. D. R. Cashman

Dorothy Castle

Dr Malcolm Cooper

Dr Michael Dillon

David Doran

Professor M. Foster Farley

Dr G. S. P. Freeman-Grenville

Dr Paul Garner

Alastair Gray

Bridget Hadaway

Derek Heater

Patrick Keeley

Donald Lindsay

Dr R. D. Lobban

Professor P. Marshall

Professor Michael C. Meyer

Professor Roger Morgan

Dr Ron Nettler

Dr A. A. Powell

The Revd D. T. W. Price

Professor Patrick Quinlivan

Dr T. Raychaudhuri

John Stokes

Peter Teed

Professor Alan Ward

Dr J. N. Westwood

Professor Cyril Williams

Professor Malcolm E. Yapp

Foreword

Most of us are curious about our ancestry, which has at least in part made us who we are. It is not surprising that we are no less inquisitive about our collective past, which has created modern nation states and the global problems which vex us today. Our knowledge of the past is 'a key to understanding the world in which we live. In this and its companion volume, *World History: from earliest times to 1800*, the *Oxford Illustrated Encyclopedia* sets out to provide a valuable historical guide and dictionary, with brief entries about people, events, countries, trends, movements, and dates. This volume covers the period from 1800 to the present day. Its emphasis is on political history in the widest sense. War, said to be the continuation of politics by other means, is included, also religion and economics, which have so often had political overtones. Scientific, technological, literary, and artistic history are dealt with in other volumes. Adlai Stevenson is in, but not Robert Louis Stevenson or George Stephenson, designer of *The Rocket*. Benjamin Disraeli and Charles Dickens both wrote novels, but the former is in and not the latter. The entries are not confined to the Western world. The coverage is global; it includes people and events from every continent.

In compiling these volumes, the editors were faced with the question of how best to divide the subject between two volumes (for it was clear that it was too large for a single volume of the series): should the division be chronological or alphabetical? The wish to ensure that every volume of the series was self-contained, and that those interested in one particular period alone need only commit themselves to one volume, favoured the chronological divide. The year 1800 was chosen, not because anything of world-shaking significance occurred in that year, but because one must draw a line somewhere, and a good round number is as satisfactory as anything else. It reminds the reader that these volumes cover the history of the world, and not just any particular area of it: 1789 may seem an obvious dividing line for France and Continental Europe, but not particularly so for Britain, where 1688 or 1832 might make better sense. As for the USA, 1776 would be the obvious date. Look across at the other side of the world, and ask what one would do about China—1644, the Manchu conquest, or 1839, the First Opium War? There is, moreover, a broad sense in which 1800 is not an entirely arbitrary date. The agricultural and industrial revolutions which transformed society, first in Britain, then Europe, then the world, and which constitute the biggest engines of change in human society, began before 1800, but their full effects were only felt afterwards, likewise the revolution in communications, which was one of their by-products.

It is strange to remember that as late as 1834 it took Sir Robert Peel, a future Prime Minister of Britain, as long to get from Rome to London when summoned to form a government as it had taken a high functionary of the Roman empire seventeen centuries earlier. Despite all the continuity, the world with which this volume deals is so different to that of the previous volume as to justify the division, even though it was made for other reasons.

In the foreword to the first history volume Dr Harry Judge states the problem of overlap. Many topics and people straddle the 1800 boundary. In most cases our solution has been to give the history down to 1800 in the first volume and continue it in the second, but there are a few exceptions such as Napoleon, for whom an entry is given in both. It is worth repeating here that Australia, too, has been treated as an exception, the entries on its early colonization being in this volume and not in the previous one, as strict chronology would dictate. The object of editors is to make the volumes as easy as they can for purpose of reference. It is certain that some exclusions will be criticized and that the impression given in short entries may disappoint experts in the field. It is to be hoped that critics will recognize the problems of space and remember that the volume is a reference book for the general reader and not a work of history in the conventional sense.

The unusually wide range of topics covered by this volume has required the editors to consult a large number of experts, many of them from the University of Oxford, and I am most grateful for all the advice that they have provided during the preparation of this volume.

ROBERT BLAKE

A User's Guide

This book is designed for easy use, but the following notes may be helpful to the reader.

ALPHABETICAL ARRANGEMENT The entries are arranged in a simple letter-by-letter alphabetical order of their headings (ignoring the spaces between words) up to the first comma (thus **détente** comes before **de Valera**). When two entry headings are the same up to the first comma, then the entries are placed in alphabetical order according to what follows after the comma (thus **Booth, Charles** comes before **Booth, John Wilkes**). Names beginning with 'St' are placed as though spelt 'Saint', so **St Laurent, Louis Stephen** follows directly after **Saigo Takamori**.

ENTRY HEADINGS Entries are usually placed after the key word in the title (the surname, for instance, in a group of names, or the location of a battle, for example, **Roosevelt, Theodore**; **Waterloo, battle of**). The entry heading appears in the singular unless the plural form is the more common usage. Monarchs and rulers are identified by their English names and regnal numbers, and are listed in numerical order (thus **William I** of Prussia, **William II** of Prussia, **William IV** of Great Britain). When an entry covers a family of several related individuals, important names are highlighted in bold type within an entry.

ALTERNATIVE NAMES For countries that have changed their names during the period covered by this volume, the main entry will be found under the most recent name, with a cross-reference, indicated by an asterisk (*), to guide the reader to the entry (thus **Persia** *Iran). For individuals with titles, the main entry appears under the name by which he or she is most commonly known, with cross-references for alternative forms (thus **Artois, Charles, comte d'** *Charles X). For Chinese names the pinyin system has been used (thus a reader looking up Chou En-lai is directed to Zhou Enlai). There are three major exceptions: Chiang Kai-shek, Kuomintang, and Sun Yat-sen. Examples of pinyin use that may not be familiar include Beijing (not Peking) and Tianjin (not Tientsin).

CROSS-REFERENCES An asterisk (*) in front of a word denotes a cross-reference and indicates the entry heading to which attention is being drawn. Cross-references in the text appear only in places where reference is likely to amplify or increase understanding of the entry being read. They are not given automatically in all cases where a separate entry can be found, so if you come across a name or a term about which you would like to know more, it is worth looking for an entry in its alphabetical place even if no cross-reference is marked.

ILLUSTRATIONS Pictures and diagrams usually occur on the same page as the entries to which they relate or on a facing page. The picture captions supplement the information given in the text and indicate in bold type the title of the relevant entry. Where a picture shows a work of art of which only one original exists (for example, a painting, sculpture, drawing, or manuscript), the caption concludes, wherever possible, by locating the work of art. Where multiple copies of an original exist (for example, printed books, engravings, or woodcuts), no location has been given, although rare items have sometimes been given a location. There are maps, charts, and line drawings to illustrate military campaigns, extents of empires, nations, and confederations, technological progress, and other topics. The time-charts to be found on the endpapers provide easy-to-read information on major historical events from 1800 onwards.

WEIGHTS AND MEASURES Both metric measures and their non-metric equivalents are used throughout (thus a measure of distance is given first in kilometres and then in miles). Large measures are generally rounded off, partly for the sake of simplicity and occasionally to reflect differences of opinion as to a precise measurement.

RELATIONSHIP TO OTHER VOLUMES This volume of world history from 1800 to the present day is Volume 4 of a series, the *Oxford Illustrated Encyclopedia*, which will consist of eight thematic volumes (and an index). It is published with a companion history volume (Volume 3) which covers world history from earliest times to 1800. Each book is self-contained and is designed for use on its own. Readers are advised that under this chronological arrangement the history of countries, for example, is begun in Volume 3 and completed in Volume 4. Further information is contained in the foreword to this volume, which offers a fuller explanation of the book's scope and organization.

The titles in the series are:

1 *The Physical World*
2 *The Natural World*
3 *World History: from earliest times to 1800*
4 *World History: from 1800 to the present day*
5 *The Arts*
6 *The World of Technology*
7 *Human Society*
8 *The Universe*

A

Abd el-Krim (1881–1963), Moroccan Berber resistance leader. In 1921 he roused the Rif Berbers, and defeated a Spanish army of 20,000. He held out until 1925, when a joint Franco-Spanish force took him prisoner. He was exiled to Réunion until 1947, when he was given permission to go to France. On the way he escaped to Cairo, where he set up the Maghrib Bureau, or Liberation Committee of the Arab West. After Moroccan independence (1956), he refused to return as long as French troops remained on African soil.

Abdication crisis, the renunciation of the British throne by *Edward VIII in 1936. The king let it be known that he wished to marry Mrs Wallis Simpson, a divorcee, which would have required legislative sanction from the United Kingdom Parliament and from all the *dominions. The British government strongly opposed the king's wish, as did representatives of the dominions. Edward chose to abdicate, making a farewell broadcast to the nation, and commending his brother, the Duke of York, who succeeded him as *George VI.

Abdul Hamid II (1842–1918), Ottoman sultan (1876–1909). He succeeded his brother Murad V and ruled until his deposition following the 1908 *Young Turk Revolution. His war with Russia (1877–8) was resolved by the Treaty of San Stefano (1878), subsequently modified by the Congress of *Berlin (1878). An autocratic ruler, he suspended Parliament and the constitution, and ruled the empire until his deposition (1909). He was noted for his exploitation of the religious feelings of his Muslim subjects and, due to internal unrest within the empire, his suppression of his non-Muslim subjects, notably the *Armenians.

Edward VIII, on an informal holiday with Mrs Wallis Simpson (*left*) in Salzburg, Austria, 1936, shortly before the **Abdication crisis** came to a head. The couple's movements, largely unreported by the British press, were covered in detail by foreign news media.

Abdullah, Sheikh Muhammad (1905–82), Kashmiri Muslim leader. He began his career as an activist in the Kashmir Muslim Conference in the 1930s, agitating against the arbitrary rule of the Hindu Dogra Maharaja of Kashmir. Later, as the leader of the Muslim National Conference, he established close links with the Indian National *Congress and Jawaharlal *Nehru. In 1947, when the Maharaja delayed his decision regarding accesion to India or Pakistan, and tribal groups from the North-West Frontier with Pakistani military support invaded *Kashmir, Abdullah approved of the decision to join India. As Chief Minister of the now Indian state of Kashmir, he worked closely with Nehru until he was deposed from his office and imprisoned on suspicion that he was planning a separatist movement. He was later released by Mrs *Gandhi and in his last years again co-operated with the Indian government and Congress.

Aberdeen, George Hamilton Gordon, 4th Earl of (1784–1860), British statesman. He was Foreign Secretary during 1828–30 and again from 1841 to 1846, when he concluded the *Webster–Ashburton and *Oregon Boundary treaties which settled boundary disputes between the USA and Canada. As a leader of those Conservatives who campaigned for *free trade, he supported Sir Robert *Peel in repealing the *Corn Laws (1846). As Prime Minister (1852–5) of the 'Aberdeen Coalition', he reluctantly involved his country in the *Crimean War and was subsequently blamed for its mismanagement. He resigned in 1855.

abolitionists, militant opponents of slavery in 19th-century USA. In the first two decades of the 19th century, there was only a handful of individual abolitionists, but thereafter, fired by religious revivalism, the abolition movement became a strong political force. Prominent as writers and orators were the Boston newspaper-owner William Lloyd Garrison, the author Harriet Beecher Stowe (whose anti-slavery novel *Uncle Tom's Cabin* sold 1.5 million copies within a year of its publication in 1852), and the ex-slave Frederick Douglass. The abolitionist cause at first found little support in Congress or the main political parties, except among a few individuals such as Charles *Sumner, but it played an increasing part in precipitating the political division which led to the *American Civil War.

Aborigine (Australian), a member of the Australoid people found in Australia and Tasmania. Prior to European settlement Australian Aborigines had sophisticated social organization, myths, and sacred rituals. Some 300 different languages were spoken. The Aboriginal population in 1788 has been estimated at about 300,000, but recent research suggests that it may have been much higher. Within a hundred years the population had declined to about 50,000, mainly as a result of loss of land, adoption of European habits, such as drinking alchohol, effects of European diseases, declining birthrates, and violence between Europeans and Aborigines. Aborigine reserves were created in central and northern Australia in the 1930s, but since World War II Aboriginal groups have emerged seeking to preserve their cultural heritage. On the northern reserves they have made demands for a share from the mining companies in the mineral exploitation, and in Western Australia they have formed and joined trade unions on the sheep stations. Legally they have been Australian citizens since 1948, but there has been considerable variation in state and federal procedures requiring Aborigines to register and vote.

Abushiri Revolt (August 1888–9), Arab revolt against German traders on the East African coast north of *Zanzibar. The Arab leader Abushiri (Abu Bashir ibn Salim al-Harthi) united local hostility to German colonization at Pangani when in August 1888 the Germans hauled down the Sultan of Zanzibar's flag and hoisted their own. British and German interference in the slave, ivory, and rubber trades, and their conduct in mosques, had already caused resentment. Abushiri's resistance spread inland to the Usambara mountains, and to Lake Victoria. The revolt was crushed by the German explorer and administrator Hermann von Wissmann in 1889.

Abyssinian Campaigns (1935–41), conflicts between Italy, Abyssinia (*Ethiopia), and later Britain. War broke out from Italy's unfulfilled ambition of 1894–6 to link *Eritrea with *Somalia, and from *Mussolini's aim to provide colonies to absorb Italy's surplus unemployed population. In 1934 and 1935 incidents, possibly contrived, took place at Walwal and elsewhere. On 3 October 1935 an Italian army attacked the Ethiopian forces from the north and east. Eventually the Ethiopians mustered 40,000 men, but they were helpless against the highly trained troops and modern weapons of the Italians. During the Italian occupation (1936–41), fighting continued. In 1940 the Italians occupied British Somaliland, but in 1941 British troops evicted the Italians entirely from Eritrea, Ethiopia, and Somalia in a four-month campaign with support from Ethiopian nationalists.

Acheh War (1873–1903), conflict in north Sumatra between the Dutch and the Achehnese. Trade rivalry and attempts by the sultan of Acheh to obtain foreign assistance against Dutch domination of north Sumatra caused the dispatch of an abortive Dutch expeditionary force in 1871. Although a larger force, sent later in the year, captured the sultan's capital, the Dutch met with fierce resistance in the interior, organized by the local religious leaders (*ulema*). The war was brought to an end between 1898 and January 1903 by military 'pacification'

Abyssinian chieftains receive homage on their arrival in Addis Ababa to undertake military service in the **Abyssinian Campaigns** for their emperor, Haile Selassie. Their obsolete rifles typify the futility of their struggle against superior Italian technology.

and concessions to the *ulema*, who were permitted to carry on their religious duties provided they kept out of politics. Anti-Dutch sentiments persisted, and the region was the first to rise against the colonial power when the Japanese invaded *Indonesia in 1942.

Acheson, Dean (Gooderham) (1893–71), US politician. He served as Assistant Secretary of State, Under-Secretary, and Secretary of State (1941–53), urging international control of nuclear power in the Acheson–Lilienthal Report of 1946, formulating plans for *NATO, implementing the *Marshall Plan, and the *Truman Doctrine of US support for nations threatened by communism.

Action Française, an extreme right-wing group in France during the first half of the 20th century, and also the name of the newspaper published to promote its views. Founded by the poet and political journalist Charles Maurras, it aimed at overthrowing the parliamentary republic and restoring the monarchy. Strongly nationalist, its relationship with royalist pretenders and the papacy was not always good. It became discredited for its overt *fascism and association with the *Vichy government in 1940–4.

Act of Union (1801), an Act to abolish the Irish Parliament. Following rebellion in *Ireland (1798) the British Prime Minister William *Pitt resolved that Ireland should be united with the rest of Britain under a single Parliament. The Act had to be passed by the British Parliament in London and the Irish Parliament in Dublin, where it was deeply resented. Pitt promised that the reward would be *Catholic emancipation, although in the event he was to find this impossible to achieve.

Adams, John Quincy (1767–1848), sixth President of the USA (1825–9). The eldest son of John Adams, second President of the USA, he was elected Federalist Senator for Massachusetts (1803–8). After five years as Minister to Russia (1809–14), Adams was one of the five commissioners sent to Ghent to negotiate the end of the *War of 1812. He was Minister in London (1815–17) and then became Secretary of State under Monroe, when he helped to shape the *Monroe Doctrine. He succeeded as President in 1825, but little of note was achieved during his term, due in part to persistent opposition from supporters of Andrew *Jackson to his attempts to extend federal powers. After leaving office in 1829, Adams entered the House of Representatives (1831), where he served until his death, taking a prominent part in the anti-slavery campaign.

Addams, Jane (1860–1935), US social worker and reformer. With her friend Ellen Grates Starr, she opened Hull House in Chicago in 1889, a pioneer settlement house for workers and immigrants on the model of *Toynbee Hall in London. A pioneer of the new discipline of sociology, she had considerable influence over the planning of neighbourhood welfare institutions throughout the country. She was a leader of the *women's suffrage movement and an active pacifist.

Addington, Henry, 1st Viscount Sidmouth (1757–1844), British statesman. He entered Parliament in 1783, and succeeded William *Pitt the Younger as Prime

Minister in 1801. His peace treaty with France (Amiens, 1802) won him some popularity, but when the conflict was resumed in the following year it became clear that he lacked the qualities of a war leader. He resigned in 1804 and later, as Lord Sidmouth, held other cabinet posts. As Home Secretary (1812–21), he introduced repressive legislation in an attempt to suppress the *Luddites and other protest groups.

Aden, a port commanding the entrance to the Red Sea. In 1839 Aden was captured from the Abdali Sultan of Lahej by a British expedition and annexed to British India. It became a free port in 1850 and enjoyed commercial prosperity as an entrepôt for the East African trade and as a station on the route from Europe to the East, especially after the opening of the Suez Canal in 1869. In 1937 Aden became a crown colony and in 1963 part of the South Arabian Federation of Arab Emirates. In the civil war of 1965–7, British forces attempted to keep the peace, but when Britain withdrew its sponsorship of the Federation, Aden became part of the People's Republic of *South Yemen.

Adenauer, Konrad (1876–1967), German statesman. He became Mayor of Cologne in 1917, but because of his opposition to *Nazism he was removed from this post in 1933 and subsequently twice arrested. In 1945 he again became mayor, but was removed by the British authorities for alleged inefficiency. In the same year he helped to create the *Christian Democratic Party. When the *German Federal Republic was created in 1949, he became the first Chancellor (1949–63). During his period in office a sound democratic system of government was established; friendship with the USA and France was secured; and the West German people started to enjoy the fruits of the so-called 'economic miracle' of the *Erhard years. However, his critics accused him of being too autocratic in manner and too little concerned about the possibility of German reunification.

Adowa, battle of (1 March 1896), a decisive defeat of the Italians by the Ethiopian Emperor *Menelik. Italy had established a protectorate in *Ethiopia in 1889. In 1895 there was a rebellion, and at Adowa an Italian force of 10,000 was routed, losing 4,500 dead and 300 prisoners. In the resulting Treaty of Addis Ababa the Italians recognized the independence of Ethiopia and restricted themselves to the colony of Eritrea. The battle ensured Ethiopian survival as an independent kingdom in Africa after the era of imperial partition.

Adrianople, Treaty of (1829), a peace treaty between Russia and the *Ottoman empire. It terminated the war between them (1828–9) and gave Russia minor territorial gains in Europe, including access to the mouth of the Danube, and substantial gains in Asia Minor. The treaty also confirmed the autonomy of *Serbia, promised autonomy for Greece, and guaranteed free passage for merchant ships through the Dardanelles.

Afars and Issas   *Djibouti.

Afghani, Jamal al-Din al- (1839–97), Muslim revivalist of Iranian origin. He advocated social and political reforms within Muslim countries and Pan-Islamism. Afghani was active as a teacher in Egypt during the 1870s, edited a newspaper (*al-Urwa al-Wuthqa*, 'the Unbreakable Link') in Paris during the 1880s, and played a part in the protest in Iran in 1891–2, which forced the government to cancel its concessions for the production and sale of tobacco. His writings influenced later Muslim thinkers of the Islamic modernist school.

Afghanistan, a country in south-central Asia. In the 19th and early 20th centuries it was the focal point of conflicting Russian and British interests. A British attempt to replace the Kabul ruler *Dost Muhammad was repulsed in the First *Anglo-Afghan War, but Afghan foreign policy came under British control in 1879 by the Treaty of Gandamak, when Britain gained control of the Khyber Pass, an important route between India and Central Asia, thus alienating the *Pathan tribes. In 1880 Abdurrahman Khan became amir. Under him a strong central government was established, and his heirs achieved some modernization and social reform. In 1953 General Mohammad Daoud Khan seized power and was Prime Minister until 1963, during which time he obtained economic and military assistance from the Soviet Union. There were border disputes with Pakistan, but it was Daoud's policy to maintain 'non-alignment' between the two super-power blocs. In 1964 Afghanistan became a parliamentary democracy, but a military coup in 1973 overthrew the monarchy and Daoud reasserted control, proclaiming himself President. In 1977 he issued a constitution for a one-party state and was re-elected President. Within a year, however, he had been assassinated and the Democratic Republic of Afghanistan proclaimed, headed by a revolutionary council, whose first President was Nur Mohammad Taraki. The new regime embarked on reforms, but there was tension and rural unrest. In February 1979 the US ambassador was killed and one month later Taraki was assassinated by supporters of the deputy Prime Minister, Hafijullah Amin, who then sought US support. In December 1979 Soviet troops entered the country. Amin was killed and replaced by Babrak Karmal. Since then anti-government guerrilla forces have waged war against Afghan troops armed and supported by Soviet tanks, aircraft, and equipment. Some three million refugees have fled to Iran and Pakistan. In 1987 the Soviet Union began to seek ways to disengage from the conflict.

Afghan Wars   *Anglo-Afghan Wars.

Africa, the second largest continent. By 1800 the coastline had been explored and in places lightly settled by Europeans, while much of the northern third had been penetrated by the religion and culture of *Islam. During the 19th century the interior was gradually opened up to explorers, traders, and Christian *missionaries in an extensive programme of colonization. Imperialist sentiments and the desire to exploit the continent's natural resources produced a series of military campaigns against the local states and tribes. After World War I Germany's former colonial empire was divided among the victorious Allies. After 1945 the rise of African nationalism accelerated the process of decolonization, most of the black countries becoming independent between 1957 and 1975, sometimes as a result of peaceful negotiation and sometimes through armed rebellion. In the south, small white élites held on to political power, but elsewhere the original inhabitants assumed re-

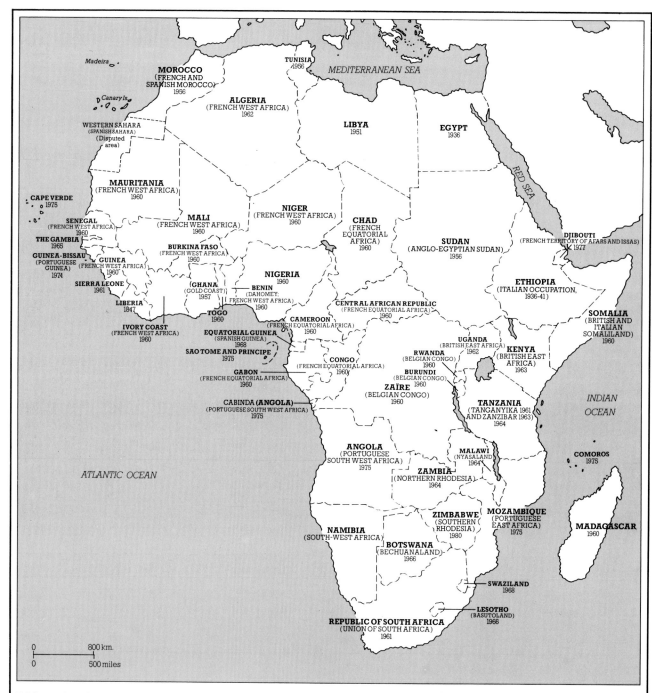

African decolonization

Agitation for political independence was strong in Islamic North Africa before World War II, but south of the Sahara the first black African independent state to emerge was Ghana (1957). Within decades, white control was confined to South Africa and Namibia.

sponsibility for their own government. The artificial boundaries produced by the colonial experience, the rapidity of the transition to home rule, and the under-developed state of many of the local economies have produced political, social, and economic problems of varying severity all over the continent, and many of the new nations remain unstable and politically impoverished.

African National Congress (ANC), a South African political party. It was established in Blomfontein in 1912 as the South African Native National Congress by a Zulu Methodist minister, J. W. Dube. In 1914 he led a deputation to Britain protesting against the Native Land Act (1913), which restricted the purchase of land by Black Africans. In 1926 the ANC established a united front with representatives of the Indian community, which aimed to create a racially integrated, democratic southern Africa. It sought to achieve racial equality by non-violent means, as practised by *Gandhi in India, and from 1952 until 1967 was led by the Natal chieftain Albert *Luthuli. Together with the more militant break-away movement, the *Pan-African Congress (PAC), it

THE PEOPLE SHALL GOVERN!
26 JUNE: SOUTH AFRICA FREEDOM DAY

A poster published by the **African National Congress**, bearing the slogans of the banned South African organization's freedom charter.

was declared illegal by the South African government in 1960. Confronted by Afrikaner intransigence on racial issues, the ANC saw itself forced into a campaign of violence. Maintaining that *apartheid should be abolished, and every South African have the vote, it formed a liberation army, 'Umkhonto Wesizwe' (Spear of the Nation). In 1962 its leader, Nelson *Mandela, was arrested and a number of its executive members were detained. Mandela and some of his colleagues were convicted of sabotage and jailed for life. The exiled wing of the ANC has maintained a campaign of violence, which escalated during the 1980s.

Afrikaner, or Boer, the name generally given to the white Afrikaans-speaking population of South Africa. It is used particularly to refer to the descendants of the families which emigrated from the Netherlands, Germany, and France before 1806, that is, before Britain seized the Cape Colony. The Afrikaans language and adherence to the Christian Calvinist tradition, out of which arose the concept of *apartheid, are unifying factors.

Aga Khan (Turkish, *aga*, 'master', *khan*, 'ruler'), title borne by leaders of the Nizari or eastern branch of the Ismaili sect of Shiite Islam. The first Aga Khan, Hassan Ali Shah of Kirman (d. 1881), fled to Afghanistan and Sind after leading an unsuccessful revolt in Iran in 1838.

Winning British favour he settled in Bombay, where in 1866 the Arnold judgment gave him control of the affairs of the Indian Khoja community. His grandson, Sultan Muhammad Shah (1877-1957), played an active part in Indian politics, attempting to secure Muslim support for British rule, particularly as President of the All-India *Muslim League (1913). He was leader of the Muslim delegation to the *Round Table Conference in 1930-2. He was succeeded by his grandson, Prince Karim.

Aguinaldo, Emilio (1869-1964), Filipino nationalist leader. He became active in the nationalist movement in the early 1890s and led an armed uprising against Spanish rule (1895-6) before going into exile in Hong Kong. He returned with American assistance during the *Spanish-American War (1898) and organized another guerrilla campaign, but, after the US victory, his nationalist aspirations resulted in war with American forces (1899-1901). Finally accepting US rule, he waged a peaceful campaign for independence for the next four decades before collaborating with the Japanese during World War II. Briefly imprisoned by the Americans in 1945, he retired from active politics after his release.

agricultural improvements, the means of achieving increased agricultural production. Although the 18th century saw many improvements in agriculture, methods in the 19th and 20th centuries enabled world cereal production to rise dramatically. This was achieved by the use of fertilizers and chemicals, by improvements in seed strains and the elimination of plant diseases, and by

Air Force

Fokker Dr.I triplane fighter
(German)

DH4 bomber (British)

b) Junker Ju 52/3m
(German)

c) Hawker Hurricane fighter
(British)

d) MIG-15 fighter
(Soviet)

e) Bell AH-1G Huey Cobra gunship
(US)

f) Boeing G-3A Sentry AWACS
(USAF & NATO)

In sketch (a) a German triplane fighter is attacking a British bomber (460 lb. bomb-load) in the last year of World War I. The Junker Ju 52/3m (b) was developed by Germany in the 1930s and adapted in the Spanish Civil War as a bomber. In World War II it served the Luftwaffe for troop-transport and bombing. Behind the Hurricane (c) are radar aerials, developed 1938–40, and a vital asset in the battle of Britain. The Soviet MIG 15 (d) was first used in the Korean War, and the US helicopter gunship (e), equipped with rockets, fought in the Vietnam War. Boeing AWACS (f) are used for surveillance and as flying communications centres.

mechanization. Sodium nitrate and guano (sea-bird excrement imported from South America and containing phosphates and nitrates) were first used in Britain in the 1830s. Although fertilizers are expensive and environmentally controversial, they remain the main reason for increased productivity in the USA and Europe. New strains of cereals have been steadily developed over the period, especially since 1950, which has witnessed the 'green revolution' in Asia, resulting, for example, in India becoming self-sufficient in cereal production by 1985. Mechanization began with the invention of machines such as the first reaper (1826) and the reaper and binder (1886). It was the latter particularly that enabled the great wheat prairies of the USA and Canada to be opened up in the later years of the 19th century. Steam threshing machines were invented early in the century, and a massive steam tractor appeared in the USA in 1889. The diesel tractor was developed early in the 20th century, while other machines included deep-ploughs, combine-harvesters, corn-pickers, and balers.

airforce, the armed service concerned with attack and defence in the air. At first in World War I aircraft were used to locate targets for artillery on the *Western Front, but from 1916 onwards they were developed for bombing, while rival fighter aircraft engaged in aerial gunfights both in France and in the *Mesopotamian campaign, where aircraft were also invaluable for reconnaissance. Airships were also developed, especially by Germany, which used Zeppelin airships for bombing attacks against civilian targets. After the disastrous crashes of the R.101 and the *Hindenburg* in the 1930s, however, airships lost popularity. Very rapid development in aircraft design between the wars meant that World War II began with both sides possessing formidable bomber and fighter capability. During the war dive-bombing techniques as well as heavily armed bombers for massed high altitude air raids (*bombing offensives) were developed, while the invention of radar assisted defenders in locating attacking aircraft. Large troop-carrying planes were also developed, together with the helicopter, which became a key weapon in later wars in Korea, Vietnam, and Afghanistan. The high costs of increasingly sophisticated aircraft since World War II have made it difficult for developing countries to compete in building and maintaining an efficient air force.

air transport, the movement of passengers and goods by air. Nineteenth-century air transport, on a very limited scale, was by hot-air balloon, glider, and airship, first using steam- and then petrol engines to drive a propeller. An integral part of the *transport revolution, airships continued to be developed, especially by Germany in World War I, and by the USA, until the disaster of the *Hindenburg* airship (1937). Development of the aeroplane since the first flight of the Wright brothers in 1903 was stimulated by both World Wars. Passenger and freight traffic, using land and sea-planes, was developing between the wars, but has seen massive growth since 1945. In 1970 Jumbo jets (Boeing 747), carrying 400 passengers, resulted in reduced fares, while in 1976 *Concorde* cut the trans-Atlantic crossing to under four hours. Helicopters were developed for military purposes during World War II and used in the *Korean War and the *Vietnam War; they also became increasingly used as taxis, in rescue operations, and for supplying isolated areas.

Alamein, El, battle of (October–November 1942), a critical battle in Egypt in World War II. In June 1942, the British took up a defensive position in Egypt. One flank rested on the Mediterranean at El Alamein and the other on the salt marshes of the Qattara Depression. In August, General *Montgomery was appointed to command the defending 8th Army. He launched an offensive in which, after a heavy artillery preparation, about 1,200 tanks advanced, followed by infantry, against the German Afrika Korps commanded by General *Rommel. Rommel was handicapped by a grave fuel shortage and had only about 500 tanks. The outnumbered Germans never regained the initiative. Rommel managed to withdraw most of his men back into Libya, but this battle marked the beginning of the end of the *North African Campaign for Germany.

Alamo, the, a mission fort in San Antonio, Texas, and scene of a siege during the *Texas Revolution against Mexico of 1836. A Mexican army of 3,000 led by *Santa Anna besieged the fort held by fewer than 200 men, under the joint command of William B. Travis and James Bowie. The siege lasted from 24 February to 6 March, when the Mexicans finally breached the walls. Travis, Bowie, Davy *Crockett, and all their men were killed. The defence of the Alamo became the symbol of Texan resistance.

Alanbrooke, Alan Francis Brooke, 1st Viscount (1883–1963), British field-marshal. He served with distinction during World War I, and in the 1930s was noted as an artillery expert. In World War II he was a corps commander during the withdrawal from *Dunkirk. Later, as Chief of the Imperial General Staff and Chairman of the Chiefs of Staff Committee (1941–6), he represented the service chiefs in discussions with Churchill. As Churchill's chief adviser on military strategy, he accompanied him to all his conferences with Roosevelt and Stalin.

A 19th-century lantern slide depicting the last stand by the defenders of the **Alamo**. A buckskin-clad frontiersman, wearing the familiar coonskin cap of Davy Crockett, fights as the Mexicans overwhelm the mission. Women and children only among the Texans were spared.

Åland Islands, a province of *Finland, consisting of over 6,000 islands in the Gulf of Bothnia. The islands were Swedish until 1809, when, together with Finland, they became a Russian grand duchy. After the collapse of the Russian empire in 1917 Finland declared the independence of the islands, and Swedish troops, who were occupying the archipelago, were ejected. The future of the islands was referred to the *League of Nations (1921), which upheld the islands' autonomy as a part of Finland.

Alaska Purchase (1867), the purchase by the USA of Alaska from Russia for $7,200,000 (less than five cents a hectare) arranged by William H. *Seward. It remained an unorganized territory until 1884. Despite extensive copper and gold discoveries there was little population growth. Matters changed with World War II, with the impact of the American military presence and the discovery of oil, and in the *Cold War the area began to play an important strategic role.

Albania, a country in south-eastern Europe. A part of the Ottoman empire from the 15th century, it was noted for the military dictatorship of Ali Pasha (c.1744-1822), whose court was described by the English poet Byron in *Childe Harold*. Nationalist resistance was crushed in 1831, but discontent persisted and a national league was created during the *Russo-Turkish War of 1877-8. It became an independent state as a result of the *Balkan Wars in 1912, and after a brief period as a republic became a monarchy under King *Zog in 1928. Invaded by Italy in 1939, it became a communist state under Enver Hoxha after World War II. Under the strong influence of the Soviet Union until a rift in 1958, it became closely aligned with China until*Mao Zedong's death in 1976. Albania left the *Warsaw Pact in 1968, and generally has remained isolated in its Stalinist policy and outlook.

Albert Francis Charles Augustus Emmanuel (1819-61), Prince-Consort of England, husband of Queen *Victoria. The younger son of the Duke of Saxe-Coburg-Gotha, he married Victoria, his first cousin, in 1840. At first meeting with mistrust and prejudice, both because he was not British and because the queen would let him take no part in state affairs, he gradually exerted his influence and that of his own adviser, Baron Stockmar, over Victoria. Albert took a keen interest in industry, agriculture, and the arts, and presided over the Royal Commission that raised the money for the *Great Exhibition of 1851. His diplomatic skills were shown in December 1861, when he moderated the government's hostile reaction to the USA over the *Trent Affair. His death from typhoid was a great shock to Victoria, who withdrew from public life for several years.

Alessandri, Arturo (1868-1950), Chilean statesman. In 1920 he was elected President on a liberal policy, but, finding his attempts at reform blocked, he went into voluntary exile in 1924. The following year he was brought back by the army when a new constitution was adopted. He extended the suffrage, separated church and state while guaranteeing religious liberty, and made primary education compulsory. He resigned again in October 1925 and went to Italy. On his return he was re-elected President (1932-8). By now the economy was experiencing the Great *Depression. He reorganized

the nitrate industry, developed schools, and improved conditions in agriculture and industry.

Alexander I (1777-1825), Emperor of Russia (1801-25). The son of Paul I (in whose murder he may indirectly have assisted), he set out to reform Russia and correct many of the injustices of the preceding reign. His private committee (Neglasny Komitet) introduced plans for public education, but his reliance on the nobility made it impossible for him to abolish serfdom. His adviser, M. Speransky, then pressed for a more liberal constitution, but the nobles secured his fall from office in 1812. At first a supporter of the coalition against *Napoleon, his defeat by the latter at *Austerlitz (1805) and Friedland (1807) resulted in the Treaties of *Tilsit and in his support of the *Continental System against the British. His wars with Persia (1804-13) and Turkey (1806-12) brought territorial gains, including the acquisition of Georgia. His armies helped to defeat Napoleon's *grande armée* at *Leipzig, after its retreat from Moscow (1812). Under the influence of the Russian novelist and mystic Baroness Juliane Krüdener he became a mystic. In an effort to uphold Christian morality in Europe he formed a *Holy Alliance of European monarchs which only those of Austria and Prussia joined, and became increasingly conservative in his domestic policies. The constitution he gave to *Poland scarcely disguised the rule of the military there. He supported *Metternich in suppressing liberal and national movements, and gave no help to the Greeks in rebellion against the *Ottoman Turks, although they were Orthodox Christians like himself. He was reported to have died suddenly while on a tour of the Crimea, but rumour persisted that he had escaped to Siberia and become a hermit.

Alexander I (1888-1934), King of Yugoslavia (1921-34). Of the Karageorgević dynasty of Serbia, he tried to overcome the ethnic, religious, and regional rivalries in his country by means of a personal dictatorship (1929), supported by the army. In the interest of greater unity, he changed the name of his kingdom, which consisted of Serbs, Croats, and Slovenes, to *'Yugoslavia' in 1929. In 1931 some civil rights were restored, but they proved insufficient to quell rising political and separatist dissent, aggravated by economic depression. He was planning to restore parliamentary government when he was assassinated by a Croatian terrorist.

Alexander II (1818-81), Emperor of Russia (1855-81). Known as the 'Tsar Liberator', he was the eldest son of *Nicholas I and succeeded to the throne when the *Crimean War had revealed Russia's backwardness. His Emancipation Act of 1861 freed millions of serfs and led to an overhaul of Russia's archaic administrative institutions. Measures of reform, however, did not disguise his belief in the need to maintain autocratic rule and his commitment to military strength, as witnessed by the introduction of universal conscription in 1874. His reign saw great territorial gains in the Caucasus, Central Asia, and the Far East, to off-set the sale of *Alaska to the USA (1867). The growth of secret revolutionary societies such as the *nihilists and *populists, culminating in an assassination attempt in 1862, completed his conversion to conservatism. After further assassination attempts, he was mortally wounded (1881) by a bomb, thrown by a member of the People's Will Movement.

Alexander III (1845-94), Emperor of Russia (1881-94). Following the assassination of his father *Alexander II he rejected all plans of liberal reform, suppressing Russian *nihilists and *populists, extending the powers of nominated landed proprietors over the peasantry, and strengthening the role of landowners in local government. A zealous Orthodox Christian, he was unsympathetic to non-Orthodox Christians and did nothing to prevent the anti-Semitic *pogroms of 1881 and later years. Autocratic in attitude, he was, however, genuinely interested in the principles of administration and his reign saw the abolition of the poll tax, the creation of a Peasant Land Bank, and tentative moves towards legalization of trade unions. Under his Minister of Finance Sergei *Witte the reign also saw large European investment in Russian railways and industry. Alexander's concept of *naradnost* (belief in the Russian people) led to the Russian language being imposed as the single language of education throughout the empire. Although he resented the loss of the Russian Balkans imposed by the Congress of *Berlin, he nevertheless continued to support Bismarck's League of the *Three Emperors, the Dreikaiserbund, until 1890, when the aggressive attitudes of the new German emperor *William II led to its replacement by an alliance with France.

Alexander of Tunis, Harold Rupert Leofric George Alexander, 1st Earl (1891–1969), British field-marshal. He served in the Irish Guards in World War I, and later in Latvia and India. In World War II he commanded the rearguard at *Dunkirk and the retreat in Burma. As commander-in-chief, Middle East, in 1942, he turned the tide against *Rommel. In 1943 he was deputy to *Eisenhower, clearing Tunisia for the Allies. He then led the invasion of Sicily and commanded in Italy until the end of the war. He was governor-general of Canada (1946–52).

Algeciras Conference (1906), an international meeting in Algeciras, Spain, held at Germany's request. Its treaty regulated French and Spanish intervention in Moroccan internal affairs and reaffirmed the authority of the sultan. It was a humiliation for Germany, which failed to obtain support for its hardline attitude towards France except from Austro-Hungary. Britain, Russia, Italy, and the USA took the side of France.

Algeria, a North African country. Conquered by France in the 1830s and formally annexed in 1842, Algeria was 'attached' to metropolitan France and heavily settled by European Christians. The refusal of the European settlers to grant equal rights to the native population led to increasing instability, and in 1954 a war of national independence broke out which was characterized by atrocities on both sides. In 1962, in spite of considerable resistance in both France and white Algeria, President *de Gaulle negotiated an end to hostilities in the *Évian Agreement, and Algeria was granted independence as the result of a referendum. In 1965 a coup established a left-wing government under Colonel Houari *Boumedienne and afterwards serious border disputes broke out with Tunisia, Morocco, and Mauritania. After Boumedienne's death in 1978, his successor Benjedid Chadli

Riot police in Algiers in 1961 confront a nationalist crowd demonstrating for a free **Algeria**. A banner calls for talks between France's President de Gaulle and Ferhat Abbas, president of the Algerian provisional government.

relaxed his repressive domestic policies and began to normalize Algeria's external relations.

Ali Pasha, Mehmed Emin (1815–71), Ottoman statesman and reformer. After service in the Foreign Ministry he became Grand Vizier in 1852. He became one of the leading statesmen of the *Tanzimat reform movement, and was responsible for the Hatt-i Humayun reform edict of 1856. This guaranteed Christians security of life and property, opened civil offices to all subjects, abolished torture, and allowed acquisition of property by foreigners. He believed in autocratic rule and opposed the granting of a parliamentary constitution.

Allenby, Edmund Henry Hynman, 1st Viscount (1861–1936), British field-marshal. In World War I he served in France before his appointment as commander of the British expeditionary forces in Egypt and Palestine in 1917. He defeated the Ottoman forces in Palestine and Syria, capturing Jerusalem, and ending Turkish resistance after the battle of Megiddo (18–21 September 1918) and the fall of Damascus and Aleppo. He was appointed Special High Commissioner for Egypt and the Sudan (1919–25), and in 1922 persuaded the British government to end its protectorate over *Egypt.

Allende (Gossens), Salvador (1908–73), Chilean statesman. As President of Chile (1970–3), he was the first avowed Marxist to win a Latin American presidency in a free election. Having bid for the office unsuccessfully on two previous occasions (1958 and 1964), Allende's 1970 victory was brought about by a coalition of leftist parties. During his brief tenure in office he set the country on a socialist path and in the process incurred the antipathy of the Chilean military establishment. Under General *Pinochet a military coup (which enjoyed some indirect support from the USA) overthrew him in 1973. Allende died in the fighting.

Alsace-Lorraine, a French region west of the Rhine. Alsace and the eastern part of Lorraine were ceded to Germany after the *Franco-Prussian War (1871) and held in common by all the German states. Rich in both coal and iron-ore, Lorraine enabled Germany to expand its naval and military power. The subsequent policy of Germanization of the region was resented by French nationalists, and the province was restored to France by the Treaty of *Versailles. In 1940 Nazi troops occupied the region and it reverted to Germany. In 1945 French and US troops recovered Alsace-Lorraine for France.

Ambedkar, Bhimrao Ramji (1893–1956), Indian leader of the Untouchables (Hindus held to defile members of a caste on touch). He led the agitation for their constitutional rights in the 1930s and when *Gandhi went on a fast against the provision of separate electorates for the Untouchables, agreed to the Poona Pact (1934) providing reserved seats for them in the legislatures. As the leader and founder of the Scheduled Castes Federation, he opposed the Indian National *Congress, but joined that party after independence. As a leading constitutional lawyer, he played a major role in formulating and drafting the Indian constitution.

American Civil War (1861–5), a war between the Northern (Union) and Southern (*Confederacy) states of the USA. It was officially known as the War of the Rebellion and usually called the War between the States in the South. Economic divergence between the industrialized North and the agricultural, slave-based economy of the South was transformed into political rivalry by the activity of the *abolitionists, and above all by the dispute over the expansion of slavery into the western territories. By the late 1850s, all efforts at compromise had failed and the slide into violence had begun with John *Brown's dramatic armed descent on Harper's Ferry (1859). South Carolina seceded from the Union in December 1860 in the wake of Abraham *Lincoln's victory in the presidential election of that year. When the war began with the bombardment of *Fort Sumter (1861), the newly established Southern Confederacy increased to eleven states under the presidency of Jefferson *Davis.

The war itself is best considered as three simultaneous campaigns. At sea, the North held the upper hand, but the blockade imposed in 1861 took a long time to become effective. Virtually no cotton was exported. Massive naval expansion produced a blockade which helped to cripple the Confederate war effort. On land a series of engagements took place in the *Virginia Campaigns, where the close proximity of the Union and Confederacy capitals, Washington and Richmond, good defensive terrain, and the military genius of General *Lee enabled the Confederacy to keep superior Union forces at bay for much of the war. In the more spacious western regions, after a series of abortive starts, the North managed to split the Confederacy in the *Vicksburg Campaign, by gaining control of the Mississippi. From here General *Grant moved through Tennessee in the *Chattanooga Campaign, opening the way for the drive by *Sherman through Georgia to the sea. This ruthless strategy, together with Lee's surrender to Grant at *Appomattox, brought the war to an end in April 1865. Over 600,000 soldiers died in the Civil War. While the immediate results were the salvation of the union and the abolition of slavery, the challenges of revitalizing the South and promoting racial justice and equality persisted.

American Colonization Society, founded in the USA in 1817 in order to resettle in Africa free-born Africans and emancipated slaves. In 1821 the society bought the site of the future Monrovia, *Liberia, which it controlled until Liberia declared its independence in 1847. After 1840 the society declined and was dissolved in 1912.

American Federation of Labor (AFL), federation of North American labour unions, mainly of skilled workers, founded in 1886. From its formation until his retirement in 1924, it was decisively shaped by its president Samuel Gompers. After mass disorders culminating in the *Haymarket Square Riot and the subsequent eclipse of the *Knights of Labor, Gompers wanted a cohesive nonradical organization of skilled workers committed to collective bargaining for better wages and conditions. However, growing numbers of semi-skilled workers in mass-production industries found their champion in John L. Lewis, leader of the more militant United Mine Workers. When he failed to convince the AFL of the need to promote industry-wide unions in steel, automobiles, and chemicals, Lewis formed (1935) the Committee (later the Congress) of Industrial Organizations (CIO), its members seceding from the AFL. In

American Civil War (1861–5)

Often called the first really modern war, the American Civil War marked a milestone in military strategy. The deployment of the long-range rifle shattered the effectiveness of traditional attack, while the use of steam power lent speed of movement to both armies. The political objective of the North was to prevent the Southern states from seceding from the Union: the Confederate states, apprehensive of centralized government, sought to retain their independence, especially over their rights to buttress their economy with slave labour.

1955 the rival organizations were reconciled as the AFL-CIO under George Meany and Walter Reuther with a total of fifteen million members. This body has remained the recognized voice of organized labour in the USA and Canada.

American System, a programme of US economic reforms propounded in the 1820s by the politician Henry *Clay. He advocated a combination of a protective tariff, a national bank, and a system of internal improvements to expand the US domestic market and lessen dependence upon overseas sources. A nationalist policy, intended as a binding force in the face of demands for increased *states' rights, the American System inspired much of the economic policy of the *Whig Party.

Amin, Idi (1926–), Ugandan head of state. Possessed of only rudimentary education, Amin rose through the ranks of the army to become its commander. In 1971 he overthrew President *Obote and seized power. His rule was characterized by the advancing of narrow tribal interests, the expulsion of non-Africans (most notably Ugandan Asians), and violence on a huge scale. He was overthrown with Tanzanian assistance in 1979.

Amritsar massacre (13 April 1919). Indian discontent against the British had been mounting as a result of the *Rowlatt Act. The massacre in Amritsar followed the killing, three days before, of five Englishmen and the beating of an Englishwoman. Gurkha troops under the command of Brigadier R. H. Dyer fired on an unarmed crowd gathered in the Jallianwala Bagh, an enclosed park, killing 379 and wounding over 1,200. Mounting agitation throughout India followed, and Dyer was given an official, if belated, censure.

anarchism, the belief that government and law should be abolished. The French social theorist Pierre Joseph *Proudhon first outlined the concept that equality and justice should be achieved through the abolition of the state and the substitution of free agreements between individuals. Groups of anarchists tried to find popular support in many European states in the 1860s and 1870s. They were hostile to *Marxism on the grounds that a seizure of state-power by the workers would only perpetuate oppression. The Russian anarchist Mikhail *Bakunin founded a *Social Democratic Alliance (1868) which attempted to wrest control of the workers *International from *Marx. Anarchists switched between strategies of spontaneous mutual association and violent acts against representatives of the state. The Presidents of France and Italy, the King of Italy, and the Empress of Austria were killed by anarchists between 1894 and 1901. Subsequently they tried to mobilize mass working-class support behind the Russian General Strike, which was a central feature of the *Russian Revolutions of 1905 and 1917. Their influence in Europe declined after the revolution in Russia and the rise of totalitarian states elsewhere. They were active in the *Spanish Civil War, and in the latter half of the 20th century anarchism has attracted urban *terrorists.

Andean Common Market, a regional economic grouping comprising Colombia, Peru, Bolivia, and Venezuela. Formally established by the Cartagena Agreement of 1969, the ACM was an attempt to enhance the competitive edge of the member states in their economic relations with the more developed economies of the Latin American region. The two major goals—reduction of trade barriers and the stimulation of industrial development—have encountered some measure of success, but the group has not realized its full potential.

Anderson, Elizabeth Garrett (1836–1917), British physician. Until her time medical schools would not admit women students, but she discovered that the Society of Apothecaries, by its constitution, could not prevent her taking its examinations. She qualified in 1865 and this entitled her to practise as a doctor. In 1866 she helped to found the Marylebone Dispensary for Women and Children (the present Elizabeth Garrett Hospital).

A photograph taken in 1889 of Elizabeth Garrett **Anderson**, pioneer of women doctors.

She lectured on medicine (1875–97) at the London School of Medicine for Women and was the first woman elected to the British Medical Association (1873).

Andersonville Prison, a prisoner-of-war camp used by the *Confederacy during the *American Civil War. It was notorious as a result of the high death rate among its inmates. Andersonville Prison had been hurriedly established in early 1864 and suffered from the general shortage of food, clothing, and medical supplies in the war-stricken South. By the time of its capture by Union (Northern) forces, nearly half the prisoners had died from disease, and as a result of the ensuing outcry the ex-commandant, Captain Henry Wirz, was tried and executed for murder. Subsequent investigation revealed the catastrophe to have been the product less of deliberate barbarity than of the collapse of the Confederate military machine.

Andover scandal, an event leading to improvement in workhouse conditions in Britain in the 1840s. The *Poor Law of 1834 had resulted in a worsening of the conditions in workhouses to discourage all but the really needy from applying for admission. In 1845 able-bodied labourers at the workhouse at Andover, Hampshire, were found to be eating the gristle and marrow from bones they were crushing to make manure. A committee was set up to investigate and the result was the establishment of a new Poor Law Board, directly responsible to Parliament, and some improvement in workhouse conditions.

Andrada e Silva, José Bonifacio (1763–1838), Brazilian scientist and statesman. In 1821 he gave his support to the regent Pedro, who was left in charge when his father *John VI returned to Portugal. By mid-1822 leading Brazilians were determined that their country should become independent and in December *Pedro I was crowned Emperor of Brazil, with Andrada appointed as Prime Minister. With his two brothers he drew up a draft constitution, but antagonism developed with the emperor, and Andrada was exiled. In 1831 he was invited to return to Brazil by the emperor to become the tutor of his son. When Pedro I abdicated in April 1831 in favour of the boy, Pedro II, Andrada was confirmed as tutor by the council of regency. He held this office for two years, but was then arrested for 'political intrigue' and again left the country.

Andrassy, Julius, Count (1823–90), Hungarian statesman. One of the radical nationalist leaders of the unsuccessful Hungarian *Revolution of 1848, he rose to prominence with Francis *Deák in the negotiations leading up to the *Ausgleich (Compromise) of 1867. By now a moderate, he served as Hungary's first Prime Minister (1867–71). From 1871 to 1879 he was Foreign Minister of the Austro-Hungarian empire, during which time he limited Russian influence in the *Balkan States.

Anglican Communion, a world-wide family of Protestant Christian churches whose origins were in Reformation England in the 16th century. In 1784 the first diocese of the American Episcopalian Church was formed, followed by others. From the 1830s, there was a considerable religious revival in Britain. John Henry *Newman and others led the Oxford Movement, which sought

British troops storm the Afghan citadel of Ghazna during the First **Anglo-Afghan War**. This coloured lithograph of 1842 is taken from a painting based on an original drawing by a British army officer and shows the final confrontation between the Afghan defenders and the British forces. (National Army Museum, London)

to restore historical continuity with the Roman Catholic Church and introduce more ceremony into Anglican worship. This 'High Church' movement was countered by the 'Low Church' or Evangelical movement, which was more Protestant in its views and very active in *missionary work and social reform. The Anglican Church remains the legally established church in England, but was disestablished in Scotland (1690), Ireland (1871), and Wales (1920). The majority of Anglicans now live outside Britain, and the churches of the Anglican Communion embrace many variations in practice and ritual, for example in the ordination of women to the priesthood in some countries.

Anglo-Afghan Wars, a series of wars between Afghan rulers and British India. The first occurred (1838–42) when Britain, concerned about Russian influence in *Afghanistan, sent an army to replace *Dost Muhammad with a pro-British king, Shah Shuja al-Mulk. Resistance to Shuja's rule culminated in an uprising (1841) which led to the destruction of the British Indian forces in Kabul during their withdrawal to Jalalabad (1842). Kabul was reoccupied the same year, but British forces were withdrawn from Afghanistan. The second (1878–80) was also fought to exclude Russian influence. By the Treaty of Gandamak (1879) Britain acquired territory and the right to maintain a Resident in Kabul, but in September of the same year the Resident, Sir Louis Cavagnari, was killed in Kabul and further campaigns were fought before the British withdrawal was accomplished. The third war was fought in 1919, when the new amir of Afghanistan, Amanullah, attacked British India and, although repulsed, secured the independence of Afghanistan through the Treaty of Rawalpindi (1919).

Anglo-Burmese Wars (1824–6, 1852, 1885), conflicts between British India and Burma. In 1824 a threatened Burmese invasion of Bengal led to a British counter-invasion, which captured Rangoon and forced the cession to Britain of Arakan and Tenasserim, the payment of a large indemnity, and the renunciation of Burmese claims to Assam. After a period of relative harmony, hostile treatment of British traders led to a second invasion in 1852, as a result of which Rangoon and the Irrawaddy delta was annexed. In 1885, the alleged francophile tendencies of King Thibaw (1878–85) provoked a third invasion which captured the royal capital at Mandalay and led to Thibaw's exile. Upper Burma became a province of British India, although guerrilla resistance to British rule was not suppressed for another five years.

Anglo-Japanese Alliance (1902), diplomatic agreement between Britain and Japan. It improved Britain's international position and consolidated Japan's position in north-east Asia at a time of increasing rivalry with Russia. The two powers agreed to remain neutral in any war fought by the other to preserve the *status quo* and to join the other in any war fought against two powers. Britain and Japan began to drift apart after World War I, and when the *Washington Conference was summoned in 1921, Britain decided not to renew the alliance, which ended in 1923.

Anglo-Maori Wars, a complex series of conflicts following the colonization of New Zealand. In the mid-1840s there were rebellions under the Maori chiefs Hone Heke and Te Rauparaha. In 1860 the *Taranaki Wars began, but Wiremu Tamihana, a leader of the *Kingitanga unity movement, negotiated an uneasy truce in 1861.

A skirmish in the New Zealand bush during the
Anglo-Maori Wars depicted by Orlando Norie. Maori
warriors, armed with spears, sticks, and stones, as well as
guns, defend a stockade against British troops. The Maori
fought a tenacious guerrilla campaign in defence of their
tribal lands. (National Army Museum, London)

Governor Browne was replaced by Sir George *Grey in
an attempt to secure peace. Grey and his advisers were
reluctant to see the Kingitanga consolidated, for fear that
British authority could not be asserted throughout New
Zealand, and that land purchases would be halted.
Fighting resumed in Taranaki in May 1863 and in July
the Waikato was invaded. Fighting with the Kingitanga
stopped in 1865 but was sustained by resistance from the
*Pai Marire (1864–5) and from Titokowaru in Taranaki
and *Te Kooti on the east coast (1868). London recalled
Grey and the British regiments that year, but the pursuit
of Titokowaru and Te Kooti, masters of guerrilla warfare,
was carried on by settler militia and Maori auxiliaries.
The last engagement was in 1872, after which Maori
resistance gradually subsided.

Angola, a country in south-western Africa north of
Namibia. The coastal strip was colonized by the Portuguese in the 16th century, but it was not until the 19th
century that, following wars with the Ovimbundu, Ambo,
Humbo, and Kuvale, they began to exploit the mineral
reserves of the hinterland. In 1951 Angola became an
Overseas Province of Portugal. In 1954 a nationalist
movement emerged, demanding independence. The Portuguese at first refused, but finally agreed in 1975 after
a protracted guerrilla war, and 400,000 Portuguese were
repatriated. Almost total economic collapse followed.
Internal fighting was continued between guerrilla
factions. The ruling Marxist party, the Popular Movement
for the Liberation of Angola (MPLA), is supported by
Cuba and East Germany, and its opponent, the National
Union for the Total Independence of Angola (UNITA),
by *South Africa and the USA. Punitive South African
raids have taken place from time to time, aimed at
Namibian resistance forces operating from Angola.

Anguilla, the most northerly of the Leeward Islands in
the West Indies. A British colony since 1650, Anguilla
formed part of the Federation of the *West Indies (1958–
62) and subsequently received associated state status with
St Kitts and Nevis. Anguilla declared independence in

1967 and two years later was occupied by British troops,
who reduced the island to colonial status once again. It
is now a British dependency with full self-government.

Annam, central region of *Vietnam. The Annamese
state split into two states in the 16th century, and was
reunited in 1802 as the empire of Vietnam by the
Annamese general Nguyen Anh with assistance from
French mercenaries. He ruled as the Emperor Gia-Long.
In 1807 a protectorate was enforced on Cambodia which
led to frequent wars with Siam. Gia-Long's xenophobia
and massacre of Catholics provoked French military
intervention in 1858, and, by 1884, the seizure of the
southern part of the country (also known as Cochin
China) and the establishment of protectorates over the
centre (Annam) and north (Tonkin). The imperial court
at Hué remained, but had only nominal power except,
briefly, under *Bao Dai in the post-1945 period.

Anschluss (German, 'connection'), specifically applied
to Hitler's annexation of Austria. The *German Second
empire did not include Austrian Germans, who remained
in Austro-Hungary. In 1934 a coup by Austrian Nazis
failed to achieve union with Germany. In February 1938
Hitler summoned Kurt von Schuschnigg, the Austrian
Chancellor, to Berchtesgaden and demanded the admission of Nazis into his cabinet. Schuschnigg attempted
to call a plebiscite on Austrian independence, failed, and
was forced to resign. German troops entered Vienna and
on 13 March 1938 the Anschluss was proclaimed. The
majority of Austrians welcomed the union. The ban on
an Anschluss, laid down in the Treaties of *Versailles
and St Germain (1919), was reiterated when the Allied
Powers recognized the second Austrian republic in 1946.

Antarctica, the ice-covered continent centred upon the
South Pole. Fabian von Bellingshausen, a Russian, was
probably the first person to sight the continent (1820).
There were significant Russian, British, French, and
American scientific and geographical expeditions to Antarctica during the 19th century. Many countries were
involved in overland exploration during the first decades

New Year's Day celebrations in **Antarctica**, 1842, from a
painting by John Edward Davis. Sailors and scientists of
the British expedition of 1839–43 were led by James Clark
Ross, and here enjoy a respite from the hardships of
Antarctic exploration in its early days. (Scott Polar
Research Institute)

of the 20th century. Ernest Shackleton's British party sledged to within 156 km. (97 miles) of the South Pole in 1909. Roald Amundsen's Norwegian party reached it in 1911. Soon after, Robert Falcon Scott's British party reached it (1912), but all members died on the return journey. The main territorial claims, not recognized by the USA or the Soviet Union, have been those made by Britain (British Antarctic Territory) in 1908, New Zealand (Ross Dependency) in 1923, Australia (Australian Antarctic Territory) in 1933, France (Adélie Land) in 1938, Norway (Queen Maud Land) in 1939, Chile (Antarctic Peninsula) in 1940, and Argentina (Antarctic Peninsula) in 1942. The Antarctic Treaty, signed in 1959, preserves Antarctica for peaceful purposes, pledges international scientific co-operation, and prohibits nuclear explosions and the disposal of radioactive waste.

Anthracite strike, a strike by the United Mine Workers of America, called in May 1902 in a bid for higher wages, shorter hours, and union recognition. The employers refused to arbitrate, and President Theodore *Roosevelt appointed a commission to mediate, which led the union to call off the strike. In March 1903 the commission gave the miners a 10 per cent wage increase but refused to recognize the union. The intransigent behaviour of the employers created public support for the federal government's intervention and the strike signalled an important extension of federal economic responsibilities.

Anti-Comintern Pact (25 November 1936), an agreement between Germany and Japan ostensibly to collaborate against international communism (the *Comintern). Italy signed the pact (1937), followed by other nations in 1941. It was in reality a union of aggressor states, first apparent in Japan's invasion of China in 1937.

Anti-Corn Law League, a movement to bring about the repeal of the duties on imported grain in Britain known as the *Corn Laws. Founded in Manchester in 1839 under Richard *Cobden and John *Bright, the League conducted a remarkably successful campaign. It organized mass meetings, circulated pamphlets, and sought to influence Members of Parliament. A combination of bad harvests, trade depression, and the *Irish famine strengthened the League's position and in 1846 the Prime Minister, Sir Robert *Peel, was persuaded to abolish the Corn Laws. The expected slump in agriculture did not take place.

Antietam (Sharpsburg), battle of (17 September 1862), a battle in the *American Civil War, fought in Maryland. After his victory at the second battle of Bull Run, General *Lee invaded the North, but with only 30,000 men under his immediate command was attacked by a Union (Northern) army under General George *McClellan at Sharpsburg on the Antietam Creek. Although the Confederates were badly mauled, they held their positions and were able to make an orderly retreat on the following day. The casualties of 23,000 (divided almost equally between the two sides) were the worst of any single day of the war. While the Confederate invasion was repulsed, McClellan missed his chance to destroy Lee's army and bring the war to an early end. The political advantages of the victory, however, were such

Heavy casualties were suffered by both sides at the battle of **Antietam**, depicted here in a Kurz and Allison lithograph. Union troops (*left*) are shown advancing across Burnside Bridge to attack Confederate artillery during the decisive battle which halted Lee's northward march.

as to deter Britain and France from diplomatic intervention and to provide *Lincoln with the opportunity to issue his Emancipation Proclamations (2 September 1862 and 1 January 1863), the executive order abolishing slavery in the areas under Confederate control.

Anti-Masonic Party, a US political party of the 1820s and 1830s opposed to Freemasons. Formed in 1826 in the wake of the disappearance of William Morgan, a New York bricklayer alleged to have divulged lodge secrets, the Anti-Masonic Party was the product of hysteria, cleverly played upon by local politicians. It played an influential part in the politics of New York and surrounding states, and drew sufficient *Whig support away from Henry *Clay in the 1832 presidential election to help sweep President *Jackson back into office. The Anti-Masonic Party was the first to hold a nominating convention, but otherwise made no lasting political contribution.

anti-Semitism, hostility towards Jews. In the late 19th and early 20th centuries it was strongly evident in France, Germany, Poland, Russia, and elsewhere, many Jewish emigrants fleeing from persecution or *pogroms in southeast Europe to Britain and the USA. After World War I early Nazi propaganda in Germany encouraged anti-Semitism, alleging Jewish responsibility for the nation's defeat. By 1933 Jewish persecution was active throughout the country. The 'final solution' which Hitler worked for was to be a *Holocaust or extermination of the entire Jewish race, and some estimated four million Jews were killed in *concentration camps between 1941 and 1945. Anti-Semitism has been a strong feature of society within the Soviet Union and Eastern Europe, especially since World War II. Tension between the Arab people (who are also Semitic) and *Zionist Jews since 1948 has been both territorial and religious.

Anti-Trust laws, US laws restricting business monopolies. After twenty-five years' agitation against monopolies, the *Congress passed the Sherman Anti-Trust Act (1890) that declared illegal 'every contract, combination, or conspiracy in restraint of trade'. The Clayton Anti-

Trust Act (1914), amended by the Robinson–Patman Act (1936), prohibited discrimination among customers and mergers of firms that would lessen competition. After World War II there was a further growth in giant multi-national corporations and the Celler–Kefauver Antimerger Act (1950) was intended to prevent oligarchic tactics, such as elimination of price competition, as being against the public interest.

Antonescu, Ion (1882–1946), Romanian military leader and fascist dictator. He became (1937) Chief of Staff and Defence Minister. In 1940 he became Prime Minister and assumed dictatorial powers. He forced the abdication of King *Carol in the same year, and supported the Axis Powers. His participation in the Nazi invasion of the Soviet Union resulted, in 1944, in the fall of his regime as the Red Army entered Romania. In 1946 he was executed as a war criminal.

ANZAC, an acronym derived from the initials of the Australian and New Zealand Army Corps, which fought during World War I. Originally it was applied to those members of the Corps who took part in the *Gallipoli campaign. The name came to be applied to all Australian and New Zealand servicemen. Anzac Day (25 April), commemorating the landing (and later contributions to other campaigns), has been observed since 1916.

ANZUS, an acronym given to a tripartite Pacific security treaty between Australia, New Zealand, and the USA, signed at San Francisco in 1951. Known also as the Pacific Security Treaty, it recognizes that an armed attack in the Pacific Area on any of the Parties would be dangerous to peace and safety, and declares that it would act to meet the common danger, in accordance with its constitutional processes. Following New Zealand's anti-nuclear policy, which included the banning of nuclear-armed ships from its ports, the USA suspended its security obligations to New Zealand (1986).

apartheid (Afrikaans, 'separateness'), a racial policy in South Africa. It involves a strict segregation of black from white, in land ownership, residence, marriage and other social intercourse, work, education, religion, and sport. As a word it was first used politically in 1943, but as a concept it goes back to the rigid segregation practised by the settlers since the 17th century. From 1948 onwards, it has been expressed in statutes, in job reservation and trade union separation, and in the absence of parliamentary representation. In accordance with it *Bantustans have been created, depriving the Bantu-speaking peoples of South African citizenship for an illusory independence. Since 1985 certain restrictions have been mitigated, by creating subordinate parliamentary chambers for Indians and Coloureds (people of mixed descent), by relaxations for sport, by abolishing the Pass Laws and the Separate Amenities Act, and by modifying the Group Areas Act.

appeasement, a term used in a derogatory sense to describe the efforts by the British Prime Minister, Neville *Chamberlain, and his French counterpart, Édouard *Daladier, to satisfy the demands (1936–9) of the *Axis Powers. Their policy of appeasement enabled Hitler to occupy the *Rhineland, to annex Austria, and to acquire the Sudetenland in Czechoslovakia after the *Munich Pact of 1938. Appeasement ended when Hitler, in direct

South African **apartheid** in its strictest form prohibits racial intermingling in many public places. Signs in Afrikaans and English delineate the areas of separation, in this case a beach reserved for whites only.

contravention of assurances given at Munich, invaded the rest of Czechoslovakia in March 1939. A policy of 'guarantees' was then instituted, by which Britain and France pledged themselves to protect Romania, Greece, and Poland should they be attacked by Germany or Italy. The German invasion of Poland five months later signalled the outbreak of World War II.

Appomattox, a village in Virginia, USA, scene of the surrender of the *Confederacy Army of Northern Virginia to the Union Army of the Potomac on 9 April 1865 at the end of the *American Civil War. Having been forced to evacuate Petersburg and Richmond a week before, General *Lee found himself almost surrounded by greatly superior forces and decided that further resistance was pointless. He was granted generous surrender terms by his victorious opponent, General *Grant. The surrender terminated Confederate resistance in the east and marked the effective end of the war.

Arabi Pasha, the English name of Ahmad Urabi Pasha al-Misri (1839–1911), Egyptian nationalist leader. A conscript in the Egyptian army, he rose to the rank of colonel in the Egyptian–Ethiopian War (1875–6). In 1879 he took part in an officers' revolt against the Turkish governor of

Egypt, and led a further revolt in 1881. In 1882, when Britain and France intervened at the request of Khedive Tawfiq, bombarding the city of Alexandria, he organized a nationalist resistance movement. The British defeated him at *Tel-el-Kebir, and exiled him to Sri Lanka. He returned to Egypt in 1901.

Arab League, an organization of Arab states, founded in Cairo in March 1945. The original members were Lebanon, Egypt, Iraq, Syria, Transjordan (now Jordan), Saudi Arabia, Yemen, and representatives of the Palestine Arabs; it was subsequently joined by Sudan and Libya. The objects of the League were to protect the independence and integrity of member states. It embodied Syrian and Lebanese hopes of Arab aid in consolidating their freedom from French rule, and confirmed feelings of Arab solidarity over Palestine. The League developed into a loose co-ordinating body which arranged after 1948 the economic boycott of Israel. Egypt was expelled following the *Camp David Accord, and readmitted in 1987.

Arbenz Guzmán, Jacobo (1913–71), Guatemalan statesman. He served as Secretary of Defence before assuming the Presidency (1951–4), as the candidate of the Revolutionary Action Party. Comprehensive agrarian reform laws made possible the expropriation of large estates owned by Guatemalan nationals and the United Fruit Company of the USA. When his administration was judged to be communist by the Roman Catholic Church and the US government, the *Eisenhower administration began sending arms to Guatemala's neighbours. These arms enabled the Guatemalan exile Carlos Castillo Armas to lead a successful counter-revolution which deposed Arbenz in 1954.

Arctic, the ice-covered ocean centred around the North Pole. The search for a north-west and a north-east passage from Europe to the Orient gave impetus to arctic explorations from the 16th century onwards. The British geographer Sir John Barrow promoted explorations in the early 19th century, while an attempt by Sir John Franklin (1845) to find the north-west passage led to his disappearance and ultimate confirmation of his death. The forty or more search parties sent out after him brought back valuable information about the Arctic regions. In 1850 the British arctic explorer Robert McClure completed a west–east crossing, but the first continuous voyage remained unachieved. In 1878–9 the Swedish Baron Nordenskjöld undertook the first traverse of the north-east passage from Norway to the Bering Strait, but the north-west passage was not completed until the voyage of the Norwegian Roald Amundsen in 1903–6.

Ardennes Campaign (also called battle of the Bulge) (16–26 December 1944), the last serious German counter offensive against Allied armies advancing into Germany in World War II (*Normandy Campaign). It resulted from a decision by Hitler to make an attack through hilly, wooded country and thereby take the US forces by surprise. Last-ditch resistance at several points, notably at Bastogne, held the Germans up long enough for the Allies to recover and prevent the Germans reaching their objective of Antwerp.

Argentina, a South American country occupying much of the southern part of the continent. Colonized by the

Pampa Indians of **Argentina** outside a store painted by E. E. Vidal in 1820. The store displays riding tackle, *bolas* (a missile used for hunting, and for winding round and entangling cattle), jaguar skins, and plumes from the rhea, the large flightless bird of the South American grasslands.

Spanish, the independence of the 'United Provinces of South America' was declared at the Congress of Tucuman in 1816. Divisional differences produced a series of conflicts between unitarios (centralists) and federales (federalists) which characterized much of the 19th century. The lack of political or constitutional legitimacy saw the emergence of the age of the *caudillos until the promulgation of the National Constitution in 1853. The second half of the 19th century witnessed a demographic and agricultural revolution. The fertile plains (pampas) in the interior were transformed by means of foreign and domestic capital, while immigrant workers (principally from Spain and Italy), an extensive railway network, and the introduction of steamships and refrigeration vastly increased the export of cattle and grain. The influx of immigrants between 1870 and 1914 contributed to an increase in the national population from 1.2 million in 1852 to 8 million in 1914. Argentina's export-orientated economy proved vulnerable to the fluctuations of the international market, and the Great *Depression saw a drop of 40 per cent in the nation's exports. The military coup of 1930 saw the emergence of the armed forces as the arbiter of Argentinian politics. The failure of civilian democratic government and of achieving sustained economic growth has led to frequent military intervention.

This was true even in the case of Peronism, the populist movement created with the support of trade unions by Juan Domingo *Peron (1946–55). Peron was re-elected as President in 1973 after an eighteen-year exile. His death in 1974 was followed by another period of military dictatorship (1976–83) in a particularly bitter and tragic period of authoritarian rule, as a result of which an estimated 20,000 Argentinians lost their lives in the 'dirty war' between the state and political terrorism. In 1982 the armed forces suffered a humiliating defeat in the war with Britain over the *Falkland (Malvinas) Islands, and in 1983 a civilian administration was elected under President Raul Alfonsin of the Radical Party. The process of redemocratization in Argentina faces severe problems, most notably a virtually bankrupt economy and the political sensitivity of the armed forces to reform.

Armenia, a region which comprised what is now north-eastern Turkey and the Armenian Soviet Socialist Republic. In 1828 north-east Armenia was ceded by the *Ottoman Turks to Russia. Agitation for independence developed in both Russian and Turkish Armenia, leading to a series of large-scale massacres that culminated in the deportation by the *Young Turk government of all Turkish Armenians to Syria and Palestine (1915), in which over one million died. A short-lived independent Transcaucasian Federal Republic was created in 1918. The separate republic of Armenia lasted until 1920 when, following the battle of Kars, Turkish Armenia was renounced, while the remainder of Armenia proclaimed itself a Soviet Republic, joining the USSR in 1936. There were further Turkish massacres and mass deportations until by the Treaty of Lausanne (1923) Turkish Armenia was absorbed into the new republic of Turkey.

arms control, the attempts made to reduce and control armed forces and weapons. Efforts had been made to bring about arms control as well as *disarmament before and after World War I (for example, the Washington Naval Agreement (1922), which limited the warships of the major powers). In 1952 a Disarmament Commission of the United Nations was set up, and helped to bring about the Non-Proliferation Treaty (1968), covering nuclear weapons. Most of the important negotiations on arms control, however, have taken place in direct talks between the USA, the Soviet Union, and other powers. A *Nuclear Test-Ban Treaty limiting nuclear testing was signed by the USA, the Soviet Union, and Britain in 1963, while direct *Strategic Arms Limitations Talks (SALT) between the USA and the Soviet Union (1969–79) led to some limitation of strategic nuclear weapons. Member states of *NATO and the *Warsaw Pact have met in Vienna since 1973 in the Mutual and Balanced Force Reduction (MBFR) talks, concerned with limiting conventional ground forces in central Europe. The final act of the *Helsinki Conference of 1975 included provision for 'confidence-building measures' (such as notification of military manoeuvres), which are also a form of arms control. Soviet and US negotiations on strategic nuclear weapons re-started in 1982 under the title of START (Strategic Arms Reduction Talks), but these were suspended in 1983. In 1987 a treaty on eliminating intermediate-range nuclear forces (INF) was signed between the superpowers, while talks continued on the reduction of long- and short-range nuclear weapons and of conventional forces, as well as the elimination of chemical weapons.

army, an organized force of men armed for fighting on land. In the late 18th century European armies were mainly of mercenaries recruited (often under pressure) and trained by a professional officer class. The first conscript armies were recruited in France to fight the *Revolutionary and *Napoleonic Wars. During the 19th century most European countries adopted a system of conscription of young men to train and serve for about two years. (Britain only enforced conscription in 1916–18, and again between 1939 and 1959.) In 1800 military battles followed a strict and formal pattern. Infantry were lined into well-drilled rows, firing muskets and advancing with bayonets, backed up by field guns. Cavalry divisions, armed with sabres, provided mobility. European armies played an essential role in 19th- and early 20th-century *imperialism, their superior fire-power enabling them to dominate the peoples of Africa and Asia. The *American Civil War 1861–5 saw large armies of the Union (the North) and the Confederacy (the South) engaged in a struggle in which railways were crucial for movement of troops, and new infantry weapons, such as the breach-loading rifle and the repeating carbine, were developed. By the time of the *Franco-Prussian War heavy artillery was developing, but infantry and cavalry tactics remained little changed until World War I, when motor transport and heavier artillery developed. Even then, armies were slow to adapt to armoured vehicles, and the massed infantry attacks of its battles still used rifle, bayonet, and hand-grenade as their basic weapons, now pitched against machine-guns. By World War II armies were fully motorized, and tanks played a major part in the *North African Campaign and at the *Eastern Front. This mobility required large back-up fuel and maintenance services. Basic infantry tactics still remained essential, even though the rifle was largely replaced by the repeating 'sten' gun, especially in the jungle warfare of the *Burma Campaign. They remained so for the later campaigns in Korea, Vietnam, and the Falklands. In the balance of power between communist and capitalist countries, large armies of *NATO and the *Warsaw Pact continue to face one another in Europe, armed with conventional weapons but also with missiles, some of which have nuclear warheads.

Arnhem, battle of (September 1944), battle in Holland in World War II. Parachutists of the 1st Allied Airborne Division (British, US, Polish) were dropped in an attempt to capture key bridges over the Lower Rhine to enable the Allied armies to advance more rapidly into Germany. The attempt failed, with 7,000 casualties. German units blocked the path of Allied divisions which were attempting to reach and reinforce the airborne troops.

Arthur, Chester A(lan) (1830–86), twenty-first President (1881–5) of the USA. Born in Vermont, he campaigned for the abolition of slavery, while helping to organize his state's *Republican Party. In 1871 he became Collector of the Port of New York, one of the most lucrative sources of political patronage. Caught in the cross-fire of the self-styled *'Stalwarts' and 'Half Breeds' (Republican reformers) over the *spoils system during the presidency of *Hayes, he was, nevertheless, nominated as James *Garfield's running mate in the election of 1880. He became President when Garfield was assassinated in office. Arthur repudiated the spoils system, and supported the *Pendleton Act, which reformed the civil services. He failed to secure re-nomination in 1884.

Army

a) Riflemen
*c.*1810 (British)

b) Cavalry
*c.*1855 (British)

c) Field-gun
*c.*1864 (US Gatling)

d) Field-gun
*c.*1917 (French long-range Filloux)

e) Tank
*c.*1941 (German PZkpfw III)

f) Mobile anti-aircraft missile launcher
*c.*1982 (British Rapier)

Sketch (a) is of British riflemen in the Peninsular War, armed with muzzle-loading flintlocks. Breach-loading rifles (Enfields) were first issued to British infantry in 1842. Mobility was provided by cavalry (sketch (b)), armed with sabre and sword, armies only slowly adapting to motorized fighting after World War I, with the development of the tank (e). Artillery field-guns (c) and (d) developed rapidly in the second half of the 19th century as the armaments industry grew, while mobile launchers (f) were used from the 1960s onwards in Vietnam, the Falklands, and Afghanistan.

Artigas, José Gervasio (1764–1850), national hero of Uruguay. He led the Uruguyan movement for independence from Spain during the years 1811–13 and maintained this in the face of the territorial ambitions of Argentina in 1814. Uruguay also had to contend with Portuguese expansionists from Brazil, and Portuguese troops captured Montevideo in 1817. Artigas was unable to dislodge them. He conducted guerrilla warfare against them for three years but in 1820 was forced to retreat to Argentina, and never returned to Uruguay.

Artois, Charles, Comte d' *Charles X.

Arusha Declaration (1967), a major policy statement by President Nyerere of *Tanzania. The text was agreed by the executive of the political party TANU (Tanganyika African National Union) and proposed that TANU implement a socialist programme by which the major means of production would be placed under the collective ownership of the farmers and workers of the country. No party member would be allowed more than one salary or to own more than one house, nor any capitalist stocks and shares. Banks were nationalized, followed by large industrial and insurance companies, as well as the larger trading firms. Nyerere was deeply committed to the concept of *ujamaa*, which sees all land and natural resources as belonging to the people within their village communities, and following the declaration there emerged many farm collectives. The policy was moderated after 1977 to allow some private investment.

Arya Samaj, a Hindu reform movement. Founded in 1875 by Swami Dayananda Saraswati (c.1825–83), it appealed to the authority of the *Vedas* in support of programmes of social reform and education. Its supporters, such as Lala Rajpat Rai, were prominent in political movements opposed to British rule in the Punjab, and their activities aggravated Hindu relations with Sikhs and Muslims.

Asante (Ashanti), the largest and most prestigious of Ghana chiefdoms. In 1807 the Asante occupied Fanti coastal territory. Following the British abolition of the slave trade they fought the British between 1824 and 1831, and again in 1874, when *Wolseley took and burned Kumasi, the Asante capital. Further troubles (1895–6) ended in the establishment of a Protectorate and the exile of Asantehene *Prempeh I, and in 1901 Britain annexed the country. In 1924 Prempeh was allowed to return and an Asante Confederacy Council was set up in 1935 as an organ of local government, the Asantehene being head. In that year the Golden Stool, symbol of the soul of the Asante people, was restored to Kumasi.

Ashley, William Henry (1778–1838), US fur-trader and politician. In 1822 he founded the Rocky Mountain Fur Company to develop the fur trade in the far west of North America. Between 1822 and 1826 he organized expeditions across the Rockies by the South Pass, penetrating to Great Salt Lake and the Green River valley, opening up rich fur areas and the route to be followed by later settlers travelling to Oregon. Instead of using the trading forts he introduced annual meetings of fur traders, the first being held in Green River, in 1825. The system revolutionized the fur trade and in two years made him a

'The First Day of the Yam Custom' celebrated among the **Asante** people of Ghana; a detail from a 19th-century painting by Edward Bowdich depicting the British mission to the kingdom of King Osei Bonsu (c.1800–24). As monarch, he established a strong, centralized state with an efficient network of communications.

private fortune. He retired from trade in 1827 and devoted the rest of his life to politics, being elected to the House of Representatives in 1831.

Asquith, Herbert Henry, 1st Earl of Oxford and Asquith (1852–1928), British statesman, Liberal Prime Minister (1908–16). He served as Home Secretary (1892–5) and in 1905 joined the government of *Campbell-Bannerman as Chancellor of the Exchequer. He introduced three skilful budgets, the third setting up Old Age Pensions, and he supported other important social legislation such as the abolition of sweatshops and the establishment of labour exchanges. When Campbell-Bannerman fell ill (April 1908) Asquith became Prime Minister, supporting *Lloyd George in his fight for the People's Budget and the creation of the *National Insurance scheme of 1911. Other important legislation included the *Parliament Act (1911) and an Act to pay Members of Parliament. The later years of his ministry were beset with industrial unrest (*Tonypandy) and violence in parts of *Ireland over his *Home Rule Bill. The Bill to disestablish the Anglican Church in *Wales provoked much hostility before being passed. In 1915 he formed a coalition government with the Conservatives, but in the conduct of World War I he was too detached to provide dynamic leadership. Discontent grew and in 1916 *Lloyd George displaced him. The division in the Liberal Party between his supporters and those of Lloyd George lasted until 1926, when Asquith resigned the leadership of the party.

Association of South-East Asian Nations (ASEAN), a regional organization formed by Indonesia, Malaysia, the Philippines, Singapore, and Thailand through the Bangkok Declaration of 1967. Brunei joined the organization in 1984. Although ASEAN has aimed to accelerate economic growth, its main success has been in the promotion of diplomatic collaboration over such matters as post-1979 Vietnamese occupation of *Kampuchea, and policies towards the super-powers. It has facilitated exchange of administrative and cultural resources and co-operation in transport and communication.

Astor, John Jacob (1763–1848), US fur trader and financier. Emigrating from Germany in 1779, he worked with his brother, a maker of musical instruments in London, until 1783, when he set sail for North America. He entered the American fur trade and by 1800 had established the beginnings of a commercial empire, with chartered ships plying both the Atlantic and the Pacific. His American Fur Company, formed in 1808, dominated the fur trade in the prairies and mountains within a decade. In 1834 he sold his interest in the fur trade and spent his remaining years managing his highly profitable property holdings.

Atatürk, Mustafa Kemal (Turkish, 'Father of the Turks', 1881–1938), founder of modern Turkey. An Ottoman officer, he distinguished himself during World War I. In May 1919 he was appointed inspector-general of the 9th Army in Samsun, Anatolia, and organized Turkish resistance to the proposed *Versailles Peace Settlement for the *Ottoman empire. The defeat of Greek forces in 1922 paved the way for the recognition of Turkey's independence at Lausanne (1923), the abolition of the sultanate, the establishment of the republic (1923),

and the abolition (1924) of the caliphate (the temporal and spiritual leadership of the Muslim community). As first President of the republic (1923–38) Atatürk defined the principles of the state in the so-called six arrows of Kemalism: republicanism, nationalism, populism, statism, secularism, and revolution. His policies involved a rejection of the *Islamic past and the creation of a secular Turkish state over which he ruled until his death.

Atkinson, Sir Harry Albert (1831–92), New Zealand statesman. A pioneer farmer, he served in the national Parliament and in various ministries for nearly thirty years. He was Premier, 1876–7, 1883, and 1887–90. Though generally representing the conservative rural interest in matters such as land tax and retrenchment of government spending, he was also noted for his advocacy of radical measures such as a national insurance scheme, leasehold tenure of crown land, proportional representation, and abolition of plural voting.

Atlantic, battle of the, the name given to a succession of sea-operations in World War II. They took place in the Atlantic, the Caribbean, and northern European waters

Turkish leader Mustafa Kemal **Atatürk** photographed in 1923 with his wife Latifeh Hanoum, a year after their marriage. One of Atatürk's later reforms was to abolish traditional Turkish costume.

and involved both submarine blockades and attacks on Allied shipping. German U-boats, sometimes assisted by Italian submarines, were the main weapon of attack, but aircraft and surface raiders also participated. About 2,800 Allied, mainly British, merchant ships were lost, placing the Allies in a critical situation. After summer 1943, with the introduction of better radar, the provision of long-distance aircraft and of escort carriers, and the breaking of German codes, the situation eased, although technical innovations subsequently increased the U-boats' effectiveness. It was only the capture of their bases by Allied land forces in 1944 that finally put an end to the threat.

Atlantic Charter, a joint declaration of principles to guide a post-World War II peace settlement. It resulted from a meeting at sea between *Churchill and F. D. *Roosevelt on 14 August 1941. It stipulated freely chosen governments, free trade, freedom of the seas, and disarmament of current aggressor states, and it condemned territorial changes made against the wishes of local populations. A renunciation of territorial ambitions on the part of Britain and the USA was also prominent. In the following month other states fighting the *Axis Powers, including the USSR, declared their support for these principles. The Atlantic Charter provided the ideological base for the *United Nations Organization.

Attlee, Clement Richard, 1st Earl Attlee (1883–1967), British statesman. He was successively a lawyer, a social worker, and a university lecturer before entering politics, becoming a Labour Member of Parliament in 1922. He served in the government of Ramsay *Mac-Donald, and in 1935 succeeded George *Lansbury as leader of the Labour Party. During World War II he served in the government of Winston *Churchill and was Deputy Prime Minister (1942–5). He was Prime Minister 1945–51, during which time his two Labour ministries had to face many post-war problems, while also extending the *Welfare State. He was defeated in the 1951 election and accepted an earldom in 1955.

Attlee ministries (1945–51), Labour governments in Britain led by Clement *Attlee. To the surprise of many, the Labour Party easily won the general election of 1945 against the Conservative Party led by Winston *Churchill. Despite a war debt of $20,000 million and severe fiscal difficulties, the government embarked on an economic and social reform programme advocated by J. M. *Keynes. It implemented the *Beveridge Report of 1942 through the creation of a *Welfare State, supported by a policy of full employment. The National Insurance Act of 1946 introduced the National Health Service, a free medical service financed from general taxation, and the extension of *national insurance to the entire adult population. Public ownership was extended, the Bank of England was nationalized, as were key industries and services, such as gas, coal, and railways. A full-employment policy was vigorously pursued through the relocation of industry, and the wartime policy of subsidizing agriculture was continued. The economic stability of the country was underpinned by the international agreements reached at the *Bretton Woods Conference. The powers of the House of *Lords were further reduced by the Parliament Act of 1946. The process of decolonization began with the granting of independence to India and Pakistan (1947), as well as to Burma, while British withdrawal from Palestine

Clement **Attlee** campaigning during the 1945 general election which brought the Labour Party to power in Britain. Himself the son of a prosperous lawyer, the wartime Deputy Prime Minister is seen addressing constituents outside flats in Limehouse, the impoverished London constituency he had represented in Parliament since 1922.

allowed the creation of Israel (1948). In 1949, with the *Cold War at its height, Britain helped to form *NATO. The second ministry, following an election in February 1950, had a smaller majority. At home it faced fierce opposition in its attempts to nationalize the steel industry, while entry into the *Korean War necessitated increased rearmament. The Festival of Britain in the summer of 1951 encouraged a sense of optimism in the future of the nation, but it did not prevent Labour from losing the election to the Conservatives in October.

Auchinleck, Sir Claude John Eyre (1884–1981), British field-marshal. He served with distinction in World War I. He commanded the land forces at Narvik in the ineffectual Norwegian campaign in April–May 1940, was commander-in-chief in India (1940–1) and from mid-1941 he commanded in North Africa. He led the advance in Libya, but was driven back by stronger German forces in 1942. When *Tobruk surrendered, he took personal command of the troops, establishing the key defensive line at El *Alamein. Churchill then replaced him with *Montgomery and he returned to India as commander-in-chief.

Aung San (1914–47), Burmese nationalist leader. Head of the pre-war *Dobama Asiayone* (We Burmans' Association), whose members took the title of *Thakin* ('lord'), Aung San first achieved prominence as an organizer of the student strike in Rangoon in 1936. After a period of secret military training under the Japanese he returned to Burma in 1942 and became the leader of the Japanese-sponsored Burma National Army which defected to the Allies in the closing weeks of the Pacific War. As head of the Anti-Fascist People's Freedom League, he led the post-war Council of Ministers, and, in January 1947, negotiated a promise of full self-government from the British, but on 19 July 1947 he and six of his colleagues were assassinated at the behest of a political rival, U Saw.

Auriol, Vincent (1884–1966), French statesman. First elected as a Socialist Deputy in 1914, he served as Minister of Finance (1936) and Minister of Justice (1938). During World War II he spent two years in internment before escaping to join General *de Gaulle and the *Free French in Britain. After the liberation he served in de Gaulle's brief government (1945–6) and then played an active part in the formation of the Fourth Republic, being elected its first President (1947–53).

Ausgleich (1867; German, 'compromise'), a constitutional compromise between Hungary and the *Austrian empire following the defeat of Austria in Italy and Germany. It was drawn up by Francis *Deák, and ratified by the Austrian emperor *Francis Joseph, granting Hungary its own parliament and constitution but retaining Francis Joseph as King of Hungary. A dual monarchy, the *Austro-Hungarian empire, was created, in which the Magyars were permitted to dominate their subject peoples, and the Austrians the remaining seventeen provinces of the empire.

Austerlitz, battle of (2 December 1805), fought by Austria and Russia against France, near the town of Austerlitz in Moravia. Alexander I of Russia persuaded *Francis I of Austria to attack before Russian reinforcements arrived. Their complicated plan to encircle the French allowed *Napoleon to split their army and defeat each half. It was a decisive battle; the Russian army had to withdraw all troops from Austria, and Austria signed the Treaty of Pressburg (1805), in which it recognized Napoleon as King of Italy, and ceded considerable territories in northern Italy, the Alpine regions, and on the Adriatic coast.

Australia, an island continent in the south-west Pacific. It was first inhabited by *Aborigines thought to have migrated from south-east Asia c.50,000–40,000 years ago. Although the first known European discoveries of the continent were those made in the early 17th century, there may have been earlier Portuguese discoveries. It was visited by an Englishman, William Dampier, in 1688 and 1699. Captain James Cook claimed British possession of the eastern part of the continent in 1770, naming it New South Wales. The British penal colony of New South Wales was founded in 1788. Immigration of free settlers from 1820 onwards aided the colony's development, as did exploration, which opened pastures for the wool industry. *Squatter settlement of much of eastern Australia led to conflict with the Aborigines, resulting in events such as the *Myall Creek Massacre (1838). Van Diemen's Land, settled in 1803, became a separate colony in 1825. Moreton Bay, founded as a *penal settlement in 1824, became the colony of Queensland in 1859. The colony of Western Australia was founded in 1829. The Port Phillip District, settled illegally in 1834, became

the colony of Victoria in 1851. South Australia, founded as a province in 1834, became a crown colony in 1842. All of the colonies except Western Australia were granted responsible government during the 1850s. The *gold rushes of the 1850s and 1860s brought many changes. The *White Australia Policy can be traced back to that period. Demands for land to be opened for *selectors increased. Western Australia, granted responsible government in 1890, developed more slowly than the other colonies. The colonies federated as self-governing states, becoming the Commonwealth of Australia in 1901.

Australia, Commonwealth of, the official title of the Australian nation. It comprises the six states (formerly colonies), which are New South Wales, Queensland, South Australia, Tasmania, Victoria, and Western Australia, together with the Northern Territory, and the Australian Capital Territory. The Commonwealth of Australia was inaugurated on 1 January 1901, with powers distributed between the Commonwealth and state governments, and with the crown through its representative, the governor-general, retaining (until 1931) overall responsibility for defence and foreign affairs. State legislators would have full responsibility for internal state affairs. *Barton, who had been prominent in the federation movement, was the first Prime Minister. The Northern Territory was transferred from South Australia to the Commonwealth in 1911. In the same year land was transferred to the Commonwealth from New South Wales, for the creation of the Australian Capital Territory, Canberra. (Jervis Bay was added to the Australian Capital Territory in 1915.) The Commonwealth Parliament met in Melbourne until 1927, when it was transferred to Canberra. In the 1930s reserves were established for the *Aborigines, and in 1981 the Pitjantjara Aborigines were granted freehold titles to land in Southern Australia. Australia fought with the Allies in both *World Wars and with the USA in *Vietnam. After World War II ties with Britain diminished, and Australia joined the *ANZUS and *SEATO powers. The Labor government of the 1970s and 1980s strengthened trade ties with the non-communist Far East, but a deteriorating economy in the 1980s led to labour unrest.

Australian federal movement (1890–1900), a movement to seek federation of the six Australian colonies and (initially) New Zealand. Two pressure groups for federation were the Australian Natives' Association and the Australasian Federation League. In 1889 the six Australian colonies and New Zealand agreed to send delegates to a federal conference in Melbourne in 1890. It was decided to hold a full convention the following year in Parliament House, Sydney, at which a draft constitution was drawn up. A second convention was held in 1893 in the small New South Wales town of Corowa convened by the Australian Natives' Association, at which it was proposed that a national referendum be held. By now indifference had increased in New Zealand as its trade shifted to Britain with the invention of refrigeration. In Australia too, opposition to New Zealand's participation was growing, and plans for the federation were dropped. In 1895 the Premiers of the six Australian colonies met to reconsider the draft constitution drawn up in 1891. New Zealand was now excluded. Ten delegates from each colony were chosen (elected by the people, except in the case of Western Australia) and the Australian Federation

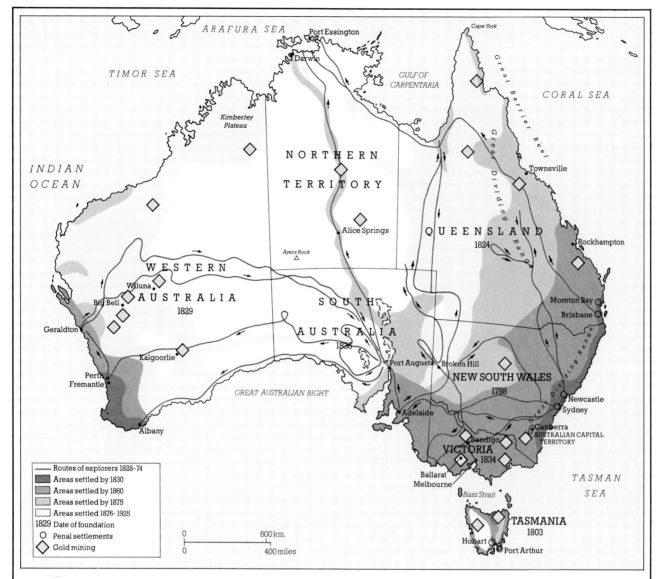

Australia

After explorers crossed the continent in the mid-19th century, white settlers gradually spread into the interior of Australia. Later, crossings west–east were achieved, with telegraph linking South and Western Australia, which became one of the world's major wheat producers.

Federation on a national scale was endorsed by popular referenda in 1898–9, and in 1901 the six colonies became states of the Commonwealth of Australia. Aborigine reserves were established in the 1930s, mostly in the Federal North Territories.

Convention first met in Adelaide (March 1897). It was agreed that a referendum should be held. It met again in January 1898, when after much compromise a proposed constitution was agreed. The first referendum failed. The second (held in 1899 after amendments were made) passed in all colonies and was given royal assent in 1900, the Commonwealth of *Australia coming into being on 1 January 1901.

Austria, a country in central Europe. The unification of Germany by Prussia in 1866–71 excluded the *Austrian empire from German affairs and destroyed hopes for the creation of a union of all the German-speaking peoples. Austria was forced to make concessions to the Hungarians by forming the *Austro-Hungarian empire. Austrian diplomats, however, retained links with the new *German Second empire, and tried to gain German support for

their ambitions against Russia in the Balkans through an alliance system. During World War I the Austrian Imperial Army was virtually under German military control. Defeat and revolution destroyed the monarchy in 1918, and the first Austrian republic which followed it was only a rump of the former state. This was destabilized by the Nazis, who in 1934 murdered *Dolfuss and attempted a coup. They were more successful in achieving *Anschluss in 1938, when Hitler's army invaded the country without opposition. Defeated in World War II, Austria was invaded by Soviet troops, and divided into separate occupation zones, each controlled by an Allied Power. In 1955 a treaty between the Allies and Austria restored full sovereignty to the country. The treaty prohibited the possession of major offensive weapons and required Austria to pay heavy reparations to the USSR, as well as to give assurances that it would ally itself with neither East nor West Germany, nor restore the Habsburgs. It has remained neutral, democratic, and increasingly prosperous under a series of socialist regimes.

Austrian empire (1806–67), those territories and peoples from whom the Habsburg emperors in Vienna demanded allegiance. Following the dissolution of the Holy Roman Empire (1806) Emperor Francis II continued to rule as *Francis I (1804–35), Emperor of Austria and of the hereditary Habsburg lands of Bohemia, Hungary, Croatia and Transylvania, Galicia (once a province of Poland), and much of northern Italy (Venetia and Lombardy). He ruled by means of a large bureaucracy, a loyal army, the Roman Catholic Church, and an elaborate police force. His chief minister was Chancellor *Metternich. Nationalist feelings were emerging, and during the reign of his successor Ferdinand I (1835–48), liberal agitation for reform developed. Vienna was becoming rapidly industrialized and in March 1848, at a time of economic depression, riots in the capital led to Metternich's resignation. The emperor abolished censorship and promised a constitution. This, published in April, was not democratic enough for radical leaders, who organized a popular protest on 15 May 1848. The emperor fled to Innsbruck and later abdicated. His 18-year-old nephew *Francis Joseph succeeded. There were movements for independence among all the peoples of the empire, including the Hungarians led by *Kossuth, the Czechs, Slovaks, Serbs, Croats, Romanians, and Italians. A *Pan-Slav conference met (1848) in Prague. But the opposition to the government in Vienna was divided and the Prime Minister, *Schwarzenberg and Francis Joseph were able to regain control. The army crushed the reform movements in Prague and Vienna and with the help of Russia, subjugated Budapest. Alexander Bach, the new Minister of the Interior, greatly strengthened the centralized bureaucracy, and the empire regained some stability, until its defeat by France and Piedmont at *Magenta and *Solferino, which ended Austrian rule in Italy. In an effort to appease nationalist feeling the emperor proposed a new federal constitution, but it came too late and after a further defeat at *Sadowa he agreed to the *Ausgleich (Compromise) of 1867 and the creation of the *Austro-Hungarian empire.

Austro-Hungarian empire (Dual Monarchy), the Habsburg monarchy between 1867 and 1919. Following Austria's defeat by Prussia (1866) Francis Joseph, the Austrian emperor, realized that Austria's future lay along the Danube and into the Balkans. Before any such expansion could occur, the differences between the two dominating peoples in his empire, Germans and Hungarians, had to be overcome. By the *Augsleich or Compromise of 1867 Austria and Hungary became autonomous states under a common sovereign. Each had its own parliament to control internal affairs: foreign policy, war, and finance were decided by common ministers. The dualist system came under increasing pressure from the other subject nations; in Hungary there was constant friction with the Croatians, Serbs, Slovaks, and Romanians (52 per cent of the population). The Czechs of *Bohemia–Moravia resented the German-speaking government in Vienna, and found a potent advocate for Czech independence in Tomas *Masaryk. *Bosnia and Hercegovina, formally annexed in

Vienna, capital of the **Austrian empire**, in flames during the 1848 Revolution. The imperial army under Prince Alfred Windischgrätz besieged the rebels within the city and its artillery fire set their headquarters ablaze. (Herresgeschichtliches Museum, Vienna)

1908, developed a strong Serbian nationalist movement, and the failure to resolve nationalist aspirations within the empire was one of the main causes of World War I. After the death of Francis Joseph (1916) his successor Charles promised constitutional reforms, but the Allies gave their support to the emergent nations and the Austro-Hungarian empire was finally dissolved by the *Versailles Peace Settlement.

Austro-Prussian War (June–August 1866), a war fought between Prussia, allied with Italy, and Austria, allied with Bavaria and other, smaller German states. War had become inevitable after *Bismarck challenged Austria's supremacy in the *German Confederation. Hostilities finally broke out when Bismarck, having gained France's neutrality and the support of Italy, proposed that the German Confederation should be abolished. Prussian troops forced the Austrians out of Schleswig-Holstein, but the Austrians defeated the Italian army at Custozza. However, the Prussian army, better trained and equipped, crushed the main Austrian army at *Sadowa. Seven weeks later the Austrians signed the Treaty of Prague, by which the German Confederation was dissolved. Austria ceded Venetia to Italy, while Prussia annexed the smaller states into the new North German Confederation. Austria, excluded from its territories in the south and from political influence to the north, turned towards the east, accepting the Hungarian *Ausgleich and forming the *Austro-Hungarian empire.

automation, a term applied to automatic performance and control of mechanical and manufacturing processes. During the 19th century a number of machines such as looms and lathes became increasingly self-regulating. At the same time transfer-machines were developed, whereby a series of machine-tools, each doing one operation automatically, became linked in a continuous production line by mechanical devices transferring components from one operation to the next. Early control devices for such systems were pneumatic or hydraulic, but in the early 20th century, with the development of electrical devices and time-switches, more processes became automatically controlled, and a number of basic industries such as oil-refining, chemicals, and food-processing were increasingly automated. The development of computers since World War II has enabled more sophisticated automation to be used in manufacturing industries, for example in iron and steel production, electricity supply, and motor manufacture, and has resulted in a reduction in the number of workers employed. From the 1960s onwards many office and banking procedures have been computer-automated resulting in the greatly increasing speed at which information can be processed.

Awami League, political party in East Pakistan. It was founded in 1952 as the Jinnah Awami Muslim League by H. S. Suhrawardy, although it existed informally before that date. It was renamed the Awami League under pressure from its East Bengal leader, Maulana Abdul Hamid Bhashani, who left the party in 1957 to form the National Awami Party. During the 1960s the Awami League grew rapidly under Sheikh Mujibur Rahman, who succeeded Suhrawardy as leader and in 1970 won a majority, completely dominating East Pakistan, which became *Bangladesh in December 1971. In August 1975 the Awami League was disbanded with other political parties. It was later re-formed and became the largest opposition party in Bangladesh.

Axis Powers, an alliance of fascist states fighting with Germany during *World War II. The term was used in an agreement (October 1936) between Hitler and Mussolini proclaiming the creation of a Rome-Berlin 'axis round which all European states can also assemble'. Japan joined the coalition on signing the *Anti-Comintern Pact (November 1936). A full military and political alliance between Germany and Italy (the Pact of Steel) followed in 1939. The Tripartite Pact between the three powers in 1940 cemented the alliance, and, by subsequently joining it, Hungary, Romania, and Bulgaria, as well as the Nazi-created states of Slovakia and Croatia, became members.

Ayub Khan, Muhammad (1907–74), military leader and President of Pakistan (1958–69). A Pathan from the Hazara district, he was a professional soldier who, when the state of Pakistan was created (1947), assumed command of military forces in East Pakistan (now Bangladesh). He was appointed commander-in-chief of the Pakistan Army in 1951, Minister of Defence (1954–6), and Chief Martial Law Administrator after the 1958 military coup. For the next ten years he ruled Pakistan as President, pursuing a policy of rapid economic growth, modest land reform, and restricted political life through 'basic democracies', introducing Pakistan's second constitution in 1962. In March 1968 he suffered a serious illness and thereafter lost political control, being replaced in March 1969 by General Yahya Khan.

Azikiwe, (Benjamin) Nnamdi (1904–), Nigerian statesman, first President of *Nigeria (1963–6). An Ibo (Igbo), he founded the National Council of Nigeria and the Cameroons (NCNC) and exerted a strong influence throughout the 1940s on emerging Nigerian nationalism. He held a number of political posts before becoming the first governor-general of Nigeria (1960–3). He was deposed by a military coup (1966), which ousted the civilian government, but remained leader of the Nigerian People's Party and was a member of the Nigerian Council of State (1979–83).

Automation in the production-line assembly of motor vehicles is exemplified by the use of computer-controlled robots. Robot welders such as these, each programmed to assemble a particular part of a car body, require minimal human supervision, thereby reducing the overall size of the workforce.

B

Baader–Meinhof gang, byname of the West German anarchist terrorist group, Red Army Faction. Its leaders were Andreas Baader (1943-77) and Ulrike Meinhof (1934-76). The group set itself to oppose the capitalist organization of German society and the presence of US armed forces by engaging in murders, bombings, and kidnappings. The leaders were arrested in 1972, and their trial and deaths (by suicide) received considerable publicity. The group continued its terrorist activities in the 1980s, forming a number of splinter cells.

Ba'athism, the doctrines of an Arab political party, the Ba'ath or Renaissance Party. Founded in Syria in 1943 by Michel Aflaq and Salah al-Din al-Bitar, it pursues a policy of Arab nationalism, unity, and socialism. The Ba'ath took power in Syria in 1963 and in Iraq in 1968, but it became divided between its Syrian and Iraqi wings and between its civilian and military elements.

Bábism, the doctrines of a Muslim messianic Shiite sect. Founded in 1844 by the Persian Sayyid Ali Muhammad of Shiraz (1819-50) known as the *Báb ed-Din* (the gate or intermediary between man and God), who declared himself to be the long-awaited Mahdi. For inciting insurrection the Báb was arrested in 1848 by the government and executed in 1850, his remains being interred (1909) on Mt Carmel, Palestine. In 1863 Baha'ullah and his son Abdul Baha declared themselves the new leaders, and their followers became knows as the *Baha'is.

Baden-Powell, Robert Stephenson Smyth Baden-Powell, 1st Baron (1857-1941), British soldier and founder of the Boy Scout movement (1907). As a soldier he had served in India and Afghanistan, and was the successful defender of Mafeking in the Second *Boer War. His knowledge of the skill of gaining information about hostile territory was fundamental to the teachings of the Scout movement. Training in self-reliance and a code of moral conduct were to be the hallmarks of the movement. By the end of World War I the movement was developing on an international scale. With his sister, Agnes, he founded the Girl Guide movement (1910).

Badoglio, Pietro (1871-1956), Italian general and Prime Minister. By 1925 he was chief of staff; Mussolini appointed him governor of Libya (1929) and sent him (1935) to rescue the faltering Italian campaign in *Ethiopia. He soon captured Addis Ababa and became governor. When leading Italians in 1943 decided that Mussolini should be deposed, it was he who was chosen to head the new non-fascist government. He made peace with the advancing Allies, declared war against Germany, but resigned soon afterwards.

Bagehot, Walter (1826-77), British economist, journalist, and man of letters. He was joint editor of the *National Review* and later editor of *The Economist* (1860-77), which he made a journal of international importance. Bagehot is best known for his book *The English Constitution* (1867), in which he applied a rigorous analysis to his country's political system in order to distinguish between the realities of power which he saw as lying in the cabinet and the House of Commons, and its formal trappings, which he saw as the crown and the House of Lords.

Baghdad Pact *Central Treaty Organization.

Baha'ism, a monotheistic religion founded by the Persian religious leader Baha'ullah (Splendour of God) (1817-92) as a branch of *Bábism. The central tenet of the Baha'i faith is that the essence of all religions is one. Its quest is for the general peace of all mankind, whose unification it regards as necessary and inevitable. Its governing body is centred in Haifa in Israel. The Baha'is have been subject to persecution by orthodox Shiites, especially in Iran after the Islamic revolution in 1978.

Bahrain, a sheikhdom consisting of a group of islands in the Persian Gulf. Iran, which ruled Bahrain from 1602 to 1783, was expelled by the al-Khalifas, who still reign. British political control dates from 1820. Oil was discovered in 1932, when the Bahrain National Oil Company was formed. After the withdrawal of Britain in 1971 and the abandonment by Iran of its claims, the country joined the Arab League. Tension between Shiite and Sunni communities increased, leading to the suspension of the national Assembly in 1975. Together with other members of the Gulf Co-operation Council (Saudi Arabia, Kuwait, Qatar, and Oman), Bahrain has repeatedly called for an end to the *Iran-Iraq War.

Baker, Sir Samuel White (1821-93), British explorer who traced the Nile tributaries in Ethiopia (1861-2). In 1864, despite the opposition of Arab slave traders, he located the Nile source in Lake Albert Nyanza. Khedive Ismail of Egypt sent him to put down the slave trade (1869-73), which he attempted by establishing Egyptian protectorates in the present *Uganda and opening the lake areas to commerce. Although he set up a skeletal administration, it was not strong enough to be effective.

Bakunin, Mikhail (1814-76), Russian revolutionary, leading exponent of *anarchism and founder-member of the Russian *Populist Movement. He served in the emperor's army until his dismissal in 1835. After taking part in the *Revolutions of 1848 he was exiled to Siberia. He escaped in 1861 and went to London, which was used as a headquarters for militant anarchists and communists. The first *International Workingmen's Association, founded in 1864, was marred by the conflict between *Marx and Bakunin, who called for violent means to destroy the existing political and social order, splitting the two factions for years to come.

Balaklava, battle of (25 October 1854), an inconclusive battle during the *Crimean War. Following their defeat on the River Alma, the Russians retreated to Sevastopol, which was then besieged by British, French, and Turkish troops, supplied from the small port of Balaklava. Russian forces moved down into the Balaklava plain, where they were met by the British cavalry division under Lord Lucan. Lord Raglan, British commander-in-chief, sent orders to the Light Brigade under Lord Cardigan to 'prevent the enemy carrying away the guns'. It is assumed this order referred to guns on the Vorontsov Heights, but

A contemporary coloured lithograph depicting an incident during the battle of **Balaklava** (1854). The Scots Greys, supported by Enniskillens, come to the aid of other British cavalry, forcing their way through massed Russian troop formations. (National Army Museum, London)

Cardigan understood it to require a direct frontal charge down the valley. Fired on by guns from both flanks and to the front, he and a few dragoons reached the Russian line before retreating. In this 'Charge of the Light Brigade' 247 men were killed out of a force of 673. The Russians failed to capture Balaklava, but they held on to the Vorontsov Heights and thus cut the paved supply-road from the port to the besieging allied forces above Sevastopol.

Baldwin, Robert (1804–58), Canadian statesman. Born in York (renamed Toronto in 1834), he was elected to the Assembly of Upper Canada (*Upper and Lower Canada) in 1829, and became one of the leaders of the campaign for reformed government which would give more say to elected representatives. After the Act of Union (1840) he was elected for Canada West and in 1842–3 led a reformist ministry with *La Fontaine. The *Durham Report had recommended the development of 'responsible government' (responsible to an executive or cabinet dependent on the votes of a majority in the elected legislature), and in 1848 Lord *Elgin was appointed governor-general and instructed to implement this recommendation. Baldwin and *La Fontaine formed a second coalition ministry (1848–51) which was to be based on such parliamentary procedures. It is sometimes called the 'Great Ministry' for its outstanding reformist accomplishments.

Baldwin, Stanley, 1st Earl Baldwin of Bewdley (1867–1947), British statesman. A Conservative Member of Parliament (1908–37), he was a member of *Lloyd George's coalition (1918–22) but led the Conservative rebellion against him. He was Chancellor of the Exchequer under *Bonar Law and was chosen as Prime Minister in preference to Curzon when Law resigned in 1923. He lost the 1923 election in an attempt to introduce tariffs but returned to office in 1924. His premiership was marked by the return to the *gold standard, the *General Strike, Neville *Chamberlain's social legislation, and the Trades Dispute Act of 1927. He lost the 1929 election, but served under Ramsay *Macdonald in the coalition caused by the 1931 crisis, succeeding him as Prime Minister in 1935. His last ministry witnessed the *Abdication Crisis, which he handled skilfully. In 1935 he approved the Hoare-Laval pact which allowed fascist Italy to annex Ethiopia. Although international relations continued to deteriorate with the German occupation of the Rhineland and the outbreak of the *Spanish Civil War, Baldwin opposed demands for rearmament, believing that the public would not support it. He resigned the premiership in 1937.

Balewa, Alhaji Sir Abubakar Tafawa (1912–66), Nigerian statesman. He entered politics in 1946, and became a member of the Central Legislative Council in 1947. The first Prime Minister of the Federation of Nigeria, he retained his office when the country became independent (1960) until he was killed in an army coup. He was a founder and deputy president general of the country's largest political party, the Northern People's Congress.

Balfour, Arthur James, 1st Earl (1848–1930), British statesman, nephew of Lord *Salisbury. As Chief Secretary of Ireland (1887–91) he was an opponent of *Home Rule and earned from the Irish the nickname 'Bloody Balfour'. His premiership of 1902–5 was undermined by his vacillation over tariff reform. His Education Act (1902) established a national system of secondary education. He created a Committee of Imperial Defence (1904), and helped to establish the *entente cordiale (1904) with France. The Conservatives were crushingly defeated in the 1906 general election. Balfour then used the House of Lords, described by Lloyd George as 'Mr Balfour's poodle', to attempt to block contentious Liberal legislation. He resigned the leadership of the Conservative Party in 1911. As Foreign Secretary in Lloyd George's war cabinet, he is associated with the *Balfour Declaration (1917) promising the Jews a national home in Palestine. In the 1920s he supported the cause of *dominion status. The Statute of *Westminster owed much to his inspiration.

Balfour Declaration (2 November 1917), a declaration by Britain in favour of a Jewish national home in Palestine. It took the form of a letter from Lord *Balfour (British Foreign Secretary) to Lord Rothschild, a prominent *Zionist, announcing the support of the British government for the establishment of a national home for the Jewish people in Palestine without prejudice to the civil and religious rights of the non-Jewish peoples of Palestine or the rights and political status of Jews in other countries. The Declaration subsequently formed the basis of the mandate given to Britain for Palestine and of British policy in that country until 1947.

Bali, an island east of Java. A Hindu enclave with strong military traditions, Bali suffered three Dutch expeditions (1846, 1848, and 1849), which established control in the north coast areas, but the rulers in the south and east only submitted after the *prang puputan* (final battles) at Den Pasar, Paměcutan, and Klungkung in 1906 and 1908, when the white-clad Hindu *rajas* and their retainers allowed themselves to be ritually slaughtered by the Dutch forces. Occupied by the Japanese during World War II, Bali became a part of *Indonesia after 1945.

Balkan States, the countries of the Balkan peninsula, an area in south-east Europe. It now includes Albania, continental Greece, Bulgaria, European Turkey, most of Yugoslavia, and south-east Romania. Ruled by the Turks as part of the *Ottoman empire for nearly 500 years, the subject nations largely retained their languages and religions, but suffered periodic persecution at the hands of the Turks. In the 19th century, the grip of Turkey weakened, and increasingly Russia and Austria quarrelled over the gains to be made there. The Balkan League of 1912 was formed to counter Turkish rule in the area and led to the outbreak of the *Balkan Wars. After Serbia's success in the wars, Austria's hostility towards *Pan-Slavism contributed to the outbreak of World War I. As a result of the *Versailles Settlement frontiers were re-drawn and attempts made to introduce democratic government. These failed and authoritarian regimes emerged in a majority of the states between the wars. The Balkan Pact of 1934 sought to unify the countries by a non-aggression treaty and guarantees of the Balkan frontiers. Since 1945 the states have varied in their allegiance between Soviet and Western politics. A second Balkan Pact between Yugoslavia, Greece, and Turkey was concluded in 1954, which provided for common military assistance in the face of aggression.

Balkan Wars (1912-13), two short wars, fought between Serbia, Montenegro, Greece, Romania, Turkey, and Bulgaria for the possession of remaining European ter-

ritories of the *Ottoman empire. In 1912 Greece, Serbia, Bulgaria, and Montenegro formed the Balkan League; officially to demand better treatment for Christians in Turkish Macedonia, in reality to seize the remaining Turkish territory in Europe while Turkey was embroiled in a war with Italy. In October 1912 the League armies captured all but Constantinople (now Istanbul). European ambassadors intervened to re-draw the Balkans map to the advantage of Bulgaria and detriment of Serbia in the Treaty of London (May 1913). A month later, Bulgaria launched a pre-emptive attack on the Serbs and Greeks, who coveted Bulgaria's gains, but was defeated. In the Treaty of Bucharest (August 1913) Greece and Serbia partitioned Macedonia, and Romania gained part of Bulgaria. Albania, which had been under Turkish suzerainty, was made an independent Muslim principality. A 'big Serbia' now presented a considerable threat to Austro-Hungary. Russia promised to support Serbia in its nationalist struggle, and Germany to give military aid to Austro-Hungary. The assassination of the Austrian archduke at Sarajevo (1914) gave Austro-Hungary the pretext to invade Serbia, leading to the outbreak of World War I six weeks later.

Ballance, John (1839-93), New Zealand statesman. In the 1890 elections the Liberals emerged as the first party in New Zealand politics, lacking organization but broadly united on a programme of radical reform, and Ballance became Premier (1891-3). His strong and able cabinet introduced graduated land taxes to break up large estates, new leasehold tenures, and new labour regulations. Ballance and his Liberal successors established the tradition of using the state to regulate the economy and protect poorer groups, laying the foundations of basic stability in New Zealand society.

Balmaceda, José Manuel (1840-91), Chilean statesman. He was first elected to the Chilean Congress as a Liberal in 1864. As leader of the anti-clerical group he was sent to Argentina in 1878 to persuade that country not to enter the War of the *Pacific. He became a member of the cabinet of President Santa Maria (1881-6) and himself was then elected President (1886-91). Chile was experiencing rapid economic expansion with both the copper and nitrate industries booming. Balmaceda instituted wide reforms and a large public works programme. Despite national prosperity, however, tension arose between President and Congress, which Balmaceda increasingly ignored. In January 1891 this resulted in civil war. Balmaceda took refuge in the Argentinian Embassy, where, rather than face trial, he shot himself.

Baltic States, *Finland and the formerly independent republics of Latvia, Lithuania, and Estonia, now constituent republics of the Soviet Union. In the 19th century the Baltic states were under Russian hegemony, with a ruling mercantile class, notably in Estonia, of German or Jewish origin. In Finland, the publication (1835) of the *Kalevala* epic heralded a nationalist revival that spread to every Baltic state. During World War I Estonia, Latvia, and Lithuania were occupied by the Germans, who ruled them through puppet regimes. After Germany's collapse (1918) the Soviet Union attempted to recover the states, but with Allied and German aid, independent governments were established. In 1939 Latvia and Estonia concluded a mutual non-aggression pact with Germany, while Lithuania, having lost (1938) Memel to Germany, made efforts to draw closer to Poland. In September and October 1939 the Soviet government concluded treaties with Estonia and Latvia, allowing Soviet naval and air bases on their territories, while a Soviet–Lithuanian mutual assistance pact (October 1939) allowed Russia the right to occupy stations of military importance there. The Finnish government rejected similar demands, and suffered a military invasion by Russia in the *Finnish–Russian War which forced it (1940) to cede its eastern territories to the Soviet Union. As a result of the *Nazi–Soviet Pact (1939), Estonia, Latvia, and Lithuania were incorporated (1940) in the Soviet Union. When Germany reneged on the Nazi–Soviet Pact, its army invaded all three states. During Nazi occupation (1941–4) the Jewish minorities in these three Baltic states were largely exterminated. Latvia, Lithuania, and Estonia were retaken by Soviet forces in 1944, and integrated into the Soviet Union, while Finland retained independence and a status of neutrality.

Bandaranaike, S(olomon) W(est) R(idgeway) D(ias) (1899–1959), Sinhalese statesman. He formed the Maha Sinhala Party in the 1920s. In 1931 he was elected to the new State Council and after independence he assumed ministerial power. In 1952 he founded the Sri Lanka Freedom Party, which was the leading partner in the coalition which won the 1956 elections, attracting

Mr and Mrs **Bandaranaike** on arrival in New York in November 1956 where the Sinhalese Prime Minister was to attend the United Nations General Assembly.

President Sukarno of Indonesia at the rostrum addressing the opening session of the **Bandung Conference** in April 1955. Among the delegates was India's Prime Minister Nehru (left).

left-wing and Buddhist support. As Prime Minister (1956–9) Bandaranaike pursued a policy of promoting the Sinhalese language, Buddhism, socialism, and neutrality. His policy alienated the Tamils. After his assassination in September 1959 by a dissident Buddhist monk, his widow, Mrs Sirimavo Bandaranaike (1916–), succeeded him as Prime Minister (1960–77).

Bandung Conference (1955), a conference of Asian and African states at Bandung in Java, Indonesia. Organized on the initiative of President *Sukarno and other leaders of the Non-Aligned Movement, the Bandung Conference brought together twenty-nine states in an attempt to form a non-aligned bloc opposed to colonialism and the 'imperialism' of the superpowers. The five principles of non-aggression, respect for sovereignty, non-interference in internal affairs, equality, and peaceful co-existence were adopted, but the subsequent emergence of the non-aligned movement was hamstrung by the deterioration of relations between India and China, and by the conflicting forces set loose by decolonization.

Bangladesh, a predominantly Muslim country of the Indian subcontinent. The state was established in 1971 in the eastern part of Bengal from territories which had previously formed the eastern part of Pakistan. Evidence of discontent in East Pakistan first appeared in the 1952 Bengali-language agitation and became much stronger after the 1965 *Indo-Pakistan war. In 1966 the *Awami League put forward a demand for greater autonomy which it proposed to implement after its victory in the 1970 elections. In March 1971, when this demand was rejected by the military government of Pakistan, civil war began, leading to a massive exodus of refugees to India. India sent help to the East Pakistan guerrillas (the Mukti Bahini). In the war of December 1971, Indian troops defeated the Pakistan forces in East Pakistan. The independence of Bangladesh was proclaimed in 1971, and recognized by Pakistan in 1974.

Bantu *Nguni.

Bantustan, term used to describe 'homelands' reserved for black Africans in the Republic of South Africa. The 'Separate Development Self-Governing Areas' were set up under the Bantu Homelands Constitution Act (1971). This granted, first self-government, and then independence to African territories within the republic; their non-residents automatically becoming citizens of the new states. Transkei, chiefly Xhosa people, became independent in 1976; Bophuthatswana, chiefly Tswana, in 1977; Venda in 1979; and Ciskei in 1981. The territories are not unitary, and for this reason, and because many of their citizens live in South Africa, they cannot be recognized in international law. Only South Africa recognizes them. Each Bantustan has a democratic constitution and a legislative assembly with similar but not identical systems. Overcrowded and undeveloped, they are likely to remain dependent on South Africa for the foreseeable future.

Bao Dai (1913–), emperor of Vietnam (1926–45). His initial aim to reform Vietnam did not receive French colonial support. During World War II he collaborated with the Japanese and in 1945 he was forced to abdicate by the *Vietminh. In 1949 he renounced his title and returned to Saigon as head of the state of Vietnam within the French Union. In 1955, after the partition of Vietnam at the *Geneva Conference, he was once again deposed when power in the new republic of South Vietnam passed to *Ngo Din Diem.

Baptist Church, an Evangelical Christian movement distinguished by its stress on adult baptism and the autonomy of local congregations. In both Britain and the USA Baptist churches grew in the late 18th century. Baptist missionaries first went to India in 1792 and in the 19th century were active all over the world including in Russia. In 1813 the Baptist Union of Great Britain was organized and Baptist churches became very popular, especially in towns. In the USA in 1845, the Southern Baptist Convention was formed and split from the Northern (later National) Convention. Black Baptist Churches grew after 1865 and have contributed significantly to black culture. Baptists constitute the largest Protestant group in the Soviet Union and many ministers have been imprisoned.

Barbary Wars   *Tripolitan War.

Baring, Evelyn   *Cromer, 1st Earl of.

Baring crisis (July 1890), a financial crisis in *Argentina. The London merchant bank of Baring Brothers was the country's financial agent in Europe, where a crisis of confidence occurred over the inflationary policy of President Juarez Celman (1886–90). The President gave way to his deputy Carlos Pellegrini, who had to stabilize the currency and adopt the *gold standard before London would give any more credit. One result was heavy urban unemployment, although the refrigerated beef and corn industries continued to expand.

Barnardo, Thomas John (1845–1905), British doctor and social reformer who founded homes for destitute children. Born in Dublin of a Spanish Protestant family, he moved to London in 1862 and subsequently qualified as a doctor of medicine. He became concerned over the plight of children living in deprivation in the East End of London. In 1870 he established his first home for destitute boys and a similar institution for girls. By the time of his death nearly 60,000 children had been cared for in 'Dr Barnardo's Homes', which still flourish.

A group of destitute children from the East End of London, c.1880. The homes set up by Dr **Barnardo** for such waifs constituted an early attempt to deal with this aspect of urban deprivation.

Barth, Heinrich (1821–65), German explorer and geographer. The most influential of 19th-century European observers of West African life, Barth was a member of a British-sponsored expedition which left Tripoli in 1850 to explore the hinterland to the south. He assumed command of the expedition in northern Nigeria and subsequently explored the area around Lake Chad and Cameroon, stayed six months in Timbuktu, and recrossed the Sahara, before returning to London in 1855. He published a five-volume account of his explorations, *Travels and Discoveries in North and Central Africa* (1857–8).

Barton, Clara (1821–1912), US humanitarian, founder of the American Red Cross. She organized supplies and nursing in army camps and on the battlefields during the American Civil War (1861–65). In 1881 she established the American Red Cross, and successfully campaigned to extend Red Cross relief internationally to calamities not caused by war, such as famines and floods.

Barton, Sir Edmund (1849–1920), Australian statesman and jurist. He held high political offices in New South Wales for most of the 1880s and 1890s. Barton with Parkes and Deakin was an acknowledged leader of the *Australian federation movement. He was the Commonwealth of *Australia's first Prime Minister (1901–3), leading a ministry which sought to protect Australian industry.

Baruch, Bernard (Mannes) (1870–1965), US industrialist and financier. The respected adviser of presidents from Wilson to Eisenhower, he preferred to be an 'eminence grise' than to run for elective office. In World War I he served on the Council of National Defense and was the successful Chairman of the War Industries Board. Between the World Wars he was a member of various presidential conferences on capital, agriculture, transportation, and labour. In the 1940s he acted as special adviser on war mobilization and post-war planning. He was appointed to the UN Atomic Energy Commission, which proposed (1946) a World Atomic Authority with full control over the manufacture of atomic bombs throughout the world; this, however, was rejected by the Soviet Union.

Basutoland *Lesotho.

Batista y Zaldívar, Fulgencio (1901-73), Cuban statesman. He was President of Cuba (1933-44, 1952-9), having come to national prominence in 1933 when, as a sergeant in the army, he led a successful revolt against President Gerardo Machado y Morales. He established a strong, efficient government, but increasingly used terrorist methods to achieve his aims. He amassed fortunes for himself and his associates, and the dictatorial excesses of his second term abetted *Castro's revolution, which drove Batista from power in 1959.

Batlle y Ordóñez, José (1856-1929), Uruguayan statesman. He was President of Uruguay (1903-7, 1911-15), and initiated legislation to increase public welfare. He believed that the Swiss Bundesrat or federal council was well suited to his own country's needs and during his second term he tried to have the office of president eliminated altogether. His political opponents compromised by agreeing to an executive branch in which power was shared between a president and a nine-man council. This decentralization of power placed Uruguay on a unique path in the 20th century.

Battenberg, Prince Louis (1854-1921), British admiral. Of Polish-German descent, he became a naturalized British subject in 1868 and joined the navy, becoming First Sea Lord in 1912 in the critical period before the outbreak of World War I. His decision, criticized by some, not to disperse the naval squadrons gathered for exercises at Portsmouth at the time of the assassination of the Archduke *Francis Ferdinand at Sarajevo in 1914, assisted Britain's readiness for war. Anti-German hysteria in the early months of the war forced his resignation in October 1914. He became a marquis in 1917, giving up his German titles, and adopting the equivalent English name of Mountbatten. He married Princess Alice, granddaughter of Queen Victoria, in 1884. The younger of their two sons was Lord Louis Mountbatten, later Earl *Mountbatten of Burma, and one grandson was Prince Philip, Duke of Edinburgh.

Bavaria, the largest state of the Federal Republic of Germany. Towards the end of the 18th century it came under French influence, supporting Napoleon in the invasion of Austria (1800). Although obliged (1801) by Napoleon to cede to France its territories on the Rhine (the Palatinate), it was enlarged and proclaimed a kingdom in 1805. In 1806 it was incorporated into Napoleon's *Confederation of the Rhine. Shortly before the battle of *Leipzig it deserted Napoleon, joining the *Quadruple Alliance and was rewarded with the restoration of the Rhineland Palatinate. The Wittelsbach king Maximilian I gave Bavaria a liberal constitution in 1818, and it became one of the three most powerful states in the *German Confederation, holding the balance of power between Austria and Prussia. In 1866 it was defeated by the Prussians, but later joined Prussia in the *Franco-Prussian War. Ludwig II secured sovereign rights for Bavaria within the new *German Second empire, but withdrew from public life. Ludwig III came to the throne in 1913, to be deposed at the end of World War I, when his kingdom was declared a republic. A communist government was set up by the *Spartakist Kurt Eisner, and after the latter's assassination, a short-lived Communist Soviet Republic was proclaimed (1920-1). On its collapse, Bavaria became the centre of right-wing politics with the first attempted Nazi revolution in a *Munich beer-hall (1923). In 1948 Bavaria became a state, minus the Rhenish Palatinate, in the Federal Republic of Germany.

Bay of Pigs, an incident in Cuba in the area of that name. There in 1961 a small force of *CIA-trained Cuban exiles from Miami was landed from US ships in an attempt to overthrow the Marxist regime of Fidel *Castro. The invaders were swiftly crushed and rounded up by Castro's troops, and the incident was a grave blow to the prestige of the USA and of President *Kennedy. It strengthened the Castro regime and tightened Cuba's links with the Soviet Union.

Bazaine, Achille François (1811-88), French general. During the *Franco-Prussian War he was appointed commander-in-chief, but was reluctant to give battle. He withdrew at Metz with 176,000 French troops and capitulated to Bismarck on 27 October 1870. Convicted of treason (1873), he was sentenced to death, but this was commuted to twenty years' imprisonment. After one year he escaped to Italy and then Spain, where he died.

Beale, Dorothea (1831-1906), pioneer, together with her friend Frances Mary Buss (1827-94), in higher education for women in Britain. In 1858 she was appointed principal of the recently established Cheltenham Ladies' College, a position she was to hold until her death. She founded (1885) St Hilda's College, Cheltenham, for women teachers and lent her support to the establishment of St Hilda's Hall (later College), Oxford, in 1893. She was also an enthusiastic advocate of women's suffrage.

Beatty, David, 1st Earl (1871-1936), British admiral. He earned rapid promotion for his daring leadership in campaigns in Egypt and the Sudan, and in the *Boxer Rising in China. Winston Churchill, First Lord of the Admiralty, secured for him in 1913 command of the battlecruiser squadrons. Beatty gained minor victories over German cruisers off Heligoland (1914) and the Dogger Bank (1915), and played a major role in the battle of *Jutland. He was commander-in-chief of the Grand Fleet (1916-19) and First Sea Lord (1919-27).

Beauregard, Pierre Gustave Toutant (1818-93), US general. He served as an engineer during the *Mexican-American War and was appointed superintendent of West Point in 1860, but he resigned at the outbreak of the *American Civil War to join the *Confederacy. As commander at Charleston, he ordered the first shot of the

war against the Union-held *Fort Sumter. Beauregard was the field commander in the Confederate victory at the first battle of Bull Run (1861) before being promoted to full general and sent to the western theatre, where, after the battle of Shiloh (1862), he commanded the Army of Tennessee. Ill-health and bad relations with President Jefferson *Davis limited his influence for much of the remainder of the war, although he was back in the field when Confederate forces finally surrendered to General *Sherman in 1865.

Beaverbrook, William Maxwell Aitken, Baron (1879–1964), British financier, statesman, and newspaper owner. In 1910 he became a Conservative Member of Parliament and in 1916 took an important part in overthrowing *Asquith and manoeuvring *Lloyd George into the premiership. By 1918 he owned the *Evening Standard*, the *Sunday Express*, and *Daily Express*, with a record world circulation. Through these newspapers he supported the Hoare-*Laval pact and Chamberlain's *appeasement of Hitler at *Munich (1938). In 1940 he became Minister of Aircraft Production and a member of Churchill's war cabinet, and it was in no small measure due to his efforts in producing fighter aircraft that the battle of *Britain was won.

Bechuanaland   *Botswana.

Beecher, Lyman (1775–1863), US temperance reformer and Presbyterian clergyman. In 1832 he was appointed first president of the Lane Theological Seminary in Cincinnati, where his daughter, Harriet Beecher Stowe, author of *Uncle Tom's Cabin* (1852), married Calvin Ellis Stowe, professor of biblical literature. Lane Seminary students were among the first *abolitionists. Beecher became a target for attack by conservative Presbyterians and had to face charges, of which he was finally acquitted, of slander, heresy, and hypocrisy.

Begin, Menachem (Wolfovitch) (1913–), Israeli statesman. Active in the *Zionist movement throughout the 1930s, he was sent with the Polish army-in-exile to Palestine (1942), where he joined the militant *Irgun Zvai Leumi. On the creation of *Israel (1948) the Irgun regrouped as the Herut (Freedom) Party and elected Begin as its head. He was leader of the Opposition in the Knesset (Parliament) until 1967, when he joined the National Unity government. In 1970 he served as joint chairman of the Likud (Unity) coalition, and after its electoral victory in 1977 became Prime Minister (1977–83). He negotiated a peace treaty with President *Sadat of Egypt at *Camp David, but remained opposed to the establishment of a Palestinian state.

Beit, Sir Alfred (1853–1906), South African financier and philanthropist. Of German origin, he settled in Kimberley as a diamond merchant in 1875, and became a close friend of Cecil *Rhodes. His interest in gold greatly contributed to the development of the Rand and the British South African Company, and later of *Rhodesia. He made benefactions to scholarship and the arts.

Belaúnde, Terry Fernando (1912–), Peruvian statesman. He was elected to the Chamber of Deputies (1945–8). In 1956 he helped to found the moderate Popular Action Party and in 1963 was elected President with the support of both Popular Action and Christian Democrats in opposition to *Haya de la Torre, the candidate for APRA (Alianja Popular Revolucionaria Americana). His first term of office (1963–8) is remembered for its social, educational, and land reforms, as well as for industrial development and the construction of a vast highway system across the Andes. He was a strong supporter of the US Alliance for Progress programme, but his economic policies resulted in high inflation. He was deposed by the army and fled to the USA. He returned briefly to Peru in 1970, but was deported. He returned again in 1976 and was again President (1980–5).

Belgium, a country in north-west Europe. Until invaded by France in 1795 it was the Austrian Netherlands ruled from Vienna. In 1815 it became one of the provinces of the kingdom of the *Netherlands, but in 1830 it separated from the Netherlands following a national revolution, and Prince Leopold of Saxe-Coburg was elected king. After an unsuccessful Dutch invasion, an international treaty was drawn up guaranteeing Belgian neutrality in 1839, a treaty that was to be recognized in 1870 but not in 1914, when Germany's invasion precipitated Britain's entry into World War I. Leopold II (1865–1909) headed an international Association of the Congo (1876), following the exploration of the River Congo by H. M. *Stanley. This association was recognized at the Berlin Conference (1884) as the *Congo Free State, with Leopold as its unrestrained sovereign. As the Congo was opened for trade, appalling atrocities against Africans were committed, leading to its transfer from Leopold's personal control to the Belgian Parliament (1908). In World War I the country was occupied by the Germans, against whom Albert I (1908–34) led the Belgian army on the *Western Front. When Germany invaded again in 1940 Leopold III (1901–83) at once surrendered. However, a government-in-exile in London continued the war, organizing a strong resistance movement. After the war Leopold was forced to abdicate (1951) in favour of his son Baudouin (1930–). Independence was granted to the Congo in June 1960, but was immediately followed by violence and bloodshed (*Congo Crisis). The main task since the war has been to unite the Flemish-speaking northerners with the French-speaking Walloons of the south. In 1977 the Pact of Egmont, introduced by the Prime Minister, Leo Tindemans, recognized three semi-autonomous regions: that of the Flemings in the north, the Walloons in the south, and Brussels.

Belize, a country on the Caribbean coast of Central America. The British settled there in the 17th century, proclaiming the area (as British Honduras) a crown colony in 1862. Subject to the jurisdiction of the governor of Jamaica, the colony sustained itself with little direct support from the British government. Grudging acceptance by its Latin American neighbours in the 19th century led to treaties recognizing its permanent boundaries. In 1964 the colony gained complete internal self-government. It adopted the name Belize in 1973, and in 1981 became an independent state within the *Commonwealth of Nations. Guatemala, which bounds it on the west and south, has always claimed the territory on the basis of old Spanish treaties.

Bello, Alhaji Sir Ahmadu (1906–66), Nigerian statesman. He became leader of the Northern People's

Congress, and, in 1952, the first elected minister in Northern Nigeria, and in 1954 Premier. When Nigeria became independent in 1960, his party combined with *Azikiwe's National Council of Nigeria and the Cameroons (NCNC) to control the federal Parliament. Bello's deputy in the NPC, Abubakar Tafawa *Balewa, became federal Prime Minister, while Bello himself remained to lead the party in the north. In 1966, when the army seized power, Bello was among the political leaders who were assassinated.

Ben Bella, Ahmed (1916–), Algerian revolutionary leader. He served in the French army in World War II, and in 1947 became a leader of the secret military wing of the Algerian nationalist movement. He organized revolutionary activities and was imprisoned by the French (1950–2). He then founded and directed the National Liberation Front (FLN), which began the Algerian war with France. In 1956, when he was on board a Moroccan airliner he was seized and interned in France. In 1962 he was freed under the *Évian Agreements; he became Prime Minister of Algeria (1962–5) and was elected the first President of the Algerian Republic in 1963. In 1965 his government was overthrown in a military coup by Colonel Houari *Boumédienne. He was kept in prison until July 1979, and under house arrest until 1980, when he was freed unconditionally.

Beneš, Edvard (1884–1948), Czechoslovak president (1935–8; 1946–8). He was, with Tomáš *Masaryk, a founder of modern Czechoslovakia (1918). As leader of the Czech National Socialist Party, he was the country's foreign minister from 1918 until he succeeded Masaryk as President in 1935. In an attempt to keep the balance of power in Eastern Europe, he formed, in 1921, the *Little Entente with Yugoslavia to enforce observance of the *Versailles Peace Settlement by Hungary and prevent a restoration of the Habsburg King Charles. He strongly supported the *League of Nations, helping to admit the Soviet Union. In 1934 he and the Greek jurist Nikolaos Politis drafted the abortive Geneva Protocol for the pacific settlement of international disputes. Exiled during *World War II, he returned as President in 1945. Refusing to sign Klement *Gottwald's communist constitution, he resigned in 1948.

The Algerian leader **Ben Bella** (*front row, second from right*) at a conference of the Organization of African Unity (OAU) held at Dakar, Senegal, in August 1963.

Israeli Prime Minister David **Ben-Gurion** photographed at his desk in 1949.

Ben-Gurion, David (1886–1973), Israeli statesman. Born in Russian Poland, he migrated to Palestine in 1906 and quickly entered politics. He was one of the organizers of the Labour Party (Mapai) and of the Jewish Federation of Labour (Histadrut), which he served as General Secretary (1921–35). As Chairman of the Jewish Agency (1935–48) he was the leading figure in the Jewish community in Palestine. As Israel's first Prime Minister (1948–53, 1955–63), and Minister of Defence, he played the largest part in shaping Israel during its formative years. In 1965 he was expelled from the Labour Party and formed a new party known as Rafi.

Benin, a country in West Africa formerly known as Dahomey. It was ruled by kings of Yoruba origin until the French occupied it in 1892. It was constituted a territory of French West Africa (1904) under the name of Dahomey. It became an independent republic within the *French Community in 1960. Since then periods of civilian government have alternated with military rule. In 1972 it was declared a Marxist–Leninist state, and its name was altered (1975) to Benin. Under the leadership of Mathieu Kerekou, Benin has achieved greater domestic stability and international standing.

Benin, kingdom of, an African kingdom in what is now Nigeria. Once one of the principal states of West Africa, its extent declined in the 19th century. Continuing slave-trading and the use of human sacrifice in religious rituals

precipitated a British military expedition in 1897, which was massacred, whereupon a British force razed Benin city. The kingdom of Benin was incorporated into the new protectorate of southern *Nigeria in 1900.

Bentham, Jeremy (1748–1832), British philosopher, the founder of the *utilitarian school of ethics and political thought. He promoted the idea that the morality of an action could be measured by its effects on people: the greatest happiness of the greatest number should be the goal, and human institutions judged by the extent to which they contributed to that happiness. He supported much humanitarian reform and provided the inspiration for the founding of London University.

Bentinck, Lord William Cavendish (1774–1839), British statesman. After serving in Flanders and Italy, he was posted to India as governor of Madras (1803–7). He was recalled to Britain after a mutiny at Velore for which, by his prohibition of sepoy beards and turbans, he was held responsible. After serving in the *Peninsular War he returned to India as governor-general of Bengal (1827–33). He was appointed the first governor-general of all India (1833–5). A liberal reformer, his administration substituted English for Persian and Sanskrit in the courts, brought about many educational reforms, suppressed the practice of ritual strangling (*thug), and abolished suttee, whereby a widow was burned on her dead husband's pyre.

Bentinck, William Henry Cavendish, 3rd Duke of Portland (1738–1809), British statesman. As leader of the Whig Party he was briefly Prime Minister at the end of the American War of Independence in 1783. Later he supported the government of William *Pitt in its opposition to the French Revolution. He became Pitt's Home Secretary (1794–1801) and greatly assisted in the

passing of the *Act of Union in 1801. After Pitt's death (January 1805) and the failure of the so-called 'Ministry of All the Talents' (1805–7), he was persuaded (1807) to take office again as Prime Minister. Then an old man, he failed to prevent internal dissension in his government, which led to the duel between *Canning and *Castlereagh, on news of which he resigned.

Beria, Lavrenti Pavlovich (1899–1953), Soviet politician. Born in Georgia, he joined the Communist Party in 1917, and became head of the secret police (*CHEKA) in that province (1921). He came to Moscow in 1938 to take charge of the secret police as head of Internal Affairs and to organize Soviet *prison camps. During World War II he was a major figure in armaments production. After Stalin's death he became a victim of the ensuing struggle for power. In July 1953 he was arrested on charges of conspiracy. He was tried in secret and executed.

Berlin, Congress of (1878), a conference of European powers. It revised the Treaty of San Stefano (1878) which had ended the war between the *Ottoman empire and Russia (1877–8). Under the chairmanship of the German chancellor, Otto von *Bismarck, the congress limited Russian naval expansion; gave Montenegro, Serbia, and Romania independence; allowed Austro-Hungary to occupy Bosnia and Hercegovina; reduced *Bulgaria to one-third of its size; and placed Cyprus under temporary occupation by the British. The congress left Russian nationalists and *Pan-Slavs dissatisfied, and the aspirations of Greece, Ser-

The German Chancellor Otto von Bismarck, presiding over the **Congress of Berlin** (1878), stands to address the leaders of Europe's other great powers. Among them is Benjamin Disraeli (Lord Beaconsfield), the British Prime Minister (*seated fourth from left*).

bia, and Bulgaria unfulfilled. Bismarck's handling of the congress antagonized Russia, and the claim of *Disraeli, that it had achieved 'peace with honour', proved unfounded.

Berlin Airlift (1948-9), a measure undertaken by the US and British governments to counter the Soviet blockade of Berlin. In June 1948 the USA, Britain, and France announced a currency reform in their zones of occupied Germany. The Soviet Union, fearing this was a prelude to the unification of these zones, retaliated by closing all land and water communication routes from the western zones to Berlin. The western Allies in turn responded by supplying their sectors of Berlin with all necessities by cargo aircraft. The siege lasted until May 1949, when the Russians reopened the surface routes. The blockade confirmed the division of Berlin, and ultimately of Germany, into two administrative units.

Berlin Wall, a barrier between East and West Berlin. It was built by the *German Democratic Republic in August 1961 in order to stem the flow of refugees from East Germany to the West: over three million had emigrated between 1945 and 1961. The Wall is heavily guarded and many people, especially in the 1960s, have been killed or wounded while attempting to cross.

Bermuda, a cluster of some 150 islands in the west Atlantic, a British colony. First settled by the Virginia Company, they have the oldest parliament in the New World, dating to 1620. A flourishing slave economy existed until 1834 when slavery was abolished. By the 20th century

Graffiti on the **Berlin Wall**, together with tributes to escapers from the East, political symbols, and slogans of protest, cover the Western side of the stark barrier which has divided the city of Berlin since 1961.

two-thirds of the population was of African or Indian descent. Strategically important to the British navy, Bermuda grew rich on trade and tourism, with close ties with the USA. US naval and air-bases were granted in 1940. Universal adult suffrage was introduced in 1944 and the present constitution in 1967, granting the colony considerable self-government. Political activity developed in the 1960s and there were sporadic and bitter race-riots in the 1970s.

Bernadotte, Folke, Count (1895-1948), Swedish international mediator. The nephew of Gustav V of Sweden, he entered the Swedish army as a young man. During World War II he worked for the Swedish Red Cross and in 1948 was appointed as United Nations mediator to supervise the implementation of the partition of *Palestine and the creation of *Israel. He was murdered by Israeli terrorists.

Besant, Annie (1847-1933), British social reformer and theosophist. She became a Fabian, a trade-union organizer (including the *match girls' strike of 1888), and a propagandist for birth control. She became a leading exponent of the religious movement of theosophy, and founded the Hindu University in India, helping to form, in 1916, the All India Home Rule League. She was President of the Indian National *Congress 1918-19, one of only three Britons to have held this office. She published numerous books on theosophy.

Betancourt, Rómulo (1908-), Venezuelan statesman. An avowed democrat, he served as President from 1959 to 1964, presiding over a period of redemocratization following a long period of military juntas. He initiated a modest programme of agrarian reform, increased the taxes paid by the foreign oil companies, and secured a series of benefits for organized labour. Attacked by the right-wing

supporters of his predecessor and by the radical socialists, he turned the presidential office over to a freely elected successor in 1964.

Bethmann-Hollweg, Theobald von (1856–1921), German statesman. He was Prussian Minister of the Interior, then Secretary of State in the Imperial Ministry of the Interior. As Chancellor (1909–17) he instituted a number of electoral reforms and gave greater autonomy to *Alsace-Lorraine. He greatly increased the German peacetime army, believing that Germany would be forced into a war by its neighbours. He hoped to retain British neutrality and, in 1912, worked successfully with Britain to keep peace in the *Balkan States. In July 1914 he and the Kaiser, *William II, convinced of the need for a short, preventive war, promised unconditional support to *Austro-Hungary. In 1917 he opposed unrestricted submarine warfare, rightly foreseeing the entry of the USA into the war. He retired in July and Hindenburg and Ludendorff then set up a military dictatorship.

Bevan, Aneurin (1897–1960), British politician. He led the Welsh miners in the 1926 *General Strike. He was elected Independent Labour Member of Parliament for Ebbw Vale in 1929, joining the more moderate Labour Party in 1931. His left-wing views and fiery personality made him a rebellious member of the Party. A founder and editor of *Tribune* magazine, he was one of Winston *Churchill's most constructive critics during World War II. As Minister of Health (1945–51), he was responsible for a considerable programme of house-building and for the creation, amid bitter controversy with the medical profession, of the National *Health Service. In 1951 he became Minister of Labour, but resigned when charges were imposed for some medical services. The Bevanite group was then formed within the Party. As shadow minister for foreign affairs in opposition from 1957, he is remembered for his opposition to unilateral nuclear disarmament. Often in conflict with his own party, he was nevertheless elected deputy leader in 1959.

Beveridge, William Henry, 1st Baron Beveridge (1879–1963), British economist and social reformer. At the invitation of Winston *Churchill he entered (1908) the Board of Trade and published his notable report, *Unemployment*, in 1909. In it he argued that the regulation of society by an interventionist state would strengthen rather than weaken the free market economy. He was instrumental in drafting the Labour Exchanges Act (1909) and the National Insurance Act (1911). In 1941 he was commissioned by the government to chair an inquiry into the social services and produced the report *Social Insurance and Allied Services* (1942). This was to become the foundation of the British *Welfare State and the blueprint for much social legislation from 1944 to 1948.

Bevin, Ernest (1881–1951), British trade union leader and politician. He was General Secretary of the Transport and General Workers' Union (1922–40), Minister of Labour and National Service (1940–5), and Foreign Secretary (1945–51). Bevin played a major role in the *General Strike of 1926 and the crisis of 1931. His influence on Labour Party politics in the 1930s was considerable, but he did not enter Parliament until invited to join Winston *Churchill's war-time coalition government in 1940. His trade-union background was invaluable in

Aneurin **Bevan**, as Labour Minister of Health, visits a hospital in Lancashire in April 1949, on the first day of the National Health Service in Britain.

mobilizing the labour force during World War II. As Foreign Secretary he took decisive action to extricate Britain from *Palestine in 1948 and to involve the USA in post-war West European affairs through the *Marshall Plan and *NATO.

Bhave, Vinoba (1895–1982), Indian leader. A follower of *Gandhi from 1916, he was active in attempts to revitalize Indian village life. Imprisoned by the British (1940–4) for defying wartime regulations, Bhave was, after Gandhi's assassination (1948), widely regarded as the leading exponent of Gandhism. He founded (1948) the Sarvodaya Samaj to work among refugees. In 1951 he began the *Bhoodan or land-gift movement, and led the Shanti Sena movement for conflict resolution and economic and social reform.

Bhoodan, a movement in India begun in 1951 by Vinoba *Bhave with the object of acquiring land for redistribution to landless villagers. At first the object was to acquire individual plots, but from the late 1950s an attempt was made to transfer ownership of entire villages to village councils. The movement had a measure of success in Bihar state.

Bhutto, Zulfikar Ali (1928–79), Pakistani statesman. In 1958 he joined *Ayub's first military government as Minister of Fuel and Power and subsequently became Foreign Minister (1963). Dismissed from Ayub's cabinet in 1967 he formed the Pakistan People's Party with a policy of Islam, democracy, socialism, and populism. In the elections of 1970 the PPP secured the largest share of the vote in West Pakistan and, after the military government was discredited by the loss of Bangladesh, Bhutto became President (1971), but stepped down in 1973 to become Prime Minister. Bhutto concluded the Simla agreement with India in 1972; recognized Bangladesh in 1974, and cultivated China; and formulated a new constitution and an ambitious economic programme, whose failure contributed to his ejection in a military coup in 1977. Bhutto was subsequently hanged on the charge of complicity in a political murder. His daughter, Benazir

Zulfikar Ali **Bhutto**, President of Pakistan (1971–3) and subsequently Prime Minister until 1977, seen here at a meeting of the United Nations Security Council in 1971.

Bhutto (1955–), leader of the Pakistan People's Party, has remained active in her party's opposition to the regime of General *Zia.

Biafra, the name of an abortive *Nigerian secessionist state (1967–70) in the south-east of the country, inhabited principally by Ibo people. It seceded after mounting antagonism between the eastern region and the western and northern regions, Colonel Ojukwu declaring the east independent. Civil war followed. Gabon, Ivory Coast, Tanzania, and Zambia recognized Biafra, while Britain and the USSR supported the federal government. When Ojukwu fled to the Ivory Coast General Effiong capitulated in Lagos in 1970 and Biafra ceased to exist.

The very young and the old were among the first to suffer from the famine caused by the fighting in **Biafra**. Severe hardship resulted from the collapse of the breakaway regime's economy during the civil war.

Bidault, Georges (1899–1982), French statesman and journalist. After serving in World War I he became professor of history in Paris. During World War II he became a distinguished leader of the French *resistance movement. He was a founder-member and leader (1949) of the Mouvement Républicaine Populaire. Bidault was Foreign Minister in several administrations of the Fourth Republic (1944, 1947, 1953–4) and Prime Minister (1946, 1949–50, 1958). He subsequently became bitterly opposed to *Algerian independence: he became President of the National Resistance Council in 1962, was charged with plotting against the state, and went into exile in Brazil. He returned to France in 1968.

Bigge Inquiry (1819–21), a British government inquiry into New South Wales, Australia. It was conducted by John Bigge to inquire into the future potential of the penal colony as a free settlement. He visited New South Wales and Van Diemen's Land and produced three official reports in 1822 and 1823. He recommended limited constitutional government and the establishment of Van Diemen's Land as a separate colony. Other recommendations, many of which were implemented, included the encouragement of the pastoral industry. Bigge was critical of Governor *Macquarie, especially in his treatment of convicts and *emancipists, which Bigge saw as being excessively lenient, and also of his expensive programme of public works.

Bikini Atoll, an atoll in the *Marshall Islands, west central Pacific. It was the site for twenty-three US nuclear bomb tests (1946–58). Despite expectations that it would be fit again for human habitation in 1968, the atoll remains too contaminated for the return of the Bikinians, who have been relocated on surrounding islands.

Biko, Steve (1956–77), student leader in South Africa. A medical student at the University of Natal, he was co-founder and president of the all-black South African Students Association, whose aim was to raise black consciousness. Active in the Black People's Convention, he was banned and then arrested on numerous occasions (1973–6). In some ways he was of greatest significance in his death in prison at the age of twenty-one, which made him a symbol of heroism in black South African townships and beyond. Following disclosures about his maltreatment in prison, the South African government prohibited numerous black organizations and detained newspaper editors, thus provoking international anger.

Billy the Kid (or William H. Bonney) (1859–81), US outlaw. He arrived in New Mexico in 1868. A frequenter of saloons, he moved effortlessly into robbery and murder. In 1878 he became prominent in a cattle war, killing the local sheriff, Jim Brady. The territorial governor, Lew Wallace, was unable to persuade Billy to cease his activities. Sheriff Pat Garrett captured him in 1880, but he escaped, only to be shot by Garrett at Fort Sumner, New Mexico.

Birch, John, Society, US political organization. Named after John Birch, a Baptist missionary killed by Chinese communists in 1945, it was formed in 1958 in Massachusetts with the aim of exposing the 'communist conspiracy' which was allegedly infiltrating the highest federal offices. Accused by its detractors of using smear

A South African mourns the death of Steve **Biko** as she holds aloft a wreath and portrait of the student leader on the steps of the Pretoria Old Synagogue. The inquest into Biko's death in prison was held in the building during November 1977.

tactics associated with *McCarthyism, it favoured *states' rights and a reduction in federal powers. It gained most support in California and the southern states and bitterly attacked Chief Justice *Warren for his part in forcing the southern states into *desegregation.

Birla, Indian commercial and industrial family of the Marwari or merchant caste. It is one of the two (with the Tatas) greatest Indian industrial families. The best known member of the family was Ghanshyam Das Birla, who became *Gandhi's principal financial backer, paying most of the cost of the *ashram* (retreat), the Harijan organizations, the peasant uplift campaign, and the national language movement, as well as supporting many other Gandhian welfare projects. It was at Birla House, New Delhi, that Gandhi was killed. Like most Marwaris the Birlas were devout Hindus.

Bishop, Maurice   *Grenada.

Bismarck, Otto von (1815–98), German statesman, known as the 'Iron Chancellor'. A Brandenburg nobleman, he entered the Prussian Parliament as an ultraroyalist and an opponent of democracy. During the *Revolutions of 1848 he opposed demands for constitutional reform and in 1851, as Prussian member of the Federal German Diet at Frankfurt, dominated by Austria, he demanded equal rights for Prussia. After a brief period as ambassador to St Petersburg (1859) and Paris (1862) he was made minister-president of Prussia (1862–90). He enlarged and reorganized the Prussian army. In 1864, in partnership with Austria, he led the German states in the defeat of Denmark, acquiring *Schleswig-Holstein, whose *Kiel Canal became of strategic importance to Germany. In 1866 he provoked a confrontation with Austria, known as the *Seven Weeks War, from which he emerged victorious. He then annexed Hanover and united most of the other German states in the North German Confederation, of which he became Chancellor. He instigated the

*Franco-Prussian War (1870–1), wresting *Alsace and Lorraine from France, capturing the French emperor, *Napoleon III, and subjecting Paris to a long and terrible siege. He then proclaimed the King of Prussia, *William I, as emperor of a *German Second empire in the French Palace of Versailles. At home, he introduced a common currency, a central bank, a single code of law, and various administrative reforms for the new empire. He unsuccessfully sought to weaken the power of the Catholic Church (the so-called *Kulturkampf), but successfully introduced the Prussian school system, with its government inspectors, into the empire. He kept the German Parliament (Reichstag) weak and the executive strong. He dealt severely with socialist supporters. In an effort to keep the working class away from the socialists and to hold trade-unionists in check, he introduced the first industrial welfare scheme in history, a series of *social security laws (1883–7) to provide for sickness, accident, and old age benefits. In foreign affairs, as Chancellor, he initiated the *Three Emperors League (Dreikaiserbund) and the later *Triple Alliance. He presided with great success over the Congress of *Berlin (1878) and the Berlin Conference on Africa (1884). As a result of his economic nationalism and protective tariffs, German industry and commerce flourished and new colonies were acquired overseas. The death of William I showed the weakness of Bismarck's position, dependent as it was on the royal will and not on

A 'wanted, dead or alive' poster offering a substantial reward for the notorious outlaw **Billy the Kid**.

REWARD

($5,000.00)

Reward for the capture, dead or alive, of one Wm. Wright, better known as

"BILLY THE KID"

Age, 18. Height, 5 feet, 3 inches. Weight, 125 lbs. Light hair, blue eyes and even features. He is the leader of the worst band of desperadoes the Territory has ever had to deal with. The above reward will be paid for his capture or positive proof of his death.

JIM DALTON, Sheriff.

DEAD OR ALIVE!
"BILLY THE KID"

Bismarck, the master of politics, sitting comfortably astride the world with France beneath his sword in this 19th-century caricature.

Chief Black Hawk, leader of the Sauk and Fox Indians in the **Black Hawk War**, wearing strings of wampum (shell beads) and a medallion. This portrait by Charles Bird King was painted in 1837, a year before the chief's death.

popular, democratic support. *William II saw Bismarck as a rival for power, and forced his resignation (1890). Bismarck spent the rest of his years in retirement.

Black-and-Tans, an auxiliary force of the Royal Irish Constabulary. The demands of the Irish Republicans for a free *Ireland led in 1919 to violence against the Royal Irish Constabulary, an armed British police force. Many of the policemen resigned, so the British government in 1920 reinforced the RIC with British ex-soldiers. Their distinctive temporary uniforms gave them their nickname of Black-and-Tans. They adopted a policy of harsh reprisals against republicans, many people being killed in raids and property destroyed. Public opinion in Britain and the USA was shocked and the Black-and-Tans were withdrawn after the Anglo-Irish truce in 1921.

Black Hand, symbol and name for a number of secret societies which flourished in the 19th and early 20th centuries. It was the name adopted by a Serbian terrorist organization, founded in 1911 by Colonel Dimitrijevic largely from army officers, to liberate Serbs still under Habsburg or Turkish rule. It organized the assassination at Sarajevo of Archduke *Francis Ferdinand (1914), an event which contributed to the outbreak of World War I. The name and symbol were adopted by organizations controlled by the *Mafia in the USA and Italy, which used intimidation and murder to gain their ends.

Black Hawk War (1832), a North American Indian war. Between the *Louisiana Purchase and the 1830s, there was steady pressure to remove the remaining Indians east of the Mississippi to the new territory, and Indian land rights were eaten away by a series of enforced treaties. In 1831, the Sauk and Fox Indians, led by Chief Black Hawk, were forced by the local militia to retreat across the Mississippi into Missouri. In the following year, threatened by famine and hostile Sioux, the Indians recrossed the river to plant corn. When they refused to comply with the local military commander's order to leave, a brief war broke out in which the starving Indians were gradually driven back, before being trapped and massacred near the mouth of the Red Axe River in early August. Black Hawk's defeat and death allowed the final alienation of Indian land rights east of the Mississippi.

Black Muslim Movement, a black nationalist organization in the USA. It seeks to unite black Americans under *Islam to secure their emancipation from white rule. Founded in 1930 and led by Elijah Muhammad from 1934 to 1975, it expanded greatly in the 1950s when *Malcolm X became one of its spokesmen, and by the 1960s, at the height of the *Black Power movement, it probably had over 100,000 members. With the assassination of Malcolm X in 1965, it lost some of its influence to the Black Panthers, but it continued to establish separate black enterprises and to provide a source of inspiration for thousands of black Americans.

Black Power Movement, a term used among black people in the USA in the mid-1960s. The movement aimed at a more militant approach towards securing *civil rights, and stressed the need for action by blacks alone, rather than in alliance with white liberals. Many blacks felt that the civil rights movement had done little to alter their lives, and under such leaders as Stokeley Carmichael they proposed that black Americans should concentrate in their own communities to establish their own political and economic power. In 1966 a Student Non-Violent Co-ordinating Committee (SNCC) was formed by Carmichael to activate black college students, and at the same time the *Black Muslim Movement was advocating Islam as the black salvation. Others, like the Black Panthers, emphasized violence and militancy, but all were concerned to stress the value of black culture and all things black. The riots in the cities in the middle and late 1960s seemed to herald new waves of black militancy, but the intensity of the Black Power Movement tended to decline in the early 1970s, when many blacks began co-operating with white organizations against the *Vietnam War.

Black September, Palestinian terrorist organization. It emerged after the defeat of the Palestinian guerrilla organizations in Jordan in September 1970, from which event it took its name. It was claimed to be an independent organization, but was apparently a cover for al-Fatah operations, the most atrocious of which was the massacre of Israeli athletes at the Munich Olympics in September 1972. Shortly after that event the organization became inactive.

Blackshirt (Italian, *camicia nera*), the colloquial name given to the *Squadre d'Azione* (Action Squad), the national combat groups, founded in Italy in 1919. Organized along paramilitary lines, they wore black shirts and patrolled cities to fight socialism and communism by violent means. In 1921 they were incorporated into the *Fascist Party as a national militia. The term also applies to the *SS in Nazi Germany.

Blackwell, Elizabeth (1821–1910), US physician. She was the first woman to gain a degree in medicine in the USA. Born in Bristol, England, she emigrated with her family to the USA in 1832. After her father's death she supported her family by teaching, and began studying medicine privately. Rejected by various medical schools, she was finally accepted by the Geneva Medical College, New York, graduating in 1849. She practised in New York but later lived in England, becoming professor of gynaecology at the London School of Medicine for Women (1875–1907).

Blaine, James Gillespie (1830–93), US politician. He was Secretary of State to President *Garfield (1881), and to President William Henry *Harrison (1889–91). As leader of the so-called 'Half Breeds' Republicans (those committed to a conciliatory policy towards the South and to civil-service reform), he helped three lesser men (Hayes, Garfield, Harrison) attain the presidency but was denied the prize himself in the 1884 election against *Cleveland. He aroused suspicion on account of his transactions with railway companies, to whom he owed his moderate wealth.

Blanc, Louis (1811–82), French politician and historian. In 1839 he published *The Organization of Labour* in which he outlined his ideal of a new social order based on the principle 'from each according to his abilities, to each according to his needs'. In 1848 he headed a commission of workers' delegates to find solutions to problems of exploitation and unemployment. The suppression of the workers' revolt later that year forced him to flee to Britain

and he did not return until 1871. He was elected a Deputy of the National Assembly and did not join the *Paris Commune in 1871, but tried instead to obtain an amnesty for those implicated in the rising. His advocacy of the control of industry by working men with the support of the state through social workshops (*ateliers sociaux*) influenced later leftist reformers, notably Ferdinand Lassalle (1825–64) and other German socialists.

Blanco, Antonio Guzmán   *Guzmán Blanco.

Blanqui, Louis Auguste (1805–81), French radical thinker and revolutionary leader. Although a member of the *Carbonari, he was decorated by *Louis-Philippe for his part in the 1830 July revolution which had deposed Charles X. Realizing that the only beneficiaries of the revolution had been bourgeois oligarchs, he took to conspiracy against them, launching an attack on the Paris Hotel de Ville in 1839. Sentenced to death, his sentence was later commuted to life imprisonment. A brief period of freedom allowed him to lead the republicans in the *Revolution of 1848. He remained in prison until 1859, was re-arrested in 1861, and escaped to Belgium in 1865, where he organized the extremist republican opposition to *Napoleon III in whose deposition he was instrumental. He was imprisoned in 1871, after attempting to overthrow the French provisional government. His influence over the *Paris Commune was considerable, and his followers vainly offered their hostages in exchange for Blanqui. He died in 1881, two years after being finally released from prison.

Bligh, William (1754–1817), British admiral. He accompanied Captain Cook on his third voyage (1776–9). On a further visit to the South Pacific islands in 1788, his

Captain William **Bligh**, depicted against a Polynesian setting; a contemporary engraving from a painting by J. Russell. Bligh and his companions survived over six weeks adrift in the Pacific in an open boat following the mutiny on the *Bounty*, eventually reaching Timor in June 1789. (National Maritime Museum, London)

irascible temper and overbearing conduct provoked the *Bounty* Mutiny. Returning to Britain, he served under Nelson at Copenhagen (1801) and in 1805 was appointed governor of New South Wales. Conflict with the New South Wales Corps culminated in the *Rum Rebellion of 1808. Settling in England in 1810, he was promoted to the rank of vice-admiral.

Blitzkrieg, German term meaning 'lightning war'. An Anglicized version, 'the Blitz', was coined by the British public to describe the German air assault on British cities in 1940. As a military concept, it was employed by the Germans in World War II and was especially successful in the campaigns against Poland, France, and Greece. It employed fast-moving tanks and motorized infantry, supported by dive-bombers, to throw superior but slower enemy forces off balance and thereby win crushing victories rapidly and with small expenditure of men and materials. After 1941, because Germany's enemies were better prepared and because new battlefields in the Soviet Union and Africa were less suited to the technique, Blitzkrieg tactics were no longer decisive.

Blood River, battle of (16 December 1838), fought between *Voortrekkers and *Zulus, led by *Dingaan,

near a tributary of the Buffalo River, subsequently called Blood River after its waters reddened with the blood of some 3,000 Zulus, killed to avenge the slaughter of about 500 *Boers earlier in the year. The Zulu defeat enabled the Boers to establish the Republic of *Natal.

Blücher, Gebhard Leberecht von (1742–1819), Prussian field-marshal. Forced to surrender to the French in 1806, he helped to re-create his country's opposition to *Napoleon, and was commander-in-chief of the armies in their victory at *Leipzig in 1813. The following year he led the invasion of France, gaining a major victory at Laon, which led to the overthrow of Napoleon. He retired to Silesia, only to be recalled when Napoleon returned. His intervention at a late stage of the battle of *Waterloo was decisive.

Blum, Léon (1872–1950), French politician and writer. An established journalist and critic, he was first drawn to politics by the *Dreyfus affair. He brought about the coalition of radical socialists, socialists, and communists which won power in 1936. As France's first Socialist Prime Minister, his government granted workers a forty-hour week, paid holidays, and collective bargaining, resulting in considerable hostility from industrialists. Radicals re-

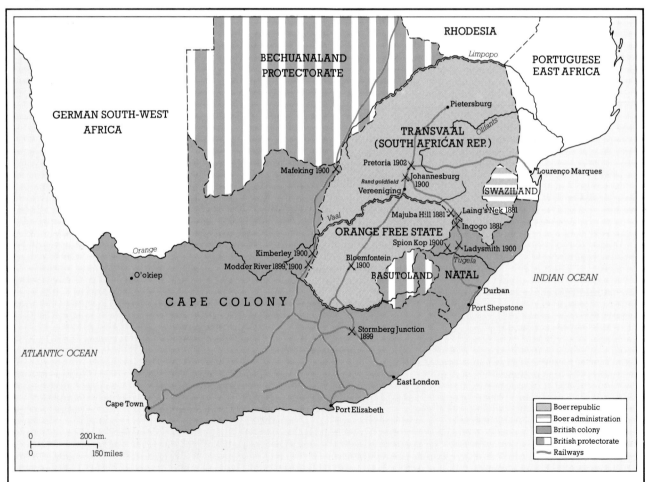

Boer Wars

The discovery of diamonds (1867) stimulated an influx of British settlers into the Boer Orange Free State and the Transvaal, where the First Boer War (1881) resulted in a British defeat. The Witwatersrand goldrush (1886) and

British imperialist ambitions led to the Second War (1899–1902). The Act of Union (1909) was an uneasy compromise, preserving white domination over black Africans and Indians.

fused to support intervention in the *Spanish Civil War while communists withdrew their support for his failure to intervene. His government fell. He was arrested in 1940, and charged with causing France's defeat, but his skilful defence obliged the authorities to call off his trial (1942). He was interned in a German concentration camp (1943–5), and returned briefly to power as the Prime Minister of a caretaker government in 1946–7.

Boer *Afrikaner.

Boer Wars (1880–1, 1899–1902) (the South African, or Anglo-Boer, Wars; the First and Second Wars of Freedom). They were fought between Britain and Transvaal and between Britain and Transvaal and the Orange Free State. The first arose from the British annexation of the Transvaal in 1877 and the incompetent administration that followed. In 1880 it was thought that the *Gladstone government would grant independence, or at least self-government; when hopes were dashed, *Kruger, Joubert, and *Pretorius took power as a triumvirate. British disasters at the battles of Laing's Nek, Ingogo, and Majuba Hill, forced peace upon Gladstone, who granted self-government. The second war (1899–1902) was caused by multiple grievances. The Boers, under the leadership of Kruger, resented the imperialist policies of Joseph *Chamberlain, which they feared would deprive the Transvaal of its independence. The refusal of political rights to *Uitlanders aggravated the situation, as did the aggressive attitude of Lord *Milner, British High Commissioner. For Britain, control of the Rand goldfield was all-important. In 1896 the Transvaal and the Orange Free State formed a military alliance. The Boers, equipped by Germany, never mustered more than 88,000 men, but defeated Britain in numerous initial engagements, for example, Spion Kop. British garrisons were besieged in Ladysmith, Kimberley, and Mafeking. In 1900 the British, under *Kitchener and Roberts, landed with reinforcements. The Boers were gradually defeated, despite the brilliant defence of the commandos. Kitchener adopted a scorched-earth policy, interning the civil population in *concentration camps, and systematically destroying farms. Peace was offered in 1901, but terms that included the loss of Boer independence were not agreed until the Peace of *Vereeniging in 1902.

Bohemia–Moravia, a region of Czechoslovakia. The independent state of Bohemia (of which Moravia was a province) passed under Austrian rule in 1526. In 1848, a Slav Congress demanding greater autonomy assembled in Prague under the leadership of *Palacký. Austrian domination was forcibly restored in 1849, and Moravia was made into a separate crown land. Concessions made to the Czechs by Vienna after 1867 served only to disconcert the Germans living in Bohemia. Independence as part of the republic of *Czechoslovakia, incorporating Bohemia, Moravia, Slovakia, and Austrian Silesia, was achieved after the collapse of the *Austro-Hungarian empire (1918). Now it was the German minority of the Sudetenland who felt oppressed by the Czechs. In 1938, having earmarked Bohemia and Moravia for German colonization, Hitler invaded the *Sudetenland and annexed the rest of Bohemia–Moravia in the following year. A lasting shift of population was effected by the expulsion by the Czechoslovak government of three million Germans, mainly from Bohemia and Moravia, after World War II.

Bolívar and South American independence

Spanish control in South America was weakened when Napoleon forced Ferdinand VII to abdicate (1808). In Caracas Simón Bolívar and other young Venezuelan aristocrats, inspired by the French Revolution, established themselves at Angostura, from where, augmented by British mercenaries, they defeated Spanish troops in New Granada and Peru. Here, in 1824, Bolívar met up with José de San Martín, who had marched across the Andes, to liberate Chile. In 1826 Bolívar proclaimed the Republic of Gran Colombia, but this soon broke up. In 1825 Upper Peru took the name of Bolivia in his honour.

Bolívar, Simón (1783–1830), South American soldier and statesman, and leader in the *Spanish-South American War of Independence. Inspired by European rationalists, he vowed to liberate Hispanic America. Participating in South American republican risings from 1812, Bolívar's crucial victory at Boyacá (1819) secured Colombia's independence from Spain, and two years later at the battle of Carabobo (June 1821) his defeat of the Spanish royalists achieved the same for Venezuela. He then marched an army to Ecuador and drove the Spanish from Quito before meeting up with José de *San Martín at Guayaquil. The two independence leaders disagreed at the Guayaquil conference over the future of South America and ultimately San Martín resigned his command and allowed Bolívar to drive the Spanish army out of Peru, the last colonial stronghold on the continent. With the independence of South America assured, Bolívar accepted the presidency of the Confederation of Gran Colombia (Venezuela, Colombia, Ecuador, and

Panama). Unable to prevent the break-up of the confederation into three independent nations, he resigned the presidency in April 1830.

Bolivia, a land-locked country of central South America. Conquered by the Spanish conquistadores (1538), independence was won under José de Sucre, at the battle of Ayacucho (1824). A National Assembly declared Upper Peru independent, naming it after Simón *Bolívar. A short-lived Peru-Bolivian Confederation was formed (1825–39). In 1842 control of the Atacara coast region, where rich guano nitrate deposits were found, was challenged by Chile, finally being lost in 1884 in the disastrous War of the *Pacific. A series of military dictatorships followed (1839–80), succeeded by more liberal regimes, with Liberal and Republican Parties alternating. In 1930 a popular revolution elected a reforming President Daniel Salamanca. In 1936, following the disastrous *Chaco War, military rule returned. In 1952 the Bolivian National Revolution overthrew the dictatorship of the junta, and *Paz Estensorro, leader of the MNR (Movimento Nacionalista Revolucionario) Party returned from exile and was installed as President. Tin mines were nationalized, adult suffrage introduced, and a bold programme of social reforms begun. Paz was re-elected in 1960 but overthrown in 1964 by a military coup. In 1967 a communist revolutionary movement, led by Ché *Guevara, was defeated. Military regimes followed each other quickly. Not all were right-wing, and that of General Juan José Torres (1970–1) sought to replace Congress by workers' soviets. Democratic elections were restored in 1978, when the first woman President, Lydia Guelier Tejada, briefly held office. There was a new military coup in 1980 and a state of political tension in the country continued until 1982, when civilian rule was restored.

Bolshevik (Russian, 'a member of the majority'), a term used to describe the wing of the Social Democratic Party in Russia which, from 1903, and under the leadership of *Lenin, favoured revolutionary tactics. It rejected co-operation with moderate reformers and favoured the instigation of a revolution by a small, dictatorial party prepared to control the working class. Their opponents, the Mensheviks ('members of the minority'), led by *Martov and *Plekhanov, favoured a loosely organized mass labour party, in which workers had more influence, and which was prepared to collaborate with the liberal bourgeoisie against the Tzarist autocracy. After the abortive *Russian Revolution of 1905 Bolshevik leaders fled abroad, having made little appeal to the peasantry, and it was the Mensheviks led by *Kerensky who joined the Provisional Government, following the February *Russian Revolution in 1917. The infiltration by Bolsheviks into *Soviets and factory committees contributed to the success of the October Revolution. During the *Russian Civil War the Bolsheviks succeeded in seizing control of the country from other revolutionary groups. In 1918 they changed their name to the Russian Communist Party. The Mensheviks were formally suppressed in 1922.

bombing offensives (World War II), attacks by bomber aircraft on military and civilian targets. As part of his *Blitzkrieg tactics, Hitler deployed dive-bombers in the offensives in Poland (1939) and western Europe (1940). In August 1940 the first major German offensive was launched against Britain, a series of daylight attacks by bombers, many of which were destroyed by British fighter aircraft in the battle of *Britain. A German night-bombing offensive on civilian targets then began which lasted until May 1941, London being attacked on fifty-seven consecutive nights and badly burned while large numbers of incendiary bombs were also dropped over other cities. In 1944 the Allied air offensive was intensified. The development of radar to intercept aircraft and direct gunfire revolutionized the Allied bombing offensive. Increasing resources were made available to the British Bomber Command under Air Marshal Sir Arthur Harris, and daylight incendiary bombing by the US Air Force, combined with British night-bombing, endeavoured to obliterate key German cities, one of the biggest such raids being against *Dresden. Meanwhile the bulk of German bombing power was turned to the Eastern Front, where fighter-bombers supported the army, and besieged cities such as Leningrad and Stalingrad were regularly attacked. Pilotless bombers (V1s) and rocket missiles (V2s), launched at England during 1944 and 1945, did relatively little damage. In the Far East a massive incendiary bomber offensive was launched against Japanese cities in October 1944. On 9 March 1945 much of Tokyo was destroyed. Finally, a US aircraft dropped an atom bomb (6 August 1945) on *Hiroshima and, three days later, a second one on *Nagasaki.

Bonaparte *Napoleon.

Bonar Law, Andrew *Law, (Andrew) Bonar.

Bonhoeffer, Dietrich (1906–45), German Lutheran

Blowing the Tzarist government sky-high: a fictional ideal depicted on the cover of the first issue of the **Bolshevik** revolutionary magazine *Voron* ('Raven'), 1905–6.

The **bombing offensives** by the German airforce against London and other British cities were popularly known as the 'blitz'. This 1941 painting by Henry Carr shows the burning of the church of St Clement Danes in central London. The dropping of incendiaries, as well as of high-explosive bombs, caused widespread fires in London and inflicted considerable damage on the city in 1940–1.

theologian. An active opponent of Nazism, he signed (1934) the Barmen Declaration in protest against attempts by German Christians to synthesize Nazism with Christianity. He was forbidden by the government to teach, and in 1937 his seminary at Finkenwalde was closed. In 1942 he tried to form a link between the Germans opposed to Hitler and the British government. Arrested in 1943, he was executed in 1945.

Bonus Army, an assemblage of ex-servicemen in the USA. It was the popular name given to the so-called Bonus Expeditionary Force (BEF), a group of about 20,000 World War I veterans, roused by poverty in the Great *Depression, who marched on Washington in the spring of 1932. Under their leader, Walter F. Waters, they demanded immediate payment of a war pension, or bonus, voted for them by Congress in 1924 but not to be paid until 1945. When the Senate rejected an enabling Bill, the Secretary of War, Patrick Hurley, had General Douglas *MacArthur use federal troops to break up the various encampments and raze them, dispersing the BEF.

Booth, Charles (1840–1916), British social researcher. As the author of *Life and Labour of the People in London* (1891–1903) he presented an exhaustive study of poverty in London, showing its extent, causes, and location. Aided by Beatrice *Webb, his methods, based on observation and on searches into public records, pioneered an approach to social studies which has been influential ever since. His special interest in the problems of old age accelerated the Old Age Pensions Act (1908).

Booth, John Wilkes (1838–65), US assassin of President *Lincoln. Brother of the tragic actor Edwin Booth, and sympathizer with the *Confederacy, he participated during the closing stages of the *American Civil War in a small conspiracy to overthrow the victorious Lincoln gov-

ernment. On 14 April 1865 he mortally wounded Lincoln in Ford's Theater in Washington and escaped to Virginia, but was discovered and killed on 26 April. Four of his fellow conspirators were hanged.

Booth, William (1829–1912), British religious leader and founder of the *Salvation Army (1878). Originally a Methodist preacher, he, assisted by his wife, Catherine Booth, preached in the streets, and made singing, uniforms, and bands a part of his evangelical mission. He used his organizational gifts to inspire similar missions in other parts of the world. National concern over the poor and the aged was increased by his *In Darkest England and the Way Out* (1890), which showed that one-tenth of the population of England and Wales was living in abject poverty.

Borden, Sir Robert Laird (1854–1937), Canadian statesman. He was chosen leader of the Conservative Party in 1901. In the general election of 1911 he defeated the Liberals and succeeded *Laurier as Prime Minister of Canada. Knighted in 1914, he remained in office throughout World War I, leading a coalition government after 1917 and joining the imperial war cabinet. He retired from political life in 1920, but remained active in public affairs until his death.

Bormann, Martin (1900–c.1945), German Nazi leader. He was briefly imprisoned for his part in a political murder in 1924, and then rewarded by appointment to *Hitler's personal staff in 1928. After the departure of *Hess in 1941 he headed the Party chancery. His intimacy with Hitler enabled him to wield great power unobtrusively. He was an extremist on racial questions, and was also behind the offensive against the Churches in 1942. He was sentenced to death *in absentia* at the *Nuremberg trials; in 1973, after identification of a skeleton exhumed in Berlin, the West German government declared that he had committed suicide after Hitler's death (1945).

Borodino, battle of (7 September 1812), fought between Russia and France, about 110 km. (70 miles west of Moscow. Here *Kutuzov chose to take his stand against *Napoleon's army. The Russian position was centred upon a well-fortified hill. After twelve hours of fierce combat, a terrific artillery bombardment and a decisive cavalry charge split the Russian forces. They were forced to withdraw and Napoleon, claiming victory, marched on an undefended Moscow. Over 80,000 men were lost in the most bloody battle of the *Napoleonic Wars.

Bosch, Juan (1909–), Dominican statesman. He founded the leftist Partido Revolucionario Dominicano (PRD) in 1939, and was exiled during the dictatorship of Rafael *Trujillo. After the latter's assassination he returned (1961) to the Dominican Republic and was elected President (1962–3) in the first free elections for nearly forty years. He introduced sweeping liberal and democratic reforms, but after nine months in office was overthrown by rightist military leaders with the backing of the Church, of landowners, and of industrialists. His supporters launched their revolt in 1965, a movement which prompted a military intervention by the USA. In 1966 he was defeated for the presidency by Joaquin Balaguer, who had heavy US backing. Since then he has remained active in politics and as a writer.

Bose, Subhas Chandra (1897-1945), Indian nationalist politician. With Jawaharlal *Nehru he founded the Indian Independence League in 1928. He became President of the Indian *National Congress Party (1938-9) but quarrelled with other leaders. He escaped from virtual house arrest (1941), went to Germany but failed to secure Nazi support and in 1943 went to Japan and Singapore. There he assumed command of the Indian National Army, recruited from Indian prisoners-of-war, and formed a provisional Indian government. He was killed in an aircrash.

Bosnia and Hercegovina, one of the six constituent republics in modern Yugoslavia. Part of the *Ottoman empire until 1878, the rise of *Pan-Slavonic nationalism provoked revolts in 1821, 1831, and 1837. A revolt in 1875 brought Austrian occupation, which was confirmed at the Congress of *Berlin in 1878 and consolidated by formal annexation into the *Austro-Hungarian empire in 1908. This provoked protest from Serbia and Russia. An international crisis only subsided when Germany threatened to intervene. Serbs continued to protest and to indulge in terrorist activity, culminating in the assassination of the Archduke *Francis Ferdinand and his wife in the capital Sarajevo in 1914. This sparked off World War I, after which Bosnia was integrated into the new Kingdom of Serbs, Croats, and Slovenes, later renamed *Yugoslavia. During World War II the two provinces were incorporated into the German puppet state of Croatia, and were the scene of much fighting by the Yugoslav partisans.

Botany Bay, an inlet on the eastern coast of Australia. Captain James Cook, the first white person to discover it, landed there in 1770 and named it Stingray Bay. The botanist Sir Joseph Banks observed many plants unknown to him, and its name was changed to Botany Bay. The British government instructed Captain Arthur Phillip to prepare a fleet for transportation to Botany Bay. The first fleet, consisting of convicts, marines, and some civilians, arrived at Botany Bay in January 1788, but Phillip decided instead to establish a settlement further north in Port Jackson. During the era of *convict transportation, the name was sometimes used as a synonym for New South Wales.

Botha, Louis (1862-1919), Boer general and statesman. He was the first Prime Minister of the Union of *South Africa. The son of a *Voortrekker, he was elected to the Natal Volksraad (parliament) in 1897. In the *Boer War he rose rapidly, and his successes at Spion Kop and elsewhere gained him promotion to general. After the Peace of *Vereeniging (1902), he worked tirelessly for reconciliation with Britain. In 1910 he became Prime Minister, and in 1911 he established the South African Party. In 1915 some of his followers turned against him in an Afrikaner rebellion. He suppressed it, and then led a successful campaign against the Germans in *South-West Africa.

Botswana, a land-locked country in southern Africa. Formerly known as Bechuanaland, British missionaries visited the southern Tswana people in 1801, and in 1817 the London Missionary Society settled at Kuruman. David *Livingstone and other missionaries operated from here during the second quarter of the 19th century. In 1885 the British protectorate of Bechuanaland was

General Louis **Botha** in uniform. During the Boer Wars he skilfully led guerrilla resistance to superior British forces in South Africa.

declared, to be administered from Mafeking. The success of the cattle industry led the Union of South Africa to seek to incorporate Botswana, along with Basutoland (Lesotho) and Swaziland, but this was rejected by the British government in 1935; no transfer would be tolerated until the inhabitants had been consulted and an agreement reached. The dominant tribe was the Ngwato, whose chief Seretse *Khama was banned from the country from 1948 until 1956 for marrying an Englishwoman. By now a nationalist movement had begun, which culminated in a democratic constitution in 1965 followed by independence on 30 September 1966, as the republic of Botswana, with Seretse Khama as President. He was succeeded on his death in 1980 by the vice-president Quett Masire. The country retains economic links with South Africa, although since 1980 it has moved closer to Zimbabwe.

Boulanger, Georges Ernest (1837-91), French general and politician. He won increasing popular support for his campaign for revenge on Germany after the *Franco-Prussian War (1870-1). In 1886 he became Minister of War but forfeited the support of moderate republicans who feared that he might provoke another war with Germany. Forced from his ministry in 1887, he became the focus of opposition to the government and won a series of by-elections. He failed to seize this opportunity to make himself President, and his popularity waned. The government prepared to have him tried for treason but he fled into exile. Two years later he committed suicide in Brussels.

Boumédienne, Houari (1925–78), Algerian statesman. In the early 1950s he joined a group of expatriate Algerian nationalists in Cairo which included *Ben Bella, and in 1955 he joined resistance forces in *Algeria operating against the French. He became chief-of-staff of the exiled National Liberation Front in Tunisia (1960–2). In March 1962 his forces occupied Algiers for Ben Bella after which a peace treaty was signed with France. He displaced Ben Bella in a coup in 1965, ruling until his death in 1978. He had close ties with the Communist bloc, but also maintained friendly relations with Western countries. He died in office in December 1978.

Bounty mutiny (1789), a British mutiny which occurred near the Tongan Islands on HMS *Bounty*, under the command of Captain *Bligh. Some of the crew, resenting Bligh's harsh exercise of authority and insults, rebelled under the leadership of Fletcher Christian. Bligh and eighteen others were cast off in a small, open boat with no chart. Thanks to Bligh's navigational skill and resource, they covered a distance of 5,822 km. (3,618 miles), arriving in Timor about six weeks later. Bligh was exonerated at a court martial in England. Some of the mutineers surrendered and others were captured and court martialled in England. Fletcher Christian and some of the other mutineers, with a number of Tahitian men and women, settled on Pitcairn Island in 1790. Their descendants moved to Norfolk Island in 1856.

Bourguiba, Habib Ali (1903–), Tunisian statesman. A staunch nationalist, he was imprisoned at different times by the French and during World War II by the Germans. He negotiated the agreement which led to Tunisian autonomy (1954) and when Tunisia became independent (1956), he was elected Prime Minister. In 1957 he deposed the Bey of Tunis, abolished the monarchy, and was himself chosen President of the Republic by the constituent As-

Tunisian nationalists acclaim Habib **Bourguiba** as he leaves France for a triumphant return to Tunisia in 1955, following his ratification of the agreement with France that opened the way for his country's independence. The successful negotiations secured Bourguiba's position as undisputed Tunisian leader.

sembly, and President for life in 1975. A moderate, Bourguiba faced riots in 1978 and 1980. After 1981 he democratized the National Assembly of his one-party state, and recognized the right of opposition by forging a coalition alliance. He was deposed in 1987.

Boxer Rising (1899–1900), a popular anti-western movement in China. The secret society of Righteous and Harmonious Fists, which was opposed to foreign expansion and the Manchu court, claimed that by

A Chinese view of Western reaction to the **Boxer Rising**. This ink and colour sketch shows warships of the Allied navies bombarding Chinese forts at Taku at the mouth of the Tianjin River, 1900. (British Museum, London)

training (including ritual boxing) its members could become immune to bullets. The movement began in Shandong province and had its roots in rural poverty and unemployment, blamed partly on western imports. It was pushed westwards and missionaries, Chinese Christians, and people handling foreign goods were attacked. The movement was backed by the empress dowager *Cixi and some provincial governors. In 1900 the Boxers besieged the foreign legations in Beijing for two months until they were relieved by an international force which occupied and looted the capital; Cixi and the emperor fled in disguise. The foreign powers launched punitive raids in the Beijing region and negotiated heavy reparations in the Boxer Protocol (1901). The rising greatly increased foreign interference in China, and further reduced the authority of the *Qing dynasty.

Boycott, Charles Cunningham (1832–97), English land agent in Ireland. When, at the direction of the Land League, Irish tenants on the estate of Lord Erne in County Mayo asked for rent reductions and refused to pay their full rents, Boycott ordered their eviction (1880). *Parnell urged everyone to refuse all communication with Boycott and to ostracize his family. The policy was successful and Boycott was forced to leave. The practice of non-communication became known as boycotting.

Bradlaugh, Charles (1833–91), British social reformer. A republican and keen supporter of reform movements, he was tried, with Annie *Besant, in 1877–8 for printing a pamphlet on birth control. The charge failed and contraceptives could thereafter be openly advertised. When returned as Member of Parliament for Northampton in 1880 his refusal to take the Bible oath of allegiance to the crown was backed by his voters and led eventually in 1886 to British Members of Parliament having the right to affirm rather than to swear allegiance.

Bradley, Omar Nelson (1893–1981), US general. In World War II he commanded a corps in the *North African and Sicilian campaigns. He commanded US land forces in the *Normandy Campaign, and later, following the *Ardennes Campaign, went beyond Eisenhower's orders to link up with the Soviet forces on the Elbe in 1945. He was instrumental in building up *NATO, formulating US global defence strategy in the post-war years, and in committing US troops to fight in the *Korean War.

Brahmo Samaj (Hindu, 'Society of God'), Indian religious movement. It was a development of a Hindu social reform movement founded in Bengal in 1828 by Ram Mohan *Roy and revived as a purely religious movement in 1842 by Maharshi Devendranath Tagore (1817–1905). Following the latter's repudiation of the vedic scriptures in 1850 the movement divided between a religious group, the Adi Brahmo Samaj, and the social reformers, Brahmo Samaj of India (under Keshab Chandra Sen (1838–84)). The latter sponsored a temperance movement and campaigned for women's education and social rights. Brahmo Samaj had a powerful influence on 20th-century Hindu society.

Brandreth, Jeremiah   *Pentrich Rising.

Brandt, Willy (Herbert Ernst Karl Frahm) (1913–), West German statesman. As a young Social Democrat he had to flee (1932) from the *Gestapo and assumed the name of Willy Brandt, living in Norway. As mayor of West Berlin (1957–66), he resisted Soviet demands that Berlin become a demilitarized free city (1958) and successfully survived the crisis arising out of the building of the *Berlin Wall in 1961. In 1964 he became Chairman of the Social Democratic Party, an office he held until 1987. He was elected Federal Chancellor in 1969. His main achievement was one of *détente* or *Ostpolitik towards eastern Europe. In 1970 he negotiated an agreement with the Soviet Union accepting the *de facto* frontiers of Europe, making a second agreement on the status of Berlin in 1971. In 1971 he also signed a non-aggression agreement with the USSR and Poland, accepting the Oder–Neisse boundary; in 1972 he negotiated the agreement with the *German Democratic Republic which recognized the latter's existence and established diplomatic relations between the two nations. In 1974 he resigned as Chancellor over a spy scandal in his office, but accepted an invitation to chair the Independent Commission on International Development Issues which published its findings in 1980 and is known as the *Brandt Report.

Brandt Report (*North–South: A Programme for Survival*, 1980), report by an international commission on the state of the world economy. Convened by the United Nations, it met from 1977 to 1979 under the chairmanship of Willy *Brandt. It recommended urgent improvement in the trade relations between the rich northern hemisphere and poor southern for the sake of both. Governments in the north have been reluctant to accept the recommendations. Members of the commission therefore reconvened to produce a second report, *Common Crisis North–South: Co-operation for World Recovery* (1983), which perceived 'far greater dangers than three years ago', forecasting 'conflict and catastrophe' unless the imbalances in international finance could be solved.

Brauchitsch, Walter von (1881–1948), German field-marshal. As commander-in-chief of the German army (1938–41), he carried out the occupation of Austria (*Anschluss) and Czechoslovakia (*Sudetenland) and conducted the successful campaigns against Poland, the Netherlands, and France. He was relieved of his command by Hitler as a scapegoat for the German failure to capture Moscow.

Brazil, the largest country in South America. By 1800 the prosperity of the colony had outstripped that of Portugal. As a result of the *Napoleonic Wars, the Portuguese court was transferred to Rio de Janeiro, which was transformed into the centre of the Portuguese empire. When John VI returned to Lisbon in 1821, his son Pedro remained behind as regent. In 1822 he became Emperor Pedro I of Brazil in an almost bloodless coup, and established an independent empire which lasted until the abdication of his son Pedro II in 1889. Brazil's neo-colonial economy based upon agricultural exports such as coffee and wild rubber produced upon the fazenda (estate), and dependent on slave labour, remained virtually intact until the downfall of the country's two predominant institutions–slavery (1888), and the monarchy (1889). In 1891 Brazil became a republic with a federal constitution. The fraudulent elections of 1930 and the effects of the Great *Depression prompted the intervention of the military and the appointment of Getúlio *Vargas as provisional president. Vargas was to remain in power until he was deposed in 1945. He remained a powerful force in international politics until his suicide in 1954. Vargas' successor, Juscelino *Kubitschek (1956–61) embarked upon an ambitious expansion of the economy, including the construction of a futuristic capital city at Brazilia, intended to encourage development of the interior. President Joao Goulart (1961–4) had to face the consequent inflation and severe balance-of-payments deficit. In rural areas peasant leagues mobilized behind the cause of radical land reform. Faced with these threats, Brazil's landowners and industrialists backed the military coup of 1964 and the creation of a series of authoritarian regimes which sought to attract foreign investment. Recent governments have attempted a slow and faltering process of redemocratization whilst promoting rapid industrialization, a policy which has increased, rather than reduced the inequalities of income distribution.

Brazzaville Conference (1944), a meeting between leaders from French West and Equatorial Africa and General *de Gaulle as head of Free France. The African leaders for the first time publicly called for reforms in French colonial rule, and were given an assurance by de Gaulle that these would be implemented. Independence was still firmly ruled out.

Breckinridge, John Cabell (1821–75), US politician and general. He served as a Democrat member of the House of Representatives (1851–5), before being elected as *Buchanan's Vice-President in 1856. He presided over the Senate during the pre-war political crisis with noted impartiality, despite his strong belief in slavery and *states' rights. When the Democratic Party split in 1860, he ran for President against Abraham *Lincoln as the candidate of the Southern Democrats. From November 1861, he saw extensive service as a major-general in the army of the Confederacy Party before becoming Secretary of State for War under Jefferson *Davis in 1865.

Brest-Litovsk, Treaty of (1918), an agreement between Soviet Russia, Germany, and Austro-Hungary, signed in the town of that name in Poland. The conference opened in December 1917 in order to end Soviet participation in World War I. *Trotsky skilfully prolonged discussions in the hope of Allied help for the *Russian Revolution or of a socialist uprising of German and Austro-Hungarian workers. Neither happened. *Lenin capitulated and ordered his delegates to accept the German terms. By the treaty, Russia surrendered nearly half of its European territory: Finland, the Baltic provinces, Belorussia, Poland, the Ukraine, and parts of the Caucasus. The German armistice in the west (November 1918) annulled the treaty, but at *Versailles Russia only regained the Ukraine.

Bretton Woods Conference (1944), a United Nations monetary and financial conference. Representatives from forty-four nations met at Bretton Woods, New Hampshire, USA, to consider the stabilization of world currencies and the establishment of credit for international trade in the post-war world. They drew up a project for an International Bank for Reconstruction and Development (*World Bank) which would make long-term capital available to states urgently needing such aid, and a plan for an *International Monetary Fund (IMF) to finance short-term imbalances in international trade and pay-

ments. The Conference also hoped to see an international financial system with stable exchange rates, with exchange controls and discriminatory tariffs being ended as soon as possible. The Bank and the Fund continue as specialized agencies of the United Nations.

Brezhnev, Leonid Ilyich (1906–82), Soviet statesman. He was President of the Praesidium of the Supreme Soviet (i.e. titular head of state) (1960–4). As First Secretary of the Communist Party, he replaced *Khrushchev (1964). Through these two offices he came to exercise effective control over Soviet policy, though initially he shared power with *Kosygin. Brezhnev's period in power was marked by the intensified persecution of dissidents at home and attempted *détente followed by renewed *Cold War in foreign affairs. He was largely responsible for the decision to invade *Czechoslovakia in 1968, maintaining the doctrine that one socialist state may interfere in the affairs of another if the continuance of socialism is at risk.

Briand, Aristide (1862–1932), French statesman. He was eleven times Premier, and Foreign Minister in fourteen successive governments. He entered Parliament in 1903, a strong socialist and an impressive orator. In 1905 he took a leading part in the separation of church from state and by 1909 had become Premier. In the 1920s he was a powerful advocate of peace and international co-operation, and supported the League of Nations. The cabinet he headed in 1921 fell because of his criticism of France's harsh treatment of Germany after the Treaty of *Versailles. Working closely with Austen *Chamberlain and *Stresemann, the British and German Foreign Ministers, his greatest achievements were the *Locarno Pact (1925) and the *Kellogg–Briand Pact (1929).

Bright, John (1811–89), British politician. An active supporter of radical causes in Victorian England, he became a founder-member of the *Anti-Corn Law League and was associated with its leader, Richard *Cobden in the movement for *free trade. Bright advocated the abolition of the East India Company, supported the Union (the North) in the *American Civil War, was prominent in the campaign which led to the *Reform Act of 1867, and helped William Gladstone to prepare his policy of Irish land reform. In old age he was a strong opponent of *Home Rule.

Britain, battle of (August–October 1940), a series of air battles between Britain and Germany fought over Britain. After the fall of France, German aircraft launched a *bombing offensive against British coastal shipping with the aim of attracting and then destroying British fighter aircraft, as a prelude to a general invasion of Britain. This was in July and August 1940, and resulted in heavy German dive-bomber losses. Then attacks were made on southern England, but German losses were again heavy. In late August and early September mass bomber attacks on British aircraft factories, installations, and fighter airfields were made; these caused heavy British losses, but Hitler ordered the offensive to be diverted to British cities just as the British Fighter Command was exhausting its reserves of machines and pilots. Hitler's priority of the day bombing of London gave time for Fighter Command to recover, so that German losses again rose. On 1 October day-bombing of major cities was replaced by night-bombing, but by this time it was clear that German losses

At the height of the **battle of Britain**, the duels fought
between the British Royal Air Force (RAF) and the German
Luftwaffe could be seen in south-east England as a
bewildering pattern of vapour trails. The painter Paul
Nash, official war artist to the Air Ministry, here depicts
German bombers in formation being attacked by RAF
fighters above an airfield protected by barrage balloons.
(Imperial War Museum, London)

were so high that the attempt to destroy British air power
had failed. Consequently Hitler on 12 October postponed
indefinitely his plan to invade Britain. Though heavily
outnumbered by the Germans, the British lost 900 aircraft
against 1,700 German losses. Radar, used by the British
for the first time in battle, made a significant contribution.

Britain, Great, the countries of England, *Wales, and
*Scotland, and small adjacent islands including the
Channel Islands, linked together as a political and
administrative unit. It is thus the larger part of the
United Kingdom of Great Britain, which after the Irish
*Act of Union (1801) included all of *Ireland, after 1921
including only *Northern Ireland. In 1979 there were
referenda in both Wales and Scotland for an extension
of home rule. That in Wales produced a large majority
against devolution. A bill to implement Scottish devolution failed in the House of Commons in March 1979,
although pressure for a separate Scottish Assembly revived
in 1987. During the reign of Queen *Victoria, colonial
expansion of the *British empire reached its height. The
*dominions and colonies gradually gained independence
and for the most part elected to join the *Commonwealth
of Nations. During *World War I and *World War II
Britain fought against Germany and its allies. Britain is

a constitutional monarchy, with, since 1969, full adult
suffrage for all over 18. Since 1832 the power of the
House of *Commons has steadily increased against that
of the monarch and the House of *Lords. Since 1967 gas
and oil from offshore wells have been commercially
produced, creating a major impact on the nation's
economy. In 1973 Britain became a member of the
*European Community.

British empire, a term used to describe lands throughout the world linked by a common allegiance to the British
crown. In 1800, although Britain had lost its thirteen
American colonies, it still retained Newfoundland, thinly
populated parts of Canada, many West Indian islands,
and other islands useful for trading purposes. It held Gibraltar from Spain and in 1788 had created a convict settlement in New South Wales, Australia. During the
*Napoleonic Wars Britain acquired further islands, for
example Malta, Mauritius, the Maldives, and also Ceylon
and Cape Colony, which was particularly valuable for
fresh food supplies for ships on the way to the East. Most
of these belonged to the *East India Company, which was
steadily developing and exploiting its trade monopoly in
India and beyond. All such acquisitions were seen as part
of the development of British commerce, as was to be the
seizure of Hong Kong in 1841. From the 1820s, a new
colonial movement began, with British families taking pas-
sages abroad to develop British settlements. In 1857 the
*Indian Mutiny obliged the British government to take
over from the East India Company the administration of
that vast sub-continent; in January 1877 Queen Victoria
was proclaimed Empress of India. New tropical colonies
were competed for in the *'Scramble for Africa' and in

the Pacific. In 1884 an Imperial Federation League was formed, seeking some form of political federation between Britain and its colonies. The scheme soon foundered, being rejected by the colonial Premiers when they gathered in London for the two Colonial Conferences of 1887 and 1897. Strategically, the key area was seen to be southern Africa, and it was the dream of Cecil *Rhodes and Alfred *Milner to create a single Cape-to-Cairo British dominion, linked by a railway, and acting as the pivot of the whole empire, a dream which faded with the Second *Boer War. Another result of the Boer War was the creation of the permanent Committee of Imperial Defence (1902), whose function was to be the co-ordination of the defence of the empire, and which was to continue until 1938. The empire reached its zenith c.1920, when German and Ottoman *mandates were acquired, and over 600

million people were ruled from London. In the later 19th century movements for home-rule had begun in all the white colonies, starting in Canada, but spreading to Australasia and South Africa, such moves resulting in 1931 in *dominion status for these lands. Although the Indian National *Congress had been founded in 1885, success by the non-white peoples of the empire for similar self-government proved more difficult. It was only after 1945 that the process of decolonization began, which by 1964 was largely complete.

British Expeditionary Force (BEF), a term applied to British army contingents sent to France at the outbreak of World War I. Following the army reforms of Richard *Haldane a territorial reserve army had been created. This was immediately mobilized when war was declared

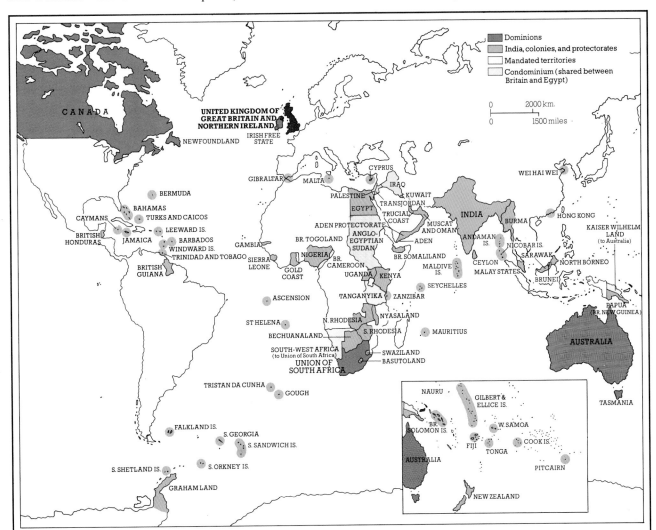

British empire (1923)

The British empire, established over a period of three centuries, resulted primarily from commercial and political motives. At its height, during the late 19th and early 20th centuries, it comprised about one quarter of the world's area and population. It acquired pre-eminence over its Dutch, Portuguese, French, and Belgian rivals through its command of the seas, and sustained its dominance through the flexibility of its rule, which encouraged the establishment of a regular civil service and relatively efficient colonial administrations. A pattern of devolution for its white colonies was inaugurated by the British North America Act of 1867, while the crown colonies, with their large indigenous populations, were ruled by a British governor and consultative councils, which delegated powers to local rulers. Nationalist agitation against economic disparities forced Britain to concede independence to most of its remaining colonies after World War II. Since then, most of the empire's former territories have elected to remain within the Commonwealth of Nations.

on 4 August and, together with regular troops, sent to France under Sir John *French. Here, as the Germans advanced into France, the BEF moved up the German eastern flank into Belgium before being halted and defeated at the battle of Mons (23-24 August). From here they steadily retreated to Ypres, where they took part in the first battle of Ypres (20 October-17 November). It is estimated that by the end of November, survivors from the original force averaged no more than one officer and thirty men per battalion of approximately 600 men. An expeditionary force was again mobilized and sent to France in September 1939.

British Honduras *Belize.

British North America Act (1867), a British Act of Parliament establishing the *dominion of Canada. As the *American Civil War drew to a close there were increasing fears in British North America of US expansionist ambitions. In 1864 representatives from United Canada joined others from New Brunswick, Prince Edward Island, Nova Scotia, and Newfoundland to discuss federation. In 1867 proposals were agreed, although Prince Edward Island and Newfoundland would not ratify them. The British Parliament passed an Act in July 1867 uniting the colonies of New Brunswick and Nova Scotia with the province of Canada, which itself was to be divided into the two provinces of Quebec (Canada East) and Ontario (Canada West), thus creating 'one Dominion under the name of Canada'. The Act formed the basis of the Canadian Constitution until the Constitution of Canada Act of 1982. The new dominion retained the status of colony, but with a system of responsible government (a cabinet government responsible to the legislature and not direct to the governor general). Provision was made for other colonies and territories later to seek admission, as was to be the case.

A contemporary engraving of Sir James **Brooke**, the redoubtable 'White Raja' of Sarawak.

British Raj (Hindi, 'rule'), a term for the British Indian empire, particularly during the period of crown rule from 1858 to 1947. Created gradually and haphazardly as a by-product of the *East India Company's trading objectives, the Raj's heyday was the half-century following the *Indian Mutiny (1857), which had abruptly ended Company rule. It was an age of *imperialism, symbolized by the proclamation of Queen Victoria as Empress of India (January 1877), and the viceroyalty of Lord *Curzon (1899-1905) over an empire 'on which the sun never sets'. The Indian National *Congress, which initiated nationalist criticism of the Raj, and eventually succeeded it, was founded in 1885. Its influence extended over the subcontinent, although more than 500 *Princely States, bound by treaty to the crown, preserved control over their domestic affairs. Control over the directly ruled territories (about three-quarters of the total area) was exercised by a Secretary of State in the British cabinet, and a Viceroy-in-Council in India. The administration was staffed by the ICS (Indian Civil Service), which was open in later years to Indians. The Indian army, with British officers in controlling positions until the 1920s and recruited from British and Indian ranks, ensured the Raj's security in conjunction with a British army garrison. The Raj ended in 1947 when Britain transferred power to the new states of India and Pakistan. British personnel withdrew, but Western modes of thought, channelled through the educational system, made a lasting contribution to the subcontinent's character.

Brooke, Sir James (1803-68), British adventurer and ruler of *Sarawak (1841-68). Arriving in Borneo in 1839 he helped one of the Brunei princes to put down a revolt and was rewarded with the governorship of Kuching in 1841. He established himself as an independent ruler (the 'White Raja') governing as a benevolent autocrat and extending his rule over much of Sarawak. Renowned for his legal reforms (which successfully adapted local custom) he resisted external attacks by Chinese opponents in 1857. Sarawak was effectively ruled by the Brooke family until the Japanese occupation of 1942-5.

Brougham, Henry Peter, 1st Baron Brougham and Vaux (1778-1868), British lawyer and statesman. A notable legal reformer, he, as Attorney-General, successfully defended Queen Caroline at her trial in 1820. An enthusiast for education, in 1828 he helped to found London University. As Lord Chancellor (1830-4) he was responsible for the setting up of the Central Criminal Court and the Judicial Committee of the Privy Council. He also helped to secure the passage of the 1832 *Reform Bill through the House of Lords and the Act of 1833 abolishing slavery in the British empire. However, his somewhat autocratic and eccentric behaviour made him enemies and, although he lived on until 1868, he never again held high office.

Brown, John (1800-59), US *abolitionist. Fired by a mixture of religious fanaticism and a violent hatred of slavery, Brown was responsible for the Pottawatomie massacre, in which five pro-slavery men were murdered. He rapidly emerged as one of the leading figures in the violent local struggle which was making 'Bleeding Kansas' into a national issue. His most dramatic gesture came in October 1859 when, at the head of a party of about twenty, he seized the federal arsenal at Harper's Ferry, Virginia, in

The militant abolitionist John **Brown** leaving prison on his way to the gallows in December 1859. This painting, by T. Hovenden, shows him bidding farewell to black slaves, to whose liberation he had dedicated himself. (Metropolitan Museum of Art, New York)

the belief that he could precipitate a slave uprising. The arsenal was recaptured by soldiers two days later, and Brown was hanged for treason and murder.

Brownshirt, member of an early Nazi paramilitary organization, the *Sturmabteilung* or SA ('assault division'). The Brownshirts, recruited from various rough elements of society, were founded by Adolf *Hitler in Munich in 1921. Fitted out in brown uniforms reminiscent of Mussolini's *blackshirts, they figured prominently in organized marches and rallies. Their methods of violent intimidation of political opponents and of Jews played a key role in Hitler's rise to power. From 1931 the SA was led by a radical anti-capitalist, Ernst Röhm. By 1933 it numbered some two million, double the size of the army, which was hostile to them. Röhm's ambition was that the SA should achieve parity with the army and the Nazi Party, and serve as the vehicle for a Nazi revolution in state and society. For Hitler the main consideration was to ensure the loyalty to his regime of the German establishment, and in particular of the German officer corps. Therefore, he had more than seventy members of the SA summarily executed by the *SS on the *'Night of the Long Knives', after which the revolutionary period of Nazism may be said to have ended.

Brown v. Board of Education of Topeka (1954), a US Supreme Court case. The Board of Education in Topeka (West Kansas) had established separate schools for white and black children, in accordance with the Supreme Court decision of *Plessy* v. *Fergusson* of 1896. The Board's policy was challenged by the National Association for the Advancement of Colored People (NAACP) and the case brought before the US Supreme Court, where it was argued by a black lawyer, Thurgood Marshall, who was himself to be the first black justice to be appointed to the Court in 1967. The Court found unanimously that racial segregation in schools violated the *Fourteenth Amendment, thus reversing the decision of 1896 and opening the way for *desegregation not only in schools but in other public facilities.

Bruce, Stanley Melbourne, Viscount Bruce of Melbourne (1883–1967), Australian statesman. A member of the House of Representatives, he represented the Nationalists and the United Australia Party. He became Prime Minister and Minister for External Affairs in the so-called Bruce–Page government. His government's policies were summed up in the slogan 'Men, Money, and Markets'. He served in the British War Cabinet and Pacific War Council (1942–5). He chaired the World Food Council (1947–51) and the British Finance Corporation for Industry (1947–57).

Brunei, country in north Borneo. By 1800 the Brunei sultanate had been reduced to *Sarawak and *Sabah. Control of Sarawak was lost to Sir James *Brooke and his successors after 1841, and in 1888 further incursions drove the sultan to accept a British protectorate, which in 1906 was extended through the appointment of a British resident. The Brunei economy was revolutionized by the discovery of substantial onshore oil deposits in 1929 and offshore oil and gas fields in the early 1960s. Partly because of these natural resources, the sultanate resisted pressure to join the newly formed Federation of *Malaysia in 1963, achieving internal self-government in 1971 and full independence from Britain in 1984.

Brüning, Heinrich (1885–1970), German statesman. As leader of the Weimar Republic's Catholic Centre Party, he was Chancellor and Foreign Minister, 1930–2. He attempted to solve Germany's economic problems by unpopular deflationary measures such as higher taxation, cuts in government expenditure, and by trying to reduce *reparation payments. But after the elections of 1930 he lost majority support in the Reichstag and ruled by emergency decrees. He was forced to resign in 1932 by President Hindenburg, whose confidence he had lost. He escaped the 1934 purge and became a lecturer at Harvard University (1939–52).

Brusilov, Aleksky (1853–1926), Russian general. He won a brilliant campaign against Austro-Hungary (1916) in south-west Russia, which, although it cost Russia at least a million lives, forced Germany to divert troops from the *Somme and encouraged Romania to join the Allies. After the fall of the Russian emperor he sided with the *Bolsheviks and directed the war against Poland.

Bryan, William Jennings (1860–1925), US politician. Elected to Congress as a Democrat, his celebrated *free-silver speech, in which he attacked *McKinley's endorsement of the *gold standard ('You shall not crucify mankind upon a cross of gold') was delivered at the Democratic Convention in 1896 and won him the presidential nomination, which he obtained again in 1900 and 1908. He supported Woodrow *Wilson, who made him Secretary of State, but Bryan resigned over Wilson's note to Germany after the sinking of the *Lusitania in World War I. In 1925 he appeared for the prosecution in the celebrated *Scopes Trial. He won the case, but died five days after the trial was concluded.

Buchanan, James (1791–1868), fifteenth President of the USA (1857–61). He served as a Democratic Senator (1835–45) and, as Secretary of State under Polk (1845–9), played a central role in the diplomatic events surrounding the *Mexican–American War, and the settlement of the *Oregon Boundary Dispute (1846). As minister to London (1853–6), he was one of the authors of the Ostend Manifesto, which backed the American claim to Cuba. As a Northerner acceptable to the South, Buchanan won the Democratic presidential nomination, and subsequently the presidency in 1856. Hard-working but limited in his vision, he consistently leaned towards the pro-slavery side in the developing dispute over slavery in the territories. He endorsed the candidacy of the Southern Democrat *Breckinridge in 1860, but supported the Union (the North) in the *American Civil War which followed.

Buddhism, the religion derived from the teachings of Buddha. The last two centuries have demonstrated the resilience of Buddhism and its ability to communicate across cultural barriers. It has had to contend with the breakdown of monarchal patronage, communist revolutions, Western technology, and commercialism. In turn it has claimed its teaching to be in tune with science and psychology while at the same time its ancient meditation techniques have maintained their appeal.

Attempts to revive Buddhism in India are indebted to the impetus of the Theosophical Society, the zeal for education reforms by the Mahabodhi Society, the spread of neo-Buddhism, particularly among the outcastes by *Ambedkar and, in recent times, the presence of Tibetan Buddhist refugees.

The Scriptures have been edited on more than one occasion from the Council called by King Rama I of Thailand in 1788 to the Sixth World Council of Buddhism at Rangoon in 1954–6. In Thailand, Buddhism continues to enjoy royal patronage, but the work of the *sangha* is also seen as an important factor in social development by neighbouring Buddhist socialist states in spite of conflict. In Sri Lanka, there have been efforts for over a century to restore its position as a leading Theravada country. Buddhism has survived even in China, while in Japan, although new religions have flourished, the Pure Land sects of Mahayana Buddhism remain popular. Like Zen, they are also represented in the USA and Europe.

Buffalo Bill *'Wild West'.

Buganda *Uganda.

Bukharin, Nikolai Ivanovich (1888–1938), Russian Bolshevik leader and theoretician. A member of the Social Democratic Party, he played an active part in the *Russian Revolution of 1917 and became editor of the Party newspaper, *Pravda* (Truth). He opposed *Lenin's withdrawal from World War I (1918), arguing in favour of promoting a European revolution. After Lenin's death

Marxist theoretician and economist Nikolai **Bukharin** (*centre front*) with Bolshevik soldiers whose zeal he hoped would inspire a general communist revolution throughout Europe.

(1924) he was a member of the *Politburo and of the *Comintern. In the 1920s he supported the *New Economic Policy, arguing that industrialization required a healthy agricultural base, and opposed *collectivization. He lost favour with *Stalin and was arrested as a 'Trotskyite' (1937). In 1938 he was put on trial, together with other prominent Bolsheviks, accused of wanting to restore bourgeois capitalism and of joining with *Trotsky in treasonable conspiracy. He was convicted and executed.

Bulganin, Nikolai Alekandrovich (1895–1975), Soviet military leader and politician. He joined the Communist Party in 1917 and served in the secret police or *CHEKA. He held various Party posts in Moscow during 1931–41 and helped organize the defence of the city during World War II. He became a Marshal of the USSR in 1945, succeeding Stalin as Minister of Defence in 1946. He was Chairman of the Council of Ministers (1955–8), during which time he shared power with *Khrushchev, who replaced him. He lost his membership of the Central Committee in 1958.

Bulgaria, a country in south-east Europe. Bulgarian nationalism in the 19th century led to a series of insurrections against the Ottoman Turks culminating in 1876, when several thousand Bulgars were massacred. Russia gave its support to Bulgaria, and war between Russia and Turkey followed. This was ended by the Treaty of San Stefano (March 1878) which created a practically independent Bulgaria covering three-fifths of the Balkan Peninsula. Britain, however, now feared that the new state would become a puppet of Russia. The Treaty of *Berlin (1878) therefore split the country into Bulgaria and Eastern Roumelia, which remained nominally under Turkish rule. In 1879 a democratic constituent assembly elected the German Prince, Alexander of Battenburg, as ruling prince and in 1885 Alexander incorporated Eastern Roumelia into Bulgaria. For this he was kidnapped by Russian officers and forced to abdicate. His successor was another German prince, Ferdinand of Saxe-Coburg (1887–1918). Taking advantage of the *Young Turk movement Ferdinand formally proclaimed full independence from Turkish rule in 1908, and was crowned king. Participation in

World War I on the side of Germany led to invasion by the Allies (1916), and the loss of territory through the *Versailles Peace Settlement. Between 1919 and 1923 Bulgaria was virtually a peasant-dictatorship under Alexander Stamboliyski, the leader of the Agrarian Union. He was murdered and an attempt by communists under *Dimitrov to seize power followed. Military and political instability persisted until 1935, when an authoritarian government was set up by Boris III (1918–43). World War II saw co-operation with Nazi Germany, followed by invasion by the Soviet Union. In 1946 the monarchy was abolished and a communist state proclaimed. Since then Bulgaria has been one of the most consistently pro-Soviet members of the *Warsaw Pact countries.

Bulge, battle of the   *Ardennes.

Bülow, Bernard, Prince von (1849–1920), German statesman. He served in the *Franco-Prussian War and the German Foreign Service before becoming German Foreign Minister (1897–1900) and then Chancellor (1900–9) under *William II. In domestic policies he was a cautious conservative, but in foreign affairs his policies were to support the emperor's wish for German imperial expansion. Following the *Boer War, when William openly supported the Boers, von Bülow improved relations with Britain, who suggested in 1900 that Germany might assist to support the decaying regime of Abdul Aziz in Morocco. France was also interested and following the *entente* with Britain (1904) the latter supported its claim. At first von Bülow retaliated by sending the emperor on a provocative visit to Tangier (1905), when Franco-German tension developed. He then, however, helped to convene the *Algeciras Conference (1906) and in 1909 agreed that France be the protector of *Morocco. He supported the Austrian annexation of *Bosnia–Hercegovina, a move which was to help precipitate *World War I. Von Bülow retired when he lost the support of the Reichstag in 1909.

Bunche, Ralph (1904–71), US administrator and diplomat. He was professor of political science at Harvard University when, during World War II, he served with the joint chiefs-of-staff and the State Department. In 1946 he joined the secretariat of the United Nations and served on the UN Palestine Commission in 1947. After Count *Bernadotte was assassinated in 1948, he carried on negotiations between the warring Arabs and Jews with such skill that he was able to arrange an armistice between them. For this achievement he was awarded (1950) the Nobel Peace Prize, the first awarded to a black American. He served as Director of the Trusteeship Division of the UN (1948–54).

Burger, Warren E. (1907–), Chief Justice of the US Supreme Court. In 1955 he was appointed judge in the Court of Appeal in the District of Columbia. In May 1969 he was appointed by President Nixon as Chief Justice of the Supreme Court, to succeed Chief Justice *Warren. In 1971 the Court supported a policy of *busing, although to little avail, as it recognized in 1974 in its judgment on *Milliken* v. *Bradley*, which accepted the reality of racial segregation by housing. In the 1978 Bakke case it supported 'positive discrimination' in favour of disadvantaged candidates for university admission, i.e. blacks or hispanics, even though it also ruled that in this particular case a rejected white candidate, Allan Bakke,

be admitted. In 1974 Burger wrote a judgment for the case of *United States* v. *Richard M. Nixon*, in which he confirmed that the Supreme Court and not the President was the final arbiter of the US Constitution.

Burkina Faso, an inland country of western Africa. Known as Haute-Volta (Upper Volta) until 1984, it was a French protectorate from 1898, originally attached to Sudan (now Mali) and later partitioned between the Ivory Coast, Sudan, and Niger. In 1958 it became an autonomous republic within the *French Community, and independent in 1960. Following a military coup in 1970, a new constitution was adopted in 1977. Since then there have been a series of military governments. It is now ruled by the National Revolutionary Council, whose president Thomas Saukara, was assassinated in 1987.

Burma, a country in south-east Asia. Having emerged in the mid-18th century from civil war under the Konbaung dynasty, Burma was invaded by the British in 1824–6, 1852, and 1885. The first two *Anglo-Burmese Wars led to the ceding of territory and the third resulted in the deposition of King Thibaw and the establishment of Upper Burma as a province of British India. The *dyarchy of 1935 led to the granting of a measure of internal self-government in 1937, and increased pressure for full independence from the nationalist Dobama Asiayone (Thakin) party. After the Japanese invasion of 1942, a government was set up under Ba Maw, and the Burma National Army formed under *Aung San. This force defected to the Allies during the final campaign of the war. Full independence was gained in 1948, Burma electing to remain outside the *Commonwealth of Nations. Civil war erupted, with challenges to central government authority by the Christian Karens and the Chin, Kayah, and Kachin hill tribes. U Nu's government succumbed to an army coup in 1962, led by *Ne Win, which established an authoritarian state based on quasi-socialist and Buddhist principles, and continued to maintain a policy of neutrality and limited foreign contact.

Burma Campaigns (World War II) (January 1942–August 1945). In 1942 two Japanese divisions advanced into Burma, accompanied by the Burma National Army of *Aung San, capturing Rangoon, and forcing the British garrison to begin the long evacuation west. The Japanese reached Lashio at the southern end of the 'Burma Road', thus cutting off the supply link from India to Nationalist China. They captured Mandalay (May 1942) and the British forces under General *Alexander withdrew to the Indian frontier. During 1943 there were attempts to reassert control over the Arakan, but these failed, although *Wingate with his Chindit units organized effective guerrilla activity behind Japanese lines, where an originally pro-Japanese population was becoming increasingly disillusioned. Early in the spring of 1944 heavy fighting took place in defence of Imphal, when an attempted Japanese invasion of Assam/Northern India was deflected in a series of bloody battles, of which Kohima was the most important. In October a three-pronged offensive was launched by British, Commonwealth, US, and Chinese Nationalist troops, and in January 1945 the 'Burma Road' was re-opened. By now a discontented Aung San had contacted *Mountbatten and in March his troops joined the Allies. Rangoon was finally captured on 1 May 1945 by an Indian division.

These portraits by Louis de Sanges of Sir Richard **Burton** and his wife Isabel were presented to the couple on their marriage in 1861. Burton was by then a world-renowned traveller and ethnologist. After his death Lady Burton destroyed many of her husband's papers, an act much regretted by later scholars. (Burton Collection, London)

Burns, John Elliott (1858–1943), British trade union leader and politician. He had been a factory worker as a child, and was largely self-educated. A radical socialist, the 1889 *London Dockers' Strike owes much of its success to his leadership. Burns was one of the first Labour representatives to be elected to Parliament (1892), but he fell out with Keir *Hardie and turned his back on socialism. As a supporter of the Liberal Party he became president of the Local Government Board (1905–14) and introduced the first Town Planning Act (1909). He was president of the Board of Trade in 1914 but resigned from the cabinet in protest against Britain's entry into World War I.

Burr, Aaron (1756–1836), US Vice-President and political adventurer. He served with distinction in the War of Independence (1775–83) and then served in the Senate (1791–7). Allying himself with the Democratic-Republicans (*Democratic Party), Burr won the same number of electoral college votes as Thomas *Jefferson in the presidential election of 1800, but was defeated in the House of Representatives and became Vice-President. His rivalry with Alexander Hamilton, who in 1804 thwarted Burr's ambition to become governor of New York, led to a duel in which Hamilton was killed and which effectively stopped Burr's public career. He became an adventurer, and was involved in a conspiracy allegedly intended to set up a separate confederacy in the west, allied to Spain, which in 1807 led to his trial (and acquittal) on a charge of treason.

Burton, Sir Richard Francis (1821–90), British explorer and writer. He joined the Indian Army in 1842 and while employed in military intelligence, he claimed to have learnt thirty-five languages. He travelled widely; in 1853, disguised as a Pathan, he went on a pilgrimage to Mecca. With J. H. Speke he travelled to uncharted

east central Africa in search of the source of the Nile. He drafted over eighty volumes on the sociology and anthropology of the countries he visited, most of which were published.

Burundi, a country in East Africa. It was ruled as a monarchy in the 19th century by *Bami* (kings) of the Tutsi tribe, who dominated a population of Bahutu. Germany annexed it as part of German East Africa in the 1890s and from 1914 it was administered by Belgium, which obtained a League of Nations *mandate and ruled it as a part of Ruanda-Urundi. In 1962 it became independent and in 1964 its union with Ruanda was dissolved. Burundi became a republic after a coup in 1966, but tribal rivalries and violence obstructed the evolution of central government.

bushranger, law-breaker who lived in the Australian bush. The term came into use in the early 19th century and the first bushrangers were escaped convicts such as

A pair of masked **bushrangers** load their weapons, ready to ambush and rob the approaching mail coach. Such hazards were not infrequent in the Australian territory of New South Wales during the 1850s.

John *Donahoe. They often operated in well-organized gangs and attacked both white settlers and *Aborigines. Bushranging was prevalent in Van Diemen's Land in the 1820s and 1830s. The main period of bushranging in the south-eastern colonies (where it was most common) was during the 1850s and 1860s.

busing, an educational policy introduced in the USA in the 1960s. Children were taken by bus from black, white, or hispanic neighbourhoods, usually to suburban schools, in order to secure racially integrated schooling. The *desegregation movement had mainly affected the southern states, where busing was first introduced, against strong opposition from white families. *De facto* segregation also existed in many northern cities, since the central areas were often inhabited entirely by blacks. In 1971 the Supreme Court approved the principle of busing. In 1972 Congress ordered that further schemes should be delayed. Busing remained a controversial issue and its use steadily declined as a means of racial integration.

Bustamante, Anastasio (1780–1853), Mexican statesman. As President of Mexico (1830–2, 1837–41) Bustamante posed as a champion of constitutionalism while violating Mexico's constitutions of 1824 and 1836. His regime was troubled by revolution and conflict with the French, who blockaded Vera Cruz (1838) as a means of obtaining compensation for damages suffered by French nationals.

Bustamante, Sir (William) Alexander (1884–1977), Jamaican statesman. He was a labour leader and founder of the Jamaican Labour Party, and became his country's first Prime Minister (1962–7) after independence from Britain in 1962. During this time he initiated an ambitious

A portrait painted in 1894 by G. F. Watts of Josephine **Butler**, whose campaigns on behalf of prostitutes, or what were then termed 'fallen women', earned her both the admiration and derision of her contemporaries. (National Portrait Gallery, London)

The introduction of compulsory **busing** of pupils to achieve racially mixed schooling provoked opposition from some white parents in the USA. Here demonstrators in Pontiac, Michigan, protest as school buses pick up local children.

five-year plan which embraced major public works projects, agrarian reform, and social welfare.

Butler, Josephine Elizabeth (Grey) (1828–1906), British social reformer. She is best known for her successful campaign against the Contagious Diseases Acts, whereby women operating as prostitutes were under regular police and medical inspection, which frequently exposed them to brutality and injustice. She was a keen supporter of higher education and social and political advancement for women. Settling with her husband in Liverpool in 1866, she devoted herself to the rescue and rehabilitation of prostitutes. She was also active in attempts to suppress the procuring of young girls for prostitution both in Britain and on the European mainland.

Butler of Saffron Walden, R(ichard) A(usten), Baron (1902–82), British statesman. He entered Parliament as a Conservative Member in 1929. During 1941–5 he was President of the Board of Education and was responsible for the Education Act of 1944, which laid down the framework for the post-war English free secondary education system and introduced the '11-plus' examination for the selection of grammar school children. He was an important influence in persuading the Conservative Party to accept the principles of the *Welfare State. Butler held several ministerial posts between 1951 and 1964, including Chancellor of the Exchequer (1951–5), but was defeated in the contest for the leadership of the Conservative Party by Harold *Macmillan in 1957 and again by Sir Alec *Douglas-Home in 1963. He became Master of Trinity College, Cambridge, and a life peer in 1965.

C

Cabral, Amilcar (1924–73), Guinean revolutionary. He founded a clandestine liberation organization against Portuguese rule. From 1963 to 1973 he led a successful guerrilla campaign which had gained control of much of the interior before he was assassinated, supposedly by a Portuguese agent. In the following year Portuguese Guinea became independent as *Guinea Bissau.

CACM   *Central American Common Market.

Cadbury, George (1839–1922), British businessman and social reformer. A Quaker, he was part owner, with his brother **Richard** (1835–99) of the cocoa and chocolate firm of Cadbury. They built the Bournville garden city estate near Birmingham, England, in 1885, for their employees. His concern for adult education and for the welfare of his workers set new standards in management. He and his wife, **Elizabeth Cadbury** (1858–1935), herself a noted social worker and philanthropist, were influential in the improvement of housing and education, and in peace movements.

Caetano, Marcello José das Neves Alves (1904–81), Portuguese statesman. As Minister for the Colonies in 1944 he drafted the law which integrated overseas territories with metropolitan Portugal. He was Prime Minister from 1968 to 1974. He was ousted from power by General Spinola in 1974 in a *putsch* which brought to an end half a century of dictatorship in Portugal, established by Caetano's predecessor, *Salazar.

Cairo Conference (22–26 November 1943), a World War II meeting, attended by *Roosevelt, *Churchill, and *Chiang Kai-shek, to decide on post-war policy for the Far East. Unconditional surrender by Japan was its prerequisite; Manchuria was to be returned to China, and Korea to its own people. At a second conference Roosevelt and Churchill met President *Inönü of Turkey, and confirmed that country's independence. The *Teheran Conference was held immediately afterwards.

Calhoun, John Caldwell (1782–1850), US statesman. He was elected to Congress in 1811. As a leader of the 'War Hawks', Calhoun committed the USA to the *War of 1812. He served as Secretary of War under President *Monroe (1817–25), and as Vice-President to both John Quincy Adams (1825–9) and *Jackson (1829–32). The leading advocate of *states' rights, he was the main architect of the theory of *nullification which led to the *Nullification crisis of 1832–3. Calhoun was the spokesman of Southern interests who saw a North–South confrontation as inevitable. He served briefly as Secretary of State under *Tyler (1844–5) before returning to the Senate.

Callaghan ministry, British Labour government (1976–9) with James Callaghan (1912–) as Prime Minister. During this ministry relations with the rest of the *European Economic Community remained cool:

some members of the cabinet were opposed to Britain's continued membership. Domestically the government could not command a majority in the House of Commons. An agreement was therefore entered into with the Liberal Party—the 'Lib–Lab Pact' (1977–8). Partly to meet Liberal interests devolution bills were introduced for Scotland and Wales, though they were rejected in referenda (1979). The government's position became weakened by widespread strikes in the so-called 'winter of discontent' (1978–9) in protest at attempts to restrain wages and it was defeated in the House of Commons on the devolution issue. The Conservatives won the election with a large majority, under Margaret *Thatcher.

Calles, Plutarco Elías (1877–1945), Mexican statesman. He achieved prominence as a military leader during the *Mexican Revolution. As President of Mexico (1924–8), Calles implemented Mexico's constitution (1917) by supporting agrarian reform, organized labour, economic nationalism, and education. During 1928–34, although not in office himself, he continued to exert a powerful influence. This period, known as the *maximato* or chieftainship, was not as successful as the Calles administration itself and caused his reputation to suffer.

Cambodia   *Kampuchea.

Cameroon, a country in West Africa believed to be the original home of the *Bantu. About 1810 King Mbwé-Mbwé walled his capital, Fomban, against the *Fulani empire. Other peoples set up small kingdoms. Germans began trading c.1860, and signed protectorate treaties in 1884. The German Protectorate of Kamerun was confirmed by the Franco-German Treaty of 1911. In 1916 Anglo-French forces occupied it, and from 1919 it was administered under *League of Nations (later UN) trusteeship, divided into British and French *mandates. In 1960 the French Cameroun became an independent republic, to be joined in 1961 by part of the British Cameroons, the remainder becoming part of Nigeria. The French and British territories in 1972 merged as the United Republic of Cameroon.

Campaign for Nuclear Disarmament (CND), a British pressure group pledged to nuclear disarmament, and to the abandonment of British nuclear weapons. CND was created in 1958 with the philosopher Bertrand Russell as President. Frustration at the lack of progress led to the creation of a splinter-group, the Committee of 100, led by Russell and pledged to civil disobedience. From 1963 to 1980 CND was in eclipse. It revived in 1980–4 mainly as a protest against the deployment of US cruise missiles at Greenham Common. Similar protest movements developed in the USA, France, Germany, and Australasia, as well as in some communist countries, notably Romania.

Campbell-Bannerman, Sir Henry (1836–1908), British statesman. He was Prime Minister of a Liberal government (1905–8). As Secretary of State for War (1895) he secured the removal of the Duke of Cambridge as army commander-in-chief but failed to introduce any far-reaching army reforms. His brief Premiership ended in 1908 with his resignation and death, but it included the grant of self-government for the *Transvaal and *Orange Free State in South Africa, support of the

'Protest and survive' was one of many slogans coined by the **Campaign for Nuclear Disarmament** (CND). This CND demonstration in London in 1983 was typical of others held in Britain and elsewhere as the peace movement revived after a decline in the 1970s.

important 1906 Trade Disputes Act, the army reforms of *Haldane, and the Anglo-Russian *entente cordiale* in 1907.

Camp David Accord (1978), a Middle East peace agreement. It was named after the official country house of the US President in Maryland, where President *Carter met President *Sadat of Egypt and Prime

United States President Jimmy Carter (*centre*), the 'matchmaker' between Egypt and Israel, celebrates the **Camp David Accord** between Egypt's President Sadat (*left*) and Israel's Prime Minister Begin (*right*). The hopes for a gradual reconciliation engendered by the 1978 agreement were never to be realized.

Minister *Begin of Israel to negotiate a settlement of the disputes between the two countries. Peace was made between Egypt and Israel after some thirty years of conflict, and provisions were agreed for an Israeli withdrawal from Egyptian territory. This agreement did not bring about peace with the other Arab countries. Instead it led increasingly to Egypt being isolated from its Arab neighbours.

Canada, a federation of ten North American provinces, the Yukon Territory, and the Northwest Territories. At the end of the American War of Independence (1783), British North America consisted of the maritime colonies of Newfoundland, Nova Scotia, Prince Edward Island, and New Brunswick, and the former French colony of *Quebec. To the north and west were areas administered by the Hudson's Bay Company. In 1791 Quebec was divided into *Upper and Lower Canada, but following the Act of Union of 1840 the two were reunited to form the Province of Canada. Two frontier agreements were made with the USA: the *Webster–Ashburton Treaty (1842) and a treaty ending the *Oregon Boundary dispute (1846). Fears of US expansion led to the British North America Act (1867), creating the Dominion of Canada. The new dominion acquired full responsibility for home affairs. In 1870 the Hudson's Bay Company's lands around the Red River were formed into the Province of Manitoba, while the Northwest Territories passed from control of the Company to the federal government. In 1873 Prince Edward Island joined the Confederation, British Columbia, including Vancouver Island, having done so in 1871. This had been on the promise of a *Canadian Pacific Railway, which was completed in 1885, enabling prairie wheat to flow east for export. Britain gave Canada title to the arctic islands

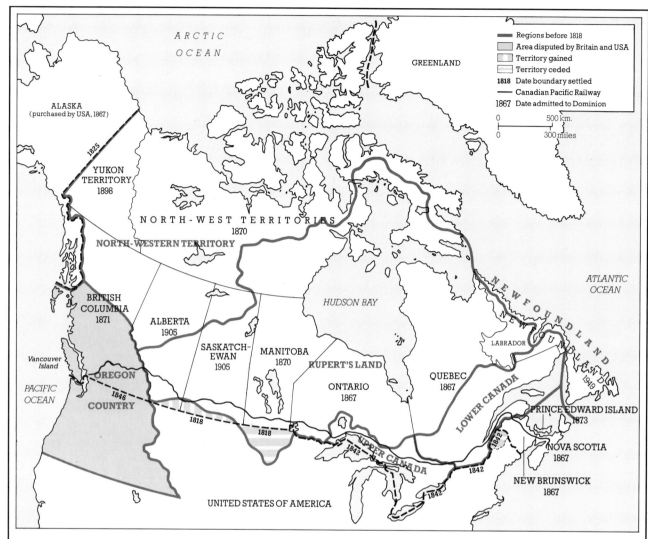

Legend:
- Regions before 1818
- Area disputed by Britain and USA
- Territory gained
- Territory ceded
- **1818** Date boundary settled
- Canadian Pacific Railway
- 1867 Date admitted to Dominion

0 — 500 km.
0 — 300 miles

Canada

Following the American War of Independence (1775–83) many loyalists to the British crown came north into the British colonies of Quebec and Nova Scotia. As the 19th century progressed, Canada evolved from colonial to dominion status (1867), establishing complete national sovereignty in 1982. The 20th century has seen an influx of immigrants from central and southern Europe to add to the earlier settlers of mainly French and British descent, the majority of residents of Quebec remaining Roman Catholic and French-speaking. Those descendants of the country's earlier inhabitants, the Indians and Inuits (Eskimos) who have not been attracted to the industrial south live in scattered settlements.

in 1880. In 1896 the Yukon boomed briefly with the Klondike *gold rush. In 1905 Alberta and Saskatchewan became federated provinces. Newfoundland joined the dominion in 1949. The Hudson's Bay Company gradually ceded all the lands for which it was responsible, but as a corporation it has retained a significant place in the Canadian economy. As the provinces developed, so did their strength *vis-à-vis* the central federal government, a strongly centralized political system being resisted. In 1982 the British Parliament accepted the 'patriation' of the British North America Act to Canada, establishing the complete national sovereignty of Canada, although it retained allegiance to the British crown as well as membership of the *Commonwealth of Nations.

Canadian Pacific Railway, the first transcontinental railway in Canada. Proposed in the 1840s, the idea was revived in 1871 on condition that British Columbia entered the new Confederation of Canada. Preparations for its undertaking led to the *Pacific Scandal of 1873, when the Prime Minister, Sir John A. *Macdonald, was forced to resign. His return to office in 1878 brought the railway its charter: the line was completed in 1885.

canals, artificial waterways to carry ships or barges. In the late 18th and early 19th centuries hundreds of miles of canals were built by pick and shovel in Britain, the USA, and Europe to carry raw materials, and manufactured goods. Narrow barges were at first towed along a footpath by horses, but later had their own steam or motor engine. Wider barges on rivers and broad canals used sail or were towed by a steam tug. By 1830 a canal network covered Britain, particularly the midlands. In the USA, Erie Canal (1817–25) linked Buffalo on Lake Erie with Albany on the River Hudson and from there to New York, which then became the focal point of westward development. In the second half of the 19th century canals in Britain and the USA gave way to rail, but on the continent of Europe a canal boom continued into the 20th century, linking all the major rivers and using large barges.

Canning, Charles John, 1st Earl Canning (1812–62), British statesman. The son of George *Canning, he was governor-general of India at the time of the *Indian Mutiny, and played a notable part in the work of reconciliation which followed it. He was subsequently first viceroy of India (1858–62), and was known as 'Clemency Canning' for his policy of no retribution.

Canning, George (1770–1827), British statesman. Entering Parliament in 1794, he became known as a firm opponent of revolutionary France. In 1807 he was appointed Foreign Secretary and was responsible for ordering the destruction of the Danish fleet at the second battle of *Copenhagen (1807) to prevent it falling into Napoleon's hands, and for the decision to wage the *Peninsular War. After a quarrel with Lord *Castlereagh, which ended in a duel, he held no important office for some years. However, following Castlereagh's death in 1822 he again became Foreign Secretary. He openly supported the *Monroe Doctrine and recognized the independence of Spain's South American colonies. He arranged the Anglo–French–Russian agreement that resulted in Greek independence from Turkey. He was Prime Minister for a few months before his death in August 1827, but had to rely on Whig support, since his policies, notably his advocacy of *Catholic emancipation, had antagonized his Tory colleagues.

Cape Province, formerly Cape Colony in South Africa. The territory was inhabited by San (Bushmen), Khoikhoi (Hottentots), and Bantu-speaking Africans when the Dutch and French built up their trading settlements there from the 16th century onwards. In 1779 the first of the *Xhosa wars broke out. From 1795 to 1803 and from 1806 to 1815 the British held the Cape and this became permanent under the Treaty of Vienna (1815). Relations between the British settlers and the *Boers over the anglicization of the courts and schools, the official use of English, control of farmland, and the emancipation of slaves quickly deteriorated. When the status of crown colony was granted in 1853, the franchise for the elected assembly was for males, whose property qualified them to vote, regardless of colour. In the 1890s the attempts by Cecil *Rhodes (as Prime Minister of Cape Colony) to unite the Cape with the three Dutch republics, under British rule, led to the Second *Boer War (1899–1902). By the Act of Union (1910) Cape Colony joined the Union of South Africa as Cape Province, and the franchise was reduced, although partially extended again in 1956.

capitalism, a system of economic organization under which the means of production, distribution, and exchange are privately owned and directed by individuals or corporations. This system developed gradually in West European countries between the 16th and 19th centuries. Capitalist methods were used in banking and commerce before being applied to industrial production at the time of the *Industrial Revolution. In its most developed form capitalism, which is based on the principle that economic

decisions should be taken by private individuals, restricts the role of the state in economic policy to the minimum. It thus stands for *free trade. Capitalist industry has at times sought the protection of the state against foreign competition: this is known as protectionism. In the 20th century capitalist societies have been modified in various ways: often a capitalist economy is accompanied by the development of a *welfare state and is therefore known as 'welfare capitalism' as in western Europe, and in addition capitalism is combined with a degree of government intervention, as for instance in F. D. Roosevelt's *New Deal or as advocated by J. M. *Keynes. Another development is the 'mixed economy', in which the production of certain goods or services is nationalized, while the rest of the economy remains in private ownership. A trend in 20th-century capitalism, particularly since World War II, has been the growth of multi-national companies operating across national frontiers, often controlling greater economic resources than small- or medium-sized states.

Capone, Al(fonso) (1899–1947), Italian-American gangster. Also known as 'Scarface', he was the most flamboyant and widely publicized criminal of the *prohibition era. In 1925 he took over Chicago's South Side gang from Johnny Torrio, and dominated the city's underworld, dealing in bootleg liquor, extortion, white slavery, and other rackets, and controlling the corrupt administration of Mayor Bill Thompsen. His war on other syndicates, culminating in the St Valentine's Day Massacre of 1929 against the North Side gang, went

The gangster Al **Capone**, pictured in a Chicago police photograph of 1929, just three months after the St Valentine's Day Massacre in which members of his gang, disguised as policemen, had shot their opponents.

unchecked until his indictment for federal income tax evasion in 1931 led to a prison sentence. Physically and mentally broken by syphilis, he was released in 1939.

Caporetto, battle of (24 October 1917). A battle fought north of Trieste when Austro-Hungarian and German forces overwhelmed the Italian army, many of whom surrendered or fled. General Cadorna withdrew his demoralized troops north of Venice, where his new line held, eventually strengthened by British and French reinforcements. Italy was effectively out of the war, and a German offensive for March 1918 on the *Western Front could now be planned.

Caprivi, Leo, Graf von (1831–99), Prussian army officer and statesman. Chosen by *William II to succeed *Bismarck as Chancellor (1890–4) he had to face the consequences of the break-up of Bismarck's coalition in the Reichstag. He surrendered *Zanzibar to Britain in exchange for Heligoland (1890). He favoured conciliation with the working classes, socialists, and Roman Catholics, and pleased industrialists with lower grain imports, thereby laying the foundation for German trade expansion. He resigned in 1894, having displeased the agrarians by encouraging industry above agriculture, and the militarists through the reduction of military service, and finding it increasingly difficult to work with the wilful and politically active, William II.

Carbonari (Italian, 'charcoal burners'), secret revolutionary society formed in Italy, and active in France, and the Iberian Peninsula. It was formed in the kingdom of Naples during the reign of Joachim Murat (1808–15) and its members plotted to free the country from foreign rule. The society was influential in the revolt in Naples in 1820 which resulted in the granting of a constitution to the kingdom of the Two *Sicilies. Similar revolts took place in Spain and Portugal (1820), Piedmont (1821), Romagna and Parma (1831), all in turn being suppressed by government troops. It was supplanted in Italy by the more broadly-based *Young Italy movement. Meanwhile the French movement after mutinies in 1821–2 also declined.

Cárdenas, Lázaro (1895–1970), Mexican statesman. As President of Mexico (1934–40), he carried the *Mexican Revolution to the left during his six-year administration. He distributed more land than all of his predecessors combined, encouraged organized labour through support of the Confederación de Trabajadores de Mexico (CTM), and nationalized the property of the foreign-owned oil companies in 1938. Himself a mestizo (of mixed American Indian and European descent), he won the support of the Indian and Mexican working classes.

Cardwell, Edward, Viscount (1813–86), British statesman. A supporter of Sir Robert *Peel, he served as Secretary to the Treasury (1845–6) and as Secretary for War in *Gladstone's ministry of 1868–74. British incompetence in the *Crimean War (1853–6) and the efficiency of the German army in the European wars of the 1860s were the background to Cardwell's military reforms. These included subordination of the commander-in-chief to the Secretary of State for War, short service enlistment for six years in the army, and six years in the Reserves, and abolition of the purchase of commissions. This reorganized army, though much smaller than those of other large European powers, proved efficient in dealing with later 19th-century imperial crises, but inadequate in the Second *Boer War, after which a more thoroughgoing reform was carried out by *Haldane.

Caribbean Community and Common Market (CARICOM), an organization formed in 1973 to promote unity among the many small nations of the Caribbean. The main purpose of the organization is to promote the economic integration of these countries by means of a Caribbean Common Market, replacing the former Caribbean Free Trade Association (CARIFTA). Member nations also co-operate on other projects in areas such as health, education, and agricultural development.

Carlist, a conservative who supported the claims of Don Carlos (1788–1855) and his descendants to the throne of Spain. Don Carlos's religious orthodoxy and belief in the divine right of kings made him the natural leader of these traditionalists. In 1830, after Ferdinand VII had excluded his brother Don Carlos from the succession, preferring his daughter, Isabella, they formed the Carlist Party. Three years later, on the death of Ferdinand, they proclaimed Don Carlos as Charles V and a civil war followed. This lasted six years until Don Carlos fled to France. His two eldest sons continued the Carlist claims until they were forced to surrender them after an unsuccessful rising in 1860. The third son, John (1822–87), briefly advanced his claims before renouncing them in favour of his son, Charles, Don Carlos VII in 1868. In the same year a revolution overthrew *Isabella II and the Carlists asserted his claim to the throne. Although routed by the new king Amadeus in 1872, the creation of a federal republic the following year gave the party renewed hope. But in 1874–5 Isabella's son was restored as Alfonso XII. After the abdication of his son Alfonso XIII (1886–1941) in 1931 and the establishment of a republic the Carlists emerged as a strong force with popular support. In the *Spanish Civil War the Carlists sided with the nationalists, and for many years obstructed Franco's aim to restore the Bourbon dynasty. In 1969 Franco overcame Carlist objections and named the grandson of Alfonso XIII, Juan Carlos, as his successor.

Carlyle, Thomas (1795–1881), Scottish historian and essayist. He was for a time a teacher, contributed to the *Edinburgh Review*, and wrote a *Life of Schiller* (1824). In 1826 he married Jane Welsh, who was to become a noted letter writer and the constant support of her husband. In 1837 Carlyle published *The French Revolution*, in which he expressed the view, to be repeated in his later works, that great men shape the course of history. He was a critic of 19th-century materialism and a sympathizer with the sufferings of the poor, whose plight he described in *Chartism* (1839) and *Past and Present* (1843).

Carnegie, Andrew (1835–1919), Scottish-born US industrialist and philanthropist. The son of a weaver, he emigrated in 1848, and rose to become, at 18, personal assistant to Thomas A. Scott of the Pennsylvania Railroad. He invested in iron and steel production and launched his own steel company in 1872. He achieved control of steel production, ensuring greater and faster production than his rivals by technological improvement, and by

Thomas **Carlyle** (*second from right*) is shown with the theologian Frederick Denison Maurice observing labourers digging up part of Heath Street, Hampstead in Ford Madox Brown's painting *Work* (1852–63). Carlyle highlighted the sufferings of the poor in much of his writing. (Manchester City Art Galleries)

demanding accountable management. In 1901 he sold out to the US Steel Corporation for the sum of $447 million, of which $250 million came to Carnegie himself. His belief that the rich should act as trustees of their wealth for the public good was set out in his essay *The Gospel of Wealth* (1889), and applied in the philanthropic distribution of over $350 million.

Carnot, Lazare Nicolas Marguerite (1753–1823), French general and military tactician. He entered the French army in 1784 and two years later published his influential *Essay on the Use of Machines in Warfare*. In 1791 he was elected to the National Assembly and voted for the execution of Louis XVI. His reorganization of recruitment and administration of the army was mainly responsible for the successes of the revolutionary armies. He was a member of the Directory between 1795 and 1797, when he accepted the royalist victory in the elections rather than the republican coup which crushed them. He fled to Germany, falsely accused of treason. He returned to become Minister of War (1800) and continued with his administrative reforms for a year under *Napoleon Bonaparte, but resigned in 1801. He served Napoleon during the *Hundred Days of 1815, as Minister of the Interior. He brilliantly defended Antwerp,

but went into exile in Germany after the second restoration.

Carol I (1839–1914), first King of Romania (1881–1914). A German-born prince and Prussian officer, he was elected in 1866 to succeed Alexander John Cuza as Prince of Romania. His pro-German sympathies made him unpopular during the *Franco–Prussian War, but skill in manipulating politicians and elections saved him from abdication. As a result of his military leadership in the *Russo–Turkish War, he gained full independence for Romania at the Congress of *Berlin and declared a Romanian kingdom in 1881. In 1883 he concluded a secret alliance with Germany and Austro-Hungary. He reformed the Romanian constitution and the monetary system, army, and network of communication. Romanian oilfields first began to be exploited in his reign, but the problem of peasant land-hunger was not solved, and there was a Romanian peasant revolt in 1907.

Carol II (1893–1953), King of Romania. The great-nephew of *Carol I, he was exiled in 1925 for his scandalous domestic life. In 1930 he returned as king, and established a royal dictatorship inspired by intense admiration of *Mussolini. In 1940 he was forced to cede large parts of his kingdom to the *Axis Powers, and to abdicate in favour of his son, Michael.

Caroline Islands, a chain of 963 islands in the north-west Pacific, inhabited for over 2,000 years. Europeans maintained contact with the islands from the 16th century. From 1885 the Carolines were controlled by the

Spanish, then, after 1899, the Germans, and, after 1914, the Japanese, who used the islands as a base for operations during World War II. In 1947 the Carolines, together with the Mariana and Marshall Islands, became part of the US Trust Territory of the Pacific.

carpetbaggers (US), Northerners who moved into the post-Civil War American South. In the wake of the *Reconstruction Acts of 1867, large numbers of Northern entrepreneurs, educators, and missionaries arrived in the South to share in the rebuilding of the former states of the *Confederacy. Some carpetbaggers (so called because it was said that they could transport their entire assets in a carpetbag) hoped to help the black ex-slave population, but others were interested only in making a quick profit. Some of the more politically active were influential in the Republican Party in the South, and were elected to various offices, including state governorships. Following the election of Rutherford *Hayes to the Presidency (1876-80) and the restoration of home rule to the South in 1877, carpetbagger influence waned.

Carranza, Venustiano (1859-1920), Mexican statesman. As President of Mexico (1917-20), he played a minor role in the revolution against Porfirio *Díaz but a major role in shaping the course of the *Mexican revolution from 1913 to 1920. A voice for moderation during the violent decade of revolutionary politics, he defeated his rival 'Pancho' *Villa and reluctantly accepted the leftist constitution of 1917. During his administration he implemented the revolutionary provisions only when forced to do so. Driven from office before his presidential term expired, he was assassinated in the village of Tlaxcalantongo on his way into exile.

Carson, Edward Henry, Baron (1854-1935), Anglo-Irish statesman. Elected to the British Parliament in 1892, he became Solicitor General (1900-5) in the Conservative government. He was determined to preserve Ireland's constitutional relationship with Britain. He opposed the third *Home Rule Bill (1912) and organized a private army of *Ulster Volunteers, threatening that Ulster would set up a separate provisional government if the Bill proceeded. In 1914 he reluctantly agreed to Home Rule for southern Ireland but insisted that *Northern Ireland, including the predominantly Catholic counties of Tyrone and Fermanagh, should remain under the British crown. He continued an inflammatory campaign against Home Rule after the war. Although he reluctantly accepted the Anglo-Irish Treaty (1921), he never ceased to speak for the interests of Ulster.

Carson, Kit (1809-68), US frontiersman and guide. He worked with *Frémont along the Oregon Trail, played an important role in the seizure of California from Mexico in 1847, and guided many groups of settlers west during the *gold rush of 1849. He served as a US Indian agent and, during the *American Civil War, was responsible for Union (Northern) scouts in the western theatre. One of the most accomplished of the *Mountain Men, Carson was the subject of many of the legends of the early days of the American West.

Carter presidency (1977-81), the term of office of James (Jimmy) Earl Carter (1924-) as thirty-ninth President of the USA. His Southern Baptist Christian background and his disassociation from the US political establishment, which had suffered from the revelations over the *Watergate Scandal, helped him to win the Democratic nomination and election of 1976, with Walter Mondale for Vice-President. Initially regimes, such as many in South America, which failed to respect the basic human rights agreed on at the *Helsinki Conference were to be deprived of US aid, but this policy was soon abandoned. Carter's measures to pardon draft-dodgers (young men imprisoned for evading conscription in the *Vietnam War) and to introduce administrative and economic reforms were popular. Although Congress had a Democratic majority Carter was not always able to secure its support. He failed to obtain approval for his energy policy, which sought to reduce oil consumption, while the Senate in 1979 refused to ratify the agreement on *Strategic Arms Limitation (SALT II). In foreign affairs the administration achieved the *Camp David Accord between Israel and Egypt and the transference of the *Panama Canal to Panama. The President's reputation was harmed by his failure to resolve the *Iran Hostage Crisis. Although renominated by the Democrats in the 1980 election he was defeated by the Republican Ronald *Reagan.

Cartier, Sir George-Etienne (1814-73), French-Canadian statesman. His involvement in 1837 in the *Papineau Rebellion forced him into brief exile in the USA. In 1848 he was elected as a Conservative to the Canadian Legislative Assembly, holding a seat there, and later in the Canadian House of Commons, until his death. From 1857 to 1862 he was leader of the French-Canadian section of the government in the Macdonald-Cartier administration and after Confederation (1867) served as Minister of Militia in the first dominion government. One of the Fathers of Confederation and a promoter of improved relations between English and French Canada, Cartier was a central figure in the political campaign for unification, wielding immense influence as a result of his hold over the electors of French Canada.

Casablanca Conference (14-24 January 1943), a meeting in Morocco between *Churchill and F. D. *Roosevelt to determine Allied strategy for the continuation of World War II. Plans were made to increase bombing of Germany, invade Sicily, and transfer British forces to the Far East after the collapse of Germany. Both leaders expressed their determination to continue the war until Germany agreed to unconditional surrender.

Casement, Roger David (1864-1916), Irish patriot. As a British consular official, he won respect for exposing cases of ill-treatment of native labour in Africa, particularly the Upper Congo, and in South America, and was awarded a knighthood by the British government. He retired from the consular service in 1913. An Ulster Protestant, he supported Irish independence and went to the USA and to Germany in 1914 to seek help for an Irish uprising. His attempt to recruit Irish prisoners-of-war in Germany to fight against the British in Ireland failed, nor would the Germans provide him with troops. Casement, however, was landed on the Irish coast in County Kerry from a German submarine in 1916, hoping to secure a postponement of the *Easter Rising. He was arrested, tried, and executed for treason. His request to be buried in Ireland, rejected at the time, was fulfilled in 1965.

Casey, Richard Gardiner, Baron (1890–1976), Australian diplomat and statesman. He was a United Australia Party Member of the House of Representatives (1931–40) and held several portfolios. He was Australia's first Minister to the USA (1940–2), joined the British war cabinet (1942–3), and was governor of Bengal (1944–6). On returning to the House of Representatives (1949–60), representing the Liberal Party, Casey held various portfolios including that of External Affairs (1951–60). He was governor-general of the Commonwealth of Australia (1965–9).

Castilla, Ramón (1797–1867), Peruvian statesman. He began his long political career during the wars for independence against Spain. As President of his country (1845–51, 1855–62), he encouraged railway development and telegraphic communication, and supported the commercial use of the guano (the nitrogen-rich droppings of fish-eating seabirds) as a fertilizer. He developed the nitrate industries by establishing government monopolies and leasing them to private individuals. He abolished slavery, and freed the Peruvian Indian from tribute payments.

Castle Hill Uprising (1804), a convict rebellion at Castle Hill, a settlement in New South Wales, Australia. Several hundred convicts, many of them Irish nationalists, captured Castle Hill as part of a plan to gain control over other settlements. The rebellion was crushed by the New South Wales Corps, martial law was proclaimed, and the ringleaders hanged, flogged, or deported.

Castlereagh, Robert Stewart, Viscount (1769–1822), British statesman. He played a leading part in the coalition that defeated Napoleonic France and in the post-war settlement at the Congress of *Vienna. He entered the Irish House of Commons in 1790 and secured the passage of the *Act of Union which united Britain and Ireland. He resigned when George III rejected a plan for *Catholic emancipation. As Secretary for War (1807–9), he was not a success, and the attack upon his policies by George *Canning led to a duel between them. Appointed Foreign Secretary in 1812, he devoted his energies to the overthrow of Napoleon and the maintenance of the balance of power in Europe. At home, he was the focus of the hostility aroused by the British government's repressive measures. He committed suicide when overwork had disturbed the balance of his mind.

Castro (Ruz), Fidel (1927–), Cuban revolutionary and statesman. Son of an immigrant sugar planter, he joined the Cuban People's Party in 1947, and led a revolution in Santiago in 1953, for which he was imprisoned. His self-defence at this trial, known by its concluding words, *History Will Absolve Me*, was to become his major policy statement at the time. Exiled in 1955 he went to Mexico and in 1956 landed on the Cuban coast with eighty-two men, including Ché *Guevara, but only twelve men survived the landing. He conducted successful guerrilla operations from the Sierra Maestra mountains, and in December 1958 led a march on Havana. The dictator, General *Batista, fled, and on 1 January 1959 Castro proclaimed the Cuban Revolution, ordering the arrest and execution of many of Batista's supporters. Castro declared himself Prime Minister and, unable to establish diplomatic or commercial agreements

Cuban statesman Fidel **Castro**, who used his considerable gifts as an orator to win support for a communist revolution in Cuba. He is seen here in his uniform as commander-in-chief of the armed forces.

with the USA, negotiated credit, arms, and food supplies with the Soviet Union. He expropriated foreign industry, and collectivized agriculture. The USA cancelled all trade agreements (1960), and from 1961 Castro was openly aligned with the Soviet Union, emerging more and more strongly as a Marxist. The abortive US and Cuban invasion (April 1961) of the *'Bay of Pigs' boosted his popularity, as did his successful survival of the *Cuban Missile Crisis (October 1962) and of several assassination plots. A keen promoter of revolution in other Latin American countries, and of liberation movements in Africa, he has achieved considerable status in the Third World through his leadership of the Non-Aligned Movement.

Catholic emancipation, the granting of full political and civil liberties to British and Irish Roman Catholics. Partial religious toleration had been achieved in Britain by the late 17th century, but by the late 18th century many reformists were agitating for total religious freedom. In Ireland, where a majority were Catholics, concessions were made from 1778 onwards, culminating in the Relief Act of 1793, passed by the Irish Parliament and giving liberty of religious practice and the right to vote in elections, but not to sit in Parliament or hold public office. William *Pitt had become convinced of the need for full Catholic emancipation by 1798, and promises were made to the Irish Parliament when it agreed to the *Act of Union in 1801. Protestant landlords as well as George III resisted emancipation and Pitt resigned. Various attempts were made, for example in 1807, to ease restrictions, but all failed. Daniel *O'Connell took up the cause for emancipation and founded the Catholic Association in 1823, dedicated to peaceful agitation. In 1828 O'Connell won a parliamentary election for County Clare, but as a Catholic could not take his seat. The Prime Minister, *Wellington, reluctantly introduced a Relief Bill to avoid civil war. The 1829 Act removed most civil restrictions; the only one to survive to the present is that no British monarch may be a Roman Catholic.

Cato Street Conspiracy (1820), a plot to assassinate members of the British government. Under the leadership of Arthur Thistlewood, a revolutionary extremist, the

A contemporary print illustrates the violent end of the **Cato Street conspiracy**. A police officer named Smithers was stabbed as the conspirators' headquarters were broken into by police on the night of 23 February 1820; the would-be assassins were divided between fight and flight.

conspirators planned to murder Lord *Castlereagh and other ministers while they were at dinner, as a prelude to a general uprising. However, government spies revealed the plot and possibly also provoked the conspirators to take action. They were arrested at a house in Cato Street, off the Edgware Road, in London. Convicted of high treason, Thistlewood and four others were executed, the rest being sentenced to transportation for life.

cattle trails, routes along which cattle in the USA were driven to the nearest railhead for despatch to market. When the *American Civil War ended many millions of wild cattle roamed the Texas range. Enterprising *cow-

Edith **Cavell**, to whom some two hundred Allied soldiers owed their escape.

boys would round up herds, brand them, and then drive them along cattle trails to a railhead at such places as Abilene and, later, Dodge City. Two early trails from Texas were the Chisholm and the Shawnee trails. From the mid-1880s encroachments of *homesteaders, the growth of large cattle ranches fenced by barbed wire, and the steady extension of railroads all contributed to the end of open-range ranching and made long-distance cattle drives uneconomical.

caudillo (Spanish, 'leader', 'hero'), military dictator in a Spanish-speaking country. In Latin American politics caudillos have tended to circumvent constitutions, and rule by military force. A product of the weakness of formal political structures and the prevalence of family and dynastic connections in politics they have nevertheless sometimes been able to bring about order, at least until they in turn have fallen.

Cavell, Edith (1865–1915), English nurse. The daughter of a Norfolk vicar, in 1906 she helped establish a training school for nurses at the Berkendael Medical Institute in Brussels. Left in charge of the Institute after it had become a Red Cross hospital in World War I she nursed German and Allied soldiers alike. She believed it was her duty to help British, French, and Belgian soldiers to escape to neutral Holland. Unable to conceal these activities she was arrested, courtmartialled, and executed (1915). Her last words were, 'I realize that patriotism is not enough. I must have no hatred or bitterness towards anyone.'

Cavour, Camillo Benso, Count (1810–61), Piedmontese statesman. A leading agriculturalist, financier, and industrialist, he became a believer in the need for Italian unity and independence. In 1847 he founded the newspaper Il *Risorgimento (Italian, 'resurgence'), which advocated constitutional reforms. The *Revolutions of 1848 in Italy made it clear that unification would have to come through the action of the strongest Italian state, Piedmont. Elected to the first Parliament of *Piedmont, he became Prime Minister in 1852, and quickly established Piedmont as a model of economic and military progress. A series of treaties with Britain, Belgium, and France encouraged free trade. The entry of Piedmont into the *Crimean War gave it an international voice and the chance of alliances essential to end Austrian control in Italy. His secret negotiations at Plombières in 1858 with *Napoleon III resulted in the promise of Savoy and Nice to France as the price of French support which led to victory against the Austrians at *Magenta and *Solferino the following year. The unexpected truce between the emperors of France and Austria at *Villafranca, whereby Venetia was to remain an Austrian province precipitated Cavour's brief resignation. He returned to office at the beginning of 1860, when French support was again forthcoming, and master-minded the unification of all of northern Italy under Victor Emanuel II in 1859. He made use of *Garibaldi's expedition to Sicily and Naples the following year to bring those states also into a united Italy. Appointed Italy's first Premier in February 1861, Cavour died four months later, still negotiating to secure complete Italian unification with the inclusion of Venetia and the papal states.

CENTO *Central Treaty Organization.

Count Camillo **Cavour**, first Prime Minister of Italy, the nation his statesmanship helped to unify.

Jean Bedel Bokassa, emperor of the **Central African Republic**, at his coronation at Bangui in December 1977. The occasion was supposedly modelled on the coronation of Napoleon Bonaparte, whom Bokassa greatly admired, but its extravagant imperial trappings practically bankrupted the country.

Central African Federation (1953–63), a short-lived African federation, comprising the self-governing colony of Southern Rhodesia (*Zimbabwe) and the British protectorates of Northern Rhodesia (*Zambia) and Nyasaland (*Malawi). In the 1920s and 1930s Europeans in both Rhodesias had pressed for union, but Britain had rejected the proposal because of its responsibilities towards Africans in Northern Rhodesia and Nyasaland. In 1953 the Conservative government in Britain allowed economic arguments to prevail, and a federal constitution was devised by which the federal government handled external affairs, defence, currency, intercolonial relations, and federal taxes. Riots and demonstrations by African nationalists followed (1960–1), and in 1962 Britain accepted in principle Nyasaland's right to secede. A meeting of the four concerned governments at the Victoria Falls Conference agreed to dissolve the Federation, which came officially to an end in 1963. Nyasaland and Northern Rhodesia became independent. Southern *Rhodesia refused to hand political control over to its African majority, and in 1965 the white government made a unilateral declaration of independence (UDI) from Britain. It was not until 1980 that the ensuing political impasse was ended, with the creation of the republic of Zimbabwe.

Central African Republic, a country in Central Africa. Formerly the French colony of Ubangi Shari, it formed part of *French Equatorial Africa. In 1958 it became a republic within the *French Community, and fully independent in 1960. In 1976 its president, Jean Bedel Bokassa, declared it an empire, and himself emperor. Following allegations of atrocities, he was

deposed in 1979, and the country reverted to a republic. Political instability persisted, and in 1981 General Kolingba seized power from the civilian government.

Central America, the land mass comprising Panama, Costa Rica, Nicaragua, El Salvador, Honduras, Guatemala, and Belize (British Honduras), together with four Mexican states. It connects the North and South

Central America

Central America, for long part of the Spanish American empire as the captaincy-general of Guatemala, gained independence in 1821. When the United Provinces of Central America (1823–38) collapsed, the various provinces declared separate independence, except for British Honduras (Belize), which Britain had already seized from Spain in 1789. Early US attempts to build a canal (across Nicaragua) were thwarted by Britain (1855–7), but the successful Panama Canal was opened in 1914. US intervention has taken place in Guatemala (1954), El Salvador (1980), and Nicaragua (1983).

American continents by the Isthmus of Panama. Independence from Spain came to Central America in 1821, most of the area being briefly annexed (1821-2) to the Mexican empire of Agustin de *Iturbide. From 1823 to 1838 it experimented with political confederation within the United Provinces of Central America (Costa Rica, Guatemala, Honduras, Nicaragua, and El Salvador), but this soon fell victim to rivalries between liberals and conservatives, and to regional jealousies. By 1839 the political unity had ended. Military *caudillos dominated the remainder of the 19th century. The US adventurer William Walker invaded (1855-7) Nicaragua. The British occupied (1848) San Juan del Norte (Greytown) and the Bay Islands of Honduras to gain control of the Mosquito coast and to block US plans to build an inter-oceanic canal, while the French applied diplomatic pressure to secure canal rights throughout the region. In 1951 the Organization of Central American States was formed to help solve common problems. Since then the Economic Commission for Latin America, an organ of the United Nations, has encouraged co-operation concerning production, tariffs, and trade between member countries of the Latin American Free Trade Association and the *Central American Common Market.

Central American Common Market (ODECA or CACM), an economic organization comprising Guatemala, Honduras, El Salvador, Nicaragua, and Costa Rica. Beginning with a treaty signed by all five countries in 1960 the CACM sought to reduce trade barriers, stimulate exports, and encourage industrialization by means of regional co-operation. With a permanent secretariat at Guatemala City, its aim was co-operation with the member countries of the Latin American Free Trade Association. During the 1970s, however, it lost impetus, owing to war, upheaval, international recession, and ideological differences among member states. By the mid-1980s it was in a state of suspension.

Central Intelligence Agency (CIA), a US government agency. It was established by Congress in 1947 and is responsible to the President through the National Security Council. Its work consists of gathering and evaluating foreign intelligence, undertaking counter-intelligence operations overseas, and organizing secret political intervention and psychological warfare operations in foreign areas. The CIA has acquired immense power and influence, employing thousands of agents overseas, and it disposes of a large budget which is not subjected to congressional scrutiny. During the 1980s it has been actively involved in Nicaragua, in Afghanistan, and in Iran.

Central Pacific Railroad, a US railway forming the western part of the first transcontinental route. The Central Pacific Railway was built eastward from Sacramento to meet the Union Pacific coming west. With a larger federal subsidy available for the company building the most track, the CPR was constructed rapidly over the most difficult territory, crossing the Sierras, passing along the Humboldt River through Nevada into Utah, where at Promontory Point on 10 May 1869 a golden spike was driven to mark the joining with the Union Pacific and the completion of the east-west rail link. Although the Union Pacific won the extra subsidy, the organizers of the CPR, Collis P. Huntington and Leland Stanford, were able to realize sufficient from the sale of millions of acres of land grants to cover the entire cost ($90 million) of construction.

Central Treaty Organization (CENTO) (1955-79), a mutual security organization composed of representatives of Britain, Turkey, Iran, Pakistan, and Iraq. In 1956 the USA became an associate member. Formed as a result of the Baghdad Pact (1955), it was designed in part as a defence against the Soviet Union and to consolidate the influence of Britain in the Arab world. Following the withdrawal of Iraq (1958), its headquarters were moved to Ankara. It became inactive after the withdrawal of Turkey, Pakistan, and Iran in 1979.

Cetshwayo (often spelt Cetewayo) (1826-84), King of the *Zulus (1873-9). He first took part in raids against European settlers in 1838. When his father Mpande died in 1872 fighting broke out: six of his half-brothers were killed and two exiled to enable him to ascend the throne. His installation was performed by Sir Theophilus Shepstone. Cetshwayo was angered by Shepstone's support of Boer claims to his territory, and increased his army to defend it. This led to the *Zulu War (1879), in which he defeated the British. The British captured his capital Ulundi eight months later. Cetshwayo was deposed and sent to London. In 1883 an attempt by the British to restore him failed; he was attacked by an old enemy and fled to a native reserve.

Chaco War (1932-5), a conflict between Paraguay and Bolivia. The Gran Chaco, an extensive lowland plain, had been an object of dispute between the two countries since the early 19th century, but Bolivia's final loss of its Pacific coast in 1929 (the *Tacna-Arica settlement) prompted it to push its claims to the Chaco. Border clashes in the late 1920s led to outright war in 1932. Bolivia had the larger army and superior military equipment, but the Aymará and Quechua Indian conscripts from the Andean highlands did not fare well in the low, humid Chaco. The Paraguayans, although poorly trained and equipped, were fighting closer to home and were accustomed to the tropics. The Paraguayan colonel José Félix Estigarribia drove the Bolivians west across the Chaco and forced his enemies to sue for peace in 1935. Paraguay gained most of the disputed territory, but the price was immense for both countries. More than 50,000 Bolivians and 35,000 Paraguayans had lost their lives. Economic stagnation was to plague both combatants for years to come.

Chad, an inland country in north central Africa. By the early 1890s much of Chad had fallen under the control of the Sudanese conqueror, *Rabeh. French expeditions advanced into the region, and French sovereignty was recognized by the European powers. After *Fashoda (1898) France declared a protectorate, and in 1908 Chad became part of French Equatorial Africa, though control was complete only in 1912. In 1920 Chad became a colony under French administration, its rich mineral deposits being rapidly exploited. In 1940 Chad was the first colony to declare for the *Free French. It became autonomous within the *French Community in 1958, and a fully independent republic in 1960, with François Tombalbaye as the first President. Since then the country has struggled to maintain unity between the Arabic-speaking Muslim peoples of the north and the more economically developed south and west. In 1977 Libya invaded and civil war continued until 1987.

Chadwick, Sir Edwin (1800-90), British public health reformer. A friend and disciple of Jeremy *Bentham, he was the architect of the *Poor Law Amendment Act (1834). His report for the royal commission set up in 1833 to investigate the conditions of work of factory children resulted in the passing of the Ten Hours Act. In 1840, concerned at the number of people driven into pauperism by the death of the breadwinner during the numerous outbreaks of cholera, he conducted on behalf of the Poor Law Commissioners an *Inquiry into the Sanitary Condition of the Labouring Population*, published in 1842. As a result of this and subsequent agitation, an Act of 1848 gave municipalities powers to set up local boards of health, subject to Public Health Commissioners, among them Chadwick himself. During his term of office as Commissioner of the Board of Health (1848-54), he persuaded urban authorities to undertake major water, drainage, and slum clearance schemes to reduce *disease.

Chamberlain, Arthur Neville (1869-1940), British statesman. Son of Joseph *Chamberlain, he first entered Parliament in 1918. As Minister of Health (1923 and 1924-9), he was responsible for the reform of the *Poor Law, the promotion of council-house building, and the systematizing of local government. A skilful Chancellor of the Exchequer (1931-7), he steered the economy back towards prosperity with a policy of low interest rates and easy credit. As Prime Minister (1937-40) his hope for a large programme of social reform was ended by the necessity for rearmament, which began in 1937. His policy, largely popular at the time but later termed *'appeasement' by his critics, was to accommodate the European dictators in order to avoid war. At his three meetings with Hitler, at Berchtesgaden, at Godesberg, and at *Munich, he made increasing concessions. He did not in fact save *Czechoslovakia from German invasion (March 1939). When Germany invaded Poland later in the year, Chamberlain had little choice but to declare war. In May 1940, following the routing of British forces in Norway, his own party rebelled against him and he

was forced to resign the Premiership in favour of Winston *Churchill, whom he whole-heartedly supported, until his death later in the year.

Chamberlain, Joseph (1835-1914), British statesman. In local politics he won distinction as a Liberal mayor of Birmingham (1873-6), pioneering municipal reform. Elected to Parliament (1876), he organized national Liberal Associations throughout the country. These helped to win the election of 1880, when he joined the cabinet. Unable to support Gladstone's *Home Rule policy for Ireland, he left the Liberals to join the Conservatives as a Liberal-Unionist (1887). As Colonial Secretary (1895-1903) he distanced himself from the *Jameson Raid, but supported *Milner's policies in South Africa which precipitated the Second *Boer War, and encouraged the formation of the Commonwealth of *Australia. A committed imperialist, he came increasingly to regard a trade policy of protection as essential to the British economy, resigning in 1903 to campaign for an end to *free trade and the introduction of tariffs to encourage trade within the empire (imperial preference).

Chamberlain, Sir Austen (1863-1937), British statesman. Son of Joseph *Chamberlain, he entered Parliament in 1892 as a Liberal-Unionist. He was Chancellor of the Exchequer (1903-05) and Secretary of State for India (1915-17), resigning over alleged blunders in the *Mesopotamia Campaign. He became Chancellor of the Exchequer again in 1919 and leader of the Conservative Party in 1921, but loyalty to *Lloyd George led to his resignation in 1922. Returned to favour, he was Foreign Secretary (1924-9), playing a major part in securing the *Locarno Treaties.

Chapultepec Conference (1945), an Inter-American Conference on the Problems of War and Peace, held in Mexico City. The Act of Chapultepec (1945), adopted by twenty republics, resolved to undertake joint action in repelling any aggression against an American state. This was formalized by the Inter-American Treaty of Reciprocal Assistance (the Rio Treaty, 1947), and constituted a significant step in the history of *Pan-Americanism.

Charge of the Light Brigade   *Balaklava.

Charles X (1757-1836), King of France (1824-30). As the Comte d'Artois, the dissolute and reactionary brother of Louis XVI, he was ordered by the king to leave France in 1789 and became the leader of the exiled royalists. He returned to France in 1814 and during the reign of his next brother, *Louis XVIII, led the ultra-royalist party. After his succession he was initially popular, but his proclamation to rule by divine right and his choice of ministers who did not reflect liberal majorities in Parliament soon led to unrest. The defeat of an unpopular ministry in June 1830 prompted him to issue the July Ordinances, which established rigid control of the press, dissolved the newly elected chamber, and restricted suffrage. These measures enraged the populace and he was forced, in the *July Revolution, to abdicate. After the succession of *Louis-Philippe, he returned to Britain.

Charles XIV (Jean Baptiste Jules Bernadotte) (1763-1844), King of Sweden and Norway (1818-44). A

Britain's Prime Minister Neville **Chamberlain** returns from Munich in September 1938. He holds the paper 'signed by Herr Hitler' which, as he announces, promises 'peace with honour'.

Charles XIV, formerly Jean Baptiste Jules Bernadotte, after his election in 1810 as crown prince Charles John of Sweden. Abandoning his Bonapartist loyalties, he henceforth devoted himself and his adopted country to the defeat of Napoleon and to union with Norway.

supporter of the French Revolution, he served brilliantly under *Napoleon Bonaparte in the Italian Campaign. At one time a rival to Napoleon, he nevertheless supported the latter when he proclaimed the empire in 1804. He fought at Austerlitz and Wagram and became governor of Hanover before being invited (1810) by the Swedish Riksdag (Parliament) to succeed the senile, childless Charles XIII. He accepted, becoming a member of the Lutheran Church. As crown prince he allied Sweden with Britain and Russia and played an important part in the defeat of Napoleon at the battle of Leipzig (1813). Having invaded Denmark, he obtained Danish agreement at the Treaty of Kiel (1814) for the transfer of *Norway to Sweden. He succeeded Charles XIII in 1818. Autocratic in style and opposed to demands for a free press

and more liberal government, he nevertheless maintained popular support throughout his reign. He was the founder of the present Swedish dynasty.

Charter 77, a Czechoslovak human rights movement. Named after a document delivered in 1977, initially signed by 242 academics, intellectuals, and churchmen, the charter appealed to the Czech government to adjust the country's laws in conformity with the Universal Declaration of Human Rights enshrined in the United Nations covenants, and to respect in practice the agreements of the *Helsinki Conference. More signatories joined the movement, even though signing inevitably cost jobs, freedom, and (in the case of Professor Patocka, one of the Charter's first spokesmen) life.

Chartism, a popular movement in Britain for electoral and social reform (1836–48). The *Reform Act of 1832 had left the mass of the population without any voice in the country's affairs, and widespread discontent was fuelled by a slump in the economy. The Chartist movement began with the formation of the London Working Men's Association, led by William Lovett and Francis *Place, who drew up a programme of reform for the common people. In 1838 *The People's Charter* was launched at a meeting in Birmingham: it called for universal male suffrage, annual parliaments, vote by ballot, abolition of the property qualification for Members of Parliament, payment of Members of Parliament, and equal electoral districts. In 1839, the Chartists, now strongly influenced by the Irish radical Feargus *O'Connor, met in London to prepare a petition to the House of Commons. The meeting revealed deep differences of opinion and after Parliament had rejected the petition, there was uncertainty about the movement's future. During that year there were riots in Birmingham and throughout the north of England; the *Newport Rising took place

The second great demonstration of **Chartism** took the form of a procession in 1842 to present a petition signed by 3,317,702 supporters to Parliament. As the march moves towards Westminster, the huge petition is borne on a litter. (Communist Party Library)

in Monmouthshire, and several Chartist leaders were arrested and imprisoned. Reorganizing themselves, in 1842 the Chartists presented a second petition, signed by three million supporters, to Parliament, which again refused to listen to their claims. The plan for a final demonstration, to be held in London in 1848 for the purpose of presenting yet another petition, was called off after the government threatened military resistance, and the movement faded into insignificance, though many Chartists were later active in radical politics.

Chattanooga Campaign (1863), a campaign during the *American Civil War. On 9 September a Federal army under General W. Rosencrans occupied Chattanooga, a strategic communication centre for the *Confederacy and pushed on south-eastwards. Forced to concentrate at Chickamauga, Rosencrans's forces were attacked by a Confederate army under General Bragg, who successfully drove Rosencrans back to Chattanooga, where the Union army was besieged for several weeks. General Ulysses *Grant assumed direct command, and, with the help of reinforcements, broke the siege and counter-attacked, winning the decisive battle of Chattanooga and opening the way for the advance on Atlanta (1864) and later for the march of General William *Sherman to the sea.

CHEKA, Soviet secret police. An acronym for the All-Russian Extraordinary Commission for the Suppression of Counter-revolution and Sabotage, it was instituted by *Lenin (December 1917) and run by Dzerzhinski, a Pole. Lenin had always envisaged the need for terror to protect his revolution and this was its purpose. Its headquarters, the Lubyanka prison in Moscow, contained offices and places for torture and execution. In 1922 the CHEKA became the GPU or secret police and later the OGPU (United State Political Administration).

chemical warfare, the use of incendiary or toxic agents in war. Chlorine gas was first used in January 1915 by Germany against the Russians in Poland with little effect; but in April 1915 its use against the British and French in Flanders won a tactical success. To counter its effects gas masks were devised, further encumbering troops in *trench warfare. In 1917 both sides used mustard gas, which severely burned and blistered the skin and caused blindness, though seldom death. Between the wars public outcry led to the virtual outlawing of gas warfare, which held throughout World War II. The German discovery of a gas which attacked the nervous system was never used, nor were gases which interfere with the functioning of the blood. Tear gas, however, continues to be used in dispersing rioters. Modern research has led to the stock-piling of potentially ever more effective chemical weapons.

Cherokee Nation, North American Indian tribe. They originally occupied a large area south of the Ohio River and supported the British in 18th-century wars against the French and during the War of Independence. Following US independence their territories were greatly reduced and in 1827 they established the Cherokee Nation in north-west Georgia through a series of treaties with the US federal government. The discovery of gold on their land resulted in pressure from the white settlers

World War I witnessed the first large-scale use of **chemical warfare**, which was employed by both sides. This photograph illustrates the effects of a gas attack. At Bethune in France, in April 1918, during the battle of Estaires, men of the British 55th (West Lancashire) Division were assailed by German tear gas. Blinded by the fumes, casualties led one another in single file to dressing stations for treatment.

to encroach on Cherokee territory. Although their treaty rights and tribal autonomy were upheld in the Supreme Court, they fell foul of both the state authorities and Jackson's policy of removal of Indian tribes to land west of the Mississippi. In 1838 President Van Buren ordered the deportation of the remaining Cherokee to the *Oklahoma Indian Territory (*Trail of Tears). In 1906 the Cherokee finally gave up their tribal allegiance and became US citizens.

Chiang Kai-shek (or Jiang Jiehi) (1887-1975), Chinese general and statesman. He took control over the *Kuomintang in 1926 and led the *Northern Expedition (1926-8). He ruthlessly suppressed trade union and communist organizations and drove the communists out of the Kuomintang. His nationalist government, established in Nanjing in 1928, lasted until 1937 and succeeded in unifying most of *China. Major financial reforms were carried out, and communications and education were improved. Chiang promoted the New Life Movement (1934-7), which reasserted traditional Confucian values to combat communist ideas. His government was constantly at war—with provincial warlords, with the communists in their rural bases, and with the invading Japanese. In 1936 he was kidnapped at *Xi'an and was released, having agreed to co-operate with the communists in resisting the Japanese. With US support, encouraged by the advocacy of his wife, Soong Mei-ling, he put nationalist forces against the Japanese from 1937 to the end of World War II, but he lost control of the coastal regions and most of the major cities to Japan early in the conflict. Talks with *Mao Zedong failed to provide a basis for agreement in 1945 and in the ensuing *Chinese Civil War, Chiang's forces were gradually worn down until he was forced to resign as President and evacuate his remaining Kuomintang forces to *Taiwan

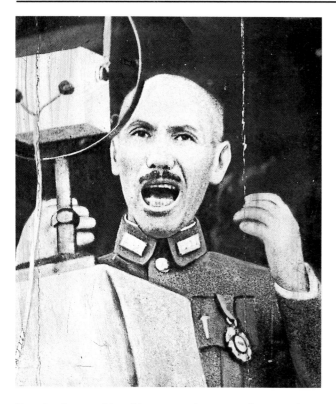

Despite the considerable personal powers of persuasion of **Chiang Kai-shek** and his access to modern propaganda media, such as radio, the Kuomintang leader never won the full support of the Chinese people.

in 1949. The administration he established there still continues as the Republic of China and Chiang was its President until his death.

Chicanos, US citizens of Mexican descent. The descendants of immigrants or of Mexicans living in the south-west of the US when the area was taken in the *Mexican–American War (1846–8), they were long an under-privileged group, but were encouraged by the *civil rights movement from the 1950s to launch Chicano organizations to secure improvements. In the 1960s Cesar Chavez wrung concessions for the California grape pickers by a series of strikes and boycotts, while in the 1970s advances were made in education, and some Mexican Americans secured high positions in government.

Chifley, Joseph Benedict (1885–1951), Australian statesman. An employee of the railways, he was active in union affairs and was dismissed during the Railway Strike of 1917. A Labor member of the federal Parliament, he held various offices, including that of Prime Minister (1945–49). He led the opposition from 1949 until his death. He successfully introduced the uniform tax scheme during the war. His attempt to nationalize the banks in the late 1940s was controversial and unsuccessful.

child labour, the employment of children in jobs which require them to work long hours. Before the *Industrial Revolution children had frequently been compelled to work from an early age, but by 1800 their employment under dirty and dangerous conditions in the new mines and factories had become a cause for concern. In 1802 the British government enacted the first laws regulating

child labour, but they proved ineffective. In 1833 a *Factory Act restricted working hours for children and provided for the appointment of inspectors. During the 19th century further factory acts and the introduction of compulsory education effectively limited child labour. Other western European countries, particularly Prussia after 1870, began to make legislation regulating the employment of children. In the USA some states passed laws to restrict the common use of child labour, but these were not always enforced. It was not until the enactment of a federal law, the Fair Labor Standards Act in 1938, that child labour could be brought to an end. In developing countries the employment of children in factories, mines, and agriculture has remained widespread.

Chile, a South American country. Chilean independence from Spain was proclaimed in 1810 by *O'Higgins; it was achieved after the South American liberator José de *San Martín crossed the Andes with an army of 3,200 men and defeated Spanish troops at the battles of Chacabuco (1817), and Maipo (1818). The discovery of rich copper deposits in the northern Atacama desert had a dramatic impact on economic life, with a railway system developing from 1851. Following war with *Bolivia and *Peru (1879–83), rich natural nitrate deposits were annexed in the north, leading to a fifty-year economic boom. By the 1920s synthetic nitrates were replacing saltpetre and dependence on copper exports placed Chile at the mercy of the world market. Political experiments after World War II failed to cope with a series of burgeoning social problems and prompted the election in 1970 of the Marxist democrat Salvador *Allende, the first avowed communist in world history to be elected President by popular vote. As the head of the Unidad Popular (a coalition of communists and socialists), Allende was faced with a majority opposition in Congress, and the hostility of the USA. He was increasingly frustrated in his attempts to implement his radical programme of

The overt hiring of **child labour** continued for some time in 19th-century London despite the introduction of laws prohibiting such exploitation. At this child market in Spitalfields, in the East End of London, uniformed police are present to keep order as a crowd of children of the poor wait to find employment.

nationalization and agrarian reform. Inflation, capital flight, and a rapidly rising balance-of-payments deficit contributed to an economic crisis in 1973. In September the army commander-in-chief Augusto *Pinochet led the military coup which cost Allende and 15,000 Chileans their lives, and prompted one tenth of the population to emigrate. The authoritarian military regime which replaced Chile's democracy has brutally suppressed all labour unions and opposition groups, and has pursued the goal of a free-market economy. Although inflation was dramatically reduced, so was demand, output, and employment. The economy has continued on a downward spiral in the 1980s with the world's highest per-capita level of external debt, the burden of which has been carried by the poorest section of society.

Chilembwe, John (c.1871–1915), *Malawi nationalist. A servant of the Baptist missionary Joseph Booth, who sent him to a Negro theological college in the USA, he became a church minister, and in 1900 established the Providence Industrial Mission in Nyasaland (now Malawi), which became a well-organized Christian community. Chilembwe protested in schools and churches against the injustices of colonial rule. In 1915 he started a rebellion but it did not gain enough African support and he was shot by the police.

China, major country of East Asia and empire until 1912. The population is predominantly ethnic Chinese (Han) with significant minorities, especially in *Tibet, *Xinjiang, and *Mongolia. After the Manchu invasion of 1644, China was ruled by the *Qing dynasty, which was at its most powerful and prosperous in the 18th century. Contact with the west precipitated crisis and decline. After the *Opium Wars, *treaty ports became the focus for both western expansion and demands for modernization. Nineteenth-century rebellions, such as the *Taiping devastated the country and undermined imperial rule in spite of the *Self-Strengthening Movement and the abortive *Hundred Days Reform. Defeat in the *Sino–Japanese War (1894–5) and the *Boxer Rising stimulated reforms, but the dynasty ended in the *Chinese Revolution of 1911. The Republic that followed *Sun Yat-sen's brief presidency degenerated into *warlord regimes after *Yuan Shikai's attempt to restore the monarchy. *Chiang Kai-shek united much of China after the *Northern Expedition and ruled from Nanjing with his nationalist *Kuomintang, but his Republic of China collapsed in the face of the Japanese invasion of 1937 and the civil war with the communists, and continued only on the island of *Taiwan after his retreat there in 1949. The *Chinese Communist Party under *Mao Zedong won the civil war, established the People's Republic of China on the mainland, and set about revolutionizing and developing China's economy and society. In the 1950s, land reform led to the *communes and the *Great Leap Forward, and urban industry was expanded and nationalized. Relations with the Soviet Union worsened and during 1966–76 the country was torn apart by the *Cultural Revolution, which ended only with Mao's death. *Deng Xiaoping and his pragmatic colleagues have since promoted a less anti-western policy with their *Four Modernizations.

China–Japan Peace and Friendship Treaty (1978), an agreement between China and Japan aimed at closer political and economic co-operation. Post-war Japanese foreign policy was characterized by a tension between dependence on the USA and popular pressure for closer relations with China. The growing western inclination of Chinese policy, the thaw in US–Chinese relations following the Nixon visit of 1972, and increasing Japanese dependence on Asia for its foreign trade improved Sino-Japanese contact, leading to the signing of the Treaty in 1978, one of the major aims of which was the establishment of closer trading links, which have since taken place.

Chinese Civil War (1927–37; 1946–9), conflicts between nationalist and communist Chinese forces. Hostilities broke out in 1927 during *Chiang Kai-shek's *Northern Expedition, with anti-leftist purges of the *Kuomintang and a series of abortive communist urban uprisings. Communist strength was thereafter most successfully established in rural areas and its supporters were able to utilize guerrilla tactics to neutralize superior nationalist strength. After a three-year campaign, Chiang finally managed to destroy the *Jiangxi Soviet established by *Mao Zedong, but after the *Long March (1934–5), the communists were able to re-establish themselves in Yan'an, in the north of the country. Hostilities between

Chinese Civil Wars and the Long March

Conflict between the nationalist Kuomintang and Chinese communists developed in 1927. By 1934 the nationalists had almost destroyed the Jiangxi Soviet, while Japan had occupied Manchuria. During 1934–5 Mao Zedong led the remnants of his army on the Long March, being joined on the way by men of the 2nd, 3rd, and 4th Front Armies. Following a truce (1937–46), civil war resumed, ending in 1949 when Chiang Kai-shek took refuge in Taiwan.

the two sides were reduced by the Japanese invasion of 1937, and, until the end of World War II in 1945, an uneasy truce was maintained as largely separate campaigns were fought against the common enemy. Violence broke out briefly immediately the war ended, resuming on a widespread basis in April 1946 after the US general George *Marshall had failed to arrange a lasting compromise settlement. During the first year of the renewed conflict, numerically superior nationalist troops made large territorial gains, including the communist capital of Yan'an. Thereafter Kuomintang organization and morale began to crumble in the face of successful guerrilla and conventional military operations by the communists, and hyperinflation and growing loss of confidence in their administration, and by the end of 1947 a successful communist counter-offensive was well under way. In November 1948 *Lin Biao completed his conquest of Manchuria, where the nationalists lost half a million men, many of whom defected to the communists. In central China the nationalists lost Shandong, and in January 1949 were defeated at the battle of Huai-Hai (near Xuzhou). Beijing fell in January, and Nanjing and Shanghai in April. The People's Republic of China was proclaimed (1 October 1949), and the communist victory was complete when the nationalist government fled from Chongqing to *Taiwan in December.

Chinese Communist Party (CCP), Chinese political party. Interest in communism was stimulated by the *Russian Revolution (1917) and the *May Fourth Movement and promoted by Li Dazhao, librarian of Beijing University, and Chen Duxiu. They were co-founders of the Chinese Communist Party at its First Congress in Shanghai in July 1921. Under *Comintern instructions, CCP members joined the *Kuomintang and worked in it for national liberation. Early activities concentrated on trade union organization in Shanghai and other large cities, but a peasant movement was already being developed by *Peng Pai. Purged by the Kuomintang in 1927 and forced out of the cities, the CCP had to rely on China's massive peasant population as its revolutionary base. It set up the *Jiangxi Soviet in southern China in 1931, and moved north under the leadership of *Mao Zedong in the *Long March (1934-5). Temporarily at peace with the Kuomintang after the *Xi'an Incident in 1936, the communists proved an effective resistance force when the Japanese invaded the country in 1937. After the end of World War II, the party's military strength and rural organization allowed it to triumph over the nationalists in the renewed civil war, and to proclaim a People's Republic in 1949. The party has ruled China since 1949. Internal arguments over economic reform and political doctrine and organization finally led to the chaos of the *Cultural Revolution (1966-76), during which the CCP appeared to turn on itself. After the death of Mao Zedong and the purge of the *Gang of Four the CCP has pursued a more stable political direction, inspired by *Deng Xiaoping.

Chinese Revolution of 1911, the overthrow of the Manchu *Qing dynasty and the establishment of a Chinese republic. After half a century of anti-Manchu risings, the imperial government began a reform movement which gave limited authority to provincial assemblies, and these became power bases for constitutional reformers and republicans. Weakened by provincial opposition to the nationalization of some major railways, the government was unable to suppress the republican *Wuchang Uprising (10 October 1911). By the end of November fifteen provinces had seceded, and on 29 December 1911 provincial delegates proclaimed a republic, with *Sun Yat-sen as provisional President. In February 1912, the last Qing emperor *Puyi was forced to abdicate and Sun stepped down to allow *Yuan Shikai to become President. The Provisional Constitution of March 1912 allowed for the institution of a democratically elected parliament, but this was ignored and eventually dissolved by Yuan Shikai after the abortive Second Revolution of 1913 which challenged his authority. Yuan had himself proclaimed emperor in 1915, but by that time central government was ineffective, and China was controlled by provincial *warlords.

Ch'ing dynasty   *Qing.

Chou En-lai   *Zhou Enlai.

Christian Church, a descriptive term for the many independent churches who believe that Jesus Christ is the Son of God. The Christian Church is divided into three main groups, the *Roman Catholic Church, the *Orthodox Churches, and the Protestant Churches, the most prominent of which are the *Anglican Communion, the *Baptist Church, the Lutheran Church, and the *Methodist Church. In the 19th century Christianity faced many challenges as a result of political, social, and scientific revolutions. There was a weakening of the close relationship between church and state, and growing scientific knowledge questioned traditional biblical accounts, most notably the creation story, which was challenged by Darwin's theory of evolution (1859). The 19th century saw great *missionary activity, particularly by the Protestant churches. There was also an increasing awareness of social deprivation, and Christian belief was an important motivator behind campaigns for the abolition of slavery, the introduction of legislation to protect workers, and the establishment of education and welfare systems. Traditional churches did not always serve growing urban areas, inspiring new Christian movements to grow, most notably the *Salvation Army. In the 20th century links between church and state were further weakened, and after the communist revolutions in Russia and China, churches were forcibly suppressed.

During the 1911 **Chinese Revolution**, republican riflemen in Chaozhou (Chao'an) in Guangdong province prepare for an offensive against imperial troops.

In 1948 the need for greater unity between the many churches was recognized with the establishment of the *World Council of Churches. Since then church membership has been declining in Western Europe but continues to grow in many developing countries.

Christian Democrats, generic term for a number of moderately conservative, mainly Roman Catholic political parties. Christian Democratic doctrine emphasizes a sense of community and moral purpose together with social reform. The parties were especially strong during the first decade after World War II and counted some of the most distinguished West European politicians as members, among them *Adenauer, *Bidault, *de Gasperi, and *Schuman. In post-war France, Italy, and the German Federal Republic Christian Democracy provided an attractive alternative to the wartime regimes.

Christian Science, a religious movement founded in the USA by Mary Baker Eddy (1821-1910), a frail woman deeply interested in medicine and the Bible. Her commitment to religious healing was deepened by her recovery from a serious illness in 1866. In 1875 she published the first of many editions of the manual *Science and Health, with Key to the Scriptures*, and in 1879 she established the Church of Christ, Scientist. This teaches that God is divine mind. Only mind is real; matter, evil, sin, disease, and death are all unreal illusions. Membership has declined in North America and Europe since 1950, but there has been considerable growth in Africa and South America. Mrs Eddy founded the international daily newspaper the *Christian Science Monitor* in 1908.

Christian Socialism, a term widely used for a form of *socialism based on Protestant Christian ideals. The term was first used in Britain in the 1840s by clergy, including Charles Kingsley, who opposed the social consequences of competitive business and unrestricted individualism, their aim being to improve the status of workers. Late 19th-century urban and industrial conditions stimulated further opposition to unrestricted capitalism, with the establishment in 1889 of the British Christian Social Union and the US Society of Christian Socialists. A belief that the established Churches were more sympathetic to the interests of capital than to the conditions of labour gave rise to the more radical Social Gospel movement. Its leaders, mostly American, studied Christ's teaching for the purpose of tracing its implications in social and economic problems.

Chulalongkorn (1853-1910), King Rama V of Siam (Thailand) (1868-1910). Only fifteen when his father Rama IV (*Mongkut) died, Chulalongkorn was represented by a regent until he reached his majority in 1873 and used the intervening years to travel and study administrative practices abroad. He then continued his father's reformist policies, undertaking a massive modernization of his country. This, together with his astute international diplomacy, in which rival British and French interests were played off against each other, helped to protect it from colonization, although he was forced to cede some territory to French Indo-China in 1907 and to British Malaya in 1909.

Churchill, Lord Randolph Henry Spencer (1849-94), British politician. Younger son of the Duke of Marlborough and father of Winston *Churchill, he was elected as Conservative Member of Parliament in 1874. He became prominent in the 1880-5 Parliament, when he and a group of young Tories became known in opposition to the Liberals as 'the Fourth Party'. Churchill emphasized the concept of Tory democracy to attract the middle and working classes, and looked on himself as the heir to *Disraeli. A gifted rhetorician, his comment in 1886 that 'Ulster will fight and Ulster will be right' became a slogan for those resisting *Home Rule for Ireland. Chancellor of the Exchequer in 1886, he resigned when the cabinet would not support him over foreign policy and cuts in military expenditure. Ill health ended his career and he died at the age of 46.

Churchill, Sir Winston Leonard Spencer (1874-1965), British statesman and war leader. The son of Lord Randolph *Churchill and of Jenny Jerome of New York, he fought at the battle of Omdurman (1898). As a journalist he covered the *Boer War, was captured, and escaped. Elected as Unionist Member of Parliament in 1900, he switched to the Liberals in 1904 as a supporter of *free trade. He served as Under-Secretary of State for the Colonies (1906-8) and in *Asquith's great reforming government from 1908. He introduced measures to improve working conditions, established Labour Exchanges, and supported *Lloyd George's Insurance Bill against unemployment in Parliament (1911). At the Admiralty from 1911 until 1915, it was largely due to him that the navy was modernized in time to meet Germany in World War I. Resigning because of the evacuation of the Dardanelles, he served briefly on the Western Front. In 1917 he became Minister of Munitions, in 1918 Minister for War and Air. Back with the Conservatives, he was Chancellor of the Exchequer from 1924 to 1929, his return to the *gold standard bringing serious economic consequences including, indirectly, the *General Strike, in which his bellicose attitude towards the trade unions was unhelpful. In the 1930s he was out of office, largely because of his extreme attitude to the India Bill, but his support for rearmament against Nazi Germany ensured his inclusion, as First Lord of the Admiralty, in *Chamberlain's wartime government. In May 1940 he be-

Winston **Churchill** at his desk in March 1944. Britain's wartime leader, only recently recovered from pneumonia, was then devoting his formidable energies to the preparation of the Allied invasion of France (June 1944). Cigar-smoking was an ever-present Churchillian passion, even in convalescence.

came Prime Minister (and Defence Minister) of a coalition government. As war leader, Churchill was superb in maintaining popular morale and close relations with the USA and the Commonwealth. Together with Roosevelt he was instrumental in drawing up the *Atlantic Charter as a buttress of the free world. Wary of Soviet expansionism, he was concerned that the USA should not concede too many of Stalin's demands as the war drew to its close. In the 1945 election he lost office, but returned as Prime Minister from 1951 to 1955. In failing health, he was preoccupied with the need for Western unity in the *Cold War, and of a 'special relationship' between Britain and the USA. He suffered a stroke in 1953, and two years later resigned the Premiership. A master of the English language, he was a notable orator and a prolific writer.

Chu Teh   *Zhu De.

CIA   *Central Intelligence Agency.

Ciano, Count Galeazzo (1903–44), Italian politician. A leading fascist, he married *Mussolini's daughter and from 1936 to 1943 was Foreign Minister. He was among those leaders who voted for the deposition of Mussolini, and for this he was tried and shot in Verona by the puppet government established by Mussolini in northern Italy. Ciano's diaries confirm that although he arranged the pre-war agreements with Germany, he soon came to resent the German connection.

civil disobedience   *passive resistance; *satyagrahi.

Civil Rights Acts (1866, 1875, 1957, 1964), legislation aimed at extending the legal and civil rights of the US black population. The first Civil Rights Act of 1866 reversed the doctrine laid down by the *Dred Scott decision of 1857 and bestowed citizenship on all persons born in the USA (except tribal Indians, not so treated until 1924). It also extended the principle of equal protection of the laws to all citizens. The provisions of the Act were reinforced by the *Fourteenth Amendment to the Constitution, but later decisions of the Supreme Court and lack of will on the part of administrators rendered them largely ineffective. For almost a century thereafter there were few effective federal attempts to protect the black population against discrimination, and in the South in particular blacks remained persecuted second-class citizens. It was only a series of legislative acts commencing with the Civil Rights Act of 1957, and culminating in the Civil Rights Act of 1964 and the Voting Rights Act of 1965, which finally gave federal agencies effective power to enforce black rights and thus opened the way to non-discrimination.

Civil War, American   *American Civil War.

Cixi (or Tz'u-hsi) (c.1834–1908), empress dowager of China (1862–1908). A Manchu, she became a concubine of the emperor Xianfeng (ruled 1851–61), giving birth to a son in 1856 who came to the throne in 1862 as the emperor Tongzhi. Cixi acted as Regent for twelve years, and after Tongzhi's death, resumed her position after the elevation of the latter's four-year-old cousin to the throne as the emperor Guangxu. She maintained her power through a combination of ruthlessness and corruption,

The empress dowager **Cixi**, who began her career as a low-ranking concubine to the emperor Xianfeng and rose to be Regent and one of the most powerful women in the history of China. Her conservative policies denied China a peaceful transition to political reform.

until the last decade of the century, when the emperor attempted to reverse her conservative policies (*Hundred Days Reform). She responded by imprisoning Guangxu, and encouraging the *Boxer Rising. Forced by foreign military forces to flee the capital, she returned in 1902, conceding some reforms, but still tried to delay the establishment of a constitutional monarchy.

Clare election (1828), an event in Ireland which led to the passing of the Roman Catholic Relief Act by the British government in 1829. In 1828 Daniel *O'Connell, an Irish lawyer, stood for election to Parliament in the County Clare constituency, winning a resounding victory over his opponent. However, O'Connell, as a Roman Catholic, could not take his seat. The Prime Minister, the Duke of Wellington, felt that if O'Connell were excluded there would be violent disorders in Ireland. Accordingly, despite furious opposition, the government pushed through a *Catholic emancipation measure allowing Catholics to sit in Parliament and hold public office.

Clarkson, Thomas (1760–1846), British philanthropist. A strong opponent of slavery, he became a founder member of the Committee for the Suppression of the Slave Trade. He collected much information about the trade and conditions on slave ships, which was published in a pamphlet and used by William *Wilberforce in his parliamentary campaign for abolition. In 1807 an Act was passed prohibiting British participation in the *slave trade. In 1823 Clarkson became a leading member of the Anti-Slavery Society, which saw its efforts rewarded with the 1833 Act abolishing slavery in the British empire.

Clausewitz, Karl von (1780–1831), Prussian general and military strategist. He served in the Rhine campaigns of 1793–4 before being admitted to the Berlin military academy (1801), where he came under the influence of the military reformer Gerhard von *Scharnhorst. Captured by the French after the battle of Jena, he returned to Prussia in 1809 and assisted Scharnhorst in his reorganization of the Prussian army. He served briefly in the Russian army (1812–14) and helped negotiate the alliance between Prussia, Russia, and Britain against Napoleon. His most famous work, *On War*, was published after his death. It sees war as a continuation of politics, the ultimate arbiter when all else fails. He supported the conception of national war and insisted that war must be conducted swiftly and ruthlessly in order to reach a clear decision in the minimum time.

Clay, Henry (1777–1852), US statesman and orator. As Speaker of the House of Representatives (1811–14) he played a central role in the agitation leading to the *War of 1812, and was one of the commissioners responsible for the negotiation of the Treaty of *Ghent which ended it. He was one of the architects of the *Missouri Compromise and won support for his *American System, a policy to improve national unity through a programme of economic legislation. His final political achievement lay in helping the passage of the *Compromise of 1850 between the opposing *Free-Soil and pro-slavery interests. His role in arranging major sectional compromises between North and South (1820, 1833, and 1850) earned him the title of 'the Great Compromiser'.

Clemençeau, Georges (1849–1929), French statesman. He entered the National Assembly (1871) as an anti-clerical republican. He fought for justice for *Dreyfus (1897) but as Minister of the Interior and Premier (1906–9) ruthlessly suppressed popular strikes and demonstrations. In 1917, with French defeatism at its peak, he formed his victory cabinet with himself as Minister of War, persuading the Allies to accept *Foch as allied commander-in-chief. Nicknamed 'The Tiger', he became chairman of the *Versailles Peace Conference of 1919, where in addition to the restoration of *Alsace-Lorraine to France, he demanded the *Saar basin and the permanent separation of the Rhine left bank from Germany, which should also pay the total cost of the war. Failing to get all these demands he lost popularity and was defeated in the presidential election of 1920.

Cleveland, (Stephen) Grover (1837–1908), twenty-second (1885–9) and twenty-fourth (1893–7) President of the USA. He became governor of New York (1883–4) and gained a reputation as a reform politician, independent of the corrupt machine politics of *Tammany Hall. He won the Democratic nomination for President in 1884 and closely defeated his republican rival *Blaine by gaining the support of many reform Republicans—the mugwumps, who voted against their party. As President, he favoured low tariffs and civil service reforms, but he had no answer for the Depression of 1893–7. He refused arguments for *free silver, relying on exorbitant loans from a bankers' consortium led by J. P. *Morgan. In foreign affairs he opposed the rising tide of imperialist sentiment, resisting US intervention in Hawaii and Cuba, but he enlarged the scope of the *Monroe Doctrine by his stand on the dispute between Britain and Venezuela over the Venezuelan-British Guiana boundary (1897), insisting that Britain should go to arbitration.

Coates, Joseph Gordon (1878–1943), New Zealand statesman. He represented the interests of the rural poor, including the *Maori. Though he was of the conservative Reform Party he advanced the New Zealand tradition of state action for the public good, developing roads and railways, the state hydro-electricity programme, afforestation, and aiding Maori farming schemes. In coalition with *Forbes during the *Depression he broke the authority of the private banks, established the Reserve Bank, and laid the foundation for economic recovery.

Cobbett, William (1763–1835), British social reformer and journalist. In 1802 he began publishing a weekly journal, the *Political Register*. He began to denounce the British government's conduct of the war with France, calling for peace and parliamentary reform. His conversion to radical reform was completed by his observations of the sufferings of the rural poor. In 1810 he was imprisoned for denouncing flogging in the army. After a period in the USA, he spent much time travelling in the English countryside, recording his impressions in his *Rural Rides* (1830). He strongly supported the *Reform Act of 1832 and was elected Member of Parliament the same year.

Cobden, Richard (1804–65), British political economist and statesman. A strong supporter of *free trade, he believed that it would help to promote international peace. In 1839, together with John *Bright, he founded the *Anti-Corn Law League and was in part responsible for the repeal of the Corn Laws in 1846. He lost his seat in Parliament in 1857 but was re-elected two years later. In 1860 he helped to negotiate a commercial treaty with France, which was based on tariff reductions and expansion of trade between the two countries. During the *American Civil War he declared his support for the Union (the North) and helped to smooth the often difficult relationships between the US and British governments. Cobden's radicalism was tempered by doubts about the extension of the franchise, a belief in minimum state interference, and a dislike of trade unions, but he retained a hatred of any kind of injustice.

Cochise (*c*.1815–74), Apache Indian chief. Noted for his courage and military prowess, he gave his word in 1860 that he would not molest US mail riders passing through his Arizona territory and in 1872 made a peace treaty with the US government. He maintained both agreements, despite hostile acts from whites and Indians.

Cochrane, Thomas, 10th Earl of Dundonald (1775–1860), British naval officer. Elected to Parliament in 1806, he conducted a campaign against naval corruption, but was himself found guilty of fraud in 1814 and courtmartialled. After fruitless attempts to clear his name, in 1817 he took command of Chile's fleet during its struggle to win freedom from Spain. He subsequently commanded the navies of Brazil (1823–5) and of Greece (1827–8) when those countries were fighting for their independence. Receiving a free pardon, he was reinstated in the Royal Navy in 1832 with the rank of rear-admiral. Cochrane was one of the first men to advocate the use of steam power in warships.

Code Napoléon (or *Code Civil*), the first modern codification of French civil law, issued between 1804 and 1810, which sought, under the direction of J. J. Cambacérès, to reorganize the French legal system. Napoleon himself presided over the commission which drafted the laws, their articles representing a compromise between revolutionary principles and the ancient Roman (i.e. civil) law which prevailed generally throughout Europe. It may be considered a triumph of individual, bourgeois rights over those established by Church and customary law, and for a long time delayed the recognition of collective (e.g. trade-union) and workers' rights. Although revised in 1904, it has remained the basis of French private law and has been adopted by many other countries. In addition to the *Code Civil*, Napoleon was responsible for the Code of Civil Procedure (1807), the Commercial Code (1808), the Code of Criminal Procedure (1811), and the Penal Code (1811).

Cod War (1972-6), the popular name for the period of antagonism between Britain and Iceland over fishing rights. The cause was Iceland's unilateral extension of its fishing limits to protect against over-fishing. Icelandic warships harassed British trawlers fishing within this new limit (1975-6), prompting protective action by British warships. A compromise agreement was reached in 1976 which allowed twenty-four British trawlers within a 320 km. (200 mile) limit. This hastened the decline of British fishing ports such as Hull and Grimsby.

Cody, William Frederick (Buffalo Bill) *'Wild West'.

Cold War, the popular term applied to the struggle between the Soviet bloc countries and the Western countries after World War II. The Soviet Union, the USA, and Britain had been wartime allies against Nazi Germany, but already before Germany was defeated they began to differ about the future of Germany and of Eastern Europe. Wartime summit meetings at *Yalta and *Potsdam had laid down certain agreements, but as communist governments seized exclusive power in Eastern Europe, and Greece and Turkey were threatened with similar take-overs, the Western Powers became increasingly alarmed. From 1946 onwards popular usage spoke of a 'Cold War' (as opposed to an atomic 'hot war') between the two sides. The Western allies took steps to defend their position with the formation of the *Truman Doctrine (1947) and the *Marshall Plan (1947) to bolster the economies of Western Europe. In 1949 *NATO was formed as a defence against possible attack. The communist bloc countered with the establishment of the Council for Mutual Aid and Assistance (*COMECON, 1949), and the *Warsaw Pact (1955). Over the following decades, the Cold War spread to every part of the world, and the USA sought *containment of Soviet advances by forming alliances in the Pacific and south-east Asia. There were repeated crises (the *Korean War, Indo-China, *Hungary, the *Cuban Missile Crisis, and the *Vietnam War), but there were also occasions when tension was reduced as both sides sought *détente. The development of a nuclear arms race from the 1950s, only slightly modified by a *Nuclear Test-Ban Treaty in 1963 and *SALT talks (1969-79), maintained tension at a high level. Tension intensified in the early 1980s with the installation of US Cruise missiles in Europe and the announcement of the US *Strategic Defense Initiative, and receded with an agreement in 1987 for limited *arms control.

collectivization, the creation of collective or communal farms, to replace private ones. The policy was ruthlessly enforced in the Soviet Union by *Stalin between 1929 and 1933 in an effort to overcome an acute grain shortage in the towns. The industrialization of the Soviet Union depended on cheap food and abundant labour. Bitter peasant resistance was overcome with brutality, but the liquidation of the *Kulaks and slaughter by peasants of their own livestock resulted in famine (1932-3). Gradually more moderate methods were substituted with the development of state farms. A modern collective farm in the Soviet Union is about 6,000 hectares (15,000 acres) in extent, nine-tenths cultivated collectively, but each family owning a small plot for its own use. Profits are shared in collective farms; in state farms workers receive wages. Since 1945 a policy of collectivization has been adopted in a number of socialist countries. The Soviet example was followed in China by *Mao Zedong in his First Five Year Plan of 1953, but was only enforced by stages. China did not copy the ruthless subordination of agriculture to industry, preferring the peasant *commune.

Collins, Michael (1890-1922), Irish patriot. A member of the *Irish Republican Brotherhood, he fought in the *Easter Rising (1916) in Dublin. Elected Member of Parliament he was one of the members of *Sinn Fein who set up the Dáil Éireann in 1919. He worked as Finance Minister in Arthur *Griffith's government and at the same time led the *Irish Republican Army. In 1921 the British government offered a reward of £10,000 for him, dead or alive. He played a large part in the negotiations that led to the Anglo-Irish truce in 1921 and the Dáil approval of the treaty in 1922. He commanded the *Irish Free State Army at the start of the Irish civil war and was killed in an ambush at Beal-na-Blath, County Cork, in August 1922.

Colombia, a country in the extreme north-west of South America. Named New Grenada by the Spanish, it re-

During the years of rigid **collectivization** in the Soviet Union, peasants were organized to work on large collective farms. This photograph shows women on their way to work in the fields, while their babies are left in the farm crèche.

mained a viceroyalty of Spain until the battle of Boyacá (1819) during the *Spanish South-American Wars of Independence, when, joined with Venezuela, it was named by Simón *Bolívar the United States of Colombia. In 1822 under his leadership New Granada, Panama, Venezuela, and Ecuador were united as the Republic of Gran Colombia, which collapsed in 1830. In 1832 a constitution for New Granada was promulgated by Francisco Santander, which was amended in 1858 to allow a confederation of nine states within the central republic now known as the Granadine Confederation. In 1863 the country was renamed the United States of Colombia. The constitution of 1886 abolished the sovereignty of the states and the presidential system of the newly named Republic of Colombia was established. The War of the Thousand Days (1899-1902), encouraged by the USA, led to the separation of Panama from Colombia (1903). Violence broke out again in 1948 and moved from urban to rural areas, precipitating a military government between 1953 and 1958. A semi-representative democracy was restored that achieved a degree of political stability, and Colombia's economy has recovered from the setbacks of the early 1970s as diversification of production and foreign investment have increased. Agriculture is the chief source of income in Colombia, but it is estimated that the country's illegal drugs trade supplies some 80 per cent of the world's cocaine market. In 1986 the Liberal Virgilio Vargas was elected President, and during the 1980s Colombia has achieved sustained economic growth and the most successful record of external debt management in the continent.

Colombian Independence War *Spanish South-American wars of independence.

Colombo Plan (for Co-operative Economic Development in south and south-east Asia), an international organization to assist the development of member countries in the Asia and Pacific regions. Based on an Australian initiative at the meeting of Commonwealth ministers in Colombo in January 1950, it was originally intended to serve Commonwealth countries of the region. The scheme was later extended to cover twenty-six countries, with the USA and Japan as major donors.

colonialism *imperialism.

Combination Acts, laws passed by the British Parliament in 1799 and 1800 in order to prevent the meeting ('combining together') of two or more people to obtain improvements in their working conditions. Flouting the law resulted in trial before a magistrate, and *trade unions were thus effectively made illegal. The laws, which were largely inspired by the fear of radical ideas spreading from France to Britain, were nevertheless unsuccessful in preventing the formation of trade unions. The Combination Acts were repealed in 1824 as a result of the skilful campaign by Francis *Place and Joseph *Hume, and were followed by an outbreak of strikes. In 1825 another Act was passed which resulted in trade union activity but limited the right to strike.

COMECON (Council for Mutual Economic Assistance), the English name for an economic organization of Soviet-bloc countries. It was established by Stalin among the communist countries of eastern Europe in 1949 to encourage interdependence in trade and production. It achieved little until 1962, when the agreements restricting the satellite countries to limited production and to economic dependency on the Soviet Union were enforced. Present members are: Bulgaria, Cuba, Czechoslovakia, German Democratic Republic, Hungary, Mongolian People's Republic, Poland, Romania, the Soviet Union, and Vietnam (Yugoslavia has associate status). Albania was expelled in 1961. In 1987 it agreed to discuss co-operation with the *European Community.

Cominform (Communist Information Bureau), an international communist organization to co-ordinate Party activities throughout Europe. Created in 1947, it assumed some of the functions of the *Internationals which had lapsed with the dissolution of the *Comintern in 1943. After the quarrel of *Tito and *Stalin in 1948 Yugoslavia was expelled. The Cominform was abolished in 1956, partly as a gesture of renewed friendship with Yugoslavia and partly to improve relations with the West.

Comintern (Communist *International), organization of national communist parties for the propagation of communist doctrine with the aim of bringing about a world revolution. It was established by *Lenin (1919) in Moscow at the Congress of the Third International with *Zinoviev as its chairman. At its second meeting in Moscow (1920), delegates from thirty-seven countries attended, and Lenin established the Twenty-one Points, which required all parties to model their structure on disciplined lines in conformity with the Soviet pattern, and to expel moderate ideologies. In 1943 *Stalin dissolved the Comintern, though in 1947 it was revived in a modified form as the *Cominform, to co-ordinate the activities of European communism. This, in turn, was dissolved in 1956.

Commons, House of, the lower chamber of the British *Parliament. Members of the House of Commons are today elected by universal adult suffrage, following a series of *Reform Acts and other legislation (1832, 1867, 1884, 1918, 1928, 1945, 1969). Although the Commons had gained considerable constitutional powers during the 17th century and had had some notable Prime Ministers such as Robert Walpole and William Pitt, it was still, at the beginning of the 19th century, no more than an equal partner with the House of *Lords. Extension of the franchise and the influence of such powerful members as Robert Peel, Lord Palmeston, Lord John Russell, and William Gladstone did much to extend its power, so that by the end of the century it was effectively regarded as the voice of the people. By the Parliament Act of 1911, the maximum duration of a Parliament became five years. The life of a Parliament is divided into sessions, usually of one year in length. As a rule, Bills likely to raise political controversy are introduced in the Commons before going to the Lords, and the Commons claim exclusive control in respect of national taxation and expenditure. Since 1911 Members have received payment. The House of Commons is presided over by an elected Speaker, who has power to maintain order in the House.

Commonwealth of Nations, an international group of nations. It consists of the United Kingdom and former members of the *British empire, all of whom are independent in every aspect of domestic and external affairs but who, for historical reasons, accept the British monarch as the symbol of the free association of its members and as such the head of the Commonwealth. The term British Commonwealth began to be used after World War I when the military help given by the *dominions to Britain had enhanced their status. Their independence, apart from the formal link of allegiance to the crown, was asserted at the Imperial Conference of 1926, and given legal authority by the Statute of *Westminster (1931). The power of independent decision by Commonwealth countries was evident in 1936 over the abdication of *Edward VIII, and in 1939 when they decided whether or not they wished to support Britain in World War II. In 1945 the British Commonwealth consisted of countries where the white population was dominant. Beginning with the granting of independence to India, Pakistan, and Burma in 1947, its composition changed and it adopted the title of Commonwealth of Nations. A minority of countries have withdrawn from the Commonwealth, notably Burma in 1947, the Republic of Ireland in 1949, Pakistan in 1972, and Fiji in 1987. *South Africa withdrew in 1961 because of hostility to its apartheid policy. In the 1950s pressure began to build up in Britain to end free immigration of Commonwealth citizens which was running at about 115,000 in 1959, mostly from the West Indies, India, and Pakistan. From 1962 onwards increasing immigration restrictions were reciprocally imposed by a series of legislative measures. Regular conferences and financial and cultural links help to maintain some degree of unity among Commonwealth members, whose population comprises a quarter of mankind.

commune, a small district of local government; in China the basic unit of agricultural organization and rural local government from 1958 to about 1978. Co-operatives were formed when the mutual aid teams that emerged during the land reform of the early 1950s were merged as part of the 'high tide of socialism' of 1955–6. During the *Great Leap Forward these co-operatives were themselves combined to form large units known as communes which were responsible for planning local farming and for running public services. Commune power was gradually devolved to production brigades after the disastrous harvests of 1959–61. In the *Four Modernizations movement after the death of Mao, communes were virtually abolished.

Commune of Paris   *Paris, Commune of.

communications revolution, an unprecedented advance in the speed of message transmission. In 1794 the French army started to use semaphore to pass messages, and a hot-air balloon to observe the enemy. In 1837 Charles Wheatstone in Britain and Samuel Morse in the USA developed an electric telegraph, the latter sending electric signals by means of a 'Morse-code'. Telegraph lines were erected between Washington and Baltimore in 1844, then across Europe, and in 1866 across the Atlantic. One of the effects of the telegraph was that it linked international banking; another was that newspapers could print up-to-date international news; thirdly, governments could exert much closer control, for example in war or in their distant colonies. With the telephone (1876) direct speech communication replaced the telegram, while radio (1899) removed the need for communicants to be linked by electric cables. Television (1926) enabled visual images to be transmitted, while satellites (from the 1970s) enabled

the whole world to watch events of supranational interest, with English becoming increasingly the language of the world. Information technology emerged in the 1980s as a computer-based means of storing and transmitting information, particularly business data.

communism, a social and political ideology which advocates that authority and property be vested in the community, each member working for the common benefit according to capacity and receiving according to needs. Perhaps the most important political force in the 20th century, communism embraces an ideology based on the overthrow, if necessary by violent means, of *capitalism. According to the theories of Karl *Marx, a communist society will emerge after the transitional period of the dictatorship of the proletariat and the preparatory stage of *socialism. In a fully communist society the state will, according to Marx, 'wither away' and all distinctions between social relations will disappear. Specifically communist parties did not emerge until after 1918, when extreme Marxists broke away from the *Social Democrats. Marx's theories were the moving force behind *Lenin and the *Bolsheviks and the establishment of the political system in the *Union of Soviet Socialist Republics. They have since been adapted to local conditions in a large number of countries, for example in China and Yugoslavia.

Communist Manifesto, the primary source of the social and economic doctrine of *communism. It was written as *Das Manifest der Kommunistischen Partei* in 1848 by Karl *Marx and Friedrich *Engels to provide a political programme that would establish a common tactic for the working-class movement. The manuscript was adopted by the German Socialist League of the Just as its manifesto. It proposed that all history had hitherto been a development of class struggles, and asserted that the industrialized proletariat would eventually establish a classless society safeguarded by social ownership. It linked *socialism directly with *communism and set out measures by which the latter could be achieved. It had no immediate impact and Marx suggested it should be shelved when the *Revolutions of 1848 failed. Since then it has been influential in all communist movements and in political philosophy.

Communist Party of India, a party that emerged in the 1920s. Originally functioning as a part of the Indian National *Congress, it was expelled for its co-operation with the war effort after the German invasion of Russia. After independence a programme of violent incursion and agrarian uprising was tried unsuccessfully in Telengana in Andhra Pradesh. The party returned to constitutional politics and formed the government in Kerala state, which was overthrown by an agitation organized by the Congress. After the Sino-Soviet rift, the party split into two and the CPI (Marxist) emerged as the more powerful wing. The latter has been the leading partner in a leftist coalition government in West Bengal, victorious in three successive elections. It is also a partner in the coalition government in Kerala elected in 1987.

Compromise of 1850, a political compromise between the North and South of the USA. Initiated by Henry *Clay, it became law in September 1850. In an attempt to resolve problems arising from slavery it provided for

Communications Revolution

1794	Semaphore first used by French army.
1822	First typesetting machine invented by William Church (USA).
1834	Charles Babbage's (UK) analytical engine (never built) anticipates the modern computer.
1837	Charles Wheatstone (UK) and Samuel Morse (USA) demonstrate message transmission by electric telegraph.
1840	Louis Daguerre (France), William Henry Fox Talbot (UK), and others pioneer photography.
	Beginning of modern postal service with Rowland Hill's Penny Post (UK).
1848	Associated Press, the first news wire service, begins in New York, and with it the emergence of a world-wide news-gathering and transmission operation.
1851	Reuters wire service begins in London.
1866	First successful laying of trans-Atlantic telegraph cable, from Ireland to Newfoundland.
1867	Christopher Latham Sholes (USA) designs the first practical typewriter (first sold commercially in 1876).
1876	Alexander Graham Bell (USA) patents the telephone.
1877	The first demonstration of effective sound recording, by Thomas Edison (USA).
1884	The first typecasting machine, the Linotype, patented by Ottmar Mergenthaler in the USA.
1894	First public cinema shows (France) by the Lumière brothers, Auguste and Louis Jean.
1896	Guglielmo Marconi first demonstrates radio transmission (wireless telegraphy) in the UK.

British Post Office officials examining Guglielmo Marconi's apparatus used to communicate by wireless telegraphy 13 km. (8 miles) across the Bristol Channel in 1897.

1897	Karl Braun (Germany) pioneers the cathode ray tube, an essential component of television technology.
	First automatic telephone exchange, designed by A. B. Strowger (USA).

A manual telephone exchange of the National Telephone Company (UK), c.1900.

1898	Valdemar Poulsen (Denmark) invents magnetic sound recording on tape.
1901	First radio transmission across the Atlantic, from Cornwall to Newfoundland, by Marconi.
1904	Offset printing developed by Ira Rubel (USA).
1907	First successful combination of recorded sound with film, demonstrated by Eugene Lauste in the UK.
1909	Colour cinematography demonstrated by George Albert Smith (UK).

1920	Regular public radio broadcasting established in Britain and the USA.
1924	Teleprinter developed in the USA.
1926	A mechanically scanning television demonstrated by John Logie Baird (UK). This sytem was superseded by an electrically scanning system based on the work of Vladimir Zworykin and Philo Farnsworth in the USA.

John Logie Baird with his first television set, in 1926.

1929	Baird arranges first public transmission of television pictures in London.
1934	Introduction of drum-scanner facsimile telegraph for the transmission by wire of documents and pictures.
1940	Chester Carlson patents the xerographic copier, forerunner of modern photocopiers (USA).
1943	The code-breaking machine Colossus, conceived by Alan Turing (UK), is the first electronic computer.
1946	ENIAC (USA) is first computer with comprehensive data processing capability.
1950s	Phototypesetting machines commercially available.
1956	First trans-Atlantic telephone cable laid from Scotland to Newfoundland.
1958	Videotape in use for television recording and playback.
1960s	Audio cassettes available for home use.
1960	First working laser, developed by Theodore Maiman (USA). Lasers make possible holograms and optical fibre technology.
1962	Telstar (USA) is the first communications space satellite. With satellite technology, instantaneous world-wide TV transmission becomes a reality.
1970s	The VCR (video cassette recorder) available for home use. Personal computers developed for use in the home or office.
1971	Invention of the microprocessor, key component of the miniaturization of computers.

Architectural plans being drawn on a microcomputer.

1979	First public videotext service (in which information from a computer data bank is obtained through telephone lines) established in Britain.
1980s	Electronic fax (facsimile) machines provide almost instantaneous document transmission.
	Compact discs developed for high-quality reproduction of music and for the storage of large quantities of information.
	World-wide spread of satellite and cable television.
	Optical fibres used for telephony.

the admission of California as a free state; the organization of the Utah and New Mexico territories with no mention of slavery; the abolition of the slave trade in the District of Columbia (Washington); and a stricter fugitive slave law. Hopes that these measures would provide an enduring solution to North–South antagonism were dashed by the passage of the *Kansas-Nebraska Act (1854) and other issues, including persistent popular interference with the return of fugitive slaves.

concentration camps (Boer War, 1900–2), camps formed in the Second *Boer War, created by Lord *Kitchener primarily to remove women and children from the hardships of his 'scorched earth' policy in the Transvaal and Cape Colony, but also to prevent civilians from assisting Boer guerrillas. Some 20,000 Boer women and children died, largely as a result of disease from unhygienic conditions. Boer indignation at what they considered 'deliberate genocide' was intense, and in London Emily Hobhouse roused public opinion against such maladministration. A commission of investigation was appointed, headed by Elizabeth *Fawcett. It identified inadequate sanitary and medical facilities and recommended changes. Lord *Milner acted on the report, and before their closure the death rate was falling.

concentration camps (Germany, 1933–45), institutions in Nazi Germany for the detention of unwanted persons. Described by *Goebbels in August 1934 as 'camps to turn anti-social members of society into useful members by the most humane means possible', they in fact came to witness some of the worst acts of torture, horror, and mass murder in the 20th century. With administration in the hands of the *SS, early inmates included trade unionists, Protestant and Catholic dissidents, communists, gypsies, and Jews (the *Holocaust). Some 200,000 had been through the camps before World War II began, when they were increased in size and number. In eastern Europe prisoners were used initially in labour battalions or in the tasks of genocide, until they too were exterminated. In camps such as Auschwitz, gas chambers

Survivors of the Nazi **concentration camp** at Mittelbaudora are sprayed with disinfectant following their liberation in 1945. Allied troops were unprepared for the horrific scenes that greeted them at such camps, where those few inmates who survived were in many cases weak and ravaged by disease.

could kill and incinerate 12,000 people daily. In the west, Belsen, Dachau, and Buchenwald (a forced labour camp where doctors conducted medical research on prisoners) were notorious. An estimated four million Jews died in the camps, as well as some half million gypsies; in addition, millions of Poles, Soviet prisoners-of-war, and other civilians perished. After the war many camp officials were tried and punished, but others escaped. Maidanek was the first camp to be liberated (by the Red Army, in July 1944). After 1953 West Germany paid $37 billion in reparations to Jewish victims of Nazism.

Concordat, agreement between the Roman Catholic Church and a secular power. One of the most important was the Concordat of 1801 between Pius VII and Napoleon I which re-established the Catholic Church in France. This provided that French archbishops and bishops should be appointed by the government, but confirmed by the pope. Church property confiscated during the Revolution was not to be restored, but the government was to provide adequate support for the clergy. This lasted until the separation of Church and State in France in 1905. Another concordat, in the form of the *Lateran Treaties of 1929, regulated the status of the papacy in Italy, which had been a source of contention since unification in 1870 abolished the temporal power of the pope. It gave the pope sovereignty over *Vatican City and restored the influence of the Catholic Church in Italy.

Condor Legion, a unit of the German airforce sent by *Hitler to aid *Franco in the *Spanish Civil War (1936) on condition that it remained under German command. It aided Franco in transporting troops from Morocco in the early days of the war, and played a major role in the bombing of rebel lines and civilian centres, notably the city of Guernica on 27 April 1937.

Confederacy, the name given collectively to the eleven southern US states which seceded from the Union of the United States in 1860–1. Seven states (Alabama, Florida, Georgia, Louisiana, Mississippi, South Carolina, and Texas) formed themselves into the Confederate States of America on 8 February 1861 at Montgomery, Alabama, with a constitution modelled on the US document but incorporating guarantees of *states' rights and the institution of slavery. Jefferson *Davis and Alexander H. Stephens were elected President and Vice-President. After the bombardment of *Fort Sumter, four further states joined the Confederacy (Arkansas, North Carolina, Tennessee, and Virginia). Although the Confederate flag contained thirteen stars, two represented Kentucky and Missouri, border states which in fact remained largely under federal control. Despite the relative weakness of its central government based at Richmond, Virginia, the Confederacy managed to sustain the civil war until its collapse in April 1865 after four years of war with most of its territory occupied, its armies defeated, and its economy in ruins.

Confederation of the Rhine (1806–15), a grouping of middle and south German states. After Napoleon's victory at Austerlitz (1805) he announced the creation of a Confederation of the Rhine, whose members were obliged to abdicate from the old Holy Roman Empire, which was then declared dissolved. After the defeat of Prussia at Jena (1806) other princely states and cities joined. Napoleon had annexed for France all the left bank of the Rhine, but the new Confederation gradually extended from the Rhine to the Elbe. It was at first welcomed by the German people as a step towards unity, but it was really a barrier against Prussian and Austrian power, and as the *Continental System began to result in economic hardship, it became less popular. It contributed a contingent to Napoleon's campaigns of 1813. After his defeat at Leipzig, however, the Confederation broke up; one by one the German states and cities made peace and supported the *Quadruple Alliance of Prussia, Britain, Russia, and Austria. A new *German Confederation was to emerge from the Congress of *Vienna.

Congo   *Zaïre.

Congo crisis (1960–5), political disturbances in the Congo Republic (now *Zaïre) following independence from Belgium. The sudden decision by Belgium to grant independence to its vast colony along the Congo was taken in January 1960. A single state was to be created, governed from Léopoldville (Kinshasa). Fighting began between tribes during parliamentary elections in May and further fighting occurred at independence (30 June). The Congolese troops of the Force Publique (armed police) mutinied against their Belgian officers. Europeans and their property were attacked, and Belgian refugees fled. In the rich mining province of Katanga, Moise *Tshombé, supported by Belgian troops and white mercenaries, proclaimed an independent republic. The government appealed to the United Nations for troops to restore order, and the UN Secretary-General *Hammarskjöld despatched a peace-keeping force to replace the Belgians. A military coup brought the army commander, Colonel *Mobutu, to power with a government which excluded the radical Prime Minister, Patrice *Lumumba. In 1961 Lumumba was killed, allegedly by 'hostile tribesmen', and Hammarskjöld died in an air crash on a visit to the Congo. The fighting continued and independent regimes were established at different times in Katanga, Stanleyville, and Kasai. In November 1965 the Congolese army under Mobutu staged a second coup, and Mobutu declared himself President.

Congo People's Republic, a country in western Africa, formerly called Congo (Brazzaville) after the French explorer de Brazza. He made the first of the series of treaties that brought it under French control in 1880. In 1888 it was united with Gabon, but was later separated from it as the Moyen Congo (Middle Congo). It was absorbed with Chad into French Equatorial Africa (1910–58). It became a member of the French Community as a constituent republic in 1958, and fully independent in 1960. In the 1960s and 1970s it suffered much from unstable governments, which alternated between civilian and military rule. Some measure of stability has been achieved by the regime of Colonel Denis Sassou-Nquesso, which has been in power since 1979. Although the Congo is a one-party Marxist state, it maintains links with Western nations, and is particularly dependent on France for economic assistance.

Congress, Indian National, the principal Indian political party. It was founded in 1885 as an annual meeting of educated Indians desiring a greater share in government in co-operation with Britain. Later, divisions

A refugee camp under United Nations protection in Elisabethville (Lubumbashi), Zaïre, 1961. The shanty-crammed camp was home to almost 30,000 people, forced from their homes by tribal fighting in the **Congo crisis**.

emerged between moderates and extremists, led by B. G. Tilak, and Congress split temporarily in 1907. Tilak died in 1920 and under the leadership of M. K. *Gandhi Congress developed a powerful central organization, an elaborate branch organization in provinces and districts, and acquired a mass membership. It began to conduct major political campaigns for self-rule and independence. In 1937 it easily won the elections held under the Government of India Act (1935) in a majority of provinces. In 1939 it withdrew from government, and many of its leaders were imprisoned during the 1941 'Quit India' campaign. In 1945–7 Congress negotiated with Britain for Indian independence. Under Jawaharlal *Nehru it continued to dominate independent India. After his death a struggle ensued between the Congress Old Guard (the Syndicate) and younger, more radical elements of whom Mrs Indira *Gandhi assumed the leadership. In 1969 it split between these two factions but was quickly rebuilt under Mrs Gandhi's leadership. In 1977 it was heavily defeated by the Janata (People's) Alliance Party, led by Morarji Desai (1896–), who became Prime Minister (1977–9). In 1978 Mrs Gandhi formed a new party, the 'real' Indian National Congress, or Congress (I) (for Indira). In 1979 she led this faction to victory in elections and again became Prime Minister in 1980. After her assassination in October 1984 the splits between factions

Delegates to the first **Indian National Congress**, 1885. From this initial meeting the Congress Party evolved.

largely healed and leadership of the Congress (I) Party passed to her son Rajiv Gandhi (1944–), who became Prime Minister.

Congress of Industrial Organizations *American Federation of Labor.

Congress of the USA,
the legislative branch of the US federal government. Provided for in Article I of the US Constitution, Congress is divided into two constituent houses: the lower, the House of *Representatives, in which membership is based on the population of each state; and the upper, the *Senate, in which each state has two members. Representatives serve a two-year term and Senators a six-year term. Congressional powers include the collection of taxes and duties, the provision for common defence, general welfare, the regulation of commerce, patents and copyrights, the declaration of war, raising of armies, and maintenance of a navy, and the establishment of the post offices and federal courts. Originally, Congress was expected to hold the initiative in the federal government, but the emergence of the President as a national party leader has resulted in the continuous fluctuation in the balance of power between legislature and executive. Much of the effective work of Congress is now done in powerful standing committees dealing with major areas of policy.

Conkling, Roscoe
(1829-88), US politician. A prominent supporter of the *Whig Party, he joined the newly formed *Republican Party on its collapse and served in the House of Representatives from 1858 to 1864 as a strong supporter of President *Lincoln. After the *American

Civil War he was distinguished for his support of radical measures against the former *Confederacy and for his insistence on retaining control of federal appointments within his state of New York. He remained a believer in the *spoils system and opponent of civil service reform, over which he resigned from the Senate.

Conservative Party
(Britain), a major political party in Britain. In 1830 it was suggested in the *Quarterly Review*, a Tory journal, that a better name for the old Tory Party might be Conservative since the Party stood for the preservation of existing institutions. The idea was favoured by Sir Robert *Peel, whose *Tamworth Manifesto brought him briefly to the premiership in 1834-5 and more firmly in 1841, but when in 1846 he was converted to *free trade, the Party split. Peel's followers after a time joined the Liberals. The majority under Lord Derby and *Disraeli gradually adopted the title Conservative, though Tory continued to be used also. Between 1846 and 1874 the Conservatives were a minority party though they were in office in 1867 and passed a *Reform Bill. In 1867 they were the first party to create a national organization with the formation of the Central Office. Disraeli described the aims of the Party as: 'the preservation of our institutions, the maintenance of our Empire and the amelioration of the condition of the people'. In 1874, his government embarked on a programme of social reforms and increased the powers of central government. In 1886 those Liberals, led by Joseph *Chamberlain, who rejected Gladstone's *Home Rule policy for Ireland, allied with the Party, whose full title then became the National Union of Conservative and Unionist Associations. The Party was strongly imperalist

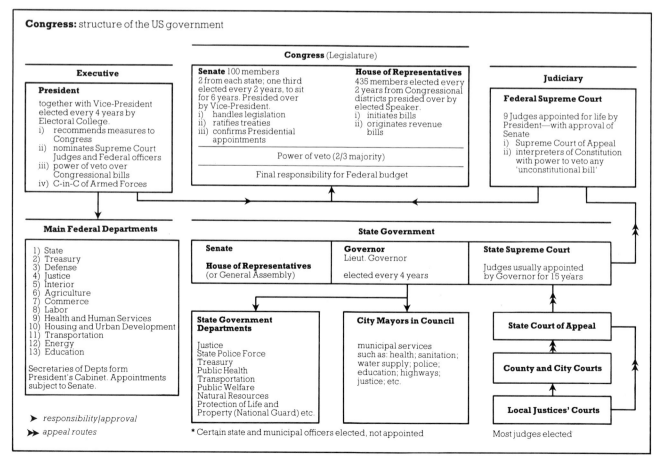

Congress: structure of the US government

*Certain state and municipal officers elected, not appointed

Most judges elected

➤ responsibility/approval
➤➤ appeal routes

throughout the first half of the 20th century, although splitting in 1903 over the issue of free trade or empire preference. From 1915 until 1945 the Party either formed the government, except for 1924 and 1929–31, or joined a *National government in coalition with the Labour Party (1931–5). Since World War II it has again been in office (1951–64, 1970–4, and since 1979). Before the 1970s the Party's policies tended to be pragmatic, accepting the basic philosophy of the *Welfare State and being prepared to adjust in response to a consensus of public opinion. Under the leadership of Margaret *Thatcher, however, it seemed to reassert the 19th-century liberal emphasis on individual free enterprise, challenging the need for state support and subsidy, while combining this with a strong assertion of state power against local authorities. With the growing crisis in *Northern Ireland since 1968 the *Ulster Unionists have more or less dissociated themselves from official Party policy. In recent years the Party has tended to return to the use of the term Tory.

Conservative Party (Canada), named in 1942 the Progressive Conservative Party of Canada, a major Canadian political party. Its origins go back to the early 19th century, when Tory supporters of the British were opposed to their critics, the 'Reformers', at the time of the *Mackenzie Rebellion in 1837. In 1864, under John A. *Macdonald, a coalition of Conservatives and Reformers was formed to work for Confederation (the union of the British colonies of British North America). When the Dominion of Canada was created in 1867, with Macdonald as the first Prime Minister, a *Liberal Party began to form and led the government from 1873 to 1878, when the Conservatives regained leadership. In 1896 they relinquished office to the Liberals under Wilfrid *Laurier, forming a coalition with the Liberals in 1917. They have since held power under Sir Robert *Borden (1911–20), Arthur Meighen (1920–1, 1926), R. B. Bennett (1930–5). John *Diefenbaker (1957–63), Joe Clark (1979–80), and Brian Mulroney (1984–).

containment, a basic principle of US foreign policy since World War II. It is aimed at the 'containment of Soviet expansionist tendencies' by the building of a circle of military pacts around the Soviet Union and its satellites. The policy was first adopted by President Truman with the creation of *NATO in 1949. This was to be a major means of containment in Europe, armed with conventional forces and nuclear devices, and stretching from the Arctic Circle to Turkey. Similar pacts in the Far East were the *ANZUS Pact of 1951 and *SEATO of 1954. In the 1960s the policy was extended to include the need to prevent Soviet participation in the affairs of states in Latin America and Africa, the most dramatic episode perhaps being the *Cuban Missile Crisis of 1962.

Continental System, economic strategy in Europe, intended to cripple Britain's economy. It was based upon the Berlin (1806) and Milan (1807) decrees of *Napoleon, which declared Britain to be in a state of blockade and forbade either neutral countries or French allies to trade with it or its colonies. At *Tilsit (1807) Russia agreed to the system and in 1808 Spain was obliged to join it. Britain responded by issuing Orders in Council which blockaded the ports of France and its allies and allowed them to trade with each other and neutral countries only if they did so via Britain. The restrictions imposed by the system

and Britain's countermeasures had serious effects on Britain and its allies. It contributed to the *War of 1812 with the USA over the right of neutral ships to trade with Europe. It gradually resulted in Napoleon losing support at home and being challenged abroad. His unsuccessful invasion of Russia in 1812 was provoked by Russian refusal to continue the system and it marked the beginning of his downfall.

Control Commissions, Allied administrations established in Germany after both World Wars. After World War I the Commission supervised German demilitarization. During World War II it was agreed by the US, British, and Soviet leaders that, after its defeat, Germany should be divided. Four zones of occupation were created in 1945, administered until 1948 by these Allies and France, the four military commanders acting as a supreme Control Council. Their responsiblity was to deal with matters relating to the whole of Germany. In practice the occupying powers administered their zones independently, while the British and US zones merged at the start of 1947. However, the Control Commission undertook significant work especially in the process of removing members of the Nazi Party from important positions. Tension between the Soviet and Western representatives led to the collapse of the system.

convict transportation, the banishment of a criminal to a penal settlement as a form of punishment. The system was used by France (to the West Indies) and in Russia (to Siberian *prison camps). It was used in Britain from the 17th century until 1868, with most of the convicts being sent to Britain's American and Australian colonies. The most common crime of Australian convicts was theft, although there were some socio-political prisoners, many of them Irish. At first, convicts were mostly used on public works. Later, the system of 'assignment', where convicts were allotted as paid servants to colonists, came to be used extensively. Various forms of probation were used, especially in Van Diemen's Land (*Tasmania) during the 1840s and early 1850s and in New South Wales from *Macquarie's time onwards. Secondary punishment included corporal punishment, solitary confinement, hard labour, confinement to segregated factories for women, and transportation to *penal settlements.

convoy system, a system whereby in war merchant vessels sail in groups under armed naval escort. In 1917 Germany's policy of unrestricted submarine (U-boat) warfare nearly defeated Britain. One ship in four leaving British ports was sunk; new construction only replaced one-tenth of lost tonnage; loss of Norwegian pit props threatened the coal industry; only six weeks' supply of wheat remained. In the face of this crisis *Lloyd George overruled the Admiralty's refusal to organize convoys, and by November 1918, 80 per cent of shipping, including foreign vessels, came in convoy. In World War II transatlantic convoys were immediately instituted in spite of a shortage of destroyers, using long-range aircraft for protection. During 1942 they were extended to the USA as the Allies were losing an average of ninety-six ships a month.

Coolidge, (John) Calvin (1872–1933), thirtieth President of the USA (1923–9). He won national fame by his firm action in face of the 1919 Boston police strike. This helped to win him the Republican nomination as Vice-President in 1920. When President *Harding died (1923) Coolidge succeeded him. Elected President in his own right (1924), he served only one full term of office, being seen as an embodiment of thrift, caution, and honesty in a decade when corruption in public life was common, even in his own administration. He showed no sympathy towards war-veterans, small farmers, miners, or textile workers, all of whom were seeking public support at that time. Foreign policy he left to his Secretaries of State, Hughes and Kellogg. Personally highly popular, he resisted pressures to stand for office again in 1928, preferring to retire into private life.

Co-operative Commonwealth Federation (CCF), a Canadian political party. It was founded in 1932, when about twelve Progressive Members of Parliament joined with supporters of the League for Social Reconstruction to form the Commonwealth Party. In 1933, at its convention, it put forward a programme for economic and social planning that would combat the depression affecting all Canada, but particularly the prairie provinces. The CCF returned members to all Canadian Parliaments after 1935, reaching a high point in 1945-9, before becoming the *New Democratic Party (NDP) with a closer relationship with organized labour in 1961.

Co-operative Movement, an organization owned by and run for the benefit of its members. First developed in many of the new industrial towns in Britain at the end of the 18th century, the Co-operative Movement was largely an attempt to offer an alternative to competitive *capitalism. In the early 19th century the social reformer Robert *Owen made several attempts to set up his own co-operative communities, but it was with the founding of the Rochdale Pioneers in 1844 that the co-operative movement in Britain really got under way. In 1864 these came together in a federation known as the Co-operative Wholesale Society. In 1869 the Co-operative Union, an advisory and educational body, was formed. The Co-operative Wholesale Society developed as a manufacturer and wholesale trader, opening its first factories and developing its own farms. The Co-operative Party was es-

Thirteen of the original members of the Rochdale Equitable Pioneers' Society, founded in 1844, marking the beginning of the modern **Co-operative Movement**.

PHOTOGRAPH OF THIRTEEN OF THE ORIGINAL MEMBERS
OF THE
ROCHDALE EQUITABLE PIONEERS' SOCIETY.

1. JAMES STANDRING.   2. JOHN BENT.   3. JAMES SMITHIES.   4. CHARLES HOWARTH.   5. DAVID BROOKS.   6. BENJ. RUDMAN.   7. JOHN SCOWCROFT.
8. JAMES MANOCK.   9. JOHN COLLIER.   10. SAMUEL ASHWORTH.   11. WILLIAM COOPER.   12. JAMES TWEEDALE.   13. JOSEPH SMITH.

tablished in 1917 to represent its members' interests in Parliament, and subsequently contested elections in alliance with the Labour Party. The movement spread rapidly to northern Europe. In the USA the first co-operatives were established at the end of the 18th and the beginning of the 19th centuries. In India and other developing countries, particularly in Africa after World War II, co-operatives have been an important factor in the growth of the economy.

Copenhagen, first battle of (1801), a naval engagement between the British and Danish fleets. The northern powers (Russia, Prussia, Denmark, and Sweden) formed a league of armed neutrality to resist the British right of search at sea. Without declaring war, a British fleet, commanded by Admiral Sir Hyde Parker, was sent to destroy the Danish fleet, anchored in Copenhagen. The British divided their fleet, *Nelson attacking the Danes from the more protected south whilst Parker attacked from the north. Despite bad weather and the loss of three ships Nelson, ignoring Parker's signal to discontinue action by fixing the telescope to his blind eye, was able to sink or take all but three of the Danish ships. The Danes agreed to an armistice and the league was disbanded. There was a second battle in 1807.

Corfu incident (31 August 1923), naval bombardment and occupation of the Greek island of Corfu by Italian troops. An Italian general and four members of his staff, engaged under international authority in determining the boundary between Greece and *Albania, had been murdered three days before. Following the bombardment by Italy in which sixteen people were killed, *Mussolini issued an ultimatum, demanding a heavy indemnity. Greece appealed to the *League of Nations, which referred the dispute to the Council of Ambassadors. The Council ordered Greece to pay 50 million lire. Under pressure from Britain and France, Italian troops withdrew. The outcome of the dispute raised serious doubts about the strength and efficiency of the League.

Corn Laws, regulations applied in Britain to the import and export of grain (mainly wheat) in order to control its supply and price. In 1815, following the end of the Napoleonic Wars, Parliament passed a law permitting the import of foreign wheat free of duty only when the domestic price reached 80 shillings per quarter (8 bushels). A sliding scale of duties was introduced in 1828 in order to alleviate the distress being caused to poorer people by the rise in the price of bread. A slump in trade in the late 1830s and a succession of bad harvests made conditions worse and strengthened the hand of the *Anti-Corn Law League. In 1846 the Corn Laws were repealed. This split the Conservative Party, but agriculture in Britain did not suffer as had been predicted. The repeal of the Corn Laws came to symbolize the success of *free trade and liberal political economy.

Cosgrave, William Thomas (1880-1965), Irish statesman. Determined to gain Irish independence from Britain, he took part in the *Easter Rising (1916). Elected to the British Parliament in 1918 as a *Sinn Fein member, he became Minister for Local Government in the provisional government of the Dáil Éireann in 1919. He reluctantly accepted the Anglo-Irish Treaty creating the *Irish Free State. He was president of the Executive

The final destruction of the Danish fleet at the first battle of **Copenhagen**. The British (*left*) pour concentrated gunfire into the trapped Danish vessels. This aquatint was dedicated to the victorious British commanders, Hyde Parker and Horatio Nelson. (National Maritime Museum, London)

Council of the Free State from 1922 to 1932, during which time the international standing of the new state was greatly enhanced. He was Opposition Leader in the Dáil Éireann (1933–44). He was the father of **Liam Cosgrave** (1920–), who in turn became leader of the Fine Gael Party in 1965 and later Taoiseach (Prime Minister) of the Republic of Ireland.

Costa Rica, a central American country. Discovered by Columbus in 1502, its upland plateau was settled by Spaniards, who created a stable agricultural community which has survived. The majority of the Indian population died from disease. It formed part of the captaincy-general of Guatemala until 1821, when it joined the independent Mexican empire (1821–3) and then the United Provinces

Distress caused by the **Corn Laws** led to 'bread riots' in British cities. Here rioters protesting at the entrance to the House of Commons in 1815 are dispersed by cavalry.

of Central America (1823–38). In 1838 it became an independent republic. A policy of isolation and stability, together with agricultural fertility, brought considerable British and US investment in the 19th century. Apart from the brief dictatorship of Federico Tinoco Granados (1917–19), Costa Rica was remarkable in the late 19th and early 20th centuries for its democratic tradition. After World War II left-wing parties emerged, including the communist. The socialist Presidents Otilio Ulate (1948–53) and José Figueres (1953–8, 1970–4), tried to disband the army, nationalize banks, and curb US investment. A new constitution, granting universal suffrage, was introduced in 1949. Political tensions in the 1970s were aggravated by economic problems and by the arrival of many fugitives from neighbouring states. President Luis Alberto Monge (1982–6) had to impose severe economic restraint.

Coughlin, Charles Edward (1891–1979), Canadian-born Roman Catholic priest. His radio broadcasts won him fame in the 1930s as the 'radio priest' first in support of the *New Deal and then, when F. D. Roosevelt refused Coughlin's plan for the coinage of silver to expand the currency, against the President and his policies with ever-mounting vituperation. Isolationist, anti-Semitic, and pro-fascist, Coughlin used his magazine *Social Justice* to promote his causes. In 1942 it was barred from the mails for violating the Espionage Act, and with full US involvement in the war his influence quickly waned.

Council of Europe, an association of West European states, independent of the *European Community. Founded in 1949, it is committed to the principles of freedom and the rule of law, and to safeguarding the political and cultural heritage of Europe. With a membership of twenty-one European democracies, the Council is served by the Committee of Ministers, the European Court of Human Rights, the European Commission of Human Rights, and the Parliamentary Assembly at Strasburg. Although without legislative powers, treaties have covered the suppression of terrorism, the legal status of migrant workers, and the protection of personal data.

country house, a large house in the country, standing in its own park or estate. In the 18th and 19th centuries new wealth from trade, banking, or industry was invested in the purchase or construction of country-house estates, and the country-house 'week-end' became a characteristic of 19th- and early 20th-century social and political Britain. After World War I some houses continued to maintain their traditions, but increasing costs and a more fragmented society reduced their numbers considerably. Some houses fell into disrepair, others became conference centres or institutions, or were given to the nation under the care of the National Trust. Others continue to be lived in by their owners, helped by revenue from tourists.

cowboy (historically) a lawless marauder. The term was first applied within the USA to some pro-British gangs during the American War of Independence, who roamed the neutral ground of Westchester county in New York state (their Revolutionary counterparts were 'skinners'). By the 1870s the term was used to describe those who herded cattle on the Great Plains. Many of them were black or Mexican; they rounded up cattle in an enclosure or rodeo, and, dividing them into herds of about 2,500 head, with a dozen cowboys for each herd, drove them along the *cattle trails to the nearest shipping points. The cattle industry spread across the Great Plains from Texas to Canada and westward to the Rocky Mountains. The introduction of barbed wire to fence in ranches rapidly encroached on the open ranges, and by 1895 railway expansion had made trail-driving uneconomical, and cowboys settled to work on the cattle ranches.

Crazy Horse (d. 1877), American Indian chief. Of the Ogala Sioux tribe, he opposed white infiltration into the mineral-rich Black Hills. He was at the first military confrontation between the Sioux and whites in 1854, and opposed Indian settlement on reservations. He was at the centre of the confederation that defeated Generals George Crook on the Rosebud River and *Custer at *Little Big Horn, both in 1876. He and his followers were starved into surrender (May 1877). Imprisoned because of a rumour that he was planning a revolt, he was reputedly stabbed to death while trying to escape.

Crédit Mobilier of America, a US finance company. It was acquired by Thomas Durant, vice-president of the *Union Pacific Railroad, to raise money for its construction. The railway had cost at least $50 million, of which $23 million had been diverted into the pockets of the promoters, who had granted construction contracts to Crédit Mobilier at exorbitant rates. Thus the Union Pacific was forced to the verge of bankruptcy while Crédit Mobilier paid excessive dividends. It was a scandal that cut at the heart of US political corruption in the so-called *Gilded Age. During the elections of 1872 fifteen leading politicians were damaged by their association with it.

Crete, an island in the eastern Mediterranean. A part of the Ottoman empire, it rose unsuccessfully in revolt against its Turkish overlords in 1866, in a pursuit of *enosis, or union with Greece. In 1897 a Greek force landed but the great powers supported Turkey. The island was declared independent, under Turkish suzerainty, but with Prince George of Greece as High Commissioner. Turkish troops were withdrawn in 1898 and after more unrest it was declared a part of Greece by the Treaty of London

in 1913. In 1941 Germany attacked Crete and, despite resistance by Greek and Allied forces, made the first successful airborne invasion in military history, followed by a bloody 12-day battle and capturing some 18,000 Allied troops.

Crimean War (1853–6), a war between Russia and Britain, France, Turkey, Piedmont, and Austria. The immediate cause was the dispute between France and Russia over the Palestinian holy places. War became inevitable after the Russians, having failed to obtain equal rights with the French, occupied territories of the *Ottoman empire in July 1853. In a bid to prevent Russian expansion in the Black Sea area and to ensure existing trade routes, a conference was convened in Vienna. Turkey was pressed by the Powers to make some concessions to placate Russia, but it refused, and declared war. In November 1853 the Russians destroyed the Turkish fleet at Sinope, in the Black Sea. This forced the hand of Britain and France, who in March 1854 declared war, expecting, with their naval supremacy, a quick victory. Austria did not join the Allies until Lord *Raglan, commander-in-chief of the ill-prepared armies, was ordered to attack the Crimea. The Allied armies were able to defeat the Russian army, skilfully led by Menschikov, at the battle of the Alma River (20 September 1854) and began bombarding the strongly armed fort of *Sevastopol. Following the battle of *Balaklava, a long winter of siege warfare ensued, aggravated by lack of fuel, clothing, and supplies for the Allied armies. Public opinion in Britain became critical of the war after reading eyewitness reports in *The Times*, sent back by the Irishman W. H. Russell, the first journalist in history to write as a war correspondent using the telegraph. Florence *Nightingale received permission to take nurses to the Crimea. Sevastopol fell on 8 September 1855; by that time the Russians, with a new emperor, *Alexander II, were already seeking peace. This was concluded at the Congress of *Paris in 1856.

Cripps, Sir (Richard) Stafford (1889–1952), British politician. He entered Parliament as a Labour Member in 1931, but was expelled from the Labour Party in 1939 because of his advocacy of a Popular Front. During World War II Cripps was Ambassador to Moscow (1940–2) and Minister for Aircraft Production (1942–5). During 1945–50 he served in *Attlee's government successively as President of the Board of Trade and Chancellor of the Exchequer. In these posts he was responsible for the policy of austerity—a programme of rationing and controls introduced to adjust Britain to its reduced economy following the withdrawal of US *Lend–Lease. He also directed a notable expansion of exports, especially after devaluation of the pound in 1949.

Crispi, Francesco (1819–1901), Italian politician. He began as a Sicilian revolutionary republican supporting *Garibaldi's invasion (1860) and ended as a monarchist, a friend of *Bismarck, and twice a dictatorial Premier. During his first ministry (1887–91) a colonial administration was formally established (1889) in the Ethiopian province of Eritrea. Italy's economic distress was aggravated by his tariff war against France, and he brutally suppressed a socialist uprising in Sicily. His foreign policy was based on friendship with Germany and adherence to Bismarck's *Triple Alliance. His second ministry (1893–6) witnessed the rout of the Italians by the Ethiopians at *Adowa (1896). Italy was obliged to sue for peace and Crispi was forced from office.

Croatia, a constituent republic of Yugoslavia. From 1809 to 1813 it was part of Napoleon's Illyrian province, during which time Croatian nationalism emerged, strongly resisting both Habsburg imperialism and Hungarian control. In 1848 a revolution reasserted Croatian independence, ending serfdom, and proclaiming all citizens equal. In the following year Austria countered by proclaiming the nation an Austrian crownland. In 1868, following the establishment of the *Austro-Hungarian empire, it was pronounced to be the autonomous Hungarian crownland of Croatia-Slovenia, apart from the coastline of Dalmatia, which was to remain an Austrian province. The Hungarian authorities tried to crush all manifestations of Croatian nationalism, with little success, and in October 1918 an independent Croatia was again proclaimed. This then joined the Kingdom of the Serbs, Croats, and Slovenes (1921), later renamed Yugoslavia. In 1941 it was once again declared an independent state under the fascist leader Ante Pavelic, whose brutal government provoked a guerrilla war. Croatia joined the new Federal Republic of Yugoslavia in 1945.

Crockett, David (Davy) (1786–1836), US pioneer and adventurer. A hunter and frontier fighter, he served two terms in the state legislature and three terms in Congress as a Jacksonian Democrat, gaining considerable fame as a result of his backwoods manner and uninhibited sense of humour. Defection to the Whigs ended his political career, and Crockett returned to the frontier, where he took up the cause of Texan independence and died at the *Alamo.

Cromer, Evelyn Baring, 1st Earl of (1841–1917), British statesman, colonial administrator, and diplomatist. A professional soldier, he became secretary to the viceroy of India. In 1879 he became Commissioner of Debt in *Egypt, rescuing the country from near bankruptcy. He was in India again in 1880–3, and then became British Agent and Consul-General in Egypt at a critical time, following the battle of *Tel-el-Kebir and Britain's occupation of the country. His ability and imposing personality made him the real and absolute ruler of Egypt until he retired in 1907.

A caricature of Davy **Crockett** as politician, published shortly after his death. The frontiersman is seen addressing the US Congress in characteristically lively fashion, to the evident amusement of some fellow Congressmen.

Crosland, (Charles) Anthony (Raven) (1918–77), British politician. He served as Labour Member of Parliament (1950–5, 1959–77). His book, *The Future of Socialism* (1956), gave an optimistic forecast of continuing economic growth which was to influence a whole generation. As Secretary of State for Education and Science (1964–7) his strongly held libertarian and egalitarian principles led to the closure of grammar schools, the establishment of a comprehensive state school system, and the growth of polytechnics. During 1965–70 and 1974–7 he held several cabinet posts, and was Foreign Secretary before his early death.

Crossman, Richard Howard Stafford (1907–74), British politician. He was assistant chief of the Psychological Warfare Division during World War II. He entered Parliament as a Labour Member in 1945. During the *Wilson administrations he was successively Minister of Housing and Local Government, Leader of the House of Commons, and Secretary of State for Social Services. His posthumous *Diaries* (1975–77) provided revealing insights into the working of government.

Cuba, a large island in the Caribbean. Colonized by the Spanish, its sugar and tobacco plantations were worked by African slaves. Slave importation ended in 1865, but slavery was not abolished until 1886. Various attempts were made by US interests to acquire the island and many Americans fought in the unsuccessful first War of Independence (1868–78). Large US investments were maintained in the sugar industry, which by now was producing one-third of the world's sugar. The second War of Independence (1895–1901) was joined by the USA (1898) after a well-orchestrated press campaign, and Cuba was occupied by US troops (1899–1901). In 1902 the Republic of Cuba was proclaimed. A series of corrupt and socially insensitive governments followed, culminating in the brutal, authoritarian regime of Gerardo Machado (1925–33), which prompted the abortive revolution of 1933–4, the island remaining under US 'protection' until 1934. Fulengio *Batista was President 1940–4 and 1952–9. Although supported by the USA, his second government was notoriously corrupt and ruthless. In 1956 Fidel *Castro initiated a guerrilla war which led to the establishment of a socialist regime (1959) under his leadership. He repulsed the invasion by Cuban exiles at Cochinos Bay, the *Bay of Pigs' (April 1961), and survived the *Cuban Missile Crisis of October 1962. The accomplishments of his one-party regime in public health, education, and housing have been considerable. Castro has maintained a high profile abroad and, although the espousal of world revolution has been tempered under pressure from Moscow, Cuban assistance to liberation movements in Latin America and Africa has been consistent. At home, after the political turbulence of the 1960s, the revolution was stabilized with the establishment of more broadly based representative assemblies at municipal, provincial, and national levels. In economic terms, the initial hopes of diversification and industrialization have not been realized, and Cuba has continued to rely on the export of sugar as well as on substantial financial subsidy from the Soviet Union. Agricultural production in the socialist state has been generally poor, and shortages and rationing are still common. Frustrations with the regime led to an exodus of 125,000 Cubans in 1980.

Cuban Missile crisis (1962), an international crisis involving the USA and the Soviet Union. It was precipitated when US leaders learned that Soviet missiles with nuclear warheads capable of hitting the USA were being secretly installed in Cuba. President *Kennedy reinforced the US naval base at Guantanamo, ordered a naval blockade against Soviet military shipments to Cuba, and demanded that the Soviet Union remove its missiles and bases from the island. There seemed a real danger of nuclear war as the rival forces were placed on full alert, and the crisis sharpened as Soviet merchant vessels thought to be carrying missiles approached the island and the blockading US forces. However, the Soviet ships were ordered by *Khrushchev to turn back, and the Soviet Union agreed to US demands to dismantle the rocket bases in return for a US pledge not to attack Cuba. An outcome of the crisis was the establishment of a direct, exclusive line of communication (the 'hot line') to be used in an emergency, between the President of the USA and the leader of the Soviet Union.

Cultural Revolution (1966–76), decade of chaos and political upheaval in China with its roots in a factional dispute over the future of Chinese socialism. Oblique criticisms of *Mao Zedong in the early 1960s prompted him to retaliate against this threat to his ideology-led position from more pragmatic and bureaucratic modernizers with ideas closer to the Soviet Union. Unable to

The cover of the *China Pictorial* showing a demonstration of 40,000 people in Hubei province in 1968, one of many held during the **Cultural Revolution** supporting its anti-intellectual and anti-Western ideals.

do so in the Communist Party, he utilized discontented students and young workers as his *Red Guards to attack local and central party officials, who were then replaced by his own supporters and often had army backing. *Liu Shaoqi, State Chairman of China since 1959 and Mao's heir-apparent, lost all his government and party posts and *Lin Biao became the designated successor. The most violent phase of the Cultural Revolution came to an end with the Ninth Party Congress in 1969, but its radical policies continued until Mao's death in 1976.

Cunard, Sir Samuel (1787–1865), Canadian shipowner. He became a successful merchant in Nova Scotia and in 1839, in partnership with George Burns and David MacIver, he successfully bid for a British government subsidy to run a steam mail-packet service between North America and the UK. With others he established the British and North American Royal Mail Steam Packet Company, later known as the Cunard Line.

Cunningham, Andrew Browne, Viscount Cunningham of Hyndhope (1883–1963), British admiral. At the beginning of World War II he was commander-in-chief, Mediterranean. Here he was faced with an Italian fleet that was numerically superior to his own. However, he asserted British domination by his air attack on the Italian base of Taranto in 1940, and at Cape Matapan in 1941, where his victory effectually neutralized the Italian fleet for the rest of the war. As First Sea Lord from 1943 he was responsible for naval strategy and attended the meetings of Allied heads of government.

Curragh incident, a mutiny at the British military centre on the Curragh plain near Dublin. In 1914 the British commander there, General Sir Arthur Paget, on the instructions of Colonel Seely, the Secretary of State for War, informed his officers that military action might be necessary against private armies in Ulster. Officers with Ulster connections were to be allowed to 'disappear' or resign. Such an action, threatening army discipline, brought about the resignation of many British army officers, as well as of Colonel Seely.

Curtin, John Joseph (1885–1945), Australian statesman. A Labor Member of the House of Representatives (1928–31, 1934–45), he led the Opposition from 1935 until 1941, and was Prime Minister from 1941 until his death. Curtin opposed conscription during World War I, but organized the defence of Australia in World War II, working closely with the USA. He introduced a limited form of conscription for overseas service and helped plan closer co-operation within the British *Commonwealth.

Curzon, George Nathaniel, 1st Marquis Curzon of Kedleston (1859–1925), British statesman. As viceroy of India (1899–1905) he achieved reforms in administration, education, and currency, and set up the North-West Frontier province (1901). He was instrumental in the partitioning of Bengal in 1901, incurring thereby the ill-feeling of the Hindus. A strong supporter of imperialism, he resigned in 1903 in a dispute with *Kitchener. *Lloyd George included him in his coalition war cabinet (1916–18). He became Foreign Secretary in 1919. Lloyd George's tendency to conduct foreign affairs himself irritated Curzon, who joined the Conservative rebellion

Lord **Curzon** with British and Indian dignitaries during a viceregal visit to Hyderabad. Seated beside the viceroy is the Nizam of Hyderabad and next to him is Curzon's American wife, Mary Leiter.

in 1922 against the coalition government. *Bonar Law became Prime Minister and made Curzon his Foreign Secretary in 1922. As Foreign Secretary he gave his name to the frontier line proposed (1920) by Lloyd George, between Poland and Russia. The broad outline of the frontier became (1939) the boundary between the Soviet and German spheres of occupied Poland. It was imposed (1945) on Poland by the Allies as the definitive frontier between itself and the Soviet Union.

Custer, George Armstrong (1839–76), US soldier. He served in most of the *American Civil War campaigns, gaining a reputation for personal courage. He personally received the Confederate white flag of truce from General Lee on 9 April 1865. In 1874 he led an expedition to look for gold in South Dakota in an area which had been agreed by treaty as a sacred hunting-ground for the Sioux and Cheyenne Indians. A *gold-rush followed and the Indians were ordered to move into a reservation by 1 January 1876 or be deemed hostile. In June Custer joined an expedition led by General Alfred Terry to round up such hostile Indians. Custer went ahead of the main force towards Mount Little Big Horn in Montana and decided to attack. Possibly he underestimated the number of Indian warriors mustered—estimated at some 3,500. He occupied a hill position, only to be surrounded. He and his entire force of 266 men were killed.

Cyprus, a large island in the eastern Mediterranean. It formed part of the Ottoman empire until 1879, when it was placed under British administration. It was formally annexed by Britain in 1914 and in 1925 declared a crown colony. From the outset there was rivalry between Greek- and Turkish-speaking communities, the former, the majority, desiring union (*enosis) with Greece. After World War II there was much civil violence in which the Greek Cypriot terrorist organization *EOKA played the leading role. In 1959 independence within the Commonwealth was granted under the presidency of Archbishop *Makarios, but by 1964 the government was in chaos and a United Nations peace-keeping force intervened. In 1974 a Greek Cypriot coup overthrew the president and Turkish forces invaded, gaining virtual control over most of the island. The Greek national government which had backed the revolt, collapsed. Talks in Geneva between

George Armstrong **Custer**, wearing the Civil War uniform of a major-general, designed by himself. He was later to carry his flamboyant command style into a new theatre of war, against the Indians with the US 7th Cavalry.

Britain, Turkey, Greece, and the two Cypriot communities failed, and, although Makarios was able to resume the presidency in 1975, the Turkish Federated State of Cyprus was formed in northern Cyprus, comprising some 35 per cent of the island, with its own president. Britain retains an important RAF base, which is also a key intelligence centre.

Czartoryski, Adam Jerzy, Prince (1770–1861), Polish statesman and nationalist leader. A cousin of the last independent king of *Poland, he worked unfailingly at the restoration of his country when Russia, Prussia, and Austria had partitioned it between them. He became the trusted adviser of the Russian Prince Alexander, who became emperor in 1801. The latter appointed him Russian Foreign Minister (1804–5). After the battle of *Leipzig (1813) he sought the re-creation of Poland from the Grand Duchy of Warsaw, formed by Napoleon. In this he was partially successful as the Polish representative at the Congress of *Vienna, which restored the kingdom of Poland, but with the Russian emperor as king. He was proclaimed President of the Provisional Government of Poland at the time of the Polish revolt of 1830, for which he was at first condemned to death but then granted exile in 1831. He became known as the 'Polish king in exile' and helped to plan the two unsuccessful Polish rebellions of 1846–9 and 1863.

After the so-called 'Prague spring' of 1968 in **Czechoslovakia** Czechs walk past Red Army tanks in the streets of the capital. Their heads are turned away, in defiance of the Soviet-led invasion that crushed the country's liberal reformist movement.

Czechoslovakia, a country in central Europe. It was created out of the northern part of the old *Austro-Hungarian empire after the latter's collapse at the end of World War I. It incorporated the Czechs of *Bohemia-Moravia in the west with the Slovaks in the east. Tomáš *Mazaryk became the republic's first President and *Beneš its Foreign Minister. Loyalty to the League of Nations, alliances with Yugoslavia and Romania (1921), France (1924), and the Soviet Union (1935) ensured a degree of stability, but danger lay in the national minorities, especially Germans and Hungarians, within its borders. In 1938, deserted by his allies, President Beneš accepted the terms dictated by Hitler at *Munich, which deprived the country of the *Sudetenland and of nearly five million inhabitants. In 1939 Hitler's troops occupied the country. During World War II a provisional government under Beneš was formed in London. After a brief period of restored independence (1945-8) under Beneš the communists under Klement *Gottwald and with the backing of the Soviet Union gained control of the government, making Czechoslovakia a satellite of the Soviet Union. In the 'Prague Spring' of 1968 an attempt by *Dubček and other liberal communist reformers to gain a degree of independence failed as Soviet troops assisted by the *Warsaw Pact armies invaded the country. Relatively industrialized and prosperous, a continuing wish for independence nevertheless remains, as is shown by the *Charter 77 movement.

D

Dahomey *Benin.

daimyo, Japanese feudal lords. For most of the Tokugawa *shogunate (1600-1868), the daimyo exercised local control over domains comprising two-thirds of Japan. The new national government at the time of the *Meiji Restoration persuaded the daimyo to surrender their titles, powers, and privileges as feudal landowners, compensating them by payment of a portion of their former revenues. This, along with the dismantling of the *samurai class of warriors who served the daimyo, helped transform Japan from a feudal to a centralized state.

Daladier, Édouard (1884-1970), French statesman. With Neville *Chamberlain he yielded to Hitler's demands at *Munich (1938) to annex the *Sudetenland of Czechoslovakia. He had served as a Radical Socialist in various ministries, was briefly Premier in 1933 and 1934 and again in 1938-40. Arrested by the *Vichy government in 1940, he was tried at Riom, together with other democratic leaders, accused of responsibility for France's military disasters. Although acquitted, he remained imprisoned in France and Germany. He was elected to the national assembly (1945-58) during the Fourth Republic.

Dalhousie, James Ramsay, 1st Marquess of (1812-60), British statesman and colonial administrator. A Conservative Member of Parliament (1837), he became governor-general of India (1848-56), when he oversaw the extension of British rule through the annexation of the Punjab (1849), of Lower Burma (1852), of Oudh (1856), and of several smaller Indian states, through the use of the so-called Doctrine of Lapse. According to this Britain annexed those states where there was no heir who was recognized by Britain. Dalhousie initiated major developments in communications, including the railway (1853), the telegraph and postal system, the opening of the Ganges canal, and in public works and industry. He removed internal trade barriers, promoted social reform through legislation against female infanticide and the suppression of human sacrifice, and fostered the development of a popular educational system in India. He introduced improved training of the Indian civil service, which was opened to all British subjects of any race.

D'Annunzio, Gabriele (1863-1938), Italian political adventurer, poet, dramatist, and novelist. He urged Italy to enter World War I and himself fought with spectacular daring in the air force 1915-18. In 1919 in defiance of the *Versailles Peace Settlement, he seized the Adriatic port of Fiume (Rijeka), imposing an authoritarian and fascist government on it until, after fifteen months, it was starved into surrender.

Dardanelles, a 61-km. (38-mile) strait between Europe and Asiatic Turkey, joining the Aegean to the Sea of Marmara, once called the Hellespont. By the 1841 London Convention the Straits were closed to all warships in time of peace. The collapse of the Ottoman empire in 1918 permitted the establishment of a new system by the *Versailles Peace Settlement (1920) under which the Straits were placed under an international commission and opened to all vessels (including warships) at all times. This arrangement was modified at Lausanne (1923) to permit the passage of warships of less than 10,000 tonnes in peace-time only and reduce the powers of the Commission. By the Montreux Convention (1936) the International Commission was abolished and control of the Straits fully restored to Turkey. The Straits were to be closed to all warships in wartime if Turkey was neutral. The Convention has remained effective despite Soviet attempts to have it revised in its favour. The Dardanelles was the scene of an unsuccessful attack on the Ottoman empire by British and French troops in 1915, with Australian and New Zealand contingents playing a major part (*Gallipoli Campaign).

Darlan, (Jean Louis Xavier) François (1881-1942), French admiral. He was the virtual creator of the French navy that entered World War II. After he became Minister of Marine in the *Vichy government in 1940 he was regarded by the British as pro-fascist. His secret order to his commanders to scuttle their vessels should the Germans attempt to take them over was not known to the British. When the Allies invaded North Africa in 1942 he was in Algiers, where he began negotiations with the Americans. He ordered the Vichy French forces to cease fire and was proclaimed Head of State in French Africa. A month later he was assassinated.

Darling, Sir Ralph (1775-1858), British military commander and colonial administrator. He was the governor of New South Wales from 1825 until 1831. A rigid disciplinarian, he faced many difficulties, largely because of continuing conflict between *emancipists and *exclusionists in the colony. In the controversial Sudds and Thompson affair (1826) Darling's harsh punishment of these two soldiers was by popular opinion held responsible for the death of Sudds. The continued agitation over instances of alleged misgovernment resulted in a British House of Commons select committee of inquiry (1835), which exonerated him.

Darrow, Clarence Seward (1857-1938), US lawyer. Known as the 'attorney for the damned', in 1894 he defended the railway leader Eugene *Debs for his part in the *Pullman Strike; although he lost, he earned a reputation for taking on controversial cases. This flair for controversy brought him to the verge of bankruptcy (1911), when he was tried, but acquitted, of conspiring to bribe jurors. He defended over fifty people charged with murder, but only once did he lose a client to the executioner. In 1925 he defended the evolutionist biology teacher in the *Scopes Trial but lost the case.

Daughters of the American Revolution, a US patriotic society. Founded in 1890 it is open to women directly descended from individuals who assisted in establishing American independence. Its stated aims are to perpetuate the spirit of the early patriots and to 'cherish, maintain, and extend the institutions of American freedom'. It encourages education and the study of American history, but it has tended to be conservative on such issues as foreign affairs and civil liberties.

Davis, Jefferson (1809-89), US statesman and president of the Southern *Confederacy (1861-5). He served in the *Black Hawk War before leaving the army in 1835 to become a Mississippi planter. He commanded the Mississippi Rifles in the *Mexican-American War. Davis served two terms in the Senate (1847-51, 1857-61) and was Secretary of War in the administration of President *Pierce (1853-7). He left the Senate when Mississippi seceded from the Union, and in 1861 was named provisional President of the Confederacy. A year later he was elected to a six-year term. His aloofness and limited political skill, as well as his interference in military affairs, aroused considerable opposition, but it is doubtful whether any other Confederate leader could have been much more effective in the difficult wartime conditions of the South.

Davitt, Michael (1846-1906), Irish nationalist and land reformer. The son of an Irish farmer who had been evicted from his holding, he opposed the British-imposed land-holding system in Ireland. In 1865 he joined the *Irish Republican Brotherhood, a movement committed to the establishment of an independent republic of Ireland. He was sentenced to fifteen years' penal servitude in 1870 for smuggling weapons for the *Fenians. Released in 1877 he helped found the Irish Land League in 1879, an organization formed to achieve land reform. With C. S. *Parnell, he sought to protect Irish peasants against evictions and high rents. He was elected a Member of Parliament in 1882 while in gaol, and again in 1892 and

Jefferson **Davis**, portrayed with the flags of the states over whose short-lived breakaway Confederacy he presided in the background. (Mississippi State Archive)

1895. The agitation which he led influenced Gladstone to introduce the 1881 Irish Land Act, guaranteeing fair rents, fixety of tenure, and freedom to sell (the *Three Fs) to tenants.

Davout, Louis Nicolas, Duke of Auerstedt (1770–1823), marshal of France. He was made a general by Napoleon after the battle of Marengo (1800) and marshal in 1804. One of Napoleon's ablest generals, his third corps played a major part at *Austerlitz, *Jena, Friedland (1807), and *Wagram (1809). He was responsible for organizing the army that invaded Russia in 1812. During the *Hundred Days Davout was Minister of War. After the restoration of Louis XVIII he was deprived of his rank and title, but they were restored two years later.

Dawes Plan (1924), an arrangement for collecting *reparations from Germany after World War I. Following the collapse of the Deutschmark and the inability of the *Weimar Republic to pay reparations, an Allied payments commission chaired by the US financier Charles G. Dawes put forward a plan whereby Germany would pay according to its abilities, on a sliding scale. To avoid a clash with France (which demanded heavy reparations and had occupied the *Ruhr to ensure collections) the experts evaded the question of determining the grand total of reparations, and scheduled annual payments instead. Germany's failure to meet these led to the Plan's collapse and its replacement by the *Young Plan.

Deák, Francis (Ferenc) (1803–76), Hungarian statesman. He entered the Hungarian Diet in 1833, becoming the leader of the moderate faction for national emancipation from the *Austrian empire. In 1848 as Minister of Justice he introduced the reforming Ten Points or 'March Law' which, together with further demands, made Hungary all but independent. He was Minister of Justice in the Hungarian ministry of 1848, but after the Austrians refused to carry out reforms, he retired. Five years later he returned to politics and was the architect of the *Ausgleich of 1867, which gave Hungary internal autonomy within an *Austro-Hungarian empire.

Deakin, Alfred (1856–1919), Australian statesman, lawyer, and journalist. He was a Member of the Victorian Legislative Assembly (1879, 1880–1900), where he held various portfolios, and was active in the *Australian federation movement. Deakin was a supporter of a *White Australia. In the new federal parliament he was a Protectionist, and then Liberal, Member of the House of Representatives (1901–13). He was Prime Minister (1903–4, 1905–8, 1909–10), and attempted to implement the so-called New Protection based on the concept of a minimum 'fair wage'. New Protection was declared unconstitutional by the Federal High Court, after the defeat of his government by Labor in 1910. Deakin led the Opposition (by then called the Liberals) until his resignation in 1913.

Debs, Eugene V(ictor) (1855–1926), US labour leader. In 1893 he founded the American Railway Union and led it in a secondary strike on behalf of the Pullman workers in 1894 (*Pullman Strike). The strike was broken by the intervention of federal troops, and Debs was imprisoned in 1895 for conspiracy. Together with Victor Berger, Morris Hillquit, and others, he formed the Socialist Party of America (1901) and stood as its presidential candidate. A leading pacifist, he was briefly imprisoned for his sedition, that is, discouraging recruitment to the US armed services in World War I.

Decatur, Stephen (1779–1820), US naval commander. He was promoted captain following his daring recapture of the frigate *Philadelphia* in the *Tripolitan War (1801–5). After the *War of 1812 he became a national hero by forcing the Bey of Algiers to sign the treaty (1815) that ended American tribute to the Barbary pirates. He was killed in a duel with a suspended naval officer.

Decembrists, members of a Russian revolutionary society, the Northern Society. A group of Russian army officers, influenced by French liberal ideas, combined to lead a revolt against the accession of *Nicholas I in 1825. Some of their supporters proclaimed their preference for a republic, others for Nicholas's eldest brother Constantine, in the hope that he would be in favour of constitutional reform and modernization. A few Guards regiments in St Petersburg (now Leningrad) refused to take an oath of allegiance to Nicholas and marched to the Senate House, where they were met by artillery fire. Betrayed by police spies, five of their leaders were executed, and 120 exiled to Siberia. The Decembrists' revolt profoundly affected Russia, leading to increased police terrorism and to the spread of revolutionary societies among the intellectuals.

Defence of the Realm Acts (DORA), legislation (1914, 1915, 1916) by the British Parliament during World War I. Under the Acts government took powers to commandeer factories and directly control all aspects of war production, making it unlawful for war-workers to move elsewhere. Left-wing agitators, especially on Clydeside, were 'deported' to other parts of the country. Strict press censorship was imposed. All Germans had already been interned but war hysteria led tribunals to harass anyone with a German name or connection (for example, the writer D. H. Lawrence) and to imprison or fine pacifists (for example, Bertrand Russell). The Act of May 1915 gave wide powers over the supply and sale of intoxicating liquor, powers which were widely resented but which nevertheless survived the war. An Emergency Powers Act of 1920 confirmed the government's power to issue regulations in times of emergency and in 1939 many such regulations were reintroduced.

de Gasperi, Alcide (1881–1954), Italian statesman. He was elected to the *Austro-Hungarian Parliament in 1911, and became Secretary-General of the Italian People's Party (1919–25). From 1929 to 1943 he was given refuge from *Mussolini's regime by the Vatican. He played an important part in creating the Christian Democrat Party as a focus for moderate opinion after the fascist era. De Gasperi was Prime Minister from 1945 to 1953, during which time he adopted a strong stand against communism and in favour of European co-operation.

de Gaulle, Charles André Joseph Marie (1890–1970), French general and statesman. He first gained a reputation as a military theorist by arguing the case for the greater mechanization of the French army. When France surrendered in 1940 he fled to Britain, from where

General **de Gaulle** returns to France in June 1944, and is here greeted by his countrymen and women as he walks through the streets of Bayeux following the town's liberation by Allied forces.

he led the *Free or Fighting French forces. He was Head of the provisional government (1944-6) and provisional President (1945-6), when he retired into private life following disagreement over the constitution adopted by the Fourth Republic. In 1947 he created the Rassemblement du Peuple Français, a party advocating strong government. Its modest success disappointed de Gaulle, who dissolved it in 1953 and again retired. He re-entered public life in 1958 at the height of the crisis in *Algeria. The Fourth Republic was dissolved and a new constitution was drawn up to strengthen the power of the President: the Fifth Republic thus came into being, with de Gaulle as President (1959-69). He conceded independence to Algeria and the African colonies. De Gaulle dominated the *European Economic Community, excluding Britain from membership. He developed an independent French nuclear deterrent and in 1966 withdrew French support from *NATO. His position was shaken by a serious uprising in Paris (May–June 1968) by students discontented by the contrast between the high expenditure on defence and that on education and the social services. They were supported by industrial workers in what became the most sustained strike in France's history. De Gaulle was forced to liberalize the higher education system and make economic concessions to the workers. In 1969, following an adverse national referendum, he resigned from office.

Delcassé, Théophile (1852–1923), French statesman. As Foreign Minister in six successive governments between 1898 and 1905, he was the principal architect of the pre-1914 European alliances. He was the key figure in negotiations which resulted in the *entente cordiale* with Britain (1904) and he paved the way for the Triple Entente with Britain and Russia (1907). In 1911 as Minister of Marine he arranged for co-operation between British and French fleets in the event of war. In 1914 he was again Foreign Minister and helped to negotiate the secret Treaty of London (1915), which persuaded Italy to fight on the side of the Allies in World War I by guaranteeing the retention of the Dodecanese Islands.

Democratic Party, a major political party in the USA. Known in its initial form as the Democratic-Republican Party, it emerged under Thomas Jefferson in the 1790s in opposition to the *Federalist Party, drawing its support from Southern planters and Northern yeoman farmers. In 1828, after a split with the National Republicans (soon called *Whigs) led by John Quincy *Adams and Henry *Clay, a new Democratic Party was formed under the leadership of Andrew *Jackson and John C. *Calhoun. Its strong organization and popular appeal kept it in power for all but two presidential terms between then and 1860, when it divided over slavery. It only returned as a major national party in the last decades of the 19th century. By then, while retaining the loyalty of the deep South, it was gaining support from the ever expanding West and from the immigrant working classes of the industrialized north-east. In the early 20th century it adopted many of the policies of the *Progressive Movement and its candidate for President, Woodrow *Wilson, was elected for two terms (1913–21). Although in eclipse in the 1920s, it re-emerged in the years of the Great *Depression, capturing Congress and the presidency: its candidate, Franklin D. *Roosevelt, is the only President to have been re-elected three times. Since then it has tended to dominate the House of Representatives, and has generally held the Senate as well. Following the *civil rights movement and *desegregation in the 1950s and 1960s it lost much of its support from the *Dixiecrat Southern states, becoming less of a coalition party and more one which favours the working classes of the big cities and the small farmers, as against business and the middle classes. The Democratic presidencies of John F. *Kennedy and Lyndon B. *Johnson saw fruitful partnership between Congress and President, although the *Vietnam War badly divided the Party in 1968. Under the Republican President *Nixon it retained control of Congress and won the presidential election for Jimmy *Carter. It lost control of the Senate in 1980, but regained it in 1986.

Deng Xiaoping (Teng Hsiao-p'ing) (1904–), Chinese statesman. He studied with *Zhou Enlai in France in the early 1920s and spent some time in the Soviet Union before returning to China and working for the communists in Shanghai and Jiangxi. During the wars of 1937–49 he rose to prominence as a political commissar, and afterwards he held the senior party position in south-west China. He moved to Beijing in 1952 and became General Secretary of the Chinese Communist Party in 1956. Since the *Great Leap Forward, Deng has been identified with the pragmatic wing of the CCP. He was discredited during the *Cultural Revolution and after one rehabilitation suffered again at the hands of the *Gang of Four. He re-emerged in 1977 as the real power behind the administration of *Hua Guofeng, and became the most prominent exponent of economic modernization and

Deng Xiaoping, the most influential Chinese leader since the death of Mao Zedong.

improved relations with the West, and effective leader of China, although he refrained from taking top party or government posts.

Denikin, Anton Ivanovich (1872–1947), Russian general and counter-revolutionary. The son of a serf, he served the Provisional Government as commander of the *Western Front in 1917. After the *October Revolution he assumed command of a 'white' army, the 'Armed Forces of the South', gaining control of a large part of southern Russia. In May 1919 Denikin launched an offensive against Moscow which the *Red Army repulsed at Orel. He retreated to the Caucasus, where in 1920 his army disintegrated and he fled to France.

Denmark, a country in northern Europe. It supported France during the Napoleonic Wars, and in 1814 was forced to cede Norway to Sweden. In 1849 a new constitution ended absolute monarchy and introduced a more representative form of government under a constitutional monarch. In 1863 Denmark incorporated *Schleswig, which its king ruled personally as a duke, but this was opposed by Prussia and Austria, whose troops invaded in 1864. Schleswig was then absorbed into the *German Second empire. After World War I north Schleswig voted to return to Denmark, which had remained neutral during the war. Despite another declaration of neutrality at the start of World War II,

the Germans occupied the country from 1940 to 1945 when all Schleswig-Holstein passed to the new German Federal Republic.

Depression, the Great (1929–33), popular term for a world economic crisis. It began in October 1929, when the New York Stock Exchange collapsed, in the so-called *Stock Market Crash. As a result US banks began to call in international loans and were unwilling to continue loans to Germany for *reparations and industrial development. In 1931 discussions took place between Germany and Austria for a customs union. In May, the French, who saw this as a first step towards a full union or *Anschluss, withdrew funds from the large bank of Kredit-Anstalt, controlled by the Rothschilds. The bank announced its inability to fulfil its obligations and soon other Austrian and German banks were having to close. Although President *Hoover in the USA negotiated a one-year moratorium on reparations, it was too late. Because Germany had been the main recipient of loans from Britain and the USA, the German collapse was soon felt in other countries. Throughout the USA and Germany members of the public began a 'run on the banks', withdrawing their personal savings, and more and more banks had to close. Farmers could not sell crops, factories and industrial concerns could not borrow and had to close, workers were thrown out of work, retail shops went bankrupt, and governments could not afford to continue unemployment benefits even where these had been available. In the colonies of the European powers and in Latin America demand for basic commodities collapsed, increasing unemployment, but also stimulating nationalist agitation. Unemployment in Germany rose to six million, in Britain to three million, and in the USA to fourteen million, where by 1932 nearly every bank was closed. In Europe, where a process of democratization since World War I had reduced class tensions, the effect everywhere was to foster political extremism. Renewed fears of a *Bolshevik uprising produced extreme right-wing, militarist regimes, inspired by fascism, not only in Italy and Germany but throughout the Balkan countries. In 1932 Franklin D. *Roosevelt was elected President of the USA, and gradually financial confidence there was

A scene typifying the hardship endured by many during the Great **Depression**: unemployed US citizens line up for free soup, coffee, and doughnuts in Chicago.

restored, but not before the *Third Reich in Germany had established itself as a means for the revitalization of the German economy.

Derby, Edward George Geoffrey Smith Stanley, 14th Earl of (1799-1869), British statesman. Entering Parliament in 1822 as a Whig, he served as Chief Secretary for Ireland (1830-3) and subsequently as Colonial Secretary (1833-4), when he introduced the successful proposals to abolish slavery in the British empire. In the later 1830s he left the Whigs and joined Sir Robert Peel's Conservative government of 1841, but resigned over the repeal of the *Corn Laws. Together with Benjamin Disraeli he led the Conservative opposition to the succeeding Whig administration. He was Prime Minister in 1852, in 1858-9, and again from 1866 to 1868, when he carried the *Reform Act of 1867 through Parliament. This act, which redistributed the parliamentary seats and more than doubled the electorate, gave the vote to many working men in the towns.

Desai, Morarji (Ranchhodji) (1896-), Indian statesman and nationalist leader. He made his reputation as Finance Minister (1946-52) and Chief Minister of Bombay (1952-6), and as Finance Minister in the Central Government (1958-63), overseeing a series of five-year plans for expanding industry, which led to a doubling of industrial output in ten years. After the death of Jawaharlal Nehru, he was a strong contender for the post of Prime Minister, but his austere and autocratic style made him too many enemies within the Congress Party. In 1977 he was the obvious candidate to lead the Janata opposition to Mrs Gandhi and led his party to victory in the election of that year. As Prime Minister (1977-9) his inflexible style handicapped him in dealing with the economic and factional problems which confronted him and he resigned in 1979.

desegregation, the name given in the USA to the movement to end the discrimination against its black citizens. Many segregation laws were passed in the Southern states after the *American Civil War, and they were supported by a Supreme Court decision in 1896 which accepted as constitutional a Louisiana law requiring separate but equal facilities for whites and blacks in trains. For the next fifty years, many Southern states continued to use the 'separate but equal' rule as an excuse for requiring segregated facilities. With the founding of the National Association for the Advancement of Colored People (NAACP) in 1909 black and white Americans began making efforts to end segregation, but they met with fierce resistance from state authorities and white organizations, especially in the South. When World War II saw over one million blacks in active military service change was inevitable, and in 1948 President Truman issued a directive calling for an end to segregation in the forces. It was only with the *civil rights movement of the 1950s and 1960s that real social reforms were made. The Supreme Court decision in 1954 against segregation in state schools (*Brown vs. *Board of Education of Topeka*) was a landmark. The efforts of Martin Luther *King, the *Freedom Riders, and others ended segregation and led to the passing of the Civil Rights Act of 1964 and the Voting Rights Act of 1965, which effectively outlawed legal segregation and ended literacy tests. There were still black ghettos in the northern cities, but the purely legal obstacles to the equality of the races was now essentially removed.

Dessalines, Jean Jacques (1758-1806), Negro emperor of Haiti. A former slave, he served under *Toussaint l'Ouverture in the wars that liberated Haiti from France. Although illiterate, he had a declaration of independence written in his name in 1804. With the defeat of the French in a war of extermination he became governor-general of Haiti, and in late 1804 had himself crowned Emperor Jacques I. The ferocity of his rule precipitated a revolt of mulattos in 1805. Dessalines was killed while trying to put down this rebellion in 1806.

Destroyer–Bases Deal (1940), a World War II agreement between F. D. Roosevelt and Churchill. Known as the Destroyer Transfer Agreement, it ensured the US transfer to Britain of fifty much-needed destroyers in exchange for leases of bases in British possessions in the West Indies, Newfoundland, and British Guiana. Being of World War I design, the destroyers became surplus to British requirements during the war, but in the early and critical stage of World War II they were an invaluable supplement to available escort vessels.

détente (French, 'relaxation'), the easing of strained relations, especially between states. It was first employed in this sense in 1908. The word is particularly associated with the 'thaw' in the *Cold War in the early 1970s and the policies of Richard *Nixon as President and Henry *Kissinger as National Security Adviser (from 1968) and Secretary of State (1973-7). The more relaxed relations were marked by the holding of the European Conference on Security and Co-operation in *Helsinki in 1972-5; the signing of the *SALT I Treaty in 1973; and the improvement in West Germany's relations with the countries of Eastern Europe following Chancellor Willy Brandt's *Ostpolitik.

de Valera, Eamon (1882-1975), Irish statesman. He devoted himself to securing independence for Ireland from Britain. He was imprisoned for his part in the *Easter Rising (1916) and would have been executed but for his American birth. After escaping from Lincoln

Eamon **de Valera** addressing an open-air meeting of pro-Irish Americans in Los Angeles, 1919.

gaol in 1919 he was active in the guerrilla fighting of 1919–21 as a member of the *Irish Republican Army. Elected as a *Sinn Fein Member of Parliament, he became president of the independent government (Dáil Éireann) set up by Sinn Fein in 1919. He did not attend the negotiations in London leading to the Anglo–Irish Treaty of 1921, and repudiated its concept of an *Irish Free State from which six Ulster counties were to be excluded. The leading opponent of *Cosgrave between 1924 and 1932, he founded *Fíanna Fáil in 1926, leading his party to victory in the 1932 election. He was president of the Executive Council of the Irish Free State from 1932 to 1937. He ended the oath of allegiance to the British crown and devised a new constitution in 1937, categorizing his country as 'a sovereign independent democratic state'. He stopped the payment of annuities to Britain and negotiated the return of naval bases held by Britain under the 1921 treaty. De Valera continued to have popular support and was twice elected President of the Republic of Ireland. His last presidency (until 1973) took him into his ninetieth year.

Dewey, George (1837–1917), US admiral. He served in the Union (Northern) navy under Farragut during the

A formal portrait photograph of the Mexican dictator Porfirio **Díaz** in uniform; he controlled his country for nearly thirty-five years.

*American Civil War. He was granted naval command of the Pacific (1897). His victory over the Spanish fleet at *Manila Bay on 1 May 1898 was not only decisive for the outcome of the *Spanish–American War but also for the future of American imperialism in the Pacific. In 1899 he made a triumphal progress through New York and was created the first ever US admiral.

Díaz, Porfirio (1830–1915), Mexican dictator of part-Indian descent. As President of Mexico he remained in control of his country for nearly thirty-five years (1877–80, 1884–1911). He began his military career by supporting *Juárez and the liberals during Mexico's War of Reform (1858–61) and during the fight against the French intervention in 1862. Responsible for the economic development and modernization of his country, he ruled in the interests of the privileged minority. The mineral resources of Mexico were largely exploited by foreigners, the *peons, or indebted labourers, lost most of their communal land, and much of the rural population was bound to debt slavery. The harsh dictatorship which he initiated prompted the *Mexican Revolution of 1910 and led to civil war (1911–18).

Diefenbaker, John George (1895–1979), Canadian statesman. He served as leader of the Progressive *Conservative Party (1956–67) and Prime Minister of Canada (1957–63). He introduced some important measures of social reform and sought to encourage economic development, but as Canada experienced increasing economic difficulties in the early 1960s he was forced to devalue the Canadian dollar. In foreign affairs he wished to reduce Canada's dependence on the USA, but his party lost the election of 1963 when he took issue with the USA over the arming with atomic warheads of missiles supplied to Canada.

Dienbienphu (1954), decisive military engagement in the *French Indo-Chinese War. In an attempt to defeat the *Vietminh guerrilla forces, French airborne troops seized and fortified the village of Dienbienphu overlooking the strategic route between Hanoi and the Laotian border in November 1953. Contrary to expectations, the Vietnamese commander General Giap was able to establish an effective siege with Chinese-supplied heavy artillery, denying the garrison of 16,500 men supply by air, and subjecting it to eight weeks of constant bombardment between March and May 1954, which finally forced its surrender. The ensuing armistice ended French rule in Indo-China within two months.

Dieppe raid (18–19 August 1942), an amphibious raid on Dieppe, Normandy, in World War II. Its aim was to destroy the German port, airfield, and radar installations and to gain experience in amphibious operations. Some 1,000 British commando and 5,000 Canadian infantry troops were involved. There was considerable confusion as landing-craft approached the two landing beaches, where they met heavy fire. The assault was a failure and the order to withdraw was given. Not only were over two-thirds of the troops lost, but German shore guns sank one destroyer and thirty-three landing-craft, and shot down 106 aircraft. Although in itself a disaster, the raid taught many lessons for later landings in North Africa, Italy, and the eventual success of the *Normandy Landings of June 1944, not least the need for careful planning.

French defenders of the isolated stronghold of **Dienbienphu** snatch a brief respite from the waves of Vietminh attacks that eventually forced their surrender. They are surrounded by the containers in which supplies were parachuted to them during the siege.

Diet (from the medieval Latin, 'a meeting for a single day'). It came to describe the estates or representative assemblies of various European countries. The Imperial Diet of the Holy Roman Empire was abolished in 1806. The *German Confederation established a federal Diet in Frankfurt with Austria holding a casting vote. Other parliamentary bodies, including those of Hungary, Bohemia, Poland, the Scandinavian countries, and Japan, have also been called Diets.

Dimitrov, Georgi (1882–1949), Bulgarian communist leader. From 1929 he was head of the Bulgarian sector of the *Comintern in Berlin. When the *Reichstag was burned (1933) he was accused with other communists of complicity. His powerful defence at his trial forced the Nazis to release him and he settled in Moscow. In 1945 he was appointed head of the communist government in *Bulgaria which led to the setting up of the Bulgarian People's Republic (1946) under his premiership, a period marked by ruthless Sovietization.

Dingaan (d. 1843), Zulu king (1828–40). He was half-brother to *Shaka, whom he murdered. At first

Georgi **Dimitrov**, the Bulgarian Premier, addressing an audience of women during the early years of communist rule in Bulgaria.

friendly to European settlers, missionaries, and the *Voortrekkers, he treacherously killed their leader Piet Retief and his followers. He attacked a white settlement near what is now Durban, but was defeated (1838) at the battle of *Blood River. He then fled, and was succeeded in 1840 by his brother Mpande. Driven into *Swaziland, he was assassinated there three years later.

Dingiswayo (d. 1817), founder of the *Zulu kingdom. In 1807 he became chief of the Mthethwa in the present northern *Natal. By conquering neighbouring *Nguni peoples he made himself paramount over all surrounding groups and established a rudimentary military state, developing trade with Mozambique. He had already designated *Shaka as his successor when he was assassinated by Zwide, chief of the Ndurande clan of the Zulu, in a rebellion against his rule.

Diponegoro (or Dipanagara) (1785–1855), Javanese prince, leader of the *Java War (1825–30) against the Dutch. This struggle gained the support of central Javanese society, many of whom saw the prince as a latter-day messiah, a Javanese 'Just King' (*Ratu Adil*), who would liberate Java from foreign influence. Fired by a mixture of Javanese mysticism and Islam, Diponegoro demanded recognition from the Dutch as protector of Islam in Java, but successful Dutch military tactics, in particular their use of mobile columns and fortified outposts (*bèntèng*), eventually forced him to the conference table, where he was treacherously arrested and exiled.

disarmament, the reduction or abolition of military forces and armaments. Attempts to achieve disarmament by international agreement began before World War I and in 1932 there was a World Disarmament Conference. In 1952 a permanent United Nations Disarmament Commission was established in Geneva. National disarmament pressure groups have tended to seek unilateral disarmament, for example the *Campaign for Nuclear Disarmament. Bilateral agreements are negotiated between two governments for both arms reduction and *arms control, while multilateral agreements are sought via international conferences or the UN Commission.

disease, infectious, a major cause of human illness. At the beginning of the 19th century the belief that infectious diseases were caused by 'miasmata', poisonous vapours given off by sewage and rubbish, led to campaigns to improve public health by suppling clean water and drainage systems. In Britain this resulted in the Public Health Act (1848) drawn up by Edwin *Chadwick, in the same year that the London doctor, John Snow, showed that cholera was carried by dirty water. In the latter half of the 19th century, Louis Pasteur, Heinrich Koch, and others began to identify the micro-organisms that caused infectious diseases and this led to further research into their prevention and cure. The incidence of malaria was reduced by killing the mosquitoes that carried the infection, while vaccination against smallpox, first described by Edward Jenner in 1798, resulted in its global eradication by 1979. In the 20th century drugs were developed that could cure many infectious diseases. In 1910 Paul Ehrlich introduced salvarsan, the first effective cure for syphilis and, following the use of sulphonamides (1935) and penicillin (1941), many life-threatening diseases can now be treated, resulting in

Infectious disease

Year	Event	Year	Event
1796	Edward Jenner carries out his first vaccination against smallpox.	1877	Lister uses a spray of phenol to reduce risk of infection during surgery.
1843	Oliver Wendell Holmes establishes contagious nature of puerperal fever (childbed fever).	1882	Robert Koch isolates the tuberculosis bacillus.
1848	First public health legislation in Britain. John Snow links cholera to unclean water.	1885	Pasteur develops vaccination against rabies. Koch identifies the cholera bacillus.
		1890	Anti-diphtheria serum developed by Emile Behring.
		1894	Alexander Yersin and Kitasato Shibasaburo independently establish the cause of bubonic plague.
		1897	Almroth Wright succeeds in controlling typhoid by inoculation.
		1897–8	Ronald Ross confirms Patrick Manson's theory that mosquitoes carry malaria.
		1900	Walter Reed identifies yellow fever virus and proves Carlos Finlay correct in linking this disease with mosquitoes (1881).
		1901	Karl Landsteiner indentifies the types of human blood. Elie Metchnikoff demonstrates how white blood cells fight disease. Paul Ehrlich introduces salvarsan to treat syphilis.
		1918–9	Influenza pandemic kills 15 million people.
		1928	Alexander Fleming discovers penicillin.
		1935	Sulphonamide drugs introduced to treat certain bacterial infections.
		1941	Penicillin first used by Howard Florey and Ernst Chain to treat patients with bacterial infections.
		1950s 1960s	Effective drug treatments become available for many infectious diseases including tuberculosis, leprosy, venereal diseases, pneumonia, bacterial meningitis, typhoid, plague, and urinary infections.
		1954	Vaccine is developed for poliomyelitis by Jonas Salk.

This cartoon from *Punch* (1852) illustrates the overcrowding and insanitary conditions in slums which enabled infectious diseases to spread.

1858	Rudolf Virchow pioneers cell pathology (histology).
1860s	Louis Pasteur develops his germ theory of disease, that diseases are transmitted by micro-organisms rather than by 'spontaneous generation'.
1865	Joseph Lister uses phenol (carbolic acid) to prevent infection of wounds.

Joseph Lister's phenol (carbolic acid) spray being used at an operation in Aberdeen, c.1880.

Ali Maow Maalin of Merka, Somalia, the world's last recorded endemic smallpox victim.

1979	Smallpox is declared 'eradicated' by the World Health Organization (WHO). (Smallpox virus and vaccine remain stored in high-security laboratories). First observation of AIDS (Acquired Immune Deficiency Syndrome).
1983	Identification of HIV retrovirus from which AIDS can result.
1986	New vaccine against hepatitis-B (causing liver disease) is produced by genetic engineering.

greater life expectancy; deaths from tuberculosis in the USA in 1900 were 110 per 100,000 people, in 1975 they were 1.2. Treatment for some virus infections, however, has proved much more difficult. There is no cure for influenza, an epidemic of which killed up to fifteen million people in 1918–19, nor for the HIV (Human Immunodeficiency Virus) which can result in AIDS (Acquired Immune Deficiency Syndrome), whose symptoms were first recognized in 1979. Cures are still needed for some tropical parasitic diseases.

Disestablishment Acts, legislation in Britain to remove the financial and other privileges of the Anglican Church. The Anglican Church had been 'established' in the reign of Elizabeth I as the only church allowed within the state, with large endowments and privileges. These came to be strongly resented by Non-Conformists in Victorian England; but proposals that all financial and other state support should be withdrawn failed. In Ireland, however, it came to be accepted as unjust that the Anglican Church should be the established church in a predominantly Roman Catholic population. It lost its privileges by Gladstone's Irish Church Disestablishment Act (1869). The Welsh also pressed for the disestablishment of the Anglican Church in Wales. Heated arguments over the financial implications of Welsh disestablishment arose in the years just before World War I, the Welsh Church Disestablishment Bill eventually becoming law in 1920.

Disraeli, Benjamin, 1st Earl of Beaconsfield (1804–81), British statesman and novelist. He gave the modern *Conservative Party its identity and provided it with a policy of imperialism and social reform. Of Jewish descent, he was baptized a Christian. He entered Parliament in 1837. By the early 1840s he had become a member of the *Young England movement of Tories who favoured an alliance between the old aristocracy and the mass of the people, in opposition to the increasingly powerful middle classes. These views provided the theme

DANGEROUS WATERS.

A satirical cartoon shows the British Prime Minister Benjamin **Disraeli** as oarsman, with Queen Victoria at the helm, in the 'dangerous waters' of imperialist policy. The statesman is shown in peril of entanglement, while long-dead British monarchs shout advice from the bank. The German emperor, a rival in the race for overseas possessions, takes a keen interest in Disraeli's plight, while the French have altogether sunk from view. Unpopular overseas ventures led to Disraeli's fall from office in 1880.

for his novels, in which he sees early Victorian England as deeply divided between rich and poor, many of whom were massed in the new northern cities. In 1846 the Prime Minister, Sir Robert Peel, repealed the *Corn Laws. This provoked fierce opposition from Disraeli and a majority of the Tories, who supported a protectionist policy, and it split the Party. For the next twenty years Disraeli led the protectionist Conservatives in the House of Commons and was Chancellor of the Exchequer in 1852, 1858–9, and 1866–8. He introduced the *Reform Bill of 1867, which enfranchised much of the urban working class, and was briefly Prime Minister in the following year. In 1874 he became Prime Minister again. He bought the largest shareholding in the *Suez Canal Company (1875), procured the title of Empress of India for Queen Victoria (1876), and averted war with Russia through his skilful diplomacy at the Congress of *Berlin (1878). At home his government passed much useful social legislation (slum clearance, public health and trade union reform, and the improvement of working conditions in factories). Economic depression and unpopular colonial wars led to a Conservative defeat in the election of 1880 and Disraeli, though in poor health, remained till his death leading his party from the House of Lords, having been created 1st Earl of Beaconsfield in 1876.

Divine, Father (c.1882–1965), black American evangelist. Born (as George Baker) in Georgia, he founded a Peace Mission movement in 1919. He opened a free employment bureau and fed the destitute in Sayville, New Jersey. His house became known as 'Heaven' and he began styling himself first Major, then Father, Divine. By the 1940s his followers numbered thousands, both black and white, and some 200 'Heavens' were established as centres for communal living. Many of his followers declared him to be the personification of God and endowed him with miraculous healing powers. His cult did not long outlast his death.

Dix, Dorothea Lynde (1802–82), US humanitarian and medical reformer. She campaigned for prison reform and improvement in the treatment of the mentally ill. During the *American Civil War Dix served as superintendent of nursing for the Union. Perhaps her greatest achievement was in persuading many states to assume direct responsibility for care of the mentally ill.

Dixiecrat, popular name in the USA for a Democrat in a Southern state opposed to desegregation. In 1948 the *States' Rights Democratic Party was founded by diehard Southern Democrats ('Dixiecrats'), opposed to President Truman's renomination as Democratic candidate for President on account of his stand on *civil rights. Its members wished each state to be able to nominate its own presidential candidate without losing the label 'Democratic'. After Truman's victory they abandoned their presidential efforts but continued to resist the civil rights programme in Congress. Many Dixiecrats moved to support the Republican Party in the 1960s and 1970s.

Djibouti, an East African country, formerly part of French Somaliland. The small enclave was created as a port c.1888 by the French and became the capital of French Somaliland (1892). Its importance results from its strategic position on the Gulf of Aden. In 1958 it was declared by France to be the Territory of the Afars and Issas, but in 1977 it was granted total independence as

Dorothea **Dix** organized nursing corps to tend American Civil War casualties in army hospitals. This ward, hung with the bunting of Northern victory, is representative of the best hospital conditions of the day.

the Republic of Djibouti under President Hassan Gouled Aptidou. Famine and wars inland have produced many economic problems, with refugees arriving in large numbers, and the President has sought to mediate between Ethiopia and Somalia.

dole *unemployment assistance.

dollar diplomacy, a term used to describe foreign policies designed to subserve US business interests. It was first applied to the policy of President *Taft, whereby investments and loans, supported and secured by federal action, financed the building of railways in China after 1909. It spread to Haiti, Honduras, and Nicaragua, where US loans were underpinned by US forces and where a US collector of customs was installed in 1911. Although the policy was disavowed by President Woodrow *Wilson, comparable acts of intervention in support of US business interests, particularly in Latin America, remained a recurrent feature of US foreign policy.

Dollfuss, Engelbert (1892-1934), Austrian statesman. As Chancellor (1932-4), his term of office was troubled by his hostility to both socialists and nationalists. In an effort to relieve the economic depression and social unrest in the country, Dollfuss secured a generous loan from the *League of Nations in 1932. Unrest and terrorism continued and in March 1933 he suspended parliamentary government. In February 1934 demonstrations by socialist workers led Dollfuss to order the bombardment of the socialist housing estate in Vienna. After fierce fighting the socialists were crushed and Dollfuss proclaimed an authoritarian constitution. By antagonizing the working classes he deprived himself of effective support against the *Nazi threat. On 25 July 1934 he was assassinated in an abortive Nazi coup.

Dominican Republic, a country in the Caribbean, the eastern part of the island of Hispaniola. Independence from Spain came in 1821, but Haiti annexed Santo Domingo in 1822. In 1843 the Dominicans revolted from Haitian domination, winning their second independence in 1844. Between 1861 and 1865 the Dominican Republic was re-annexed to Spain and fought a third war for independence (1865) under Buenaventura Báez. Anarchy, revolutions, and dictatorships followed, and by 1905 the country was bankrupt. The USA assumed fiscal control, but disorder continued and the country was occupied (1916-24) by US marines. A constitutional government was established (1924), but this was overthrown by Rafael *Trujillo, whose military dictatorship lasted from 1930 to 1961. On his assassination, President Juan Bosch established (1962-3) a democratic government, until he was deposed by a military junta. Civil war and fear of a communist take-over brought renewed US intervention (1965), and a new constitution was introduced in 1966. Since then redemocratization has steadily advanced, the Partido Reformista being returned to power in the 1986 elections. The country occupies a strategic position on major sea routes leading from Europe and the USA to the Panama Canal.

dominion, the term used between 1867 and 1947 to describe those countries from the *British empire which had achieved a degree of autonomy but which still owed allegiance to the British crown. The first country to call itself a dominion was Canada (1867), followed in 1907 by New Zealand. Australia called itself a Commonwealth (1901), South Africa a Union (1910). After World War I, in which all these countries had aided Britain, it was felt that there was a need to define their status. This came about at the Imperial Conference (1926) when they were given the general term dominion. Their power to legislate independently of the British government was confirmed and extended by the Statute of *Westminster (1931). After World War II the concept became obsolete as the *Commonwealth of Nations included countries that were republics and did not owe allegiance to the crown, though accepting the monarch as symbolic head of the Commonwealth.

domino theory, the theory that one (especially a political) event precipitates other events in causal sequence, like a row of dominoes falling over. After the defeat of the French in Vietnam at *Dienbienphu in 1954, it was argued that the loss of Vietnam to communism would have a domino effect: neighbouring countries in south-east Asia would follow, one by one. In the early 1960s this theory became generally accepted and was the main justification for the increasingly active involvement of the USA in the *Vietnam War (1964-73).

Donahoe, John (c.1806-30), Australian *bushranger. Born in Dublin, he was sentenced to transportation for life. He and two companions went bushranging in New South Wales from 1827 to 1830, when he was killed in a fight with police and soldiers in the Bringelly scrub near Campbelltown. He, together with Ned Kelly (1855-80), inspired the glorification and cult of bushranging in Australian society.

Don Pacifico affair, an international incident. In 1847 an angry crowd in Athens ransacked and burnt the house of Don Pacifico, a Portuguese moneylender who was also a Jew, injuring his wife and children. Pacifico, who had been born in Gibraltar and could therefore claim British nationality, demanded compensation from the Greek government. Insisting on Pacifico's rights as a British subject, the British Foreign Secretary, Lord Palmerston, took up the case in 1850 and decided to reinforce his entitlement to compensation by blockading Greece with the British fleet. He defended his action, which almost precipitated a war with France, with a masterly speech in Parliament.

Dost Muhammad (c.1798-1863), amir of Afghanistan. He was ruler of Kabul and Afghanistan (1826-39, 1843-63). Defeated in the first *Anglo-Afghan war, he regained power in 1843 and consolidated his rule in Afghanistan through control of Kandahar (1855), northern Afghanistan (1850-9), and Herat (1863), so establishing the territorial outlines of modern Afghanistan.

Douglas, Clifford (1879-1952), British engineer and economist. Before and during World War I he developed his theory of *Social Credit, that in every productive establishment the total cash issued in wages, salaries, and dividends was less than the collective price of the product. To remedy deficiencies of purchasing power, either subsidies should be paid to producers or additional moneys go to consumers. His ideas became fashionable in Britain (1921-2), and in the dominions, particularly in Canada and New Zealand.

Douglas-Home ministry, British Conservative government (1963–4) with Sir Alec Douglas-Home (1903–) as Prime Minister. Other important ministers included Reginald Maudling (Chancellor of the Exchequer) and R. A. *Butler (Foreign Secretary). The short ministry was notable for monetary expansion and the acceptance of the Robbins Report on higher education. The Conservatives were narrowly defeated in the general election in 1964 and Harold *Wilson became Prime Minister.

Douglass, Frederick (c.1817–95), US black *abolitionist. Born in slavery in Maryland, he made his escape to the free states in 1838. In 1841 he became an agent for the Massachusetts Anti-Slavery Society and a prominent advocate of abolition. An adviser of *Lincoln during the *American Civil War, he remained throughout his long life an advocate of full civil rights for all. From 1889 to 1891 he served as US minister to Haiti.

Dowding, Hugh Caswall Tremenheere, 1st Baron Dowding (1882–1970), British air chief marshal. On the outbreak of war in 1914 he was appointed commandant of the newly formed Royal Flying Corps and served as a pilot. In 1936 he was appointed commander-in-chief of Fighter Command. During the next three years he built up a force of Spitfire and Hurricane fighter aircraft, encouraged the key technological development of radar, and created an operations room which would be able to control his command. It was here that he fought the battle of *Britain in September–October 1940. Mentally and physically exhausted, he was replaced in November 1940 when the German Luftwaffe had abandoned its daylight *bombing offensive.

Drago, Luis Maria (1859–1921), Argentine statesman, jurist, and writer on international law. The Drago Doctrine, enunciated in 1902 and intended as a corollary to the *Monroe Doctrine, states that no country has the right to intervene militarily in a sovereign American state for the purpose of collecting debts. Drafted in response to a naval blockade (1902) of Venezuela by Britain, Italy, and Germany, the basic principles of the doctrine were accepted internationally by the Second Hague Conference in 1907.

Dreadnoughts, a class of British battleship. They were designed in response to a perceived threat from the German naval development of *Tirpitz (1898) and the first was launched in 1906. They revolutionized naval warfare. Powered by steam turbine engines, their speed of 21 knots and heavy fire power enabled them to fight outside the range of enemy torpedoes.

Dred Scott decision (1857), a US Supreme Court decision regarding slave status. Dred Scott, a slave, had in 1834 been taken by his master into Illinois (a non-slave state) and later into territory in which slavery had been forbidden. Years later, his then owner sued for Scott's freedom in a Missouri (slave state) court, claiming that because of his earlier stay in free territory he should be free. In 1857 the case was decided by the US Supreme Court. The majority of the court held that Scott, as a slave and as a black, was not a citizen of the US, nor was he entitled to use the Missouri courts. He was not free since his status was determined by the state in which he lived when the case was brought, i.e. Missouri. In the highly tense political atmosphere of the 1850s, the Dred Scott decision immediately deepened divisions over slavery, in particular because it declared unconstitutional the *Missouri Compromise of 1820 which had banned slavery from all territory north of the 36°30′ line of latitude.

Dreikaiserbund *Three Emperors' League.

Dresden raid (February 1945), one of the heaviest airraids on Germany in World War II. The main raid was on the night of 13–14 February 1945 by Britain's Bomber Command; 805 bombers attacked the city, which, because of its beauty, had until then been safe. The main raid was followed by three more in daylight by the US 8th Air Force. The Allied commander-in-chief General *Eisenhower was anxious to link up with the advancing *Red Army in south Germany, and Dresden was seen as strategically important as a communications centre, as well as being a centre of industry. The city was known to be overcrowded with some 200,000 refugees, but it was felt that the inevitably high casualties might in the end help to shorten the war. Over 30,000 buildings were flattened. Numbers of dead and wounded are still in dispute, estimates varying from 55,000 to 400,000.

Dreyfus, Alfred (1859–1935), French officer. Jewish by birth, he served in the French War Office, and in 1894 was accused and found guilty of selling military secrets to

The French officer Alfred **Dreyfus**, photographed before the time of his first trial in 1894.

Germany. He was sentenced to life imprisonment on Devil's Island in the West Indies. Doubts quickly arose about the fairness of the trial. In 1896 Colonel Picquart, chief of the intelligence section, satisfied himself that Dreyfus was the innocent victim of a spy, Major Esterhazy. When Esterhazy was pronounced innocent by a military court the storm broke (1898). Emile Zola, the novelist, encouraged by *Clemenceau, in a newspaper article headed '*J'Accuse*' accused the judges of having obeyed orders from the War Office in their verdict. The case was exploited by nationalist, militarist, and royalist elements on the one hand, and republican, socialist, and anti-clerical supporters on the other. The case against Dreyfus collapsed as the forger, Major Henry, committed suicide (1898) and the real spy Esterhazy confessed. The Supreme Court ordered the military to re-try the case (1899). A military court found Dreyfus 'guilty with extenuating circumstances', and he was pardoned by the President of the republic. It was not until 1906 that he was fully exonerated and reinstated in the army. The French political left-wing was both strengthened and unified as a result of the affair, and army influence declined. Anticlericalism gained widespread support, and in 1905 the Roman Catholic Church was disestablished in France.

Druze, a closed, tightly knit, relatively small, religious and political sect of *Islamic origin with Shiite influences. The main communities are in Syria, Lebanon, and Israel. In the 18th and 19th centuries they expanded from southern Lebanon to south-western Syria, where they drove out the inhabitants of Jabal Hawran and became known as Jabal Druze. Throughout the 19th and 20th centuries the Druze have persistently been involved in clashes with *Maronite Christians, but occasionally also with the Turks when under the *Ottoman empire. After the French *mandate was created in Syria (1920), Druze tribes rebelled (1925–7) against French social and administrative reforms. In retaliation the French bombarded Damascus city in 1925 and 1926. In 1944 the Druze of Syria became, theoretically, amalgamated under the country's central government. In *Lebanon, the Druze have held high political office in recent years, and are embroiled in that country's civil war.

Dual Monarchy *Austro-Hungarian empire.

Dubček, Alexander (1921–), Czechoslovak communist statesman. He fought with the Slovak Resistance in World War II and held several Communist Party posts between 1945 and 1968, when he became First Secretary and leader of his country. In what came to be known as the 'Prague Spring' he and other liberal members of the government set about freeing the country from rigid political and economic controls. He promised a gradual democratization of Czech political life and began to pursue a foreign policy independent of the Soviet Union. The latter organized an invasion by *Warsaw Pact forces of Czechoslovakia. Dubček, together with other leaders, was called to Moscow and forced to consent to the rescinding of key reforms. He was removed from office in 1969 and expelled from the Party in 1970.

Du Bois, William Edward Burghardt (1868–1963), US black *civil rights leader and author. Seeking a self-sufficient black society, he was a co-founder of the National Association for the Advancement of Colored People

(NAACP, 1909). His enrolment in the Communist Party earned him the Lenin Peace Prize (1961); this followed federal indictment (and acquittal) during the years of the *McCarthy witch-hunts. In 1962 he became a citizen of Ghana, where he died.

Dulles, John Foster (1888–1959), US international lawyer and statesman. He served as adviser to the US delegation at the San Francisco Conference (1945) which set up the *United Nations, and as the chief author of the Japanese Peace Treaty (1951). As Secretary of State under *Eisenhower (1953–9) he became a protagonist of the *Cold War and, advancing beyond the *Truman doctrine of *containment, he urged that the USA should prepare a nuclear arms build-up to deter Soviet aggression. He helped to prepare the *Eisenhower Doctrine of economic and military aid to halt aggression in the Middle East, and gave clear assurances that the USA was prepared to defend West Berlin against any encroachment.

Duma, an elective legislative assembly introduced in Russia by *Nicholas II in 1906 in response to popular unrest. Boycotted by the socialist parties, its efforts to introduce taxation and agrarian reforms were nullified by the reactionary groups at court which persuaded the emperor to dissolve three successive Dumas. The fourth Duma (1912–17) refused an imperial decree in February 1917 ordering its dissolution and established a provisional government. Three days later it accepted the emperor's abdication, but soon began to disintegrate.

Dumbarton Oaks Conference (1944), an international conference at Dumbarton Oaks in Washington, DC, when representatives of the USA, Britain, the Soviet Union, and China drew up proposals that served as the basis for the charter of the *United Nations formulated at the San Francisco Conference the following year. Attention at Dumbarton Oaks was focused on measures to secure 'the maintenance of international peace and security', and one of its main achievements was the planning of a *Security Council.

Dunant, Jean-Henri *Red Cross.

Dunkirk evacuation, a seaborne rescue of British and French troops in World War II (26 May–4 June 1940).

The fourth **Duma** established a provisional post-imperial government in Russia in 1917, but was short-lived. This photograph shows army representatives meeting in Petrograd.

The withdrawal from **Dunkirk**, painted by Charles Cundall. Allied troops crowding the beaches, and under German air attack, wait for boats to ferry them to the assortment of craft, military and civilian, that crossed the English Channel to evacuate what remained of the British Expeditionary Force. (Imperial War Museum, London)

German forces advancing into northern France cut off large numbers of British and French troops. General Gort, commanding the British Expeditionary Force, organized a withdrawal to the port and beaches of Dunkirk, where warships, aided by small private boats, carried off some 330,000 men—most, but not all, of the troops.

Durbar (Persian and Urdu, 'court'), term used by the Mogul emperors of India to describe their public audiences and appropriated under British rule to describe major ceremonial gatherings usually connected with some royal event. On 1 January 1877 the viceroy, Lord Lytton, held a Durbar at Delhi to proclaim the adoption by Queen Victoria of the title Empress of India, and thereafter such gatherings became features of the British Raj. The most magnificent was that attended by George V at Delhi in December 1911.

Durham Report (1838), a report on constitutional reform in British North America. In 1837 the liberal reformer John George Lambton, Earl of Durham (1792–1840), was appointed governor-general of British North America in the wake of the *Mackenzie and *Papineau Rebellions of 1837. Soon after his arrival he submitted the *Report on the Affairs of British North America*, recommending that *Upper and Lower Canada be united under a single parliament and given freedom to govern itself. It also proposed a reform of the land laws, extensive railway building to unify the country, and an end to French nationalism in Canada. Its recommendations met with initial hostility in Britain, but it became the seminal document for British policy in the white dominions for the rest of the 19th century. It was partially implemented in 1841 following the Act of Union uniting Upper and Lower Canada.

Dutch East Indies   *Indonesia.

Dutra, Eurico Gaspar (1885–1974), Brazilian statesman. He served as Minister of War before becoming President of Brazil (1946–51). Winning the presidency as the candidate of the Partido Social Democratico, Dutra had a new constitution adopted during his first year in office and promoted economic nationalism and industrialization. A political conservative, he did not repudiate the idea of state participation in the economy.

Duvalier, François (1907–71), dictator of Haiti (1957–71). His 1957 election was the first held under the rule of universal adult suffrage, but within a year after coming to office he suspended all constitutional guarantees and established a reign of terror based on the Tontons Macoutes, a notorious police and spy organization. The economy of his country declined severely, and 90 per cent of his subjects remained illiterate. By the time of his death

François **Duvalier**, known as 'Papa Doc', with his armed Tontons Macoutes henchmen. Duvalier's frail appearance belied the cruelty and repression of his dictatorship in Haiti.

in 1971 'Papa Doc', as he was called, had assured the succession of his son, Jean-Claude. In the face of popular unrest the latter was deposed and forced to flee to France in 1986.

dyarchy, the system of government introduced into British India by the 1919 Government of India Act. Powers in the Indian provinces were to be divided into a reserved area (finance, police, and justice) under the control of the governor, and a transferred area, such as education and health, under the control of ministers chosen from the elected members of a legislative council. The plan was devised by Lionel Curtis, founder of the Royal Institute of International Affairs. It was largely unsuccessful owing to the opposition of Indian nationalists.

E

East African Community (1967–77), an economic association of East African countries. It began with a declaration of intent (June 1963) between Kenya, Tanzania, and Uganda to improve trade, communications, and economic development. This was the East African Common Services Organization and provided for common currency, common market, and customs. This was developed by the Treaty of Kampala (1967) into the East African Community with headquarters in Arusha, Tanzania. It made considerable economic headway before 1971 when Uganda came under *Amin's regime. The Community broke up in 1977.

Eastern Front Campaigns (World War II, 1939–45), a series of military campaigns fought in eastern Europe. The first campaign (September 1939) followed the *Nazi–Soviet pact (1939), when Germany invaded Poland. Soviet forces entered from the east, and Poland collapsed. Finland was defeated in the *Finnish–Russian War. In June 1941 Hitler launched a surprise offensive against his one-time ally, the Soviet Union. Italy, Romania, Hungary, Finland, and Slovakia joined in the invasion. By the end of 1941 Germany had overrun Belorussia and most of the Ukraine, had besieged Leningrad, and was converging on Moscow. The Russian winter halted the German offensive, and the attack on Moscow was foiled by a Soviet counter-offensive. Britain, now allied with the Soviet Union, launched a joint British–Soviet occupation of *Iran (1941), thus providing a route for British and US supplies to the Red Army, as an alternative to ice-bound Murmansk. During 1942 *Leningrad continued to be besieged, while a massive German offensive was launched towards *Stalingrad and the oil-fields of the Caucasus. *Kursk, Kharkov, and Rostov all fell, as did the Crimea, and the oil centre of Maikop was reached. Here the Soviet line consolidated and forces were built up for a counter-offensive which began in December 1942, the relief of Stalingrad following in February 1943. The surrender of 330,000 German troops there marked a turning point in the war. A new German offensive recaptured Kharkov, but lost the massive battle of Kursk in July. The Red Army now resumed its advance and by the winter of 1943–4 it was back on the River Dnieper. In November 1943 Hitler ordered forces to be recalled from the Eastern Front to defend the Atlantic. Soviet offensives from January to May 1944 relieved Leningrad, recaptured the Crimea and Odessa, and re-entered Poland. Through the rest of the year and into 1945 the Red Army continued its advance, finally entering Germany in January 1945. By April it was linking up with advance troops of the Allied armies from the west, and on 2 May Berlin surrendered to Soviet troops. Victory on the Eastern Front had been obtained at the cost of at least twenty million lives.

Easter Rising (April 1916), an insurrection in Dublin when some 2,000 members of the Irish Volunteers and the Irish Citizen Army took up arms against British rule in Ireland. The *Irish Republican Brotherhood had

Above: The bitter fighting following the German invasion of the Soviet Union in 1941 left many thousands of casualties, both military and civilian, in the **Eastern Front Campaigns**. This Soviet photograph shows Ukrainians searching for relatives and friends among the bodies of the dead. Large-scale atrocities added to the human cost of the campaigns.

Below: One of the main Sinn Fein targets during the 1916 **Easter Rising** was the General Post Office in Dublin. The building was severely damaged during the week-long resistance, as is shown in this photograph taken after British forces had regained control.

planned the uprising, supported by the *Sinn Fein Party. A ship carrying a large consignment of arms from Germany was intercepted by the British navy. Roger *Casement of the IRB, acting as a link with Germany, was arrested soon after landing from a German U-boat. The military leaders, Pádraic Pearse and James Connolly, decided nevertheless to continue with the rebellion. The General Post Office in Dublin was seized along with other strategic buildings in the city. The Irish Republic was proclaimed on 24 April, Easter Monday, and a provisional government set up with Pearse as president. British forces forced their opponents to surrender by 29 April. The rising had little public support at first. Many Irishmen were serving in British forces during World War I. Sixteen leaders of the rebellion were executed and over 2,000 men and women imprisoned. The executions led to a change of feeling in Ireland and in the 1918 general election the Sinn Fein (Republican) Party won the majority vote.

East India Company, English, an English company controlling commercial policy and administration in India. The company gradually changed from a narrowly commercial enterprise into an instrument of colonial government; having lost its commercial monopolies by 1833, it served as Britain's administrative agent in India. Widespread risings in 1857 during the *Indian Mutiny

determined, through the *India Acts, the transfer of India from company to British government control in 1858, and the company was finally dissolved in 1873.

Ebert, Friedrich (1871–1925), German statesman. When Germany collapsed at the end of World War I he was Chancellor for one day (9 November 1918). He steered a difficult course between revolution and counter-revolution in order to give Germany a liberal, parliamentary constitution and he became the first President (1920–5) of the unpopular *Weimar Republic. He lost support from the Left for crushing the communists and from the Right for signing the Treaty of *Versailles.

Eboué, Félix (1884–1944), French colonial administrator. He was appointed governor of Martinique (1932), of Guadeloupe (1936), of Chad (1938), and then governor-general of French Equatorial Africa. In World War II he supported the *Free French Army of *de Gaulle, sending men and materials from the French African colonies to support the Allies.

Economic Community of West African States (ECOWAS), an economic grouping. It was constituted largely on the initiative of General *Gowon at Lagos in 1975 by fifteen West African countries, and later (1977) joined by Cape Verde. Its object was to provide a programme of liberalization of trade and to bring about a customs union within fifteen years. There is a common fund to promote development projects, and to compensate states that suffer losses; there are specialized commissions within it: for trade, industry, transport, and social and cultural affairs.

Ecuador, a country on the north-west coast of South America. Following the victory at Pichincha (1822) by Antonio *Sucre it joined Gran *Colombia. When this broke up (1830) it became a separate republic, whose politics reflected the tension between the conservative landowners of the interior and the more liberal, business community of the coastal plain. This led to an almost total breakdown in government (1845–60). Garcia Moreno ruthlessly re-established order as President (1860–75) and, on his assassination, there was a period of stable government under anti-clerical liberal governments. After World War I increasing poverty of the masses led to political turbulence. Although US military bases in World War II brought some economic gain, a disastrous war with Peru (1941) forced Ecuador to abandon claims on the Upper Amazon. Between 1944 and 1972 the *caudillo José Maria Velasso Ibarra alternated with the military as ruler, being elected President five times. The discovery of oil in the 1970s might have brought new prosperity, but in fact the mass of the population remained poor and illiterate, with the great *haciendas surviving intact. The election of the social democrat Jaimé Roldos Aquilera as President (1979–81) promised reform, but he died in a mysterious air-crash. His successor, Osvaldo Hurta do Larrea (1981–4), was accused of embezzlement, and President Febres Cordero (1984–) has faced military intervention, a crisis of external indebtedness, trade union unrest, and a decline in the oil price.

Eden, (Robert) Anthony, 1st Earl of Avon (1897–1977), British statesman. Foreign Secretary in 1935–8, 1940–5, and 1951–5, he was noted for his support for the *League of Nations in the 1930s, and was deputy to *Churchill (1945–55), whom he succeeded as Prime Minister (1955–7). Eden's premiership was dominated by the *Suez Crisis. He had opposed *appeasement of the dictators in the 1930s, and was determined to stand up to President *Nasser of Egypt, whom he perceived as a potential aggressor. Widespread opposition to Britain's role in the *Suez Crisis, together with his own failing health, led to his resignation.

Edo, Treaty of (1858), treaty between Japan and the USA. It extended the rights granted to the USA four years earlier by the Treaty of *Kanagawa, establishing diplomatic relations, accepting a conventional tariff, and granting US citizens extra-territorial rights in five treaty ports. Along with treaties signed with other foreign powers, this agreement opened the way to the westernization of Japan, but exposed the *shogunate to nationalist hostility which was to play an important role in its downfall in 1868.

education   *school systems.

Edward VII (1841–1910), King of Great Britain and Ireland and dependencies overseas, Emperor of India (1901–10). The eldest son of Queen *Victoria and Prince Albert, he was 59 before he succeeded to the throne on the death of his mother. As Prince of Wales he served on the Royal Commission on working-class housing (1884–5), but in general the queen excluded him from public affairs, denying him access to reports of cabinet meetings until 1892. As monarch his state visit to Paris in 1903 improved relations between Britain and France, and promoted public acceptance of the *entente cordiale. In domestic politics he was influential in 1910 through his insistence that his approval for the Parliament Bill to reform the House of *Lords must be preceded by a general election. He was succeeded in 1910 by his second son, *George V.

Edward VIII (1894–1972), King of Great Britain and Northern Ireland and of dependencies overseas, Emperor of India (1936). The eldest son of King *George V and Queen Mary, he served as a staff officer in World War I. The *Abdication Crisis (1936), provoked by his desire to marry Wallis Simpson, a divorcee, led to his abandoning the crown. Created Duke of Windsor, his only subsequent public role was as governor of the Bahamas during World War II. He settled in France, but was buried at Windsor in 1972, together with the duchess after her death in 1986.

Egypt, a north-east African country. Occupied by the French at the turn of the 19th century, Egypt was restored to the Ottoman empire in 1802 but enjoyed almost total independence under the rule of pashas (*Mehemet Ali) in Cairo. The construction of the Suez Canal in the 1860s made Egypt strategically important and in 1882 the British occupied the country in the wake of a nationalist revolt led by *Arabi Pasha. They ruled the country in all but name through the Agent and Consul-General Lord *Cromer. Egypt became a British protectorate in 1914 and received nominal independence in 1922 when Britain established a constitutional monarchy, with Sultan Ahmed as King Fuad I. Britain retained control of defence and imperial communications.

French troops explore **Egypt** during Napoleon's military occupation. This painting by Jean Charles Tardieu shows soldiers relaxing among the ruins at Syene (modern Aswan). One soldier is adding some French graffiti to the ancient stones, while his comrades barter for souvenirs with Egyptian traders. (Musée de Versailles)

In 1936 an Anglo-Egyptian treaty of alliance was signed, providing for a British garrison for twenty years, but for a gradual British withdrawal. This was interrupted by World War II. In 1948 Egyptian forces failed to defeat the emerging state of Israel, and in 1952 King *Farouk was overthrown by a group of army officers, one of whom, Colonel *Nasser, emerged as the head of the new republic. Nasser's nationalization of the *Suez Canal in 1956 provoked abortive Anglo-French military intervention, and in the same year he embarked on another unsuccessful war against Israel. Helped by Soviet military and economic aid, Nasser dominated the Arab world, although he suffered another heavy defeat at Israeli hands in the *Six-Day War of 1967. His successor, Anwar *Sadat, continued his confrontationalist policies, but after defeat in the *Yom Kippur War of 1973, he turned his back on the Soviet alliance, sought an accommodation with Israel, and strengthened his contacts with the West. This change of policy damaged Egypt's standing in the Arab world and in 1981 Sadat was assassinated by Islamic fundamentalists. His successor, President Mubarak, has followed a policy of moderation and reconciliation.

Eichmann, (Karl) Adolf (1906–62), Austrian Nazi administrator. A salesman by trade, he joined the Austrian Nazi Party in 1932 and by 1935 was in charge of the *Gestapo's anti-Jewish section in Berlin. In 1942, at

a conference in Wannsee on the 'final solution to the Jewish problem', he was appointed to organize the logistic arrangements for the dispatch of Jews to *concentration camps and promote the use of gas chambers for mass murder. Abducted (1960) by Israelis from Argentina, he was executed after trial in Israel.

Eisenhower, Dwight D(avid) (1890–1969), US general and thirty-fourth President of the USA (1953–61). In World War II he was appointed to command US forces in Europe. He was in overall command of the Allied landings in North Africa in 1942. In 1944–5, as Supreme Commander of the Allied expeditionary force, he was responsible for the planning and execution of the *Normandy landings and subsequent campaigns in Europe. His success in rolling back German forces was limited by Soviet advances from the east, and the pressure to bring US troops back home. The resultant vacuum of power in central Europe led to the *Cold War and to the need to establish *NATO. At home his popularity led to nomination as Republican presidential candidate and a sweeping electoral victory (*Eisenhower presidency).

Eisenhower Doctrine, a statement of US foreign policy issued by President Eisenhower after the *Suez crisis and approved by Congress in 1957. It proposed to offer economic aid and military advice to governments in the Middle East who felt their independence threatened and led to the USA sending 10,000 troops to Lebanon (1958) when its government, fearing a Muslim revolution, asked for assistance. Britain had also sent troops (1957) to protect Jordan, and despite Soviet protests US and British forces remained in the Middle East for some

General Dwight D. **Eisenhower** (right) confers with Lieutenant-General George S. Patton in Tunisia, 16 March 1943. Following the Allied landings in North Africa, Eisenhower was later to be supreme Allied commander after the 1944 D-Day invasion of France.

months. The Doctrine, whose assumption that Arab nationalism was Soviet-inspired came to be seen as fallacious and lapsed with the death (1959) of the US Secretary of State, John *Dulles.

Eisenhower presidency (1953–61), period of office of Dwight D. *Eisenhower as thirty-fourth President of the USA, with Richard *Nixon as Vice-President. His new 'modern Republicanism' sought reduced taxes, balanced budgets, and a decrease in federal control of the economy. The administration was embarrassed by the extreme right-wing 'witch-hunt' of Senator Joseph *McCarthy, with its anti-communist hysteria. In spite of tough talk there was a move towards reconciliation with China and a decision not to become engaged in Indo-China following the defeat of France there in 1954. A truce to the *Korean War was negotiated in July 1953. John *Dulles, as Secretary of State, held to a firm policy of *containment and deterrence by building up US nuclear power and conventional forces against possible Soviet aggression. Thus the *NATO and *ANZUS pacts of President Truman were extended by the *SEATO Pact of 1954. The *Eisenhower Doctrine (1957) committed the US to a policy of containment in the Middle East. After Dulles's death in 1959 Eisenhower took a more personal role in foreign policy, seeking to negotiate with the Soviet Union, when in May 1960 a US U-2 reconnaissance plane was shot down by the Russians over Soviet territory, an incident that destroyed his hopes.

El Alamein   *Alamein, El.

ELAS, a communist-dominated guerrilla army in Greece. The initials stand for the Greek words meaning National People's Liberation Army. It was created during World War II by the communist-controlled National Liberation Front (EAM) to fight against German occupation forces. By the time of the German defeat and withdrawal (1944–5) EAM/ELAS controlled much of Greece and opposed the restoration of the monarchy, aiming to replace it with a communist regime. A bitter civil war broke out (1946–9), which prompted US promise of support in the *Truman Doctrine (1947). Stalin's unwillingness to support the Greek communists contributed to their defeat.

Electoral College, a group of people chosen to elect a candidate to an office. Probably the oldest College is that which meets in Rome to elect a new pope, consisting of the cardinals of the Church. The idea was adapted by the framers of the American Constitution in 1787, each state appointing as many electors as it had members of Congress, these electors then meeting to choose the President of the USA. As states extended their franchise these electors came to be chosen by direct election. With the emergence of organized political parties, the holding of a national party convention to select presidential candidates developed. Candidates in each state are all now chosen beforehand by party associations and their vote is decided by their party's convention. Thus, for each state (except Maine since 1969), following a presidential election, the candidate who has won a majority of the popular vote in that state will gain all that state's electoral votes. In the event of a tied election the President is chosen by a vote in the House of Representatives.

Elgin, James Bruce, Earl (1811–63), British statesman and colonial administrator. As governor-general of British North America (1847–54) he was given the task of carrying out the recommendations made by the *Durham Report. In 1848 he implemented 'responsible' government with the formation of the Baldwin–La Fontaine ministry to be responsible to the elected legislative assembly. He introduced measures to improve education in Canada and to stabilize the economy, which was depressed by the new British policy of *free trade. After leaving Canada, he jointly led an Anglo-French force that marched into Beijing in 1860 to secure ratification of the Treaty of Tianjin. In 1862 he was appointed viceroy of India, dying in office a year later.

Elizabeth II (1926–), Queen of Great Britain and Northern Ireland and dependencies overseas, head of the *Commonwealth of Nations (1952–). As elder daughter of *George VI she became heir to the throne on the abdication in 1936 of her uncle *Edward VIII. She was trained in motor transport driving and maintenance in the Auxiliary Territorial Service (ATS) late in World War II and in 1947 married her distant cousin Philip Mountbatten, formerly Prince Philip of Greece and Denmark. Their first child and heir to the throne, Prince Charles was born in 1948. Her coronation in 1953 was the first major royal occasion to be televised. Since then she has devoted much of her reign to ceremonial functions and to tours of the Commonwealth and other countries. While strictly adhering to the convention of the British constitution, she has always held a weekly audience with her Prime Minister and shown a strong personal commitment to the Commonwealth.

On 7 February 1952 Princess Elizabeth, now **Elizabeth II**, arrives at London airport with the Duke of Edinburgh after a night flight from Kenya following news of the death of her father, George VI. Earl Mountbatten of Burma (*far left*) was among the party awaiting them.

Ellice Islands   *Tuvalu.

Ellis Island, an island in New York Bay off Manhattan Island. Long used as an arsenal and a fort, from 1892 to 1943 it served as the centre for immigration control. From 1943 until 1954 it acted as a detention centre for aliens and deportees. In 1965 it became part of the Statue of Liberty National Monument, and open to sightseers.

El Salvador, the smallest Central American country. After independence from Spain in 1821, El Salvador joined (1824) the United Provinces of *Central America, and with the break-up of that entity in 1838, became an independent republic (1839). Internal struggles between liberals and conservatives and a series of border clashes with neighbours retarded development in the 19th century. By the early 20th century the conservatives had gained ascendancy and the presidency remained within a handful of élite families as if it were their personal patrimony. El Salvador's 20th-century history has been dominated by a series of military presidents. While some of them, such as Oscar Osorio (1950–6) and José M. Lemus (1956–60), appeared mildly sympathetic to badly needed social reform, they were held in check by their more conservative military colleagues in concert with the civilian oligarchy. Fidel *Castro's Cuban revolution and leftist guerrilla activity in other Central American countries have pushed the Salvadoran army steadily to the right. Repressive measures and violations of human rights by the army during the 1970s and 1980s were documented by a number of international agencies, and have posed a large refugee problem.

emancipists, a term used in Australia during the period of *convict transportation. In a narrow sense, it referred only to those convicts who had been pardoned, conditionally or absolutely, by the governor. In a broader sense, it was applied to all ex-convicts who, having served their term of imprisonment or enforced servitude, had become free, and in some cases, wealthy. There was much conflict between emancipists and *exclusionists in New South Wales, Australia, especially during *Macquarie's governorship (1810–21). The term was also applied to members of a political group, consisting of emancipists and liberals, which campaigned for reforms during the 1820s, 1830s, and early 1840s. William Wentworth was the acknowledged leader of this group for many years. In 1835, it founded the Australian Patriotic Association.

Villagers in **El Salvador** listen to the words of a government spokesman, a local army officer, in 1983. Government and insurgents vied for rural support in El Salvador during the unsettled years of the 1980s.

Emin Pasha (Mehmed Eduard Schnitzer) (1840–92), German explorer and physician. He joined the Ottoman army in 1865, and in 1876 he served under General *Gordon in Khartoum. Gordon used him for administrative duties and diplomatic missions and in 1878 appointed him governor of the Upper Nile area of Equatoria, where he surveyed the region and suppressed slavery. Isolated when the *Mahdi controlled the Sudan, he was rescued in 1888 by H. M. *Stanley. In 1890 he was employed by the German government in East Africa. While engaged in exploration for Germany, Arab slave-raiders murdered him.

Emirates, Fulani   *Fulani empire of Sokoto.

Emmet, Robert (1778–1803), Irish nationalist. Involved in the United Irishmen movement, during 1800–2 he visited France in an attempt to win support for Irish independence. Returning to Ireland in 1803 with a small band of followers, he began an insurrection in Dublin against British rule, which ended in disaster. Emmet escaped but was subsequently captured and executed. Gallant and reckless, he was to become a potent symbol in the cause of Irish nationalism.

Ems telegram (13 July 1870), a dispatch from the Prussian king *William I to his chancellor, *Bismarck. A relative of the Prussian king, Prince Leopold of Hohenzollern-Sigmaringen, had accepted an offer to the Spanish throne. This alarmed the French, who feared Prussian influence south of the Pyrenees. Leopold withdrew his claim a few days later, but the French ambassador approached William at the German spa town of Ems, asking for an assurance that Leopold's candidacy would never be renewed. The king refused, politely but firmly, and he sent his chancellor a telegram to the effect that the crisis had passed. Bismarck, intent on provoking war with France, published a shortened version which turned the refusal into an insult. French public opinion was outraged and Napoleon III declared the *Franco-Prussian War, whose consequences were to cause the downfall of the French and the creation of the *German Second empire.

Engels, Friedrich (1820–95), German social philosopher and businessman. He was, with *Marx, one of the founders of modern *communism. A partner in the Ermen and Engels cotton plant in Manchester, England, he was converted to communism by the radical, Moses Hess. Engels believed that England, with its advanced industry and rapidly growing proletariat, would lead the world in social upheaval. In 1844 his *The Condition of the Working Class in England in 1844*, based largely on official parliamentary reports, attracted wide attention. In the same year he met Karl Marx in Paris. Together they joined the socialist League of the Just in 1847, transforming it into the Communist League, and publishing its *Communist Manifesto the following year. He participated in the revolutionary movement in Baden and later published a penetrating work on the failure of the *Revolutions of 1848 in Germany. Engels returned to England, working as a successful businessman whilst supporting Marx financially. After Marx's death he served as the foremost authority on Marxism, continuing work on *Das Kapital* from Marx's drafts, publishing the second and third volumes in 1885 and 1894.

Enghien, Louis Antoine Henri de Bourbon-Condé, duc d' (1772–1804), son of Henri, Prince of Condé. After the unsuccessful invasion of 1792, he used his military training to lead a force of exiled royalists. This force was dissolved after the peace of Luneville in 1801 and he retired to Baden. Three years later he was wrongly accused by *Napoleon of being involved in a plot to invade France. This charge led to his being kidnapped and shot on Napoleon's orders.

enosis (Greek, 'union'), a Greek-Cypriot campaign for union of *Cyprus with Greece, launched by *EOKA in the 1950s. Archbishop *Makarios's acceptance of independence from Britain without union (1960) led to renewed demands for *enosis* (1970), and its proclamation in 1974. In response Turkey invaded and partitioned the island to protect the Turkish minority.

entente cordiale (1904), friendly understanding between Britain and France. It aimed to settle territorial disputes and to encourage co-operation against perceived German pressure. Britain was to be given a free hand in Egyptian affairs and France in Morocco. Germany, concerned over this *entente*, tested its strength by provoking a crisis in Morocco in 1905, leading to the *Algeciras Conference (1906). The *entente* was extended in 1907 to include Russia and culminated in the formal alliance of Britain, France, and Russia in World War I against the Central Powers and the Ottoman empire.

Enver Pasha (1881–1922), Ottoman Turkish general and statesman. He played a prominent part in the 1908 *Young Turk revolution which restored the liberal constitution of 1876, and subsequently led a successful coup in 1913. As Minister of War (1913–18) he played the leading role in determining the entry of the Ottoman empire into World War I on the side of the Central Powers and in the conduct of Ottoman strategy during the war. In 1921 he fled to Turkistan, where he was killed leading opposition to Soviet rule.

EOKA (National Organization of Cypriot Fighters), the militant wing of the *enosis* movement in Cyprus. Colonel

Enver Pasha, accompanied by a German officer, on his way to inspect troops on the Romanian front during World War I. Enver Pasha was instrumental in promoting Turkey's wartime alliance with Germany and the other Central Powers.

Georgios Grivas (1898–1974), commander of the Greek Cypriot national guard, was its most famous leader. During 1954–9 guerrilla warfare and terrorist attacks were waged against the British forces. In 1956 *Makarios was exiled on the charge of being implicated with EOKA. After independence in 1960 the organization was revived as EOKA–B.

Equatorial Guinea, a West African country that includes Fernando Po island. Formerly a Spanish colony, it was a haunt of foreign slave-traders and merchants. Britain administered it from 1827 to 1858. The mainland was not effectively occupied by Spain until 1926. Declared independent in 1968, a reign of terror followed until President Macias Nguema was overthrown and executed (1979) by his nephew, Obiang Nguema. The new regime has pursued less repressive domestic policies with some degree of success.

Erhard, Ludwig (1897–1977), German economist and statesman, Chancellor of the German Federal Republic (1963–6). He became a Christian Democrat member of the German Federal Republic's *Bundestag*, and was Minister for Economic Affairs from 1949 to 1963, during which time he assisted in his country's 'economic miracle' (German, *Wirtschaftswunder*), which trebled the gross national product in the post-war years.

Eritrea, a province of *Ethiopia, on the Red Sea. In 1869 Italy purchased the coastal town of Assab, and in 1885 began the occupation of the rest of the province, which it declared a colony in 1889. It was from here that the Italians launched their disastrous campaign against Ethiopia in 1896, ending in their defeat at *Adowa. Under British military administration (1941–52), a plan to join the Muslim west with the Sudan and the Christian centre with Ethiopia failed. Instead, the United Nations voted to make Eritrea a federal area subject to Ethiopia. In 1962 Emperor *Haile Selassie declared it a province and the Eritrean Liberation Front (ELF) then emerged, seeking secession. Fierce fighting between the ELF and the Ethiopian regime has continued through the 1980s, in spite of drought and famine.

Ethiopia, a north-east African country (known formerly as Abyssinia). The unification of Ethiopia was begun by Tewodros II (1855–68), and was continued during the reign of *Menelik II. The country successfully repelled Italian attempts at colonization by a decisive victory at *Adowa in 1896, but was conquered by *Mussolini in 1935–6. The Ethiopian emperor *Haile Selassie was restored in 1941 after the *Abyssinian Campaigns, and in the 1950s and 1960s Ethiopia emerged as a leading African neutralist state. Haile Selassie's failure to deal with severe social and economic problems led to his deposition by a group of radical army officers in 1974. A subsequent coup brought Colonel Mengistu to power in 1977, but his centralized Marxist state was confronted by a Somali-backed guerrilla war in *Eritrea. Famine broke out on a massive scale, and despite Soviet and Cuban military assistance in the war and an international relief effort to alleviate starvation, neither problem has yet been solved.

Eugénie, Marie de Montijo (1826–1920), Empress of the French and wife of *Napoleon III. Throughout her husband's reign she contributed much to the brilliance of his court and acted as regent on three occasions. When the empire collapsed (September 1870), she fled to England. She retained a close interest in European affairs until her death in Spain in 1920.

Eureka Rebellion (1854), an armed conflict between diggers and authorities on the Ballarat gold fields of

Hand-to-hand fighting brought an end to the **Eureka Rebellion** as soldiers and police overran the Australian diggers' stockaded mining camp. This painting by B. Ireland shows the confusion as the fifteen-minute-long battle raged. (Latrabe Library, Victoria)

Australia. Gold had been found here in 1851; by 1853 over 20,000 diggers from around the world had crowded into Ballarat. Their grievances, which included the licence system and its administration, corruption among officials, lack of political representation, and limited access to land, culminated in an attack by soldiers and police on diggers who were in a stockade on the 'Eureka lead'. Thirteen diggers faced charges of high treason; one case was dropped and the others were acquitted. A royal commission led to reforms on the gold-fields.

European Coal and Steel Community   *European Community.

European Community, an organization of West European states. It came into being (1967) through the merger of the *European Economic Community (EEC), the European Atomic Energy Community (Euratom), and the European Coal and Steel Community (ECSC), and was committed to economic and political integration as envisaged by the Treaties of *Rome. Its membership is identical with that of the EEC. It operates within the framework of the European Commission (an executive body with powers of proposal), various consultative bodies, and the Council of Ministers (a decision-making body drawn from the member governments) with headquarters in Brussels. The *European Parliament, which held its first direct elections in 1979, has powers of supervision and consultation, and a measure of control over the Community's budget. The decisions of the European Court of Justice at The Hague are directly binding on its member states and are superior over any national Act of Parliament that is inconsistent with it.

European Economic Community (EEC, Common Market), an economic association of European nations set up by the Treaties of *Rome, operating within the *European Community. Its member states have agreed to co-ordinate their economic policies, and to establish common policies for agriculture, transport, the movement of capital and labour, the erection of common external tariffs, and the ultimate establishment of political unification. The world's biggest trading power, its present members (with dates of accession) are: Belgium, France, the German Federal Republic, Italy, Luxemburg, and the Netherlands (sometimes referred to as 'the Six'— 1958); Denmark, Great Britain, Ireland (1973); Greece (1981); Portugal and Spain (1986). Norway withdrew in 1972 and Greenland in 1985. From its inception the EEC provided an extension of the functional co-operation inaugurated by the European Coal and Steel Community. It owed much to the campaigning initiative of Jean *Monnet and to the detailed planning of Paul-Henri *Spaak. Preliminary meetings were held at Messina in 1955, which led to the Treaties of Rome in 1957 and the formal creation of the EEC in January 1958. Much controversy surrounded Britain's entry, which was delayed for thirteen years from initial application, mainly by the use of the French veto under President *de Gaulle. In recent years market intervention and artificial currency levels in the Community's Common Agricultural Policy (CAP), together with over-production due to high consumer prices, have meant the absorption by the CAP of about two-thirds of the Community's budget. The European Monetary System (EMS) ensures a degree of economic stability by limiting fluctuations in the exchange

rates of member states (Britain excepted) against a central rate. The Community has an agreement (the *Lomé Convention) with some sixty African, Caribbean, and Pacific countries which removes customs duties without reciprocal arrangements for most of their imports to the Community, and offers substantial aid programmes.

European Free Trade Association (EFTA), a customs union of European states. Brought into existence by the Stockholm Convention in 1959, its membership has at times consisted of Austria, Britain, Denmark, Norway, Portugal, Sweden, Switzerland, Liechtenstein, Finland, and Iceland. Unencumbered by the political implications of the *European Economic Community, it had been created by a British initiative as an alternative trade grouping. In 1973 Britain and Denmark entered the European Economic Community and left EFTA. In 1977 EFTA entered into an agreement with the EEC which established industrial free trade between the two organizations' member countries, followed in the 1980s by wider areas of co-operation.

European Parliament, one of the constituent institutions of the *European Community, meeting in Strasbourg or Luxemburg. From 1958 to 1979 it was composed of representatives drawn from the Assemblies of the member states. However, quinquennial direct elections have taken place since 1979. Treaties signed in 1970, 1975, and 1986 gave the Parliament important powers over budgetary and constitutional matters, assuming, through the single European Act (1986), a degree of sovereignty over national parliaments.

European Recovery Program   *Marshall Plan.

Evatt, Herbert Vere (1894–1965), Australian statesman. Chief Justice of New South Wales (1960-2), and a federal politician (1925-30, 1940-60), he led the Labor Opposition from 1951 until 1960. Noted for his championship of the rights of the smaller nations, and for greater independence from Britain, Evatt presided over the United Nations General Assembly (1948-9).

Évian Agreements (1962), a series of agreements negotiated at Évian-les-Bains in France. Secret negotiations between the government of General *de Gaulle and representatives of the provisional government of the Algerian Republic of *Ben Bella began in Switzerland in December 1961 and continued in March 1962 at Évian. A cease-fire commission was set up and the French government, subject to certain safeguards, agreed to the establishment of an independent Algeria following a referendum. The agreements were ratified by the French National Assembly but were violently attacked by the extremist Organization de l'Armée Secrète (OAS).

exclusionist (or exclusive), Australian settler opposed to the emancipation of ex-convicts. The name was applied in New South Wales, during the period of *convict transportation, to those people who opposed the restoration of civil rights to ex-convicts or *emancipists. The exclusives were composed for the most part of civil and military officials and of gentleman squatters and settlers who were called in derision 'Pure Merinos'.

F

Fabians, British socialists aiming at gradual social change through democratic means. The Fabian Society was founded in 1884 by a group of intellectuals who believed that new political pressures were needed to achieve social reforms. George Bernard Shaw, Beatrice and Sidney *Webb, Annie *Besant, and Hubert Bland were its leading members. The slogan of the early Fabians was 'the inevitability of gradualism'. Reforms would be secured by the patient, persistent use of argument, and propaganda through constitutional methods. It was one of the socialist societies which helped found the Labour Representation Committee, the origin of the *Labour Party, in 1900. Trade Union militancy from 1910 to 1926, and the harshness of unemployment in the 1930s, weakened the appeal of Fabian gradualism but by 1939, with moderate leaders, such as Clement *Attlee, coming to the forefront, their influence revived.

Factory Acts, laws to regulate conditions of employment of factory workers. Textile factories first developed in Britain and the USA in the late 18th century, employing many workers, especially women and children, and replacing the older 'domestic' textile industry. Conditions were sometimes dangerous and working hours were long. In Britain two early Acts of Parliament in 1802 and 1809, which aimed to protect children and apprentices, failed because they could not be enforced. The Factory Act of 1833 banned the employment of children under 9, restricted working hours of older children, and provided for the appointment of factory inspectors. Sometimes fiercely opposed by industrialists, it was fought for in Parliament by Christian philanthropists such as Lord *Shaftesbury. Legislation in Britain (1844 and 1847) extended protection of workers into mines and other industries and reduced the working day to ten hours. A Factory Act (1874) consolidated the ten-hour day and raised the age of children in employment to 10, this being further raised to 12 in 1901 and 14 in 1920. In the 20th century a complicated structure of industrial law developed. Legislation similar to the British was enacted in most European countries, particularly during the 1890s, and also in the USA, where each state developed its own factory laws. These were consolidated by a federal act, the Fair Labor Standards Act of 1938. In the early 20th century conditions of work in much of the Far East remained poor, with child labour often used and excessive hours being demanded, although in British India legislation had begun in 1881 to try to tackle these problems. It was to counter the problem of child labour and the exploitation of factory workers, particularly women, that the *International Labour Organization (ILO) was formed by the League of Nations (1919).

Fair Deal, the name given by US President *Truman to his proposed domestic programme in 1949. By it he hoped to advance beyond the *New Deal, to introduce measures on *civil rights, fair employment practices, education, health, social security, support for low-income housing, and a new farm subsidy programme. A coalition

Saudi Arabia's King **Faisal** (in dark robes) with Saudi officials and foreign visitors at Riyadh's principal mosque. The king's standing as an Islamic leader strengthened Saudi opposition to both Israeli expansion and communism.

of Republicans and conservative southern Democrats blocked most of his measures in Congress, and although he did secure some advances in housing and social security the bulk of his proposals were lost.

Faisal I (1885–1933), King of Iraq (1921–33). The son of *Hussein Ibn Ali, he commanded the northern Arab army in Jordan, Palestine, and Syria in association with T. E. *Lawrence in the Arab Revolt of 1916. In 1920 Faisal was chosen King of Syria by the Syrian National Congress but was expelled by France, the mandatory power. He was then made King of Iraq by Britain, who held the mandate for that territory. As ruler of Iraq (1922–33) he demonstrated considerable political skill in building up the institutions of the new state.

Faisal, Ibn Abdul Aziz (1905–75), King of Saudi Arabia (1964–75). Brother of King Saud ibn Abd al-Aziz, he became effective ruler of Saudi Arabia in 1958, dealing with the main consequences for Saudi Arabia of the immense increase of oil revenues. Pro-West, he worked in association with the USA while remaining inflexible in his opposition to Israel's ambitions and unyielding on Arab claims to Jerusalem. Faisal stood against the demands of radical Arab nationalism represented by Egypt under *Nasser. He was assassinated by a nephew.

Falange, the (Spanish, 'phalanx'), a Spanish political party, the Falange Española. Founded in 1933 by José António Primo de Rivera, the son of General *Primo de Rivera, its members were equally opposed to the reactionary Right and the revolutionary Left. Their manifesto of 1934 proclaimed opposition to republicanism, party politics, capitalism, Marxism, and the class war, and it proposed that Spain should become a syndicalist state on Italian *fascist lines. During the *Spanish Civil War Franco saw the potential value of the Falange provided that its aims were made acceptable to traditionalists and monarchists. The death of José António at the hands of the Republicans made it possible for Franco to adopt the movement in April 1937. After World War II it ceased to be identified with fascism.

Falkland Islands (Spanish, 'Islas Malvinas'), a group of two main islands and nearly 100 smaller ones in the South Atlantic. First visited by European explorers in the late 16th century, the Falklands were successively colonized by the French, British, and Spanish. In 1806 Spanish rule over Argentina ceased, and in 1820 the Argentinians claimed to succeed Spain in possession of the Falklands. The British objected and reclaimed the islands (1832) as a crown colony. In 1882–3 a British naval squadron occupied the islands for the protection of the seal-fisheries. Since 1833 the Argentinians have exerted their claims and have disputed possession of the Falklands by the British. This rivalry culminated in 1982 with the *Falkland (Malvinas) War. The large British naval and military force in the area was reduced following the completion of Mount Pleasant airport in 1987.

Falkland (Malvinas) War (2 April–14 June 1982), the Argentine–British war in the *Falkland Islands. Repeated attempts at negotiation for the transfer of the islands from British to Argentine rule having failed, an Argentine warship was sent by General Leopoldo Galtieri's military junta to land a party of 'scrap dealers' on South Georgia on 19 March 1982 with the intention of reclaiming the Falkland Islands. This was followed on 2 April by a full-scale military invasion. Attempts by the UN, the USA, and Peru to secure a peaceful resolution to the conflict failed, and Britain sent a task force of thirty warships with supporting aircraft and auxiliary vessels across 13,000 km (8,000 miles) of sea to recover the islands. Although all but three Latin American nations supported Argentina, the USA, in a difficult position because of close ties to both countries, sided with the British. The ten-week conflict, which claimed the lives of nearly 1,000 British and Argentine servicemen and civilians, ceased with the surrender of the Argentine forces on 14 June. The British victory contributed to the downfall of General Galtieri's government.

Fante Confederation, a loose association of small states along the Gold Coast (Ghana) in West Africa. Having migrated from the north in the 17th century, the Fante served as middlemen between the slave and gold-producing states of the African interior and European traders along the coast. The coastal states were threatened by the rise of *Asante power in Kumasi in the 19th century, and they supported the British in the Asante Wars. The Fante have played a prominent role in the affairs of Ghana since independence in 1957.

Farouk (1920–65), King of Egypt (1936–52). Son of Fuad I, whom the British had installed in 1922, he ruled autocratically, as had his father. His pro-*Axis sympathies during World War II resulted in a clash with the British, who imposed on him (1942) the Wafd leader Mustapha an-Nahas Pasha as a Premier who would support the Allies. Farouk's defeat in the Arab–Israeli conflict (1948), and the general corruption of his reign led to a military coup in 1952, headed by *Nasser. He was forced to abdicate in favour of his infant son, Fuad II, in 1952, who was deposed in 1953.

fascism (from the Italian *fasces*, the bundle of rods and axe laid before a Roman magistrate symbolizing unity and power), an extreme right-wing totalitarian political system. It arose in opposition to *communism, but adopted communist styles of propaganda, organization, and violence. The term was used by the Fascio di Combattimento in Italy in 1919. *Mussolini shaped fascism into a potent political force in Italy and *Hitler developed a more racialist brand of it in Germany. Similar movements sprang up in Spain (*Falangists), Portugal, Austria, the Balkan states, France, and South America. In Britain the National Union of Fascists under *Mosley was founded in 1932, and between 1934 and 1936 adopted a strongly *anti-Semitic character. The Public Order Act of 1936, banning private armies and political uniforms in Britain (in this case, the fascists' *black shirts) discouraged further activities. Once in power (in 1922 in Italy, in 1933 in Germany, and 1939 in Spain) fascists proceeded to establish police states. Fascism as a power in Germany and Italy was removed at the end of *World War II. It lingered for some years in Spain and Portugal, however, and has erupted and re-erupted on occasion in Central and South America. It has remained a latent, if minimal, force in almost every country in the western world. In France the Front National of Jean-Marie Le Pen made considerable gains in the 1986 elections.

Fashoda incident (18 September 1898), the culmination of a long series of clashes between Britain and France in the *'scramble for Africa'. The French objective, to occupy the sub-Saharan belt from west to east, countered the British aim of linking their possessions from the Cape to Cairo. Thus in 1896 the French dispatched a force under *Marchand from *Gabon to occupy the *Sudan, at the same time that *Kitchener was moving up the Nile to recover Khartoum. Both reached Fashoda

A contemporary British cartoon ridicules the **Fashoda incident**. The French tricolour flag is being waved from a precarious and defensive position up a palm tree while John Bull confronts his choleric French counterpart in a characteristically confident pose.

HISTORY OF THE MONTH IN CARICATURE.
(NOVEMBER.)

I.—FASHODA.

Westminster Budget. "IN A DIFFICULT POSITION." [Oct. 14, 1898.

during the summer of 1898, and as neither side desired conflict, they agreed that both French and British flags should fly over the fort. The matter was referred to London and Paris, and for a while tension between the two countries was extreme. In December the French ordered Marchand to withdraw, and this enabled an agreement to be reached whereby the Nile and Congo watersheds should demarcate the respective spheres of influence by the two countries in Africa.

Fatah, al- (Arabic, 'victory'), a militant Palestinian organization. It was founded (1962) in Kuwait to fight for the restoration of *Palestine to the Arabs. Al-Fatah assumed the leadership of the *Palestine Liberation Organization in 1969. Its guerrilla units were expelled from Jordan after the civil war in 1970, and it withdrew to southern Lebanon (Fatahland). Subsequently al-Fatah was drawn into the Lebanese imbroglio and became divided; a part was expelled from Lebanon after the Israeli invasion of 1982. Leadership remained in the hands of Yassir Arafat (1929-), who had led al-Fatah from its foundation.

Faulkner of Downpatrick (Arthur), Brian Deane Faulkner, Baron (1921–77), Northern Ireland states-man. A Unionist Member of Parliament at *Stormont (1949-73), he was Minister of Home Affairs (1959-63, 1971-2), and Prime Minister (1971-2). His negotiations with the Westminster government for constitutional changes in *Northern Ireland lost him support in his own party.

Fawcett, Dame Millicent Garrett (1847–1929), Brit-ish feminist. Sister of Elizabeth Garrett *Anderson, she was a pioneer of the movement in Britain to secure equality for women in voting, education, and careers. She was strongly supported by her husband, Henry Fawcett, a Liberal politician and academic. In 1897 she became president of the National Union of Women's Suffrage Societies, whose policy was to gain votes for women without the militancy soon to be associated with the *suffragettes. Though overshadowed by the actions of the latter, the reliance of the 'suffragists' on peaceful methods favourably influenced public opinion.

Federal Bureau of Investigation (FBI), the in-vestigative branch of the US Department of Justice. Established by Attorney-General Charles J. Bonaparte (1851-1921) in 1908, it was at first called the Bureau of Investigation. It was reorganized in 1924 when J. Edgar *Hoover was appointed as director, giving it wider powers to investigate violations of federal laws. Hoover successfully led the 1930s drive against gangsters. During World War II the FBI began spying activities against Nazi sympathizers in the USA and Latin America. The later excesses of Hoover, in particular his harassment of political dissidents and radicals such as Martin Luther *King, brought its counter-intelligence activities into disrepute. It was roundly criticized by the Senate in investigations of 1975-6.

federalism, a political system uniting separate con-stituent political entities within an overall political struc-ture. An early example of a federal state is that of Switzerland, but it was the USA, whose federal Con-stitution was devised in 1787, which became a model for many later federations, for example Canada, Australia, the German Federal Republic, and Yugoslavia. For any federal system to succeed, the balance between the powers of each constituent part and those of the central federal government needs to be agreed. In the USA before the *American Civil War, the excessive demands for *states' rights ultimately led to the break-up of the Union and to war. The term federalist was early used in the USA to describe those who believed in the need for strong central government.

Federalist Party, US political party. The first political party to emerge after the US Constitution became operative (1789), it took its name from the *Federalist Papers*, a collection of essays written by Madison, Hamilton, and Jay to influence the ratification of the Constitution by New York. The party of George Washington and John Adams, it had support in New England and the north-east generally both from commercial interests and wealthier landowners. It stood for strong central government and the firm enforcement of domestic laws, was pro-British in foreign affairs, and identified itself with the economic policies of Alexander Hamilton. The emergence of new political issues, disagreements over commercial and for-eign policy, and the narrowness of its popular appeal, gradually undermined the Party, although it continued to elect members to Congress until it finally disappeared in 1825.

Federation of Rhodesia and Nyasaland   *Central African Federation.

Feisal   *Faisal.

feminism, a movement concerning social, political, and economic rights. Its advocates have for the most part demanded equal rights for women as for men, but sometimes have asserted the right of women to separate development. Throughout the ages women had generally been subordinated to men and largely excluded from education and from the ownership of property. A move-ment for the elevation of women's status began with the French Revolution. The philosopher Antoine Nicolas Condorcet wrote an essay (1790) on the admission of women to full citizenship, and in Britain Mary Wollstonecraft published the first great document of feminism, the *Vindication of the Rights of Woman* (1792), in which she argues that educational restrictions alone keep women in a state of 'ignorance and slavish dependence'. In the USA the movement grew out of anti-slavery agitation. Abigail Adams, Mercy Otis Warren, and Emma *Willard were early campaigners for the rights of women. In 1848 the first feminist convention, led by Elizabeth Cady Stanton and Lucretia Mott, was held at Seneca Falls, New York. In Britain a series of Married Women's Property Acts from 1870 onwards increasingly allowed women to hold and manage property. *Women's suffrage, first achieved in the US state of Wyoming (1869), has become almost universal. Notable developments in the UK have been the Sex Disqualification Removal Act (1919), the Equal Pay Act (1970), and the Sex Discrimination Act (1975). The influence of books such as the French author Simone de Beauvoir's *The Second Sex* (1953) led to the formation in the 1960s of *women's liberation movements throughout the Western world, which in turn have produced much feminist writing, both

The **Fenian** ambush in Manchester, 1867. Armed Fenians attacked a police van and rescued two Irish prisoners. Such incidents alerted British politicians to the dangers inherent in an unsolved Irish problem.

scholarly and creative. The revival of Islam, with its enforced social isolation of women, has led to the establishment of segregated systems of banking, commerce, and education in Muslim communities.

Fenian, originally a member of a secret revolutionary society, named after the Fianna, the Irish armed force in legendary times. Founded as the Fenian Brotherhood in the USA by John O'Mahony and as the Irish Republican Brotherhood by James Stephens in Ireland (1858), the name was later applied to supporters of Irish republicanism. Many of its early members had been actively involved in the *Young Ireland movement. Its military wing was known as the *Irish Republican Army (IRA). Fenian invasions of Canada (1866, 1870, 1871) failed. In England the Fenians attempted to seize Chester Castle; they rescued two of their number in Manchester, and in an unsuccessful rescue of prisoners in London (1867) killed twelve people. Several Fenians were executed and hundreds imprisoned. Their exploits drew attention to Irish discontent and helped to convince *Gladstone of the urgent need to find a solution to Ireland's problems. Several Fenians became Members of Parliament at Westminster during the *Home Rule period. In the latter part of the 1860s the Fenian Brotherhood split into three sections, each in theory supporting the IRB but in practice sharply divided by personalities and policies. The organization was superseded in the USA by Clan-na-Gael, a secret society headed by John Devoy, and by other open Irish-American organizations supporting Irish republicanism.

Ferdinand VII (1784–1833), King of Spain (1808–33). He succeeded to the throne after the forced abdication of his father, Charles IV, and was in turn forced by the French to abdicate in favour of *Napoleon's brother, Joseph Bonaparte, spending the years of the *Peninsular War in prison in France. Known as 'The Desired One', he was released in 1814 and restored to the throne. He abolished the liberal constitution of 1812 and instituted his own absolutist rule, relying on the support of the Church and the army. The loss of the colonies in America

(*Spanish-American Wars of Independence) deprived the government of a major source of income, and his troops mutinied. The revolutionaries held him practically a prisoner until 1823, when French forces came to his aid. Restored to power, he carried out a bloody revenge on the insurgents.

Ferry, Jules François Camille (1832–93) French statesman. Mayor of Paris (1870–1) during its siege in the *Franco-Prussian War, his narrow escape (18 March 1871) from the *Paris Commune left him with a strong dislike of extremist politics. After serving as French ambassador in Greece, he was elected to the French Chamber of Deputies (1876–89) and was in government 1879–85, twice as Prime Minister (1880–1, 1883–5). He was responsible for much liberal legislation, extending freedom of association and of the press and legalizing trade unions. He weakened the grip of the Roman Catholic Church on education, extended higher education, created lycées for girls, and made French elementary education both free and compulsory (1881). His ministries also saw wide French colonial development in Tunisia (1881), the Congo (1884), Madagascar (1885), and Indo-China (1885). This latter lost him support and he fell from office. He narrowly failed to be elected President of the Republic in 1887.

Ferdinand VII, King of Spain, painted by Goya in 1828, after his restoration to the Spanish throne. (Museo del Prado, Madrid)

Fiji. A 19th-century painting by J. J. Wild shows the village of Ngaloa on the Fijian island of Kandavu. The idealized waterside scene contrasts sharply with Fiji's notoriety in the 1800s for cannibalism and tribal warfare.

Fíanna Fáil (Gaelic, 'soldiers of destiny'), Irish political party. Its main aim is to create a united republican Ireland, politically and economically independent of Britain. Eamon *de Valera founded the Party in 1926 from opponents of the Anglo-Irish Treaty (1921) which established the *Irish Free State. The Party won control of the government (1932). It dominated Irish politics for the following years, being out of office only for short periods. In 1973 it lost to an alliance of the Fine Gael and the Labour Party, but returned to power for a period in 1977 and again in 1987.

Fiji, a group of some 840 islands in the Melanesian archipelago of the south-west Pacific. Nineteenth-century Fiji was notorious for inter-tribal wars and cannibalism, a situation not assisted by an influx of deserting seamen, traders seeking sandalwood, and whalers. The islands became a British crown colony in 1874, the Western Pacific High Commission being set up for the pacification and control of the labour trade. By 1879 Indians began to be imported under the indenture system. By the 1950s Indians outnumbered Fijians and were dominating commercial life, while Fijians owned most of the land. The country became independent in 1970. The election of a government with an Indian majority (1987) brought ethnic tensions to a head leading to two military coups to restore indigenous Fijian control, and to the withdrawal of Fiji from the Commonwealth of Nations.

filibusters, a term used originally to describe a class of piratical adventurers or freebooters who pillaged the Spanish colonies in the 17th century. It was extended in the 19th century to such lawless adventurers from the USA as those who between 1850 and 1860 followed *Lopez in his expedition to Cuba, and Walker in his expedition to Nicaragua. Subsequently it was used of anyone who engaged in unauthorized war against foreign states. From this the term came to be used to describe speakers in the US Congress or any other assembly seeking to delay legislation by making lengthy speeches and so obstructing business.

Fillmore, Millard (1800–74), thirteenth President of the USA (1850–3). In 1834 he joined the *Whig Party and became its leader in the House of Representatives. He was elected as US vice-president to President Zachary *Taylor (1848). Two years later, on Taylor's death, he became President (1850–3). While in office, he signed the *Compromise of 1850 and approved of the naval expedition under Matthew *Perry to force trading arrangements with Japan (1853). Denied renomination in 1852 as a result of the split of his party over the issue of slavery, he remained politically active and ran as the presidential nominee of the American (*Know Nothing) Party in 1856, but managed to carry only one state (Maryland).

Fine Gael (Gaelic, 'United Ireland'), Irish political party. Founded in 1923 as Cumann na nGaedheal, it changed its name in 1933. It originated among supporters of the Anglo-Irish Treaty that created the *Irish Free State. William *Cosgrave was its leader (1935–44). Fine Gael gained power as the dominant element in a coalition in 1948, electing John Costello as its leader. This government in 1949 declared Ireland to be a republic. Since then, Fine Gael has been intermittently in power, but has required coalition support to remain so. It has advocated the concept of a united Ireland achieved by peaceful means.

Finland, a Baltic country. The Treaty of *Tilsit (1807) between Alexander I and Napoleon led to the annexation of Finland as a grand duchy of Russia until 1917. Attempts to impose the Russian language and military conscription brought discontent and the *Russian Revolution of 1917 offered opportunities for national assertion. Independence was achieved (1919) under Marshal *Mannerheim, and a democratic, republican constitution introduced. In 1920 Finland joined the League of Nations, which achieved one of its few successes in resolving the *Åland Islands dispute. After the *Nazi–Soviet Pact of 1939, Finland was invaded in the *Finnish-Russian War (1939-40). Finnish resistance excited international admiration but no practical help, and surrender entailed a considerable loss of territory (Karelia and Petsamo). When Germany invaded the Soviet Union in 1941 the Finns sought to regain these territories by fighting on the side of the *Axis Powers, but capitulated to the Soviet Union in 1944 and were burdened with a huge reparations bill. Since World War II Finland has accepted neutrality in international affairs.

Finnish–Russian War ('Winter War') (1939-40), fought between Finland and the Soviet Union. The Finnish government under General *Mannerheim had rejected Soviet demands for bases and for frontier revisions similar to those accepted by the lesser *Baltic states. Soviet armies attacked on three fronts, and at first the Finns' superior skill in manoeuvring on skis on the frozen lakes and across the Gulf of Finland, and in the forests of their country, kept the Soviet forces at bay. After fifteen weeks of fierce fighting the Soviets breached the Mannerheim Line and Finland was forced to accept peace on Stalin's terms, ceding its eastern territories and the port of Viipuri (Viborg).

First World War *World War I.

Fisher, Andrew (1862-1928), Australian statesman. He was a Member of the Queensland Legislative Assembly (1893-6, 1899-1901) and led the Labor Party (1907-15). He was federal Prime Minister and Treasurer (1908-9, 1910-13, 1914-15). His second government extended social welfare and established the Commonwealth Bank. Wartime stress, and conflict with W. M. *Hughes, led to his resignation, after which he was High Commissioner in London (1916-21).

Fisher, John Arbuthnot, 1st Baron (1841-1920), British sailor and First Sea Lord (1904-10, 1914-15). He successfully persuaded his political masters of the importance of strengthening the British navy before World War I, securing the implementation of the *Dreadnought programme. He, together with Winston *Churchill, was a prime instigator in 1915 of the *Gallipoli expedition. The failure of this attempt and the resulting strained relations with Churchill led Fisher to resign in May 1915.

FLN *Front de Libération nationale.

Foch, Ferdinand (1851-1929), Marshal of France. He fought on the *Western Front in World War I, co-ordinating the actions of Allied forces in preventing the loss of the Channel ports in 1914, commanded the French troops on the *Somme in 1916, and was appointed Allied commander-in-chief in 1918. He was famous for

Marshal Ferdinand **Foch** (*right*) photographed in Cologne while visiting the British army of occupation on the Rhine after the Allied victory over Germany in 1918. With the French commander-in-chief is the British general Sir William Robertson.

insisting that constant attack was the sole recipe for victory. Mutual dislike and lack of co-operation between the Allied generals had prevented concerted action until the German offensive in 1918 led to his appointment as Allied commander-in-chief. He achieved final victory in his July counter-offensive and received the German surrender at Compiègne on 11 November.

Food and Agriculture Organization (FAO), a specialized agency of the *United Nations organization, established in 1945. The FAO is one of the largest and most effective of the UN agencies. It has collected and disseminated facts and statistics, given advice on improvements to food distribution, provided important technical advice for increasing agricultural production, and channelled food aid through its world food programme.

Forbes, George William (1869-1947), New Zealand statesman. A Member of Parliament for the Liberal (later United) Party, Forbes succeeded the Prime Minister, J. G. *Ward, in 1930. He attempted to meet the *Depression by balancing the budget and paying 'the dole' only to men who worked on labour-intensive public works schemes. In 1931 Forbes, in coalition with the more dynamic J. G. *Coates, won the election on a policy of retrenchment, but in 1935 was swept out of office by Labour with its policy of economic expansion.

Force Acts, popular name for US Acts designed to enforce the law. Such an Act was passed in 1833 and was designed to counteract *nullification: it empowered President *Jackson to use the army and navy, if necessary, to enforce the laws of Congress. Four Enforcement Acts of Congress (1870-5) were intended to compel recognition by the South of the *Civil Rights Act of 1866, and of the *Fourteenth and Fifteenth Amendments: Congressional elections were placed under national control, and the acts of armed organizations such as the *Ku Klux Klan were declared tantamount to rebellion.

Ford, Henry (1863-1947), US industrialist and pioneer in car manufacture. In 1903 he founded the Ford Motor Company, in Detroit, which produced the classic Model T in 1908. Adapting the *mass production techniques of the conveyor belt and assembly line to car production, he was able to make large numbers of cheap cars by 1913. At a time when the average wage in manufacturing was $11 a week, he was paying his employees $5 a day and turning out one 'Tin Lizzie' every three minutes. In World War I he became a leading producer of aeroplanes, tanks, ambulances, and submarine chasers. In the early 1920s one car in two throughout the world was a Ford Model T. World War II saw him once more converting his factories to the production of war material. Among his philanthropic legacies is the Ford Foundation (established 1936), the largest philanthropic trust in the world.

The twenty-millionth **Ford** car is driven out of the River Rouge, Detroit, works by Henry Ford himself on 24 April 1931. In the front seat beside the millionaire industrialist is his son, Edsel. The car subsequently made a tour around the United States.

Ford Presidency (1974-7), the term of office of Gerald R. Ford (1913-) as the thirty-eighth President of the USA. Ford had replaced Spiro Agnew as Republican Vice-President to *Nixon after Agnew's resignation in 1973. When Nixon himself resigned Ford automatically succeeded him as President. He continued with Nixon's attempts to control inflation with some success, although unemployment continued to rise. The *Vietnam War was finally ended with an air-lift of some 237,000 troops and refugees out of the country in April 1975. Egypt and Israel were helped by his Secretary of State Henry *Kissinger to settle a territorial dispute. In the election campaign of 1976 he won the Republican nomination, although he was defeated by the Democratic candidate Jimmy *Carter.

Forrest, Nathan Bedford (1821-77), US general in the army of the *Confederacy. Although he participated with distinction in several of the large engagements in the western theatre of the *American Civil War (most notably Shiloh, Murfreesboro, and Chickamauga), he was best known for his brilliant raids behind Union (Northern) lines, tying down large numbers of troops and causing constant confusion and disruption. By the end of the war he was a lieutenant-general, although his reputation had been clouded by his involvement in the alleged massacre of Negro troops at Fort Pillow, Tennessee. After the war he became Grand Wizard of the *Ku Klux Klan but, disapproving of the lawlessness of many of its members, sought to disband the organization in 1869.

Forrest, Sir John, 1st Baron Forrest of Bunbury (1847-1918), Australian explorer and statesman. He led

several major expeditions, including a search for the lost explorer Leichardt (1869), and an expedition from Perth to Adelaide (1870). Forrest was Western Australia's first Premier (a courtesy title not conferred by the Constitution, but by usage) from 1890 until 1901. He was prominent in the *Australian federation movement, being a federal politician (Protectionist, Fusion, Liberal, and finally Nationalist) from 1901 until his death. He held various portfolios, serving mostly as Treasurer, and introduced fifteen federal budgets.

Forster, William Edward (1818–86), British statesman and educational reformer. Entering Parliament in 1861, he became Under-Secretary for the Colonies in 1865 and in 1868 vice-president of the Council. As such he, with Lord *Ripon, secured the passing of the 1870 Education Act, which laid the foundations of compulsory, government-financed elementary education. In 1880 he was appointed Chief Secretary for Ireland, incurring unpopularity because of the severe measures he used to maintain order. He resigned in 1882 and became an opponent of Gladstone's plans for *Home Rule.

Fort Sumter, military stronghold in Charleston harbour. The Confederates, having seized Federal funds and property in the South, demanded the evacuation of the Federal Fort Sumter in Charleston Harbor. Major Robert Anderson, in command, refused and General Beauregard bombarded it (12-13 April 1861) just as relief for the Federalists approached. The fall of the fort marked the beginning of the *American Civil War.

Foster, William Zebulon (1881–1961), US political leader. He joined the International Workers of the World in 1909, and after organizing the steelworkers' strike in the Chicago area in 1919, became a member of the newly formed US Communist Party. He ran as its presidential candidate in 1924, 1928, and 1932, and in 1945, after the discrediting of Earl Browder, the Party's war-time leader, became its chairman. He died in Moscow.

Four Modernizations, key aspects of China's post-Mao development. The need to modernize agriculture, industry, national defence, and science and technology was implied in a speech by *Mao in 1963, but in the *Cultural Revolution ideology was considered to be more important than economic development. After *Deng Xiaoping came to power, the Four Modernizations began to take priority. Training of scientists, engineers, and managers, and the reform of agriculture by the 'responsibility system' (the transfer of management power from the commune to the individual) are key examples.

Fourteen Points (8 January 1918), a US peace programme, contained in President Woodrow *Wilson's address to Congress. They comprised freedom of the seas, equality of trade conditions, reduction of armaments, adjustment of colonial claims, evacuation of Russian territory and of Belgium, the return to France of Alsace-Lorraine, recognition of nationalist aspirations in eastern and central Europe, freedom for subject peoples in the Turkish empire, independence for Poland, and the establishment of a 'general association of nations'. Accepted, with some reluctance, by the Allies, they became in large part the basis for the peace negotiations of the *Versailles Settlement.

Fourteenth Amendment (1868), the most important of the three *American Civil War and Reconstruction amendments to the US Constitution. Drawn up in 1866 by the Joint Committee of Fifteen, the Fourteenth Amendment extended US citizenship to all persons born or naturalized in the USA (and thus, by including ex-slaves, reversed the *Dred Scott decision). It also prohibited the states from abridging the privileges and immunities of citizens or depriving any person of life, liberty, or property without due process of law, or denying any person the equal protection of the laws. Another clause reduced the representation in Congress of states which denied the vote to blacks. The Fourteenth Amendment has caused more legal controversy than any other part of the constitution. In the late 19th century, it was used as a device to protect big business from state regulation. In the 20th century it has been the main constitutional instrument of the *civil rights movement and, in recent years, of the women's rights movement. (The Fifteenth Amendment, adopted in 1870, provided that the right to vote should not be denied on grounds of race, colour, or previous condition of servitude.)

Fox, Charles James (1749-1806), British statesman. A critic of Lord North's American policy, he favoured American independence. After North's resignation and Rockingham's death, Fox collaborated with North to bring down the Shelburne ministry in 1783. The Fox-North coalition lasted until Fox's India Bill was defeated in the House of Lords, and *George III took the opportunity to dismiss a coalition he detested. Fox spent most of the rest of his career in opposition to the Prime Minister, the Younger *Pitt. His rivalry with Pitt led him into opposition to some of the latter's reforms, but in 1801 he supported Pitt in his fight for *Catholic Emancipation. When Pitt died in 1806, Fox became a leading member of the new ministry, and through his dying efforts Parliament agreed to an anti-slavery Bill.

France, a country in western Europe. The First Republic (1792-1804), established after the fall of the Bourbon monarchy, lasted until the First Empire (1804-14) under

The raising of the Union flag at **Fort Sumter**, as depicted by the painter Edwin White. The fort's commander Major Anderson (*left*) kneels in prayer as his garrison prepares to defy the Charleston rebels.

Francis I of Austria, the last Holy Roman Emperor. A proud and reactionary ruler who despised Napoleon for his humble origins, he nevertheless gave his daughter Marie-Louise to the French emperor as his second wife.

*Napoleon I, when France became the dominant political power in Europe. After his fall the monarchy was restored (1814) and, with a brief interval in 1815, lasted until the abdication of Louis Philippe (1848). During this period, having lost influence in India and Canada, France began to create a new overseas empire in North Africa. The Second Republic, established in 1848, lasted until 1852, when *Napoleon II proclaimed the Second Empire (1852–70). It saw further expansion of the French empire, particularly in south-east Asia. The Third Republic (1870–1940) was established after the capture and exile of Napoleon III and France's defeat in the *Franco-Prussian War (1870). France took part in the Berlin Conference (1884) on Africa, and by 1914 ruled over Morocco, Tunis, Madagascar, and the huge areas of *French West Africa and *French Equatorial Africa. The Third Republic fell in 1940, following defeat by Nazi Germany. Northern France was occupied by the Germans, unoccupied France to the south was under the *Vichy government, and a *Free French government was proclaimed in London. The Fourth Republic (1946–58) was replaced by the Fifth Republic (1958–), under the strong presidency of Charles *de Gaulle (1959–69). Protracted and costly wars led to the decolonization of Indo-China (1954) and of Algeria (1962), while, from 1956, the rest of the African empire gained increasing independence. Since 1945 France has regained its position as a major European power and was a founder member of the *European Economic Community (1958). As a nuclear power it refused to sign the *Nuclear Test-ban Treaty (1963) and withdrew formally from NATO in 1966.

Francia, José Gaspar Rodriguez de (1776–1840), dictator of Paraguay. A leader of the Paraguayan move-

ment for independence from Spain (1811), he dominated the post-independence period by establishing (1814) one of the most absolute dictatorships in 19th-century Latin American history. Dogmatic, anti-clerical, and xenophobic, he placed his country in almost complete isolation from the outside world. At home his autocratic rule earned him the name *El Supremo*. As time went on, he grew more arbitrary and despotic. The extravagances of his later years were considered symptomatic of his insanity, although he held office until his death in 1840.

Francis I (1768–1835), Habsburg monarch, last Holy Roman Emperor (as Francis II, 1792–1805), and first Emperor of Austria (as Francis I, 1804–35). Following defeat at *Austerlitz he was forced to abdicate the title of Holy Roman Emperor, losing lands to Russia, Bavaria, and France by the Treaty of Schönbrunn (1809). He had little choice but to allow *Napoleon to marry his daughter, Marie-Louise, in 1810, but he changed sides with decisive effect in 1813 and played a major part in the eventual downfall of Napoleon. In 1815 at the Congress of *Vienna, he regained much of the territories lost in the war, largely as a result of the diplomatic skill of *Metternich, responsible for foreign affairs. During the last twenty years of his reign he was identified with the *Holy Alliance and its policy of repression. Denounced by the liberals he was, however, popular with the mass of his subjects.

Francis Ferdinand (1863–1914), Archduke of Austria and heir presumptive to Emperor *Francis Joseph. He aimed to transform the *Austro-Hungarian empire into a triple monarchy to include a Slavic kingdom. He was opposed by the Hungarians, who refused to make concessions to Slavs, and by extreme Slav nationalists (including Serbs), who saw no future for the emergent nations within the empire. On 28 June 1914, while on an inspection tour at Sarajevo, he and his wife were assassinated by Gavrilo Princip, a Serbian nationalist. The subsequent ultimatum by Austria to Serbia led directly to the outbreak of World War I.

Francis Joseph (1830–1916), Emperor of Austria (1848–1916), King of Hungary (1867–1916). He succeeded to the throne (aged 18) amid the *Revolutions

Archduke of Austria **Francis Ferdinand** and his wife leave Sarajevo's town hall for their car shortly before their assassination on 28 June 1914.

of 1848. He suppressed all nationalist hopes until forced to meet Hungarian aspirations in the establishment of the *Austro-Hungarian empire (1867). His foreign policy lost Habsburg lands to Italy (1859 and 1866) and led to the loss of Austrian influence over German affairs and to the ascendancy of Prussia. Seeking compensation in the *Balkan states, he aroused Slav opposition which ultimately resulted in World War I. His wife Elizabeth was assassinated by the Italian anarchist Lucheni. Opposed to social reform, Francis Joseph maintained administrative centralization and opposed the federalist aspirations of the Slavs.

Franco (Bahamonde), Francisco (1892–1975), Spanish general and head of state. A monarchist, he rose rapidly in his profession until 1931, when Alfonso XIII abdicated and was replaced by a republican government. He was temporarily out of favour, but by 1935 was chief of the General Staff. Elections in February 1936 returned a more left-wing government and the army prepared to revolt. At first he hesitated to join in the military conspiracy but in July led troops from Morocco into Spain to attack Madrid and overthrow the republic. After three years of the savage *Spanish Civil War he was victorious and became dictator of Spain (1939). In 1937 Franco adopted the *Falange, expanding it into a Spanish fascist party and banning all political opposition. During World War II he remained neutral though sympathizing with Hitler and Mussolini. His government

General **Franco** gives the fascist salute to a German guard of honour as he escorts Adolf Hitler during the Nazi leader's visit to Spain in October 1940. This photograph came from the album of the German Foreign Minister, von Ribbentrop.

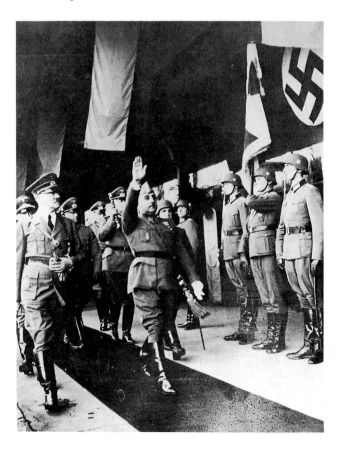

was ostracized by the new United Nations until, with the coming of the *Cold War, his hostility towards communism restored him to favour. His domestic policy became slightly more liberal, and in 1969 he named Prince Juan Carlos (1938–), grandson of Alfonso XIII, not only as his successor but as heir to the reconstituted Spanish throne. On his death Spain returned to a democratic system of government under a constitutional monarchy.

Franco-Prussian War (1870–1), between France, under *Napoleon III, and Prussia. The war itself was provoked by *Bismarck, who had skilfully isolated the French, and altered an uncontroversial message from his king (the *Ems telegram). Prussian armies advanced into France; the French forces led by *MacMahon were driven out of Alsace whilst a second French army, led by Napoleon, was forced to retire to Metz. MacMahon, marching to relieve Metz, was so comprehensively defeated by *Moltke at Sedan that Napoleon, discredited in the eyes of the French, ceased to be emperor. Bismarck refused to make peace, and in September the siege of Paris began. Hopes of a French counter-attack were dispelled when *Bazaine surrendered at Metz and Paris finally gave way in January 1871. An armistice was granted by Bismarck, and a national assembly elected to ratify the peace, but the population of Paris refused to lay down arms and in March 1871 rose in revolt and set up the *Paris Commune. The French government signed the Treaty of *Frankfurt in May and French prisoners-of-war were allowed through Prussian lines to suppress the Commune. For Prussia, the proclamation of the *German Second empire at Versailles in January was the climax of Bismarck's ambitions to unite Germany.

Frank, Anne (1929–45), a Jewish girl who became a *concentration camp victim. She was living with her family in Amsterdam when the Nazi Germans invaded in 1940. From July 1942 to April 1944 the family and four other Jews were hidden by a local family in a sealed-off back room, but were eventually betrayed. She died in Belsen. During the years of hiding Anne kept a diary of her experiences which, since its publication in 1947, has attracted a world-wide readership.

Frankfurt, Treaty of (10 May 1871), an agreement that ended the *Franco-Prussian War. By it France surrendered Strasburg, Alsace and part of Lorraine, together with the great fortresses of Metz to *Bismarck's Germany. An indemnity of five billion gold francs was imposed by Germany on France, and a German army of occupation was to remain until the indemnity had been paid. Bismarck's aim in this treaty was to ensure that France would be entirely cut off from the Rhine.

Franz Josef *Francis Joseph.

Fraser, Peter (1884–1950), New Zealand statesman. He was gaoled during World War I for opposing conscription, and upon his release joined the New Zealand Labour Party. Elected to Parliament in 1918, he became the party's deputy leader in 1933, and Prime Minister (1940–9). One of the architects of the United Nations (1945), he held a life-long commitment to equality in education, and the modern New Zealand education system is perhaps his most enduring monument.

Frederick William III (1770-1840), King of Prussia (1797-1840). After his defeat at the battle of *Jena he was forced by the Treaty of *Tilsit (1807) to surrender half his dominions by the creation of the kingdom of Westphalia and the grand duchy of Warsaw. In 1811 he joined *Napoleon in the war against Russia but, following the retreat of Napoleon from Moscow, he signed a military alliance with Russia and Austria. From 1807 onwards he supported the efforts for reform made by *Stein and *Hardenberg, and at the Congress of *Vienna he won back for Prussia Westphalia and much of the Rhineland and of Saxony. He signed the *Holy Alliance, and became progressively more reactionary during the last years of his reign.

Frederick William IV (1795-1861), King of Prussia (1840-61). A patriarchal monarch by temperament, he was the champion of a united Germany, but could not accept the degree of democracy envisaged by the Frankfurt *Diet in 1848. He therefore refused (1849) the offer of a constitutional monarchy for the *German Confederation. For Prussia he promulgated a conservative constitution allowing for a parliament, but with a restricted franchise and limited powers. This remained in force until 1918.

Freedom Riders, groups of non-violent black and white protesters in US southern states. They were volunteers from the north who in 1961 began chartering buses and riding through the southern states to challenge the segregation laws. Many Freedom Riders were arrested or brutally attacked by southern whites, but their actions did help to arouse public opinion in support of the *Civil Rights campaign.

Free French, the, a World War II organization of Frenchmen and women in exile. Led by General *de Gaulle, it continued the war against the *Axis Powers after the surrender of *Vichy France in 1940. Its headquarters were in London, where, apart from organizing forces that participated in military campaigns and co-operating with the French *resistance, it constituted a pressure group that strove to represent French interests. In 1941 its French National Committee was formed and this eventually developed into a provisional government for liberated France. The Free French army in French Equatorial Africa, led by General Leclerc (Philippe, vicomte de Hauteclocque), linked up with the British forces in Tripoli (1943), after completing an epic march of c.2,400 km. (1,500 miles) from Lake Chad. A provisional Free French government was established in Algiers, moving to Paris in 1944.

Free-Silver Movement, a movement in the 19th century in the USA for an unlimited silver coinage. Following the *gold rushes of the 1850s and 1860s, large deposits of silver were discovered in the West. Silver miners wished to see unlimited production, but in 1873 Congress refused to include the silver dollar in its list of authorized coins. A protest movement resulted, and in 1878 the silver dollar became legal tender, the US Treasury agreeing to purchase silver to turn into coins. In 1890 the Sherman Silver Purchase Act doubled the agreed issue of silver, but following a stockmarket crisis in 1893 the Act was repealed. Eastern bankers were blamed for a depressed silver market, and the Democratic Party adopted the demand for unlimited free silver in the presidential campaign of 1896. Following the 1900 election, a Republican Congress passed the Gold Standard Act, which made *gold the sole standard of currency.

Free Soil Party, a minor US political party. In 1846 David Wilmot from Pennsylvania proposed to Congress a proviso that slavery would be forbidden in the huge territories then being seized from Mexico in the *Mexican-American war. This Wilmot Proviso failed, but its disappointed supporters held a convention in 1848 where they formed a Free Soil Party advocating 'free soil, free labour, and free men'. They gained support from the small farmers as well as from anti-slave groups. The party never won more than 10 per cent of the popular vote in presidential or Congress elections and failed to prevent either the Compromise of 1850 or the *Kansas-Nebraska Act of 1854. It became absorbed after 1854 into the newly formed Republican Party.

free trade, a doctrine advocating a free flow of goods between countries to encourage mutual economic development and international harmony by the commercial interdependence of trading nations. A policy of free trade prohibits both tariffs on imports and subsidies on exports designed to protect a country's industry. The doctrine's best early statement was by Adam Smith in his *Wealth of Nations* (1776). The argument appealed to many British industrialists in the early 19th century, who came to be called the *'Manchester School'. It became increasingly government policy with the repeal of the *Corn Laws in 1846, and was fully adopted in 1860. The contrary doctrine, that of protectionism or the imposition of import tariffs to protect home industries, was advocated in the later 19th century in a number of countries, for example the USA, Germany, and Australia. In 1903 Joseph Chamberlain began a campaign in Britain for *tariff reform, which was a major factor in British politics until 1932, when a conference in Ottawa approved a system of limited tariffs between Britain and the newly created dominions, in the first instance for five years. After World War II the USA tried to reverse the trend to protection. At a conference in Geneva in 1947 a first schedule for freer world trade was drawn up, the *General Agreement on Tariffs and Trade (GATT). For over a decade after the war Britain also was a strong supporter of moves to restore freer trade. It was a founder member of the *European Free Trade Association (EFTA) in 1958, but as adverse economic conditions developed in the 1960s Britain sought entry into the *European Economic Community (EEC). In Eastern Europe a similar community, *Comecon, was established in 1949, which, since 1987, has sought co-operation with EEC countries. The highly successful growth of the Japanese economy after the war led many countries to seek tariffs against Japan and by the 1980s world economic policies were confused, with advocates supporting both free trade and protection.

Frei (Montalva), Eduardo (1911-82), Chilean statesman. He was President of Chile (1964-70). A founder-member of the Falange Nacional, later Partido Demócrata Cristiano (PDC), he was a severe critic of the US Alliance for Progress programme to aid Latin American countries. The programmes he initiated, including agrarian reform and the 'Chileanization' of the copper industry (whose controlling interest had until then been held by US companies), were ambitious, but his failure to check inflation or to redistribute wealth turned many of his supporters against him.

Frelimo War (1964–75), a war fought between *Mozambique nationalist groups united into the Mozambique Liberation Front (Frelimo) and Portuguese troops. In 1963 Frelimo recruits were sent to Algeria and Egypt for political and guerrilla training. Operations, headed by Eduardo Mondale, began in 1964. The Portuguese failed to contain the conflict, and by 1968 Samora *Machel claimed one-fifth of the country. A Portuguese resettlement programme (*aldeamentos*), and public and social works failed to satisfy the guerrillas, who were being armed by supplies from China, Czechoslovakia, and the Soviet Union. Brutal Portuguese counter-terrorism made conciliation even more impossible and Portugal conceded independence in 1974. Frelimo became the dominant political force in the new People's Republic of Mozambique.

Frémont, John Charles (1813–90), US explorer, soldier, and political leader. Frémont played an important

The US soldier, explorer, and political leader John Charles **Frémont**, in the uniform of a general of the US Army c.1850.

part in mapping the area between the upper reaches of the Mississippi and Missouri rivers before leading three major expeditions to the West between 1842 and 1846 which earned him the name of the 'Pathfinder'. He was involved controversially in the US conquest of California at the time of the *Mexican–American War. In 1856, he was the first presidential candidate of the new Republican Party, but was defeated by *Buchanan. During the *American Civil War, he commanded Union forces in the Department of the West (1861) and in western Virginia (1862) with little success. He briefly challenged Lincoln in the presidential campaign of 1864.

French, John Denton Pinkstone, 1st Earl of Ypres (1852–1925), British field-marshal. Having distinguished himself in the Sudan and the Second *Boer War (1899), he was appointed commander-in-chief of the *British Expeditionary Force in France (1914). Under instructions from Lord *Kitchener he opposed the German advance through Belgium and Flanders. He and his armies were ill-equipped for the kind of *trench warfare in which they found themselves involved, and in December 1915 French resigned in favour of Sir Douglas *Haig. At the Irish *Easter Rising in 1916 French dispatched two divisions to suppress the uprising. He served as Lord Lieutenant of Ireland (1918–21) at a time when outrages and reprisals were widespread.

French Community (1958–61), an association consisting of France, its overseas territories and departments, and various independent African states formerly part of the *French empire. Established in 1958, it was a system aimed at associating the former colonies with metropolitan France. Members of the Community were allowed considerable autonomy though they were denied control of their own higher education, currency, defence, or foreign affairs. The West African territory of Guinea immediately voted against the arrangement and became independent. Hostility to the restrictions on autonomy led to a revision of the Constitution in 1960. Nevertheless, some African states left, while continuing to retain close links with France. By 1961 the system had virtually collapsed and some of its institutions were abolished.

French empire, the colonial empire of France. In the 18th century a long rivalry with Britain ended with the loss of *Quebec and recognition of British supremacy in India. By 1815 only some West Indian Islands, French Guiana, and Senegal and Gabon were left. The 19th century witnessed a rapid growth of the empire. The conquest of Algeria began (1830), while Far Eastern possessions—Cochin China, Cambodia, and New Caledonia—were added. In the 'Scramble for Africa', Tunisia became a protectorate (1881), and by 1912 French Somaliland, Sahara, Morocco, Dahomey, Senegal, Guinea, Congo, Ivory Coast, and Madagascar were gained, making the empire twenty times the size of France itself. Britain frustrated French aspirations in Egypt and the Sudan, and rivalry at *Fashoda (1898) nearly caused war until the *entente cordiale brought agreement. After World War I Togoland and the Cameroons, former German colonies, became French *mandates, as did Syria and Lebanon (1923). Defeat in World War II and short-lived post-war governments prevented urgent reforms, thereby causing the loss of both Far Eastern and African empires. In Indo-China the communist leader, *Ho Chi Minh, established his Vietnamese republic (1945) which France refused to recognize. Open warfare (1946–54) ended with the French capitulation at *Dienbienphu and the consequent independence of Cambodia, Laos, and Vietnam. In Algeria almost the entire French army failed to quell an Arab rising (1954). By 1958 *de Gaulle realized that independence was inevitable, it followed in 1962. In 1946 the empire was formed into the French Union, which was replaced in 1958 by the *French Community.

French Equatorial Africa, a former French federation in west central Africa. It included the present republics of the Congo, Gabon, Central Africa, and Chad, all originally French colonies. To them was attached (1920) the League of Nations Mandated Territory of Cameroon, now the Federal Republic of Cameroon. The Federation was formed mainly through the efforts of the Franco-Italian empire-builder, Savorgnan de Brazza (1852–1905). Proclaimed in 1908, it was administered centrally from Brazzaville until its constituents became autonomous republics within the French Community in 1958. The member states formed (1959) a loose association called the Union of Central African Republics.

French Foreign Legion, a French volunteer armed force consisting chiefly of foreigners. In 1831 *Louis-Philippe reorganized a light infantry legion in Algeria as the *régiment étranger*, the foreign legion. It fought in numerous 19th-century wars and in both World Wars. Following Algeria's independence in 1962 the legion was transferred to France. No questions are asked about the origin or past of the recruits, whose oath binds them absolutely to the regiment whose unofficial motto is *legio patria nostra* ('the legion is our fatherland').

French Guiana, a French possession on the Caribbean coast of South America. The first French settlement was made at Cayenne Island at the beginning of the 17th century. The rivalry with the British, the Dutch, and the Portuguese was intense until 1817, when French possession was finally secured. Immigration schemes were ineffective and the colony was populated by former prisoners from Devil's Island and other off-shore convict settlements. In 1946 it became an overseas department of France and in 1968 the launch site of the European Space Agency was established there.

French Indo-China, former French colonial empire in south-east Asia. Having gained early influence in the area through assisting Gia-Long in establishing the Vietnamese empire in the early 19th century, the French colonized the area between the late 1850s and 1890s, using the term Indo-China to designate the final union of their colonies and dependencies within Annam, Cambodia, Cochin-China, Laos, and Tonkin. Nationalist movements aiming particularly at the formation of an independent and united Vietnam sprang up between the wars, and French influence in the area was fatally undermined in the early 1940s by the collaboration of the *Vichy colonial administration with the Japanese. The *Vietminh resistance movement became active during the war consolidating a peasant base and resisting attempts by the French to reassert their control after 1945. A protracted guerrilla war eventually brought France to defeat at *Dienbienphu in 1954. In the same year the

*Geneva Conference formally ended French control, transferring power to national governments in Cambodia, Laos, and North and South Vietnam.

French Indo-China War (1946–54), a conflict fought between French colonial forces and *Vietminh forces largely in the Tonkin area of northern Vietnam. The Vietminh began active guerrilla operations during the Japanese occupation of World War II and in September 1945 their leader, *Ho Chi Minh, proclaimed a Vietnamese Republic in Hanoi. The French opposed independence, and launched a military offensive. Ho Chi Minh was forced to flee Hanoi and begin a guerrilla war in December 1946. By 1950, foreign communist aid had increased Vietminh strength to the point where the French were forced into defensive lines around the Red River delta, but Vietminh attempts to win the war failed in 1951. Guerrilla operations continued until an ill-advised French attempt to seek a decisive engagement led to the encirclement and defeat of their forces at *Dienbienphu in 1954. The war, and French rule in Indo-China, were formally terminated at the *Geneva Conference in April–July of that year.

French West Africa, the former federation of French overseas territories in West Africa. In the late 19th century, France sought to extend its colonial interests inland from existing trading settlements on the Atlantic coast. Substantial military force had to be used to overcome local Islamic states, and in 1895 the formation of Afrique Occidentale Française was proclaimed including the present republics of Mauritania, Senegal, Mali, Burkina Faso, Guinea, the Ivory Coast, Niger, and Benin. The neighbouring colony of Togo was captured from Germany in World War I and partitioned as a mandate of the League of Nations between Britain and France. French West Africa supported the *Vichy government from 1940 to 1942, when it transferred its allegiance to the *Free French cause. In 1958 the constituent areas became autonomous republics within the French community, with the exception of Guinea, which, following a referendum, voted for immediate independence. Full independence was granted throughout the area in 1960.

Freyberg, Bernard Cyril, 1st Baron Freyberg (1889–1963), New Zealand general. A professional soldier, he was appointed commander-in-chief of the New Zealand Expeditionary Forces (November 1939). He commanded the unsuccessful Commonwealth expedition to Greece and Crete (June 1941). He took an active part in the North African and Italian campaigns, where his New Zealand Division greatly distinguished itself, although suffering heavy losses. From 1945 until 1952 he was governor-general of New Zealand.

Front de Libération nationale (FLN), Algerian radical Muslim independence movement. It was formed in 1954 as the political expression of the ALN (Armée de Libération Nationale) when the Algerian war of independence broke out. In spite of differences of opinion between military, political, and religious leaders, the movement hung together, and brought its military leader, *Ben Bella, to power successfully as the first president of Algeria in 1962 following President de Gaulle's successful national referendum on the *Évian Agreement. The

The British philanthropist Elizabeth **Fry** reading to inmates of Newgate prison in London. To her right is a party of her reformist supporters, while among the prisoners are a nursing mother, young children, and women concealing liquor, snuff, and playing cards.

principal policies of the party were independence, economic development in a socialist state, non-alignment, and brotherly relations with other Arab states.

Fry, Elizabeth (Gurney) (1780–1845), British philanthropist and prison reformer. The wife of a London Quaker, Joseph Fry, she subsequently became recognized as a preacher in the Society of Friends. After a visit to Newgate prison in 1813, horrified by what she found there, she began to press for more humane treatment for women prisoners. Her unflagging determination resulted in eventual reform. She visited other European countries to advocate improvement in prison conditions and in the treatment of the insane. She also founded hostels for the homeless.

Fuchs, Klaus (1911–), German-born British physicist and spy. In 1943 he became a member of the team which developed the atomic bomb at Los Alamos. On his return to Britain he worked at the Atomic Energy Research Establishment at Harwell. From 1943 Fuchs passed vital secret information about the atomic bomb and plans for developing a hydrogen bomb to Soviet secret agents. In 1950 he was arrested on the evidence obtained by the USA from confessed communist agents. Fuchs pleaded guilty and was imprisoned. He was released in 1959 and went to the German Democratic Republic, where he became director of the Institute for Nuclear Physics.

Fugitive Slave Acts, US legislation providing for the return of escaped slaves to their masters. After the abolition of slavery in the northern states of the USA, these 'free states' became lax in enforcing the first Fugitive Slave Act of 1793. The second Act, part of the *Compromise of 1850, introduced more stringent regulations, specifically aimed at the *Underground Railroad. Unpopular in the North, it added fuel to the slavery controversy, and 'liberty laws' passed by free states to thwart the Act drove the South further towards rebellion. Fugitive slave legislation was finally repealed in 1864.

Fulani empire of Sokoto, West African Islamic empire. In the late 18th century the Fulani came into contact with the nominally Muslim Hausa states. One of their clerics, *Uthman dan Fodio (1754–1817), had built up a community of scholars at Degel in the Hausa state of Gobir, whose new sultan in 1802 enslaved Uthman's followers. A quarrel developed, Uthman was proclaimed 'commander of the faithful', and in the ensuing *jihad* (holy war) all the Hausa states collapsed. By 1810 Uthman had created a vast empire, to be administered by emirs in accordance with Koranic law. High standards of public morality replaced the corruption of the Hausa states and widespread education was achieved. In 1815 he retired, appointing his son Muhammed Bello his successor and suzerain over all the emirates. Bello had built the city of Sokoto, of which he became the sultan, and he considerably extended the empire, establishing control of west Bornu and pushing down into the *Yoruba empire of Oyo. Although losing some of its high ideals, the Fulani empire of Sokoto continued under Bello's successors. In the late 19th century British penetration of the empire increased. Kano and Sokoto were sacked in 1903, when the empire ended, although the emirates survived under the system of indirect rule instituted by the first High Commissioner, Frederick *Lugard.

Fulbright, James William (1905–), US politician. An early enemy of *isolationism, he was Chairman of the Senate Foreign Relations Committee from 1959 to 1974. He was an active critic of US foreign policy in the 1960s and 1970s, attacking particularly the US involvement in the *Vietnam War, and urging that Congress should have more control over the President's powers to make war. A Rhodes Scholar himself, in 1946 he sponsored the Fulbright Act, which provided funds for the exchange of students and teachers between the USA and other countries.

Fundamentalist Churches, Protestant religious bodies which stress traditional Christian doctrines, especially the literal truth of the Bible. Fundamentalism developed in the 1920s in opposition to modern theological teachings, and its adherents have been particularly powerful in the USA, especially among the various Baptist groups. Fundamentalists were and are particularly noted for their hostility to the theory of evolution, as shown by the prosecution of J. T. *Scopes in Tennessee. Several US fundamentalists have acquired fame and influence as preachers and presenters on television churches and as leaders of the so-called 'Moral Majority' movement. Fundamentalist churches are probably increasing in membership throughout the world, and among them the 'House Churches' in Britain are growing rapidly.

G

Gabon, a country in equatorial West Africa. In 1839 the French made it a naval base to suppress slave trade. Thus a French colony developed, exploiting the rare woods, gold, diamonds, other minerals, and oil. The country became autonomous within the French Community in 1958 and fully independent in 1960. Almost entirely on the basis of its natural resources it has had one of the fastest economic growth rates in Africa. After early years of political instability, there has been considerable support for the presidency of Omar Bongo, and his one-party constitution (1961, revised 1981).

Gadsden Purchase (1853–4), US acquisition of Mexican territory. Following the *Mexican–American War and under pressure to construct a transcontinental railway across the American south-west, the administration of President *Pierce sent Senator James Gadsden to negotiate the necessary redefinition of the Mexico–US border. In the resulting transaction, Mexico was paid $10 million for ceding a strip of territory 76,767 sq. km. (29,640 sq. miles) in the Mesilla Valley, south of the Gila River. The area completed the present borders of the mainland USA.

Gaitskell, Hugh Todd Naylor (1906–63), British politician. He entered Parliament in 1945, holding several government posts dealing with economic affairs (1945–51), including Chancellor of the Exchequer (1950–1). He became leader of the Labour Party (1955–63). He represented the moderate right-wing of his party and believed in the welfare legislation of 1944–51 and the need for a balance between private and state finance. Gaitskell was particularly vigorous in his opposition to the government over the *Suez crisis and in resisting the unilateralists within his own party.

Gallatin, Abraham Alfonse Albert (1761–1849), US statesman and ethnologist. After serving in Congress, he became Secretary of the Treasury (1801–14), a post he filled with considerable success, carrying out an extensive programme of financial reforms and economies. As one of the US commissioners he helped to negotiate the Treaty of *Ghent ending the *War of 1812. He undertook ethnological work on the American Indians and founded the American Ethnological Society.

Gallipoli Campaign (1915–16), an unsuccessful Allied attempt to force a passage through the Dardanelles during *World War I. Its main aims were to force Turkey out of the war, and to open a safe sea route to Russia. A naval expedition, launched in February and March 1915, failed. A military expedition (relying mainly upon British, Australian, and New Zealand troops), with some naval support, was then attempted. The first landings, on the Gallipoli peninsula and on the Asian mainland opposite, were made in April 1915. Turkish resistance was strong and, although further landings were made, fighting on the peninsula reached a stalemate. The Australian casualties on Gallipoli were 8,587 killed and 19,367 wounded. The Allied troops were withdrawn. Winston *Churchill, who was largely responsible for the campaign, was blamed for its failure.

Gambetta, Leon (1838–82), French statesman. After the defeat at *Sedan and the collapse of the Second Empire he played a principal role in proclaiming the Third Republic and was Minister of War and Minister of the Interior in the provisional government. He opposed the armistice with Prussia, was elected to the National Assembly for Alsace, and left the Assembly when Alsace was ceded to Prussia. He was soon re-elected and in the 1870s led the Republican Party in its campaign to secure

'V' beach at **Gallipoli**, as seen from the steamer *River Clyde*. Rope hoists and small craft are being used to offload stores from supply ships, while specially fitted lighters ferry fresh horses and mules to the Allied troops encamped on the shore.

the nation for republicans. He died soon after the fall of his brief ministry of 1881-2.

Gambia, a small West African country. The beginning of the colony was the building of a fort by the British at Banjul in 1816, as a base against the slave trade. Renamed Bathurst, the new town was placed under *Sierra Leone in 1821. Gambia became a British colony in 1843. The Soninki-Marabout Wars in neighbouring Senegal caused serious disturbances and were ended by Anglo-French intervention in 1889. A British Protectorate over the interior was proclaimed in 1893. Gambia became an independent member of the Commonwealth in 1965, and a republic in 1970, with Sir Dawda Kairaba Jawana the country's first president. In 1981 Gambia and Senegal formed a limited confederation, Senegambia, for defence, economic, and foreign policy purposes.

Gamelin, Maurice Gustave (1872-1958), French general. As a staff officer in World War I he helped to plan the successful battle of the *Marne (1914) and served with distinction through the war. In World War II as commander-in-chief of the Allied forces, he was unprepared for the German thrust through the Ardennes which resulted in disaster for his forces. In mid-May 1940 he was replaced by General *Weygand.

Gandhi, Indira (1917-84), Indian stateswoman. The daughter of Jawaharlal *Nehru, in 1939 she joined the Indian National *Congress Party, and spent over a year in prison for her wartime activities. Her first years in politics were spent as an aide to her father when he was

Indira **Gandhi** with supporters in 1978, while in opposition following her defeat in elections in 1977. An astute politician, she formed her own breakaway party, Congress (I), in order to regain power.

Prime Minister, serving as President of Congress (1959-60). She became Minister for Broadcasting and Information in Lal Bahadur Shastri's cabinet, and in 1966 was chosen to succeed Shastri as Prime Minister. During the following years she was engaged in a protracted struggle with the older leadership of the Congress, but with the aid of the Congress left wing defeated them in 1969-70. After the successful *Indo-Pakistan War of 1971 her popularity stood high, but it waned during the 1970s. When threatened with the loss of her position through a court case for illegal electoral activities, she declared a state of emergency (1975-7) and governed India dictatorially, assisted by favourites such as her younger son, Sanjay. After her defeat in the 1977 elections by Morarji *Desai her career seemed finished, but in 1979 her faction of the Congress Party was re-elected to power and she ruled until her assassination in 1984 by a Sikh extremist. Her elder son, Rajiv Gandhi (1944-), succeeded her as Prime Minister.

Gandhi, Mohandas Karamchand (1869-1948), Indian national and spiritual leader. Born into a family of Hindu Bania (merchant) caste in Porbandar, he was educated in India and Britain, qualifying as a barrister in London. He practised law briefly in India, but moved to South Africa (1893-1914), where he became a successful lawyer. There he developed his technique of *satyagraha ('truth-force' or non-violent resistance). His campaign for equal rights for Indians in South Africa met with partial success. Returning to India, he formed political connections through campaigns for workers' and peasants' rights (1916-18). Subsequently he led Indian nationalists in a series of confrontations with the *British Raj, including the agitation against the *Rowlatt Act (1919). From 1920 he dominated the Indian National *Congress, supporting the *Khilafat movement and initiating the decision by Congress to promote a non-co-operation movement (1920-2), suffering frequent imprisonment by the British. The *Salt March to Dandi (1930) was followed by a campaign of *civil disobedience until 1934, individual *satyagraha*, 1940-1, and the 'Quit India' campaign of 1942. As independence for India drew near, he co-operated with the British despite his opposition to the partition of the sub-continent. In political terms Gandhi's main achievement was to turn the small, upper-middle-class Indian National Congress movement into a mass movement by adopting a political style calculated to appeal to ordinary Hindus (using symbols such as the loin cloth and the spinning-wheel) and by creating a network of alliances with political brokers at lower levels. In social and economic terms he stressed simplicity and self-reliance as in village India, the elevation of the status of the Untouchables (*harijans*), and communal harmony. In intellectual terms his emphasis was upon the force of truth and non-violence (*ahimsa*) in the struggle against evil. His acceptance of partition and concern over the treatment of Muslims in India made him enemies among extremist Hindus. One such, Nathuram Godse, assassinated him in Delhi. Widely revered before and after his death, he was known as the Mahatma (Sanscrit, 'Great Soul').

Gang of Four, four radical Chinese leaders. Jiang Qing, *Mao's fourth wife, Wang Hongwen, Yao Wenyuan, and Zhang Chungqiao all rose to prominence during the *Cultural Revolution, with a power base in Shanghai.

Mohandas **Gandhi** enjoying the affectionate greeting of his grand-daughters in New Delhi. At the time of India's independence in 1947, Gandhi was a figure of international standing, and the inspiration for the country's advance to statehood.

They occupied powerful positions in the Politburo after the Tenth Party Congress of 1973. After the death of Mao in 1976 they are alleged to have planned to seize power, and in 1980 were found guilty of plotting against the state. They have been blamed for the excesses of the Cultural Revolution.

Garcia Moreno, Gabriel (1821–75), President of Ecuador (1861–5, 1869–75). An extreme conservative, his goal was to convert Ecuador into the leading theocratic state of Latin America. His 1861 constitution, which accorded wide powers to the president, and his 1863 Concordat with the Vatican almost succeeded in doing so. The Catholic Church enjoyed greater power and privilege during his two administrations than it ever had before. A sound administrator, he put Ecuador on a stable financial basis and introduced material reforms. He was assassinated in 1875.

Garcia y Iñigues, Calixto (1836–98), Cuban nationalist. He led his country's preliminary struggles for independence from Spain. A leader during the Ten Years' War (1868–78), and the Little War (1879–80), his military efforts enjoyed scant success and resulted in his prolonged imprisonment. Cuban troops under his command supported US forces during the *Spanish-American War, but Garcia, shortly before his death, rejected his erstwhile allies, fearing that Cuba had simply exchanged one master for another.

Garfield, James Abram (1831–81), twentieth President (1881) of the USA. He served in the *American Civil War, retiring as a major-general, and then entered national politics as a Republican Congressman (1863-80). One of the politicians smeared with the *Crédit Mobilier scandal, he was rescued by James G. *Blaine. During the Republicans' feud between the self-styled 'HalfBreeds' (members of that wing of the Republican Party that favoured a conciliatory policy towards the South, and advocated civil service reforms) and 'Stalwarts' (their Republican opponents), he emerged as the compromise candidate to fight, and win, the presidential

election of 1880. His assassination within months of taking office by a disappointed office-seeker sealed the fate of the Stalwarts and ensured support for civil service reform in the *Pendleton Act.

Garibaldi, Giuseppe (1807–82), Italian leader and military commander of the *Risorgimento. In 1834 he led an unsuccessful republican plot in Genoa in support of *Mazzini. He fled, and spent the next twelve years in exile in South America, where he became a master of guerrilla warfare and formed his famous 'Redshirts'. He returned to Italy to take part in the *Revolutions of 1848, and formed a volunteer army to defend the short-lived Roman Republic against French forces intervening for Pope Pius IX. His resistance and his gallant retreat when Rome fell in June 1849 made him a popular hero. He was invited by *Cavour to help defeat the Austrians in northern Italy. When this was achieved (1859) the 'Thousand' captured Sicily (1860) and then Naples, capital of the kingdom of the Two *Sicilies, handing the whole of southern Italy to *Victor Emanuel II four months later. In 1862 Garibaldi, in an attempt to secure the Papal States, led his forces unsuccessfully against Rome. In 1867 he was defeated by French and papal forces at Mentana, while attempting once more to capture Rome.

Garvey, Marcus Moziah (1887–1940), Jamaican black leader. He organized the Universal Negro Improvement and Conservation Association (UNICA) in the USA in 1914 to encourage racial pride and black unity with a slogan 'Africa for the Africans at home and abroad', promising repatriation of US blacks to a new

Jamaican-born black separatist leader Marcus **Garvey** in 1922. At the height of his fame, he proclaimed himself head of an 'empire of Africa' for which he created imperial trappings and decorations. Behind the flamboyance of Garvey's Harlem parades, however, lay a sincere belief in the ideals of black nationalism and self-help.

African republic (to be created out of former German colonies). He established four branches of UNICA in South Africa in 1921, which encouraged the growth of black movements there in the 1930s. Deeply resented by William *Du Bois, Garvey's followers clashed with more moderate blacks in the 1920s. Although personally honest and sincere, he mismanaged his movement's finances and was convicted (1923) of attempted fraud.

GATT *General Agreement on Tariffs and Trade.

gaucho, a horseman of South America, often Indian or mestizo. Early in the 19th century the gauchos took part in the *Spanish South-American Wars of Independence, and later were prominent on the Argentine pampas in the development of the cattle industry. By the late 19th century the pastoral economy had given way to more intensive land cultivation in fenced-off estancias (estates), forcing many gauchos to become farmhands or peons.

General Agreement on Tariffs and Trade (GATT), an international trade agreement. Established by the *United Nations in 1948, it aimed to promote international trade by removing obstacles and trade barriers, to lay down maximum tariff rates, and to provide a forum for the discussion of trading policies. GATT promoted the postwar expansion of world trade, but the poorer countries felt that its terms favoured the developed countries, a criticism that led to the founding of *UNCTAD in 1964, and to an agreement in 1965 that developing countries should not be expected to offer reciprocity when negotiating with developed countries.

By the 1980s there were increasing demands for some general modification of the GATT agreements.

General Assembly, United Nations, the assembly for all member countries of the *United Nations. It is responsible for the UN's budget and may discuss and make recommendations on any question concerning the work of the UN. Its peace-keeping powers were strengthened by a resolution in 1950 which gave it the right to step in, when the *Security Council had failed to act, and to make recommendations, including the use of force. The General Assembly holds a regular session each year, but there are special sessions when either the Security Council or a majority of members call for them, and emergency sessions can be called at twenty-four hours' notice if peace is threatened. Such emergency sessions were called over the Soviet action in *Hungary in 1956 and over the *Suez Crisis in the same year. In the early years the USA and the Western Powers normally had a majority in the Assembly against the communist bloc, but as more Third World countries became independent, the numerical balance shifted to favour the developing nations. The Assembly has been particularly opposed to any remnants of colonialism.

General Strike (1926), a British trade union strike. It was in support of the National Union of Mineworkers whose members were under threat from mine owners of longer hours and lower wages because of trading difficulties. The owners had locked out the miners from the pits to try to compel acceptance. The General Council of the Trades Union Congress responded by calling workers out on strike in certain key industries such as the railways, the docks, and electricity and gas supply. This began on 4 May 1926 and ended nine days later. Irresolute trade union leadership, skilful government

During the 1926 **General Strike** normal food distribution links (including the railways) were halted. Food was brought in to London by road convoys under army escort.

handling of information to the public, and help by troops and volunteers to keep vital services running, all led to the collapse of the strike. It was followed in 1927 by a Trade Union Act, restricting trade union privileges.

Geneva Conference (1954), conference held in Switzerland to negotiate an end to the *French Indo-Chinese War. Planned by the wartime Allies to settle the future of *Korea and Indo-China, rapid progress was made on the latter after the French defeat at *Dienbienphu. The resulting armistice provided for the withdrawal of French troops and the partition of Vietnam, with the north under the control of *Ho Chi Minh's *Vietminh and the south under Saigon. Intended as a prelude to reunification through general elections, the Conference actually resulted in the emergence of two antagonistic regimes which were not to be united until Hanoi's victory in the *Vietnam War in 1975.

Geneva Conventions, a series of international agreements on the more humane treatment of victims of war. The first (1864) was the direct result of the work of the Swiss Henri Dunant, founder of the *Red Cross. It laid down basic rules for the proper treatment of wounded soldiers and prisoners-of-war, as well as for the protection of medical personnel. It was amended and extended in the second convention (1906) insisting that all modern facilities for treating the sick and wounded must be available. World War I led to the third convention (1929) by which the USA and representatives of forty-six other nations agreed on rules about the treatment and rights of prisoners-of-war. Because of the failure of some belligerents in World War II to abide by these conventions, the fourth convention (1949) extended and codified existing provisions for four groups of victims— the sick and wounded, shipwrecked sailors, prisoners-of-war, and civilians in territory occupied by an enemy.

George II (1890–1947), King of the Hellenes (1922–3, 1935–47). He came to the throne when General Palstiras deposed his father (1922), but the continuing unpopularity of the Greek royal family caused him to leave Greece (1923). In 1935 a plebiscite favouring the monarchy enabled him to return. His position was difficult, for real power lay with the dictatorial General *Metaxas. In April 1941 Hitler attacked Greece, driving him into exile again. With strong backing from Britain, the king returned to Greece in 1946, but the Greek communists who had resisted Hitler now waged civil war against the restoration of the monarchy. He died in 1947 and US support for his brother Paul brought the civil war to an end in 1949.

George III (1738–1820), King of Great Britain and Ireland and of dependencies overseas, King of Hanover (1760–1820). He succeeded to the throne on the death of his grandfather George II. He disliked the domination of the government by a few powerful Whig families. He was against making major concessions to the demands of the American colonists, having an abhorrence of the American aim of independence. He suffered from porphyria, a metabolic disease which causes mental disturbances. Increasing reliance on *Pitt the Younger reduced his political influence. When the king refused to consider *Catholic emancipation in exchange for Pitt's *Act of Union, Pitt resigned (1801). In 1811 increasing

senility and the onset of deafness and blindness brought about the *regency of the profligate Prince of Wales, which lasted until he succeeded as *George IV.

George IV (1762–1830), King of Great Britain and Ireland and of dependencies overseas, King of Hanover (1820–30). As regent (1811–20) and later king, he led a dissolute life and was largely responsible for the decline in power and prestige of the British monarchy in the early 19th century. The eldest son of *George III, he cultivated the friendship of Charles James *Fox and other Whigs. In 1785 he secretly and illegally married a Roman Catholic widow, Maria Fitzherbert (1756–1837). Ten years later he reluctantly married Caroline of Brunswick, and separated from her immediately after the birth of their only child, Princess Charlotte. George III became increasingly senile at the end of 1810 and in the following year the prince was appointed regent. He gave his support to the Tories, but soon quarrelled with them too, leaving himself without a large personal following in Parliament. His reign saw the passage of the *Catholic Emancipation Act (1829). His attempt to divorce Caroline for adultery in 1820 only increased his unpopularity. He was a leader of taste, fashion, and the arts, and gave his name to the *Regency period. He was succeeded by his brother *William IV.

George V (1865–1936), King of Great Britain and Ireland (from 1920, Northern Ireland) and dependencies overseas, Emperor of India (1910–36). The son of *Edward VII, he insisted that a general election should precede any reform of the House of *Lords (1911). He brought together party leaders at the Buckingham Palace Conference (1914) to discuss Irish *Home Rule. His acceptance of Ramsay *MacDonald as Prime Minister of a minority government in 1924, and of a *National government in 1931, contained an element of personal choice. He was succeeded by *Edward VIII.

George VI (1895–1952), King of Great Britain and Northern Ireland and dependencies overseas (1936–52), Emperor of India until 1947. He succeeded his brother, *Edward VIII, after the *Abdication Crisis. His preference for Lord Halifax rather than Winston *Churchill as Prime Minister in 1940 had no effect, but he strongly supported Churchill throughout World War II. Likewise he gave his support to Clement *Attlee and his government (1945–50) in the policy of granting Indian independence. He and his wife, Elizabeth Bowes-Lyon, will be remembered for sustaining public morale during the German bombing offensive of British cities. He was succeeded by his elder daughter, *Elizabeth II.

German Confederation (1815–66), an alliance of German sovereign states. At the Congress of *Vienna (1815) the thirty-eight German states formed a loose grouping to protect themselves against French ambitions. Austria and Prussia lay partly within and partly outside the Confederation. The Austrian chancellor *Metternich was the architect of the Confederation and exercised a dominant influence in it through the Federal Diet at Frankfurt, whose members were instructed delegates of state governments. As the rival power to Austria in Germany, Prussia tried to increase its influence over other states by founding a federal customs union or *Zollverein. In the *Revolutions of 1848 a new constituent assembly

German Confederation

The German Confederation of 1815 spanned thirty-eight states, of which the rival powers of Austria and Prussia were the greatest.

--- German Confederation of 1815
— North German Confederation of 1866
◻ Prussia 1815
◻ Acquired by Prussia 1866-7
◻ Austrian empire

first five years the republic had to pay heavy *reparations to the Soviet Union, and Soviet troops were used to put down disorder in 1953. In 1954, however, the republic proclaimed itself a sovereign state; in 1955 it became a founder-member of the *Warsaw Pact; and in 1956 it formed the National People's Army. Walter *Ulbricht (1893–1973) was General Secretary of the Socialist Unity Party (1946–71) and Chairman of the Council of State (1960–71). In 1972 the German Federal Republic, as part of the policy of *Ostpolitik, established diplomatic relations with the republic. Admission to the UN followed in 1973, after which the republic was universally recognized. Although economic recovery from World War II was slower than in the west, East Germany is today one of the major industrial nations of the world.

German Second empire (Reich) (1871–1918), a continental and overseas empire ruled by Prussia. (The First Reich was the Holy Roman Empire, which ended in 1806.) It replaced the *German Confederation and the short-lived Northern German Confederation (1866–70). It was created by *Bismarck following the *Franco-Prussian War, by the union of twenty-five German states under the Hohenzollern King of Prussia, now Emperor William I. An alliance was formed with *Austro-Hungary in 1879 and German economic investment took place in south-east Europe. In 1884 Bismarck presided over a conference of European colonial powers in Berlin, to allocate territories in Africa. In the same year Karl Peters founded the Society for German Colonization, and Bismarck was prepared to claim three areas of Africa: German South-West Africa, bordering on Cape Colony; the Cameroons and Togoland, where Britain had long monopolized the trade; and German East Africa, thus threatening British interests in Zanzibar. Northern New

was elected to Frankfurt, and tried to establish a constitutional German monarchy, but in 1849 the Austrian emperor refused the crown of a united Germany because it would loosen his authority in Hungary, while the Prussian king, *Frederick William IV, refused it because the constitution was too liberal. The pre-1848 Confederation was restored, with *Bismarck as one of Prussia's delegates. In 1866 Bismarck proposed that the German Confederation be reorganized to exclude Austria. When Austria opposed this, Bismarck declared the Confederation dissolved and went to war against Austria. In 1867, after Prussia's victory over Austria in the *Austro-Prussian War (1866), the twenty-one secondary governments above the River Main federated into the North German Confederation (Norddeutscher Bund), with its capital in Berlin and its leadership vested in Prussia. Executive authority rested in a presidency in accordance with the hereditary rights of the rulers of Prussia. The federation's constitution served as a model for that of the *German Second empire, which replaced it after the defeat of France in the *Franco-Prussian War (1871).

German Democratic Republic (East Germany), an East European country. It emerged in 1949 from the Soviet zone of occupation of Germany. Its frontier with Poland on the Oder–Neisse line, agreed at the *Potsdam Conference, was confirmed by the Treaty of Zgorzelec in 1950. Its capital is East Berlin, but the existence of West Berlin, politically a part of the German Federal Republic but separated from it by 150 km. (93 miles) has caused serious problems (*Berlin airlift, *Berlin Wall). In the

William I of Prussia is proclaimed first emperor of the **German Second empire**, on 18 January 1871 in this painting by Anton von Werner. Bismarck (in the white uniform) witnesses the triumph of Prussian policy at the ceremony held in the Hall of Mirrors at Versailles, France, following Prussia's defeat of France in the Franco-Prussian War. (Staatsbibliotek, Berlin)

Guinea and the Bismarck Archipelago in the Pacific were also claimed. With the accession of William II (1888) colonial activity, especially in the Far East, increased. In 1898 Germany leased the Chinese province of Shandong and purchased the Caroline and Mariana Islands from Spain. In 1899 Samoa was partitioned between Germany and the USA. Potential friction with Britain was averted by a mutual agreement in 1900, following its intervention to crush the *Boxer Rising. In that year von *Bülow became Chancellor (1900–9). The growth of German industry had now made it the greatest industrial power in Europe, and inevitably the search for new markets led to tension with other colonial powers. The expansion of the German navy under von *Tirpitz led to rivalry with the British navy, while competition with France in Africa led to a crisis over *Morocco (1905). In a second Moroccan Crisis (1911), an international war came close. The assassination at *Sarajevo caught the empire unawares. After some debate it was decided that the 1879 alliance with *Austro-Hungary must be honoured even if it meant war against Russia and France. During *World War I most German African territories were conquered and at the *Versailles Peace Settlement Germany was stripped of its overseas empire, which became mandated territories, administered by the victorious powers on behalf of the *League of Nations. At the end of the war the emperor abdicated and the *Weimar Republic was created.

Germany, a country in central Europe. The alliance of 400 separate German states of the Holy Roman Empire (962–1806) had been reduced by the end of the *Napoleonic Wars to thirty-eight. At the Congress of *Vienna these were formed into a loose grouping, the *German Confederation, under Austrian leadership. The Confederation was dissolved as a result of the *Austro-Prussian War (1866), and in 1867 all northern Germany formed a new North German Confederation under Prussian leadership. This was in turn dissolved in 1871, and the new *German Second Empire proclaimed. After Germany's defeat in World War I, the *Weimar Republic was instituted, to be replaced in 1933 by the *Third Reich under Adolf *Hitler. Since the end of World War II the country has been divided into the *German Federal Republic of *Germany (West Germany) and the *German Democratic Republic (East Germany).

Germany, Federal Republic of (West Germany), a country in north-west Europe. It was created in 1949 from the British, French, and US zones of occupation. It became a sovereign state in 1955, when ambassadors were exchanged with world powers, including the Soviet Union. It consists of eleven Länder or states, each of which has wide powers over its domestic affairs. Konrad *Adenauer, as Chancellor (1949–63), was determined to see eventual reunification of Germany and refused to recognize the legal existence of the *German Democratic Republic (East Germany). A crisis developed over Berlin in 1958, when the Soviet Union demanded the withdrawal of Western troops and, in 1961, when it authorized the erection of the *Berlin Wall. The Berlin situation began to ease in 1971, during the chancellorship of the socialist Willy *Brandt (1969–74) with his policy of *Ostpolitik. This resulted in treaties with the Soviet Union (1970), Poland (1970), Czechoslovakia (1973), and one of mutual recognition and co-operation with the *German Demo-

cratic Republic (1972), with membership of the UN following in 1973. Economic recovery was assisted after the war by the *Marshall Plan. The challenge of rebuilding shattered cities and of absorbing many millions of refugees from eastern Europe was successfully met, as was that of re-creating systems of social welfare and health provision. The Federal Republic joined *NATO in 1955, when both army and airforce were reconstituted; large numbers of US and British troops have remained stationed there. In 1957 it signed the Treaty of *Rome, becoming a founder-member of the *European Economic Community in 1958. In recent years, although the pace of economic growth has slackened, the economy has remained one of the strongest in the world, under a stable democratic regime.

Geronimo (c.1829–1909), Apache Indian chief. He led his people in resistance to white settlement in Arizona until 1886. For twelve years, after 4,000 Apache had been moved to the inhospitable reservation at San Carlos, Arizona, he waged war against US troops conducting brutal raids. In 1886 he surrendered on the promise that his braves could live peacefully in exile in Florida. Instead, they were imprisoned and moved to Oklahoma.

Gestapo, the Nazi secret police or *Geheime Staatspolizei*. In 1933 Hermann *Goering reorganized the Prussian plain-clothed political police as the Gestapo. In 1934 control of the force passed to *Himmler, who had restructured the police in the other German states, and headed the *SS or *Schutzstaffel*. The Gestapo was effectively absorbed into the SS and in 1939 was merged with the SD or *Sicherheitsdienst* (Security Service), the intelligence branch of the SS, in a Reich Security Central Office under Reinhard *Heydrich. The powers of these organizations were vast: any person suspected of disloyalty to the regime could be summarily executed. The SS and the Gestapo controlled the *concentration camps and set up similar agencies in every occupied country.

Gettysburg, battle of (1–3 July 1863), a battle in the *American Civil War. On 1 July elements of the Army

Geronimo is given star billing on a poster for a Wild West show. The caption to his picture declares the Apache Indian chief to be, in the words of General Miles, 'the worst Indian that ever lived'—celebrated because his capture cost the US government the sum of a million dollars.

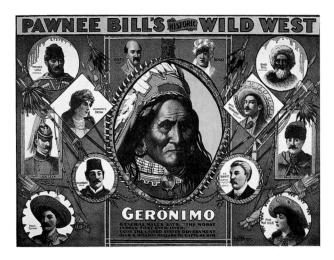

of Northern Virginia under *Lee and the Union Army of the Potomac under Meade came into contact west of Gettysburg, Pennsylvania. Although early Confederate attacks were repulsed, the arrival of reinforcements forced the Union troops to retreat back through the town. By the following day, however, fresh Union troops in strong defensive positions on Cemetery Ridge repelled Confederate attacks. On the third day, Pickett's charge against the centre of the Union line was defeated with heavy losses, and Lee was forced to abandon his invasion of the North. He lost 20,000 men from a force of 70,000 and Meade 23,000 from one of 93,000. The retreat from Gettysburg, together with the simultaneous surrender of *Vicksburg in the west, marked the turning point of the war, although over a year and a half of heavy fighting would follow before Lee was finally forced to surrender.

Ghana, a West African country. In 1800 there were British and Danish trading forts along the so-called Gold Coast and the British Colonial Office bought out the Danes in 1850 and gave some protection to the *Fante Confederation. Inland the area was dominated through the 19th century by the *Asante Confederacy. Britain occupied the capital Kumasi after wars with the Asante in 1824 and 1874, when the colony of the Gold Coast was established. Further wars against the Asante followed in 1896 and 1900. After 1920 economic growth based on mining and the cocoa industry, combined with high standards of mission schooling, produced a sophisticated people demanding home rule. Following World War II, in which many Ghanaians served, there were serious riots in Accra (1948) leading to constitutional discussions. In 1957 the Gold Coast and British *Togoland to the east were combined to become the independent Republic of Ghana, under the leadership of Kwame *Nkrumah, the first British African colony to be granted independence. Nkrumah transformed the country into a one-party state. Economic problems and resentment over political repression and mismanagement led to his overthrow by the army in 1966. Since his fall continuing economic and political problems have unbalanced Ghana. After a succession of coups, a group of junior officers under Flight-Lieutenant Jerry Rawlings took power in 1979, executed three former heads of state, and installed a civilian government. When this failed, Rawlings again seized power (December 1982), suspending the constitution and establishing a Provisional National Defence Council, with himself as Chairman.

Ghent, Treaty of (1814), a treaty ending the *War of 1812 between Britain and the USA. Negotiations to end the war, which was not popular in either country, began in August 1814 in the Belgian city of Ghent. British territorial demands obstructed progress for several months, but general war-weariness eventually brought a settlement. The resulting treaty did not address the issues which had caused the war, but provided for the release of all prisoners, restoration of all conquered territory, and the appointment of a commission to settle the north-eastern boundary dispute. Other questions, including naval forces on the Great Lakes and fishing rights, were left to future settlement.

Gibraltar, a town and rocky headland at the southern tip of Spain. It was important as a naval base during the two World Wars and still remains a British de-

pendency with the support of the inhabitants, some of whom are Italian, Maltese, and Portuguese in origin. In 1969 the border was closed by Spain, which claims possession of Gibraltar, but following an agreement signed in Brussels in 1984, was reopened in 1985.

Gilbert Islands *Kiribati.

'Gilded Age', a term used by writers in the USA in the 1870s and 1880s. It was coined by Mark Twain as the title of his novel *The Gilded Age* (1873), which satirizes the gross materialism of the decade following the *American Civil War, with its currency inflation and loose business and political morals. The decade had seen massive economic expansion by the victorious North across the whole continent, but at considerable cost, with unrestrained speculation in railway construction, currency manipulation, and many business failures. Later writers satirized the lawlessness of frontier life, the corruption of presidential elections, and the ruthless business methods of the *'robber barons'.

Giolitti, Giovanni (1842–1928), Italian statesman. He dominated Italian politics from 1900 to 1920, being five times Premier. A skilled conciliator and liberal, he introduced universal male suffrage for those over thirty (1912) and did not oppose the growth of trade unionism. He agreed to the conquest of Libya and argued against Italy's entry into World War I (1915). At first he gave some support to the fascists but later attacked the introduction of the corporate state.

Gladstone, William Ewart (1809–98), British statesman. He was the outstanding figure in British political life in the Victorian era and four times Prime Minister (1868–74, 1880–5, 1886, 1892–4). He entered Parliament as a Tory in 1832. A firm supporter of *free trade, he had to resign, together with the Prime Minister, Sir Robert *Peel, after the repeal of the *Corn Laws in 1846. He became Liberal Chancellor of the Exchequer during 1859–66, introducing an important series of budgets which cut tariffs and restrained government expenditure. Chosen leader of the Liberal Party in 1867, he took office as Prime Minister in 1868. His first administration passed some notable measures, including the disestablishment of the Church of Ireland, a secret ballot at elections, reforms in the legal position of trade unions, in the legal system itself, and in education. He also arranged the settlement of outstanding disputes with the USA through international arbitration. Defeated by Disraeli in 1874, Gladstone became Prime Minister once more in 1880 following his *Midlothian Campaign, and secured the passage of the third *Reform Act in 1884. However, he incurred much unpopularity over the fall of Khartoum and resigned in June 1885. When he took office again in the following year Gladstone introduced a bill to give Ireland *Home Rule. In doing so he split the Liberal Party and his government was defeated. Becoming Prime Minister again in 1892, he made a further attempt with a new Home Rule bill in 1893, but it was rejected by the House of Lords. In 1894 Gladstone resigned for the last time.

Goa, a district on the west coast of India, formerly a Portuguese territory. It was taken from the Portuguese by India in 1961 and now, together with two other

Joseph **Goebbels** in a characteristic posture at a Nazi gathering in Berlin's Lustgarten in August 1934. He had been appointed Hitler's Enlightenment and Propaganda Minister the year before.

former Portuguese territories, Daman and the island of Diu, annexed by India in 1962, forms a Union Territory of India.

Gobineau, Joseph Arthur, comte de (1816-82), French diplomat and scholar, the intellectual founder of 'racism'. His most famous book, *Essay on the Inequality of Human Races* (1853-5), put forward the thesis that the races are innately unequal and that the white Aryan race is not only the purest but also superior to all others. His writings were to have a sinister influence on the German *Nazi theorists, for whom they became a justification for *anti-Semitism.

Goderich, Frederick John Robinson, Viscount and 1st Earl of Ripon (1782-1859), British statesman, Prime Minister (1827-8). He first entered Parliament in 1806 and held a succession of offices, including those of President of the Board of Trade (1818-23), Chancellor of the Exchequer (1823-7), and Secretary for War and the Colonies (1827). As Viscount Goderich he succeeded George Canning as Prime Minister in August 1827, but, ill-suited to the task, resigned the following January. He was created Earl of Ripon in 1833.

Goebbels, (Paul) Joseph (1897-1945), German Nazi propagandist. Rejected by the army because of a club foot, he joined the *Nazi Party and founded a new paper for party propaganda, *Der Angriff* ('The Attack'), from now on exploiting his considerable gifts of oratory and

manipulation of the masses to further the Nazi cause. His brilliantly staged parades and mass meetings helped *Hitler to power. In 1933 he became Hitler's Enlightenment and Propaganda Minister, giving him control over the press, radio, and all aspects of culture until 1945. After Germany's defeat at *Stalingrad he was entrusted with the implementation of 'total war' within Germany. Faced with the advancing Soviet army, he committed suicide in Berlin with Hitler, first killing his wife and six children.

Goering, Hermann Wilhelm (1893-1946), German Nazi leader. In World War I he gained the highest award for bravery in the air. He joined the *Nazis in 1922, commanded their *Brownshirt paramilitary organization, and fled the country after being wounded in *Hitler's unsuccessful *Munich 'beer-hall' putsch. In 1934 he became commander of the German air force, and was responsible for the German rearmament programme. Until 1936 Goering headed the *Gestapo, which he had founded. He was then entrusted by Hitler with the execution of the four-year economic plan and directed the German economy until 1943. In 1937 he became Minister for Foreign Affairs and in 1938 Hitler's first Deputy. Increasingly dependent on narcotics, he was deprived by Hitler of all authority in 1943 and finally dismissed (1945), after unauthorized attempts to make peace with the Western Allies. Sentenced to death at the Nuremberg trials, he committed suicide in his cell by swallowing poison.

Gokhale, Gopal Krishna (1866-1915), Indian nationalist politician. The leader of the moderate faction in Congress, he became prominent in the Indian legislative Council established under the *Morley-Minto Reforms in 1910, specializing in finance. He also founded in 1905 the Servants of India Society, an austere organization dedicated to the service of India.

gold rushes, sudden influxes of people to newly discovered gold fields. The most famous gold rush was to California, where in 1848 gold was found by a Swiss settler, J. A. Sutter. As news spread, adventurers from

Towns of the American West such as Helena, Montana, sprang into being following nearby strikes in the **gold rushes**. The gold strike at Last Chance Gulch (July 1864) created Helena, today the state capital of Montana. This photograph of Bridge Street in 1865 shows newly built stores and ox-drawn wagons loaded with supplies for the mining camps.

all over the world made for California. Hard-drinkers and gamblers, the 'forty-niners' created an archetypal saloon society, where more fortunes were made from speculation in land and goods than from gold, as the city of San Francisco boomed. The second great rush was to Australia, where gold was first found near Bathurst in New South Wales in 1851 and later in Victoria at Bendigo and Ballarat, the richest alluvial gold field ever known. A ten-year boom brought diggers back across the Pacific from the declining California field, as well as from Britain, where Cornish tin-mining was declining. The population of Victoria rose from 97,000 to 540,000 in the decade 1851-60. Later rushes were to New Zealand (1860), to North Australia, Alaska, Siberia, and South Africa (1880s), and to Klondike in Canada and Kalgoorlie in West Australia (1890s). The most important was probably to Witwatersrand, South Africa, in 1886, where the influx of loose-living miners (*uitlanders or outlanders) precipitated political tensions which led to the Second *Boer War.

gold standard, a currency system in which the basic monetary unit of a country was defined in terms of a fixed quantity of gold. Paper money was convertible into gold on demand, gold could be freely imported and exported, and exchange rates between countries were determined by their currency values in gold. In 1821 Britain became the first country to introduce an official gold standard. It was followed some fifty years later by France, Germany, and the USA, and by 1900 the major countries had adopted the gold standard. Its main advantage was that any country's trade deficit would be automatically corrected. Most countries were unable to maintain the gold standard during World War I because gold could no longer be easily moved about. Britain returned to the gold standard in 1925, but abandoned it in 1931 because of the Great *Depression, and other countries were soon obliged to follow its example.

Gómez, Juan Vicente (1856-1935), Venezuelan statesman. During his twenty-seven-year rule as President (1908-35) he established an absolute dictatorship. The foreign investment that he attracted to Venezuela enabled him to build extensive railways, highways, and other public works. Rich petroleum discoveries (1918) in the Lake Maracaibo basin provided a budgetary surplus which not only enabled Gómez to pay off the foreign debt but also assured him a favourable reputation abroad. Because of the brutal nature of the dictatorship, this reputation was not shared at home. When he died in office the city of Caracas marched in celebration.

Gomulka, Wladyslaw (1905-82), Polish politician. He was Secretary-General during the crucial period 1943-9 when the Polish United People's Party was being formed. Gomulka's attempted defiance of Stalinism led to his dismissal and imprisonment (1951). He was restored to power (1956) on the intervention of Khrushchev, after Polish and Soviet frontier troops had exchanged fire in the wake of the Poznan workers' trial, in a Soviet attempt at compromise with Poland. He helped to sustain a degree of post-Stalin liberalism, but resigned in 1970 following popular disturbances against increases in food prices.

Good Neighbor Policy, the popular phrase applied

Mikhail **Gorbachev** (left) and US President Ronald Reagan, flanked by their interpreters, during a press conference at the summit meeting in Geneva, 1985.

to the Latin American policy of the early administration of President F. D. *Roosevelt. It was implemented by the withdrawal of US marines from Latin American countries and the abrogation of the Platt Amendment, which had given the US government a quasi-protectorate over Cuba. The Montevideo Conference (1933) declared that 'no state has the right to intervene in the internal or external affairs of another'.

Gorbachev, Mikhail S. (1931-), Soviet leader, the youngest Soviet leader to take power since Stalin. A Communist Party member since 1952, he was elected to the Central Committee in 1979 and to the Politburo in the following year. On the death of Konstantin Chernenko in March 1985 he became the Soviet leader, exercising control from his position as General Secretary of the Communist Party of the Soviet Union (CPSU). Gorbachev's efforts to carry out *perestroika*, the economic and social reform of Soviet society, has led to a gradual process of liberalization and the introduction of high technology to the Soviet Union. Together with his foreign minister, Eduard Shevardnadze (1928-), he negotiated (1987) an *arms control treaty with the West, to reduce nuclear forces in Europe. On the domestic front he has encouraged a greater degree of *glasnost* or openness and accountability in the face of inefficiency and corruption, and has introduced stringent laws against alcohol abuse.

Gordon, Charles George (1833-85), British general and administrator. In 1860 at the request of the Chinese government he entered the Chinese service and turned a small army of 3,500 peasants into a hard-fighting unit to defend Shanghai against *Taiping rebels. For this he was called 'Chinese Gordon'. In 1873 the governor of Egypt, Khedive Ismail, appointed him Governor of Equatoria (an area of the Upper Nile), where he brought peace and order. In 1880 he resigned, but returned to the *Sudan in 1884 at the British government's request to evacuate Egyptian forces from Khartoum, threatened by an army of the rebellious Muhammad Ahmed, calling himself the *Mahdi. In March 1884 the Mahdi besieged Khartoum and, despite Gordon's leadership, the city was captured on 26 January 1885. Gordon and his staff were killed two days before a relief expedition, dispatched belatedly from Britain, reached the garrison. Gordon's death stirred British public indignation and contributed to the collapse of the *Gladstone government.

Gottwald, Klement (1896-1953), Czechoslovak politician. He was a founder-member of the Czechoslovak Communist Party in 1921, becoming General Secretary in 1927. In protest at the *Munich Agreement (1938) Gottwald went to the Soviet Union. After World War II he returned to Czechoslovakia. He was Prime Minister in a coalition government in 1946-8 and, after the communist coup in 1948, President (1948-53) in succession to *Beneš. He dominated the country through purges, forced labour camps, and show trials, culminating in the Slansky trial and the execution (1952) of leading communists. He acquiesced in Stalin's plan of reducing Czechoslovakia's industries to satellite-status within the *Comecon economy.

Gowon, Yakubu (1934-), Nigerian statesman and soldier. He was a colonel in the Nigerian army at the time of the military coup of January 1966. Following a second coup in July he was invited to lead a new government. In a new constitution he divided Nigeria into a federation of twelve states, to replace the federal republic of four regions. The eastern Ibo region rejected the constitution and declared itself the state of Biafra, under General Ojukwu. In the *Biafra War which followed Gowon did not take field command, and helped to reconcile the defeated Ibo people after the war ended in 1970. He was largely responsible for the creation of the *Economic Community of West African States. By 1975 he was emerging as an international figure, but within Nigeria corruption was rife and he was deposed by the army in July 1975.

Graf Spee *Plate, battle of the River.

Gramsci, Antonio (1891-1937), Sardinian political theorist and founder of the Italian Communist Party (1921). He argued during the Turin general strike of 1920 that the workers should take over and run the factories, as a starting-point of a new communist society. He was captured (1926) and imprisoned by *Mussolini for eleven years, which caused his death. His *Prison Notebooks*, published in 1947, became influential in left-wing European circles during the 1970s and 1980s.

Granger Movement, a movement begun in 1867 as a social and educational association of mid-western farmers in the USA. It was more properly known as the National Grange of the Patrons of Husbandry. It believed in the importance of the family farm and opposed what it saw as a drift toward economic monopolies, particularly the abuse of freight rates. It never formed a major political party, but its supporters in a number of states did produce local legislation to fix maximum railway rates. These were validated by the US Supreme Court, which accepted the important new presumption that public regulation of private property was legitimate if the property provided public service. After the mid-1870s the movement declined.

Grant, Ulysses Simpson (1822-85), US general and eighteenth President of the USA (1869-77). He entered the *American Civil War as a colonel of volunteers in support of the Union (Northern states), and success

TO THE GRAND ARMY OF THE REPUBLIC
This print of
OUR OLD COMMANDER
GENERAL U. S. GRANT

US general Ulysses S. **Grant** encourages Union troops as they go into action during the American Civil War: a contemporary coloured lithograph by Currier and Ives. (Library of Congress, Washington)

brought him rapid promotion. He was active in most of the early engagements in the western theatre, winning the nickname 'Unconditional Surrender' for his capture of Fort Donelson (12 February 1862). As a major-general, he captured *Vicksburg in 1863 and was again successful at *Chattanooga before being promoted to the supreme command of all the Union forces in February 1864. Basing himself in Virginia, he maintained relentless pressure on *Lee in a bloody year-long campaign in which Lee was finally forced to abandon Richmond and surrender at *Appomattox, bringing the war to an end. Grant served briefly as Secretary of War (1867-8) before being twice elected as Republican President of the USA. His Presidency saw the collapse of the *reconstruction programme in the South. One of the greatest of all American soldiers, Grant was also one of the least successful of American presidents.

Grattan, Henry (1746-1820), Irish statesman. A champion of Irish independence, he entered the Irish Parliament in 1775, and led the movement to repeal Poynings' Law, which made all Irish legislation subject to the approval of the British Parliament. Grattan strongly opposed the *Act of Union (1801) which merged the British and Irish Parliaments. In 1806 he became member for Dublin in the British House of Commons and devoted the rest of his life to the cause of *Catholic emancipation.

The Crystal Palace, seen here in a painting by Louis Haghe, was the main feature of the **Great Exhibition** of 1851. Sponsored by Prince Albert to 'wed high art with mechanical skill', it was the first prefabricated building, 95 per cent of which was built of glass. Critics of the Exhibition had expressed the fear that to allow thousands of people and their leaders to come together was to invite massive riots. In the event, the Exhibition was an unprecedented success.

Graziani, Rodolfo (1882–1955), Italian general and colonial administrator. He was governor of Italian Somaliland (1935) and, after his success in the Ethiopian War, was made viceroy of Ethiopia. He commanded the Italian forces in Libya that invaded Egypt in September 1940 and reached Sidi Barrani, where he was routed by the British under *Wavell. In Mussolini's rump government of 1943–4 he was defence minister. Arrested in 1945 for collaboration with the Germans after the Italian armistice (1943), he was released in 1950 and became active in the Italian Neo-Fascist Party.

Greater East Asia Co-Prosperity Sphere, pseudo-political and economic union of Japanese-dominated Asian and Pacific territories during World War II. Announced in the aftermath of Japan's dramatic conquests of 1941–2, some nationalist leaders (for example, Indonesia's *Sukarno and Burma's *Aung San) collaborated with the Japanese for tactical reasons, but the hardships wrought by the latter (principally through their requisitioning of supplies and use of forced labour) soon disabused the local populations about Japan's intentions. By the end of the war, the Co-Prosperity Sphere had become an object of hatred and ridicule, referred to as a sphere of Co-Poverty and Co-Suffering.

Great Exhibition, the international exhibition held in Hyde Park, London, in 1851. Conceived by Henry Cole, it was keenly supported by *Albert, the Prince Consort. It was housed in the Crystal Palace, a vast glass and iron structure designed by Joseph Paxton, and its aim was to demonstrate to the world the industrial supremacy and material prosperity of Britain. The effect of the exhibition was to create a consciously Victorian visual style and, through the newly established School of Design, to encourage industrial design.

Great Leap Forward (1958), Chinese drive for industrial and agricultural expansion through 'backyard' industries in the countryside and increased production quotas to be reached by the people's devotion to patriotic and socialist ideals. Massive increases in the quantity of production were announced, but quality and distribution posed serious problems. In agriculture, *communes became almost universal, but disastrous harvests and poor products discredited the Leap, and its most important advocate, *Mao, took a back seat until the late 1960s. The *Cultural Revolution can be seen partly as his attempt to reintroduce radical policies.

Great Trek, the, the movement northwards in the 1830s by Boers to escape from British administration in the Cape Colony. From 1835 onwards parties of Voortrekkers reached *Natal, where in 1837 *Zulu resistance provoked them to kill some 3,000 Zulus at the battle of *Blood River in revenge for the death of their leader, Piet Retief. Natal became a British colony in 1843 and migration continued northwards into the Orange River country and the *Transvaal under Andries Hendrik Potgieter and Andried *Pretorius.

Greece, a country in south-east Europe. The *Greek War of Independence (1821–33) resulted in the establishment of an independent Greece, with Duke Otto of Bavaria as king. Otto was deposed in 1862 and a Danish prince, William, installed, taking the title George I of the Hellenes (1863–1913). A military coup established a republic (1924–35). *George II was restored in 1935 but fled into exile in 1941. Occupied by the Germans in World War II, the country suffered bitter fighting between rival factions of communists and royalists, the monarchy being restored by the British in 1946. Civil War developed in 1946 and lasted until 1949, when the communists were defeated. With the help of the *Marshall Aid programme, recovery and reconstruction began in 1949, Field-Marshal Alexandros Papagos becoming civilian Prime Minister (1952–5). In 1967 a military coup took place. Constantine II fled to Rome and government by a military junta (the 'Colonels') lasted for seven years, the monarchy being abolished in 1973. A civilian republic was established in 1974 and in the 1981 general election Andreas Papandreou became the first socialist Prime Minister.

Greek War of Independence (1821–32), the revolt by Greek subjects of the *Ottoman empire against Turkish domination. It had its origins in the nationalistic ideas of the Hetairia Philike ('Society of Friends'), who chose Alexander Ypsilanti, a Russian general, and son of the ruler of Wallachia, to lead the revolt. Links were established with Romanian peasants, Serb rebels, and Ali Pasha, the warlord of western Greece. Ypsilanti

The war between Greeks and Turks during the **Greek War of Independence** is vividly portrayed in this contemporary Greek painting. It shows a panorama of war, infantry and cavalry engagements as well as naval actions, with Greek Orthodox priests bearing arms alongside their country's soldiers in traditional *Evzone* costume. (Royal Library, Windsor Castle)

crossed into Turkish territory in March 1821, but only after his defeat in June did the Greeks rebel. Although atrocities took place on both sides, the revolt gained the popular support of the Christian world and many foreign volunteers (of whom Lord Byron, who went out in 1823, was the most celebrated) joined the Greek forces. By the end of 1821 the Greeks had achieved striking successes on land and sea and in January 1822 an assembly met to declare Greece independent. Four years later, however, *Mehemet Ali of Egypt reconquered the Peloponnese and threatened to restore Turkish control. At the Treaty of London in 1827, Britain and Russia offered to mediate and secure an autonomous Greek state. When the Turks refused, Britain, Russia, and France sent a combined fleet which destroyed the Egyptian fleet at Navarino (1827). The following year the Russian army seized Adrianople and threatened Constantinople. The Turks agreed to make peace (1829), and the Conference of London (1832) confirmed Greek independence. The following year a Bavarian prince, Otto I, was crowned King of *Greece.

Greenback Party, a minor political party in the USA, formed in 1874. During the *American Civil War the Lincoln government had issued some $400 million of paper money not backed by gold ('greenbacks'), to meet rising Union (Northern) wartime costs. After the war a Greenback Movement urged the retention of this currency. In 1874 the Greenback Party demanded the issue of additional greenbacks to stimulate the economy. However, it could not prevent conservative banking interests from obtaining a Resumption Act (1875) which provided that all greenbacks must be redeemable by the *gold standard. The Party fought unsuccessfully to have this Act repealed. It won support in Congress (1878) for

a general expansion in currency, backed by gold. After that it disappeared, currency expansionists then turning their attention to silver with a *Free Silver movement.

Green Movement, political organizations and pressure groups, formed largely in the 1970s, devoted to the protection of the environment against pollution and exploitation. A Green party has established a particularly strong position in the Federal Republic of Germany, securing 42 seats in the 1987 federal elections.

Greensboro incident (1960), an incident in the US *Civil Rights campaign. Four young black students staged a sit-in at a segregated lunch counter in Greensboro, North Carolina. This led to similar sit-ins in many other southern towns and sparked off waves of protest against the *segregation laws and practices.

Grenada, a West Indian island. It was colonized by the French, and from 1763 by the British. Universal adult suffrage was granted in 1950 when the United Labour Party, led by Matthew Gairy, emerged. The Windward Islands were granted self-government in 1956 and became a member of the West Indies Federation (1958-62) (*West Indian Independence). Following the break-up of the federation, the various Windward Islands sought separate independence. This was gained by Grenada in 1974, when Gairy became Prime Minister. He was

A village in **Grenada** witnesses (1983) the raising of the American flag as US forces take control of the island.

deposed in a bloodless coup (1979) by Maurice Bishop (1944-83), leader of a left-wing group, the New Jewel Movement, who proclaimed the People's Revolutionary Government (PRG). He encouraged closer relations with Cuba and the Soviet Union but, following a quarrel within the PRG, he was overthrown and killed by army troops led by General Austin in 1983. Military intervention by the USA prevented a Marxist revolutionary council from taking power. US troops left the island in December 1983, after the re-establishment of democratic government.

Grenville, William Wyndham, Baron (1759-1834), British statesman. He entered the House of Commons in 1782 and served as Secretary of State for Foreign Affairs (1791-1801) under his cousin William Pitt the Younger. After Pitt's death he formed the so-called 'Ministry of all the Talents' (February 1806-March 1807). Attempts to end the *Napoleonic War failed, but a bill for the abolition of the British overseas slave trade succeeded, following a resolution introduced by Charles James *Fox. A bill to emancipate Roman Catholics, however, was rejected by George III, and Grenville resigned from politics.

Grey, Charles, 2nd Earl Grey (1764-1845), British statesman. Entering the House of Commons in 1786, he became an advocate of electoral reform. After he had held office briefly as Foreign Secretary in the coalition government of 1806-7, his liberal views kept him politically isolated for some years. As leader of the Whigs, he was Prime Minister (1830-4) in the government which eventually secured the passage of the first great parliamentary *Reform Act (1832). In 1833 his government passed important factory legislation and the Act abolishing slavery throughout the British empire. He retired in 1834.

Grey, Sir Edward, Viscount Grey of Fallodon (1862-1933), British politician. He was Foreign Secretary from 1905 to 1916, negotiating the Triple Entente (1907), which brought Britain, France, and Russia together, and in 1914 persuading a reluctant British cabinet to go to war, because Germany had violated Belgian neutrality.

Grey, Sir George (1812-98), British colonial administrator. He served as governor of South Australia (1841-5) and of New Zealand (1845-53, and 1861-8), where he oversaw the introduction of representative government and gained a reputation for suppressing rebellion and promoting the 'amalgamation' of the *Maori people into settler society. After the onset of the *Taranaki War, he failed in negotiations with the quasi-nationalist *Kingitanga, and invaded the Waikato in 1863. He ended his governorship in disfavour with the Colonial Office, as the war dragged on. He was Premier of an undistinguished New Zealand ministry (1877-9).

Griffith, Arthur (1872-1922), Irish statesman. The editor of *United Irishman* then *Sinn Fein*, his initial goal was an independent Irish parliament under a dual monarchy of England and Ireland. He helped to found the *Sinn Fein Party. Following the emergence of the militant *Ulster Volunteer Movement in 1912, Sinn Fein became a militant political force and Griffith took part in gun-running for the Irish Volunteers in 1914. He opposed Irish participation in World War I, but did not take part in the *Easter Rising of 1916. Imprisoned by the British (1916-18), he was elected to the Westminster Parliament in 1918, and became instrumental in forming the Dáil Éireann in 1919. When the Irish Republic was declared in 1919 he became Vice-President. He led the Irish delegation that signed the Anglo-Irish Treaty (1921) establishing the *Irish Free State.

Guam, a Pacific island, east of the Philippines. Discovered by Magellan in 1521, it was ruled by Spain between 1688 and 1898, when together with the Philippines it was ceded to the USA after the *Spanish–American War. Between 1941 and 1944 it was in Japanese hands. It has since been a major US naval and air base, and remains one of the most important centres for US strategic intelligence.

Guatemala, a country in Central America. Its population is largely descended from Maya Indians. In 1821 it declared itself independent from Spain and became part of the short-lived Mexican empire of *Iturbide. When that collapsed (1823), Guatemala helped to found the United Provinces of Central America (1823-38). Strong opposition to federation, led by Rafael Carrera, resulted in its collapse, Guatemala declaring itself an independent republic with Carrera its first President (1839-65). His successors as President became increasingly despotic. A left-wing government under Jacobo

Arbenz (1951–4) instituted social reforms, before being forced to resign, following US intervention through the *Central Intelligence Agency. Ten years of disorder were followed by the peaceful election of Julio Cesar Mendez Montenegro as President (1966) on a moderate platform. But military intervention recurred, and during the 1970s and early 1980s violent suppression through the violation of human rights occurred. In 1985 civilian elections were restored, and President Vinico Cerezo elected, ending a dispute over *Belize and restoring diplomatic relations with Britain.

Guderian, Heinz (1888–1954), German general and tank expert. A proponent of the *Blitzkrieg tactics, he used tanks in large formations in the conquest of Poland (1939)and of France (1940). As commander-in-chief of the Panzer (tank) forces, he played a leading role in the German victories of 1940–1, but was dismissed when he disagreed with Hitler's order to stand fast in the 1941–2 Soviet counter-offensive outside Moscow. In 1944 he became chief-of-staff to the German Army High Command, but in March 1945 was again dismissed, this time for advocating peace with the Western Allies.

guerrilla (Spanish, 'little war'), a person taking part in irregular fighting by small groups acting independently. The term was coined during the *Peninsular War (1807–14) to describe the Spanish partisans fighting the armies of Napoleon. From Spain the use of the word spread to South America and thence to the USA. John S. Mosby used guerrilla warfare tactics during the *American Civil War to confuse the Federal army, and in the Arabian desert T. E. *Lawrence's cavalry troops were able to contain superior Turkish forces. *Mao Zedong, a leading proponent of guerrilla warfare, conducted a large-scale guerrilla campaign during the 1920s and 1930s against the Kuomintang and Japanese in China. It was during *World War II, however, that guerrillas became most prevalent, when *resistance movements were formed to harass the Japanese and Germans. In post-war years they have become associated with revolutionary movements like those in South America under *Guevara, as well as in Asia, the Middle East, and Africa, with acts of *terrorism spreading to urban areas.

Guevara, Ernesto 'Che' (1928–67), South American revolutionary and political leader. An Argentine by birth, he joined the pro-communist regime in Guatemala, and when this was overthrown (1954) he fled to Mexico. Here he met Fidel *Castro and helped him prepare the guerrilla force which landed in Cuba in 1956. Shortly after Castro's victory Guevara was given a cabinet position and placed in charge of Cuban economic policy. He played a major role in the transfer of Cuba's traditional economic ties from the USA to the communist bloc. A *guerrilla warfare strategist rather than an administrator, he moved to Bolivia (1967) in an attempt to persuade Bolivian peasants and tin-miners to take up arms against the military government. The attempt ended in failure as Guevara was captured and executed shortly thereafter. His refusal to commit himself to either capitalism or orthodox communism turned him into an archetypal figurehead for radical students of the 1960s and early 1970s.

Guiana   *French Guiana, *Guyana, *Surinam.

'Che' **Guevara**, a member of the revolutionary force that landed in Cuba in 1956: out of the 82 men who struggled ashore, only 12, including himself and Fidel Castro, survived the counter-attack by Batista's soldiers.

Guild Socialism, a British labour movement. Founded in 1906 by Samuel Hobson, it called for revolutionary change in the organization of British industry. Like the *syndicalists in France, it wished workers to be given control of industry, organized through monopolistic guilds authorized by the state. After World War I it established a National Guilds League, but this split when many members joined the newly formed Communist Party. Ideological differences later weakened their influence and the movement had ended by 1923.

Guinea, a country on the west coast of Africa, formerly a French colony. From about 1879 most of eastern Guinea became a part of the empire of *Samori Touré, who fought against the French. In 1904 Guinea was made part of French West Africa, and it remained a French colony until 1958, when a popular vote rejected membership of the French Community, and Ahmed Sékou *Touré became first President. His presidency was characterized by severe unrest and repression, and almost complete isolation from the outside world, although before his death in 1984 a degree of liberalization was introduced. This trend has continued under the military regime of President Lansana Conté.

Guinea-Bissau, a country in West Africa, formerly Portuguese Guinea. Part of the Portuguese Cape Verde Islands, it became a separate colony in 1879. Its boundaries were fixed by the 1886 convention with France. The struggle against colonial rule intensified in the 1960s, led by Amilcar *Cabral, and in 1974 Portugal formally recognized its independence. In 1977 an unsuccessful attempt was made to unite with Cape Verde (a newly formed republic of islands to the west). In 1980 a military coup established a revolutionary council which became (1984) a council of state, and a Parliament was established.

Guizot, Francois Pierre Guillaume (1787–1874), French historian and statesman. Entering official service in 1815. he lost office in 1822 and led the liberal opposition to *Charles X's government. Involved in the *July Revolution of 1830, he returned to official service and, in 1833, introduced a national system of primary education. For the next eighteen years he served *Louis Philippe. A moderate monarchist, he succeeded *Thiers as leader of the government in 1840. His passive but immovable resistance to change finally led to his downfall in the *Revolution of 1848.

gunboat diplomacy, diplomacy supported by the threatened use of force by one country in order to impose its will on another. The term is used specifically with reference to the 19th century when, in furtherance of their own interest, the great maritime nations, notably Britain, employed their naval power to coerce the rulers of small or weak countries. It was used by the British Foreign Secretary, Lord *Palmerston, during the *Opium Wars. In 1882 a British fleet bombarded Alexandria in order to crush a nationalist movement. During the uprising by the *Boxers in China in 1900 the European powers combined their forces in order to protect their interests and punish the rebels. Gunboat diplomacy was also employed by the USA in the Philippines and has been used to enforce US policies in Latin America.

Guyana, a country on the north-east coast of South America. British rule was formally secured in 1831 when three colonies, Essequibo, Demerara, and Berbice (named from the three rivers) were consolidated to form the crown colony of British Guiana. Boundary problems with neighbours dominated the 19th century. During World War II the lease of military and naval bases to the USA proved useful to the Allied war effort. Britain granted independence to the colony in 1966 and Guyana became a nominally co-operative republic in 1970. Its Prime Minister, Forbes Burnham, became executive President (1980–6) with supreme authority under an authoritarian constitution. He was succeeded by Desmond Hoyte.

Guzmán Blanco, Antonio (1829–99), Venezuelan statesman. Appointed to negotiate loans from London bankers in 1870, he seized power and two years later had himself elected President. He was absolute ruler of Venezuela (1870–89). He fostered railroad construction, public education, and free trade. An efficient administrator, he reformed the civil service and instituted public works. His extravagance gradually alienated the Venezuelan populace and in 1888, while visiting Paris, he was deposed by Juan Paúl.

Haakon VII (1872–1957), King of Norway (1905–57). Formerly Prince Charles of Denmark, he was elected by the Norwegian Storting or parliament to the throne in 1905. In April 1940 he was driven out by the German invasion. Refusing the suggestion of the government of Vidkun *Quisling to abdicate, he continued the struggle from London. He returned to Norway in 1945. He dispensed with much of the regal pomp attached to the monarchy, and became known as the 'people's king'.

Habsburg (or Hapsburg), the most prominent European royal dynasty from the 15th to the 20th centuries. It had for the most part acquired its extensive territories throughout Europe by judicious marriages, the head of its House holding the title of Holy Roman Emperor until 1806. As rulers of the *Austrian empire they were increasingly rivals of the kingdom of *Prussia, surviving the *Revolutions of 1848, but being obliged, following the *Austro-Prussian War of 1866 to make concessions to Hungarian nationalism with the formation of the *Austro-Hungarian empire. The emperor *Francis Joseph II came increasingly to clash with Russian ambitions in the *Balkans, turning more and more to the *German Second empire, with whom he formed an alliance in 1879. Nationalist aspirations led eventually to the disintegration of his empire during World War I. The last Habsburg monarch, Emperor Charles I of Austria (Charles IV of Hungary), renounced his title in November 1918 and was later deposed.

hacienda, a large estate with a dwelling-house, originally given by monarchs in Latin America as a reward for services done. Such estates are known as *estancias* in Argentina and *fazendas* in Brazil. Nineteenth-century land laws made possible the further concentration of land in the hands of a few, and, with population increase, exacerbated tension in rural areas. The first major eruption of violence, calling for the break-up of the haciendas, occurred in Mexico in 1910. Most Latin American countries have experienced similar demands during the 20th century, and they have remained a major political issue.

Haganah, a Jewish defence force in Palestine. It was established in 1920 first as an independent, armed organization and then under the control of the Histadrut to defend Jewish settlements. During the 1936–9 Arab rebellion it was considerably expanded. It gained a general staff and was put under control of the Jewish Agency, acquiring new duties of organizing illegal Jewish immigration and preparing for the fight against Britain, who held the *mandate over Palestine. In 1941 the Palmah (assault platoons) were formed. In 1948 Haganah provided the nucleus of the Israeli Defence Force, formed to protect the newly created state of Israel.

Haig, Douglas, 1st Earl (1861–1928), British field-marshal. After being chief-of-staff in India (1909) he commanded the 1st Army Corps in Flanders at Ypres and Loos and succeeded Sir John French as commander-in-chief. His strategy of attrition while being prepared to accept huge casualties on the *Somme (1916) and at *Passchendaele (1917) was much criticized. His conduct of the final campaign (1918) ended the war more quickly than *Foch expected. After the war he devoted himself to working tirelessly for ex-servicemen, and instituted the 'Poppy Day' appeal associated with his name.

Haile Selassie (1892–1975), Emperor of Ethiopia (1930–74). Baptized a Coptic Christian under the name of Tafari Makonnen, in 1907 he was named regent and heir apparent by a council of notables. Crowned king in 1928 and emperor in 1930, in 1931 he promulgated a constitution, with limited powers for a Parliament, which proved abortive. From 1935 his personal rule was interrupted by the *Abyssinian War and Italian colonial occupation. He was forced to seek exile in Britain and regained power in 1941 with British aid. In spite of efforts to modernize Ethiopia, he lost touch with the social problems of his country, and in 1974 he was deposed by a committee of left-wing army officers.

Haiti, a country in the Caribbean, the western third of the island of Hispaniola. French rule was challenged in 1791 by a slave insurrection led by *Toussaint l'Ouverture. The country declared its independence (1804) and *Dessalines was proclaimed emperor. After his assassination (1806) a separate kingdom was set up in the north, while the south and west became republican. The country was re-united in 1820 as an independent republic. Haiti and the eastern part of the island (later the *Dominican Republic) were united from 1822 to 1844. In 1859 it became a republic whose anarchic history has been exacerbated by the mulatto–black hostility. The USA, fearing that its investments were jeopardized and that Germany might seize Haiti, landed its marines (1915) and did not withdraw them until 1934. The country was dominated by President François *Duvalier (1957–71), and by his son and successor, Jean Claude (1971–86). When the latter was exiled to France, a council assumed power.

Haldane, Richard Burdon, Viscount Haldane of Cloan (1856–1928), British politician. As Secretary for War (1905–12) he showed great organizational skill in the reforms of the British army. Recognizing the growing danger from German militarism, he used his knowledge of the German army to redevelop the military organization in Britain to meet the requirements of modern warfare. A small expeditionary force ready for instant action was formed with a Territorial Army as a reserve, and an Imperial General Staff to organize military planning on an improved basis. Haldane was sent on a mission to Berlin in 1912 to secure a reduction in naval armaments, but failed. He became Lord Chancellor in 1912 but Asquith dismissed him in 1915, partly because of a press campaign, unjustified, that he was pro-German. He was briefly Lord Chancellor again in 1924.

Halder, Franz (1884–1972), German general. As Nazi chief-of-staff from 1938 he was responsible for the planning of the *Blitzkrieg campaigns of World War II. His early dislike of *Hitler changed to acquiescence after the latter's controversial military decisions had seemed to be proven right by events. However, he opposed Hitler's

Haile Selassie (in light-coloured uniform), then almost 80, attending the first All-African Trade Fair in Nairobi, Kenya, in 1972. In front of the Ethiopian emperor is Jomo Kenyatta, Kenya's President, and directly behind is Ugandan dictator Idi Amin.

decision to strike against *Stalingrad in 1942, and was dismissed. After the *July Plot against Hitler he was sent to a concentration camp. He was freed in 1945.

Halifax, Edward Frederick Lindley Wood, 1st Earl of (1881–1959), British politician. From 1925 to 1931 he was governor-general and viceroy of India (as Lord Irwin), and was closely involved in that country's struggle for independence. Halifax, who favoured *dominion status for the sub-continent, ordered the imprisonment of *Gandhi after the *Salt March. As a member of *Chamberlain's government, he visited Germany and met Hitler. An advocate of *appeasement, Halifax accepted the post of Foreign Secretary in 1938 on *Eden's resignation. He accepted, *de facto*, the *Anschluss of Austria and the dismemberment of *Czechoslovakia after the *Munich Pact. Halifax refused an invitation to Moscow, thus losing the chance of agreement with the Soviet Union, and leaving the door open for Hitler and Stalin to draw up the *Nazi–Soviet pact. During World War II he was British ambassador to the USA.

Halsey, William Frederick (1882–1959), US admiral. In 1941, commanding the Pacific Fleet aircraft carriers, he and his fleet were out of harbour when the Japanese attacked *Pearl Harbor. In 1942 he led a spectacular raid against the Marshall and Gilbert Islands and during the campaign of the *Solomon Islands he took command of the South Pacific area. As commander of the 3rd Fleet at the battle of *Leyte Gulf (1944), he sank a number of Japanese aircraft carriers and in 1945 led the seaborne *bombing offensive against Japan.

Hammarskjöld, Dag Hjalmar Agne Carl (1905-61), Swedish diplomat and Secretary-General of the *United Nations (1953-61). In 1953 he was elected UN Secretary-General as successor to Trygve *Lie. He was re-elected in 1957. Under him, the UN established an emergency force to help maintain order in the Middle East after the *Suez crisis, and UN observation forces were sent to Laos and Lebanon. He initiated and directed (1960-1) the UN's involvement in the *Congo crisis, making controversial use of Article 99 of the UN Charter, which he believed allowed the Secretary-General to exercise initiative independent of the *Security Council or *General Assembly. While in the Congo he was killed in an aeroplane crash over Northern Rhodesia.

Hampton Roads Conference (3 February 1865), an abortive conference to negotiate an end to the *American Civil War. At a meeting on a Union (Northern) steamer moored in Hampton Roads, Virginia, Confederate demands, put forward by the Southern President Jefferson *Davis, that the *Confederacy be treated as a sovereign state foundered on the refusal of President *Lincoln to negotiate on any other terms but reunion and abolition of slavery. The conference broke up without result, and the war carried on until the Confederate surrender two months later at *Appomattox.

Hanna, Marcus Alonzo (1837-1904), US businessman and politician. One of the most powerful political organizers of modern times, with the help of unprecedently high campaign contributions from big business he achieved William *McKinley's defeat of *Bryan in the presidential election of 1896. He supported labour's right to organize and, briefly, opposed US overseas expansion. His major contribution to the Republicans was the introduction of novel sales techniques in elections, paid for by levies on business firms, and the deft use of patronage.

Hanover, a former German kingdom. During the *Napoleonic Wars it became part of the kingdom of Westphalia, but in 1815 it was restored to the British crown, to which the Elector of Hanover had succeeded in 1714 as George I. It also became a kingdom in its own right within the new *German Confederation in 1815. Succession in Hanover, unlike Britain, was governed by the Salic Law, which forbids succession through the female line; thus, when *Victoria became queen in 1837, her uncle Ernest Augustus became King of Hanover. He revoked the liberal constitution of 1815, but was forced to restore it in 1848. In 1866 Hanover was annexed by Prussia, whom it had opposed in the *Austro-Prussian War. The kingdom was dissolved and it became a province of Prussia within the North German Confederation. After World War II Hanover was incorporated into adjoining territories to form the state of Lower Saxony.

Hara Takashi (or Hara Kei) (1856-1921), Japanese statesman. Leader of the Seiyukai (Friends of Constitutional Government) Party and a strong advocate of government by political party rather than by interest groups, he became the first commoner to hold the post of Prime Minister (1918-22). Attempts to build links with the business community brought him under suspicion of corruption and he failed either to prevent the breakdown of civil order or stop military intervention in the *Russian Civil War. He planned to end the military administration of Taiwan and Korea but was assassinated by a right-wing fanatic.

Hardenberg, Karl August, Prince (1750-1822), Prussian statesman and reformer. In 1810 he was appointed Chancellor of Prussia and continued the domestic reforms inaugurated by *Stein. These included the improvement of Prussia's military system, the abolition of serfdom and of the privileges of the nobles, the encouragement of municipalities, the reform of education, and civic equality for Jews. In 1813 he persuaded *Frederick William III to join the coalition against Napoleon. He represented Prussia at the Congress of *Vienna where he achieved substantial gains for his country.

Hardie, (James) Keir (1856-1915), British politician. He gained experience of leadership in the National Union of Mineworkers; this cost him his job and he was black-listed by the coal-owners. In 1888 he broke with the Liberal Party and was elected a Member of Parliament for the Scottish Labour Party in 1892, becoming chairman of the *Independent Labour Party in 1893. In 1900 he linked it with other socialist organizations to form the Labour Representation Committee. Hardie became leader in the House of Commons of the first Labour group of MPs (1906). An outspoken pacifist and chief adviser (from 1903) to the women's *suffragette movement, Hardie gained popular support from his pursuit of improvement in working-class conditions and from his strongly practical Christian beliefs.

British socialist Keir **Hardie** makes a vigorous point during a speech in Trafalgar Square, London. He was addressing an anti-war demonstration during the early part of World War I.

Harding, Warren Gamaliel (1865-1923), twenty-ninth President of the USA (1921-3). He was the tool of the ambitious lawyer Harry Daugherty, who helped him win the office of lieutenant-governor of Ohio (1904-5) and Senator (1915-21), eventually promoting him as the successful compromise Republican candidate for President in 1920. Instructed to straddle the issue on whether or not the USA should join the *League of Nations, Harding, in his campaign, pledged a 'return to normalcy'. His fondness for his self-seeking friends, the 'Ohio Gang', whom he took into office, resulted in the worst political scandals since the 1870s, notably the *Teapot Dome scandal. Harding died suddenly before the worst revelations of his administration's incompetence and corruption.

Harmsworth, Alfred Charles William, 1st Viscount Northcliffe (1865-1922), British journalist and newspaper proprietor. In 1887 he used his savings to form a general publishing business with his brother, Harold Harmsworth (1868-1940). In 1894 the two brothers acquired the *Evening News* and two years later founded the *Daily Mail*. This opened a new epoch in Fleet Street by presenting news to the public in a concise, interesting style, using advertisements and competitions, and financing schemes of enterprise and exploration. In 1903 the *Daily Mirror* was founded, in 1905 the *Observer* came under Harmsworth's control, and in 1908 he became chief proprietor of *The Times*. He was appointed to head the British War Mission to the USA in May 1917, soon after America entered World War I.

Harrison, Benjamin (1833-1901), twenty-third President of the USA (1889-93), grandson of President William Henry *Harrison. During his Republican administration US interests helped to stimulate the *Pan-American movement, while Pacific imperialist interests were advanced in Hawaii and Samoa. At home Congress passed the Sherman *Anti-Trust Act, the *McKinley Tariff Act, and the Sherman Silver Purchase Act. Harrison honoured promises to implement the Pendleton Act for reform of the civil service, at the cost of some popularity within the Republican Party, and became personally involved in opening *Oklahoma Territory to settlement in 1889.

Harrison, William Henry (1773-1841), ninth President of the USA (1841). While governor of the Indiana Territory (1801-12), he defeated the Indian leader *Tecumseh at *Tippecanoe (1811), and as a major-general in the *War of 1812, he recaptured Detroit and established US supremacy in the west by defeating a British and Indian force at the battle of the *Thames (1813). He served in the House of Representatives (1816-19) and the Senate (1825-8), and was elected President in 1840, running with John Tyler on the Whig ticket under the slogan 'Tippecanoe and Tyler too', but died one month after his inauguration.

Hartford Convention (1814-15), US political conference, held by *Federalist supporters to consider the problems of New England in the *War of 1812. Dominated by moderates rather than extremists, the Convention adopted the establishment of an inter-state defence machinery independent of federal government provision, the prohibition of all embargoes lasting more than sixty days, as well as a series of constitutional amendments. The Treaty of *Ghent brought an abrupt end to its deliberations, and the adverse publicity which it attracted accelerated the decline of the *Federalists.

Hasan, Muhammad Abdille Sayyid (1864-1920), Somali nationalist leader, known to the British as the 'Mad Mullah'. After a visit to Mecca he joined the Salihiya, a militant and puritanical Islamic fraternity. He travelled to the Sudan, Kenya, and Palestine before coming to Berbera on the Gulf of Aden. He believed that Christian colonization was destructive of Islamic faith in Somaliland and in 1899 he proclaimed a *jihad* (holy war) on all colonial powers. Between 1900 and 1904 four major expeditions by the British, Italians, and Ethiopians failed to defeat him. After a truce (1904-20) he resumed war again and was routed and killed by a British attack in 1920.

Havelock, Sir Henry (1795-1857), British general. He spent almost his entire career in India, where he took part in the first *Anglo-Afghan war (1839), the *Sikh wars (1843-9), and the *Indian Mutiny, in which he led the troops which re-took Kanpur, where the garrison had been massacred. In 1857 he relieved *Lucknow, where he died.

Hawaii, a state of the USA comprising a chain of islands in the North Pacific. The populated islands, inhabited by Polynesians and ruled by kings, were first named by Captain Cook (1778) after his patron, the Earl of Sandwich. Americans entered the islands from the 1820s, and helped evolve a written language and the first constitution (1839). By 1893 a number of the new settlers wanted US annexation and were powerful enough to overthrow the Hawaiian monarchy under Queen Liliuokalani, though not to persuade the USA into annexation until 1898. The increasing interest in the mid-Pacific led the USA to declare Hawaii an organized territory (1900), and to install its chief Pacific naval base in *Pearl Harbor. Hawaii became the fiftieth state in 1959.

Hawley-Smoot Tariff Act (1930), US legislation directed against imported goods and materials. Drafted before the Wall Street crash of 1929, it was endorsed by President *Hoover in the belief that it would help the hard-pressed farmers if increased tariffs were imposed. The Act aroused deep resentment abroad. Within two years, twenty-five countries had established retaliatory tariffs, and foreign trade, already declining, slumped even further.

Haya de la Torre, Victor Raúl (1895-1979), Peruvian statesman. He founded and led the Alianza Popular Revolucionaria Americana (APRA), known as the Aprista Party (1924), which became the spearhead of radical dissent in Peru. He advocated social and economic reform, nationalization of land and industry, and an end to US domination of South American economies. After the *Leguía regime fell he urged his APRA followers to overthrow the army-backed conservative oligarchy. He stood for President in 1931, but ballots were rigged and Colonel Sánchez Cerro was proclaimed victor. He was imprisoned 1931-3 and, after the latter's assassination, was in hiding in Peru (1935-45), becoming widely known through his writings. In 1945 the Aprista Party took the name Partido del Pueblo (People's Party) and supported José Luis Bustamante as President, but when he was overthrown in 1948 Haya took asylum in the Colombian Embassy in Lima until 1954, when he went into exile in Mexico until 1957. He contested the 1962 Presidential election, but the army intervened and Terry *Belaúnde was declared the winner. In 1979 Haya de la Torre drafted the new constitution which restored parliamentary democracy.

Hayes, Rutherford B(irchard) (1822–93), nineteenth President of the USA (1877–81). An Ohio Whig who turned Republican, he sat in the House of Representatives (1865–7). He became President by one electoral vote, following the disputed, fraudulent election of 1876. Congress had decided to concede the victory to him, rather than to the Democratic candidate Tilden, who had the majority of popular votes, on condition that Hayes ended the process of radical reconstruction (*Reconstruction Acts) in the South by withdrawing Federal (Northern) troops. Senator Roscoe *Conkling, who had connived at Hayes's election, tried to undermine his authority by obstructing Hayes's attempts to reform the civil service. This led to bitter division in Republican ranks between 'Stalwarts' (professional politicians, led by

The **Haymarket Square riot**, in a drawing by Thore de Thulstrup made at the time from sketches and photographs. A bomb, reputedly thrown by an anarchist, explodes among the Chicago police as they break up the disturbance, while gunfire adds to the confusion.

Conkling) and 'Half Breeds' (reformers led by James G. *Blaine). Although Hayes ousted some of Conkling's supporters, including Chester A. *Arthur, he was firm in his resolve not to seek a second term in 1880.

Haymarket Square riot (1886), outbreak of violence in Chicago, USA. A protest at the McCormick Harvester Works culminated in a riot at which 100 people were wounded and several died. Eight anarchists were convicted of incitement to murder, and four were hanged. The international sympathy aroused for the accused led Governor John P. Altgeld to pardon the survivors on the grounds of judicial prejudice and mass hysteria.

Hay Treaties (1901, 1903), US treaties concerning the construction of a Central American canal linking the Atlantic and the Pacific. Negotiated by US Secretary of State John Milton Hay, the Hay–Pauncefote treaty (1901) nullified the Clayton–Bulwer Treaty of 1850, which had prevented British or US acquisition of territory in Central America. The Hay–Herrán Treaty of 1903 leased the USA a canal zone from Colombia. Agreement broke down and was followed by a revolt in *Panama (then a department of Colombia), undertaken with US connivance. Independence of Panama (1903) was followed by a new treaty, the Hay–Bunau–Varilla treaty, which granted the USA a larger zone in perpetuity. President Roosevelt (rather than Secretary Hay) has, however, been held more responsible for this treaty.

health services, provision of hospital, medical, and dental services. During the 19th century there was

A 1930s photograph of US newspaper magnate William Randolph **Hearst** (*centre*) entertaining guests at his castle at San Simeon, California. Late in life his debts threatened him with ruin, but he recouped his fortune by sales of his art treasures, including castles dismantled in Europe and transported to the USA.

startling progress in medical science, but also an increased awareness of health hazards and the need for improved urban public health. In the late 19th and early 20th centuries the development of public health-service hospitals and clinics became one of the main provisions of the *welfare state, the British National Health Service being introduced in 1946. In the Soviet Union, East European republics, Cuba, and other communist regimes state health services alone are officially available, while in Britain and many western countries health services through privately financed insurance schemes are an alternative to state services. In the USA most health facilities are so funded, apart from *Medicare and Medicaid. Rising pharmaceutical costs, increased surgical skills, and higher life-expectancy are putting ever greater strains on public health services.

Hearst, William Randolph (1863–1951), US newspaper publisher and journalist. His exaggerated accounts of Cuba's struggle for independence from Spain was popularly believed to have brought on the *Spanish–American War (1898). He opposed US entry into World War I and was unremittingly hostile to the League of Nations. From newspapers he branched into magazines and films, amassing a colossal fortune. But his own incursions into politics, for example as candidate for mayor of New York, were consistently unsuccessful.

Heath ministry, British Conservative government (1970–4) with Edward Heath (1916–) as Prime Minister. It was during this ministry that Britain became a member of the *European Economic Community (January 1973), a move to which Heath was personally deeply committed. Meanwhile, the troubles in *Northern Ireland worsened; Brian *Faulkner resigned and direct rule from London

was introduced. In domestic and economic affairs the Heath ministry was beset with difficulties. The problems of inflation and balance of payments were serious and were exacerbated by the great increase in oil prices by *OPEC in 1973. However, attempts to restrain wage rises led to strikes in the coal, power, and transport industries in the winter of 1973–4. After a national stoppage in the coal industry, Heath called an election to try to strengthen his position, but was defeated. Harold *Wilson became Prime Minister. Heath was replaced as party leader by Margaret *Thatcher in 1975.

Hejaz, Red Sea coastal plain of the Arabian peninsula. Under the *Ottoman empire the Hejaz was opened up through the improvement of communications: first the construction of the *Suez Canal, then the opening of the Pilgrim Railway (1908), which linked Medina with Damascus. Following the fall of the Ottoman empire and the 1916 Arab Revolt, King *Hussein ibn Ali abdicated (1924) in favour of his son, Ali, who also abdicated (1925) in the face of a *Wahhabi invasion. Ibn Saud, the sultan of Najd, assumed the title of King of Hejaz (1926), uniting all districts under his control to form the kingdom of Saudi Arabia in 1932.

Heligoland, a small island in the North Sea. Originally the home of Frisian seamen, it was Danish from 1714 until seized by the British navy (1807). It was ceded to

Britain (1815) and held until exchanged with Germany for *Zanzibar and Pemba (1890). Germany developed it into a naval base of great strategic importance. Under the terms of the *Versailles Peace Settlement its naval installations were demolished (1920-2). They were rebuilt by the Nazis and again demolished (1947). It was returned to the Federal Republic of Germany (1952).

Helsinki Conference (1973-5), meetings by political leaders of thirty-five nations at the European Conference on Security and Co-operation, held in Helsinki and later in Geneva. The conference was proposed by the Soviet Union with the motive of securing agreement to the permanence of the post-1945 frontiers, of furthering economic and technical co-operation, and of reducing East-West tension. The conference produced the Helsinki Final Act containing a list of agreements concerning technical cooperation and human rights. All signatories to the agreement, including the leaders of Soviet-bloc countries, agreed to respect 'freedom of thought, conscience, religion, and belief'. As a result of the continued persecution of human rights activists in the Soviet bloc, protest groups were formed and *Charter 77 was established in Czechoslovakia. Periodic reviews of the implementation of the agreements have been made.

Herero Wars (1904-8), campaigns by German colonialists against the Herero people in German South-West Africa. A Herero rebellion resulted in the near-extermination of the population by the Germans, their numbers falling from over 100,000 to 15,000. The survivors were resettled in the inhospitable desert land of contemporary Hereroland in *Namibia.

Herrin massacre (22 June 1922), a clash between unionized strikers and non-unionized miners in the USA. It occurred in Herrin, Illinois, where the employers had attempted to break a strike in a local mine by importing non-union men. Striking miners forced these to stop working, promised them safe-conduct, marched them from the mine, and then opened fire, killing around twenty-five men. A grand jury returned 214 indictments for murder and related offences but local feeling prevented convictions.

Hertzog, James Barry Munuik (1866-1942), South African statesman. In the Second *Boer War he was a brilliant guerrilla leader. In 1910 he joined the first Union of South Africa cabinet. In 1912 he opposed *Botha and in 1914 he formed the *National Party, aiming to achieve South African independence and oppose support for Britain in World War I. From 1924 to 1929, as Prime Minister, he made Afrikaans an official language, instituted the first Union flag, and was a protectionist. In 1933 he formed a coalition with J. C. *Smuts, and in 1934 they united the Nationalist and South African Parties as the *United Party. In racial affairs he was a strict segregationist. Although he won the 1938 election, his opposition to joining Britain in World War II brought about his downfall (1939).

Herzen, Alexander Ivanovich (1812-70), Russian author and revolutionary. He was exiled to Viatka in 1835 after being suspected of sympathy with the *Decembrists. He returned to Moscow in 1842, where he became a leader of the 'Westernizers' who believed that Russia must adopt the free institutions and secular thought of western Europe in order to progress. He left Russia in 1847. An exile in London, Geneva, and Paris, he wrote prolifically, supporting both moderate reform and radical revolution in turn and influencing Russian liberals and communists alike.

Herzl, Theodor (1860-1904), Hungarian Jewish writer and founder of modern *Zionism. Sent to Paris as correspondent of a Vienna newspaper to cover the *Dreyfus trial (1894), the *anti-Semitism he witnessed there confirmed his resolve that the only solution to the Jewish problem was the establishment of a Jewish national state. In the pamphlet *Der Judenstaat* (1896) he set out his aims, and convened the first Zionist Congress at Basle (1897). His arguments were forcibly underlined by a series of *pogroms against Jews in Russia. He approached rulers, statesmen, and financiers in many countries, hoping for their support for a Jewish national home in *Palestine, but died before his efforts could be realized.

Hess, (Walther Richard) Rudolf (1894-1987), German Nazi leader. An early member of the German Nazi Party, sharing imprisonment with *Hitler after the *Munich 'beer-hall putsch', he was Hitler's deputy as party leader and Minister of State. In 1941, secretly and of his own volition, he parachuted into Scotland to negotiate peace between Britain and Germany. He was imprisoned by the British for the duration of the war and then, for life, by the Allies at the *Nuremberg Trials. From 1966 he was the sole inmate of Spandau Prison in Berlin, where he committed suicide.

Hesse, a state within the Federal Republic of Germany. In the 19th century the name was used for both the Grand Duchy of Hesse-Darmstadt, and the electorate of Hesse-Kassel. Hesse-Kassel allied with Austria in 1866, whereupon Prussian troops invaded and the electorate was annexed by Prussia, who imposed reforms. In 1815 Hesse-Darmstadt regained territories on the Rhine (Rheinhessen) from France, joined the *Zollverein (1828), and saw considerable economic growth. After 1848, under the reactionary Chief Minister Baron Dalwigk, Roman Catholicism was encouraged to oppose the ambitions of Protestant Prussia. In 1871 it became one of the states of the new *German Second empire, after which it was simply termed Hesse. Louis IV (1877-92) married a daughter of Queen Victoria; through their daughter, who became the last empress of Russia, the disease of haemophilia was transmitted to the imperial family of Russia. The last grand duke abdicated in 1918. In 1945 the territory amalgamated with parts of Hesse-Kassel to form the new state of Hesse.

Heydrich, Reinhard (1904-42), German Nazi police official. He joined the *SS in 1931, and in 1934 became deputy head of the *Gestapo. He played a leading part in several of the darkest episodes of Nazi history, and from 1941 administered the Czechoslovak territory of Bohemia-Moravia, his inhumanity and his numerous executions earning him the names the 'Hangman of Europe' and 'the beast'. He was assassinated by Czech nationalists in 1942. The Germans retaliated with one of the most extreme reigns of terror in World War II, set to paralyse Czech opposition both in Prague and in the rural areas. Civilians were indiscriminately executed and the entire male population of the village of Lidice murdered.

Hidalgo y Costilla, Miguel (1753-1811), Mexican priest and freedom fighter. He inspired (16 September 1810) the Mexican War for Independence from Spain with his *Grito de Dolores* ('Cry of Dolores'), proclaimed from the pulpit of his parish church at Dolores. Within a few months he had amassed an army of almost 80,000, but lacking the military skill necessary to complement his well-developed social ideology, the movement was defeated after much bloodshed at the hands of the royalist army. Captured in 1811, he was tried by the Inquisition and referred to the secular authorities for execution.

highland clearances, deliberate removal of Scottish 'crofter' peasants by landlords. In the later 18th century, Scottish society in the Highlands suffered severely with the collapse of the system of chiefs and fighting clans. Subsistence farming could not sustain an increasing population and this was aggravated by the policy of many major landowners of clearing their land for sheep farming by expulsion of crofters and the burning of their cottages. The potato famine during the *Hungry Forties aggravated the problem and in the 1880s, after the arrival of the railway, sheep were replaced by deer. In 1882 there were outbreaks of violence, the 'Crofters War', which was investigated by a Royal Commission. In 1885 the crofters voted for the first time in a general election, and an Act of Parliament in 1886 gave them some security of tenure. Yet depopulation steadily continued. Many Scottish highlanders emigrated throughout the British empire.

Hill, James Jerome (1838-1916), Canadian-born US railway builder. He opened the Great Northern Railway line to Seattle, Washington, in 1893, building it without government subsidies. He gained control of the Northern Pacific and the Chicago, Burlington, and Quincy lines. His holding company, the Northern Securities Company, was outlawed by the US Supreme Court in 1904, but he remained a potent financial force.

Hill, Sir Rowland (1795-1879), British administrator and inventor. He was the originator of the penny postage-stamp system, subsequently adopted throughout the world. In 1837 he published a pamphlet, *Post Office Reform*, in which he proposed that a uniform, low rate of postage, prepaid by the sender, should be introduced. In it he described his invention of the postage stamp, 'a bit of paper just large enought to bear the stamp, and covered at the back with a glutinous wash'. He adopted the notion from Charles Knight's proposal in 1834 that the postage of newspapers should be collected by means of uniformly stamped wrappers. Despite bureaucratic opposition, his proposals were put into effect in 1840.

Himmler, Heinrich (1900-45), German Nazi police chief. He began life as a poultry farmer in Bavaria; as an early member of the Nazi Party he took part in the *Munich 'beer-hall putsch' (1923). He became chief of the *SS in 1929. With the help of *Heydrich he founded the SD (security service) in 1932. In 1936 he became chief of all the police services, including the *Gestapo, and as head of the Reich administration from 1939 extended his field of repression to occupied countries.

The penny post devised by Rowland **Hill** resulted in a rapid increase in letter-sending. This is London's first letter-box, designed by A. E. Cowper, and sited on the corner of Fleet Street and Farringdon Street in 1855.

From 1943 he was Interior Minister, and commander of the reserve army. From a position of supreme power he was able to terrorize his own party and all German-occupied Europe. Although personally nauseated by the sight of blood, he established and oversaw *concentration camps in which he directed the systematic genocide of Jews. He ruthlessly put down the conspiracy against Hitler in the *July Plot of 1944, but a few months later was himself secretly negotiating German surrender to the Allies, hoping to save himself thereby. Hitler expelled him from the Party, and Himmler attempted to escape. He was caught (1945) by British troops and committed suicide by swallowing poison.

Hindenburg, Paul von (1847-1934), German general and statesman. He fought at the battle of Königgratz (*Sadowa) and in the *Franco-Prussian War (1870-1) and retired in 1911. He was recalled to active service at the outbreak of *World War I and crushed the Russians at Tannenberg in east Prussia (August 1914). In 1916 he became chief of the general staff. After the failure of Germany's offensive (1918) he advised the need to sue for peace. After the war he came to tolerate the *Weimar Republic and in 1925 was elected as President in succession to *Ebert. Re-elected (1932), he did not oppose the rise of *Hitler, but appointed him as Chancellor (January 1933) on the advice of Franz von *Papen.

Hinduism, one of the world's major religions. The many reform movements of the 19th century varied in their response to the catalyst of Western influence. The worship of the *Brahmo Samaj founded in 1828 by Ram Mohan *Roy (1772-1833) was patterned on Unitarianism with an emphasis on rationality and a distaste for idolatry. In 1843 Devendranath Tagore (1817-1905) amalgamated his own association, the Tatwabodhini Sabha, with the Brahmo Samaj, instituting doctrinal and social reforms. His protégé Keshab Chandra Sen (1838-84), after initial success, caused divisions within the movement. In Maharashtra, the Prerthana Samaj led by Mahadev Govind Ranade (1842-1901) engaged in the reform of Hindu religion and society. Dayananda Saraswati (1824-83), who founded the Arya Samaj in Bombay in 1875, was another reformer with many noble aims but aggressive in his promotion of Hinduism and the authority of the Vedas. The Samaj became more politically involved and suffered grievously during the partition of India. In 1897 Swami *Vivekananda established the Ramakrishna Mission, a philanthropic religious and educational movement named after the mystic Sri Ramakrishna (1836-86). The Theosophical Society had already established its International Headquarters in Varanasi in 1882. The foremost reformer of the 20th century has been Mohandas *Gandhi, champion of the deprived and the untouchables, revered and reviled for his advocacy of non-co-operation, *ahimsa* (non-violence), *satyagraha* (truth-force), and *swaraj* (self-rule). Hinduism, like other religions, today contends with increasing secularization.

Hindu Mahasabha, a Hindu communal organization. It was first established in the Punjab before 1914, and became active during the 1920s under the leadership of Pandit Mohan Malaviya (1861-1946) and Lala Rajpat Rai (1865-1928), when it campaigned for social reform and for the reconversion of Hindus from Islam. Its attitude towards Hindu–Muslim relations strained the relations of the Mahasabha with *Congress and in 1937, under the leadership of V. D. Sarvarkar, it broke away from Congress. After independence the party declined in importance as the Jana Sangh became the leading exponent of Hindu communal ideas.

Hirohito (1901-), Emperor of Japan (1926-). The eldest son of Crown Prince Yoshihito (later the Taisho Emperor), Hirohito was appointed Regent in 1921 and, after surviving an assassination attempt, succeeded to the throne in 1926, initiating the Showa era. Although he did not approve of military expansion, he had little opportunity to exercise his full technical sovereignty, allowing the political triumph of *Tojo and the militarists. He continued to follow his counsellors' advice not to weaken the throne by becoming involved in politics until 1945 when, convinced of the need to end World War II, he intervened to force the armed services to accept unconditional surrender. Saved from trial as a war criminal by *MacArthur, Hirohito renounced his divinity but retained the monarchy, albeit as a symbol without governmental power, in the new constitution of 1947.

Hiroshima, Japanese city in southern Honshu. Hitherto largely undamaged by the US bombing campaign, Hiroshima became the target of the first atomic bomb attack on 6 August 1945, which resulted in the virtual obliteration of the city centre and the deaths of about one-third of the population of 300,000. The attack on Hiroshima, together with that on *Nagasaki three days later, led directly to Japan's unconditional surrender and the end of World War II.

Hiss case (1949-50), a legal case in the USA. A State Department official, Alger Hiss (1904-), was found guilty of perjury for having denied on oath the charge that he had passed secret documents to Whittaker Chambers, a self-confessed Communist Party courier. Hiss maintained his innocence, and high government

The scene after the first atomic bomb attack at **Hiroshima** was one of near-total devastation. One of the few city-centre structures to survive in any recognizable form was this cinema, blasted and blackened by fire.

officials testified for him, but he was sentenced to five years in prison. His controversial trial was a symbol of the fears aroused by the *Cold War; many people believed that the *Federal Bureau of Investigation had tampered with evidence so as to obtain his conviction. He was released in 1954 and returned to private life.

Hitler, Adolf (1889-1945), German dictator. He was born in Austria, the illegitimate son of Anna Schicklgruber and Alois Hitler. He volunteered for the Bavarian army at the start of World War I, became a corporal, twice won the Iron Cross medal for bravery, and was gassed. After demobilization he joined a small nationalist group, the German Workers' Party, which later became the National Socialist German Workers (or *Nazi) Party, and discovered a talent for demagoguery. In Vienna he had imbibed the prevailing *anti-Semitism and this, with tirades against the *Versailles Peace Settlement and against Marxism, fell on fertile ground in a Germany humiliated by defeat. In 1921 he became leader of the Nazis and in 1923 staged an abortive uprising, the *Munich 'beer-hall putsch'. During the months shared in prison with Rudolf *Hess he dictated *Mein Kampf*, a political manifesto in which he spelt out Germany's need to rearm, strive for economic self-sufficiency, suppress trade unionism and communism, and exterminate its Jewish minority. The Great *Depression beginning in 1929 brought him a flood of adherents so that, aided by violence against political enemies, his Nazi Party flourished. After the failure of three successive Chancellors, President *Hindenburg appointed Hitler head of the government (1933). As a result of the *Reichstag fire, Hitler established his one-party dictatorship, and the following year eliminated his rivals in the *'Night of the Long Knives'. On the death of Hindenburg he assumed the title of President and 'Führer of the German Reich'. He began rearmament in contravention of the Versailles Treaty, reoccupied the *Rhineland in 1936, and took the first steps in his intended expansion of his *Third Reich: the *Anschluss with Austria in 1938 and the piecemeal acquisition of Czechoslovakia, beginning with the *Sudetenland. He concluded the *Nazi–Soviet non-aggression pact with Stalin in order to invade Poland, but broke this when he attacked the Soviet Union in June 1941. His invasion of Poland had precipitated *World War II. Against the advice of his military experts he pursued 'intuitive' tactics and at first won massive victories; in 1941 he took direct military control of the armed forces. As the tide of war turned against him, he intensified the mass assassination that culminated in the Jewish *holocaust. He escaped the *July plot to kill him, and undertook a vicious purge of all involved. In 1945, as the Soviet army entered Berlin, he went through a marriage ceremony with his mistress, Eva Braun. All evidence suggests that both committed suicide and had their bodies cremated in an underground bunker.

Hitler Youth, a *Nazi agency to train young Germans. In 1931 Baldur von *Schirach was appointed Youth Leader of the Nazi Party. In 1936 *Hitler outlawed all other youth organizations and announced that all German non-Jewish boys and girls should join the Jungvolk (Young Folk) at the age of 10, when they would be trained in out-of-school activities, including sports and camping, and receive Nazi indoctrination. At 14 the boys were to enter the Hitler Youth proper, where they would

Adolf **Hitler** addressing a Nazi rally in Berlin during World War II. Dramatic hand gestures characterized Hitler's oratory which he used to great effect at mass rallies, a key component of the Nazi propaganda machine.

be subject to semi-military discipline, out-door activities, and heavy Nazi propaganda, and girls the League of German Maidens, where they would learn motherhood and domestic duties. At 18 they would join the armed forces or the labour service. By 1936 3.6 million members had been recruited, and by 1938 7.7 million, but efforts to enrol every boy and girl were failing, so that in March 1939 a conscription order was issued. There was continuous tension between the Hitler Youth and the schools, some of which tried to retain a liberal tradition, and this survived into the war years.

Members of the **Hitler Youth** at a German railway station. The rucksacks, shorts, and walking boots of these young Germans preparing to set off for camp recall the more innocent Scout movement. But the Hitler Youth had a sinister aim: the indoctrination of impressionable minds with Nazi philosophy.

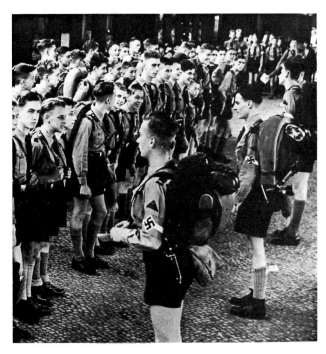

Ho Chi Minh (b. Nguyen Tat Thanh, also called Nguyen Ai Quoc) (1890–1969), Vietnamese statesman. In 1917 he moved to Paris where he became active in left-wing politics. He travelled to the Soviet Union and in 1924 went to Guangzhou in southern China as a *Comintern agent. In 1930 he presided over the formation of the Vietnamese Communist Party, re-named the Indo-Chinese Communist Party, and played a key role in its development. After a failed conspiracy in 1940, he took refuge in China and was imprisoned by the nationalist regime of *Chiang Kai-shek. In 1943 he returned to the north of Vietnam to found the *Vietminh guerrilla movement to fight the Japanese occupying forces, adopting the name Ho Chi Minh ('he who enlightens'). In 1945 after the Japanese surrender, he proclaimed the Democratic Republic of Vietnam but was forced back into guerrilla war after the return of French colonial forces. The *Geneva Conference (1954) accepted the Vietminh triumph over the French and left Ho in control of North Vietnam. From 1963 he committed his forces on an ever-increasing scale to the communist struggle in South Vietnam (*Vietnam War) and until his death he remained unswervingly committed to the reunification of Vietnam under communism, subordinating social and economic reform to the needs of the military struggle with the USA and the Saigon government.

Hofer, Andreas *Tyrol.

Hohenzollern, formerly a Prussian province, now part of the state of Baden-Württemberg in the Federal Republic of Germany. It gave its name to a dynasty which steadily gained power in Germany from the 11th century. In 1871 William I of Prussia took the title Emperor William I of the German Empire. His grandson *William II abdicated in 1918. A member of a second

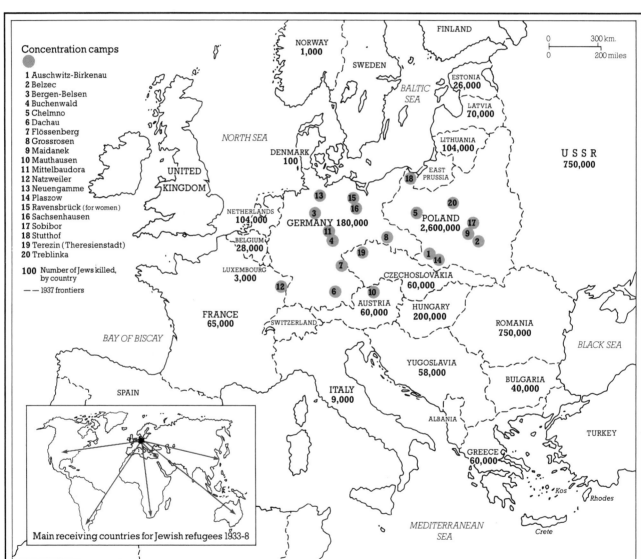

The Holocaust

The Nazi determination to rid Europe of Jews was rooted in the anti-Semitism of Central and Eastern Europe. From the earliest years of the Third Reich Jews were persecuted, and from 1941 onwards a programme of extermination began, with certain camps in Eastern Europe equipped with gas-chambers for systematic slaughter. It is estimated that four million Jews died in these camps. Perhaps a further million died in ghettos by starvation and disease, and over a million were shot by mobile killing squads (Einsatzgruppen).

Ho Chi Minh relaxing in the company of Vietnamese children. His determination to unify Vietnam under North Vietnamese communist rule cost the country dear during its long-drawn-out civil war. Ho himself did not live to see the triumphant culmination of his policy.

branch, Prince Charles Hohenzollern-Sigmaringen, was elected Prince of *Romania in 1866, becoming King Carol in 1881. His brother Leopold was offered the throne of Spain in 1870 and turned it down, but not before Bismarck had used the incident in the *Ems telegram.

Holland *Netherlands.

Holland, Sir Sidney George (1893-1961), New Zealand statesman. As leader of the National Party from 1940 and Prime Minister (1949-57), he was noted for his staunch support of private enterprise, for his vigorous handling of the 1951 waterfront strike, and for the abolition of the Legislative Council—the Upper House of the New Zealand Parliament. In the tradition of pragmatic conservatism Holland retained and even strengthened much of the welfare state and economic regulatory machinery put in place under Labour.

Holocaust, the, the term used to describe the ordeal of the Jews in *Nazi Europe from 1933 to 1945. Conventionally it is divided into two periods, before and after 1941. In the first period various *anti-Semitic measures were taken in Germany, and later Austria. In Germany, after the Nuremberg Laws (1935) Jews lost citizenship rights, the right to hold public office, practise professions, inter-marry with Germans, or use public education. Their property and businesses were registered and sometimes sequestrated. Continual acts of violence were perpetrated against them, and official propaganda encouraged 'true' Germans to hate and fear them. As intended, the result was mass emigration, halving the half-million German and Austrian Jewish population. The second phase, that of World War II from 1941, spread to Nazi-occupied Europe, and involved forced labour, massed shootings, and *concentration camps, the latter being the basis of the Nazi 'final solution' of the so-called Jewish problem through mass extermination in gas chambers. During the Holocaust an estimated six million Jews died. Out of a population of three million Jews in Poland, less than half a million remained in 1945.

Holt, Harold Edward (1908-67), Australian statesman. He represented the United Australia Party and then the Liberal Party, holding a series of portfolios from 1939.

After *Menzies retired in 1966, Holt became Prime Minister. His term of office coincided with Australia's increasing and controversial involvement in the *Vietnam War. He disappeared while swimming near Portsea in Victoria and was drowned.

Holy Alliance (1815), a loose alliance of European powers pledged to uphold the principles of the Christian religion. It was proclaimed at the Congress of *Vienna (1815) by the emperors of Austria and Russia, and the king of Prussia. All other European leaders were invited to join, except the pope and the Ottoman sultan. The restored French king Louis XVIII did so, as did most others; Britain did not. As a diplomatic instrument it was short-lived and never effective and it became associated with repressive and autocratic regimes.

Holyoake, Sir Keith Jacka (1904-83), New Zealand statesman. A farmer active in agricultural organizations in the 1930s and 1940s, he entered Parliament in 1932, becoming leader of the National Party and Prime Minister in 1957 and 1960-72. An able politician in the tradition of pragmatic conservatism, he led New Zealand skilfully in the decades of growing racial tension. He served a term as governor-general of New Zealand after his retirement from politics.

Home Guard, a World War II military force raised in Britain. In 1908 the Territorial Force, a home defence organization, had been created, which became the Territorial Army in 1921. The Home Guard, initially known as the Local Defence Volunteers, existed from 1940 to 1944. In 1942 enrolment in the force became compulsory for sections of the civilian population. About a million men served in their spare time, and in its first vital year it possessed considerably more men than firearms. It was never put to the test, but it helped British morale in 1940-1.

Home Rule, Irish, a movement for the re-establishment of an Irish parliament responsible for internal affairs. An association, founded in 1870 by Isaac Butt, sought to

A war artist's view of the **Home Guard**. This gently sardonic sketch by Edward Ardizzone is captioned: 'We are arrested by ferocious Home Guards'. The possibility of an invasion of Britain by Hitler's forces remained a very real threat during the first years of World War II. (Imperial War Museum, London)

Irish **Home Rule** excited strong passions on all sides. This postcard, supporting the parliamentary union of Britain and Ireland, was published in Belfast, which was to become the capital of Northern Ireland.

repeal the *Act of Union (1801) between Britain and Ireland. This became a serious possibility when Charles *Parnell persuaded the Liberals under *Gladstone to introduce Home Rule Bills. The first (1886) was defeated in the House of Commons. It provided for an Irish Parliament at Dublin, with no Irish representation at Westminster; it ignored the problem of Ulster where a predominantly Protestant, pro-British population had been settled since the early 17th century. Gladstone's second Bill (1893) was also defeated. The third Bill (1912), introduced by Asquith, was passed by Parliament but its operation was postponed when war broke out in Europe in 1914. It left unresolved the question of how much of Ulster was to be excluded from the Act. When World War I ended the political situation in Ireland was greatly changed. The *Easter Rising in 1916 and the sweeping majority for *Sinn Fein in the 1918 general election were followed by unrest and guerrilla warfare. Lloyd George was Prime Minister when the fourth Home Rule Bill (1920) was introduced in the Westminster Parliament. The Bill provided for parliaments in Dublin and Belfast linked by a Federal Council of Ireland. The Northern Ireland Parliament was set up in 1920 while fighting continued in Ireland. Following the Anglo-Irish truce the *Irish Free State was set up; the new state had a vague *dominion status at odds with the independence claimed by Dáil Éireann in 1919. The Anglo-Irish agreement was approved by sixty-four votes to fifty-seven in the Dáil. The majority group wanted peace and partial independence, the minority group, headed by Eamon *de Valera, desired the immediate independence of all Ireland and the setting up of a republic.

Homestead Act (1862), a US Act to encourage migration west. The Homestead Act gave any citizen who was head of a family and over 21 years of age 65 hectares (160 acres) of surveyed public land for a nominal fee. Complete ownership could be attained after five years of continuous residence or by the payment of $1.25 per acre after a six-month period. While some 15,000 such homesteads were created during the *American Civil War years, speculation and the development of mechanized large-scale agriculture reduced the effectiveness of the Act in later years.

Homestead strike (1892), a US labour dispute. It was the bitter climax of deteriorating relations between the Carnegie Steel Company at Homestead, outside Pittsburgh, run by Henry Clay Frick, and the Amalgamated Association of Iron and Steel Workers, who had refused to accept a disadvantageous new contract and ordered a strike. When Frick imported 300 *Pinkerton detectives to protect the plant and the non-union workers, they were repulsed in an armed battle in which several people were killed. The state governor introduced state militia to restore order, and the strike failed. The union collapsed after the anarchist Alexander Berkman tried to kill Frick, and unionism in the industry was seriously weakened until the 1930s.

Honduras, a Central American country. Honduras was attached administratively to the captaincy-general of Guatemala throughout the Spanish colonial period. When independence came in 1821 it briefly became part of the empire of Agustin de *Iturbide before joining the United Provinces of Central America (1825–38). Separate independent status dates from 1838, when the union broke up. An uninterrupted succession of *caudillos dominated the remainder of the 19th century. Improvement in the political process came slowly in the 20th century. Military dictators continued to be more prominent than civilian presidents, but the election in 1957 of Ramón Villeda Morales gave hope for the future. This optimism proved premature as the Honduran army overthrew him before he could implement the reform programme he had pushed through the congress. Military entrenchment was further solidified as Honduras fought a border war with El Salvador in 1969. The military has controlled the country's political life, directly or indirectly, ever since.

Hong Kong, British crown colony south-east of Guangzhou (Canton) on the coast of China, consisting of Hong Kong Island, Kowloon, and the New Territories (an area comprising some mainland territory and many small islands). Hong Kong Island was occupied by the British in 1841 during the First *Opium War and was formally ceded by China in the Treaty of *Nanjing in the following year. The mainland peninsula of Kowloon was added by the Treaty of Beijing in 1860, and in 1898 the colony's hinterland was extended when the New Territories were leased from China for 99 years. Hong Kong grew as a trading centre, attracting both Europeans and Chinese. After two weeks' fighting it surrendered to the Japanese on 25 December 1941. Reoccupied by the British in 1945, there was an influx of refugees and capital, especially from Shanghai, following the communist victory in China. The United Nations embargo on trade with China during the Korean War stimulated the development of industry and financial institutions, and in the 1970s and 1980s Hong Kong became an important international economic and business centre. With the end of the lease on the New Territories approaching, Britain agreed in 1984 to transfer sovereignty of the entire colony to China in 1997. China undertook not to alter Hong Kong's existing economic and social structure for fifty years, but there remains some unease within Hong Kong.

Hoover, Herbert (Clark) (1874–1964), thirty-first President of the USA (1929–33). A successful mining engineer and businessman, Hoover earned a reputation as a humanitarian, organizing the production and distribution of foodstuffs in the USA and Europe, during and after World War I. As Secretary of Commerce (1921-8), he persuaded large firms to adopt standardization of production goods and a system of planned economy. Esteemed as a moderate liberal, he received the Republican nomination for President and easily defeated his Democratic rival, Alfred E. Smith, in 1928. However, his presidency was marked by his failure to prevent the Great *Depression, following the *Stock Market crash of 1929. He ran for re-election in 1932, but was overwhelmingly defeated by Franklin D. *Roosevelt. Long after his electoral rout he became a respected elder statesman, co-ordinator of the European Food Program (1947), and chairman of two executive reorganization commissions (1947-9, 1953-5). Many of his recommendations were adopted, including the establishment of the Department of Health, Education, and Welfare.

Hoover, J(ohn) Edgar (1895-1972), US administrator. In 1924 he was appointed director of the Bureau of Investigation now the *Federal Bureau of Investigation with the task of raising its standards after the disrepute of the *Harding years. He achieved this by vigorous selection and training of personnel, and the creation of a scientific crime detection laboratory and the FBI National Academy. In the 1930s his widely publicized entrapment of certain criminals, while not destroying syndicate crime, earned the FBI a reputation of integrity, but in his late years a persistent interest in the sex lives of various public figures clouded his supposed political impartiality, and his antipathy to *civil rights activities earned widespread criticism.

Hopkins, Harry (Lloyd) (1890-1946), US administrator and public servant. When Franklin D. *Roosevelt was governor of New York state he made Hopkins his adviser on social and welfare policies. This continued throughout the years of the *New Deal. Hopkins' record in public social service was, perhaps, without equal: head of New York's Temporary Emergency Relief Administration (1931); the Federal Emergency Relief Administration (1933); the *Works Projects Administration (1933); and Secretary of Commerce (1938-40). He was Roosevelt's manager when he ran for a third term as President in 1940. Before and after US entry into the war Hopkins served as Roosevelt's untitled second-in-command. He played a pivotal role in the San Francisco conference of 1945, which launched the charter

FBI chief J. Edgar **Hoover** at the Senate Crime Investigation Committee in Washington in 1951, where he explained his solution to the US crime problem. The hearings were televised and attracted much publicity.

of the *United Nations, and in the last 'big three' conference at *Potsdam.

Horthy de Nagybánya, Nikolaus (1868-1957), regent of Hungary. He commanded the Austro-Hungarian fleet in World War I. In 1919 he was asked by the opposition to organize an army to overthrow Béla *Kun's communist regime. In January 1920 the Hungarian Parliament voted to restore the monarchy, electing Horthy regent. This post he retained, but thwarted all efforts of Charles IV, the deposed King of Hungary, to support the *Habsburg claim. He sought to maintain the established social order and ruled virtually as dictator. He agreed to Hungary joining Germany in World War II and declared war on the Soviet Union, but in 1944 he unsuccessfully sought a separate peace with the Allies. He was imprisoned by the Germans (1944) and released by the Allies (1945).

Houphouét-Boigny, Félix (1905-), African statesman. In 1944 he was a co-founder of the Syndicat Agricole Africain, formed to protect Africans against European agriculturalists. He represented the *Ivory Coast in the French Assembly (1945-59), and in 1946 formed the Parti Démocratique de la Côte d'Ivoire. At first allied with the Communist Party, he broke with it in 1950, and co-operated with the French to build up the economy of his country. When Ivory Coast was offered independence in 1958, he campaigned successfully for self-government within the *French Community. He became President of the Ivory Coast in 1960 in a one-party state, and his international policies have been recognizably moderate.

Houston, Samuel (1793-1863), US military leader and statesman. Houston lived for a time with the Cherokee Indians and became a popular hero as a result of his exploits, while serving under Andrew *Jackson, against the Creek Indians in 1814. In 1835-6, as military leader of the Texan insurgents, he defeated and captured *Santa Anna at the Battle of *San Jacinto, thus securing Texan independence. He was twice President of the *Texan Republic (1836-8, 1841-4), and after Texas joined the United States served in the Senate from 1847 to 1859. In 1859 his local popularity won him election as governor of his adopted state, but when Texas joined the *Confederacy in 1861, Houston, a strong supporter of the Union, was deposed.

Hua Guofeng (or Hua Kuo-feng) (1920-), Chinese statesman. He served for twelve years with the 8th Route Army before rising through the provincial bureaucracy to become deputy governor of Hunan province. The leading provincial official to survive the *Cultural Revolution, Hua won a succession of key posts between 1968 and 1975, becoming acting Premier after the death of *Zhou Enlai in 1976. He succeeded *Mao Zedong as chairman of the Central Committee, having defeated a challenge from the *Gang of Four. His appointment only disguised the power struggle in which *Deng Xiaoping was emerging as the victor. Hua resigned as Premier in 1980 and as chairman in 1981.

Huerta, Victoriano (1854-1916), Mexican statesman and general. Appointed (1912) as commander of the federal forces, he became President of Mexico (1913-14) by leading a coup against Francisco *Madero. He instituted a ruthless dictatorship in which torture and assassination of his political opponents became commonplace. He was forced to resign under insurgent military pressure, supported not only by his Mexican opponents but also by the government of Woodrow *Wilson in the USA. An attempted return to power in 1915 ended unsuccessfully when he was arrested by US agents while attempting to cross the US–Mexican border.

Huggins, Godfrey (Martin), 1st Viscount Malvern (1883–1971), Rhodesian statesman. As leader, first of the Reform Party and then of the United Party, he was Prime Minister of Southern Rhodesia (1933–53). He served as Prime Minister of the *Central African Federation from 1953 until 1956. As a white Rhodesian he was saddened in his later years to see the Federation break up (1963), and the emergence of racial intolerance.

Hughes, Charles Evans (1862–1948), US jurist and statesman. He fought, and lost, the presidential election of 1916 as Republican candidate against Woodrow *Wilson. As Harding's Secretary of State (1921–5), he organized the successful Washington conference on naval limitation (1921–2). His career was crowned by his years as Chief Justice of the US Supreme Court (1930–41), during which he enhanced the efficiency of the federal court system, and gave firm support to the freedoms guaranteed to citizens against state actions under the First Amendment. He was largely instrumental in defending the Supreme Court against a plan of F. D. *Roosevelt (1937) to 'pack' it by adding judges.

Hughes, William M(orris) (1862–1952), Australian statesman. He became a Labor Member of the House of Representatives in 1901, where he served for over fifty years. He was Prime Minister of the Commonwealth of Australia (1915–24), and represented Labor until it split in 1916. In 1917 he helped to form the Nationalist Party, which he led until 1923, when the Country Party, now in coalition with the Nationalists, refused to serve with him. In 1929, he gave his support to the Labor Party over a matter of industrial relations and the Nationalists expelled him. The United Australia Party, which he led from 1941 until 1943, was transformed by *Menzies into the Liberal Party in 1944 and Hughes became a back-bench member.

Hukbalahap, Filipino peasant resistance movement with roots in the pre-war *barangay* (village) and tenant organizations in central Luzon. Led by Luis Taruc, the movement developed during World War II into the Anti-Japanese People's Army, a left-wing guerrilla organization which was as much opposed to the Filipino landlord élite and their US backers as to the Japanese. Active against the latter from 1943, the 'Huks' controlled most of central Luzon by the end of the war, but were denied parliamentary representation and went into open rebellion against the Manila government until all but destroyed by government forces between 1950 and 1954.

Hull, Cordell (1871–1955), US statesman. As Secretary of State (1933–44), he achieved a progressive tariff by the Reciprocal Trade Agreements Act of 1934. His *Good Neighbor policy resulted in the US withdrawal of marines from Haiti (1934) and the cancellation of the *Platt amendment. He worked steadily for modification of the *isolationist Neutrality Acts (1935–7). F. D. Roosevelt, however, found him too cautious for his purposes and continually by-passed him in planning wartime policies.

Hume, Joseph (1777–1855), British radical politician and advocate of social reform. Associated with the radical reform group led by Francis *Place, he entered Parliament in 1812, re-entered as a Radical MP in 1818, and soon made a name for himself by his ruthless scrutiny of government expenditure. He played an important part in securing the repeal of the *Combination Acts (1824) and campaigned for the abolition of flogging in the army.

'Hundred Days' (20 March–28 June 1815), name given to the period between *Napoleon's return from the island of Elba and the date of the second restoration of *Louis XVIII. Napoleon landed at Cannes on 1 March while the European powers were meeting at the Congress of *Vienna. He won great popular acclaim as he moved north through Grenoble and Lyons. He arrived in Paris on 20 March, less than twenty-four hours after Louis had fled. Napoleon's attempt to win over moderate royalist opinion to a more liberal conception of his empire failed. Moreover, he failed to persuade the Allies of his peaceful intentions, and had to prepare to defend France against a hastily reconstituted 'Grand Alliance'. By the end of April he had only raised a total strength of 105,000 troops, the Allies having a force of almost 130,000 men. Nevertheless, Napoleon took the offensive and forced the Prussians to retreat at Ligny. Two days later, on 18 June, Napoleon was defeated at *Waterloo. He returned to Paris and on 22 June abdicated for the second time. Six days later Louis XVIII was restored to power.

Hundred Days Reform (1898), Chinese reform campaign. Inspired by *Kang Youwei, and supported by Liang Qichao (1873–1921), it attempted to reform the *Qing state. Kang utilized official disenchantment with the measures of the *Self-Strengthening Movement and concern at renewed foreign intrusions in the wake of the *Sino-Japanese War of 1894–5 to have extensive reforms, based on western thinking, adopted by Emperor Guangxu. These included a constitutional monarchy and modernization of the civil service. After a period of 103 days the reform programme was destroyed by a conservative backlash, the empress dowager *Cixi launching a palace coup in which the emperor was imprisoned, the reforms rescinded, and the reformers themselves exiled, dismissed, or put to death. Many of these reforms were finally implemented between the *Russo-Japanese War of 1904–5 and the *Revolution of 1911.

Hundred Flowers Movement (1956–7), Chinese political and intellectual debate. Drawing its name from a slogan from Chinese classical history, 'let a hundred flowers bloom and a hundred schools of thought contend', the campaign was initiated by *Mao Zedong and others in the wake of Khrushchev's denunciation of Stalin. Mao argued that self-criticism would benefit China's development. After some hesitation, denunciation of the Communist Party and its institutions appeared in the press and there was social unrest. The party reacted by attacking its critics and exiling many to distant areas of the country in the Anti-Rightist Campaign.

Children in a Budapest street take over a gun abandoned by the Soviet army during the **Hungarian Revolution**. Soon, however, the Russians were back in Hungary, to put down the uprising.

Hungarian Revolution (1956), a revolt in Hungary. It was provoked by the presence in the country of Soviet troops, the repressive nature of the government led by Erno Gerö, and the general atmosphere of de-Stalinization created at the *Twentieth Congress of the CPSU. Initial demonstrations in Budapest led to the arrival of Soviet tanks in the city, which served only to exacerbate discontent, Hungarian soldiers joining the uprising. Soviet forces were then withdrawn. Imre *Nagy became Prime Minister, appointed non-communists to his coalition, announced Hungary's withdrawal from the *Warsaw Pact, and sought a neutral status for the country. This was unacceptable to the Soviet Union. Powerful, mainly Soviet but some Hungarian, forces attacked Budapest. Resistance in the capital was soon overcome. Nagy was replaced by János *Kádár, while 190,000 Hungarians fled into exile. The Soviet Union reneged on its pledge of safe conduct, handing Nagy and other prominent figures over to the new Hungarian regime, which executed them in secret.

Hungary, a country in central Europe. In the 19th century Magyar nationalism was antagonized by the repressive policies of *Metternich, leading to rebellion under *Kossuth in 1848. The Austrians, with Russian help, reasserted control. After defeat by *Prussia the Austrians compromised with the Magyars in 1867, setting up the *Austro-Hungarian empire, or Dual Monarchy, which was first and foremost an alliance of Magyars and Austrian Germans against the Slav nationalities. Defeat in World War I led to revolution and independence, first under Károlyi's democratic republic, then briefly under Béla *Kun's communist regime. Dictatorship followed in 1920 under *Horthy, and lasted until 1944. Allied to the *Axis Powers in World War II, defeat brought Soviet domination and a communist one-party system. This was resented and, briefly, in 1956, the *Hungarian Revolution saw resistance to the Soviet Union. Though brutally crushed, the Hungarians have since, under the leadership of János *Kádár, been able to liberalize their regime, while remaining in the Soviet bloc.

Hungry Forties, a period in the early 1840s in Britain when the country experienced an economic depression, which caused much misery among the poor. In 1839 there was a serious slump in trade, leading to a steep increase in unemployment, accompanied by a bad harvest. The bad harvests were repeated in the two following years and the sufferings of the people, in a rapidly increasing population, were made worse by the fact that the *Corn Laws seemed to keep the price of bread artificially high. In 1845 potato blight appeared in England, spreading to Ireland later in the year and ruining a large part of the crop. The potato blight returned in 1846, bringing the *Irish Famine.

Hunt, Henry ('Orator') (1773-1835), British political reformer. He advocated, among other things, full adult suffrage and secret ballots. An outstanding public speaker, in August 1819 he addressed the crowd at the great meeting at St Peter's Fields (*Peterloo Massacre), Manchester. For this he was subsequently sentenced to two years' imprisonment. During 1830-5 he was Radical Member of Parliament for Preston.

Huskisson, William (1770-1830), British statesman. He encouraged the incipient movement towards *free trade in Britain. A supporter of William *Pitt the Younger, he held a number of minor posts before being appointed President of the Board of Trade in 1823. He reduced duties on a wide range of articles, removed restrictions on colonial trade, and proposed a relaxation of the *Corn Laws. He was killed by a train at the opening of the Liverpool and Manchester Railway.

Hussein, ibn Ali (1856-1931), Arab political leader. A member of the Hashemite family, he was sharif of Mecca and leader of the 1916 Arab revolt. In 1916 he assumed the title of King of the Arab Countries, but the Allies only recognized him as King of the *Hejaz. As ruler of the Hejaz (1916-24) he came into conflict in 1919 with Ibn *Saud, the original ruler of Najd. He abdicated in favour of his son Ali in October 1924. His son Abdullah became ruler of Trans-Jordan, and another son, *Faisal I, founded the royal line of Iraq.

Hyderabad, one of the largest and most important *Princely States in south-central India. Its rise to dominance was the achievement of the Mogul viceroy, Asaf Jah Nizam ul-Mulk, who in 1724 established virtually independent rule. His successors ruled until 1948 as nizams (governors) of Hyderabad. Absolute power was short-lived, however, for inability to challenge European expansion soon dictated co-operation with the British, whose protection was extended to Hyderabad in 1798 in return for the upkeep of *East India Company troops. Although internal control was retained, British influence became dominant in the 19th century. On British withdrawal from India in 1947, the nizam acceded to the Indian Union. In 1956 the territories of Hyderabad were divided among the new linguistically based states.

I

Ibn Saud *Saud.

Iceland, an island country in the North Atlantic. Under the rule of Denmark since 1380, a nationalist movement achieved the restoration of the Althing or parliament in 1845. Iceland acquired limited autonomy in 1874 and independence in 1918, although it shared its king with Denmark till 1943. It became an independent republic in 1944. An Allied base during *World War II, it joined the *United Nations and *NATO (1949), and has since engaged in sometimes violent dispute with Britain over fishing limits, resulting in the *'Cod War' of 1972-6.

Iguala, Plan of (1821), constitutional guarantees for an independent Mexico. The plan, proclaimed in the Mexican town of Iguala by the Creole leader *Iturbide, with the support of the guerrilla leader, Vincent Guerrero, provided that independent Mexico would be organized as a constitutional monarchy under Ferdinand VII or another European prince, that Roman Catholicism would be the state religion, and that any person, regardless of race, could hold office. The Viceroy Apodaca was deposed and his successor confirmed the guarantees by the convention of Córdoba, but the Spanish government rejected them. The Plan was discarded when Iturbide proclaimed himself emperor (1822).

Ikeda Hayato (1899-1965), Japanese statesman. He entered the government tax service and rose by 1945 to become head of the National Tax Bureau. Having served as Vice-Minister of Finance in the *Yoshida cabinet of 1947, he was elected to the House of Representatives in 1949 and became successively Minister of Finance and Minister of International Trade. Serving in a succession of high ministerial posts throughout the 1950s, Ikeda became Prime Minister (1960-4) and devoted himself to sustaining Japanese economic growth through a broadening of international trading connections.

IMF *International Monetary Fund.

imperialism, the policy of extending one country's influence over other, less developed and less powerful countries. The *Industrial Revolution introduced a new form of imperialism as European countries competed throughout the world both for raw materials and for markets. In the late 19th century imperial ambitions were motivated in part by the need for commercial expansion, the desire for military glory, and diplomatic advantage. Imperialism generally assumed a racial, intellectual, and spiritual superiority on the part of the newcomers. The effects of imperialism, while in some measure beneficial to the indigenous population, often meant the breakdown of traditional forms of life, the disruption of native civilization, and the imposition of new religious beliefs and social values. The dreams of imperialism faded in the 1920s as anti-imperialist movements developed, and from the 1940s colonies gained their independence. The French overseas territories became, with France, the *French Community, and the *British empire formed the *Commonwealth of Nations. In post-war years the phenomenon of neo-imperialism has emerged, in which their critics claim that the developed countries, including the USA and the Soviet Union, largely control the economic development of the Third World through restrictive trading practices, monopolies, and development loans.

Imperial Preference *tariff reform.

incident of 26 February 1936, attempted military coup in Japan. Young extremists of the Imperial Way faction (Kodo-ha) had been active within the Japanese army since the late 1920s, intent on using violent means to overthrow the conservative civilian government and set Japan on a course of military expansion, particularly in China. Their activities culminated in the attempted coup in which several prominent politicians were murdered (the Prime Minister Okada Keisuke only escaping through a case of mistaken identity) and much of central Tokyo seized. The revolt was put down on 29 February and most of its leaders executed, after which leadership of the military expansionist cause passed to the more moderate Control faction (Tosei-ha).

income tax, a tax levied on the income of an individual. It was imposed on the incomes of the propertied classes for the first time in Britain in December 1798 by William *Pitt the Younger, to help finance the war with France. It was temporarily abolished by Parliament against the government's wishes in 1816. It was revived in 1842 by *Peel in exchange for a reduction in customs and excise duties. In the 1850s *Gladstone planned to abolish it gradually, but was in fact obliged to raise it to finance the Crimean War, and by the 1860s it had become accepted as a permanent necessity on higher incomes. In the 20th century tax liability tended to reach further down the social scale so that by World War II most full-time employees were taxed. Income taxes were introduced in some European countries during the 19th century and in the British dominions from 1891. In the USA federal income tax was imposed in 1862 to help finance the American Civil War. An attempt by the federal government to reintroduce it in 1894 was declared unconstitutional by the Supreme Court, and a federal tax did not become effective until 1913. By then most individual US states had their own income tax.

Independent Labour Party (ILP), British socialist organization. It was founded at Bradford in 1893 under the leadership of Keir *Hardie. Its aim was to achieve equality in society by the application of socialist doctrines. The ILP was one of the constituent groups of the Labour Representation Committee (1900), which in 1906 became the *Labour Party. A split developed between the ILP and the Labour Party between the two World Wars. The sympathy of the ILP for communism, its pacifism, and its theoretical approach to politics were regarded as electoral liabilities by leading Labour politicians; from 1939 its influence declined.

India, the greater part of the subcontinent of South Asia. By 1800, the English *East India Company had emerged as the paramount power in India. The *Maratha Wars (1803-18) and the annexation of the *Punjab

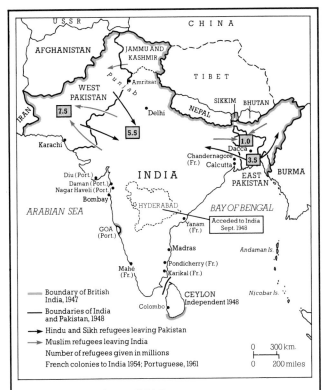

Indian subcontinent (1948)

The Indian subcontinent was steadily subjected to British control in the first half of the 19th century. Independence in 1947 created Pakistan (largely Muslim) and the Union of India (largely Hindu). The latter is now a federal republic of twenty-five states and six Union territories, organized on an ethnic basis. In 1971 East Pakistan became an independent country, as Bangladesh.

(1848-9) extended its power to all parts of the subcontinent. A mutiny of the Company's Indian soldiers (*Indian Mutiny) followed by a rebellion of the civil population in several parts of India in 1857 led to the assumption of power directly by the British government. Administrative consolidation, new systems of land tenure, modern irrigation, and transport and communication systems, initiated in the 1830s, were developed further.

The development of western-style higher education began in 1817 with the establishment of the Hindu College, Calcutta. The new western-educated middle classes were attracted by the ideology of nationalism and liberal democracy. Initially enthusiastic about British rule, they became increasingly critical, for example in social reform movements like the *Brahmo Samaj and the *Arya Samaj. The Indian National *Congress, established in 1885, provided an all-India forum for political activity. The government, anxious to ensure co-operation of politicized Indians, provided for limited association of representative Indians within the legislatures by the Councils Act of 1909, promised 'progressive realization of responsible government' in 1917, and transferred some responsibilities to elected ministers in the provinces by the Government of *India Act in 1919. Agitation organized by Mohandras *Gandhi against a bill for suppression of sedition led to the notorious massacre at *Amritsar. The campaign of *satyagraha and *non-co-operation launched by Gandhi was aimed

at achieving *swaraj (self-government) and had the support of the *Khilafat movement. The Civil Disobedience Movement (1930-4) demanding independence, and the 'Quit India' Movement, which followed the arrest of Gandhi and other leaders in 1942, consolidated the popular support for the Congress. After World War II, the British opened negotiations for transfer of power. Ever since the 1880s, politicized Muslims were anxious to protect their interests against possible encroachment by a Hindu majority. The *Muslim League, founded in 1905, co-operated with the Congress in 1916 and the Khilafat agitation, but after 1937 emphasized the Muslims' separate aspirations and demanded a separate Muslim homeland, Pakistan, in 1940. Under M. A. *Jinnah's leadership the League gained the support of the majority of Muslims. The demand for *Pakistan was conceded and a separate state created in 1947 comprising the Muslim majority areas in north-western and eastern India.

The two states of India and Pakistan fell out over the accession of Kashmir to India and have fought three wars, the last leading to the secession of East Pakistan (1971) as *Bangladesh. In foreign policy, Pakistan has been closely associated with the USA, especially since the Soviet intervention in Afghanistan, while India, despite its emphasis on non-alignment, has had a close relationship with the Soviet Union since the 1971 war. Despite a measure of economic growth, both nations have problems of mass poverty, a high rate of population growth, and several ethnic groups within their borders aspiring towards autonomy.

India, Union of, a country comprising the territories of the British Indian Provinces minus the Muslim majority areas in the north-west and the east. It is one of the two (now, with *Bangladesh, three) successor states of Britain's Indian empire. Established in 1947, the Union opted to remain within the *Commonwealth even though it adopted a republican constitution. The *Princely States within the boundaries of the Indian Union plus *Kashmir all acceded to the Union, though pressure had to be used in some instances, especially Travancore-Cochin and *Hyderabad. Eventually the Princely States were integrated or set up as separate states. The French voluntarily surrendered their few possessions in India, while the Portuguese territories agitating for accession were integrated through military action. The semi-autonomous state of Sikkim was absorbed into India through political pressure but without bloodshed. *Pakistan's claims over Kashmir, the bulk of which is formally integrated with India, remain a source of dispute. India is a federation of 25 states and 6 Union territories organized primarily on a linguistic basis. Since independence it has had three wars with Pakistan and one with China, and the relationship with *Sri Lanka is strained by the Indian Tamils' support for the Sri Lankan Tamils' movement for autonomy. The Sikh demand for autonomy and their terrorist action remain intractable problems in the Punjab. India's first Prime Minister was Jawaharlal *Nehru (1947-64), who initiated a policy of planned economic growth and non-alignment. Indira *Gandhi, his daughter, became Prime Minister in 1966. After splitting the *Congress Party and experimenting with autocratic rule (1975-7) she suffered electoral defeat. She returned to power (1979) and was assassinated by a Sikh (1984). Her son, Rajiv Gandhi (1944-), succeeded her.

India Acts, British parliamentary Acts for the government of India. The Act for the Better Government of India (1858) replaced rule by the English *East India Company by that of the crown. The viceroy would be assisted by a Council, which from 1861 was to have Indian as well as European members. The India Act (1909) allowed Indians a share in the work of legislative councils (*Morley–Minto Reforms). The Government of India Act (1919) following the *Montagu–Chelmsford proposals, established a two-chamber legislature at the centre, enlarged provincial legislatures, and gave both an elected majority. Central government remained under the control of the viceroy's Executive Council, but in the provinces a measure of self-government was conceded through the system known as *dyarchy. The Government of India Act (1935) separated Burma from India, and provided for provincial autonomy in British India, a federation of Indian princes, and for a dual system of government at the centre based on the principle of dyarchy. The provisions of this Act were never fully implemented.

Indian Mutiny (1857–8), an uprising against British rule in India. It began as a mutiny of Indian sepoys in the army of the English *East India Company, commencing at Meerut on 10 May 1857, and spreading rapidly to Delhi and including most regiments of the Bengal army as well as a large section of the civil population in Uttar Pradesh and Madhya Pradesh. The immediate cause was the soldiers' refusal to handle new cartridges apparently greased with pig and cow fat (an outrage to Muslims and Hindus respectively). The rapid introduction of European civilization coupled with harsh land policies carried out by governor-general Dalhousie and his successor Lord Canning were contributing factors to its rapid spread into the civilian population, to become a full-scale civil rebellion. The mutineers seized Delhi. The rebels restored the former Mogul Emperor Bahadur Shah II to his throne, whereupon the movement spread to *Lucknow, which was besieged, and to Cawnpore (now Kanpur), where the massacre of the British garrison is believed to have been instigated by Tantia Topi, a Maratha Brahman who became the military leader of the rebels. The recapture of Delhi by forces from the Punjab on 14 September 1857 broke the back of the mutiny. The fighting was marked by atrocities on both sides. The rebels were ruthlessly crushed in 1858 in a series of campaigns by Sir Hugh Rose in central India. These included the defeat of Lakshmi Bai, the Rani of Jhansi, who was killed in battle. Bahadur Shah was exiled and many civilians executed without trial. Tantia Topi became a fugitive, but was betrayed and executed. Following the restoration of British control, the East India Company's rule was replaced by that of the crown.

Indian National Congress *Congress, Indian National.

Indian reservations, land set aside in the USA for the occupancy and use of Indian tribes. The reservations were first created by a policy inaugurated in 1786. President *Jackson first practised removal to reservations on a large scale after Congress passed the Indian Removal Act of 1830. This sent the Creek, *Seminole, Chickasaw, Choctaw, and *Cherokee tribes to an Indian territory in modern *Oklahoma. In all some 200 reservations were set up in over forty states. All but a handful proved economically unviable, so furthering the Indians' poverty.

A cavalry encounter between British-officered troopers of the unit Hodson's Horse, raised by Lt W. S. R. Hodson, and their opponents at Rhotuch during the **Indian Mutiny**. A lithograph after Atkinson's *Campaign in India 1857–8*. (National Army Museum, London)

Indo-Chinese War (20 October–22 November 1962), a border skirmish between India and China in the Himalayan region, which China claimed had been wrongly given to India by the *McMahon decision in 1914. Chinese forces began an offensive across the McMahon Line into India. Indian forces retreated and Assam appeared to be at the mercy of China, when the latter announced a cease-fire and withdrew to the Tibetan side of the Line, while retaining parts of Ladakh in Kashmir. Some of the border areas are still disputed.

Indonesia, south-east Asian country. Partly administered by the Dutch East India Company from the 17th century, the islands were formed into the Netherlands-Indies in 1914. By the 1920s, indigenous political movements were demanding complete independence. Prominent here was *Sukarno's Indonesian Nationalist Party (*Partai Nasionalis Indonesia*), banned by the Dutch in the 1930s. The Japanese occupation of 1942–5 strengthened nationalist sentiments, and, taking advantage of the Japanese defeat in 1945, Sukarno proclaimed Indonesian independence and set up a republican government. Dutch attempts to reassert control were met with popular opposition (the *Indonesian Revolution), which resulted in the transfer of power in 1949. By 1957 parliamentary democracy had given way to the semi-dictatorship or 'Guided Democracy' of President Sukarno, a regime based on the original 1945 constitution, with a strong executive and special powers reserved for the army and bureaucracy. Sukarno's popularity began to wane after 1963, with the army and right-wing Muslim landlords becoming increasingly concerned about the influence of communists in government. Rampant inflation and peasant unrest brought the country to the brink of collapse in 1965–6 when the army under General *Suharto took advantage of a bungled coup by leftist officers to carry out a bloody purge of the Communist Party (*PKI) and depose Sukarno (1967). Despite his initial success in rebuilding the economy and restoring credit with its Western capitalist backers, Suharto's regime has become increasingly authoritarian and repressive, moving ruthlessly against domestic political opponents, particularly members of fundamentalist Islamic groups.

Indonesian Revolution (1945–9), nationalist struggle for independence from Dutch rule in *Indonesia. In 1945, *Sukarno proclaimed Indonesia's independence. Attempts by the Dutch to re-establish their pre-war colonial administration led to sporadic fighting which was temporarily brought to an end by a compromise agreement signed in 1946. This provided for the establishment of a United States of Indonesia tied to the Netherlands under a federal constitution. But the nationalists refused to accept this, forcing the Dutch to launch a new offensive which recaptured most of the estate areas and ended in a cease-fire in 1947. A second Dutch 'police action' a year later, which captured the republican capital and took Sukarno and most of his cabinet prisoners, increased international pressure and forced the Dutch to convene a conference at The Hague in 1963. As a result, all of the Dutch East Indies, with the exception of western New Guinea, were transferred to the new state of Indonesia in 1949. Western New Guinea (now Irian Jaya) came under Indonesian administration in 1963.

Indo-Pakistan War (September 1965), a border conflict between India and Pakistan. The main cause of the war was an attempt by Pakistan to assist Muslim opponents of Indian rule in Kashmir. Fighting spread to the Punjab, which was the scene of major tank battles. A UN cease-fire was accepted and by the Tashkent Declaration of 11 January 1966, a troop withdrawal was agreed. A brief renewal of frontier fighting occurred in 1971, at a time when *Bangladesh was seeking independence from Pakistan.

industrialization, the process of change from a basic agrarian economy to an industrialized one. It was first experienced by Britain in the *Industrial Revolution and at much the same time in the New England states of the USA, from which it spread along the eastern seaboard and, after the American Civil War, across the continent. Belgium was the first continental European country to experience industrialization, which then spread to north-east France and, particularly after 1870, to Germany, where its growth was so rapid that by 1900 German industrial production had surpassed that of Britain. During those thirty years all industrialized nations saw rapid development and expansion in such heavy industries as iron and steel, chemicals, engineering, and shipbuilding. Japan was the first non-European power to become industrialized, which it had done by the end of

The process of **industrialization** began in 19th-century Britain, and the population of cities grew to meet the demands of factories for labour. This detail from a view of Sheffield, painted by William Ibbit in 1854, shows the mixture of housing, factories, and forms of transport that characterized industrial cities. Sheffield had been a centre of cutlery manufacture since the Middle Ages, and expanded as a result of new technologies that enabled goods to be mass produced. (Sheffield City Museum)

the 19th century, while many others, for example India and China, have become increasingly industrialized in the 20th century. The Soviet Union saw industrialization on a massive scale under Stalin. A country is described as 'industrialized' when there are high levels of manufacturing and service industries.

Industrial Revolution (*c.*1750–*c.*1850), the term used to describe the change in the organization of manufacturing industry which transformed first Britain, then other countries, from rural to urban economies. The process began in England as a result of a combination of economic, political, and social factors, including internal peace and availability of capital. Preceded by major changes in agriculture, which freed workers for the factories, it was caused by the rise of modern industrial methods, with *steam power replacing the use of muscle, wind, and water power, the growth of factories, and the mass production of manufactured goods. The textile industry was the prime example of industrialization and created a demand for machines, and for tools for their manufacture, which stimulated further mechanization. Improved transport was needed, provided by canals, roads, and railways, and the skills acquired were exported to other countries. It made Britain the richest and most powerful nation in the world by the middle of the 19th century.

Simultaneously it radically changed the face of British society, throwing up large cities, particularly in the Midlands, the North, Scotland, and South Wales, as the population shifted from the countryside, and causing a series of social and economic problems, the result of low wages, slum housing, and the use of child labour.

The term Second Industrial Revolution has been used to describe the *Technological Revolution since World War II.

Inkerman, battle of (5 November 1854). It took place during the *Crimean War near *Sevastopol when the Russian army, led by Menshikov, launched a surprise attack on a portion of the besieging French and British armies. The attack was finally repulsed, but the French and British preparations for an assault were destroyed. The siege of Sevastopol continued, the city finally capitulating in September 1855.

Inönü, Ismet (1884–1973), Turkish soldier and statesman. He served against the Greeks during the Turkish war of independence (1919–22). He was chosen by *Atatürk as first Prime Minister of the Turkish republic (1923–37) and in 1938 succeeded him as President, remaining in power until the Democrat victory of 1950. Inönü remained leader of the Republican People's Party and served again as Prime Minister (1961–5) in the aftermath of the 1960 military coup.

International Brigades, international groups of volunteers in the *Spanish Civil War. They were largely communist, on the side of the republic against *Franco. Organized by the *Comintern, their members were largely working people together with a number of intellectuals and writers, such as the English poet W. H. Auden and the writer George Orwell. At no time were there more than 20,000 in the Brigades. They fought mainly in the defence of Madrid (1936) and in the battle of the River Ebro (1938).

International Court of Justice, a judicial court of the United Nations which replaced the Cour Permanente de Justice in 1945 and meets at The Hague. The General Assembly of the UN accepted (1948) the Universal Declaration of Human Rights. By 1966 two international covenants—that on civil and political rights and that on economic, social, and cultural rights—were promulgated. Appeals to the court, based on these covenants, are enforceable only if the nation concerned has previously agreed to be bound by its decisions.

International Labour Organization (ILO), an agency founded in 1919 to improve labour and living standards throughout the world. Affiliated to the *League of Nations until 1945, it has sought to improve labour conditions, promote a higher standard of living, and further social justice. Affiliated since 1946 to the United Nations, it has become increasingly concerned with human rights and the provision of technical assistance to developing countries.

International Monetary Fund (IMF), an international agency linked to the *United Nations. Proposed at the *Bretton Woods Conference in 1944 and constituted in 1946, it was designed to assist the expansion of world trade by securing international financial co-operation and stabilizing exchange rates. Member countries subscribe funds in accordance with their wealth; these provide a reserve on which they may draw (on certain conditions) to meet foreign obligations during periods of economic difficulty.

Internationals, associations formed to unite socialist and communist organizations throughout the world. There were four Internationals. The First (1864), at which *Marx was a leading figure, met in London but was riven by disputes between Marxists and *anarchists. By 1872 it had become clear that divisions were irreconcilable and it was disbanded (1876). The Second,

Men of the British battalion of the **International Brigades** in Spain. Their leader, Sam Wild (*centre front*), had served in the Royal Navy and was a longtime communist who joined the Brigades in 1936. He commanded the British battalion from February 1938 until its return to Britain in December of that year.

or Socialist (1889), International aimed at uniting the numerous new socialist parties that had sprung up in Europe. With headquarters in Brussels, it was better organized and by 1912 it contained representatives from all European countries and also from the USA, Canada, and Japan. It did not survive the outbreak of World War I, when its plan to prevent war by general strike and revolution was swamped by a wave of nationalism in all countries. The Third, usually known as the Communist International or *Comintern (1919), was founded by *Lenin and the *Bolsheviks to promote world revolution and a world communist state. It drew up the Twenty-One Points of pure communist doctrine to be accepted by all seeking membership. This resulted in splits between communist parties, which accepted the Points, and socialist parties, which did not. The Comintern increasingly became an instrument of the Soviet Union's foreign policy. In 1943 Stalin disbanded it. The Fourth International (1938), of comparatively little importance, was founded by *Trotsky and his followers in opposition to *Stalin. After Trotsky's assassination (1940) it was controlled by two Belgian communists, Pablo and Germain, whose bitter disagreements had by 1953 ended any effective action.

Invergordon mutiny (1931), a mutiny by sailors of the British Atlantic Fleet at the naval port on Cromarty Firth, Scotland. Severe pay cuts imposed by the *National Government led the ratings to refuse to go on duty. The cuts were slightly revised but foreign holders of sterling were alarmed; an Act suspending the *gold standard was rushed through Parliament, but the value of the pound fell by more than a quarter. The mutiny ended and the ratings' ringleaders were discharged from the navy.

Iqbal, Muhammad (1876–1938), Indian philosopher, poet, and political leader. He took an active part in politics in the Punjab and was President of the *Muslim League in 1930 when he advanced the idea of a separate Muslim state in north-west India, the beginning of the concept of Pakistan.

Iran (formerly Persia), a country in south-west Asia. Trade between Muslim countries and European powers had developed throughout the 19th century and both Russia and Britain were anxious to increase their influence over the *Qajar dynasty in Iran. In 1906 Muzaffar al-Din granted a constitution; his successor sought to suppress the *Majles* (Parliament) which had been granted, but was himself deposed. In 1901 oil concessions were granted to foreign companies to exploit what is estimated as one-tenth of the world's oil reserves. In 1909 the Anglo-Persian Oil Company (later BP) was founded and southern Iran came within Britain's sphere of influence, while Russia dominated northern Iran. Following the *Russian Revolution of 1917 British troops invaded Russia from Iran; at the end of this 'war of intervention' an Iranian officer, Reza Khan, emerged and seized power (1921), backed by the British. In 1924 he deposed the Qajar dynasty and proclaimed himself as *Reza Shah Pahlavi. In World War II Iran was occupied by British and Soviet forces and was used as a route for sending supplies to the Soviet Union. The Shah abdicated (1941) and was replaced by his son *Muhammad Reza Shah Pahlavi. It took him twenty years to establish political supremacy, during which time one of his Prime Ministers,

*Mussadegh, nationalized the Anglo-Iranian Oil Company. In 1961 the Shah initiated a land-reform scheme and a programme of modernization, the so-called 'White Revolution' (1963–71). The secularization of the state led *Islamic leaders such as *Khomeini into exile (1964), while popular discontent with secular Western, especially US, influence was masked by ever-rising oil revenues, which financed military repression, as well as industrialization. Riots in 1978 were followed by the imposition of martial law. Khomeini co-ordinated a rebellion from his exile in France. The fall and exile of the Shah in 1979 was followed by the return of Khomeini and the establishment of an Islamic Republic which proved strong enough to sustain the *Iran Hostage Crisis of 1979–81 and to fight the long and costly *Iran–Iraq War (1980–   ).

Irangate scandal   *Reagan presidency.

Iran Hostage Crisis (4 November 1979–20 January 1981), a prolonged crisis between *Iran and the USA. Followers of the Ayatollah Khomeini alleged US complicity in military plots to restore the Shah, *Muhammad Reza Pahlavi, and seized the US Embassy in the Iranian capital, Teheran, taking sixty-six US citizens hostage. All efforts of President *Carter to free the hostages failed, including economic measures and an abortive rescue bid by US helicopters in April 1980. The crisis dragged on until 20 January 1981, when Algeria successfully mediated, and the hostages were freed. It seriously weakened Carter's bid for presidential re-election in November 1980, and he lost to Ronald *Reagan.

Iran–Iraq War (Gulf War) (1980–   ), a border dispute between *Iran and *Iraq which developed into a war of international proportions. In 1980 President Saddam Hussein of Iraq abrogated the 1975 agreement granting Iran some 518 sq. km. (200 sq. miles) of border area to the north of the Shatt-al-Arab waterway in return for assurances by Iran to cease military assistance to the Kurdish minority in Iraq, which was fighting for independence. Calling for a revision of the agreement to the demarcation of the border along Shatt-al-Arab, a

Iraqi soldiers celebrate their initial success in the **Iran–Iraq War**, 1980. The Iraqi forces hoped for a swift victory against an enemy disorganized by revolution, but Iran's resistance halted Iraq's offensive and the conflict stiffened into a grinding war of attrition.

return to Arab ownership of the three islands in the Strait of Hormuz (seized by Iran in 1971), and for the granting of autonomy to minorities inside Iran, the Iraqi army engaged in a border skirmish in a disputed but relatively unimportant area, and followed this by an armoured assault into Iran's vital oil-producing region. The Iraqi offensive met strong Iranian resistance, and Iran has since recaptured territory from the Iraqis. In 1985 Iraqi planes destroyed a partially constructed nuclear power plant in Bushehr, followed by bombing of civilian targets which in turn led to Iranian shelling of Basra and Baghdad. The war, which is estimated to have cost up to 1.5 million lives, entered a new phase in 1987. Iran increased hostilities against commercial shipping in and around the Gulf, resulting in naval escorts being sent to the area by the USA and other nations to protect merchant ships against both Iranian and Iraqi attack, and to engage in counter-attack.

Iraq (ancient Mesopotamia), a country in the Middle East bordering on the Persian Gulf. Following the British *Mesopotamian Campaign in World War I, the country was occupied by Britain, who was then granted responsibility under a League of Nations *mandate (1920-32). In 1921 Britain offered to recognize amir Ahd Allah Faisal, son of *Hussein, sharif of Mecca, as King Faisal. British influence remained strong until the fall of the monarchy in 1958. Further political rivalries ended with the 1968 coup, which led to rapid economic and social modernization paid for by oil revenues and guided by the general principles of the *Ba'ath Socialist Party. A heterogeneous society, of many ethnic and religious groupings, Iraq has long been troubled by periodic struggles for independence for its *Kurds. It has often been isolated in Arab affairs by its assertiveness in foreign policy, though the long and bloody *Iran–Iraq War launched against Khomeini's Iran by President Saddam Hussein in 1980 received financial support from formerly critical monarchist Arab states.

Ireland, an island to the west of Great Britain. As a result of the *Act of Union (1801), Ireland lost its parliament and became subject to direct rule from London. In the 1840s, the failure of the potato crop resulted in the *Irish Famine. Continuing social and economic problems produced resentment against British rule and made the campaign for *Home Rule the dominant issue in domestic politics in the second half of the 19th century. The granting of Home Rule was delayed by the outbreak of World War I, and armed resistance to British rule finally broke out in the *Easter Rising of 1916. In 1920 the Government of Ireland Act provided for two Irish parliaments, one (*Stormont) for six of the counties of Ulster in the north and one for the remaining twenty-six counties of Ireland. The Anglo-Irish Treaty of 1921 suspended part of the 1920 Act: while *Northern Ireland remained part of the United Kingdom, the twenty-six counties gained separate dominion status as the *Irish Free State and in 1949 attained full independence as the Republic of *Ireland.

Ireland, Republic of, a western European country. After years of intermittent fighting, the Anglo-Irish Treaty of December 1921, concluded by Lloyd George with the *Sinn Fein leaders, gave separate *dominion status to Ireland (as the Irish Free State) with the exception of

six of the counties of Ulster, which formed the state of *Northern Ireland. Irish republicans led by *de Valera rejected the agreement and fought a civil war against the Irish Free State forces, but were defeated in 1923. After the *Fíanna Fáil Party victory in the election of 1932, de Valera began to sever the Irish Free State's remaining connections with Great Britain. In 1937 a new constitution established it as a sovereign state with an elected president; the power of the British crown was ended and the office of governor-general abolished. The title of Irish Free State was replaced by Ireland; in Irish, Eire. An agreement in 1938 ended the British occupation of certain naval bases in Ireland. Having remained neutral in World War II, Ireland left the *Commonwealth of Nations and was recognized as an independent republic in 1949. De Valera was elected president in 1959. He was succeeded as Taoiseach (prime minister) by Sean Lemass (1959–66) and Jack Lynch (1966–73). In 1973 Ireland joined the European Community and a *Fine Gael–Labour coalition led by Liam Cosgrave came to power. Subsequent governments have been controlled alternately by the Fíanna Fáil and the Fine Gael–Labour coalition. In November 1985 Ireland signed the Anglo-Irish Accord (the Hillsborough Agreement) giving the republic a consultative role in the government of Northern Ireland. The agreement thus ensured a role for the republic on behalf of the nationalist minority in the north. In the 1980s its economy has sharply declined, leading to unemployment and renewed emigration.

Irigoyen, Hipólito (1850–1933), Argentine statesman. A leader of the Radical Party, he was elected to the presidency (1916–22). He actively supported organized labour until a series of strikes in 1918 and 1919 threatened economic paralysis. He then turned against the union movement with the same enthusiasm that he had shown in previously supporting it. Sitting out one term, Irigoyen was elected for a second time in 1928. It was a bad period in Argentine history when corruption, continued labour unrest, and large budget deficits were all exacerbated by the Great *Depression. The Argentine military stepped in to overthrow Irigoyen in 1930.

Irgun (Hebrew, 'Irgun Zvai Leumi', National Military Organization, byname ETZEL), an underground *Zionist terrorist group active (1937–48) in Palestine against Arabs and later Britons. Under the leadership of Menachem *Begin from 1944, it carried out massacres of Arabs during the 1947–8 war, notably at Dir Yassin (9 April 1948), and blew up the King David Hotel in Jerusalem (22 July 1946), with the loss of ninety-one lives.

Irish Famine (1845–51), period of famine and unrest in Ireland. In 1845 blight affected the potato in Ireland and the crop failed, thus depriving the Irish of their staple food. Farmers could not pay their rents; often they were evicted and their cottages destroyed. Committees to organize relief works for such unemployed persons, together with soup kitchens, were set up, and, especially in the western counties, large numbers sought refuge in workhouses. Deaths from starvation were aggravated by an epidemic of typhus, from which some 350,000 died in the year 1846–7. The corn harvest in 1847 was good and, although the blight recurred, the worst of the famine was over. It is estimated that one million people died in

Ireland of starvation in the five years 1846–51 and another million emigrated to America or elsewhere.

Irish Free State   *Ireland, Republic of.

Irish Republican Army (IRA), terrorist organization fighting for a unified republican Ireland. Originally created by the *Fenian Brotherhood in the USA, it was revived by *Sinn Fein in 1919 as a nationalist armed force. Its first commander in Ireland was Michael *Collins and at one time Sean McBride was chief of staff. Since its establishment the IRA has been able to rely on support from sympathizers in the Irish-American community. Bomb explosions for which the IRA was held responsible occurred in England in 1939 and hundreds of its members were imprisoned. During World War II hundreds of members were interned without trial in Ireland. In 1956 violence erupted in *Northern Ireland and the IRA performed a series of border raids. Following violence against civil rights demonstrators and nationalists by both the IRA and *Ulster Unionists, the IRA split into Provisional and Official wings (1969). The Provisional IRA (PIRA) and the Irish National Liberation Army (INLA) have in recent years staged demonstrations and hunger strikes, military attacks, assassinations, and bombings in both Northern Ireland and Britain.

Irish Republican Brotherhood (IRB), a secret organization founded in Dublin in 1858 by James Stephens (1824–1901) to secure the creation of an independent Irish republic. It was closely linked with the *Fenian

Digging for the few potatoes not devastated by the fungal blight during the **Irish Famine**. The repeated failure of the crop caused terrible suffering to the Irish peasants, since their economy depended on the potato harvest.

An **Irish Republican Army** funeral in Northern Ireland. Hooded men in paramilitary uniform fire a salute over the coffin draped with the flag of the Irish republic. The IRA has waged an unsuccessful campaign to gain legal recognition as a political rather than a criminal force in Northern Ireland.

Brotherhood in the USA and its members came to be called Fenians. The primary object of the IRB was to organize an uprising in Ireland; the Fenian Brotherhood worked to support the IRB with men, funds, and a secure base. The British government acted swiftly; IRB leaders including Stephens were arrested. The 1867 Fenian Rising, led by Thomas Kelly, was a failure. The *Home Rule League, the *Land League, the Irish Volunteers, and *Sinn Fein often appeared to supersede the IRB as political forces; but Fenians were active in all these organizations. The Home Rule Bills failed to satisfy them and in World War I the IRB led by Pádraic Pearse, sought German help for the abortive *Easter Rising.

Ironclads, name given to the first wooden battleships protected by armour-plating. As a result of the loss of French and British wooden battleships during the *Crimean War, the French government ordered the construction of five armour-plated vessels for service in the Black Sea, the first entering service in 1859. In 1862, during the *American Civil War, the first Ironclad battle, between the *Monitor and Merrimack took place. The design was quickly adopted by most nations until succeeded by steel-framed, *Dreadnought-type battleships at the beginning of the 20th century.

Iron Curtain, popular description of the frontier between East European countries dependent on the Soviet Union and Western non-communist countries. Its application to countries within the Soviet sphere of influence originates in a leading article by *Goebbels in Das Reich, February 1945. This was reported in British newspapers, and the phrase was first used by Churchill in a cable to President Truman four months later, 'I view with profound misgivings . . . the descent of an iron curtain between us and everything to the eastward.'

Irredentism (derived from Italia irredenta, Italian, 'unredeemed Italy'), Italian patriotic movement. Its members aimed at liberating all lands, mainly in the Alps and on the Adriatic, inhabited by Italians and still held by Austro-Hungary after 1866. Its activities were restrained when the Italian government entered into the *Triple Alliance with Austro-Hungary and Germany in

Isabella II, a painting by Vicente Lopez. The child queen poses with her hand resting on the crown of Spain, which she was to wear uneasily before her eventual deposition. (Museo Arte Moderno, Madrid)

1882, but they headed the campaign for Italy's intervention in World War I in 1915. The Settlement of *Versailles satisfied most of their claims.

Irwin, Baron *Halifax, 1st Earl of.

Isabella II (1830–1904), Queen of Spain (1833–70). The daughter of *Ferdinand VII, her accession was contended by her uncle, Don Carlos, and this led to the *Carlist Wars that raged until 1839. Her reign, after two unpopular regencies, was a succession of personal scandals, governmental changes, and conflicts between political factions. Isabella finally fled to France after an insurrection (1868), and was deposed. The crown, offered by the new constitutional Cortes to five successive candidates, was accepted by the sixth, the Duke of Aosta (1845–90), the second son of Victor Emanuel I of Italy. As Amadeus I he ruled from 1871 to 1873, when he abdicated and the first Spanish republic was declared.

Isandhlwana, battle of (22 January 1879), fought between Zulu and British forces. The *Zulu War had begun on 11 January when *Cetshwayo ignored the British ultimatum to disband his army of 30,000 Impis gathered in Ulundi. A British force, under Lord Chelmsford, of some 7,000 regulars, with as many African levies, advanced on Ulundi. The British were caught unawares

at Isandhlwana, with 1,600 killed in close-combat fighting. That night a force of Zulu warriors went on to attack a mission hospital at Rorke's Drift on the Buffalo River. Eighty defenders under Lieutenant Chard killed some 470 Zulus before being relieved by Chelmsford.

Islam (in Arabic scripture, 'surrender to the will of Allah'), one of the world's major religions, preached by the prophet Muhammad. Its tenets are to be found in the Koran (believed by Muslims to be the final revelations which came to the prophet from God) and the Sunna (Muhammad's words and actions recorded by his companions but not written by him), while the political and social framework is provided by the Shariah (the legal code of Islam). By 1800 Islam was the official religion of the *Ottoman empire, whose sultan was caliph. It had spread into central Asia, the Indian sub-continent, and the East Indies. Today Islam is the official religion of approximately forty-five nations with some eight hundred million believers. New communities of Muslims have become established in all European countries as well as in North and Latin America, and it has been estimated that there are some seven million in West and East Europe and twelve million in the Soviet Union. The course of African history in the 19th century was affected by Islamic movements among the *Sanusi in Libya, the Fulbe and *Fulani in West Africa, the *Mahdists in the Sudan, all influenced by the Persian teacher, Jamal al-Din al-Afghani (1838–97). With the fall in 1921 of the Ottoman empire and the discontinuation of the caliphate (the temporal and spiritual leadership of the Muslim community) in Constantinople, a tide of Muslim nationalism emerged. The early 20th century witnessed two well-organized movements for the revival of fundamental Islamic beliefs and the countering of Western influences: the Society of Muslim Brothers (al-Ikhwan al Muslimun) founded in 1928 by Hasan al-Banna (1906–49) in Egypt, which has since spread throughout the Middle East, and the Islamic Society (jamma'at-i-Islami) founded in 1941 by Syed Abu al-Mawdudi (1903–79) in India. Both of these have influenced other Islamic movements across the world, including the movement initiated by Ayatollah *Khomeini in *Iran. All emphasize the social and political reconstruction of Islamic society and of the Islamic state. The two major groupings of Islam today are the Sunni, comprising the main community in most Muslim countries, and the Shiite, centred chiefly in Iran. Both adhere to the same body of tenets, but differ in community organization and in theological and legal practices. Central to these is the Shiite belief that only the descendants of the prophet Muhammad may adopt the title and role of imam (religious leader), while the Sunni choose the latter by consensus. The tensions between Sunni and Shiite have been a major cause of social and political unrest in Middle Eastern countries during the 20th century.

isolationism, an approach to US foreign policy that advocates non-participation in alliances or in the affairs of other nations. It derives its spirit from George Washington's proclamation of neutrality in 1793, and was further confirmed by the *Monroe Doctrine (1823). It foiled Woodrow Wilson in his attempt to take the USA into the *League of Nations (1919 and 1920), and it hindered Franklin D. Roosevelt's support for Britain, France, and China before and during World War II, by ensuring passage of four restrictive Neutrality Acts. Present-day isolationists favour political and military withdrawal from overseas bases as well as the establishment of a 'fortress America' protected by military systems such as the *Strategic Defense Initiative.

Isonzo, a river in north-east Italy, the scene of fierce battles between Italians and Austrians following Italy's entry into World War I on the Allied side (1915). Some dozen battles, in which Italy had twice as many casualties as Austria, were fought along this front between May 1915 and October 1917, culminating in the Italian disaster at *Caporetto (1917).

Israel, a country in the Middle East. The modern state of Israel has developed from the *Zionist campaign for a Jewish state in *Palestine. Under the British *mandate in Palestine the Jewish community increased from about

10 per cent of the population in 1918 to about 30 per cent in 1936. In 1937 the Peel Commission recommended the partition of Palestine and the formation of Jewish and Arab states. Subsequently Britain abandoned the partition solution, but, after its referral of the Palestine problem to the United Nations in 1947, a United Nations Special Commission recommended partition and a resolution to that effect passed the General Assembly. The British mandate ended on 14 May 1948 and the independent Jewish state of Israel in Palestine was established. The creation of the state was opposed by the Palestinian Arabs supported by Syria, Lebanon, Jordan, and Egypt, but after a violent conflict Israel survived and considerably enlarged its territory at the expense of the proposed Arab state. A substantial Palestinian refugee problem was created as many Arabs were materially impelled to leave Israel-controlled territory. Further Israeli–Arab wars took place in 1956 (*Suez War), 1967

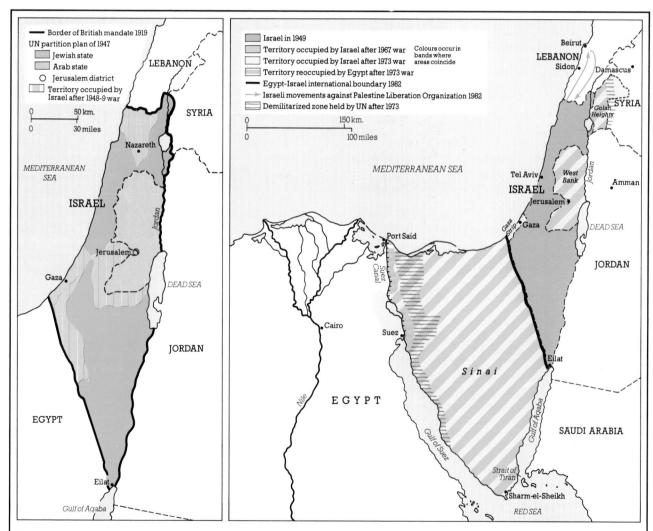

Israel

When Israel proclaimed itself a state (May 1948) the surrounding Arab states of Egypt, Jordan, Lebanon, and Syria declared war. After fierce fighting Israel agreed to a UN armistice. Israel made substantial gains, including territory beyond Nazareth. Following the Suez War (1956) a UN force was stationed in the Sinai desert and along the Gaza strip. This was withdrawn in 1967, whereupon Egypt blockaded the Gulf of Aqaba. Israel again went to war

against its Arab neighbours, occupying the Gaza strip, the Sinai desert, West Jordan, and the Golan Heights, all within six days. Jewish settlers quickly moved into these areas, often expelling Arab inhabitants. After a fourth war—the Yom Kippur War—in 1973 Israel, under pressure from both the USA and the Soviet Union, agreed to a phased withdrawal from Sinai, and UN troops were reinstated in the buffer zones.

(*Six-Day War), 1973 (*Yom Kippur War), and 1982 (Lebanon War). As a result of these wars Israel extended its occupation to include all the territory of the former British mandate. After 1948 immigration into Israel took place from over 100 different nations, especially Jews from communist and Arab countries, as well as from Europe, raising the population from about 700,000 in 1948 to 4·2 million by 1985. Despite a high inflation rate, the development of the economy has made Israel the most industrialized country in the region, greatly aided by funding from the USA and European powers.

Italian Campaign (July 1943–May 1945), a military campaign in World War II. Following the *North African Campaigns, *Montgomery and *Patton prepared British and US troops to invade Sicily. The landing was launched (July 1943) from Malta, and by the end of the month both the island's principal cities, Palermo and Catania, were captured, and on the mainland *Mussolini arrested. The German army under *Kesselring was withdrawn from Sicily and British and American forces landed in southern Italy (September 1943). An armistice was signed, ending hostilities between the Anglo-American forces and those of the new government of *Badoglio. A third surprise Allied landing on the 'heel' of Italy captured the two ports of Taranto and Brindisi, and on 13 October 1943 Italy declared war on Germany. A large and well-organized partisan force now harassed the Germans, but reinforcements successfully reached Kesselring, who took a stand at Monte Cassino (late 1943), site of the ancient monastery of St Benedict. The Allies decided to by-pass this, landing 50,000 men at Anzio (January 1944), south of Rome, but also bombing the monastery, which was finally captured (May 1944) by Polish troops. Rome fell (June 1944), and Florence was captured after bitter fighting (August 1944). The Germans consolidated in the River Po valley and fought a hard battle through the autumn of 1944. In April 1945 the Allied armies launched their final attacks, and on 2 May *Alexander accepted the surrender of the whole German army group serving in northern Italy and southern Austria.

Italy, a country in southern Europe. The Napoleonic invasion of 1796 ended in 1815, when the country reverted to a grouping consisting of Lombardy and Venetia, ruled by the *Habsburgs from Vienna; the kingdom of Piedmont Sardinia, which then consisted of most of Savoy, Piedmont, and the island of Sardinia; the Papal States, ruled by the popes in Rome; the duchies of Tuscany, Parma, and Modena, also ruled by the Habsburgs; and the Kingdom of the Two *Sicilies, now ruled by restored Bourbons from Naples. France ruled part of Savoy and Corsica, but had lost Genoa to Piedmont. Revolutionary societies, such as the *Carbonari and *Young Italy, were formed. The new forces of the *Risorgimento created hopes of independence from Austrian and French rule. Under such leaders as *Cavour, *Mazzini, and *Garibaldi, unification of Italy was finally achieved, and in 1861 *Victor Emanuel II was crowned king of Italy. In an effort to join the *'Scramble for Africa' the Italian Premier and Minister of Foreign Affairs, Francesco *Crispi, claimed (1889) the colony of *Eritrea, but the abortive bid for *Ethiopia led to a decisive defeat (1896) at the battle of *Adowa. During the Turko-Italian War (1911–12), Italy conquered north Tripoli and by 1914 had occupied much of Libya,

declaring it an integral part of the country in 1939. In World War I Italy supported the Allies, regaining Trieste and part of the Tyrol. The fascist dictator *Mussolini, determined to establish an Italian empire, successfully invaded (1935) Ethiopia, combining it with Eritrea and Italian Somaliland to form Italian East Africa. In World War II Mussolini at first allied himself with Hitler, but by 1943 the country had lost its North African empire and in the same year declared war on Germany. In 1946 the king abdicated in favour of a republic. The post-war period brought remarkable and sustained economic growth but also political instability, characterized by frequent changes of government. The Italian Communist Party has adjusted to democracy, but there have been terrorist kidnappings and outrages, notably during the 1970s by the 'Red Brigade'.

Ito Hirobumi (1841–1909), Japanese statesman. As a young *samurai of the Choshu clan, he opposed westernization before becoming aware of the benefits offered by modernization. He became one of the major political figures after the *Meiji Restoration (1868), travelling in Europe in search of a model for the *Meiji Constitution which he subsequently framed, and serving four times as Prime Minister, first in 1885. After the politics of the 1890s had shown the considerable veto power that political parties were able to exercise, and when *Yamagata Aritomo had given the armed services the power to break civilian governments, Ito formed (1901) the Seiyukai (Friends of Constitutional Government) Party. He retired from active politics soon after and exercised a moderating influence on imperial policy as first resident-general (1905–9) of the Korean protectorate. After his resignation he was assassinated by a Korean nationalist.

Iturbide, Agustín de (1783–1824), Mexican independence leader. A Creole officer in the Spanish royalist army, his decision to join the movement for independence from Spain and to proclaim the Plan of *Iguala was significant, as many other royalist officers followed his lead. With the defeat of the Spanish forces (1821) Iturbide managed to have himself proclaimed by his soldiers as Emperor Agustín I, and persuaded (1822) a hostile Congress to ratify the proclamation. On his accession he revoked the Plan of Iguala, refused to carry out his promised social reforms, and instituted a dictatorial government. A revolution led by *Santa Anna and Guadalupe Victoria overthrew the empire after eleven months. Iturbide left for Europe, but when he returned in 1824 he was arrested, tried by the Congress of Tamaulipas, and executed.

Ivory Coast, a country in West Africa. France obtained rights on the coast in 1842, establishing a colony in 1893, which in 1904 became a territory of French West Africa. In 1933 most of the territory of Upper Volta was added to the Ivory Coast, but in 1948 this area was returned to the reconstituted Upper Volta, today *Burkina Faso. The Ivory Coast became an autonomous republic within the *French Community in 1958, and achieved full independence in 1960, becoming a one-party republic governed by the moderate Democratic Party of the Ivory Coast and with Félix *Houphouét-Boigny its president. The country has an expanding economy, with large petroleum deposits and a developing industrial sector.

J

Jackson, Andrew (1767–1845), seventh President of the USA (1829–37). Trained as a lawyer, he helped draft the constitution of the state of Tennessee. After two years in the Senate (1797–8), he became judge of the Tennessee Supreme Court (1798–1804), but was forced out of politics as the result of a series of duels. On 8 January 1815 he won a victory over the British at the battle of *New Orleans, and in 1818–19 he invaded Florida during the *Seminole War, clearing the way for its inclusion in the USA. Entering national politics, Jackson rapidly became a leading figure in the Democratic Party. He narrowly lost the presidential election of 1824, but won easily in 1828 and was re-elected in 1832. During his Presidency, the new two-party system of *Democrats and *Whigs took shape. It featured the *spoils system, party-nominating conventions, and the establishment of the *kitchen cabinet. Jackson responded robustly to the *nullification challenge from South Carolina. He vetoed the renewal of the charter of the Bank of the United States, while his 'hard money' policies helped to precipitate a financial panic in 1837. Depending on his reputation as a hard-headed frontiersman, Jackson greatly increased the power and independence of the Presidency, while promoting a new style of popular democratic politics.

Jackson, Robert Houghwout (1892–1954), US lawyer and judge. He was a committed supporter of President Franklin D. Roosevelt's *New Deal policy in the 1930s, and defended New Deal laws in hearings before the Supreme Court. As a Justice of the Supreme Court, his decisions reflected his opposition to monopolies and vested interests, and he continually stressed that judges should display independence. He also acted as the chief US counsel at the *Nuremberg trials in 1945–6.

Jackson, Thomas Jonathan ('Stonewall') (1824–63), US general in the army of the Southern *Confederacy. He joined the Confederate army at the beginning of the *American Civil War and earned his nickname 'Stonewall' at the first battle of Bull Run (July 1861), when his troops held off assaults by Union (Northern) troops at a crucial phase of the engagement, paving the way for a Southern victory. He emerged as the most trusted lieutenant of General *Lee and one of the best fighting leaders on either side. He successfully immobilized a superior enemy force in the Shenandoah Valley campaign and played a major part in Confederate victories at the second battle of Bull Run (August 1862), Fredericksburg, and Chancellorsville (May 1863). During the latter, he was fatally wounded in an accident involving some of his own men.

Jaja of Opobo (d. 1891), Nigerian merchant prince. A former slave, he became head of the Anna Popple trading house at Bonny in the Niger Delta, acting as a middleman between the coastal markets and the Nigerian interior. He established (1869) his own state at Opobo on the Gulf of Guinea. From here he was able to prevent rival supplies from reaching the coast. In 1873 Jaja was

A portrait of 'Stonewall' **Jackson** by John Adams Elder. A stern disciplinarian, Jackson was Robert E. Lee's most able general. (Corcoran Gallery of Art, Washington)

recognized as King of Opobo. In the 1880s Jaja opposed increasing British influence in the area, and in 1887 was deported by Britain to the West Indies.

Jamaica, a large island in the West Indies. Captured by the British in 1655, it was a prosperous sugar colony by 1800. When slavery was abolished (1834) its economy suffered. A Negro rebellion in 1865 was ruthlessly suppressed by Governor Eyre. In 1866 it became a crown colony, and representative government gradually developed from 1884. In the 1930s there was widespread rioting, caused by racial tension and economic depression, and in 1944 self-government, based on universal adult suffrage, was granted. Economic recovery followed World War II. In 1958 Jamaica became a founding member of the Federation of the West Indies. When this collapsed, the Jamaican Labour Party (JLP) under William A. *Bustamante negotiated independence as a dominion in the *Commonwealth of Nations. Administrations have alternated between the JLP and PNP (People's National Party), whose leader Michael Manley introduced many social reforms in the 1970s. In 1980 the JLP returned to office under Edward Seaga, whose conservative economic policies failed to reverse economic decline.

James, Jesse (Woodson) (1847–82), US outlaw. He joined the Quantrills Raiders supporting the *Confederacy forces. After the *American Civil War he joined with his brother Frank (1843–1915) and the Younger brothers to form the most notorious band of outlaws in US history, specializing in bank and train robberies. The reward offered for the James brothers corrupted one of the gang, Robert Ford, who shot Jesse. Frank lived out his life in obscurity on a Missouri farm.

Jameson raid (1895–6), a conspiracy in southern Africa led by the British colonial administrator Dr Leander Storr Jameson, who planned to overthrow the South African government of Paul *Kruger. Privately financed by *Rhodes it sought to take advantage of a rebellion of the *Uitlanders against the *Transvaalers, and to further Rhodes', ambition for a united South Africa. Jameson's small band of volunteers was soon halted by the Boers, and Jameson himself captured and handed over to the British for punishment. The episode contributed to Boer dislike for Britain. Rhodes was forced to resign as Premier of Cape Colony because of his knowledge of the conspiracy. Jameson himself was tried in London and sentenced to imprisonment, but was soon released. The defeat of Jameson's raid prompted *William II to congratulate the Boers in the famous telegram to President *Kruger.

Janata Party, Indian political organization. A grouping of mostly right-wing Hindu political parties, it was based upon the Jana Sangh (People's Party, founded 1951). The so-called Janata Front was a broad coalition formed to oppose Indira *Gandhi in 1975. The Janata Party proper was formed in 1977 as a coalition under Morarji Desai; it won the 1977 election and formed a government which endured until the end of 1979.

Japan, north-east Asian country. An isolated and backward feudal country, Japan had been dominated by the Tokugawa *shogunate since the early 17th century, but in the first half of the 19th century Tokugawa power was gradually undermined by economic problems, insurrection, and the arrival of Western trading and naval expeditions, most notably those of the American Commodore *Perry (1853–4). The shogunate's failure to resist foreign penetration served as the catalyst for armed opposition, which in 1868 finally succeeded in replacing the shogunate with a new regime led formally by the emperor Meiji (*Meiji Restoration). In the succeeding decades feudalism was dismantled and a centralized state created which was dedicated to the rapid modernization of society and industrialization. Japan's new strength brought victory in the *Sino-Japanese War (1894–5) and the *Russo-Japanese War (1904–5), and established it as the dominant power in north-east Asia. Japan fought on the Allied side in World War I, but thereafter its expansionist tendencies led to a deterioration in its diplomatic position, most notable *vis-à-vis* the United States. In the inter-war period, expansionist-militarist interests gradually gained power within the country, and, after the occupation of Manchuria (1931) and the creation of *Manchukuo (1932), full-scale war with China was only a matter of time. The Sino-Japanese War finally broke out in 1937, and, having already allied itself with Germany and Italy in the *Anti-Comintern Pact, Japan finally entered World War II with a surprise attack on the US fleet at *Pearl Harbor in December 1941. Initially overrunning the colonial empires of south-east Asia at great speed, Japanese forces were eventually held and gradually driven back (*Pacific Campaigns).

In September 1945, after the dropping of two atomic bombs, Japan was forced to surrender and accept occupation. A new *Japanese Constitution was introduced, and full independence was formally returned in 1952. Japan embarked on another period of rapid industrial development, which has today left it as one of the major economic powers in the world. Its relations with China and south-east Asian countries have improved, but the large imbalance in its favour in its trade with Western nations (particularly the USA) has strained relations.

Japan, occupation of (1945–52), allied occupation of Japan after World War II. After Japan's unconditional surrender on 2 September 1945, it came under the control of the Allied forces of occupation led by General Douglas *MacArthur in his capacity as Supreme Commander of the Allied Powers (SCAP). Although technically backed by an eleven-nation Far Eastern Commission and a Four-Power Council (Britain, China, USA, and the Soviet Union), the military occupation was entirely dominated by the USA, with policy remaining in the hands of MacArthur and, after his removal in April 1951, of his successor General Matthew Ridgway. American occupation policy had two main components, the demilitarization of Japan and the establishment of democratic institutions and ideals. The first objective was achieved through the complete demobilization of the army and navy and the destruction of their installations, backed up by the peace clause of the new *Japanese Constitution. The second was much more difficult, and although the new Constitution was in operation well before the occupation was formally terminated in 1952, the impact of American-inspired reforms on Japanese socio-political institutions is still debated.

Japanese Constitution (1947), a constitution introduced during Allied occupation after World War II, with the emphasis placed on the dismantling of militarism and the extension of individual liberties. Drafted under US influence, it was finally adopted on 3 May 1947, leaving the emperor as head of state but stripping him of governing power, and vesting legislative authority in a bicameral Diet, the lower House of Representatives (originally 466, now 512 seats) being elected for four years and the upper House of Councillors (252 seats) for six (half at a time at three-year intervals). Executive power is vested in the cabinet which is headed by a Prime Minister and is responsible to the Diet. The constitution specifically renounces war but has been interpreted as allowing self-defence, and has led to the creation of the Self Defence Forces, although defence expenditure remains low by Western standards.

Japan–United States Security Treaty (1951), international defence agreement between Japan and the USA. Negotiated as part of the package of arrangements attending the formal return of Japanese independence after defeat in World War II, the treaty established the USA as the effective arbiter of Japan's defence interests, granting it a large military presence in Japanese territory. Renewal of the treaty in 1960, as a Mutual Security Treaty, with revised terms, produced a major political crisis in Japan, and when renewal next came due in 1970, both countries agreed to a process of automatic extension on the condition that revocation could be achieved on one year's notice by either.

Jaurès, Jean (1859–1914), French socialist leader. Entering parliament (1885), his campaign on behalf of *Dreyfus and against *anti-Semitism strengthened socialist support in France. In 1905 he formed the United Socialist Party, which put pressure on the radical governments in order to achieve reforms for the working class. He opposed militarism, but tried to reconcile socialist internationalism and French patriotism. He was assassinated by a French nationalist in 1914.

Java War (1825–30), war fought against the Dutch and their Javanese allies in central Java, led by *Diponegoro. The uprising proved difficult to suppress until the adoption by the Dutch of a system of rural strong points (*bèntèng*) and the use of mobile columns. Deprived of peasant support, Diponegoro was forced to negotiate with the Dutch and, when he refused to renounce his claim to the title of sultan and the status of Protector of Religion (*Panatagama*) in Java, he was arrested and banished. Thus ended the last and most serious challenge to the extension of Dutch rule in Java.

Jefferson, Thomas (1743–1826), third President of the USA (1801–9). He was a delegate to the House of Burgesses (1769–75) and to the Continental Congress (1775–6). He drafted the Declaration of Independence, was active in Virginia during the War of Independence,

The French socialist Jean **Jaurès** speaking in Paris, May 1913. He was expressing his opposition to the law that increased conscription from two to three years, in line with his belief that France should seek reconciliation with Germany rather than prepare for war. To Jaurès's right (seated, in black hat) is Pierre Renaudel, a fellow-founder of the French Socialist Party.

and was governor of the state (1779–81). A slave owner who favoured gradual emancipation, he never felt able to implement such a policy. After service as US minister to France (1785–9), he was Washington's first Secretary of State. His opposition to Alexander Hamilton's economic policies led to his resignation in 1794. He later became leader of the Democratic-Republican Party and was Vice-President under John Adams before becoming President in 1801. His administration was marked by retrenchment and reduction in the scale of government itself, but also by the *Tripolitan War, which ended tribute payment to Barbary pirates, the *Louisiana Purchase, the *Lewis and Clark expedition, and the Embargo Act in defence of US neutral rights. Jefferson, who believed in the virtues of an agrarian republic and a weak central government, has been a uniquely influential figure in the evolution of American political tradition.

Jellicoe, John Rushworth, 1st Earl (1859–1935), British admiral. He commanded the Grand Fleet at the inconclusive battle of *Jutland (1916) and then became First Sea Lord with an influence on strategic planning. He implemented the *convoy system introduced by *Lloyd George, but was dismissed from office in December 1917. After the war he was appointed governor-general of New Zealand (1920–4).

Jena, battle of (14 October 1806), a battle between the Prussian army, led by Prince Hohenlohe, and *Napoleon's French forces. Fought near the East German town of that name, the Prussians underestimated the size of the French force and were comprehensively defeated. The defeat by *Davout of the Duke of Brunswick's main Prussian army at Auerstädt on the same day left the road to Berlin unprotected. After its double defeat Prussia embarked on a series of military, political, and social reforms in order to be able to challenge the French.

Jiangxi Soviet, Chinese communist rural base formed in 1931. Under *Kuomintang attack in 1927, some communists moved to the countryside, maintaining their strength through guerrilla warfare in remote mountain regions. The group led by *Mao Zedong, which first established itself on the Hunan–Jiangxi border, merged with a group led by *Zhu De, and the First National Congress of the Chinese Soviet Republic was held in November 1931. Four nationalist 'Encirclement Campaigns' were thwarted by guerrilla tactics between December 1930 and early 1933, but a fifth, beginning in October 1933, forced the evacuation of the Soviet and the commencement of the *Long March a year later. Many communist policies, including land reform, were first tried out in the Jiangxi Soviet which, at its height, had a population of some nine million.

'Jim Crow' laws, US discriminatory statutes against blacks. 'Jim Crow', as a synonym for blacks, derives from the minstrel show of that name in about 1828. Applied to legislation, it distinguishes a body of US state laws, enacted between 1881 and 1907, that established racial segregation in respect of public transport, schools, restaurants and hotels, theatres, and penal and charitable institutions. This condition was not effectively challenged until after World War II, by which time racial barriers had been eroded.

Muhammad Ali **Jinnah** (*right*) strolling with fellow-nationalist leader Nehru in May 1946, when a unified, independent India still seemed possible. After 1947 and partition, as Pakistan broke away to become a separate state under Jinnah, relations between the two men were less cordial.

jingoism, a mood of inflated patriotism. It orginated in 1878, when Russian successes in a war against the *Ottoman empire had created at the Treaty of San Stefano a Bulgaria which Britain regarded as a threat to its Eastern interests. Strong anti-Russian feeling developed in Britain, where *Disraeli called up reserves for army service. War-fever gripped the country. The realities of *trench warfare in World War I changed the meaning of the word to that of blustering patriotism.

Jinnah, Muhammad Ali (1876–1948), founder of Pakistan. He entered politics as a strong supporter of the moderates in *Congress and as a proponent of Hindu–Muslim unity. In 1916 he was one of the principal architects of the Congress League Lucknow Pact in which Congress conceded that Muslims should have adequate legislative representation. After 1919 he became increasingly disillusioned with *Gandhi's leadership of Congress and in 1930 he went to London. Returning to India in 1934 he led the *Muslim League in the 1937 elections. Thereafter he devoted his energies to extending the hold of the Muslim League over the Muslims of British India; in 1945–6 the Muslim League won an overwhelming victory in Muslim seats, confirming Jinnah's claim to speak for Indian Muslims. He also led his party to espouse the demand for an independent Muslim state of Pakistan (Lahore 1940). In 1946–7 his

determined rejection of attempts to find a compromise led to the partitioning of India and the creation of the state of Pakistan, of which he became the first governor-general (1947–8) and President of its constituent assembly.

Jodl, Alfred (1890–1946), German Nazi general. Throughout World War II he was chief of the Armed Forces' Operations Staff, and was Hitler's closest adviser on strategic questions. He was executed as a war criminal after the *Nuremberg Trials. His diaries reveal his complicity in many of Hitler's war crimes, counselling terror *bombing offensives on British cities and signing orders to execute prisoners-of-war.

Joffre, Joseph Jacques Césaire (1852–1931), Marshal of France and French commander-in-chief on the *Western Front (1914–16). As chief of the general staff (1911) he had devised Plan XVII to meet a German invasion. Its effectiveness was proved in the first battle of the *Marne (1914), which frustrated German hopes of a swift victory. As commander-in-chief of all French armies (1915), he took responsibility for French unpreparedness at Verdun (1916) and resigned.

Johnson, Andrew (1808–75), seventeenth President of the USA (1865–9). He was the only southern senator to support the Union in the *American Civil War and was appointed military governor of Tennessee. Having been elected as Vice-President to Abraham *Lincoln, in 1864,

A caricature of Marshal **Joffre**, French commander-in-chief on the Western Front until he was stripped of the power of direct command in the third year of World War I, and resigned.

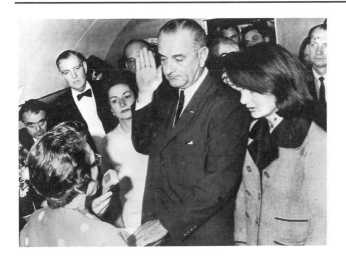

The tragic start to the **Johnson presidency**. Lyndon B. Johnson is sworn in as US President on board a crowded presidential jet on 22 November 1963. Beside Johnson (*right*) is Jacqueline Kennedy, widow of President Kennedy, who had been assassinated in Dallas only hours before. Johnson's wife, Claudia Alta (nicknamed Lady Bird), stands behind the new President on his right.

he became President as a result of Lincoln's assassination in April 1865. His *reconstruction policy, which failed to protect the interests of the ex-slaves, brought him into bitter conflict with the Republican majority in Congress, and his vetoes of several reconstruction measures were over ridden by two-thirds majorities in Congress. His dismissal of his Secretary of War, Edwin Stanton (in defiance of a Tenure of Office Act) led to his impeachment (a US legal procedure for removing officers of state before their term of office expires), and Johnson only survived by a single vote in the Senate (1868). He returned to the Senate in 1875 but died soon after.

Johnson presidency (1963–8), term of office of Lyndon B. Johnson (1908–73) as the thirty-sixth President of the USA. As a Democrat Johnson had represented Texas in Congress (1937–61). As Vice-President to John F. *Kennedy when the latter was assassinated, he was immediately sworn in as President. Johnson acted decisively to restore confidence and pressed Congress to pass the former President's welfare legislation, especially the *civil rights proposals. He won a sweeping victory in the presidential election of 1964, with Hubert Humphrey as Vice-President. The administration introduced an ambitious programme of social and economic reform. It took his considerable negotiating skills to persuade Congress to support his measures, which included medical aid for the aged (*Medicare) through a health insurance scheme, housing and urban development, increased spending on education, and federal projects for conservation. In spite of these achievements, urban tension increased. Martin Luther *King and *Malcolm X were assassinated and there were serious race riots in many cities. The USA's increasing involvement in the *Vietnam War overshadowed all domestic reforms, and led Johnson on an increasingly unpopular course involving conscription and high casualties. By 1968 this had forced Johnson to announce that he would not seek re-election.

John VI (1769–1826), King of Portugal (1816–26). The son of Maria I and Peter III, he took over the control of the government in 1792 from his mother, who had become insane, and assumed the title of regent in 1799. A repressive monarch, he was submissive to Napoleon, who nevertheless forced him into exile in *Brazil in 1807. In 1816 he was recognized as King of Portugal but continued to live in Brazil until 1822, when he returned to accept the role of a 'constitutional monarch'. In the same year he overcame a rebellion by his son Dom Miguel. In 1825 he recognized his other son, Dom Pedro, as emperor of an independent Brazil.

John XXIII (Angelo Giuseppe Roncalli, 1881–1963), Pope (1958–63). During his pontificate he made energetic efforts to liberalize *Roman Catholic policy, especially on social questions. Particularly notable were his encyclicals, *Mater et Magistra*, on the need to help the poor, and *Pacem in Terris*, on the need for international peace. He also summoned the Second Ecumenical Vatican Council (1962–5) to revitalize the life of the Church by bringing up to date its teaching, discipline, and organization, with the unity of all Christians as its ultimate goal. He was succeeded by *Paul VI.

Jordan, Middle Eastern country (correctly the Hashemite Kingdom of Jordan). The region was part of the *Ottoman empire until 1918, when it came under the government of King Faisal in Damascus. In 1920 Transjordan, as it was then called, was made part of the British *mandate of Palestine. In 1921 Britain recognized Abdullah ibn *Hussein as ruler of the territory and gave him British advisers, a subsidy, and assistance in creating a security force. In 1946 the country was given full independence as the Hashemite Kingdom of Jordan, with Abdullah ibn Hussein as king. In 1948–9 the state was

Pope **John XXIII**, photographed in the Vatican by Douglas Glass during the first year of his pontificate.

considerably enlarged when Palestinian territories on the West Bank, including the Old City of Jerusalem, were added. As a result of the *Six-Day War in 1967, these West Bank territories passed under Israeli occupation. The king was assassinated in 1951, his son Talal was deposed in 1952 as mentally unstable, and since 1952 Jordan has been ruled by Talal's son, Hussein (1935–). Palestinian refugees from territory under Israeli occupation established a commando force (*fedayeen*) in Jordan to raid Israel. Hostility from the Palestinian refugees from the West Bank to the moderate policies of Hussein erupted in 1970 between the guerrillas and the government. The mainly Bedouin regiments loyal to the king broke up the military bases of al-*Fatah, and the *Palestine Liberation Organization moved its forces (1971) to Lebanon and Syria.

Joseph (*c*.1840–1904), chief of a group of American Nez Percé Indians. He defied the efforts of the US government to move his people from their traditional Oregon lands, and was one of the leaders in a campaign (1877) against overwhelming military odds. He was finally defeated during a retreat to Canada. He spent his later years on a reservation, devoting his life to the welfare of his people.

Joséphine, Marie Rose Tascher de la Pagerie (1763–1814), Empress of the French (1804–9) and wife of *Napoleon I. Briefly imprisoned during the Reign of Terror, Joséphine married Napoleon in 1796, two years after her first husband, the Vicomte de Beauharnais, had

Chief **Joseph** of the Nez Percé Indians, photographed in 1903 by Edward S. Curtis.

been executed by the Jacobins. In 1810 Napoleon had the marriage annulled because of her alleged sterility, taking Marie Louise, daughter of *Francis I, as his wife.

Juárez, Benito (1806–72), Mexican statesman. His Plan of Ayutla (1854), calling for a constituent assembly within a federal constitution, paved the way for the War of Reform (1858–61) which exiled *Santa Anna and established a liberal government under Juárez. He was unable to prevent France's attempt to establish a Mexican empire (1864–7) under *Maximilian but was re-elected President on the emperor's assassination. He reduced the privileges of the army and Church, but his reforms were curtailed by division among the liberals.

Judaism, one of the world's major religions, practised by Jews, with an estimated world population of 13 million, centred in the USA (5.7 million), in Israel (3.5 million), and in the Soviet Union (1.5 million). Its beliefs are enshrined in the Hebrew Scriptures (the Torah) and in its oral traditions (the Mishna and the Talmud). Over the centuries Jews had spread to Mediterranean countries (the Sephardi), to Holland, Germany, the Baltic states, and into Central and Eastern Europe (the Ashkenazi). They lived in ghettos or legally enforced residence areas, worshipping in their homes and synagogues, guided by rabbis. By the 18th century pressures were growing within the Jewish community for the granting of equal rights and the abandonment of the ghetto. In Berlin the philosopher Moses Mendelssohn (1729–86) campaigned for Jewish emancipation. From his work there developed the Haskalah, or Jewish Enlightenment, which spread into Eastern Europe in the early 19th century. Jewish religious reform, as well as racial *anti-Semitism, and *pogroms led to large-scale emigration, especially to the USA. Legal limitations to Jews were abolished in most European countries during the course of the 19th century. Three strands of Judaism emerged: Reform, Orthodox, and Conservative. Followers of Mendelssohn sought to reconcile Judaism with contemporary Europe by means of a Reform Movement, which among other things allowed the language of the country to be used in the synagogue. The movement came to Britain and the USA in the 1840s, gaining many middle-class adherents. In reaction were Orthodox Jews, seeking to reject secular culture and to preserve ancient practices, and supporters of Hasidism, an influential, mystical movement that stressed the development of a personal spiritual life and attacked any manifestations of modernity. Between these emerged Conservative Judaism. Its forerunner was Zacharias Frankel (1801–75), whose 'Positive Historicism' sought to harmonize Jewish tradition with modern knowledge. When *Zionism emerged in the later 19th century, both Reform and Conservative Judaism supported it, although a majority of Orthodox Jews were and have remained suspicious of the concept of a Jewish state. During the 20th century Jews were decimated by the *Holocaust, but succeeded in establishing the state of *Israel (1948). Most Jews outside Israel have continued their assimiliation into the population of their country while preserving some traditions of the Jewish community.

July Plot (20 July 1944), a plot to assassinate Adolf *Hitler. Disenchanted by the *Nazi regime in Germany, an increasing number of senior army officers believed that Hitler had to be assassinated and an alternative government, prepared to negotiate peace terms with the Allies, established. Plans were made in late 1943 and there had been a number of unsuccessful attempts before that of July 1944. The plot was carried out by Count Berthold von Stauffenberg, who left a bomb at Hitler's headquarters at Rastenburg. The bomb exploded, killing four people, but not Hitler. Stauffenberg, believing he had succeeded, flew to Berlin, where the plotters aimed to seize the Supreme Command headquarters. Before this, however, news came that Hitler had survived. A counter-move resulted in the arrest of some 200 plotters, including Stauffenberg himself, Generals Beck, Olbricht, von Tresckow, and later Friedrich Fromm. They were shot, hanged, or in some cases strangled. Field-Marshal *Rommel was implicated and obliged to commit suicide. The regime used the occasion to execute several prominent protesters such as Dietrich *Bonhoeffer.

July Revolution (1830), a revolt in France. It began when *Charles X issued his ordinances of 25 July, which suspended the liberty of the press, dissolved the new chamber, reduced the electorate, and allowed him to rule by decree. His opponents erected barricades in Paris and after five days of bitter street fighting Charles was forced to abdicate. The duc d'Orléans, *Louis-Philippe, was invited to become 'King of the French', a title which replaced the more traditional 'King of France'. His accession marked the victory of constitutional liberal forces over arbitrary and absolutist rule.

Justo, Agustín Pedro (1876–1943), Argentine statesman. He participated in the conservative military coup which overthrew President Hipólito *Irigoyen in 1930 and was rewarded with the presidency (1932–8). Faced by the effects of revolution, high unemployment, and the economic decline caused by the Great *Depression, his regime was autocratic, outlawing the Communist Party in 1936. But he supported *Pan-American co-operation and closer links with Britain. He lost the 1937 presidential election to Roberto Ortiz, having been instrumental in

The **July Revolution** of 1830, celebrated allegorically in this detail from a contemporary painting by Eugene Delacroix, *Liberty at the Barricades*. (Musée du Louvre, Paris)

Among the eleven ships lost by the German fleet at the battle of Jutland in 1916 was the battle cruiser *Seydlitz*, seen here in flames after being raked by shells from British battleships. Survivors of the *Seydlitz* are preparing to abandon the stricken warship. In the same encounter, the British sank the German flagship *Lützow*.

ending the *Chaco Wars. A supporter of the Allies in World War II, he enlisted in the Brazilian army in 1942 and was killed.

Jutland, battle of (31 May 1916), a battle fought off the coast of Jutland in the North Sea between Britain and Germany. It was the only major battle fought at sea in World War I and began between two forces of battle cruisers, the British under *Beatty and the German under von Hipper. Suffering heavy losses, Beatty sailed to join the main British North Sea Fleet under *Jellicoe, which now engaged the German High Seas Fleet under *Scheer. Battle began at 6 p.m. at long range (approximately 14 km or 9 miles), but as the Germans headed for home in the night, they collided with the British fleet, several ships sinking in the ensuing chaos. Both sides claimed victory. The British lost 14 ships, including 3 battle cruisers; the Germans lost 11 ships, including 1 battleship and 1 battle cruiser; but the British retained control of the North Sea, the German fleet staying inside the Baltic for the rest of the war.

Kádár, János (1912-), Hungarian statesman. He joined the illegal Communist Party in Budapest in 1931, being often arrested. He helped to organize *resistance movements during World War II, after which he was appointed Deputy Chief of Police (1945) and then Minister of the Interior (1949). Imprisoned (1951-4) during the *Rakosi regime, he joined the short-lived government of Imre *Nagy, who had pledged liberalization. The *Hungarian Revolution which followed (October 1956) resulted in the fall and execution of Nagy and harsh Soviet military control. Kádár survived, becoming First Secretary of the Communist Party of Hungary (General Secretary since 1958); he was installed by the Soviet Union to curb revolt through repressive measures and became the effective ruler of Hungary. While remaining loyal to Moscow in foreign affairs, recent more liberal policies at home have allowed for an increasingly diversified economy and a higher standard of living.

Kamenev, Lev Borisovich (1883-1936), Soviet communist leader. He joined the Social Democratic Party (1901) and sided with the *Bolshevik faction when the party split in 1903. For ordering Bolshevik deputies in the Duma to oppose World War I he was exiled to Siberia. In 1917 he presided over the Second All-Russian Congress of Soviets. On *Lenin's death Kamenev, *Zinoviev, and *Stalin formed a triumvirate to exclude Kamenev's brother-in-law, *Trotsky, from power. In 1936 he was accused of complicity in the murder of Kirov in the first public show trial of Stalin's great purge, and was shot.

kamikaze (Japanese, 'Divine Wind'), a Japanese aircraft laden with explosives and suicidally crashed on a target by the pilot. It was the name chosen in World War II by the Japanese naval command for the unorthodox tactics adopted in 1944 against the advancing Allied naval forces. At first volunteers were used, but the practice soon became compulsory. Off *Okinawa in 1945, when these tactics came closest to repelling an Allied attack, over 300 kamikaze pilots died in one action.

Kampuchea (or Cambodia), south-east Asian country. In the 19th century Cambodia slipped into decline as Thai power rose to the west and Vietnamese power to the east. Continuing foreign domination forced Cambodia to seek French protection in 1863, and from 1884 it was treated as part of *French Indo-China, although allowed to retain its royal dynasty. After Japanese occupation in World War II, King Norodom *Sihanouk achieved independence within the French Union (1949) and full independence (1953). Sihanouk abdicated in 1955 to form a broad-based coalition government. Cambodia was drawn into the *Vietnam War in the 1960s and US suspicions of Sihanouk's relations with communist forces led to his overthrow by the army under Lon Nol in 1970 following a US bombing offensive (1969-70) and invasion. The Lon Nol regime then came under heavy pressure from the communist *Khmer Rouge. Following the fall of Phnom Penh in 1975 the Khmer Rouge under *Pol Pot launched a bloody reign of terror which is estimated to have resulted in as many as 2 million deaths or nearly a third of the population. Border tensions led to an invasion of Cambodia (now renamed Democratic Kampuchea) by Vietnam in 1978, and the overthrow of the Pol Pot regime two weeks later. The Vietnamese installed a client regime under the ex-Khmer Rouge cadre, Heng Samrin, but Khmer Rouge forces operating from bases across the Thai border keep up an increasingly destructive guerrilla war and hundreds of thousands of Cambodians have been forced to flee the country.

Kanagawa, Treaty of (also called Perry Convention) (31 March 1854), treaty between Japan and the USA. After three years of negotiation, the American Commodore *Perry came to an agreement with the Tokugawa *shogunate, opening two ports to US vessels, allowing the appointment of a consul, and guaranteeing better treatment for shipwrecked sailors. The Treaty of Kanagawa was followed within two years by similar agreements with Britain, Russia, and the Netherlands, and in 1858 by the more wide-ranging Treaty of *Edo with the USA, and marked the beginning of regular political and economic intercourse with the Western nations.

Kanakas, a term used indiscriminately by Europeans to describe Pacific Islanders. Kanakas, mainly from the New Hebrides and the Solomon Islands, were brought to Australia between 1863 and 1904 as cheap labour. Theoretically, they voluntarily entered contracts for fixed terms. In practice, they were subjected to abuses which included kidnap, slavery, and murder. Their entry to Australia was banned in 1904. Most of those in Australia were deported back to the islands from 1906 onwards, as part of the *White Australia policy.

Kang Youwei (or K'ang Yu-wei) (1858-1927), Chinese philosopher and political reformer. A scholar, whose utopian work *Da Tong Shu* (*One World Philosophy*, 1900), portrayed Confucius as a reformer, he believed that China's crisis could only be solved through the modernization of institutions along modified western lines. In 1898, at a time when foreign intervention presented particular dangers, he persuaded the emperor Guangxu to adopt his policies. The resulting *Hundred Days Reform was brought to a premature end by the empress dowager *Cixi's conservative coup, and Kang spent the next fifteen years in exile. He remained a monarchist in spite of the republican trend and spent his last years trying unsuccessfully to engineer an imperial restoration.

Kansas–Nebraska Act (1854), an Act of the US Congress concerning slavery. Following the *Mexican–American War, the *Compromise of 1850 had allowed *squatters in New Mexico and Utah to decide by referendum whether they would enter the Union as 'free' or 'slave' states. This was contrary to the earlier *Missouri Compromise. The Act of 1854 declared that in Kansas and Nebraska a decision on slavery would also be allowed, by holding a referendum. Tensions erupted between pro- and anti-slavery groups, which in Kansas led to violence (1855-7). Those who deplored the Act formed a new political organization, the *Republican Party, pledged to oppose slavery in the Territories. Kansas was to be admitted as a free state in 1861, and Nebraska in 1867.

Kapp putsch (March 1920), the attempt by Wolfgang Kapp (1858-1922), a right-wing Prussian landowner and politician, to overthrow the *Weimar Republic and restore the German monarchy. Aided by elements in the army, including *Ludendorff, and the unofficial 'free corps' which the new government was trying to disband, Kapp's forces seized Berlin, planning to set up a rival government with himself as Chancellor. The *putsch* was defeated by a general strike of the Berlin workers and the refusal of civil servants to obey his orders.

Kara George, Petrović (1766-1817), Serbian revolutionary leader, founder of the dynasty of Karageorgević. The son of a peasant, in 1804 he became leader of a Serbian revolt against the Turkish army, and played a major role in forcing the Turks out of *Serbia. Four years later he was proclaimed leader of Serbia, but his ruthless and autocratic rule led to unrest. In 1813 his army was defeated by the Turks and he fled. He returned in 1817 only to be murdered, some say by his rival, Milos *Obrenović. His son, Alexander, ruled Serbia as prince from 1843 to 1858, but was displaced by an Obrenović. Alexander's son, Peter, became King of Serbia in 1903; his grandson became *Alexander I of Yugoslavia.

Károlyi, Mihály, Count (1875-1955), Hungarian statesman. Of liberal views, he favoured a less pro-German policy for the *Austro-Hungarian empire and equal rights for all nations within it. There was no hope of achieving this until the empire collapsed (November 1918), when Hungary proclaimed itself a republic with Károlyi as President. When in March 1920 he learned that Hungary must cede territory to Romania, Czechoslovakia, and Yugoslavia he resigned and was replaced by Béla *Kun's communist regime.

Kasavubu, Joseph (1910-69), Congolese statesman. He became the first President (1960-5) of the Republic of the Congo (now *Zaïre). He was a member of undercover nationalist associations to free the Congo of the Belgians. In 1955 he became President of Abako (Alliance des Bakongo), a cultural association of the Bakongo tribe, and turned it into a powerful political organization. On independence in 1960 he became Head of State. His Abako party formed a coalition with *Lumumba's party, and then ousted him as premier. In 1965 he himself was deposed from the Presidency by *Mobutu in a bloodless military coup.

Kashmir, a former state on the border of India, since 1947 disputed between India and Pakistan. The state, exposed successively to Hindu and Muslim rule, was annexed (1819) to the expanding Sikh kingdom. After the first Sikh War the territory was acquired by Gulab Singh, then Hindu raja of the Jammu region. It was a *Princely State for the rest of the British period. The Maharaja, a Hindu ruling over a predominantly Muslim population, initially hoped to remain independent in 1947, but eventually acceded to the Indian Union. War between India and Pakistan (1948-9) over Kashmir ended when a United Nations peace-keeping force imposed a temporary cease-fire line which divided the Indian Union state of Kashmir (including Jammu) from Pakistani-backed Azad Kashmir. Kashmir remains divided by this line. Conflicts between India and Pakistan over Kashmir flared up again in 1965 and 1971.

Kato Komei (or Kato Takaaki) (1860-1926), Japanese statesman. He served as ambassador to Britain (1909) and Foreign Minister (1914-15), but was forced to resign after his presentation of the *Twenty-One Demands to China. He reorganized and led the conservative Kenseikai, and as Prime Minister (1924-5) pursued a moderate foreign policy while introducing universal manhood suffrage, cutting expenditure, and reducing the size of the army. He also introduced the stringent Peace Preservation Law to balance the possibly destabilizing effects of manhood suffrage. His cabinet was called the 'Mitsubishi government' because both he and his foreign minister Shidehara Kijuro had marriage ties with the Mitsubishi *zaibatsu.

Katyn massacre, a massacre in Katyn forest in the western USSR. In 1943 the German army claimed to have discovered a mass grave of some 4,500 Polish officers, part of a group of 15,000 Poles who had disappeared from Soviet captivity in 1940 and whose fate remained unknown. Each victim had a German bullet in the base of his skull, and it is assumed by historians in the West that the officers had been massacred in the early era of close *Nazi-Soviet collaboration. The Soviet Union has consistently denied involvement in the massacre. The incident resulted in a breach between the exiled Polish government of General *Sikorski in London and the Soviet Union and led to the agreement at Teheran (1943) that the post-war Polish–Soviet border should revert to the so-called *Curzon Line (1920).

Kaunda, Kenneth David (1924-), Zambian statesman. At first a schoolmaster, he joined (1949) the

Zambian leader Kenneth **Kaunda** addressing the sixth conference of the non-aligned nations at Colombo, Sri Lanka, in 1976.

*African National Congress (ANC). In 1959 he became its President and led opposition to the *Central African Federation, instituting a campaign of 'positive non-violent action'. For this he was imprisoned by the British, and the movement banned. Released (1960), he was elected President of the newly formed United National Independence Party (UNIP), which had become the leading party when independence was granted in 1964, Kaunda being elected first President of the new republic. During his presidency education expanded, and the government made efforts to diversify the economy to release Zambia from its dependence on copper. Ethnic differences, the Rhodesian and Angolan conflicts, and the collapse of copper prices engendered unrest and political violence, which led him to institute a one-party state (1973). Later, with the civil war in *Angola, he assumed emergency powers. In spite of these difficulties he was re-elected President in 1978 and again in 1983.

Kefauver, Carey Estes (1903–63), US politician. A state Senator (1949–63), he came to national prominence in the early 1950s when, as chairman of a US Senate committee investigating organized crime, he exposed nation-wide gambling and crime syndicates which had infiltrated legitimate business and gained control of local politics. The evidence of corruption among federal tax officials led to several dismissals and the resignation of the commissioner of Internal Revenue. Kefauver won the Democratic Party's nomination for Vice-President (1956), but President *Eisenhower (Republican) was re-elected.

Keitel, Wilhelm Bodewin Johann Gustav (1882–1946), German field-marshal. He was chief-of-staff of the High Command of the German armed forces (1938–45). He handled the armistice negotiations with France in 1940, and ratified the unconditional surrender of Germany in 1945. He was a close adviser of *Hitler, and bore some of the responsibility for repressive measures taken by the army in occupied territory. He was hanged after trial at *Nuremberg.

Kellogg–Briand Pact, or Pact of Paris (1928), a multilateral agreement condemning war. It grew out of a proposal by the French Premier, Aristide *Briand, to the US government for a treaty outlawing war between the two countries. The US Secretary of State, Frank B. Kellogg, countered with a suggestion of a multilateral treaty of the same character. In August 1928 fifteen nations signed an agreement committing themselves to peace; the US ratified it in 1929, followed by a further forty-six nations. The failure of the Pact to provide measures of enforcement nullified its contribution to international order.

Kemal, Mustafa *Atatürk.

Kennedy, John Fitzgerald (1917–63), thirty-fifth President of the USA (1961–3). After service in the US Navy in World War II, he became a Democratic member of the House of Representatives and subsequently a Senator. In 1960 he won the Democratic nomination and defeated Vice-President Nixon in the closest presidential election since 1884. Soon after his inaugural address ('ask not what your country can do for you—ask what you can do for your country'), Kennedy brought a new spirit of hope and enthusiasm to the office. Although Congress

John F. **Kennedy** at a press conference early in his Presidential campaign. Nomination by the Democrats and his subsequent election victory over Republican Richard M. Nixon took Kennedy to the Presidency in 1960, as the youngest man and the first Roman Catholic ever to hold the office.

gave support to his foreign aid proposals and space programme, it was reluctant to accept his domestic programme known as the 'New Frontier' proposals for *civil rights and social reform. In foreign affairs he recovered from the abortive *Bay of Pigs incident in Cuba to resist Khrushchev over Berlin in 1961, and again over the *Cuban Missile Crisis in 1962. He helped to secure a *Nuclear Test-Ban Treaty in 1963. He became increasingly involved in Vietnam, by despatching more and more 'military advisers' and then US troops into combat-readiness there. Kennedy established (1961) the Alliance for Progress to provide economic assistance to Latin America. In November 1963 he was assassinated while visiting Dallas, Texas. The Warren Commission, appointed by his presidential successor, Lyndon B. Johnson, concluded that he had been killed by Lee Harvey *Oswald. John F. Kennedy was a member of a noted political family. His brother Robert F. Kennedy (1925–60) was Attorney-General (1961–64), and was a candidate for the Democratic nomination in 1968, but as his support was growing, he also was assassinated. His brother Edward M. Kennedy (1932–) is a Senator and an influential figure in the Democratic Party.

Kenya, a country in East Africa. The Masai pastoral people came into the area in the 18th century from the north, but during the 19th century they were largely displaced by the agricultural Kikuyu, who steadily advanced from the south. British coastal trade began in the 1840s, and in 1887 the British East African Association (a trading company) secured a lease of coastal strip from the Sultan of Zanzibar. The British East Africa Protectorate was established in 1896, when thousands of Indians were brought in to build railways. The British crown colony of Kenya was created in 1920. By then a great area of the 'White Highlands' had been reserved for white settlement, while 'Native Reserves' were established to separate the two communities. During the 1920s there was considerable immigration from Britain, and a development of African political movements, demanding a greater share in the government of the country. Kikuyu nationalism developed steadily, led by Jomo *Kenyatta. From this tension grew the Kenya Africa Union, and the militant *'Mau Mau' movement (1952–7). An election in 1961 led to the two African political parties, the Kenya African National Union (KANU) and the Kenya African Democratic Union (KADU), joining the government. Independence was achieved in 1963, and in the following year Kenya became a republic with Kenyatta as president. Under him, Kenya remained generally stable, but after his death in 1978 opposition to his successor, Daniel Arap Moi, mounted, culminating in a bloody attempted coup in 1982. Elections in 1983 saw the return of comparative stability with Moi still President.

Kenyatta, Jomo (c.1892–1978), Kenyan statesman. He visited England in 1928 as Secretary of the Kenya Central Association, campaigning for land reforms and political rights for Africans. He remained in Britain from 1932 to 1946, taking part with Kwame *Nkrumah in the *Pan-African Conference at Manchester (1945). He returned to Kenya in 1946, and became President of the Kenya African Union. In 1953 he was convicted and imprisoned for managing the *Mau Mau rebellion, a charge he steadfastly denied. Released in 1961, he shortly afterwards entered Parliament as leader of the Kenya African National Union (KANU) and won a decisive victory for his party at the 1963 elections. He led his country to independence in 1963 and served as its first President from 1964 to his death in 1978. Once in power, he reconciled Asians and Europeans by liberal policies and economic common sense, but he was intolerant of dissent and outlawed opposition parties in 1969.

Kerensky, Alexander Feodorovich (1881–1970), Russian revolutionary. He was a representative of the

Kenyan President Jomo **Kenyatta**, one of Africa's most respected statesmen. He is wearing a ceremonial leopard skin cap and holding a fly whisk, a traditional symbol of authority.

moderate Labour Party in the Fourth Duma (1912) and joined the Socialist Revolutionary Party during the *Russian Revolution. After the emperor's abdication in March (February, old style), he was made Minister of War in the Provisional Government of Prince Lvov, succeeding him as Premier four months later. Determined to continue the war against Germany, he failed to implement agrarian and economic reforms, and his government was overthrown by the *Bolsheviks in the October Revolution. He escaped to Paris, where he continued as an active propagandist against the Soviet regime.

Kesselring, Albrecht (1885–1960), German field-marshal. He commanded the *bombing offensive over Poland, the Netherlands, and France before commencing the battle of *Britain, when he was hampered by interference from *Goering and *Hitler. Posted to the Mediterranean soon after, from 1943 to 1945 he commanded all German forces in Italy, and in 1945 in the West. Condemned to death as a war criminal in 1947, he had his sentence commuted to life imprisonment and was freed in 1952.

Keynes, John Maynard, Baron (1883–1946), British economist. He achieved national prominence when, in *The Economic Consequences of the Peace* (1919), he criticized the damaging effects on the international economy of the vindictive *reparations policy towards Germany. *Free trade, supported by some Liberals, was regarded by Keynes as an unwanted Victorian relic. Keynes did not support state *socialism, but argued that governments had greater financial responsibilities than balancing their budgets and leaving problems to be solved by market forces. In a depression governments should increase, not decrease, expenditure. This would 'prime the pump' of economic activity. His advocacy of full employment influenced William *Beveridge's report, and the policies of governments in the immediate post-war years. Keynes believed that governments should consider deficit financing, relying on international arrangements to right the balance in the long run. Keynes played a major role in the *Bretton Woods conference in 1944, which resulted in the setting up of the *International Monetary Fund and the *World Bank. After World War II his views were raised almost to the level of orthodoxy and underpinned the foundation of the British *welfare state. In more recent years the validity of 'Keynesian economics' has been questioned and in part abandoned.

KGB (Russian abbreviation, Committee of State Security). Formed in 1953, it is responsible for external espionage, internal counter-intelligence, and internal 'crimes against the state'. The most famous chairman of the KGB has been Yuri Andropov (1967–82), who was Soviet leader (1982–4). He made KGB operations more sophisticated, especially against internal dissidents.

Khalifa, Abdallah (Muhammad al-Ta'a'ishi) (c.1846–99), the successor of the Sudanese *Mahdi. In 1883 the Mahdi made him commander of his army, and he was largely responsible for the victory over the British at Khartoum in 1885. When the Mahdi died he eliminated the remaining Egyptian garrisons, and waged war on *Ethiopia until 1889. After his defeat by *Kitchener at Omdurman he fled to Kordofan, where he died in battle.

Khama, Sir Seretse (1921–80), Botswana statesman. He was educated in South Africa and Britain, where his marriage to an Englishwoman prevented him from becoming chief. In 1956 he returned to *Botswana and established the Bechuanaland Democratic Party. The party won the 1965 elections, and Seretse Khama became his country's leader. Botswana became independent in 1966 with Seretse as its first President. A strong believer in multiracial democracy, he strengthened the economy of Botswana and achieved universal free education.

Khilafat Movement, an Indian Muslim movement. It aimed to rouse public opinion against the harsh treatment accorded to the Ottoman empire after World War I and specifically against the treatment of the Ottoman sultan and caliph (khalifa). The movement began in 1919 and, under the leadership of the Ali brothers, Muhammad Ali (1878–1931) and Shaukat Ali (1873–1938), assumed a mainly political character in alliance with the Indian National *Congress, adopting the non-co-operation programme in May 1920. The Khilafat movement had considerable support from Muslims but was extinguished in 1924 after the abolition of the caliphate by *Atatürk.

Khmer Rouge, Kampuchean communist movement. Formed to resist the right-wing, US-backed regime of Lon Nol after the latter's military coup in 1970, the Khmer Rouge, with Vietnamese assistance, first dominated the countryside and then captured the capital Phnom Penh (1975). Under *Pol Pot it began a bloody purge, liquidating nearly the entire professional élite as well as most of the government officials and Buddhist monks. The majority of the urban population were relocated on worksites in the countryside where large numbers perished. The regime has been responsible for

The Ayatollah **Khomeini** is helped down airline steps on his return to Iran from exile in France on 1 February 1979. The deposed Shah had left Iran two weeks earlier; now Khomeini took power as the religious leader of a far-reaching Islamic revolution.

Seretse **Khama** in September 1956 with his British-born wife Ruth and their two children at their London home shortly before he returned to Bechuanaland (now Botswana), thus ending his seven-year exile.

an estimated 2 million deaths in *Kampuchea, and for the dislocation of the country's infrastructure. Frontier disputes with Vietnam provoked an invasion by the latter in 1978 which led to the overthrow of the Khmer Rouge regime, although its forces have continued a guerrilla war against the Vietnamese-backed Heng Samrin regime from bases in Thailand.

Khomeini, Ruhollah (c.1900–), Iranian religious and political leader. The son and grandson of Shiite religious leaders, he was acclaimed as an *ayatollah* (Persian, from Arabic, 'token of God', i.e. major religious leader) in 1950. During the anti-government demonstrations in 1963 he spoke out against the land reforms and Westernization of *Iran by *Muhammad Reza Shah Pahlavi, and was briefly imprisoned. After exile in Iraq (1964), he settled near Paris (1978), from where he agitated for the overthrow of the Shah. Khomeini returned to Iran in 1979 and was proclaimed the religious leader of the revolution. Islamic law was once more strictly imposed, and he has enforced a return to strict fundamentalist Islamic tradition. The *Iran Hostage Crisis confirmed his anti-US policy, and the *Iran–Iraq War his military intransigence. He has supported Islamic revolutions throughout the Middle East.

Khrushchev, Nikita Sergeyevich (1894–1971), Soviet statesman. During World War II he organized resistance in the Ukraine. He was actively involved in agriculture after the war, creating and enlarging state farms to replace collectives. On the death of *Stalin he became First Secretary of the Communist Party (1953–64) and Chairman of the Council of Ministers (1958–64). In a historic speech at the *Twentieth Congress (1956) he denounced Stalin and the 'cult of personality'. At home he attempted to tackle the problem of food supply by arranging for cultivation of the 'virgin lands' of Kazakhstan. He continued the programme of partial decentralization, and introduced widespread changes in regional economic administration. He restored some legality to police procedure and closed many *prison camps. In foreign affairs he subdued both the Poles under *Gomulka and the *Hungarian Revolution. In 1962 he came close to global war in the *Cuban Missile Crisis, but agreed to the withdrawal of Soviet missiles. His ideological feud with *Mao Zedong threatened a Sino-Soviet war. However, his policy of 'peaceful coexistence' with the West did notably ease the international atmosphere. He was dismissed from his offices in 1964, largely as a result of his handling of foreign crises and the repeated failures in agricultural production.

kibbutz (Hebrew, 'gathering', 'collective'), an Israeli collective settlement, usually agricultural but sometimes also industrial. The land was originally held in the name of the Jewish people by the Jewish National Fund, and is now owned or leased at nominal fees by its members, who also manage it. The first kibbutz, Deganya, was founded in 1910, and they now number around 300 in Israel.

Kiel Canal, an artificial waterway connecting the North Sea with the Baltic. Conceived by Bismarck (1873) so as to give German ships quick access to the North Sea, it

was built 1887–95. The construction of larger battleships (1907) forced Germany to widen it. Because of its strategic importance it was internationalized in 1919, but Hitler repudiated its international status in 1936, a condition re-imposed after World War II.

King, (William Lyon) Mackenzie (1874–1950), Canadian statesman. He entered the Canadian Parliament as a Liberal in 1908 and was appointed Minister of Labour (1909–11) under Sir Wilfrid *Laurier. Chosen (in 1919) as leader of the Liberal Party of Canada, he became Prime Minister in 1921, a post he filled, except for a brief interval in 1926, until 1930. He again served as Prime Minister, 1935–48. He never had a stable majority in Parliament, and combined his support from French-Canadian Liberals with endorsements from the Progressives, the farmers' party of Western Canada. This was reflected in his policies with their assertion of Canadian sovereignty, maintenance of political unity between English- and French-speaking Liberals, and cautious extension of social and economic reform measures. He led the Canadian war effort in World War II, but delayed conscription as long as possible. After the war he promoted the Canadian role in reconstruction, the United Nations, and NATO.

King, Martin Luther, Jr (1929–68), US pastor and *civil rights leader. As a Baptist pastor in black churches

American civil-rights activist Martin Luther **King** Jr, the most charismatic and effective black leader in modern US history.

Soviet leader Nikita **Khrushchev** (*right*) applauds crowds gathered on his arrival in Czechoslovakia in 1957. With him is the Czech Communist Party leader Antonín Novotný (*left*). Khrushchev's policy of de-Stalinization was to lead to demands for greater national independence by the Soviet Union's East European allies.

in Alabama and Georgia, he won national fame by leading (1955-6) a black boycott of segregated city bus lines in Montgomery, Alabama, which led to the desegregation of that city's buses. He then organized the Southern Christian Leadership Conference, and through this launched a nation-wide civil rights campaign. A powerful orator, he urged reform through non-violent means, and was several times arrested and imprisoned. He organized (1963) a peaceful march on the Lincoln Memorial in Washington, in which some 200,000 took part. In 1964 he was awarded the Nobel Peace Prize. His campaign broadened from civil rights for the black population to a criticism of the Vietnam War and of society's neglect of the poor. He was about to organize a Poor People's March to Washington when he was assassinated in Memphis, Tennessee (4 April 1968).

Kingitanga, a *Maori movement in New Zealand intended to unify the Maori under an hereditary kingship and restrain individual chiefs from selling land. In 1858, under the guidance of Wiremu Tamihana (the king-maker), *Potatau, the first king, was recognized by tribes of central North Island. The Kingitanga sought to establish and enforce its own laws, but its more moderate leaders, including Tamihana, were willing to contemplate a defined authority under the British crown. Governor George *Grey was disinclined to recognize a movement which would hinder British authority and settlement. Independent-minded members of the Kingitanga such as Rewi Maniapoto became involved in the *Taranaki war and gave Grey grounds for invading the Waikato in 1863. Even so, for many years, government authority did not run in 'the King Country'. In 1883, the King Country chiefs admitted settlement and King *Tawhiao returned to his traditional land in lower Waikato. In the 20th century, largely under the influence of *Te Puea, the Kingitanga came to terms with government and became a focus for economic and cultural revival.

Kiribati (pronounced Kiribas), a group of islands in the central and southern Pacific, named the Gilbert Islands in the 1820s by the Russian hydrographer Krusenstern. From 1837 European traders began to live in the group, over which Britain declared a protectorate in 1892. In 1916 the group was named a crown colony, as the Gilbert and Ellice Islands. In 1942 Japanese naval forces occupied the islands, and in 1943 US marines landed and crushed Japanese resistance after fierce fighting. In 1974 the Ellice Islanders voted to secede from the colony, which became the independent nation of Kiribati (1979). The Ellice Islands became independent as *Tuvalu in 1978.

Kirk, Norman (1923-74), New Zealand statesman. A long-time Labour Party member, he entered Parliament in 1957. Making his mark quickly, he successfully challenged A. H. Nordmeyer as parliamentary party leader in 1965. After two defeats, Kirk led the party to a landslide victory in 1972. As Prime Minister he embarked on a programme of social reform. After Kirk's death in 1974, his government swiftly lost popularity and was defeated in the 1975 general election.

Kirov, Sergei Mironovich (1886-1934), Soviet revolutionary leader. A strong supporter of *Stalin, he began his revolutionary activities in Caucasia but moved to Leningrad (1928) and became a member of the

*Politburo (1930). In 1934 he was assassinated by a young party member, Leonid Nikolayev, possibly at Stalin's instigation. Stalin used Kirov's murder to launch the show trials and party purges of the late 1930s.

Kishi Nobusuke (1896-1987), Japanese statesman. A member of *Tojo's government, he was increasingly opposed to Japan's policies later in World War II. Imprisoned in 1945, he was released without trial. Elected to the Japanese House of Representatives (1953), he emerged as leader of the *Liberal Democratic Party, becoming Prime Minister in 1957. In foreign affairs he aimed to ease tensions with neighbouring Asian countries, while encouraging the US-Japanese link. His domestic policy was conservative, especially in education and over law and order. He resigned in 1960, following a riot within the Japanese Diet building, allegedly over his revised *Japan-US Security Treaty.

Kissinger, Henry Alfred (1923-), US statesman. He acted as government consultant on defence (1955-68) and was appointed by President Nixon as head of the National Security Council (1969-75) and as Secretary of State (1973-7). He was largely responsible for improved relations (*détente*) with the Soviet Union, resulting in the *Strategic Arms Limitation Treaty (SALT) of 1969. In addition, he helped to achieve a resolution of the Indo-Pakistan War (1971), rapprochement with communist China (1972), which the USA now recognized for the first time, and above all the resolution of the *Vietnam War. This he had at first accepted, supporting the bombing offensive against Cambodia (1969-70), but he changed his views and after prolonged negotiation he reached agreement for the withdrawal of US troops in January 1973. Later in that year he helped to resolve the Arab-Israeli War and restored US diplomatic relations with Egypt. After the *Watergate Scandal and President Nixon's resignation, he remained in office to advise President *Ford.

Kita Ikki (d. 1937), Japanese revolutionary and political thinker. A former socialist and member of the nationalist Kokuryukai (Black Dragon Society), Kita played a key role in the upsurge in violent right-wing militarism in the 1930s, inspiring young dissidents with his call for a revolutionary regime, headed by the military, which would nationalize wealth, sweep away existing political forms, and prepare Japan to establish leadership over all of Asia. He was executed in 1937 for alleged involvement in the *Incident of 26 February 1936.

kitchen cabinet, a popular term for unofficial advisers to a President or Prime Minister. The term was coined during the first years of Andrew *Jackson's Presidency in the USA (1829-37). In his first years of office Jackson's official cabinet contained many strong but opposed personalities, including his first Vice-President, John *Calhoun, and his Secretary for War, John Eaton. Thus while official cabinet meetings were held as seldom as possible Jackson took most of his advice from *Van Buren (later his second Vice-President and successor), John Eaton, Amos Kendall, Francis Blair (newspaper editors), and various personal friends appointed as minor government officials. After a cabinet reorganization in 1831 the President relied rather more on members of his official cabinet.

A portrait of **Kitchener** of Khartoum, perhaps the best known of all British military commanders, was the obvious choice for the early recruiting poster shown here. It urges volunteers to join the British Army for service in World War I. (Imperial War Museum, London)

Kitchener of Khartoum and of Broome, Horatio Herbert, 1st Earl (1850–1916), British general. He commanded the Anglo-Egyptian army which conquered the *Sudan (1896–8). His organization of supplies and the effective use made by his troops of the machine-gun, a recent invention, enhanced his reputation and made him a popular hero. In the Second *Boer War he was chief-of-staff to Lord Roberts, and had to curb the activities of Boer guerrilla fighters in 1900–2. The destruction of Boer farmhouses and placing of noncombatants in *concentration camps earned him criticism from Liberal politicians. Nevertheless, the peace terms of the Treaty of *Vereeniging (1902) owed much to him. He served in India as commander-in-chief and in Egypt (1911–14) before being appointed Secretary of State for War. Unlike many of his colleagues, he realized that the war would be a long one and campaigned successfully to secure volunteers. It was largely due to his determination that Britain survived the disasters of the first two years of the war. Set-backs on the *Western Front, blunders over the supply of artillery shells, and Kitchener's advice to abandon the *Dardanelles campaign, which ended disastrously, damaged his reputation. He was drowned in 1916.

Knights of Labor, a US industrial trade union, founded in 1869 at a tailors' meeting in Philadelphia. By 1879 it was organized on a national basis, with membership open to all workers. Its goals were reformist rather than radical, and included the demand for an eight-hour day. Its growth was phenomenal. In 1882 the Knights helped push through Congress the Chinese Exclusion Act, prohibiting the entry into the USA of Chinese labourers. The union was at its height in 1886 under the leadership of Terence V. Powderly, with a membership of almost a million, but declined thereafter, partly due to involvement in unsuccessful strikes and to general antipathy to labour organizations after the *Haymarket Square Riot. Factional disputes reduced its membership after the *American Federation of Labor was founded, and by 1900 it was virtually extinct.

Know-Nothings (Native American Party), a US political party, formed in New York in 1849 as a secret oath-bound society calling itself the Star Spangled Banner. The Party derived its later name from the members' standard response to questions about their activities. In reaction to increasing Irish immigration, the Party called for a twenty-one-year naturalization period and the exclusion of Catholics and foreigners from office, and began to win local and state elections in many parts of the country. In 1854, after the passing of the *Kansas–Nebraska Act, the Party abandoned secrecy and entered the national stage as the Native American Party. It was joined by seceding *Whigs under ex-President *Fillmore, but was increasingly divided over the slavery issue. It failed badly in the 1856 presidential election and quickly faded away.

Knox, Philander Chase (1853–1921), US statesman. Born in Pennsylvania, he served in the cabinets of three Presidents (*McKinley, T. *Roosevelt, and *Taft) and also served in the US Senate (1904–9, 1917–21). He initiated the policy referred to as *dollar diplomacy by his methods of protecting US financial and big business interests abroad. He was most criticized for having US marines occupy Nicaragua to save New York bankers from loss during the revolution of 1912.

Königgratz   *Sadowa.

Kolchak, Aleksander Vasileyvich (1874–1920), Russian admiral and explorer. After Russia's defeat by Japan (1905), he helped to reform the navy and explored a possible route between European Russia and the Far East. After the *Russian Revolution he became War Minister in an anti-*Bolshevik government at Omsk (October 1918), and proclaimed himself supreme ruler of Russia. With *Denikin he fought the Bolsheviks, clearing them from Siberia, but ultimately failed, owing to defections among his supporters. Betrayed to the Bolsheviks, he was shot.

Konfrontasi, diplomatic and military confrontation between *Indonesia and *Malaysia (1963–6). It centred around the formation of the Federation of Malaysia (1963) which President *Sukarno saw as a Western inspired ploy to oppose anti-colonist forces in south-east Asia. Asserting that the Federation was part of a British plot against Indonesia, Sukarno launched a guerrilla war in Malaysia's Bornean territories, Sarawak and Sabah, in April 1963, hoping for support from local Chinese communist elements. His 'confrontation' policy, however, only served to increase support for the new federal arrangements within the Malaysian states (only *Brunei, with its massive oil reserves, remaining aloof). It led to increased disaffection in the Indonesian army which ultimately contributed to his downfall. With the guerrilla forces defeated by Malaysians with British, Australian, and New Zealand help, Sukarno's successor General *Suharto ended Konfrontasi in 1966.

Koniev, Ivan Stepanovich (1897–1973), Soviet field-marshal. He joined the *Red Army and the Communist Party in 1918. Having escaped *Stalin's purge of the Red Army, he commanded several army groups in World War II. In 1945 his 1st Ukrainian Army Group advanced through Poland and Silesia and took a major part in the capture of Berlin.

Konoe Fumimaro (1891–1945), Japanese statesman. He entered politics after World War I and as a member of the upper house emerged as a leading advocate of popularly based parliamentary democracy and opponent of the military domination of government. As Prime Minister (1937–9, 1940–1), he strove unsuccessfully to control the political situation and prevent war with the USA, but in October 1941 was forced out of office by his War Minister, *Tojo Hideki. He committed suicide in December 1945 when summoned to answer charges of war crimes.

Korea, north-east Asian country. A vassal state of China since the 17th century, Korea was ruled by the Yi dynasty. In the 19th century it became the object of intense Russian and Japanese rivalry. Opened to Japanese trade in 1876, Korea was granted independence by the Treaty of *Shimonoseki in 1895, only to become a battle ground during the *Russo-Japanese War (1904–5), and finally to be annexed by Japan in 1910. After *World War II, Korea was divided into US and Soviet zones of occupation along the 38th parallel before the proclamation of the independent Korean People's Democratic Republic (*North Korea) and Republic of Korea (*South Korea) in 1948. Rival plans for unificaiton led to the invasion of the south by the communist north and the start of the *Korean War (1950–3) which saw heavy intervention by the United Nations and communist China. The restoration of peace returned the border to the pre-war line, but tension remained high until ameliorated to some extent by an agreement signed by the North and South Korean governments in July 1972, which laid foundations for possible future reunification.

Korean War (1950–3), war fought between North Korea and China on one side, and South Korea, the USA, and United Nations forces on the other. From the time of their foundation in 1948, relations between North and South Korea were soured by rival plans for unification, and on 25 June 1950 war finally broke out with a surprise North Korean attack. In the temporary absence of the Soviet representative, the Security Council asked members of the UN to furnish assistance to South Korea. On 15 September US and South Korean forces, under command of General *MacArthur, launched a counter-offensive at Inchon, and by the end of October UN forces had pushed the North Koreans all the way back to the Yalu River near the Chinese frontier. Chinese troops then entered the war on the northern side, driving south as far as the South Korean capital of Seoul by

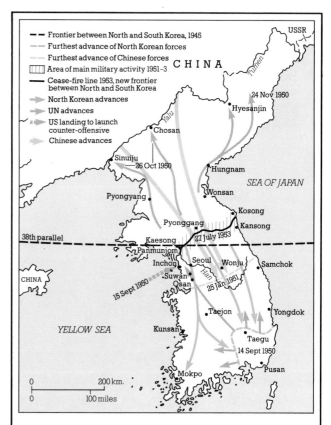

Korean War (1950–3)

In 1945 Japanese forces surrendered in North Korea to the Soviet Union and in the south to the USA. Two new nations were created, with a frontier along the 38th parallel. US troops withdrew in 1949, whereupon North Korea attempted unification by force. The United Nations sent reinforcements, mainly US servicemen, to South Korea, and the Chinese aided the North. In 1953 a ceasefire was negotiated by General Ridgway. Of more than 10,000 US soldiers captured, only 3,746 returned after the war.

January 1951. After months of fighting, the conflict stabilized in near-deadlock close to the original boundary line. Peace negotiations, undertaken in July 1951 by General M. B. Ridgway (who had succeeded MacArthur in April of that year), proved difficult, and it was not until 27 July 1953 that an armistice was signed at Panmunjom and the battle line was accepted as the boundary between North and South Korea.

Kossuth, Lajos (1802–94), Hungarian public official and revolutionary. He made his name as deputy for an absentee count in the Hungarian Diet of 1832–6. His reports on the proceedings (the debates were not officially published) were widely read and led to his arrest in 1837. Popular agitation persuaded *Metternich to release him three years later, but he intensified his demands for national independence. He led the extreme liberals in the Hungarian Diet of 1847, and in 1848 became Minister of Finance in the government of a semi-autonomous Hungary. When Austria and Hungary went to war at the end of 1848, he became the virtual dictator of the nation, declaring its independence (1849) with himself as President. Hungarian forces repulsed Austrian military intervention, but when Russian troops intervened in favour of Austria, Kossuth surrendered and fled the country, when his republic collapsed.

Kosygin, Alexei Nikolayevich (1904–81), Soviet politician. He joined the Communist Party in 1927 and became an expert in economics and industry. He was Chairman of the Council of Ministers from 1964 to 1981. During his period in office he shared power with *Brezhnev, who came to overshadow him. Kosygin achieved a notable diplomatic success in bringing the 1965–6 *Indo-Pakistan War to an end.

Kropotkin, Peter, Prince (1842–1922), Russian *anarchist. From 1872, after meeting leaders of the First *International in Switzerland, he became interested in revolutionary ideas. Imprisoned in France (1883–6) for his anarchist views he moved to Britain where he was welcomed as a scholar. In his most famous book, *Mutual Aid* (1902), he refuted the Darwinian theory of human society as essentially competitive, and in *Fields, Factories and Workshops* (1898) he outlined a social organization based upon communes of producers linked with each other through free contract. Returning to Russia (1917) he supported *Kerensky, but remained an outspoken critic of *Bolshevism.

Kruger, (Stephanus Johannes) Paul(us) (1825–1904), South African statesman. From 1864 he was Commandant-General of the South African Republic (*Transvaal) until it was annexed by the British in 1877. After it regained its independence (1881) he was elected President (1883), and re-elected in 1888, 1893, and 1898. He consistently pursued an expansionist policy, in *Bechuanaland, *Rhodesia, and *Zululand, so as to enlarge the Transvaal frontiers. He successfully defeated

A French cartoon of 1900 supports **Kruger** (*foreground*) and the Boer cause against Britain. John Bull, 'le Petit', is shown hindered by a ball and chain, vainly pursuing with a butterfly net the elusive Boer guerrilla leader Christiaan de Wet.

A warning from the **Ku Klux Klan** in the aftermath of the South's defeat in the American Civil War. This woodcut predicts the fate in store for 'those great pests of Southern society—the carpetbagger and scalawag' (a white Southerner supporting the Republicans) if found in Alabama after 4 March 1869. It appeared in a Tuscaloosa newspaper of September 1868.

the *Jameson Raid, receiving a telegram of congratulations from *William II (1896). Kruger's refusal to allow equal rights to non-Boer immigrants (*Uitlanders) was one of the causes of the Second *Boer War. When Bloemfontein and Pretoria were occupied in 1900, he retired to Utrecht, where his efforts to rouse European support for the Boers were unsuccessful.

Kubitschek, Juscelino (1902–76), Brazilian statesman. He was governor of the Province of Minas Gerais (1950–6), where he initiated a programme of industrial and agricultural development. He then served as President of Brazil (1956–61). Determined to diversify the economy and reduce unemployment, he embarked on a massive public works programme, including the creation of the new capital city of Brasília. Economic prosperity followed, but at the cost of high inflation. Brazil's national debt rose to $4 billion, while its population soared to over sixty million. He was forced into exile for three years by his successor Castel Branco.

Ku Klux Klan, a secret society founded (1866) in the southern USA after the *American Civil War to oppose *reconstruction and to maintain white supremacy. Famous for its white robes and hoods, it spread fear among blacks to prevent them voting. Its use as a cover for petty persecution alienated public opinion and led to laws in 1870 and 1871 attempting to suppress it. The Klan reappeared in Georgia in 1915 and during the 1920s spread into the north and mid-west. It was responsible for some 1,500 murders by lynching. At its height it boasted four million members and elected high federal and state officials, but it also aroused intense opposition. A series of scandals and internecine rivalries sent it into rapid decline. It survives at the local level in the southern states.

kulak (Russian, 'fist'), originally applied to money-lenders, merchants, and anyone who was acquisitive. The term became specifically applied to wealthy peasants who, as a result of the agrarian reforms of *Stolypin (1906), acquired relatively large farms and were financially able to employ labour. As a new element in rural Russia they were intended to create a stable middle class and a conservative political force. During the period of Lenin's *New Economic Policy (1921) they increasingly appeared to be a potential threat to a communist state, and Stalin's *collectivization policy (1928) inevitably aroused their opposition. Between 1929 and 1934 the great majority of farms were collectivized and the kulaks annihilated.

Kulturkampf (German, 'conflict of cultures' or beliefs), the conflict between the German government headed by *Bismarck and the Roman Catholic Church (1872–87) for the control of schools and Church appointments. Bismarck, anxious to strengthen the central power of the *German Second empire in which southern Germany, Alsace-Lorraine, and the Polish provinces were predominantly Catholic, issued the May Decrees (1873), restricting the powers of the Catholic Church and providing for the punishment of any opponents. By 1876 1,300 parishes had no priest: opponents had become martyrs. Needing Catholic support in the Reichstag, Bismarck repealed many of the anti-Church laws or let them lapse.

Kun, Béla (1886–1937), Hungarian communist leader. In World War I he was captured on the Russian front and joined the *Bolsheviks. He was sent back to Hungary to form a communist party and in March 1919 persuaded the Hungarian communists and Social Democrats to form a coalition government and to set up a communist state under his dictatorship. His Red Army overran Slovakia, but promised Soviet help was not forthcoming. In May 1919 he was defeated by a Romanian army of intervention. Kun fled the country and is assumed to have been liquidated in a Stalin purge.

Kuomintang (or Guomindang; National People's Party), Chinese political party. Originally a revolutionary league, it was organized in 1912 by Song Jiaoren and *Sun Yat-sen as a republican party along democratic lines to replace the Revolutionary Alliance which had emerged from the overthrow of the *Qing dynasty. Suppressed in 1913 by *Yuan Shikai, it was reformed in 1920 by Sun and reorganized with *Comintern assistance in 1923 in an arrangement that allowed individual communists to become members. At the party congress in 1924 it formally adopted the 'Three Principles of the People': nationalism, democracy, and 'people's livelihood'. In 1926 its rise to power began in earnest with the commencement of *Chiang Kai-shek's *Northern Campaign. The communists were purged in 1927 and the capture of Beijing in 1928 brought international recognition for its Nanjing-based Nationalist Government. It fought the *Chinese Civil War with the communists and retreated to Chongqing after the Japanese invasion of 1937. After World War II, the civil war recommenced, and by 1949 the Kuomintang's forces had been decisively defeated and forced to retreat to *Taiwan, where it still continues to form the government of the Republic of China.

Kurdestan, a mountainous area divided between Turkey, *Iran, and *Iraq, with small areas in Syria and Soviet Armenia, inhabited by Islamic Kurds. Kurdish nationalism in the *Ottoman empire developed in the late 19th century; after World War I the Treaty of Sèvres promised an independent Kurdestan, but this never materialized and there has been spasmodic armed resistance in Kurdish Turkey, notably during 1944–5, 1978–9, and 1984. In the years 1944–5 a short-lived Kurdish Republic of Mahabad was formed in Iran with Soviet help. An armed struggle for autonomy in Iraq (1958–74) resulted in a plan for limited autonomy in 1974, but fighting resumed in 1975 with military assistance from Iran. In 1979 Iran granted its Kurds limited autonomy. Guerrilla fighting in Iran, Iraq, and Turkey continues.

Kursk, battle of (5–15 July 1943), a fierce tank battle between the Red Army and Hitler's German forces around Kursk in the central European Soviet Union. Hitler had ordered the elimination of the important railway junction of Kursk. Under Field-Marshal Walter Model he concentrated 2,700 tanks and assault guns on the city, supported by over 1,000 aircraft. They were confronted by Marshal *Zhukov's Tank Army, backed by five infantry armies. Many of the large German tanks were mined and others became stuck in the mud. The Russians had more guns, tanks, and aircraft, and when they counter-attacked, the Germans were forced to retreat, losing some 70,000 men, 1,500 tanks, and 1,000 aircraft. The battle ensured that the German army would never regain the initiative on the *Eastern Front.

Kut, siege of (December 1915–April 1916), successful siege by Turkish troops in World War I. Kut-al-Amara is on the River Tigris and was garrisoned by a British imperial force under General Townshend, who had retreated there after his defeat by the Turks at Ctesiphon. Badly organized relief forces failed to break through and the garrison capitulated on 29 April 1916 after a four-month siege. Ten thousand prisoners were marched across the desert, two-thirds dying on the way, while some 23,000 troops of the relieving force were also lost. The defeat severely weakened Britain's prestige as an imperial power although Kut-al-Amara was recaptured in February 1917.

Kutuzov, Mikhail Ilarionovich, Prince of Smolensk (1745–1813), Russian field-marshal. He distinguished himself in the Russo-Turkish War (1806–12), bringing Bessarabia into Russia. He commanded the Russian armies in the wars against Napoleon and was forced to retreat after the defeat of *Borodino (7 September 1812). Kutuzov's decision to disperse in the face of the advancing *grande armée* undermined Napoleon's plans for a swift victory, and forced him to retreat from Moscow before the severe Russian winter.

Kuwait, a country on the north-west coast of the Persian Gulf. It was founded in the early 18th century by members of the Utub section of the Anaiza tribe, and has been ruled since 1756 by the al-Sabah family. In 1899 the ruler, Muvarak, made a treaty with Britain which established a *de facto* British protectorate over Kuwait, although it remained under nominal Ottoman suzerainty until 1914, when the protectorate was formalized. Kuwait became independent in 1961 when an Iraqi claim was warded off with British military assistance. Oil was discovered in 1938 and after World War II it became one of the world's largest oil producers.

L

Labor Party (Australia), the oldest surviving political party in Australia. Founded in the 1880s and 1890s, the title of the Labor groups varied from state to state until 1918, when all adopted the name Australian Labor Party. Labor governments existed briefly in 1904 and 1908-9, and at the federal general election of 1910 Labor obtained clear majorities in both houses, remaining in power until 1912. Under W. H. *Hughes (1915-17), the government established some social reforms, but split in 1916 when a majority voted against conscription. It was replaced by a Nationalist-Country Alliance, until the general election of 1929 returned it to power under J. H. Scullin (1929-31). Labor split again over policy differences during the Great *Depression. Some Labor followers combined with the Nationalist Party to form the United Australia Party under J. A. Lyons. Together with the Country Party it dominated federal and state politics until 1937, usually in coalition governments. The Labor Party was again in power 1941-9. A breakaway Labor group emerged in 1955 over the attitude of the Party to communism, a group of federal Labor members forming the new Anti-Communist Labor Party, which later became the Democratic Labor Party. The Party suffered the *Whitlam Crisis in the 1970s, but has remained in power during the 1980s despite a programme of economic austerity.

Labour Party (Britain), a major political party in Britain. Following the third *Reform Act (1884), a movement developed for direct representation of labour interests in Parliament. In 1889 a Scottish Labour Party was formed, winning three seats in 1892, including one by Keir *Hardie, who next year helped to form the *Independent Labour Party, advocating pacifism and *socialism. In 1900 a Labour Representative Committee was formed which in 1906 succeeded in winning twenty-nine seats and changed its name to the Labour Party, though still a loose federation of trade unions and socialist societies. In 1918 the Party adopted a constitution drawn up by the Fabian Sidney *Webb. Its main aims were a national minimum wage, democratic control of industry, a revolution in national finance, and surplus wealth for the common good. By 1920 Party membership was over four million. The Party now became a major force in British municipal politics, as well as gaining office with the Liberals in national elections in 1923 and 1929. The Party strongly supported war in 1939 and through leaders such as *Attlee, *Bevin, and *Morrison played a major role in Winston *Churchill's government (1940-5). In 1945 it gained office with an overall majority and continued the programme of *welfare state legislation begun during the war. It was in power (1964-70) when much social legislation was enacted, and 1974-9, when it faced grave financial and economic problems. During the 1970s and early 1980s left-wing activists pressed for a number of procedural changes; for example in the election of Party leader. From the right-wing a group of senior Party members split from the Party in the 1980s to form the *Social Democratic Party. The Party has always favoured military disarmament and in 1986 adopted a policy of unilateral nuclear *disarmament.

Labour Party (New Zealand), a major political party in New Zealand. It was formed in 1910 out of the trade union movement but was rivalled by the militant *'Red Feds'. Re-formed in 1916, the Party supported compulsory industrial arbitration and constitutional change. Its policies favoured nationalization of much industry and state leasehold of land, and these were further modified before it won office in 1935 under Michael *Savage. The first Labour government (1935-49) effected radical change, stimulating economic recovery through public works, state support for primary produce marketing, and minimum wages. It introduced *social security, including free medical care. In World War II the Labour government declared war on Germany, introduced conscription for military service, and entered strongly into collective security arrangements. The Labour Party has held office again (1957-60, 1972-5) and since 1984. It has devalued, then floated, the exchange rate, removed subsidies, and vigorously de-regulated the economy. It has implemented an anti-nuclear defence and foreign policy, and seeks improvements in health, education, and social welfare.

La Fontaine, Sir Louis-Hippolyte (1807-64), French-Canadian statesman. A member of the legislative assembly of Lower Canada (1830-7), he opposed *Papineau's Rebellion and was not in sympathy with *Mackenzie's Rebellion of 1837. An outspoken advocate of nationalism, he was arrested in 1838, but soon released, and, after the union of *Upper and Lower Canada (1841), assumed political leadership of the French-Canadian reformers. In partnership with Robert *Baldwin he twice formed the government of United Canada (1842-3, 1848-51), on the second occasion serving as Prime Minister in an administration which was notable for its reforms and its achievement of full parliamentary ('responsible') government in Canada. He left politics in 1851.

Lake Erie, battle of (10 September 1813), a naval engagement in the *War of 1812. Control of Lake Erie was of critical importance to both the USA and the

The eventual victor of the battle of **Lake Erie**, Commodore Oliver Perry, transfers from his disabled flagship *Lawrence* to the *Niagara*. He then sailed into the British line, compelling their surrender on 10 September 1813. (National Maritime Museum, London)

British, and by the late summer of 1813 each had commissioned a locally built and manned squadron of warships. Ten US vessels under Commodore Oliver Hazard Perry met six British vessels under Captain James Barclay off Put-in-Bay. After hard fighting, Perry captured the entire British squadron. His victory opened the way for a renewed US attack on Canada.

Lancaster, Joseph (1778–1838), British teacher and pioneer of mass education. As a young man he became a Quaker and in 1798 opened a school for poor children. Since he could not afford to employ other teachers, he devised a monitorial *school system whereby the children taught one another, the abler pupils passing on to the rest what they had learnt from the teacher and at the same time helping to keep order. His teaching methods were widely adopted and paved the way for universal elementary education in Britain.

Land League, an agrarian organization in Ireland. It was founded in 1879 by an ex-*Fenian, Michael *Davitt, to secure reforms in the land-holding system. With Charles *Parnell as president, it initiated the *boycotting of anyone replacing a tenant evicted because of non-payment of rent. The campaign for land reform was linked with parliamentary activity by the Irish *Home Rule Members of Parliament. The British government declared the Land League illegal and imprisoned Davitt and Parnell. Branches of the League were formed in Australia, the USA, and elsewhere. Between 1881 and 1903 British governments passed Land Acts to remove the worst features of the landlord system in Ireland.

Lansbury, George (1859–1940), British Labour politician and pacifist. As leader of Poplar Council in east London in 1921, he went to prison rather than reduce relief payments for the unemployed. Refusing to join the *National government in 1931, he became leader of the rump of the Labour Party (1931–5). His rejection of sanctions against Italy, following that country's invasion of Ethiopia in 1935, alienated his colleagues, and he resigned.

Lansdowne, Henry Charles Keith Petty-Fitzmaurice, 5th Marquis of (1845–1927), British statesman. As Foreign Secretary in the *Salisbury and then *Balfour governments (1900–5) Lansdowne negotiated the *Anglo-Japanese alliance (1902) and the *entente cordiale* (1904) with France. From 1906 Lansdowne led the Conservative Opposition in the House of Lords. His use of the veto on legislation from the House of Commons resulted in the Parliament Act (1911), which reduced the power of the *Lords to a two-year suspensory veto. He was the author of the Lansdowne letter of November 1917 in the *Daily Telegraph* advocating a negotiated peace with Germany.

Laos, a country in south-east Asia. United as Lanxang from the 14th century, Laos broke up into rival kingdoms in the 18th century and gradually fell under Siamese (Thai) domination before Siam was forced to yield its claim to France in 1893. Occupied by the Japanese during World War II, it emerged briefly as an independent constitutional monarchy (1947–53), but was undermined by guerrilla war as a result of the increasing influence of the communist *Pathet Lao as a political force in the

mid-1950s. A coalition government was established under Prince *Souvanna Phouma in 1962, but fighting broke out again soon after, continuing into the 1970s, with Laos suffering badly as a result of involvement in the Vietnam War. A ceasefire was signed in 1973 and a year later Souvanna Phouma agreed to share power in a new coalition with the Pathet Lao leader, his half-brother Souphanouvong, but by 1975 the Pathet Lao were in almost complete control of the country and on 3 December the monarchy was finally abolished and the People's Democratic Republic of Laos established, which has maintained close links with Vietnam.

Largo, Caballero Francisco (1869–1946), Spanish statesman. As a socialist he was imprisoned for life in 1917 for taking part in a general strike, but released on his election to Parliament in 1918. After the fall of *Primo de Rivera (1930) he joined the government of the Second Republic as Minister for Labour; after this collapsed he was imprisoned again (1934–5) for supporting an abortive rising, but acquitted and released. He was leader of the Popular Front, which won the elections of February 1936, but did not become Prime Minister until September 1936, two months after the outbreak of the *Spanish Civil War, when he headed a coalition of socialists, republicans, and communists. He resigned following a communist take-over in Barcelona in May 1937.

Lateran Treaties (11 February 1929), agreements between *Mussolini's government and Pius XI to regularize relations between the Vatican and the Italian government, strained since 1870 when the Papal States had been incorporated into a united Italy. By a treaty (*concordat) and financial convention the *Vatican City was recognized as a fully independent state under papal sovereignty. The concordat recognized Roman Catholicism as the sole religion of the state. The Vatican received in cash and securities a large sum in settlement of claims against the state.

La Trobe, Charles Joseph (1801–75), British colonial administrator. He became superintendent of the newly settled Port Phillip District of New South Wales in Australia in 1839. When it was separated from New South Wales (and re-named *Victoria) in 1851, he became lieutenant-governor. Almost immediately he was confronted with the *gold rushes, when the population of Victoria rose in six months from 15,000 to 80,000. He introduced, among other measures, the licence system which later became a cause of the *Eureka Rebellion.

Laurier, Sir Wilfrid (1841–1919), Canadian statesman. He entered *Quebec politics as an anti-clerical Liberal, and was elected to the Canadian House of Commons in 1874, where he held a seat until his death. He became Liberal leader in 1887 (the first French-Canadian to lead a national party), and was victorious in the elections of 1896, 1900, and 1904 remaining Prime Minister until 1911. His policies sought to avoid closer ties with the British empire and to establish greater Canadian autonomy, while advocating a removal of economic barriers with the USA. He could not, however, escape the growing disapproval of Quebec nationalists, led by his former supporter, Henri Bourassa. Nevertheless he supported Canadian entry into World War I, while opposing, as far as possible, conscription.

Laval, Pierre (1883–1945), French politician. He trained as a lawyer before entering politics as a socialist. Gradually moving to the right, in 1931–2 and 1935–6 he was Prime Minister but was best known as Foreign Minister (1934, 1935–6), when he was the co-author of the Hoare–Laval pact for the partition of Ethiopia between Italy and Ethiopia. He fell from power soon after, but after France's defeat in 1940 he became chief minister in the *Vichy government. He advocated active support for Hitler, drafting labour for Germany, authorizing a French fascist militia, and instituting a rule of terror. In 1945 he was tried and executed in France.

Law, (Andrew) Bonar (1858–1923), British politician. He became leader of the Conservative Party in 1911, and supported Ulster's resistance to *Home Rule. A tariff reformer, in 1915 he joined *Asquith's coalition as Colonial Secretary and continued under *Lloyd George, serving as Chancellor of the Exchequer (1916–19) and Lord Privy Seal (1919–21). In 1922 the Conservatives rejected the coalition government of Lloyd George and he was appointed Prime Minister. He resigned the following May for reasons of ill health.

Lawrence, John Laird Mair, 1st Baron (1811–79), British colonial administrator. As civil servant under the East India Company, he joined his brother, Sir Henry Lawrence (1806–57) on the Punjab Board in 1849, and in 1853 became chief commissioner of the Punjab, distinguishing himself by his control of that province during the *Indian Mutiny. Returning to London, he was appointed viceroy (1864) in succession to Lord Elgin. He encouraged the expansion of public works programmes in India but opposed the expansionist policies that led to the Second *Anglo-Afghan war.

Lawrence, T(homas) E(dward) (1888–1935), British soldier, scholar, and author. He worked as an archaeologist in the Near East before World War I, when he joined the Arab Bureau in Cairo. He played a major role in support of the Arab Revolt, notably with Amir *Faisal. He took part in the capture of Damascus (1918) and subsequently argued for British support of Arab claims in Syria. In 1921 Lawrence joined Churchill's new Middle Eastern Department as adviser and helped to plan the Middle East settlement of that year. He then withdrew from public life and enlisted in the ranks of the Royal Air Force under the name of John Hume Ross. In 1923 he joined the Tank Corps as T. E. Shaw, but returned to the RAF in 1925. His account of the Arab Revolt entitled *The Seven Pillars of Wisdom* (1926) has become one of the classics of English literature.

League of Nations, an organization for international co-operation. It was established in 1919 by the *Versailles Peace Settlement. A League covenant embodying the principles of collective security, arbitration of international disputes, reduction of armaments, and open diplomacy was formulated. Germany was admitted in 1926, but the US Congress failed to ratify the Treaty of Versailles, containing the covenant. Although the League, with its headquarters in Geneva, accomplished much of value in post-war economic reconstruction, it failed in its prime purpose through the refusal of member nations to put international interests before national ones. The League was powerless in the face of Italian, German,

A rubble-strewn street in Beirut, the capital of **Lebanon**, after a bomb explosion in 1979. Civil war has reduced the once prosperous and cosmopolitan city to a network of barricaded armed camps, and turned its streets into battlefields.

and Japanese expansionism. In 1946 it was replaced by the *United Nations.

Lebanon, a country in south-west Asia with a coastline on the Mediterranean Sea. Part of the Ottoman empire from the 16th century, it became a French *mandate after World War I. A Lebanese republic was set up in 1926. The country was occupied (1941–5) by *Free French forces, supported by Britain. Independence was achieved in 1945. Growing disputes between Christians and Muslims, exacerbated by the presence of Palestinian refugees, undermined the stability of the republic. Hostility between the differing Christian and Muslim groups led to protracted civil war and to the armed intervention (1976) by Syria. The activities of the *Palestine Liberation Organization brought large-scale Israeli military invasion and led to Israeli occupation (1978) of a part of southern Lebanon. A UN peace-keeping force attempted unsuccessfully to set up a buffer zone. A full military invasion (1982) by Israel led to the evacuation of the Palestinians. A massacre by the Phalangist Christian militia in Israeli-occupied West Beirut of Muslim civilians in the Chabra and Chatila refugee camps brought a redeployment of UN peace-keeping forces. Syria again intervened in 1987, but many problems remained unresolved.

Lebensraum (German, 'living-space'), *Nazi political doctrine. It claimed the need to acquire more territory in order to accommodate the expanding German nation. The term was first introduced as a political concept in the 1870s, but was given patriotic significance by *Hitler and *Goebbels. The corollary to '*Lebensraum*' was the '*Drang nach Osten*' (German, 'drive to the East'), which claimed large areas of eastern Europe for the *Third Reich as territories where the Nazi master race should subjugate and colonize the Slavic peoples.

Lebrun, Albert (1871–1950), French statesman, 17th and last President (1932–40) of the Third Republic. A moderate conservative, he pursued a respected but unspectacular parliamentary career (1920–32) until his election to the Presidency. He acquiesced in the French armistice (1940) that led to the *Vichy government of Marshal *Pétain. He was interned in Austria (1943–4) until the liberation of France, when he acknowledged *de Gaulle as head of the provisional government.

Lee, Robert Edward (1807–70), North American general in the army of the Southern *Confederacy. A member of a prominent Virginia military family, Lee served in the *Mexican–American War. While on leave from a posting in Texas, he supervised the capture of

A painting by John Adams Elder of Robert E. **Lee**, commander of the Confederate Army of Northern Virginia, and the Confederacy's most admired leader.

Lee Kuan Yew, Prime Minister of Singapore since independence. He is renowned as a skilful communicator abroad and a shrewd political tactician at home.

John *Brown at Harper's Ferry in 1859. Offered the field command of the Union (Northern) army at the outbreak of the *American Civil War, he refused, but instead became military adviser to President *Davis, and then in June 1862 Commander of the (Confederate) Army of Northern Virginia. He ended the threat of General McClellan to Richmond in the Seven Days' battle and then forced the Union Army to retreat from Virginia after the second battle of Bull Run (August 1862). Although his first invasion of the north was checked at *Antietam, he won major victories at Fredericksburg (December 1862) and Chancellorsville (May 1863) and invaded again, only to be defeated at *Gettysburg. He then held off attacks by General *Grant on Richmond for almost a year in the Wilderness and Petersburg campaigns, before being forced to surrender at *Appomattox. A master of both strategy and tactics, Lee was idolized by his men and did more than any other leader to keep the Confederate cause alive during the war.

Lee Kuan Yew (1923–), Singapore statesman, Prime Minister (1959–). In 1955, he formed the People's Action Party, a democratic socialist organization, which under his leadership has dominated Singapore politics since the late 1950s. He led Singapore as a component state of the newly formed Federation of *Malaysia in 1963, and then as a fully independent republic. Since then his policies have developed along increasingly authoritarian socialist lines and have centred on the establishment of a one-party rule and a free-market economy, tight government planning, and a hard-working population supported by an extensive social welfare system.

Leguía, Augusto Berardino (1863–1932), Peruvian statesman. As leader of the Civilian Party he was Prime Minister of Peru (1903–8) and President (1908–12, 1919–30). During his first term in office he settled frontier disputes with Bolivia and Brazil, introduced administrative reforms, and improved the public health system. He was reinstated as President by the army in 1919, introducing a new constitution in 1920. He chose largely to ignore this, governing by increasingly dictatorial methods. His second term saw rapid industrialization, but was adversely affected by the Great *Depression and fall in commodity prices. Criticized also for the *Tacna–Arica Settlement, he lost popularity and was forced from office by the military.

Leipzig, battle of (also called the 'Battle of the Nations', 16–19 October 1813), a decisive battle in the *Napoleonic Wars. It was fought just outside the city of Leipzig in Saxony, by an army under Napoleon of some 185,000 French, Saxon, and other allied German troops, against a force of some 350,000 troops from Austria, Prussia, Russia, and Sweden, under the overall command of *Schwarzenberg. Napoleon took up a defensive position and at first successfully resisted attacks by Schwarzenberg from the south and *Blücher from the north. Next day Russian and Swedish troops arrived, while Napoleon's Saxon troops deserted him. The battle raged for nine hours, but at midnight Napoleon ordered a retreat. This began in an orderly fashion until, early in the afternoon of 19 October, a bridge was mistakenly blown up, stranding the French rear-guard of 30,000 crack troops, who were captured. Following the battle French power east of the Rhine collapsed as more and more German princes deserted Napoleon, who abdicated in 1814.

Lend–Lease Act, an arrangement (1941–5) whereby the USA supplied equipment to Britain and its Allies in World War II. It was formalized by an Act passed by the US Congress allowing President F. D. *Roosevelt to lend or lease equipment and supplies to any state whose defence was considered vital to the security of the USA. About 60 per cent of the shipments went to Britain as a loan in return for British-owned military bases. About 20 per cent went to the Soviet Union.

Lenin, Vladimir Ilyich (1870–1924), Russian revolutionary statesman. Born Vladimir Ilyich Ulyanov, a formative influence on his life may have been the execution (1887) of his elder brother at the age of 19, for implication in a plot against the emperor. Lenin himself was arrested in 1895 for propagating the teachings of Karl *Marx among the workers of St Petersburg (now Leningrad), and was for a period exiled in Siberia. Living in Switzerland from 1900, he became the leader of the *Bolshevik party and took a prominent part in socialist organization and propaganda in the years preceding

Lenin (*centre*) with his fellow-revolutionary Trotsky (saluting), stands among snow-covered communist supporters in Soviet Russia in 1920.

World War I. He returned to Russia on the outbreak of the *Russian Revolution and quickly established Bolshevik control, emerging as chairman of the Council of People's Commissars and virtual dictator of the new state. He took Russia out of the war against Germany and successfully resisted counter-revolutionary forces in the *Russian Civil War (1918–21). His initial economic policy (called war communism), including nationalization of major industries and banks, and control of agriculture, was an emergency policy demanded by the civil war, after which his *New Economic Policy (NEP), permitting private production and trading in agriculture, was substituted. It came too late to avert terrible famine (1922–3). He did not live to see the marked recovery as agricultural and industrial production increased. Lenin's own outlook and character deeply affected the form that the revolution took; he set an example of austerity and impersonality which long remained a standard for the Party. Perhaps the greatest revolutionary of all time, later communist leaders have continued to look to his writings for their inspiration.

Leningrad, siege of (September 1941–January 1944), the defence of Leningrad by the Soviet army in World War II. The German army had intended to capture Leningrad in the 1941 campaign against the Soviet Union but as a result of slow progress in the Baltic area and the reluctance of Germany's Finnish ally to assist, the city held out in a siege that lasted nearly 900 days. As few preparations had been made, and as evacuation of the population was not permitted by the Soviet government, there may have been a million civilian deaths in the siege, caused mainly by starvation, cold, and disease. Over 100,000 bombs were dropped over the city, and between 150,000 and 200,000 shells fired at it. Soviet counter-attacks began early in 1943, but it was nearly a year later before the siege was completely lifted.

Leopold I (1790–65), King of the Belgians (1831–65). The son of Francis, Duke of Saxe-Coburg-Saalfeld and uncle of Queen *Victoria, he declined the throne of Greece (1830) before accepting that of the newly formed Belgium in 1831. In 1832 he married the daughter of *Louis-Philippe, King of the French. In 1839 he successfully negotiated peace with William I of Holland. He spent the remainder of his life in maintaining the independence of his kingdom and instituting reforms.

Leopold II (1835–1909), King of the Belgians (1865–1909). His reign saw considerable industrial and colonial expansion, due in large part to the wealth gleaned from the *Congo. The Berlin Colonial Conference (1884–5) had recognized Leopold as independent head of the newly created Congo Free State and he proceeded to amass great personal wealth from its rubber and ivory trade. Thanks to the report of an Englishman, Edmund Morel, his maltreatment of the Congo native population became an international scandal (1904) and he was forced (1908) to hand over the territory to his parliament.

Lesotho, a country in southern Africa, surrounded by the Republic of South Africa. It was founded as Basutoland by Moshoeshoe I in 1832, who built a stronghold on Thaba Bosigo and unified the Sotho (Basuto). After fighting both Boers and British, Moshoeshoe put himself under British protection in 1868, and until 1880 Basutoland was administered from Cape Colony. In 1884 it was restored to the direct control of the British government with the Paramount Chief as titular head. When the Union of South Africa was formed in 1910, Basutoland came under the jurisdiction of the British High Commissioner in South Africa. It was re-named Lesotho and became independent in 1966 as a constitutional monarchy, with a National Assembly (1974) which works with the hereditary chiefs.

Leticia dispute (1932–4), a border dispute between Colombia and Peru. The territory of Leticia on the upper Amazon River was ceded to Colombia by a treaty in 1924; it became an object of contention again in 1932 when Peruvian citizens seized the territory and executed Colombian officials. Colombia carried the dispute to the *League of Nations, which awarded Leticia to Colombia (1934). The active involvement of the USA with the League established a precedent, allowing the interference by an international body in territory protected by the *Monroe Doctrine. The successful reconciliation in the Leticia dispute was only one example of how the League of Nations was a striking success in Latin America.

Lewis, John Llewellyn (1880–1969), US labour leader. He worked as a miner and rose to become President of the United Mine Workers of America (1920–60). He built the union up into one of the strongest in the USA, securing considerable improvements in the wages and conditions of his members. Although he initially supported President F. D. Roosevelt, he led many bitter strikes in the 1930s and during World War II. Although not a communist, he defied the restrictive *Taft–Hartley Act by refusing to declare on oath that he was not a communist, and led his union to break away (1947) from the *American Federation of Labor.

Lewis, Sir Samuel (1843–1903), African jurist and statesman. The son of freed slaves, he trained in England as a barrister, practising in Sierra Leone, Gambia, Nigeria, and the Gold Coast. He served twice as Chief Justice of Sierra Leone. He was a Member of its Legislative Council from 1882 to 1902, and elected (1895) the first Mayor of Freetown. He was the first African to receive a knighthood (1896).

Lewis and Clark expedition (1804–6), the most important transcontinental journey in US history. The expedition was commissioned by President *Jefferson to explore the vast area acquired as a result of the *Louisiana Purchase. Commanded by Meriwether Lewis (1774–1809) and William Clark (1770–1838), it left St Louis in 1804 and sailed up the Missouri to winter in North Dakota before crossing Montana to the foothills of the Rockies. Crossing the Continental Divide at Lemhi Pass, Idaho, the expedition then moved north and in November 1805 reached the Pacific via the Clearwater, Snake, and Columbia rivers. In 1806 the expedition returned, dividing after crossing the Rockies to cover more ground and reuniting near the junction of the Yellowstone and the Missouri before returning to St Louis. Carried out under the most difficult and dangerous conditions, the explorations of Lewis and Clark produced a wealth of topographical, geological, and botanical data, and won them an unrivalled place at the head of the early explorers of the American west.

An engraving by Patrick Gass from the published journal of the **Lewis and Clark expedition** shows Captain Lewis and Lieutenant Clark holding a council with Indians, one of whom is smoking a calumet or peace pipe. (British Library, London)

Leyte Gulf, battle of (October 1944), a naval battle off the Philippines. In the campaign to recover the Philippines, US forces landed on the island of Leyte. Four Japanese naval forces converged to attack US transports, but in a series of scattered engagements 40 Japanese ships were sunk, 46 were damaged, and 405 planes destroyed. The Japanese fleet, having failed to halt the invasion, withdrew from Philippine waters.

Liaqat Ali Khan, Nawabzada (1895–1951), first Prime Minister of Pakistan (1947–51). He began his political career in British India as a Muslim leader in the United Provinces, and from 1933 became *Jinnah's right-hand man in the *Muslim League. In 1946 he became Finance Minister in the Interim Government. Between the death of Jinnah in 1948 and his own assassination in 1951 he was the most powerful politician in Pakistan, when he endeavoured to achieve a reconciliation with India via the Delhi Pact (1950).

Liberal Democratic Party, dominant Japanese post-war political party. Political alignments were slow to coalesce in post-war Japan, but in 1955 rival conservative

US troops land supplies on the Philippine island of Leyte. The amphibious landing on 20 October 1944 took place two days before the decisive naval air battle at **Leyte Gulf** which crippled the Japanese navy, ensured Allied control of the Pacific, and guaranteed the eventual recapture of the Philippines.

groups combined to form the Liberal Democratic Party, which has succeeded in holding power ever since. The Party's early leaders were *Kishi Nobusuke and his brother; *Sato Eisaku. *Tanaka Kakuei was forced to resign in 1974 as a result of a bribery scandal, and four leaders followed in the space of eight years as the party's fortunes waned before some degree of recovery was achieved under the forceful leadership of *Nakasone Yasuhiro, who has served as Prime Minister and LDP president from 1982 to 1987. The party has developed close links with business and with interest groups such as agriculture. A key feature is its structure of internal factions, which are less concerned with policy than with patronage, electoral funding, and competition for party leadership. Even so, the party appeals to a wide range of the electorate.

Liberal Party (Australia), a major political party in Australia. The original party emerged in 1910 as an alliance of various groups opposed to the Australian *Labor Party, led by *Forrest. They were known for a while as the Fusion Party, adopting the title Liberal in 1913. When Labor split in 1916 over the issue of conscription, they joined with elements of Labor to form the Nationalist Party. Shedding the Labor elements in 1922, they joined with the more right-wing Country Party, staying in power until 1929. In 1931 a new United Australia Party was formed, which was in power with the Country Party until 1941. The new Liberal Party was created in 1944 by Robert *Menzies, and a Liberal–Country coalition has alternated with Labor since then.

Liberal Party (Britain), a major political party in Britain. It emerged in the mid-19th century as the successor to the Whig Party and was the major alternative party to the *Conservatives until 1918, after which the *Labour Party supplanted it. Lord Palmerston's administration of 1854 is regarded as the first Liberal government. It was identified with *free trade and the need for civil and political liberty. *Gladstone's Liberal government (1868–74) achieved a national *school system, voting by secret ballot, the legalization of trade unions, reconstruction of the army, and reform of the judicial system. Although the Party split in 1886 over Irish *Home Rule and again in 1900 over the Second *Boer War, it won a sweeping victory in 1906 and proceeded to implement a large programme of social reform, introducing old age pensions, free school meals, national insurance against unemployment and ill-health, and a fairer taxation system; it also passed a Bill to disestablish the Anglican Church in Wales, and the third Home Rule Bill. After the outbreak of war (1914) it formed a coalition with the Conservatives (1915) and was in five coalition governments between 1916 and 1945. Since World War II it has been an opposition party of varying fortune, forming a Lib–Lab pact with the Labour government (1977–8) and the Alliance (1983–7) with the *Social Democratic Party.

Liberal Party (Canada), a major political party in Canada. Before the *Mackenzie Rebellion of 1837 a group of reformers emerged that gained a voice in the government of the Province of Canada in the mid-1850s. Following the Confederation of Canada in 1867, a Liberal Party took shape as a major political force and remained so, forming a government (1873–8) under Alexander Mackenzie. The Liberal Party has had a strong appeal for French Canadians, who have produced three distinguished Liberal Prime Ministers: Wilfrid Laurier (1896–1911), Louis St Laurent (1949–57), and Pierre Elliott Trudeau (1968–79, 1980–4), who succeeded Lester *Pearson (1963–8). It has held power for most of the 20th century until 1984. In the first half of the century its policies were less sympathetic to the idea of empire and keener on Canadian autonomy than those of the Conservatives. Its longest-serving leader, Mackenzie *King (1921–6, 1926–30, 1935–48), was a powerful influence in bringing about the Statute of *Westminster.

Liberia, a country in West Africa. It is the oldest independent republic in Africa (1847). It owes its origin to the philanthropic *American Colonization Society. US negotiations with local rulers for a settlement for the repatriation of freed slaves began in 1816. The first settlements were made in 1822, and the name Liberia was adopted in 1824. Independence was proclaimed by Joseph Jenkins Roberts, first President, in 1847. The real beginning of prosperity was in the 1920s, when the Firestone Rubber Company provided a permanent and stable market for rubber. W. V. S. *Tubman was President from 1944 until his death in 1971. With a decline in world rubber prices, the economy suffered in the 1970s and a bloody revolution in 1980 brought in the People's Redemption Council, a military government, under Master-Sergeant Samuel Doe.

Libya, a North African country. Administered by the Turks from the 16th century, Libya was annexed by Italy after a brief war in 1911–12. The Italians, however, like the Turks before them, never succeeded in asserting their full authority over the Sanussi tribesmen of the interior desert. Heavily fought over during World War II, Libya was placed under a military government by the Allies before becoming an independent monarchy in 1951 under Emir Sayyid Idris al-Sanussi, who in 1954 granted the USA military and air bases. Idris was overthrown by radical Islamic army officers in 1969, and Libya emerged as a radical socialist state under the charismatic leadership of Colonel Muammar *Qaddafi. It has used the wealth generated by exploitation of the country's rich oil resources to build up its military might and to interfere in the affairs of neighbouring states. Libyan involvement in Arab terrorist operations has blighted its relations with western states and produced armed confrontations with US forces in the Mediterranean. In April 1986, there were US air strikes against Tripoli and Benghazi.

Lie, Trygve Halvdan (1896–1968), Norwegian politician and first secretary-general of the *United Nations (1946–53). He held several ministerial posts in the Norwegian Parliament before having to flee (1940) to Britain, where he acted as Foreign Minister until 1945. He was elected secretary-general of the United Nations as a compromise candidate. When forces of the Soviet-sponsored Republic of North Korea crossed the border into South Korea (1950), Lie took the initiative in sending UN forces to restore peace, thus incurring the enmity of the Soviet Union. He later re-entered Norwegian politics.

Liebknecht, Wilhelm (1826–1900), German political leader. An early interest in *socialism led to his expulsion from Berlin in 1846 but, after the outbreak of the

*Revolutions of 1848, he returned to Germany to help set up a republic in Baden. Forced to flee, he went to England, where he spent thirteen years in close association with *Marx. He returned to Germany in 1861 and in 1863 founded the League of German Workers' clubs with August Bebel. A pacifist, he refused to vote credits for the war with France in 1870, was convicted of treason, and spent two years in prison. In 1874 he returned to the Reichstag and in 1875 helped to form the German Social Democratic Labour Party with the followers of the late Ferdinand Lassalle (1825–64) which became in 1891 the German Social Democratic Party (SPD).

Light Brigade, Charge of the   *Balaklava.

Li Hongzhang (or Li Hung-chang) (1823–1901), Chinese statesman. He formed the regional Anhui army to help suppress the *Taiping Rebellion. In 1870 he became governor-general of Zhili (or Chihli) and High Commissioner for the Northern Ocean, becoming responsible for Chinese foreign affairs until 1895. Recognizing the superiority of foreign military technology, he became the leader of the *Self-Strengthening Movement, establishing arsenals and factories and creating the Beiyang fleet (China's first modern navy). His reforms were piecemeal,

Li Hongzhang, photographed in 1874 at the beginning of his period of political prominence during which he attempted to modernize China.

however, and his prestige suffered when his forces were defeated in the *Sino-Japanese War.

Lincoln, Abraham (1809–65), sixteenth President of the USA (1861–5). Born into a poor Kentucky frontier family, Lincoln moved westwards with his parents, eventually settling in Illinois. He served as a *Whig in the state legislature and was elected to Congress in 1846. His opposition to the *Mexican–American War cost him support and he did not stand for re-election in 1848. His opposition to the *Kansas–Nebraska Act led him into the newly formed *Republican Party and he won national fame as a result of his campaign against the Democratic incumbent, S. A. Douglas, in the Illinois senatorial election in 1858. Although Douglas won the contest, Lincoln's performance established him as the national spokesman for a strong federal union in opposition to the further extension of slavery and helped him to win the Republican nomination for the 1860 presidential election. Lincoln's subsequent election on an anti-slavery programme precipitated the secession of a number of Southern states and then the *American Civil War. Through his exercise of executive authority and skilful management of those around him, Lincoln steered the Union (Northern) war effort through many storms, and although his Proclamation for the Emancipation of *Slaves (1862) was the most dramatic statement of his presidency, he consistently made the salvation of the Union his priority. He won re-election in 1864 and refused to consider a compromise peace at the Hampton Roads Conference in 1865. By the time of the surrender of Robert E. *Lee at *Appomattox, Lincoln was looking ahead to the problems of peaceful reconciliation with the defeated South, and justice for the freed slaves, but he was fatally wounded by the Southern fanatic John Wilkes *Booth in Ford's Theater, Washington. As the saviour of the Union, the emancipator of the slaves, the embodiment of the log-cabin-to-White House legend, and the eloquent champion of American democracy, Lincoln occupies a unique place in American history.

Lin Biao (or Lin Piao) (1908–71), Chinese general and statesman. Graduating from the *Kuomintang's Whampoa Military Academy in 1926, he joined the communists and rose rapidly through their military

The murder of Abraham **Lincoln** on Good Friday, 14 April 1865. John Wilkes Booth (*right*) entered the President's box at Ford's Theater in Washington and shot Lincoln in the head from close range.

command, leading an army in the *Long March of 1934–5 and operating successfully against the Japanese. He became commander of the North-West People's Liberation Army in 1945 and conquered Manchuria in 1948. He led the Chinese forces in the *Korean War (1950–3), was elevated to the rank of marshal in 1955, became Minister of Defence in 1959, and popularized the concept of the 'people's war'. He politicized the People's Liberation Army and collaborated closely with *Mao Zedong during the *Cultural Revolution. In 1969 he was formally designated as Mao's successor, but in September 1971 he apparently attempted a coup and was killed in an aircraft crash in Manchuria when trying to flee the country.

Lin Zexu (or Lin Tse-hsu) (1785–1850), Chinese imperial administrator. In 1838 he was sent to Guangzhou (Canton) as imperial commissioner to stop the illegal opium trade. He ordered foreign traders (mainly British) to surrender their opium and tried to suppress the domestic trade in the drug. Under increasing pressure from the Chinese, including the terminating of all trade, the British representative handed over 21,306 cases of opium, which Lin destroyed. His apparent success won him promotion, but when Britain reacted militarily in the *Opium War, he was disgraced and exiled, although he was rehabilitated just before his death.

Little Big Horn, battle of (25 June 1876), scene of General *Custer's last stand in South Dakota when he and 266 men of the 7th Cavalry met their deaths at the hands of larger forces of Sioux. The battle was the final move in a well-planned strategy by the Sioux leader *Crazy Horse, following the invasion of the Sioux Black Hills by white *gold rush prospectors in violation of a treaty of 1868.

Little Entente (1920–38), alliance of Czechoslovakia, Romania, and the new Kingdom of Serbs, Croats, and Slovenes (later termed *Yugoslavia). It was created by the Czech Foreign Minister Edvard *Beneš, who in August 1920 concluded treaties (extended in 1922 and 1923) with both Romania and Yugoslavia. The principal aim of the Entente was to protect the territorial integrity and independence of its members by means of a common foreign policy, which would prevent both the extension of German influence and the restoration of the Habsburgs to the throne of Hungary. France supported the Entente, concluding treaties with each of its members. In 1929 the Entente pledged itself against both Bolshevik and Hungarian (Magyar) aggression in the Danube basin, while also seeking the promotion of Danube trade. In the 1930s, however, the members gradually grew apart. Romania under *Carol II (1930–40) leaned towards Hitler's *Third Reich, Czechoslovakia signed a non-aggression treaty with the Soviet Union (1935), while in February 1934 Romania and Yugoslavia joined Greece and Turkey to form the so-called Balkan Entente. In 1937 Yugoslavia and Romania were unwilling to give Czechoslovakia a pledge of military assistance against possible aggression from Germany, and when the *Sudetenland of Czechoslovakia was annexed (September 1938), the Entente collapsed.

Little Rock, the capital of Arkansas, USA. It achieved notoriety when (1957) the state governor, Orval Faubus, called out national guards to prevent black children

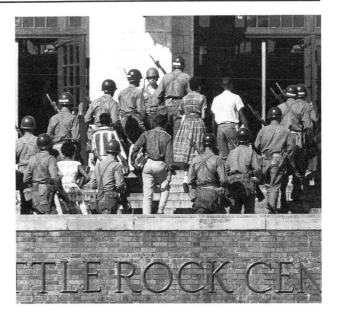

Black students at **Little Rock**, Arkansas, enter school under armed guard in 1957 at the height of the desegregation crisis.

from entering local segregated schools. A federal court injunction required the guards to be removed, and President Eisenhower sent federal troops to secure the entry of the black children to the schools. After this incident segregation in US schooling rapidly declined.

Litvinov, Maxim Maximovich (1876–1951), Soviet revolutionary politician. He joined the *Bolsheviks (1903), and from 1917 to 1918 was Soviet envoy in London. He headed delegations to the disarmament conference of the League of Nations (1927–9), signed the *Kellogg–Briand Pact (1928), and negotiated diplomatic relations with the USA (1933). He was a strong advocate of collective security against Germany, Italy, and Japan. He was dismissed (1939) before *Stalin signed the *Nazi–Soviet Pact.

Liu Shaoqi (or Liu Shao-ch'i) (1898–1969), Chinese statesman. He served as a communist trade union organizer in Guangzhou (Canton) and Shanghai before becoming a member of the Central Committee of the *Chinese Communist Party in 1927, and its chief theoretician. On the establishment of the People's Republic in 1949 he was appointed chief vice-chairman of the party. In 1959 he became chairman of the Republic, second only to *Mao Zedong in official standing, but during the *Cultural Revolution he was fiercely criticized by *Red Guards as a 'renegade, traitor, and scab' and in 1968 he was stripped of office. In 1980 he was posthumously honoured.

Liverpool, Robert Banks Jenkinson, 2nd Earl of (1770–1828), British statesman. He was first elected to Parliament in 1790. He was appointed Foreign Secretary in 1801 and helped to negotiate the Peace of Amiens with France in the following year. He was Home Secretary during 1804–6, and declined the premiership on the death of William Pitt. He became Secretary for War and the Colonies in 1809, reluctantly taking office as Prime Minister (1812–27) after the assassination of Spencer

*Perceval. After the *Napoleonic Wars his government used repressive measures to deal with popular discontent (*Peterloo Massacre), opposing both parliamentary reform and *Catholic Emancipation. Towards the end of his tenure the more liberal influences of men like *Peel and *Huskisson led him to support the introduction of some important reforms.

Livingstone, David (1813–73), British missionary doctor and explorer. Sent to Bechuanaland (now *Botswana) by the London Missionary Society, he was convinced that his duty lay in opening up the vast unexplored centre of Africa to missionaries. He crossed Africa from the Atlantic to the Indian Ocean (1852–6), and first reported the existence of the Victoria Falls. He was commissioned by the British government to explore Central Africa (1858–64). His last journey (1866–73) took him deeper into the continent in search of the sources of the Nile. In 1871 he was met at Ujiji by H. M. *Stanley, sent by the *New York Herald* to look for him. Livingstone died before reaching his goal, and his party of guides carried his embalmed body over 1,500 km. (900 miles) back to Zanzibar.

Lloyd George of Dwyfor, David, 1st Earl (1863–1945), British statesman. He was Liberal Member of Parliament for Caernarvon Boroughs from 1890 to 1945. In 1905 he was appointed President of the Board of Trade and in 1908, when *Asquith became Prime Minister, Lloyd George succeeded him as Chancellor of the Exchequer. He was responsible for the *National Insurance Act (1911), protecting some of the poorest sections of the community against the hazards of ill health and unemployment; in later life he regarded this as his greatest achievement. His budget of 1909, challenged by the House of *Lords, led to the Parliament Act (1911) reducing their powers, and provided an opportunity to use his oratorical skills. Created Minister of Munitions in 1915 his administrative drive ended the shell shortage on the *Western Front. In 1916 he became Prime Minister, replacing Asquith and forming a coalition government. He galvanized the Admiralty into accepting the *convoy system against U-boat attacks. At the *Versailles Peace Conference, fearing the consequences of French vindictive *reparations against Germany, he strove for moderation. At home, the Conservatives, disliking his individualistic style of government, left the coalition and Lloyd George resigned as Prime Minister in 1922. The Liberal Party, split between followers of Asquith and of Lloyd George, was overtaken in the 1920s by the *Labour Party as the alternative to Conservative governments. In the 1930s he opposed the *National government over the *Ottawa Agreements, but supported Britain's entry into the war in 1939.

Lobengula (*c.*1836–94), Ndebele leader. The son of *Mzilikazi, he was the second and last *Ndebele king (1870–93). After his father's death there was civil war until 1870, when he acceded to the throne. He compromised Ndebele independence by land concessions to the British (1886), and in 1888 signed the concessions which gave mining rights to *Rhodes's British South Africa Company. Later, Lobengula tried to resist the expanding European settlement and influence in his country. In 1893 he went to war against the British, but was defeated. He died as he was escaping from his capital of Bulawayo.

Locarno, Treaties of (1 December 1925), a series of international agreements. Their object was to ease tension by guaranteeing the common boundaries of Germany, Belgium, and France as specified in the *Versailles Peace Settlement in 1919. *Stresemann, as German Foreign Minister, refused to accept Germany's eastern frontier with Poland and Czechoslovakia as unalterable, but agreed that alteration must come peacefully. In the 'spirit of Locarno' Germany was invited to join the *League of Nations. In 1936, denouncing the principal Locarno treaty, *Hitler sent his troops into the demilitarized Rhineland; in 1938 he annexed the *Sudetenland in Czechoslovakia, and in 1939 invaded Poland.

Lodge, Henry Cabot (1850–1924), US politician. He represented Massachusetts in the US Senate from 1893 to 1924 and was a dominant figure in Senatorial politics. A conservative Republican, he specialized in foreign affairs and led the fight against US membership of the *League of Nations. A close friend of Theodore *Roosevelt, he supported the *gold standard and a high protective tariff, and was a bitter foe of Woodrow *Wilson's peace policy.

Lomé Convention (1975), a trade agreement reached in Lomé, the capital of Togo, between the *European Economic Community and forty-six African, Caribbean, and Pacific Ocean states, for technical co-operation and development aid. The developing countries received free access for their products into the markets of the EEC, plus aid and investment. A second agreement, 'Lomé II', was signed in 1979 by fifty-eight African, Caribbean and Pacific states and the EEC.

London Dockers' strike (1889), a major episode in the history of British trade unionism. The London dockers, objecting to low pay and casual employment, took successful strike action. They aimed to secure pay of sixpence an hour instead of fivepence, and a minimum

The **London Dockers' strike** of 1889 successfully mobilized the dock workers, here parading through east London. Among the group on the wagon are coal-heavers, exhibiting their block-and-tackle lifting gear as evidence of the hard manual labour carried out by the strikers.

working engagement of four hours. The episode signalled the beginning of 'New Unionism', characterized by the increasingly effective use of strikes.

Long, Huey (Pierce) (1893–1935), US politician. A demagogue, he rose through various political offices in Louisiana to become governor (1928–31) and Senator (1931–5). He used dictatorial methods to modernize Louisiana and had roads, bridges, and privately owned public buildings constructed, fighting the vested interests of the public utility companies. Despite initial support, Long quickly turned against Franklin D. *Roosevelt, seeing the *New Deal as a rival programme to his own 'Share Our Wealth' by the redistribution of income. His charismatic if neo-fascist appeal came close to dividing the Democratic Party, but he was assassinated before he could run for the presidency.

Long March (1934–5), the epic withdrawal of the Chinese communists from south-eastern to north-western China. By 1934 the *Jiangxi Soviet was close to collapse after repeated attacks by the *Kuomintang army. In October a force of 100,000 evacuated the area. *Mao took over the leadership of the march in January 1935. For nine months it travelled through mountainous terrain cut by several major rivers. In October Mao and 6,000 survivors reached Yan'an, having marched 9,600 km. (6,000 miles). Other groups arrived later, in all about 20,000 surviving the journey. The march established Mao as the effective leader of the Chinese Communist Party, a position he consolidated in his ten years in Yan'an.

López, Carlos Antonio (1792–1862), Paraguayan statesman. As President (1844–62), his domestic policy, although authoritarian, promoted highway and railroad construction, reorganized the country's judicial system and its military, and attempted to strengthen the economy by creating government monopolies. He abandoned the isolationist foreign policy of his predecessor, but in the process involved his country in a series of disputes with the USA, Great Britain, Argentina, and Brazil.

López, Francisco Solano (1827–70), Paraguayan statesman. Son of Carlos Antonio *López, he became President (1862–70) on his father's death. He initiated grandiose building schemes and then led his country into a disastrous war with Brazil, Argentina, and Uruguay. This war (1865–70) was one of the fiercest and bloodiest ever fought in the New World. It halved the population of Paraguay and left the country in a state of economic collapse; López himself was defeated and killed. Considered a cruel and dictatorial *caudillo in his lifetime, he afterwards came to be regarded as the champion of the rights of small countries against more powerful neighbours.

López, Narciso (1797–1851), Spanish-American field-marshal and politician. A Venezuelan by birth, he enrolled in the Spanish imperial army during the revolution against Spain. He was sent as governor to a province of Cuba, where he began to plot a revolt against Spain, hoping that with independence Cuba would become a state within the USA. With the help of pro-slavery Southern volunteers he twice led an unsuccessful military invasion against Cuba. His third expedition (1851) ended in disaster; the expected popular

uprising failed to happen and López, together with fifty volunteers, was caught by Spanish forces near Havana, and executed.

Lords, House of, the upper chamber of the British *Parliament. It is made up of the Lords spiritual (senior bishops of the Church of England), and the Lords temporal (hereditary peers and peeresses, Law Lords, and, from 1958, life peers and peeresses). In the early 19th century it was still the dominant House of Parliament, but, following the 1832 *Reform Bill, its influence gradually declined as that of the House of *Commons increased. The Parliament Act of 1911 reduced the Lords' powers to a 'suspensory veto' of two years (further reduced to one year in 1949). By it Bills can be delayed, but if passed again by the Commons, become law. The Lords' function is chiefly to revise Bills or to initiate reforms. The House also has judicial powers as the ultimate British Court of Appeal.

Louis XVIII (1755–1824), King of France (1795–1824). The brother of Louis XVI, he became titular regent after the death of the latter in 1793, and declared himself king on the death in prison of the ten-year-old Louis XVII. Known as the comte de Provence, he had fled to Koblenz, and then to England, where he led the counter-revolutionary movement. His exile ended in 1814 and with the help of *Talleyrand he returned to the throne of France and issued a constitutional charter. Many of *Napoleon's reforms in the law administration, church, and education were retained but after the assassination (1820) of his nephew the duc de Berry, he replaced moderate ministers by reactionary ones. Civil liberties were curbed, a trend which continued under his younger brother and successor, *Charles X.

Louisiana Purchase (1803), US acquisition from France of over two million sq. km. (828,000 sq. miles) of territory stretching north from the mouth of the Mississippi to its source and west to the Rockies. France had ceded Louisiana to Spain in 1762 but regained it by

Louis-Philippe I, who supported the French Revolution, visits the battlefield of Valmy (20 September 1792): a detail from a painting by Jean Baptiste Mauzaisse. As 'citizen king' from 1830, his unpopularity with the lower bourgeoisie led to his downfall. (Musée de Versailles)

treaty in 1801. Concerned at the possible closure of the Mississippi to commerce and the related threat to US security, President *Jefferson sent James *Monroe to France in 1803 to help negotiate free navigation and the purchase of New Orleans and west Florida. At war again with Britain, Napoleon was anxious not to have extensive overseas territories to defend and sold the whole of Louisiana to the US for $15 million. Although the Constitution gave no authority to purchase new territory or promise it statehood, the Senate confirmed the agreement, increasing US territory by some 140 per cent, and transforming the USA into a continental nation.

Louis-Philippe I (1773-1850), King of the French (1830-48). The son of the duc d'Orléans he, along with his father, renounced his titles and assumed the surname Égalité. On the restoration of *Louis XVIII to the French throne he recovered his estates, and was elected King of the French, the 'citizen king', after the *July Revolution in 1830. During his reign political corruption, judicial malpractices, and limited parliamentary franchise united liberals and extremists in a cry for reform. After 1840, a series of disastrous foreign ventures and alliances with reactionary European monarchies alienated the liberal opinion on which his authority had been based. His rule ended in February 1848 when, after popular riots, he agreed to abdicate, and escaped to England as 'Mr Smith'.

Lovett, William (1800-77), British radical political reformer. He became involved in working-class radical groups, and in 1836, with Francis *Place, set up the London Workingmen's Association. Lovett outlined a programme of political reform, which in 1838 was presented as the People's Charter. He was secretary of the first *Chartist Convention in 1839, but was imprisoned following violent incidents in Birmingham. He subsequently gave up all political activity.

Lowell, Francis Cabot (1775-1817), founder of the US cotton industry. A Boston merchant, in 1814 he established the first US factory to use both spinning and weaving machinery (and the first in the world to manufacture cotton cloth using power machinery enclosed in a single building) at Waltham, Massachusetts. Lowell was singular among early US industrialists for the paternalistic concern he demonstrated for his workforce and for their living and working conditions.

Lucknow, siege of (1857-8), a siege of the British garrison during the *Indian Mutiny. The abolition by the British of the Kingdom of Oudh, whose capital Lucknow had been, became one of the causes of the Mutiny. On the outbreak of hostilities the British and Indian garrison, together with women and children, were confined to the Residency, and during the ensuing five-month siege they suffered heavy casualties. Lucknow was relieved first on 26 September by troops under Sir Henry *Havelock. He in turn was besieged, and only

The charge of the Queen's Bays against hostile Indian forces surrounding **Lucknow**, in northern India, on 6 March 1858: a detail from a painting by Henry Payne. The siege had ended the previous November, but here elephants mingle with cavalry in the Indian rebel army to confront the final British effort to regain control of the garrison and free its European inhabitants.

A drawing of a **Luddite** mob, 'maddened men, armed with sword and firebrand', by Hablot K. Browne ('Phiz'). The machine-breakers were seen by some as revolutionaries, by others as misguided reactionaries, and by sympathizers as displaced workers angered by the rapid changes brought about by the Industrial Revolution.

relieved on 16 November by troops under Sir Colin Campbell. The city was not finally restored to British possession until 21 March 1858.

Luddite, a member of a 19th-century protest group of British workers, who destroyed machinery which they believed was depriving them of their livelihood. The movement began in Nottinghamshire in 1811, when framework knitters began wrecking the special type of 'wide frames' used to make poor-quality stockings, which were undercutting the wages of skilled craftsmen. The men involved claimed to be acting under the leadership of a certain 'Ned Ludd' or 'King Ludd', although it is doubtful whether such a person ever existed. The outbreaks of violence spread rapidly and by the early part of 1812 were affecting Yorkshire and Lancashire. Large groups of men stormed the cotton and woollen mills in order to attack the power looms. The government responded harshly by making machine-breaking an offence punishable by death. There were further sporadic outbreaks in 1816, but the movement subsequently died out, although the term survived.

Ludendorff, Erich (1865–1937), German general. A brilliant strategist, he helped in World War I to revise the proposals of *Schlieffen for war in the West by extending the German army's southern flank, and largely planned the battle of Tannenberg (1914). With *Hindenburg he exercised a virtual military dictatorship from 1917 and forced the resignation of the Chancellor, *Bethmann-Hollweg. He directed the war effort with Hindenburg until the final offensive failed (September 1918). He fled to Sweden, returning to attempt to overthrow the *Weimar Republic in the *Kapp putsch (1920). He joined *Hitler in the abortive *Munich 'beer-hall putsch' (1923) and sat in the Reichstag as a National Socialist (1924–8). He became a propagandist of 'total war' and of the new 'Aryan' racist dogma and wrote pamphlets accusing 'supranational powers'—the Roman Catholic Church, the Jews, and the Freemasons— of a common plot against Germany.

Lugard, Frederick Dealtry, 1st Baron (1858–1945), British colonial administrator. After early service in the army, he joined the British East Africa Company and was posted (1890) to Uganda. He persuaded the British government to assume a protectorate over Uganda in 1894. He was sent to *Nigeria, becoming High Commissioner in 1900, and by 1903 had effectively occupied northern Nigeria. He was governor of *Hong Kong (1907–12), and served as governor-general in Nigeria (1912–19), administratively amalgamating the northern and southern protectorates. He developed the doctrine of indirect rule, believing that the colonial administration should exercise its control through traditional native chiefdoms and institutions.

Lumumba, Patrice (Emergy) (1925–61), Congolese nationalist and politician. He founded the influential MNC (Mouvement National Congolais) in 1958 to bring together radical nationalists. He was accused of instigating public violence and was gaoled by the Belgians, but was released to participate in the Brussels Conference (January 1960) on the Congo. He became Prime Minister and Minister of Defence when the Congo became independent in June 1960. Sections of the army mutinied, the Belgian troops returned, and Katanga province declared its independence (*Congo Crisis). Lumumba appealed to the *United Nations, which sent a peace-keeping force. President Kasavubu, his rival in power, dismissed him and shortly afterwards he was put under arrest by Colonel Mobutu. He escaped, but was recaptured and killed.

Lusitania, British transatlantic liner, torpedoed (7 May 1915) off the Irish coast without warning by a German submarine, with the loss of 1,195 lives. The sinking, which took 128 US lives, created intense indignation throughout the USA, which until then had accepted Woodrow *Wilson's policy of neutrality. Germany refused to accept responsibility for the act, and no reparations settlement was reached. Two years later (1917), following Germany's resumption of unrestricted submarine warfare, the USA severed diplomatic relations and entered the war on the side of the Allies.

Luthuli, Albert John (1898–1967), South African political leader. A Zulu by birth, he served as a tribal chief (1936–52), and was a member of the Native Representative Council until its dissolution in 1946. In 1952 he was elected President of the *African National Congress, and became universally known as leader of non-violent opposition to *apartheid. From 1956 he suffered frequent arrests and harassment by the South African government. In 1959 the government banished him to his village and in 1960 outlawed the ANC. In 1961 he was awarded the Nobel Peace Prize.

Luxemburg, Rosa (1871–1919), Polish revolutionary socialist. A brilliant writer and orator, she defended the cause of revolution against moderates in the German Social Democratic Party. She took part in the *Russian Revolution of 1905 in Poland and argued that the mass strike, not the organized vanguard favoured by *Lenin, was the most important instrument of the proletarian revolution. She was active in the second *International,

A photograph taken in January 1919 of the revolutionary Rosa **Luxemburg**. The Polish-born founder of the German Communist Party was murdered later that month by right-wing troops during the Spartakist Revolt.

and with Karl *Liebknecht founded the *Spartakist League. A founder of the German Communist Party, she was captured by right-wing irregular troops (*Freikorps*) in Berlin, and together with Liebknecht was murdered during the Spartakist Revolt of 1919.

Lyons, Joseph Aloysius (1879–1939), Australian politician. He was a Labor politician in Tasmania from 1909 until 1929 and was Premier there from 1923 until 1928. He became a federal politician in 1929. When the Labor Party split in 1931, Lyons and others left, joining with the Nationalists to form the United Australia Party. Lyons led the Opposition from 1931 until 1932, when he became Prime Minister (1932–9). He foresaw war and embarked in 1937 on a rigorous programme of armament.

M

McCarran Act (1951), a US Act that required the registration of communist organizations and individuals, prohibited the employment of communists in defence work, and denied US entry to anyone who had belonged to a communist or fascist organization. It arose out of the fear in the early 1950s of a communist conspiracy against the USA; but in 1965 the Supreme Court ruled that individuals could refuse to admit being communists by claiming the constitutional privilege enshrined in the Fifth Amendment against self-incrimination.

McCarran–Walter Act (1952), a codification of US immigration laws. Passed over the President's veto, it maintained the quota system, whereby immigrant quotas were allocated by nationality, but it tightened up laws governing the admission and deportation of aliens, limited immigration from eastern and south-eastern Europe, removed the ban on the immigration of Asian and Pacific people, provided for selective immigration on the basis of skills, and imposed controls on US citizens abroad.

MacArthur, Douglas (1880–1964), US general. He was US army chief-of-staff (1930–5) and retired from the army in 1937 to become military adviser to the Philippines. In 1941 President F. D. Roosevelt recalled him to build up a US defence force on the islands. In 1941 Japanese troops successfully invaded the Philippines, and MacArthur transferred to Australia. He commanded the Allied counter-attack against the Japanese (July 1942–January 1943) in the Papuan campaign in New Guinea. From here (1943–4) his troops advanced towards the Philippines, which were recaptured in the spring of 1945. By now he was commander of all US army forces in the Pacific and received the Japanese surrender in Tokyo (2 September 1945). As commander of the Allied occupation forces (*Japan, occupation of), he took an active role in many reforms, as well as in the drafting of the new *Japanese constitution. Appointed UN commander in the *Korean War he led his troops into North Korea (October 1950) but was forced to retreat by an invading Chinese army. In 1951 he resumed the offensive, but tension arose with President *Truman, who believed that MacArthur was prepared to risk a full-scale atomic war, and he was dismissed in April 1951. He failed to obtain nomination for the presidential election in 1952.

Macarthur, John (1767–1834), Australian sheep breeder and viticulturist. In 1797 he purchased and had brought to New South Wales some merino sheep from the Cape. As a result of a duel, he was sent back to England in 1801, and interested manufacturers in the use of merino wool. He returned to Australia in 1805 and was influential in creating the trade in Australian wool. He was involved in the *Rum Rebellion and returned to London again until 1817. He also helped found the Australian wine trade.

McCarthy, Joseph Raymond (1908–57), US politician. A Republican Senator from Wisconsin (1947–57),

General **MacArthur** signs the document of surrender that ended the war against Japan, 2 September 1945, on board the US battleship *Missouri*. Witnessing the Allied victory ceremony was General Wainwright, recently released from his three-year captivity by the Japanese, and General Percival, who had led the British surrender at Singapore in 1942.

he launched a campaign in the early 1950s alleging that there was a large-scale communist plot to infiltrate the US government at the highest level. Despite the conclusions reached by a Senate investigating committee under Millard Tydings that the charges were a fraud, McCarthy continued to make repeated attacks on the government, the military, and public figures. The term 'McCarthyism' became synonymous with the witch-hunt that gripped the USA from 1950 to 1954. In 1953, as chairman of the Senate Permanent Subcommittee on Investigations, McCarthy conducted a series of televised hearings where his vicious questioning and unsubstantiated accusations destroyed the reputations of many of his victims. At length his methods were denounced by President *Eisenhower, and the Senate censured him for his conduct. After the 1954 election, with the Democrats again in control of Congress, McCarthy's influence declined.

Macaulay, Thomas Babington, 1st Baron (1800–59), British historian, essayist, and politician. He began his long parliamentary career in 1830, and became a prominent advocate of reform, religious toleration, and the abolition of slavery. During 1834–8 he worked in India as a member of the Supreme Council, where he played a large part in drafting the penal code which became the foundation of India's criminal law. On his return to England he wrote his partially completed *History of England* (published 1848 and 1855), which greatly influenced liberal perception of British history.

McClellan, George Brinton (1826–85), US general. Given command of the Department of Ohio at the beginning of the *American Civil War, McClellan secured a series of minor victories in West Virginia, and in November 1861 succeeded Winfield *Scott as general-in-chief of the Union (Northern) armies. While he lost this post in March 1862, he retained command of the Army of the Potomac. His attempt to capture Richmond in the Peninsular campaign proceeded too slowly and was then ruined by counter-attacks from forces of the *Confederacy in the Seven Days' battles. He checked *Lee at *Antietam, but missed the opportunity to destroy his opponent and bring the war to an end, and was subsequently removed from command. McClellan unsuccessfully ran as the Democrat presidential candidate against *Lincoln in 1864.

MacDonald, (James) Ramsay (1866–1937), British statesman and first Labour Prime Minister (1924, 1929–31). In 1906 he was elected a Labour Member of Parliament, and became leader of the Parliamentary Labour Party in 1911. At the outbreak of World War I his belief in negotiation, not war, with Germany made him unpopular, and he resigned. He was leader of the Opposition (1922–3), and then Prime Minister. His ministry of 1924, the first time Labour had formed a government, was too brief for any significant achievement. His second Labour government (1929–31) broke down through cabinet divisions over proposals to reduce unemployment benefits. MacDonald, however, was able to continue as Prime Minister of a *National Government (1931–5). A section of the Labour Party, led by George *Lansbury, refused to support the government, feeling that MacDonald, though a social reformer, had ceased to have sufficient regard for socialism. MacDonald was closely involved in the international disarmament schemes of the early 1930s. Like many others he did not discern the menace of Nazism. Failing health led to his resignation of the premiership in 1935.

Prime Minister Ramsay **Macdonald** makes a radio broadcast from the cabinet room of 10 Downing Street, August 1931. This was the year that Macdonald made his controversial decision to form a National coalition government to replace the Labour government.

Sir John **Macdonald** led the sixteen-strong Canadian delegation, the 'Fathers of Confederation', who met in London in 1866. A year later Macdonald was the first Prime Minister of the new Dominion of Canada. (House of Commons, Ottawa)

Macdonald, Sir John (Alexander) (1815–91), Canadian statesman. Elected a Tory member of the House of Assembly of United Canada in 1844, he was the leading figure in bringing about Confederation (1867) of the provinces of British North Canada as the Dominion of Canada after the passage of the *British North America Act. He became the first Prime Minister (1867–73) of the new Dominion of Canada. During his years in office, which continued from 1878 to 1891, Canada expanded territorially and experienced growth in its economy, its internal communications, and its sense of national purpose.

Machel, Samora Moises (1933–86), Mozambique statesman. He trained as a guerrilla in Algeria and became commander-in-chief of the Mozambique Liberation Front, Frelimo (Frente de Libertaçao de Moçambique) in 1966, and one of its leaders in 1969. He led the *Frelimo War against the Portuguese (1964–74), and became the first President (1975–86) of the People's Republic of Mozambique. A Marxist, he nationalized multi-national companies, and allowed his country to be used as a base for nationalist guerrilla forces from *Rhodesia and *South Africa. Nevertheless, his politics became increasingly pragmatic, accepting Portuguese aid and contact with South Africa. He died in an aircraft accident.

Machonaland *Zimbabwe.

Mackenzie, Alexander (1822–92), Canadian statesman. In 1861 he was elected to the Legislative Assembly of the Province of Canada. After the creation of the Dominion of *Canada in 1867 he became leader of the Liberal opposition in the first House of Commons. One of the dominant figures of Canada's early days of nationhood, Mackenzie defeated the Conservatives under Sir John (Alexander) *Macdonald to become the country's first Liberal Prime Minister (1873–8). During his term in office voting by ballot was introduced, the

Canadian Supreme Court formed, and the territorial government of the Northwest Territories successfully organized.

Mackenzie's Rebellion (1837), a popular uprising in Upper Canada (now part of Ontario). Growing pressure for democratic reform in Upper Canada could find no peaceful outlet after the defeat of the Reformers by the Tories in the election of 1836. The radical Reformers, led by the Scottish-born journalist and political agitator William Lyon Mackenzie (1794–1861), attempted an armed uprising on York (now Toronto) that was soon put down. Mackenzie fled to the USA where he set up a provisional government. In 1849 all those exiled as a result of this rebellion and *Papineau's Rebellion in Lower Canada, were pardoned, and in 1850 Mackenzie returned to Canada, settling in Toronto.

McKinley, William (1843–1901), twenty-fifth President of the USA (1897–1901). He served in the Union army during the *American Civil War and entered Congress as a Republican in 1876, giving his name to the tariff of 1890 which steeply raised import duties on foreign goods. With the aid of Marcus *Hanna he was elected governor of Ohio (1892–6), and his support for the *gold standard brought him the Republican nomination against *Bryan in 1896. Elected as President, he supported high tariffs on imported goods and US expansion into the Pacific, fighting the *Spanish-American War, and accepting the consequent acquisitions of Puerto Rico, Guam, and the Philippines as well as annexing Hawaii. He won re-election in 1900, again against Bryan, on the issue of prosperity, but was assassinated by the anarchist Leon Czolgosz in Buffalo. His vice-president, Theodore *Roosevelt, succeeded him.

MacMahon, Marie Edme Patrice Maurice, comte de (1808–93), French statesman and Marshal of France. Of Irish descent, he fought successfully in the Crimea and at the battles of *Magenta and *Solferino in 1859. As a general in the *Franco-Prussian War he was defeated at Worth (1870) and, with *Napoleon III, capitulated at *Sedan, but he commanded the army, that crushed the *Paris Commune in 1871. He had little sympathy with the new (Third) republic but did not support a royalist restoration either and agreed to succeed *Thiers às President (1873–9). Dislike of the recently elected Chamber of Deputies as too republican led him to dissolve it, but the electorate returned an even more republican Chamber (1877). This incident established the principle of ministerial accountability to the Chamber rather than to the President.

McMahon Line, a boundary line dividing Tibet and India. It was marked out by the British representatives led by Sir Henry McMahon at the Simla Conference (1914) between Britain, Tibet, and China. The Chinese government refused to ratify the agreement, and after the reassertion of control by China over Tibet in 1951 boundary disputes arose between India and China culminating in the *Indo-Chinese war of 1962.

Macmillan, Maurice Harold, 1st Earl of Stockton (1894–1987), British statesman. He served as Member of Parliament for Stockton-on-Tees (1924–9, 1931–64). During the 1930s he was a critic of *appeasement and of economic policy. In 1940 he joined the government of *Churchill. In 1951, as Minister for Housing and Local Government, he was responsible for the largest local authority building programme yet seen in Britain. He became Minister of Defence (1954) under Churchill, and Foreign Secretary and Chancellor of the Exchequer (1955) under *Eden, whom he succeeded as Prime Minister (1957-63) after the *Suez Crisis. He went on comfortably to win the general election of 1959. His health failed in 1963 and he resigned in October. He entered the House of Lords as the Earl of Stockton in 1984, where he criticized the economic and social policies of the *Thatcher government.

Macmillan ministries, British Conservative governments (1957-9 and 1959-63) with Harold *Macmillan as Prime Minister. Macmillan became Prime Minister on the resignation of *Eden. During the Macmillan period Britain started to taste the fruits of affluence and the majority of people agreed with the Prime Minister's comments at a speech (1957) in Bedford, borrowing the US Democratic Party's slogan, 'Most of our people have never had it so good', as confirmed by the Conservatives' success in the 1959 general election. During the Macmillan ministry life-peerages were introduced (1958), and the National Economic Development Council ('Neddy') was set up. Legislation in the form of the Commonwealth Immigration Act (1962) was passed to limit uncontrolled entry into the UK. Overseas, Macmillan enjoyed friendly relations with President *Kennedy, supporting him over the *Cuba Crisis and reaching the Nassau agreement (1962) that the US should furnish nuclear missiles for British submarines. In Africa, Britain accepted the need for independent statehood (*'Wind of Change'). However, the government was frustrated by *de Gaulle's veto on Britain's application to join the *European Economic Community (1963). Macmillan's government was weakened by public concern about an alleged Soviet espionage plot (June 1963) involving his Secretary of State for War, John Profumo, but it succeeded (July 1963) in negotiating a *Nuclear Test-Ban Treaty between the USA, the Soviet Union, and Britain.

Macquarie, Lachlan (1762-1824), Australian governor. He was appointed governor of the convict settlement of New South Wales in Australia after the *Rum Rebellion (1808) and took office in 1810. He was the last Australian governor with virtually autocratic power. In his view the colony was a settlement for convicts where free settlers had little place, and where convicts should be treated with every encouragement. The colony's nature was changing; there was conflict between *emancipists and *exclusionists; many exclusives, notably Jeffrey Hart Bent, the first judge of the Supreme Court of New South Wales, clashed with Macquarie. The *Bigge Inquiry into the condition of the convict population was held during Macquarie's term of office.

Madagascar, an island off south-east Africa. The Merina people, led by King Andrianampoinimerina (1787-1810), became dominant in the early 19th century, and in 1860 Radama II gave concessions to a French trading company. This led in 1890 to a French Protectorate, although resistance lasted until 1895. After 1945 Madagascar became an Overseas Territory of the French Republic, sending Deputies to Paris. It became

Lachlan **Macquarie**, from a portrait by Richard Read Jr, painted at the time of his appointment as governor of New South Wales, Australia in 1810. (Mitchell Library, Sydney)

a republic in 1958, and regained its independence (1960) as the Malagasy Republic, changing its name back to Madagascar in 1975. Severe social and economic problems have caused recurrent political problems in the 1970s and 1980s, including violent unrest and frequent changes of government.

Madero, Francisco Indalécio (1873-1913), Mexican statesman. President of Mexico (1911-13), he assumed leadership of the *Mexican Revolution of 1910. Thwarted in his attempt to unseat the dictator Porfirio *Díaz through legal means, he fled to the USA, from where he organized an armed movement which was unleashed on 20 November 1910. The Díaz dictatorship fell six months later and in the ensuing elections Madero won the Mexican presidency. He was unable to put into effect any of his political and social reforms because of dissent among his supporters and his own administrative inability. After putting down five insurrections against him, Madero fell victim to the sixth led by General Victoriano *Huerta in 1913. He was murdered a few days after being deposed.

Madison, James (1751-1836), fourth President of the USA (1809-17). He helped draft the Virginia State Constitution, served in the Continental Congress (1780-3), and secured Jefferson's Bill for religious freedom. He was almost certainly the most influential figure in the Constitutional Convention (1787); he also proposed the Bill of Rights (1791). He became leader of the Democratic-Republican Party and drew up the Virginia Resolves of 1798, condemning the Alien and Sedition Acts. Serving as Secretary of State under *Jefferson (1801-9), he was involved in disputes with France and Britain over America's rights to neutrality. As President (1809-17) he took the USA into the *War of 1812 against Britain. He signed Bills in 1816 incorporating the Second Bank of the USA and introducing the first protective tariff in US history.

'Mad Mullah' *Hasan, Muhammad Abdille Sayyid.

Mafia, an international secret society originating in Sicily. In its modern form the Mafia can be said to date from the period 1806-15, when, under British pressure, attempts were being made to break up the huge estates of the Sicilian feudal aristocracy. Their disbanded private armies often became brigands. They represented an alternative system of control and justice beyond the purview of the legal authorities. In 1860 the new government of *Victor Emanuel II tried to rid Sicily of them, but by now they controlled many police and government officials. In the 1880s many Sicilians emigrated to the USA and the Mafia became established in New York and Chicago. In the 1920s the fascist government in Italy brought their leaders to trial, but some escaped to the USA, where they were active during the *prohibition era, and their activities are now world-wide.

Magdala, a small fortified town in the *Ethiopian Highlands. After the British Foreign Office allegedly failed to answer a letter addressed to Queen *Victoria by Tewodros II, the latter imprisoned the British consul and his aides at Magdala. After further misunder-

standings, a British force of 32,000 landed in 1868 under Sir Robert Napier and marched inland. On 13 April it stormed Magdala almost without loss, and released the captives, only to find that Tewodros had died by his own hand. A short, punitive campaign against Ethiopia followed and John IV succeeded to the throne in 1872.

Magenta, battle of (4 June 1859), a battle fought in a town in Lombardy, between the French and the Sardinians against the Austrians. The patriotic movement of the *Risorgimento had gained from the French the offer of military support at a meeting (1859) at Plombières between *Napoleon III and *Cavour. When war broke out, the French and Sardinian forces defeated the Austrians in a disorganized fight at Magenta, which opened the way to the occupation of Lombardy and Milan. Two weeks later it was followed by the decisive battle of *Solferino.

Maginot Line, a series of defensive fortifications in France. Begun in 1929, it stretched along France's eastern frontier from Switzerland to Luxemburg. Named after the Minister of War, André Maginot, it was built because French military theorists believed that defence would predominate in the next war and because it reduced the demand for soldiers. Partly because of objections from the Belgians, who were afraid they would be left in an exposed situation, the line was not extended along the Franco-Belgian frontier to the coast; consequently it could

The storming of **Magdala** on 13 April 1868. Napier's troops, established in strength around the fortress, advance up the narrow defile to the town. Napier and his staff watch this climax to the assault, as a rocket battery fires from a nearby hill (*left*) and mortars (*right*) give artillery support. (National Army Museum, London)

Archbishop **Makarios**, robed for a service of the Greek Orthodox Church. He was primate of Cyprus from 1950 until his death, combining church leadership with a vigorous political role as head of state.

be outflanked, as indeed happened in spring 1940. However, the Maginot defences proved impregnable to frontal assault and so fulfilled their original purpose.

Mahdi, the, the Islamic spiritual and temporal saviour. According to Islamic teaching he will be sent by divine command to prepare human society for the end of earthly time by means of perfect and just government. Many have claimed to be the Mahdi at different times. Best known was Muhammad Ahmad bin Abdallah (1843-85). Of Nubian origin, he claimed descent from Muhammad. Feeling called to purify the world from wantonness and corruption, he gathered many followers and proclaimed himself Mahdi in 1881. In 1882 the Egyptian government sent expeditions against him, but by 1884, with the capture of Khartoum, he made himself master of Sudan. General *Gordon was killed in Khartoum on 30 January 1885; the Mahdi himself died, probably of typhus, five months later. Politically his struggle was carried on by the *Khalifa Abdallah until *Kitchener defeated him at Omdurman in 1898.

Mahmud II (1784-1839), Ottoman Sultan (1808-39). He came to the throne on the deposition of his brother, Mustafa IV, and continued the reforming policies of his cousin, Selim III (1789-1807). He rid himself of the Janissaries, the traditional military corps that had become unruly and inefficient, by having them massacred, and established a new, European-style army. He curbed the power of the religious classes, centralized government, and reduced provincial autonomy. He was attacked by *Mehemet Ali of Egypt, and his army defeated in the battle of Nizip (1839) in Syria.

Maine, the, US battleship destroyed by an explosion in Havana harbour in 1898 with the loss of 260 lives. The US naval inquiry found the cause of sinking to be a submarine mine, while the Spanish inquiry concluded that it was due to an explosion in the forward magazine. The truth was never known, but the episode precipitated the *Spanish-American War.

Maji-Maji, a rebellion in German East Africa (1905-7) in the south and centre of present-day *Tanzania. The African warriors believed that magic water (*maji*) could make them immune to bullets. German settlers, missionaries, and traders were murdered, and the towns of Liwale and Kilosa sacked. The Germans adopted a scorched-earth policy which ended the rebellion but greatly retarded economic development.

Makarios III (Mihail Christodoulou Mouskos) (1913-77), Greek Cypriot archbishop and statesman. Primate and archbishop of the Greek Orthodox Church in *Cyprus (1950-77), he reorganized the movement for *enosis* (the union of Cyprus with Greece). He was exiled (1956-9) by the British for allegedly supporting the *EOKA terrorist campaign of Colonel Grivas against the British and Turks. Makarios was elected President of Cyprus (1960-76). A coup by Greek officers in 1974 forced his brief exile to London, but he was reinstated in 1975 and continued in office until his death.

Malagasy   *Madagascar.

Malan, Daniel F(rançois) (1874-1959), South African statesman. Rising to prominence in the National Party in Cape Province, he was elected to Parliament in 1918. His political thinking was dominated by desire for secession from Britain and republicanism. He was Prime Minister (1948-54), and initiated the racial separation laws known as *apartheid.

Malawi, a country in south central Africa, formerly known as Nyasaland. Slave-traders from *Zanzibar raided the area frequently in the 1840s, and its desolation is described by *Livingstone in 1859. In 1875 Scottish missionaries settled, and for a while governed parts of the country. Colonial administration was instituted when Sir H. H. Johnston proclaimed the Shire Highlands a British Protectorate in 1889. This became British Central Africa in 1891, then Nyasaland from 1907 until 1964. Unwillingly a member of the *Central African Federation (1953-63), it gained independence (1964) as Malawi, with Dr Hastings Banda as first Prime Minister. When the country became a republic in 1966, he became President. A one-party state governed by the Malawi Congress Party, it has elected the President for life, but in 1978 parliamentary elections were held. Close dependency of its economy on South Africa continues to cause considerable problems.

Malaya, south-east Asian country, since 1963 part of the Federation of *Malaysia. Malaya came under increasing British influence after the foundation of settlements at Pinang (1786) and Singapore (1819), Pinang, Melaka, and Singapore were linked, as the *Straits Settlements, and ruled until 1867 from British India. In 1874 Perak accepted a British resident whose advice the sultan had to follow in all matters except religion and customs. By 1888 Britain had extended the residency system to all the west-coast states except Johore. These were linked as the Federated Malay States in 1896, and made rapid economic progress after the introduction of rubber and of western tin-mining technology. The immigration of Chinese and Tamil labourers turned Malaya into a multi-racial society with the indigenous (*bumiputra*) Malays in danger of becoming a minority within their own land. The prosperity of the Federated States led Kedah, Perlis, Kelantan, and Trengganu to accept British protection (1909) with agreement from Siam, their former suzerain. These Unfederated States were relatively loosely controlled. Johore accepted a British adviser in 1914. After the Japanese invasion and occupation during World War II, the British experimented briefly with a centralized Union of Malaya (1946), and then set up a Federation of Malaya (1948) to include Johore, Kedah, Kelantan, Labuan, Melaka, Negeri Sembilan, Pahang, Pinang, Perak, Perlis, Selangor, and Terengganu. The legacy of inter-communal conflict in 1945, and the strength of the Chinese-dominated Communist Party, which resented Malay dominance within the Federation helped to spark the armed insurrection of 1948–60 known as the *Malayan Emergency. Despite this, the Federation achieved sovereign independence in 1957 and was finally expanded into Malaysia in 1963.

Malayan Campaign (December 1941–August 1945), a military campaign in south-east Asia in World War II. After taking over military bases in Vietnam in July 1941 through an agreement with the *Vichy French administration, and securing a free passage through Thailand, whose dictator, Field Marshal *Pibul Songgram, had allied himself with Japan in 1939, Japanese troops under General *Yamashita Tomoyuki invaded northern Malaya in December 1941 while Japanese aircraft bombed Singapore. The British, Indian, and Australian troops retreated southwards, where they were taken prisoner after the *Fall of Singapore in February 1942. During the retreat a small guerrilla *resistance force was organized to conduct sabotage, operating behind Japanese lines. Known as the Malayan People's Anti-Japanese Army (MPAJA), it consisted largely of Chinese, most of whom were communists. In May 1944 Allied troops, advancing from Imphal, began the gradual reconquest of Burma, and liberated Malaya in 1945.

Malayan Emergency, communist insurgency in Malaya (1948–60). Minority Chinese resentment of Malay political dominance of the new Federation of Malaya was exploited by the, mainly Chinese, communist guerrillas who had fought against the Japanese. They initiated a series of attacks on planters and other estate owners, which between 1950 and 1953 flared up into a full-scale guerrilla war. Led by Chin Peng and supported by its own supply network (the Min Yuen), the communist guerrillas of the Malayan Races Liberation Army caused severe disruption in the early years of the campaign, but, particularly during the time of *Templer's period in charge of British and Commonwealth forces (1952–4), the insurgents were gradually defeated through the use of new jungle tactics, and the disruption of their supply network. The loyalty of the Malay and Indian population to the British, and the skilful use by the British of local leaders in the government committees, facilitated the peaceful transition to independence in 1957. By then the insurrection had been all but beaten, although the emergency was not officially ended until 1960.

Malaysia, a country in south-east Asia comprising the states of the Federation of *Malaya (Peninsular Malaysia) and *Sabah and *Sarawak (East Malaysia). Established in 1963, the Federation originally included *Singapore

but it was forced to secede in 1965 because of fears that its largely Chinese population would challenge Malay political dominance. *Brunei refused to join the Federation. The establishment of Malaysia was first suggested (1961) by Tungku Abdul *Rahman, who became its first Prime Minister (1963–70). The Federation aroused deep suspicion in Indonesia, and provoked President *Sukarno's policy of confrontation (*Konfrontasi), resulting in intermittent guerrilla war in Malaysia's Borneo territories which was only defeated with Commonwealth military assistance (1963–6). In 1969, inequalities between the politically dominant Malays and economically dominant Chinese resulted in riots in Kuala Lumpur, and parliamentary government was suspended until 1971. As a result, there was a major restructuring of political and social institutions designed to ensure Malay predominance, the New Economic Policy being launched to increase the Malay (*bumiputra*) stake in the economy.

Malcolm X (1925–65), US black leader. Born Malcolm Little, he rejected the co-operation with white liberals that had marked the *Civil Rights movement. He became a leading spokesman for the *Black Muslims in the 1950s, but was suspended by the movement's leader, Elijah Muhammad. In 1964, after conversion to orthodox Islam, he preached a brotherhood between black and white, and formed the Organization of Afro-American Unity. He was assassinated in 1965.

Maldives, a country consisting of a chain of coral islands in the Indian Ocean. The inhabitants came under Portuguese influence in the 16th century. From 1887 to 1952 the islands were a sultanate under British protection. Maldivian demands for constitutional reform began in the 1930s; internal self-government was achieved in 1948 and full independence in 1965. In 1968 the sultanate was abolished, and a republic declared. The Maldives became a full member of the *Commonwealth of Nations in 1985.

Mali, a country in West Africa. It was freed from Moroccan rule at the end of the 18th century, and divided among the Tuareg, Macina, and Ségou. France colonized it in the late 19th century. In 1946 it became an Overseas Territory of France. It was proclaimed the Sudanese Republic in 1958, an autonomous state within the *French Community. It united with Senegal as the Federation of Mali in 1959, but in 1960 Senegal withdrew and Mali became independent. A military government took over in 1968, although some degree of civilian participation has been reintroduced by General Moussa Traoré over the past decade. A brief border war broke out with *Burkina Faso in 1985, but prolonged drought in the north has caused far more severe problems.

Malthus, Thomas Robert (1766–1834), British political economist and demographer. In 1798 he published his *Essay on the Principle of Population*, in which he put forward the argument that the general standard of living could not be raised above subsistence level, since a nation's population always grows more rapidly than food supplies. Numbers are kept down only by natural disasters such as famine and war. Malthus' theories, which contradicted optimistic opinions about human progress, had considerable influence on 19th-century political thought.

Malvinas *Falkland Islands.

Manchester School, a group of economists, businessmen, and politicians who became influential in Britain in the 1840s. Based in Manchester, the centre of the cotton industry, and led by such men as John *Bright and Richard *Cobden, the group followed the *laissez-faire* philosophy of Adam Smith and David *Ricardo. They supported *free trade and political and economic freedom, and opposed any interference by the state in industry and commerce. Their influence faded in the 1860s, when many European countries began to favour state intervention in economic matters.

Manchukuo, Japanese puppet state in Manchuria (1932–45). Using the *Mukden Incident as a pretext, the Japanese seized the city of Mukden in September 1931 and within five months had extended their power over all Manchuria. Manchukuo was established as a puppet state under the notional rule of the last Chinese emperor *Puyi, but effective control remained in the hands of the Japanese army. Japanese expansion to the west was halted by the Soviet army in 1939, but the Japanese remained in control of Manchukuo, managing a partial development of its mineral resources, until the Chinese communists (with support from the Soviet Union, who removed large quantities of industrial equipment) took over at the end of World War II.

mandate, a form of international trusteeship. Mandates were devised by the *League of Nations for the administration of those colonial territories in Africa and Asia which had been the former possessions of Germany and the Ottoman empire, and which in 1919 were assigned by the League to one of the Allied nations. Marking an important innovation in international law, the mandated territories were theoretically to be supervised by the League's Permanent Mandates Commission. The latter, however, had no means of enforcing its will on the mandatory power, which was responsible for the administration, welfare, and development of the native population until considered ready for self-government. In 1946 this arrangement was replaced by the United Nations' trusteeship system for the remaining mandates.

Mandela, Nelson Rolihlahla (1918–), South African nationalist leader. He became a leader of the *African National Congress (ANC) and a member of its militant subsidiary, the Spear of the Nation. He was banned from the country (1953–5), but a year later was among those charged in a mass treason trial. The trial lasted until 1961, when all were acquitted. He continued his opposition to *apartheid by campaigning for a free, multi-racial, and democratic society. He was arrested in 1962, and imprisoned for five years. Before this sentence expired, he was charged under the Suppression of Communism Act and after a memorable trial (October 1963–June 1964), in which he conducted his own defence, he was sentenced to life imprisonment. His authority as a moderate leader of black South Africans did not diminish, though his absence from the political scene enabled a more militant generation of leaders to emerge. Offered a conditional release by the South African government, Mandela refused to compromise over the issue of apartheid. His wife, Winnie Mandela, continues to be politically active.

Nelson **Mandela**, South African black nationalist leader, at a conference in Pietermaritzburg, South Africa, in 1961. The African National Congress, of which he remains a leader, was banned by the South African government that year; in 1964 Mandela was imprisoned for life.

manifest destiny, 19th-century US political doctrine advocating territorial expansion. It was proclaimed by John O'Sullivan as 'Our manifest destiny to overspread the continent allotted by Providence for the free development of our yearly multiplying millions'. A tenet of the Democratic Party, it gained support among Whig, and later Republican, interests, and played a significant part in raising popular support for the annexation of *Texas (1845) and the *Mexican–American War (1846–8). It was later invoked by *Seward in the purchase of *Alaska (1867), and re-emerged in the 1890s with the annexation of *Hawaii and the acquisition of Spanish territories after the *Spanish–American War.

Manila Bay, battle of (1 May 1898), naval engagement during the *Spanish–American war in the Philippines, in which a US fleet under George *Dewey sank a Spanish fleet at dawn without losing a man. Dewey's objective had been to paralyse the Spanish fleet at the outset of the *Spanish–American War of 1898 over Cuba, but his overwhelming victory widened the scope of the war by opening the way for US expansion in the Pacific.

Mannerheim, Carl Gustav Emil, Baron von (1867–1951), Finnish military leader and statesman. Trained as an officer in the Tsarist army, he rose to the rank of general, and, defeating the Finnish Bolsheviks (1918), he expelled the Soviet forces from Finland. He was appointed chief of the National Defence Council (1930–9), and planned the 'Mannerheim Line', a fortified line of defence across the Karelian Isthmus to block any potential aggression by the Soviet Union. When Soviet forces attacked (1939) he resisted in the *Finnish-Russian War, and in alliance with Germany renewed the war (1941-4). In 1944 he signed an armistice with the Soviet Union. The Finnish Parliament elected Mannerheim as President (1944-6). In March 1945 he brought Finland into the war against Germany.

Manning, Henry Edward (1808-92), British cardinal. After becoming a Church of England priest in 1832, he was strongly influenced by the Oxford Movement and by the pre-Reformation Catholic Church. Received into the Roman Catholic Church in 1851 he became Archbishop of Westminster in 1865, and a cardinal in 1875. Theologically conservative, he supported the proclamation of papal infallibility (1870). He was an early supporter of *trade union rights and showed his sympathy for the Agricultural Labourers' Union in the 1870s. His mediation helped to secure a just settlement in the *London Dockers' Strike (1889); and he supported *Home Rule for Ireland.

Maori, a branch of the eastern Polynesian race living in New Zealand. The Maori migrated to New Zealand from the central Pacific from about AD 800 in several waves. By the 18th century a settled population of more than 100,000 had developed. Common language and customs, inter-marriage, and trade linked almost all the tribes, yet there was constant feuding between them. The term 'Maori' (meaning 'normal' or 'ordinary') was adopted when the tribes perceived a common identity in face of the European settlement. They responded to contact positively while attempting to control the intruders and preserve some selectivity. Hence the welcoming of traders but increasing resistance to shore-based settlement. Hence also, in 1840, the chiefs' signing of the Treaty of *Waitangi. Similarly they created the *Kingitanga to help preserve the land and a measure of independent Maori authority. By the Treaty direct purchase of land from the Maoris was forbidden, but the *Anglo-Maori Wars were mainly caused by the rapid acquisition of Maori land by the government and pressure by settlers for direct purchase. The wars were followed by participation in elections for the four Maori seats in Parliament and the establishment of village schools teaching in English. The process of putting land through the Maori Land Court, established in 1865, and replacing customary title with a crown-granted title before sale, was accompanied by the transfer of most good land from Maori possession to the settlers. Disease reduced Maori population to a low of 42,000 in 1896. After World War II the Maori migrated systematically to towns in search of wider opportunities. Their numbers increased; but threatened by rising unemployment and loss of identity, they have in recent years again demanded recognition of their language, values, and culture in national life, and the preservation of remaining Maori land.

Mao Zedong (or Mao Tse-tung) (1893-1976), Chinese revolutionary and statesman. He served in the revolutionary army during the *Chinese Revolution of 1911-12, and became involved first in the *May Fourth Movement and the *Chinese Communist Party in Beijing in 1919-21. Converted to Marxism, Mao moved to Shanghai in 1923 to become a *Kuomintang political

Mao Zedong, leader of communist China (*right*), with his Minister of Defence, Lin Biao. During the Cultural Revolution Mao encouraged a personality cult, using the Red Guards in his struggle against the Party bureaucracy.

organizer. After the Kuomintang turned on its communist allies in 1927, Mao used his experience of organizing the peasantry and his belief in their potential as a revolutionary force to establish the *Jiangxi Soviet. With *Zhu De he developed the *guerrilla tactics which were to be the secret of his success in the long civil war with the Kuomintang. In 1931 he became chairman of the Jiangxi Soviet, but following its successful blockade by Chiang Kai-shek's nationalist forces, he eventually led his followers in the *Long March (1934-5) to a new base in north-west China. Having emerged as the *de facto* leader of the Chinese Communist Party, he devoted considerable time to the theoretical writings which were to provide the ideological basis for the future communist state. Mao's well-organized guerrilla forces, capably led by such men as Zhu De and *Lin Biao, resisted the Japanese and defeated Chiang Kai-shek's nationalist forces. On 1 October 1949 he proclaimed the establishment of the People's Republic of China. Although he served as chairman of the new state from the time of its formation, Mao took little active part in administration until the mid-1950s, when he pioneered a series of reform movements, most notably the *Great Leap Forward, in an attempt to galvanize economic and political development. The rift with the Soviet Union which increasingly refused to support the Chinese communist struggle, reached its climax in the early 1960s. In 1959 Mao retired from the post of chairman of the Republic, but re-emerged in 1966 to initiate the *Cultural Revolution, a dramatic attempt to radicalize the country and prevent the revolution stagnating. Thereafter he gave his tacit support to the radical *Gang of Four, but their bid for power was stopped by his nominated successor, *Hua Guofeng, after Mao's death.

Maquis (Corsican Italian, *macchia*, 'thicket'), French *resistance movement in World War II. After the fall of France in 1940, it carried on resistance to the Nazi occupation. Supported by the French Communist Party, but not centrally controlled, its membership rose in 1943–4, and constituted a considerable hindrance to the German rear when the Allies landed in France. Its various groups, often operating independently, were co-ordinated into the Forces Françaises de l'Interieur in 1944.

Maratha Wars (1774–82, 1803–5, 1817–18) wars between the Maratha peoples and troops of the English *East India Company in India. By the late 18th century the Maratha Hindus, divided into over ninety clans, had formed an uneasy confederacy which became a significant force in northern and central India. Rivalries between chiefs were exploited by the British. In the Second War Sir Arthur Wellesley (later Duke of Wellington) won the battles of Assaye and Argaon. The Charter of the East India Company was renewed in 1813, when no further British acquisitions were envisaged, but in 1817 Company troops under Lord Hastings invaded Maratha territory to put down *Pindari robber bands supported by Maratha princes, and finally made British power dominant within the sub-continent.

Marchand, Jean-Baptiste (1863–1934), French explorer and general. He served in Africa from 1887 until 1895. In 1897 he was instructed to obtain French control of the region between the Niger and the Nile. He made a heroic journey through unexplored territory, reaching *Fashoda in 1898. Here he was confronted by an Anglo-Egyptian army led by Lord *Kitchener, who claimed the town for Egypt. Marchand's mission was withdrawn and France yielded its claim to the upper Nile region. Subsequently he served in the *Boxer Rising and in World War I.

Marcos, Ferdinand Edralin (1917–), Filipino statesman, President (1965–86). He entered Congress in 1949, subsequently becoming Senate leader in 1963. A ruthless and corrupt politician, he initially achieved some success as a reformer and identified closely with the USA, but after his election to a second term he became increasingly involved in campaigns against nationalist and communist guerrilla groups, and in 1972–3 he first declared martial law and then assumed near dictatorial powers. Although martial law was lifted in 1981 and some moves made towards the restoration of democracy, hostility to Marcos intensified after the murder of the opposition leader Benigno Aquino Jr in 1983. US support for his regime waned as a result of his failure to achieve consensus, and in February 1986 he was forced to leave the country after his attempts to retain power in a disputed election caused a popularly backed military revolt. This gave uncertain support to Corazon Aquino, widow of the slain opposition leader.

Maritime strike (1890), a shipping, mining, and shearing strike in Australia. It was a time of economic depression and workers were fighting for trade union recognition (including the 'closed shop'), while employers required freedom to make their own contracts, although other issues were also involved. The unions were defeated, some collapsing completely. This defeat has been seen as being a catalyst in the development of the labour movement and the formation of the Australian *Labor Party.

Marne, battles of (5–12 September 1914, 15 July–7 August 1918), two battles along the River Marne in east central France in World War I. The first battle marked the climax and defeat of the German plan to destroy the French forces before Russian mobilization was complete. By September the Germans were within 24 km. (15 miles) of Paris and the government moved to Bordeaux. *Joffre's successful counter-offensive has been hailed as one of the decisive battles in history. The retreating Germans dug themselves in north of the River Aisne, setting the pattern for *trench warfare on the *Western Front. The second battle ended *Ludendorff's final offensive, when, on 18 July, *Foch ordered a counter-attack.

Maronite, a member of an Eastern rite community of the Roman Catholic Church, founded by a 4th-century patriarch, Maron, with a membership of approximately 1.5 million adherents, mainly living in *Lebanon, but with important groups in Cyprus, Palestine, and Egypt. Since the 19th century a few have migrated to the Americas, Australia, and Africa. The massacre (1860) of Maronites by the *Druze brought French intervention and established French control in Lebanon and Syria. Following the dissolution of the *Ottoman empire in 1920 the Maronites in Lebanon became self-ruling under French protection. Since the establishment (1945) of an independend Lebanon, they have constituted one of the country's two major religious groups, ruling the country through a coalition of Druze, Christian, and Muslim parties. Rivalry between Maronite clans has resulted in two main political factions developing: the Phalange, founded in the 1930s by Pierre Gemayel, and the Chamounists founded by ex-President Camile Chamoun. Inter-communal feuding has given the Phalange greater control over Christian Lebanon. At the beginning of the Lebanese civil war (1975), the Phalangists attempted to extend their authority over all Lebanon, while their later aims have been to achieve autonomy. The Chamounists demand full independence.

Ferdinand **Marcos** with his wife Imelda on the occasion of her installation as governor of Manila in 1976. Marcos proposed that Imelda should stand for election to the presidency of the Philippines in order to perpetuate his hold on power. Her influence and extravagance contributed towards the unpopularity that eventually caused Marcos to be deposed.

Marquesas, the, twelve Polynesian islands in the South Pacific. From an estimated 50,000 people in 1813, foreign diseases had more than halved the population by 1842, when France annexed the islands, and by 1926 it was down to a little more than 2,000. Since then, with the development of agriculture, it has recovered to some 6,000. They are now administered as a division within French Polynesia, a French Overseas Territory.

Marshall, George Catlett (1880-1959), US general and statesman. He was army chief of staff (1939-45) responsible for enlarging the US army when his country entered World War II, and for overall strategic military planning. In late 1945 he led an abortive mission to bring about a settlement between the *Kuomintang and the communists in China. As Secretary of State (1947-9) he organized aid to Greece and Turkey, and fostered the European Recovery Program, the so-called *Marshall Plan, to promote economic recovery. Following the collapse of his economic aid plan for Eastern Europe because of the Soviet Union's hostility to it, he helped to create *NATO and supported the firm line taken by the Western Powers over the Soviet blockade of Berlin. He was Secretary of Defense (1950-1).

Marshall, John (1755-1835), US lawyer. Elected to Congress as a Federalist in 1799, he became Chief Justice (1801-35). He raised the power and prestige of the US Supreme Court, and moulded the Constitution through his interpretations, despite the fact that he frequently held opposing views to those of incumbent Presidents. Marshall established the practice whereby the Court reviewed state and federal laws and pronounced final judgment on their constitutionality. In a series of major decisions between 1810 and 1830, the Marshall Supreme Court proved the Constitution on the one hand to be a precise document that established specific powers, and on the other a living instrument to be broadly interpreted both to give the federal government power to act effectively and to limit the powers of the states.

Marshall Islands, a cluster of twenty-nine atolls and five islands in the central Pacific. The islands were named after a British captain who visited them in 1788. In 1886 the Marshall Islands became a German protectorate. After World War I the islands were administered by Japan, and after World War II they became a UN Trust Territory under US administration. From 1946 the US used *Bikini and other atolls in the group for atomic bomb tests. In 1986 they were given semi-independence in a 'compact of free association' in which the USA maintained control over military activities.

Marshall Plan (European Recovery Program), US aid programme. Passed by Congress in 1948 as the Foreign Assistance Act to aid European recovery after World War II it was named after the Secretary of State, George *Marshall. It invited the European nations to outline their requirements for economic recovery in order that material and financial aid could be used most effectively. The Soviet Union refused to participate and put pressure on its East European satellites to do likewise. To administer the plan, the Organization for European Economic Co-operation was set up, and between 1948 and 1951 some $13.5 billion was distributed. The Marshall Plan greatly contributed to the economic recovery of Europe, and bolstered international trade. In 1951 its activities were transferred to the Mutual Security Program. All activities ceased in 1956.

Martov, L. (b. Yuly Osipovich Tsederbaum) (1873-1923) Russian revolutionary and leader of the Menshevik Party in opposition to *Lenin's *Bolsheviks. He had at first co-operated with Lenin as joint editor of *Iskra* ('the Spark'), but, at the meeting of the second Russian Social Democratic Party in London (1903), he split with Lenin over the degree of revolutionary class-consciousness in the labour movement and the extent to which it had to be controlled by a small party. Mensheviks favoured a mass labour party. After 1917 he supported Lenin against the 'White' armies but continued to oppose his more dictatorial policies. He left the Soviet Union in 1920.

Marx, Karl Heinrich (1818-83), German social philosopher. The most important figure in the history of *socialism, he was thwarted in his ambition to follow an academic career at Bonn because of his radicalism. He helped to run (1842) an anti-government newspaper at Cologne, which in turn was suppressed by the censor. Moving to Paris, he met Friedrich *Engels, and from then on collaborated with him in works of political philosophy. In 1845 he went to Brussels, joined (1847) the Socialist League of the Just, later renamed the Communist League, and in conjunction with Engels wrote for it the *Communist Manifesto* (1848). The outbreak of the *Revolutions of 1848 made it possible for him briefly to return to Cologne. Expelled soon after from most European countries, he finally settled in London, where he lived, supported by Engels, for the rest of his life. He envisaged a global political and social revolution as a result of the conflict between the working classes and the capitalists, who used the state to enforce their own dominance. His goal was to unite all workers in order to achieve political power. He was a key figure in inspiring the foundation of the First *International, and was later chosen as its leader. The ideological clash between Marx and *Bakunin led to its disintegration in

Karl **Marx**, a German, spent much of his life in exile, mostly in England. He believed that power would inevitably move from industrial capitalists to the working classes. His writings inspired communist movements throughout the world in the 20th century.

Tomáš **Masaryk**, the founder of modern Czechoslovakia as a western-style democracy has been played down by communist historians in his country, but he enjoyed brief popular recognition during the 'Prague Spring' of 1968.

1876. Marx's theories were developed at length in *Das Kapital* (1867; ed. by Engels, 1885-94) and inspired the *communist movements of the 20th century.

Masaryk, Jan (1886-1948), Czechoslovak diplomat and statesman. The son of Tomáš *Masaryk, he helped in establishing the Czech republic and thereafter was mainly involved in foreign affairs. As ambassador to Britain (1925-38), he resigned in protest at his country's betrayal at *Munich (1938). On the liberation of Czechoslovakia by the Allies (1945) he became Foreign Minister, and was dismayed at the Soviet veto of Czechoslovak acceptance of US aid under the *Marshall Plan. At the request of President *Beneš, he remained in his post after the communist coup of February 1948, but he either committed suicide or was murdered three weeks later.

Masaryk, Tomáš Garrigue (1850-1937), Czechoslovak statesman. A member of the Austrian Parliament (1891-3 and 1907-14), he achieved European fame by defending Slav and Semitic minorities. During World War I he worked with *Beneš in London for Czech independence and for his country's recognition by the Allies. By their efforts Czech independence was proclaimed in Prague (1918) and he was elected President. He favoured friendly relations with Germany and Austria, and was a strong supporter of the League of Nations. He felt that the rising *Nazi menace needed a younger President and he resigned (1935) in favour of Beneš.

Massey, William Ferguson (1856-1925), New Zealand statesman. He founded the Reform Party which campaigned for freehold tenure and free enterprise. As Prime Minister (1912-25) Massey also made extensive purchases of remaining *Maori land. He was challenged by militant unionists (*Red Feds) and broke the strikes of 1912-13, having enrolled farmers as special constables— 'Massey's Cossacks'. He committed New Zealand manpower heavily in World War I, but his hold on domestic politics weakened with increasing urbanization.

mass production, a system of industrial production. It involves the division of labour advocated by Adam Smith in his book *The Wealth of Nations* (1776), and the breaking down of a process into a large number of specialized operations. Probably James Watt in Britain was the first to introduce the principles of mass production, when he designed standard and interchangeable parts for his steam-engine in the 1780s; while in the USA in 1798 Eli Whitney initiated the mass production of muskets by making machine tools to turn out the various parts of the musket. Its opponents during the 19th century, such as John Ruskin and William Morris, criticized the loss of a sense of craftsmanship which was being replaced by monotonous factory operations. It was in the automobile industry, initially in the US Ford Company in 1913, that the age of mass production was fully inaugurated. Standardized parts were brought together on a moving assembly-line to turn out standardized cars at low cost, but with high wages and profits. Mass production methods thereafter were steadily introduced into other areas of manufacturing, for example in World War II in shipbuilding and aircraft manufacture. Production was speeded up even more by increasingly sophisticated techniques of *automation.

Masurian Lakes, the scene of heavy fighting in east Prussia in World War I. Two Russian armies, urged to advance quickly in August 1914, to relieve pressure on France, were not fully prepared and the Lakes separated their attack. Prittwitz, the German commander, was indecisive, and only the despatch of *Ludendorff to the Eastern Front and the resulting battle of Tannenberg saved Germany. In February 1915 Ludendorff continued his offensive. Taking advantage of frozen swamps, he captured four Russian divisions in the region of the Lakes.

Matabeleland *Zimbabwe.

Match Girls' strike (1888), an industrial dispute in England. It involved the girls at the factory of Bryant and May in the East End of London, who complained about their low pay and the disfigurement of the jaw, nicknamed 'phossy jaw', caused by the phosphorus used in match-making. Annie *Besant, a journalist and a *Fabian, organized their strike. Demonstrations, partly to show their disfigurement, won public sympathy. Their success in gaining an increase in pay was a small-scale prelude to the growing strength of *trade unionism.

Matteotti, Giacomo (1885-1924), Italian socialist leader. A member of the Italian Chamber of Deputies, he began in 1921 to organize the United Socialist Party. He openly accused *Mussolini and his *Blackshirts of winning the 1924 parliamentary election by force, giving examples of attacks on individuals and the smashing of the printing presses belonging to opposition newspapers. Within a week he was found murdered. How far Mussolini was personally responsible is uncertain, but in January 1925 he took full responsibility for the crime, and proceeded to tighten up the fascist regime.

Mau Mau, a militant nationalist movement in Kenya. Its origins can be traced back to the Kikuyu Central Association, founded in 1920, and it was initially confined to the area of the White Highlands which Kikuyu people regarded as having been stolen from them. It imposed

A **Mau Mau** suspect is searched by a British officer following his arrest by security forces in Kenya, 1953. Police and troops pursued a vigorous campaign against the Kenyan nationalists in areas such as the Aberdare mountains.

fierce oaths on its followers. It was anti-Christian as well as anti-European. From 1952 it became more nationalist in aim and indulged in a campaign of violence, killing some 11,000 black Africans who were opposed to its brutalities and some 30 Europeans. Jomo *Kenyatta was gaoled as an alleged Mau Mau leader in 1953. In a well-organized counter-insurgency campaign the British placed more than 20,000 Kikuyu in detention camps. Widespread political and social reforms followed, leading to Kenyan independence in 1963.

Mauritania, a country in West Africa. French penetration of the interior, which forms part of the Sahara Desert, began in 1858, and in 1903 the country became a French protectorate. In 1920 Mauritania was made a territory of French West Africa. It became an autonomous republic within the *French Community in 1958, and fully independent in 1960. Following the Spanish withdrawal from the western Sahara in 1976 *Morocco and Mauritania divided between them the southern part of this territory, known as Tiris-el-Gherbia. Bitter war ensued, but in 1979 Mauritania relinquished all claims. The country's first president, Moktar Ould Daddah, was replaced by a military government in 1978, to be followed in 1980 by civilian rule.

Mauritius, an island in the Indian Ocean east of Madagascar. Discovered by the Portuguese in 1511, it was held by the Dutch (1598–1710), and the French (1710–1810). The British took it in 1810, and under their rule massive Indian immigration took place. Mauritius became an independent republic in 1960. Sugar has always been the principal crop, and since the fall of world prices in the 1980s a vigorous programme of agricultural diversification has been conducted. Politically it has maintained stability as a multi-cultural state.

Maximilian (1832–67), Austrian Archduke and Emperor of Mexico. A brother of the Austrian emperor Francis Joseph, he was persuaded by the French emperor *Napoleon III and Mexican royalists to accept the crown of the newly founded Mexican empire (1864–7). He soon lost the support of the conservatives by his liberal tendencies, and confirmed the hostility of Benito *Juárez's followers by ordering the summary execution of their leaders (1865). Maximilian's only protection was the presence of French troops, and, when these were withdrawn (1866–7), he took personal command of his soldiers. After a siege at Querétaro he was captured, tried by a court martial, and shot.

Maxton, James (1885–1946), British socialist politician. A member of the Conservative Club at Glasgow University, a speech by Philip *Snowdon in 1904 inspired him to join the *Independent Labour Party. He soon made a name for himself as an orator and a rebel. In 1916, during World War I, he urged Glasgow workers to strike and was imprisoned for a year for sedition. As a member of the ILP he represented Bridgeton in Parliament from 1922 to 1946. In 1932 Maxton, then leader of the ILP, broke with the Labour Party, dissatisfied with their policies.

May Fourth Movement, Chinese nationalist movement that began on 4 May 1919 with the student protest in Beijing at the *Versailles Settlement decision that

Giuseppe **Mazzini**, whose republican zeal fired the Italian Risorgimento. In 1831 he founded the Young Italy movement. (Museo del Risorgimento, Milan)

Japan should take over former German concessions in Shandong. In the New Culture Movement which emerged intellectuals grappled with Marxism and liberalism in their search for reforms. Socialist ideas became popular, and the movement played a major role in the revival of the *Kuomintang and the creation of the *Chinese Communist Party.

Mayhew, Henry (1812–87), British journalist. As a young man he abandoned the law for journalism and was associated with two periodicals before he helped to found *Punch* in 1841. His chief work, *London Labour and the London Poor* (1851–62), based on a series of interviews, was a remarkable piece of reporting, combining sensitive observation with shrewd economic and social analysis.

Mazzini, Giuseppe (1805–72), Italian patriot and revolutionary. The militant leader of the *Risorgimento movement for a united Italian republic, his membership of the *Carbonari led to his arrest and exile in 1830 and to his formation of the *Young Italy Society in Marseilles the following year. In 1834 he led a fruitless invasion of the Piedmontese province of Savoy from Switzerland, was sentenced to death *in absentia*, and spent a period of exile in London from 1837. During the *Revolutions of 1848, he was active in Italy. The flight from Rome of *Pius IX led to his setting up the short-lived Roman republic in 1849. This failure and abortive risings in Mantua (1852) and Milan (1853) greatly weakened his influence and he returned to London. He was considered a dangerous and irresponsible agitator by *Cavour and played an insignificant part in events in northern Italy of 1859–60.

Mboya, Tom (1926–69), Kenyan political leader. He served as treasurer of the Kenya Africa Union and was the chief trade-union organizer in Kenya in the late 1950s. In 1960, as leader of the Kenya Independence Movement, he attended a conference in London on the future of Kenya, and was instrumental in securing a constitution which would give Africans political supremacy. In 1960 he became Secretary-General of the newly formed Kenya African National Union. After Kenya gained its independence (1963) he served in various senior ministerial posts. He was assassinated in 1969.

Medicare, the provision of medical care and assistance to persons aged 65 and over in the USA. It was introduced in 1965 by President *Johnson as one of a series of reforms for his Great Society, which also included Medicaid for sick people with absolutely no personal resources. Both schemes were to be financed from social security taxes.

Mehemet Ali (c.1769–1849), Pasha (or viceroy) of *Egypt (1805–49). An Albanian by birth, he rose from the ranks to command an Ottoman army in an unsuccessful attempt to drive Napoleon from Egypt. In 1801 he returned to Egypt in command of Albanian troops and by 1811 he had overthrown the Mamelukes, who had ruled Egypt almost without interruption since 1250. Technically viceroy of the Ottoman sultan, *Mahmud II, he was effectively an independent ruler and reorganized the administration, agriculture, commerce, industry, education, and the army, employing chiefly French advisers, making Egypt the leading power in the

Mehemet Ali, mounted, poses in triumph as his army conquers Egypt. The capture of the citadel at Cairo from the Mamelukes in 1811 is depicted in the background.

eastern Mediterranean. He occupied the *Sudan (1821–3), and campaigned for the *Ottoman government in Arabia (1811–18) against the *Wahhabis, and in Greece (1822–8), as a result of which his fleet was destroyed by the combination of British, French, and Russian navies at the battle of *Navarino (1827). He took Syria (1831–3), and defeated the Ottoman troops at the battle of Nizip (1839). Threatened by united European opposition, he agreed to accept the suzerainty of the Ottoman sultan in 1841 and in return was granted a request that his family be hereditary pashas of Egypt.

Meiji Constitution, constitution of the restored imperial Japanese state. Framed by *Ito Hirobumi and modelled on the existing German form, the Meiji Constitution was gradually developed from 1884 with the institution of a European-style peerage (1884), a cabinet system (1885), and a privy council (1888), and formally completed in 1889. It was centred on a bicameral system with an elected lower house and an upper house of peers, but effective power rested with the executive as representatives of the emperor, in whom ultimate power still resided. Although political leaders always experienced difficulties in controlling the lower house (which by 1925 was elected through universal manhood suffrage), policy in the earlier period was usually dominated by a group of highly influential senior statesmen, including such men as Ito Hirobumi and *Yamagata Aritomo, and later fell under military influence. The old constitution was finally replaced by a new *Japanese Constitution, prepared under American supervision, on 3 May 1947.

Meiji Restoration, restoration of imperial rule in Japan, often defined as the overthrow of the Tokugawa *shogunate in 1868, but also sometimes considered to stretch from the overthrow of the shogunate to the formal institution of the new *Meiji Constitution in 1889. The Tokugawa shogunate was faced by increasingly severe internal problems in the first half of the 19th century, and its failure to deal effectively with foreign incursions into Japanese territory resulted in the uniting of opposition forces behind a policy of restoring the emperor to full power. Faced by a powerful alliance of regional forces, the last shogun formally surrendered his powers to the Meiji emperor Mutsuhito, who resumed formal imperial rule in January 1868, moving his capital to Tokyo a year later. Thereafter, the feudal *daimyo and *samurai systems were quickly dismantled, a western-style constitution introduced, and a policy of government-sponsored industrial development implemented, which would transform Japan into a centralized modern state.

Meir, Golda (1898–1978), Israeli stateswoman. Born as Golda Mabovitch in Kiev, Russia, she was brought up in the USA before emigrating to Palestine in 1921. She worked for several organizations before becoming (1946–8) head of the Political Department of the Jewish Agency, and involved in negotiations over the foundation

Golda **Meir**, a Jewish immigrant to Palestine, who became Prime Minister of Israel (1969–74).

of Israel. She served from 1949 to 1956 as Minister of Labour, and from 1956 to 1966 as Foreign Minister. In 1966 she became Secretary General of the Mapai Party and in 1967 helped merge it with two dissident parties into the Israel Labour Party. She was Prime Minister (1969–74) of a coalition government, and faced criticism over the nation's lack of readiness for the *Yom Kippur War. She was succeeded by General Rabin.

Melbourne, (Henry) William Lamb, 2nd Viscount (1779–1848), British statesman. He was elected to Parliament in 1806 as a Whig, becoming Chief Secretary for Ireland in 1827–8. As Home Secretary (1830–4) he dealt harshly with agrarian riots in southern England in 1830–1 and gave reluctant support to the *Reform Act of 1832. Briefly Prime Minister in 1834, he upheld the sentences passed on the *Tolpuddle Martyrs. He became Prime Minister again in 1835 and the trusted adviser of Queen *Victoria during the early years of her reign, remaining in office until 1841.

Mellon, Andrew William (1855–1937), US financier and philanthropist. He served as Secretary of the Treasury (1921–32) under *Harding, *Coolidge, and *Hoover. Mellon opposed high expenditure by federal government, working to reduce taxes. He presided over the boom of the twenties, convinced that government was an extension of big business, to be run on business lines. His tax-cuts purposely helped the rich to aid their investments which, he felt, would in time bring employment and other benefits to the less well off. In face of the accelerating depression he had no policy other than increasing retrenchment. He donated his considerable art collection together with funds, to establish the US National Gallery of Art (1937).

Mendès-France, Pierre (1907–82), French statesman. Elected as a Radical-Socialist Deputy in 1932, he was an economics minister in the government of Leon *Blum in 1938. He was imprisoned by the *Vichy government, escaped to London (1941), and joined the exiled Free French government of General *de Gaulle. After the war he was critical of France's post-war policy in *Indo-China. He became Premier in May 1954, after the disaster of *Dienbienphu, promising that France would pull out of Indo-China. He honoured this pledge, rejected the plan for a European Defence Community, prepared *Tunisia for independence, and supported claims for *Algerian independence. An austere economic policy led to his downfall in 1955. He served in the government of Guy Mollett (1956), but was unhappy with the constitution of the French Fifth Republic created by de Gaulle in 1958. He resigned from the Radical Party in 1959, after which he never had an effective power-base. He became increasingly opposed to the autocratic use of presidential power by de Gaulle and supported the bid by François *Mitterand to oppose him in 1965. He retired from political life in 1973.

Menelik II (1844–1913), Emperor of Ethiopia (1889–1913). Originally ruler of Shoa (1863–89), with Italian support he seized the throne after Emperor John IV died. He made the Treaty of Ucciali (1889) with Italy, but when he learnt that the Italian wording of the treaty made Ethiopia a protectorate of Italy, he denounced the agreement. Italy's invasion and subsequent defeat at

Melbourne, with pen and documents, stands before the young Queen Victoria at her first Privy Council, held at Kensington Palace, June 1837. Also shown in this painting by Sir David Wilkie are Russell and Palmerston (*to right and left of Melbourne*), the Duke of Wellington, and Peel (*on Wellington's left*). (The Royal Collection)

*Adowa greatly strengthened Menelik's position. He modernized Ethiopia, initiated public education, attempted to abolish slavery, and gave France a railway concession from *Djibouti, which opened up commerce. His conquests doubled the size of the country, and brought south Ethiopia into his domain.

Menon (Vengalil Krishnan) Krishna (1896–1974), Indian politician. He was a prominent spokesman in Britain of the Indian nationalist cause before 1947. He subsequently served as India's High Commissioner in London (1947–52) and as India's representative at the United Nations. As Minister of Defence (1957–62) he was heavily criticized for his autocratic decisions and for the defeat in the *Indo-Chinese War of 1962, and he was forced to resign.

Menshevik   *Bolshevik.

Menzies, Sir Robert Gordon (1894–1978), Australian statesman. He entered the Victorian Parliament in 1934, representing the Nationalist Party and later the United Australia Party. He held various offices, including that of Deputy Premier. He moved to the federal Parliament in 1934, representing the United Australia Party. A conservative and avowed anti-communist, Menzies was Prime Minister from 1939 until 1941 and leader of the Opposition from 1943 until 1949, during which time he founded the Liberal Party to replace the United Australia Party. He was again Prime Minister from 1949 until his retirement in 1966.

Mercier, Honoré (1840–94), French-Canadian statesman. He opposed Confederation (1867) of the British North American provinces and founded (1871) the Parti National to represent the French element in Canada. He served in the Canadian House of Commons from 1872 to 1874. Elected to the legislative assembly of the province of Quebec in 1879, he became leader of the Liberal Party of Quebec in 1883, and in 1886, in the wake of the *Riel Rebellion, defeated the Conservatives to become Premier of Quebec. His immense popularity was eventually undermined by charges of corruption in connection with railroad subsidies, and in 1891 he was dismissed from office and defeated in the subsequent election. The new government acquitted Mercier of the charges.

Meredith incident (1962), an episode in the US *civil rights struggle. Officials of the state of Mississippi defied a federal court order requiring the University of Mississippi to allow a black man, James Meredith, to enrol as a student. The governor made defiant statements, thus encouraging thousands of segregationists to attack the federal marshals assigned to protect Meredith as he entered the university. President *Kennedy sent in national guardsmen and regular army troops to restore peace, enforce the court's orders, and secure the entry of Meredith as the university's first black student.

Mesopotamia Campaign (World War I), a British campaign fought against the *Ottoman Turks. In 1913 Britain had acquired the Abadan oilfield of Persia (now Iran), and when war broke out in 1914 it was concerned to protect both the oilfields and the route to India. When Turkey joined the war in October 1914 British and Indian troops occupied Basra in Mesopotamia (now Iraq). They began to advance towards Baghdad, but were halted and suffered the disaster of *Kut-al-Amara. General Sir Frederick Maude recaptured Kut in February 1917, entering Baghdad on 11 March. One contingent of British troops reached the oilfields of Baku (May 1918), which it occupied until September, when the Turks reoccupied it. A further contingent moved up the River Euphrates to capture Ramadi (September 1917) and another up the River Tigris as far as Tikrit (July 1918), before advancing on Mosul. Meanwhile from Egypt General Sir Edmund Allenby was driving north into Palestine, aided by Arab partisans organized and led by T. E. *Lawrence. In December 1917 Jerusalem was occupied, from where Allenby moved north towards Damascus (October 1918). After the armistice of Mudros (30 October), British troops briefly reoccupied Baku (November 1918–August 1919), aiming to deprive the *Bolsheviks of its oil and to use it as a base in the *Russian Civil War. Britain had now occupied all Mesopotamia, and for a brief while considered the possibility of creating a single British dominion of Mesopotamia, consisting of Palestine, Jordan, Iraq, and Iran, linking Egypt with India and providing a bulwark against Bolshevism.

Metaxas, Ioannis (1871–1941), Greek general and statesman. He became chief of staff in the Greek army, but was exiled (1917) when Greece joined the Allies in World War I. He returned (1920), and, after leading a coup, was exiled again (1923–4). A strong monarchist, he held several ministerial posts from 1928 to 1936. With royal approval he dissolved Parliament and became dictator of Greece (1936–41). Under him in 1940 a united country fought against Italian invaders, but in 1941 Hitler's forces intervened and occupied Greece.

Methodist Church, a Protestant Church founded by John Wesley (1703–91). Wesley wished his followers to remain within the Church of England, but after his death Methodism rapidly became a distinct Church. In Wales the religious revival inspired by Howel Harris and Daniel Rowlands in the 18th century led to the establishment in 1811 of a dominant, Calvinist form of Methodism. The Church has suffered many secessions, especially of the Methodist New Connexion, the Primitive Methodist Church, and the Bible Christians. In the USA the Methodist Church divided into many groups, largely over attitudes towards slavery. In the 20th century the various Methodist bodies came together again, notably by the merging of the Wesleyan Methodists, the Primitive Methodists, and the United Methodists as the Methodist Church in Britain in 1932, and by the uniting of three Methodist Churches as the Methodist Church in the USA in 1939. The World Methodist Council, founded in 1881, provides a link between the forty million Methodists in the world.

Metternich Winneburg, Clemens Wenzel Nepomuk Lothar, Prince (1773–1859), Austrian statesman. At *Napoleon's request he became Austrian ambassador in Paris (1806), and took part in all the major negotiations with the emperor. Following Napoleon's victory at Wagram, he returned to Vienna as Chancellor and Foreign Minister. As such he negotiated the marriage of Napoleon to Princess Marie-Louise. Following Napoleon's defeat at the battle of *Leipzig he signed the Treaty of Paris (April 1814) with *Louis XVIII and dominated the Congress of *Vienna, which followed. From 1821 to 1848 he was Austrian Court Chancellor and Chancellor of State. His 'system', based

A portrait of Prince **Metternich** by Sir Thomas Lawrence. Metternich was the architect of the restoration of conservatism in Europe following the Napoleonic Wars. (The Royal Collection)

upon co-operation between the great powers and the suppression of revolutionary movements, created stability at the cost of liberal reform in Europe. He sought to maintain the *Austrian empire, the majority of whose subjects were non-German speaking, against the forces of nationalism by the use of police despotism. In the *Revolutions of 1848 he fled to England, returning in 1851 to live in retirement in his castle on the Rhine.

Mexican–American War (1846–8), a conflict between the USA and Mexico. Hostilities between the two countries began shortly after the USA annexed (1845) the Mexican state of *Texas and sought to expand the boundaries of the state to include still more territory. In the ensuing war General Stephen Kearny took over the New Mexico territory and Captain John *Frémont annexed the California territory almost without a fight. In northern Mexico stiffer opposition was encountered as General Zachary Taylor invaded Mexico across the Rio Grande and defeated General Antonio López de *Santa Anna in the bloody battle of Buena Vista (22–23 February 1847). The fiercest fighting occurred in central Mexico. General Winfield Scott's order of a mortar bombardment of Vera Cruz resulted in the deaths of hundreds of civilians. The US army then moved inland to Mexico City, where hotly contested engagements were fought at Molino del Ray and Chapultepec Hill (12–13 September 1847). The US capture of the capital city (1847) occasioned the Mexican surrender. The Treaty of Guadalupe Hidalgo (1848) ended hostilities. By the terms of the treaty the USA confirmed its claim to Texas and gained control of the area which would later become the states of New Mexico, Arizona, and California (where gold had recently been discovered). The USA agreed to pay Mexico US 15 million dollars in return.

Mexican Revolution (1910–40), period of political and social reform in *Mexico. The roots of the revolution can be traced to the conflicts and tensions generated by demographic, economic, and social changes which occurred during the rule of President Porfirio *Díaz, known as the Porfiriato (1876–1911). The regime became increasingly centralized and authoritarian, favouring Mexico's traditional and newly emerging élites, but failing to incorporate growing urban middle-class and labour groups into national politics. In 1910 Francisco *Madero, the leader of the Anti-Re-electionist movement, received an enthusiastic response to his call to arms to overthrow the dictator. Díaz resigned in May 1911, and Madero was elected President, but he failed to satisfy either his radical supporters or his Porfirian enemies, and was assassinated in a counter-revolutionary coup led by General Victoriano *Huerta in 1913. Huerta was defeated by an arms embargo, diplomatic hostility from the USA, and a coalition of revolutionary factions led by Emiliano

*Zapata, Pancho *Villa, Venustiano *Carranza, and Alvaro Obregon. The victorious revolutionaries split into Constitutionalists (Carranza and Obregon), who sought to reform the 1857 Liberal Constitution, and Conventionists (Zapata and Villa) who wished to implement the radical proposals of the convention of Aguascalientes (1914). The civil war which ensued was protracted and bitter. In February 1917 the reformed Constitution was promulgated. However, the document was largely ignored, and Carranza's procrastination prompted his overthrow and assassination in 1920. Mexico's post-revolutionary leaders faced the difficult tasks of economic regeneration and the reconstitution of central political authority, but were hampered by strong opposition from the Catholic Church. Tension culminated in the so-called War of the Cristeros (1928–30), when thousands of Christian peasants arose in protest against the new 'godless' state, and were finally defeated at the battle of Reforma (1930). When President Avila Camacho (1940–6) was elected, a period of consolidation and reconciliation marked the end of the revolution and the beginning of a period of industrial development.

Mexico, a Central American country. As part of New Spain, Mexico had been under Spanish rule since 1521. Inspired by French revolutionary ideas, an independence movement developed, led by two priests, Miguel *Hidalgo y Costilla and José María *Morelos y Pavón, both of whom were captured and shot by the Spanish authorities (1810 and 1814). In 1821 Augustín de *Iturbide briefly

US troops led by Colonel Doniphan attack the Bishop's Palace at Monterey, California, 1846, during the **Mexican–American War**. Monterey, with the rest of California, fell into US hands in this, the first year of the war.

created an independent Mexican empire, which included the captaincy-general of *Guatemala. Following his exile (1823), the first Mexican constitution was proclaimed (1824), based on the US constitution. Two parties, the Federalist and the Centralists, quickly appeared, and in 1833 the liberal federalist Antonio López de *Santa Anna emerged as President. He was not able to prevent the declaration of independence of *Texas (1836) nor the *Mexican–American War (1846–8), which resulted in the loss of huge territories, added to by the *Gadsden Purchase of Arizona in 1853. A period of reform followed, and a new constitution (1857) was promulgated. But economic difficulties and French imperialist dreams resulted in the imposition by troops of *Napoleon III of the Habsburg prince *Maximilian as emperor (1864–7). When French troops were withdrawn, Maximilian was defeated, captured, and shot. There followed the long dictatorship of Porfirio *Díaz (1876–1910) and then the prolonged *Mexican Revolution (1910–40). Under President Miguel Alemán (1946–52), the process of reconciliation begun by his predecessor, Avilo Camacho, continued. Since then democratic governments have continued to follow moderate policies, while seeking further to modernize the economy, bolstered by oil revenues. The presidency of Miguel de la Madrid Hurtado (1983–) was faced by a fall in oil prices, a massive national debt, and one of the fastest growing birth-rates in the world. Yet Mexico continued to enjoy the advantages of a strong manufacturing base, self-sufficiency in oil and natural gas, and large capitalist investment in a modernized agricultural system.

Midlothian Campaigns (1879–80), a series of electioneering speeches by William *Gladstone to mass audiences in Britain, marking a new phase in party electioneering. Queen Victoria and *Disraeli, the Conservative Prime Minister, regarded the tactic as unconstitutional, but it helped the Liberals to win a large majority in the general election of 1880. The campaigns recognized the importance of the new mass electorate created by the 1867 *Reform Act, and of the growing influence of newspaper reports of political speeches. Before setting out by train from London to his Edinburgh constituency of Midlothian Gladstone addressed the crowds and repeated this at stations where the train stopped. The climax came in speeches at Edinburgh and other Scottish cities in which he strongly criticized the government. He gained the Midlothian seat from the Conservative, Lord Dalkeith.

Mihailovich, Draza (1893–1946), Yugoslav partisan leader. He was a Serbian army officer who, after the fall of Yugoslavia in World War II, organized royalist partisans against German forces. He structured his forces into bands (četa, pronounced cheta), and these became known as chetniks. Their relations with the communist partisans of *Tito were uneasy. The Allies withdrew their support because chetnik forces were reported to be collaborating with the Germans against Tito. In 1944 Peter II of Yugoslavia withdrew his support from Mihailovich, who continued to fight on the borders of Serbia and Bosnia. After Tito gained power, Mihailovich was tried and shot for collaboration and war crimes.

Mikoyan, Anastas Ivanovich (1895–1978), Soviet politician. He joined the Communist Party in 1915,

taking part in the *Russian Revolution and fighting in the *Russian Civil War. He held several ministerial posts (1926–55), mainly in the field of trade. He associated himself with *Khrushchev's denunciation of *Stalin and became his negotiator in Soviet relations with discontented East European states. He was President of the Praesidium of the Supreme Soviet (1964–5).

Mill, John Stuart (1806–73), Scottish philosopher and economist. The son of James Mill, the historian and advocate of utilitarianism, he was subjected by his father to a rigorous system of education from an early age. He became a leading exponent of *Bentham's utilitarian philosophy, though also in his later years a strong critic of it. In *Utilitarianism* (1863) he expounded the view that Bentham's doctrine of the greatest happiness of the greatest number should be achieved through legislation. Mill is perhaps best remembered now for his championship of individual liberty against interference by the state in *On Liberty* (1859) and for his advocacy of equality for women—his *The Subjection of Women* (1869) remains a classic exposition of the case for female emancipation.

Milner, Alfred, Viscount (1854–1925), British statesman and colonial administrator. As private secretary to G. J. Goschen when Chancellor of the Exchequer, he developed a capacity for finance, which he later applied in Egypt (1889–92) and as Chairman of the Board of Inland Revenue (1892–7). As High Commissioner for *South Africa (1897–1905), his policies precipitated the Second *Boer War, but he made preparations for peace in the settled areas by schemes of reorganization and reform, by which stability was given to the new South Africa. He joined Lloyd George's War Cabinet in 1916 and was Secretary of State for the Colonies (1919–21).

MI5 and MI6 (Military Intelligence), the popular names for the Security Service and the Secret Intelligence Service in Britain. The sphere of MI5, founded in 1909, covers internal security and counterintelligence on British territory while MI6, formed in 1912, covers all areas outside the United Kingdom. MI6, whose successful operations in co-operation with *resistance movements overseas in World War II contributed considerably to the outcome of the war, has since then received adverse public exposure through disclosures that some of its employees, notably intellectuals recruited in Cambridge in the 1930s, such as Philby, Burgess, Maclean, and Blunt, were double agents whose final allegiance lay with the communist bloc. Some accusations have been made about MI5's extra-constitutional role during the years of the Labour government after 1974. A persistent campaign to liberalize the provisions of the Official Secrets Acts (1911, 1920) seeks to reduce the potential abuse of the law by governments anxious to suppress criticism and to keep secret matters which should be publicly known.

Mindszenty, József (b. József Pehm) (1892–1975), Hungarian prelate. He was imprisoned by the Hungarian puppet government (1944–5) and, after its collapse in 1945, became Archbishop and Primate of Hungary. After visiting the USA he was arrested and charged with treason and currency offences. At his trial in 1948 he expressed his hostility to communism and his support of the Habsburg monarchy. He was sentenced to penal servitude for life, commuted to house detention. He was freed at the time of the *Hungarian Revolution. On the return of Soviet forces, Mindszenty sought refuge in the US Legation in Budapest. He stayed there until 1971, when he went to Rome, finally settling in Vienna.

missionaries, Christian, persons working within organized groups whose aim is to propagate Christianity. In the early 19th century a new missionary spirit was aroused in Great Britain, the USA, and Germany, through the growing evangelical movements within the Protestant Church. In the wake of *imperialism, Roman Catholic and Protestant missionaries went to India, Africa, China, and to the South Seas from Europe and the USA, and US mission societies took their faith to American Indians, Negroes (both slave and free), and Inuit (Eskimos) in the Arctic. Missionaries provided medical and educational services as well as spiritual teaching; among noted missionary doctors and nurses have been David *Livingstone, Albert *Schweitzer, and Mother Teresa of Calcutta. In 1981 the International Missionary Council became part of the World Council of Churches, and the trend to co-operate in relief and welfare work with members of other faiths has been encouraged. Other faiths, particularly *Islam, have also sought to proselytize by sending out representatives to convert people to their beliefs.

Missouri Compromise (1820-1), a series of legislative measures, passed by the US Congress to end controversy over the extension of slavery in the territories beyond existing state boundaries. It was agreed that Maine would enter the Union of the United States as a free state and Missouri as a slave state and that slavery would be prohibited elsewhere in the *Louisiana Purchase north of 36° 30′. This held out to the South the prospect of Florida and Arkansas being admitted as slave states, while securing the greater proportion of unsettled territory to the free North. The Compromise of 1820 temporarily laid the issue of slavery to rest, but the drawing of precise geographical lines between slave and non-slave areas led to fresh divisions.

Mitterrand, François (1916-), French statesman. A leader of the French *resistance movement during World War II, he was elected Deputy in the French National Assembly in 1946. He served in all the governments of the French Fourth Republic. Seeking to build a coalition between French parties of the Left—Radical, Socialist, and Communist—he founded (1965) the Federation of Democratic and Socialist Left, when he stood for President against *de Gaulle, winning seven million votes. He stood unsuccessfully again in 1974 against Giscard d'Éstaing, and in 1981 was finally elected President of France. Early measures to decentralize government, raise basic wages, increase social benefits, and nationalize key industries were followed by economic crisis and a reversal of some policies. A committed supporter of both nuclear power and a nuclear bomb for France, he has advocated a strong foreign policy.

Mizuno Tadakuni (1793–1851), Japanese statesman and reformer. As Chief Senior Councillor to the Tokugawa *shogunate, Mizuno responded to the crisis engendered by famine, insurrection, and foreign pressure in the 1830s by instituting a comprehensive programme of social, political, and economic measures known collectively as the Tempo reform. He introduced strict price controls, abolished restrictive merchant guilds, and attempted to bring outlying lands under direct government control through a land requisition scheme. Opposition to the last measure led to Mizuno being driven from office in 1845.

Mobutu, Sese Seko (1930-), African statesman. President of Zaïre (1965-). He joined the Force Publique (Belgian Congo Army) in 1949. In 1960 he attended the Brussels Round Table Conference on Congolese independence. He was then appointed chief-of-staff of the Force Publique, the nucleus of the new Congolese Army, and in 1961 commander-in-chief. Following five years of chaos and civil war (*Congo crisis), he seized office as President in 1965. While maintaining order by harsh military discipline, he appointed civilians as ministers and struggled to give stability to the economy. As part of the programme of Africanization, he changed the name of the country in 1971 from Congo to Zaïre. While Mobutu has remained in power for over two decades and has achieved some degree of political reform, his regime has persistently been shaken by violent opposition, including rebel invasions from neighbouring Angola in 1977-8.

Mollet, Guy (1905-75), French statesman. He entered politics in 1946 and became General Secretary of the Socialist Party (1946-69). He was Prime Minister in 1956-7. He participated in the events leading to the *Suez crisis, and during his ministry the war in *Algeria intensified.

Molly Maguires, a secret US organization of Irish Americans (c.1865-75). Its name was based on an anti-landlord organization in Ireland. It dominated the eastern Pennsylvania coalfields, campaigning against

President **Mobutu** of Zaïre visiting the town of Lubumbashi, 1978.

anti-union mine-owners and managers. Resorting to murder and intimidation of the police, it was broken after infiltration by a *Pinkerton detective, who produced evidence which hanged twenty members and imprisoned fourteen others.

Molotov, Vyacheslav Mikhailovich (b. V. M. Skryabin) (1890-1986), Soviet leader. He joined the Communist Party at 16 and as a student in Kazan was exiled by the Tzarist regime. He played a prominent part in the establishment of the official communist paper *Pravda*, and was its editor during the *Russian Revolution. Working closely with both *Lenin and *Stalin, he was instrumental in the compulsory nationalization of factories and workshops under the *Bolsheviks. For the next forty years he remained at the heart of the Soviet political élite. He took a leading part in the liquidation of the Mensheviks, and in 1926 put down the *Zinoviev opposition. In 1939 he was the Soviet signatory in the *Nazi-Soviet Pact and, after Hitler's invasion of Russia (1941), signed the Anglo-Soviet Treaty (1942) against his former allies. At the *Yalta and *Potsdam conferences in 1945 he was Stalin's closest adviser. Surviving Stalin, he was expelled from all his posts by *Khrushchev, who appointed him ambassador to Outer Mongolia. In 1984 he was rehabilitated within the Soviet Communist Party.

Moltke, Helmuth, Graf von (1800-91), Prussian field marshal. As chief of the Prussian General Staff (1857), he was the first European general to appreciate the revolutionary effect of railways on warfare. Men and equipment could be deployed over wider areas, and troops could be maintained in the field throughout the

Von **Moltke**, in Prussian army uniform. A brilliant military strategist, he made the Prussian army the most mobile and adaptable in Europe.

year. Staff officers needed to be better trained, as orders rather than detailed instructions were called for on vast battle fronts. His ideas were proved in wars against Denmark (1864), Austria (1866), and France (1870).

Mongkut (1804-68), King Rama IV of Siam (*Thailand) (1851-68). The rightful heir to the throne of Siam on the death of Rama II in 1824, Mongkut remained a Buddhist monk for a further 27 years following the usurpation of his elder brother Pra Nang (Rama III). When he finally did ascend the throne in 1851, he rejected the policy of isolationism and maintained friendly relations with the French and British through a series of judicious concessions. He made a crucial contribution to the maintenance of Siamese independence at a time when the rest of South-East Asia was falling under European influence, and his enlightened policies were continued by his successor, *Chulalongkorn.

Mongolia, central Asian region. Mongolia remained part of the Chinese empire until the fall of the *Qing dynasty in 1911, although Russia mounted an increasingly strong challenge for the area in later years. While Inner Mongolia remained in Chinese hands, Outer Mongolia seized independence in 1911 and reasserted it after brief Chinese and White Russian occupations in 1919-21. Outer Mongolia became communist in 1924 and has remained so ever since, traditionally following a policy of alliance with the Soviet Union.

Monitor v. Merrimac (March 1862), a naval battle in the *American Civil War, the first between *ironclads. The troops of the *Confederacy captured the scuttled Union (Northern) frigate *Merrimac* at the beginning of the war and converted her into an ironclad. In the early stages of the battle of Hampton Roads, the *Merrimac* (now renamed *Virginia*) easily defeated a blockading Union squadron. She was then confronted by the *Monitor*, a revolutionary warship designed around a large revolving armoured turret. In the resulting five-hour duel neither vessel was able to inflict major damage on the other, but the Confederate threat to the blockade was countered. Neither vessel played any further part in the war but they had altered the nature of naval combat.

Monnet, Jean (1888-1979), French economist and administrator. In 1947 he devised and became commissioner-general of a plan which bore his name, whose object was to restore the French economy by means of centralized planning. Monnet was an internationalist and campaigned for European unification, working out the details of the *Schuman Plan and later becoming the first President of the High Authority of the *European Coal and Steel Community (1952-5).

Monroe, James (1758-1831), fifth President of the USA (1817-25). A member of the Virginia legislature (1782-3), the Continental Congress (1783-6), and the Senate (1790-4), Monroe emerged as a firm supporter of *states' rights and an opponent of the *Federalists. After serving as minister to France (1794-6) and Governor of Virginia (1799-1802), his political fortunes were transformed by the electoral triumph of the Democratic Republicans under Thomas *Jefferson. He helped to negotiate the *Louisiana Purchase and served as minister to England, before becoming Secretary of War (1814-

Monitor v. Merrimack. This drawing of the early stages of the first ironclad encounter on 9 March 1862 shows the larger *Merrimack* (*left, background*) exchanging fire with the turreted *Monitor* in Hampton Roads, Virginia. The historic duel is watched impotently by conventional warships. (Library of Congress, Washington)

15) in the *Madison administration. With Republican power well entrenched, Monroe easily won the presidential election of 1816 and repeated the triumph in 1820. His tenure of office saw the temporary settlement of the slavery issue in the *Missouri Compromise, the settlement of the Canadian boundary dispute, successful expansion into Florida, and the enunciation of the *Monroe Doctrine.

Monroe Doctrine, US foreign policy declaration warning European powers against further colonization in the New World and against European intervention in the governments of the American hemisphere, and disclaiming any intention of the USA to take any part in the political affairs of Europe. The background of the doctrine, spelt out in President *Monroe's annual message to Congress in 1823, was to be found in the threat of intervention by the *Holy Alliance to restore Spain's South American colonies, and in the aggressive attitude of Russia on the north-west coast of America. The doctrine was infrequently invoked in the 19th century, but after the development of territorial interests in Central America and the Caribbean it became a tenet of US foreign policy. During the early 20th century it developed into a policy whereby the USA regarded itself as the policeman of North and South America and this consistently complicated relations with Latin American countries.

Montagu–Chelmsford Proposals (1918), constitutional proposals for British India. They were made in the Report on Indian Constitutional Reforms (1918) by Edwin Montagu (1879-1924), Secretary of State for India, and Frederick John Thesiger, 3rd Baron Chelmsford (1868-1933), governor-general of India (1916-21). They followed the promise of responsible government made in 1917, and envisaged a gradual progress towards devolution of power and Indian self-government. In some ways the Government of India Act (1919) which followed can be seen as a measure to shore up imperial authority by containing opposition; it was regarded by Indian nationalists as inadequate and not a full implementation of the proposals.

Montenegro, one of six constituent republics of *Yugoslavia. The only southern Slavic nation to remain outside the Ottoman empire, it expanded its territory, often with encouragement from Russia. Under Nicholas Petrovic Njegos (1910-18) it engaged in the *Balkan War, and greatly extended its territory. Deposing Nicholas in 1918, it was absorbed into Serbia in 1919 and united with the new kingdom of *Yugoslavia from 1929. It was the scene of bitter fighting in World War II.

Montgomery of Alamein, Bernard Law Montgomery, 1st Viscount (1887-1976), British field-marshal. He served with distinction in World War I, and in World War II commanded the 8th Army (1942-4) in the *North African and *Italian campaigns. He led his troops at El *Alamein, one of the most decisive victories of the war, enabling the Allies to begin the advance that removed the Germans from North Africa. In 1944 he commanded the British Commonwealth armies in *Normandy with considerable success. His idea of an attack on *Arnhem failed, but he played a major role in beating the German counter-offensive in the *Ardennes. He held various senior military posts after the war.

Moore, Sir John (1761-1809), British soldier. He joined the army in 1776 and served with distinction during the war with France. In 1808 he was appointed commander of the British army in Portugal with orders to drive the French from the Iberian Peninsula. He advanced into northern Spain but was forced to retreat to the coast by overwhelming numbers of French troops. Moore was killed in the successful rearguard action at Corunna which enabled the British army to be safely evacuated.

Morelos y Pavón, José María (1765-1815), Mexican mestizo priest and leader of the revolution against Spain. Joining the insurrection (1810) he led a successful campaign in the south, and assumed command of his

General **Montgomery**, photographed in 1943 after his successful North African campaign and shortly before his leading role in the Allied D-Day invasion. The sight of 'Monty's' preferred headgear, a soft beret, was familiar to everyone in the 8th Army.

country's struggle after the capture and execution (1810) of Miguel *Hidalgo y Costilla. From 1812 to 1815 he kept the royalist army on the run with guerrilla warfare. He sponsored the Congress of Chilpancingo (1813), which formally declared Mexican independence and adopted Mexico's first constitution with Morelos as head of government. Before its administrative, social, and fiscal reforms could be adopted, Morelos was captured by Spanish forces and shot.

Moreno, Mariano (1778–1811), Argentine revolutionary. His study of the enlightenment led him to challenge Spanish mercantilist policies in the Rio de la Plata region. In his *Memorial of the Landowners*, he argued for free trade in terms similar to those of Adam Smith. Although trade restrictions were eased, Moreno became the secretary of the first revolutionary governing junta in 1810, which deposed the Spanish viceroy. Dispatched to Europe to secure assistance for the independence movement, he died aboard ship.

Morgan, John Pierpont (1837–1913), US financier and philanthropist. A member of a distinguished family of financiers, he came to New York from Boston in 1857. He established his own company there in 1860, becoming increasingly interested in railways. Morgan believed in the merits of centralization, working to eliminate wasteful competition, but his banking opponents saw him as a monopolist. In 1893 and 1907 his personal weight was sufficient to stabilize critically unbalanced financial markets. He was perhaps the leading private art collector of his day.

Morínigo, Higinio (1897–1985), Paraguayan military dictator. He was an army officer during the *Chaco War and subsequently became Minister of War. When President José Félix Estigarribia was killed in an air crash (1940) Morínigo became President (1940–8). He suspended the constitution and established a harsh military dictatorship. Housing and public health were improved during his tenure, and, in 1946, political exiles were brought into a more democratic cabinet. When these were ousted on suspicion of treason (1947), civil

The US financier J. P. **Morgan** (*centre*) with his son, John Pierpoint Morgan Jr (1867–1943); their companion is Louisa P. Satterlee.

war broke out and a broadly based opposition, led by the Colorado Party, forced his retirement and exile to Argentina (1948).

Morley–Minto reforms (1909), name for constitutional changes in British India, introduced to increase Indian participation in the legislature. They were embodied in the Indian Councils Act (1909) following discussions between John Morley, Secretary of State for India (1905–14), and Lord Minto, viceroy (1905–10). The reforms included the admission of Indians to the Secretary of State's council, to the viceroy's executive council, and to the executive councils of Bombay and Madras, and the introduction of an elected element into legislative councils with provision for separate electorates for Muslims. The reforms were regarded by Indian nationalists as too cautious, and the provision of separate electorates for Muslims was resented by Hindus.

Mormons, members of a Christian religious movement, also known as the Church of Jesus Christ of Latter-day Saints. The Mormon religion was founded in 1830 in New York state by Joseph *Smith and was based on the revelation to him of the Book of Mormon. This was a mixture of religious concepts and mythical history, including the belief that American Indians were the lost tribe of Israel. Rapidly gathering followers, Smith moved westwards, but local hostility, largely occasioned by the Mormon espousal of polygamy, led to Smith's lynching in Illinois in 1846. Under the leadership of Brigham *Young the group made its final westward move to Utah. After many hardships, the Mormons created a self-contained economy centred around Salt Lake City, although continued difficulties with polygamy prevented Utah becoming a state until 1896. Mormons dominate the administration of Utah.

Moro, Aldo (1916–78), Italian statesman. He entered Parliament as a Christian Democrat in 1946. As Minister of Justice (1955–7) he reformed the prison system. He was Foreign Minister in several governments (1965–74), and Prime Minister (1963–8, 1974–6). Moro was kidnapped in 1978 by the *Red Brigade, who demanded the release of imprisoned terrorists for his return. The government's refusal to accede led to Moro's murder.

Morocco, a North African country. An independent sultanate since the Middle Ages, Morocco had lapsed into endemic disorder by the 19th century and became the target for French and Spanish imperial ambitions. In the early 20th century, German opposition to French expansionism produced serious international crises in 1905 and 1911 which almost resulted in war. In 1912 it was divided between a French protectorate, a Spanish protectorate, and the international Zone of Tangier. Rif rebels under *Abd el-Krim fought the Spanish and French occupying powers in the 1920s, and Morocco became an independent monarchy under Muhammad V in 1956 when it absorbed Tangier. Muhammad was succeeded by his son Hassan II in 1961, but opposition sparked the suspension of parliamentary government in 1965, and royal authority has been maintained in the face of abortive military coups in the early 1970s and intermittent republican opposition. In 1980 a new constitution proclaimed the kingdom of Morocco to be a constitutional monarchy. Since the mid-1970s Morocco has been involved in an inconclusive desert war in the former Spanish Sahara with the local nationalist movement Polisario.

Morris, William Richard *Nuffield, Viscount.

Morrison of Lambeth, Herbert Stanley Morrison, Baron (1888–1965), British politician. As leader of the London County Council (1934–40), he unified the transport system under public ownership and created a 'green belt' around the metropolis. Morrison was Minister of Supply and then Home Secretary in *Churchill's coalition government during World War II. He drafted the programme of nationalization and social services in the 1945 election. In 1945 he was deputy prime minister, but was defeated by *Gaitskell in the election for leadership of the Labour Party in 1955.

Mosley, Oswald Ernald, Sir (1896–1980), British political leader. He was a Member of Parliament successively as Conservative (1918–22), Independent (1922–4), and Labour (1925–31). He formed a progressive socialist movement, the New Party (1931) advocating state intervention. Calling for a dictatorial system of government, he formed the National Union of Fascists in 1932. *Anti-Semitic and *fascist in character, its blackshirted followers staged violent marches and rallies in the East End of London. Mosley was interned during 1940–3. In 1947 he founded the 'Union Movement', whose theme was European unity.

Mossadeq, Muhammad *Mussadegh.

motor vehicle, a vehicle propelled by an engine that is driven along roads. In France in 1769 Nicholas-Joseph Cugnot demonstrated the first self-propelled road vehicle, driven by a steam engine. In 1801 Richard Trevithick devised a steam-powered road vehicle in England and by 1828 Sir Goldworth Gurney built and operated a steam carriage between London and Bath. The origins of the modern motor industry can be linked to the development in Germany of the internal combustion engine (1876) by Nicholas Otto and its subsequent use by Karl Benz (1885) and Gottlieb Daimler (1886) to power wheeled vehicles. Thereafter the industry developed quickly in Europe and the USA, so that by 1914 motor-buses, lorries, and taxis were being manufactured, while the introduction of *mass production techniques by Henry *Ford in 1913 reduced the cost of motor cars. World War I stimulated the industry, troop movements becoming increasingly motorizied, and motorized weapons, such as the tank, being developed. After World War II, during which motor vehicles played a crucial role, the industry continued to develop not only in Europe and the USA but also in Japan, which had become a leading manufacturer by the 1970s.

Mountain Men, US fur trappers and traders, who explored and developed the Rocky Mountains between the 1820s and 1840s. They caught public attention through their exploits and occupy an important position in the frontier legend. Their living conditions were harsh, and only a handful, such as Kit *Carson, Jedediah *Smith, and Thomas Fitzpatrick, survived long enough to return to a more settled existence after a decline in beaverskin prices in the 1840s.

An idealized view of **Mountain Men**, carrying their possessions through the wilderness: detail from *Mountain Jack and Wandering Miner* painted c.1850 by E. Hall Martin. Their lives may have seemed romantic but they endured many hardships and only the toughest survived. (Oakland Museum, California)

Mountbatten, Louis Francis Albert Victor Nicholas, 1st Earl Mountbatten of Burma (1900–79), British admiral and administrator. After service in World War I as a midshipman in *Beatty's flagships, he accompanied the Prince of Wales on two empire tours. In 1940–1 he commanded a destroyer flotilla that was badly bombed in the battle of Crete. He became Chief of Combined Operations in 1942 and did much for the subsequent landings in North Africa, Italy, and Normandy. In 1943 he was appointed Supreme Allied Commander, South-East Asia, where he restored the morale and capacity of the hard-hit and neglected Commonwealth forces fighting the Japanese in the *Burma Campaigns. In 1947 he became the last viceroy of *India, charged with the transfer of sovereignty from the British crown. This transfer was promptly effected, although marred by inter-communal massacres. At the invitation of the new Indian government, he stayed on until 1948 as the first governor-general. Resuming his naval career, he rose to Chief of the Defence Staff (1959–65), in which capacity he supervised the merging of the service ministries into a unified Ministry of Defence. Active in retirement, he criticized reliance on nuclear weapons. He was assassinated in 1979 by the *Irish Republican Army while on a holiday in Ireland.

Mozambique, a south-east African country. A Portuguese colony initially dependent on the slave trade, its African resistance movements were suppressed in the 19th century. In 1964 the Marxist guerrilla group *Frelimo, was formed. By the mid-1970s Portuguese authority had reached the point of collapse, and in 1975 an independent People's Republic was established under the Frelimo leader Samora *Machel. Support for the guerrilla campaigns in Rhodesia and South Africa led to repeated military incursions by troops of those countries, and the establishment of a stable government within the framework of a one-party Marxist state was further hindered by the weak state of Mozambique's agricultural economy. In 1984 Mozambique and South Africa signed a non-aggression pact, the Nkomati Accord, but South African support for anti-government guerrillas persists.

Robert **Mugabe**, Zimbabwe's first Prime Minister, speaking to the seventh non-aligned summit meeting in New Delhi, India, 1983. Mugabe took an active part in calling for stronger moves against South Africa, Zimbabwe's southern neighbour.

The coronation of **Muhammad Reza Shah Pahlavi**, in 1967, took place only after his third wife, Farah Diba, had provided him with an heir. The Shah sits on the magnificent Peacock Throne, with his young son and Farah (officially designated Empress of Iran) beside him: his dynasty and government were to remain in power until 1979.

Mugabe, Robert Gabriel (1924–), African statesman. In 1963 he helped form the Zimbabwe African National Union (ZANU), breaking away from Joshua *Nkomo's Zimbabwe African People's Union (ZAPU). He was imprisoned in 1964 for 'subversive speech', during which time he was elected leader of ZANU. He was freed in 1975, and, with Nkomo, led the guerrillas of the Zimbabwe Patriotic Front against Ian Smith's regime. When the war ended he won a landslide victory in elections held under British supervision in 1980, and became Prime Minister (1980–). His contest with Nkomo now sharpened, Nkomo maintaining the supremacy of Parliament, while Mugabe openly declared a Marxist one-party state as his objective. ZANU added PF (Patriotic Front) to its title, and held a congress in 1984 which set up a ninety-member central committee with a fifteen-member politburo to supervise both the party and Zimbabwe.

Muhammad Ali *Mehemet Ali.

Muhammad Reza Shah Pahlavi (1919–1980), Shah of Iran (1941–79). The son of *Reza Shah he succeeded on the abdication of his father. After the fall of *Mussadegh in 1953 he gained supreme power and with the aid of greatly increased oil revenues, embarked upon a policy of rapid social reform and economic development, while maintaining a regime of harsh repression towards his opponents. In 1962 he introduced a land reform programme to break landlord power. In 1979 he was deposed by a revolution led by the Islamic clergy, notably Ayatollah *Khomeini, whose supporters were bitterly opposed to the pro-western regime of the Shah. He died in exile in Egypt.

Mujibar Rahman, Sheikh (1920–75), Bangladeshi statesman. Popularly known as Sheikh Mujib, he came into political prominence as co-founder and general secretary of the *Awami League and as a champion of the Bengalis of East Pakistan, who he feared were being dominated by West Pakistan. He was imprisoned in 1954 under the rule of *Ayub Khan, and again in 1966, but as leader of the Awami League after the death of Suhrawardy, he became the leading politician of East Pakistan. He was released after the fall of Ayub and led his party to victory in the 1970 elections. In the conflict between East and West Pakistan that followed the elections, he was again arrested (1971), but was released in 1972 to head the government of East Pakistan, renamed the People's Republic of *Bangladesh, confirming his leadership at the elections of 1973. In 1975 he became Bangladesh's first President. His attempts at establishing a parliamentary democracy having failed, he assumed dictatorial powers under the new constitution which established a one-party Awami League government. In 1975 he and his family were murdered in an army coup.

Mukden incident (18 September 1931), Japanese seizure of the Manchurian city of Mukden. A detachment of the Japanese Guandong army, stationed in Manchuria in accordance with treaty rights, used an allegedly Chinese-inspired explosion on the South Manchurian Railway as an excuse to occupy the city of Mukden (now Shenyang). Acting without reference to their own government, and in the face of condemnation from the League of Nations, Japanese military authorities then went on to occupy all of Manchuria before the end of 1931, establishing the state of *Manchukuo. Japan, labelled an aggressor by the League of Nations, withdrew its membership.

Muldoon, Sir Robert (1921–), New Zealand statesman. He entered New Zealand politics in 1960 as a member of the National Party. He held various offices before being elected leader of his party in 1974 and winning a general election in 1975. He was Prime Minister and Minister of Finance (1975–84) at a difficult time for New Zealand, which was faced by grave economic problems. Oil prices had risen steeply in 1973 and its traditional market in Britain for farm and dairy produce had been threatened by the latter's entry into the *European Economic Community. He wished to provide economic incentives for industrial growth, but also imposed freezes in prices and wages, leading to conflict with trade unions. He served on the Board of Governors of the World Bank (1979–80). The National Party lost power in 1984 and was replaced by the Labour Party, under the leadership of David Lange.

Munich 'beer-hall' putsch (8 November 1923), an abortive rebellion by German *Nazis. In a beer-hall in Munich a meeting of right-wing politicians, denouncing the *Weimar Republic and calling for the restitution of the Bavarian monarchy, was interrupted by a group of Nazi Party members led by Adolf *Hitler. In a fierce speech Hitler won support for a plan to 'march on Berlin' and there instal the right-wing military leader General *Ludendorff as dictator. With a unit of *brownshirts (SA), he kidnapped the leader of the Bavarian government and declared a revolution. Next day a march on the centre of Munich by some 3,000 Nazis was

Sheikh **Mujibar Rahman** (at the microphone), first President of Bangladesh, with Bengali supporters. Leader of the East Pakistan secessionist movement, the Sheikh was freed from a Pakistani prison to head the new state's first government.

met by police gunfire, sixteen demonstrators and three policemen being killed in the riot which followed. Many were arrested. Ludendorff was released, but Hitler was sentenced to five years in prison, serving only nine months, during which he dictated the first volume of his autobiography and manifesto *Mein Kampf* (1925) to his fellow prisoner, Rudolf *Hess.

Munich Pact (29 September 1938), agreement between Britain, France, Germany, and Italy concerning Czechoslovakia. *Hitler had long demanded protection for the German-speaking *Sudetenland and shown readiness to risk war to attain his end. To avert conflict at all costs the British Prime Minister, *Chamberlain, had met Hitler at Berchtesgaden (15 September), and again at Bad Godesberg (23 September), by which time Hitler had extended his demands. He now stipulated the immediate annexation by Germany of the Bohemian Sudetenland and demanded that Germans elsewhere in Czechoslovakia should be given the right to join the *Third Reich. In a final effort Chamberlain appealed to *Mussolini, who organized a conference at Munich where he, Chamberlain, and Hitler were joined by *Daladier, the French Premier. No Czech or Soviet representative was invited. Hitler gained most of what he wanted and on 1 October German troops occupied the greater part of Czechoslovakia. As part of the agreement, Poland and Hungary occupied areas of Moravia, Slovakia, and Ruthenia. What remained of Czechoslovakia fell under German influence and *Beneš, the Czech President, left the country. Germany, which now dominated the entire Danubian area, emerged as the strongest power on the mainland of Europe.

Murat, Joachim (1767–1815), Marshal of France, King of Naples (1808–15). He distinguished himself in the

Benito **Mussolini** reviews fascist troops in Naples, 30 October 1922, before his march on Rome. Either side of him are (*left*) Lieutenant Balbo and General de Bono, and (*right*) General de Vecchi and Michele Bianchi.

Italian campaigns and was promoted general on the field at Aboukir (1799). He married Napoleon's sister, Caroline, in 1800. He was made King of Naples in succession to Joseph Bonaparte in 1808, where he undertook important reforms. He took part in Napoleon's Russian campaign in 1812 but abandoned the emperor during the retreat from Moscow. In order to safeguard his own throne he intrigued unsuccessfully with Austria and England. In 1815, however, Ferdinand IV was restored to Naples as Ferdinand I, King of the Two Sicilies, and Murat escaped to France. He joined Napoleon during the *Hundred Days but was defeated by the Austrians at Tolentino in Italy. In October he was captured, handed over to Ferdinand, and shot.

Muslim League, political party founded in 1905 to represent the separate interests of the Indian Muslims who felt threatened by the prospects of a Hindu majority in any future democratic system. The radical nationalist elements in the League forged a pact with the *Congress in 1916 on the basis of separate electorates and reserved seats in Muslim minority provinces. A section of the League co-operated with the Congress in the *non-co-operation movement. In the provincial elections (1937), the League captured very few Muslim seats, but it succeeded in convincing the Muslim masses that the elected Congress ministries were oppressing Muslims. In 1940 it put forward the demand for an autonomous Muslim homeland, Pakistan, interpreted by its leader, M. A. *Jinnah, as an independent state during the transfer of power negotiations. He called for a Direct Action Day in August 1946. Mass rioting followed, whereupon the British and the Congress agreed to partition. The League was virtually wiped out at the first elections in Pakistan.

Mussadegh, Muhammad (1880–1967), Iranian political leader. An Iranian landowner and politician, in 1950 he led the democratic-nationalist opposition to the policies of Muhammad *Reza Shah in Parliament. A militant nationalist, he forced (1951) the nationalization of the Anglo-Iranian Oil Company, and after rioting in Abadan, was appointed Prime Minister (1951–3). He ruled with left-wing support until he was dismissed by the Shah in 1953.

Mussolini, Benito (1883–1945), Italian dictator. After a turbulent career as a school teacher, he became a leading socialist journalist. During World War I he resigned from the Socialist Party and advocated Italian military support for the *Austro-Hungarian empire. Called up, he became an army corporal and was wounded. He returned to journalism, bitterly opposing the *Versailles Peace Settlement. He organized radical right-wing groups which were merged into the *Fascist Party. Widespread violence by his supporters, the weakness of democratic politicians, and the connivance of the king, who feared a communist revolution, enabled him to take power in 1922 after the so-called 'march on Rome'. Violence, including murder, against political enemies and, at first, popular esteem as 'Il Duce', enabled him to consolidate his position. A brilliant orator, his skill, in conditions of strict censorship, of presenting himself as all-powerful meant that his incompetence was long unnoticed. The *Vatican State was set up by the Lateran Treaty (1929). His quest for a new Italian empire led to his annexation (1936) of *Ethiopia, and Albania (1939). *Hitler, one of his early admirers and imitators, became his ally and then a resented senior partner in the *Axis. Having entered *World War II at the most favourable moment (1940), he nevertheless was unable to avoid a series of military defeats. He was deposed by hitherto acquiescent fascist leaders in 1943, but was rescued by German paratroopers and established a puppet government in the small town of Salo in north Italy. In 1945 he was captured by Italian partisans, who shot him.

Mutesa I (d. 1884), Kabaka (King) of Buganda (now in Uganda) (1857-84). An autocratic monarch, he furthered his country's wealth by opening up trade, often in slaves, with Arab merchants. He strengthened Buganda's army and improved its bureaucracy. He subdued Bunyoro, the leading state in south Uganda. Although a Muslim, he welcomed Christian missionaries. They were followed by the British East Africa Company, causing tensions that were unresolved when he died suddenly.

Mutesa II, Sir Edward Frederick (1924-69), last Kabaka of Buganda (1939-63) and President of *Uganda (1963-6). He assumed office after a regency, aged 18. Progressive in spirit, he nevertheless backed the protectorate government in suppressing Buganda nationalist risings in 1945 and 1949. In 1953, fearing the loss of Buganda's independence, he claimed the right of his kingdom to secede from the Ugandan protectorate. This was denied him by the Ugandan High Court, and he was deported by the British. In 1955 he returned as a constitutional monarch, and in 1963 was elected first President of Uganda. He disagreed with the left-wing policies of the Prime Minister, Dr Milton *Obote, who deposed him in 1966.

MVD (Ministry for Internal Affairs), Soviet police organization that, together with the MGB (Ministry of State Security), replaced the NKVD (People's Commissariat of Internal Affairs) in 1946. The MVD controlled all police forces and administered forced *prison camps. During the last years of *Stalin's rule it became a significant factor in the Soviet economy, one of its most notorious chiefs being Lavrenti *Beria. The powers of the

Mutesa I of Buganda, a painting of the youthful ruler by Lady Stanley (Dorothy Tennant) from a photograph of the king taken by H. M. Stanley. (Private Collection)

A drawing of the **Myall Creek massacre** by Hablot K. Browne ('Phiz'), June 1838. It shows aborigines in New South Wales being dragged to their death by convicts.

MGB were extended to supervise and control police agencies throughout the Soviet bloc, and to eliminate all anti-Soviet, anti-communist opposition in the satellite countries. Both agencies were drastically reduced and decentralized between 1953 and 1960, when they were replaced by the *KGB.

Myall Creek massacre (1838), an incident in which, in retaliation for an alleged 'outrage', white station hands killed twenty-eight Aborigines at Myall Creek in New South Wales, Australia. As a result, eleven whites were tried for murder. They were acquitted, but seven of them were retried, found guilty, and hanged. There was much public protest. Prosecution of whites for such an incident was unusual; the initial incident was not, and Aborigines continued to be indiscriminately killed as settlers moved into other parts of the continent.

Mzilikazi (c.1796-1868), *Ndebele leader. The first great ruler of the Ndebele, he united his people into a nation under his leadership. He became a war leader under *Shaka, King of the Zulu, but rebelled in 1822. He led his people away from Zululand to what is now the western Transvaal, and then settled with his subjects in the area of Bulawayo. In 1837 he fled north, subduing the Shona, and ruling the Ndebele until his death.

N

Nagasaki, Japanese city in Kyushu. On 9 August 1945, three days after the first atomic bomb attack on *Hiroshima, Nagasaki became the next target. The hilly terrain protected the population of 230,000 from the full effects of the explosion, but 40,000 people were killed and tremendous destruction caused. On the following day Japan offered to surrender and the ceasefire began on 15 August, the official surrender finally being signed on 2 September.

Nagy, Imre (1896–1958), Hungarian statesman. He took part in the *Russian Revolution and the *Russian Civil War. As Hungarian Minister of Agriculture (1945–6) he was responsible for major land reforms and helped in the communist take-over in Hungary. Prime Minister (1953–5), he became popular because of his policy of liberalization and de-collectivization. Denounced for *Titoism, he was removed from power (1955). Shortly before the outbreak of the *Hungarian Revolution he was reappointed. After the collapse of the Revolution, he was seized by Soviet authorities and handed over to János *Kádár, who had him tried in secret and executed.

Nakasone Yasuhiro (1918–), Japanese statesman. He was elected to the House of Representatives in 1947. As a member of the ruling *Liberal Democratic Party, he held an intermittent series of cabinet posts from the 1960s. Nakasone succeeded Suzuki as Prime Minister and LDP president in October 1982 and held power, until his resignation in 1987, despite opposition to his strong style of leadership and the damage done to his party by the involvement of *Tanaka in the Lockheed corruption scandal. His domestic policies were based on a package of administrative, fiscal, and educational reforms, while internationally he was committed to close ties with the USA and greater involvement in world affairs.

Namibia, a territory in southern Africa. German missionaries went there in the 19th century and in 1884 the German protectorate of South-West Africa was established. In 1915, during World War I, it was captured by South African forces, and in 1920 became a *League of Nations mandated territory under South Africa. In 1946 the *United Nations refused to allow it to be incorporated into South Africa and ended the mandate (1964), renaming the territory Namibia. In 1971 the International Court of Justice at The Hague ruled that continued South African occupation was illegal and the UN has recognized a black nationalist group, the *South West Africa People's Organization (SWAPO), as the legitimate representative of the people of Namibia. Although a National Assembly for internal government was established by South Africa in 1979, the dispute continues, with SWAPO guerrillas fighting South African units.

Nana Sahib (or Brahmin Dhundu Panth) (*c.*1820–59), Hindu leader. On the outbreak of the *Indian Mutiny in Cawnpore (now Kanpur) (1857), he reluctantly joined the rebels and accepted the surrender of the British garrison under Sir Hugh Wheeler, promising safe conduct to its people. His hostility to Britain was attributed to the rejection of his claim to the title and pension granted to his adoptive father, Baji Rao, the former peshwa (hereditary prime minister). A reluctant recruit to the Mutiny, he subsequently fled to Nepal and his fate is uncertain, it being thought that he died in the jungle.

Nanjing, Treaty of (1842), treaty between Britain and China that ended the First *Opium War . The first *Unequal Treaty, it ceded Hong Kong to Britain, broke the Chinese monopoly on trade, and opened the *treaty ports of Xiamen (Amoy), Guangzhou (Canton), Fuzhou (Foochow), Ningbo (Ningpo), and Shanghai to foreign trade. Further treaties extended trade and residence privileges to other nations and set up the framework for Western economic expansion in China.

Naoroji, Dadabhai (1825–1917), Indian nationalist leader. He was the first Indian to be elected to the British House of Commons, serving as Liberal Member of Parliament for Central Finsbury (1892–5). His campaign against the drain of wealth from India to Britain, defined in his classic study *Poverty and Un-British Rule in India* (1901), stimulated economic nationalism in the subcontinent. Active in promoting Indian social and political causes, he was a founder of the Indian National *Congress, serving as its President (1886, 1893, and 1906).

Napier, Robert Cornelis, 1st Baron of Magdala (1810–90), British field-marshal and civil engineer. He served with distinction in the *Sikh Wars and during the *Indian Mutiny, but made his reputation as the engineer chiefly responsible for the programme of public works in the Punjab, 1849–56. He led an expedition to Ethiopia

A portrait of **Napier** of Magdala by Lowes Dickinson.

in which he captured *Magdala (1868) and compelled the release of British captives. In 1870 he became commander-in-chief in India.

Napoleon I (Napoléon Bonaparte) (1769-1821), Emperor of the French (1804-14). Born in Ajaccio, he was a Corsican of Italian descent. He was educated in military schools in France and served in the French Revolutionary army. By the age of 26 he was a general, and placed in supreme command of the campaign against Sardinia and Austria in Italy (1796-7), which resulted in the creation of the French-controlled Cisalpine Republic in northern Italy. In 1798 he led an army to Egypt, intending to create a French empire overseas and to threaten the British overland route to India. *Nelson, by destroying the French fleet at the battle of the Nile (1798), prevented this plan. Bonaparte returned to France (1799) and joined a conspiracy which overthrew the Directory and dissolved the First Republic. Elected First Consul for ten years, he became the supreme ruler of France. During the next four years he began his reorganization of the French legal system (*Code Napoléon), the administration, the Church, and education.

With the Treaties of Lunéville (1801) with Austria, and Amiens (1802) with Britain, France now became paramount in Europe. In 1803 Britain again declared war on France, and Napoleon prepared to invade it. The ruthless execution (1804) of the duc d'*Enghien on suspicion of conspiracy provoked criticism throughout Europe. In the same year Napoleon crowned himself Emperor of the French. He created an imperial court and nobility around himself, while at the same time restricting the liberal provisions of the earlier revolutionary constitution. In 1804-5, a European coalition against Napoleon was formed and he launched his armies

The coronation of **Napoleon I**, 2 December 1804, depicted here in Jacques-Louis David's painting. The ceremony took place in the cathedral of Notre Dame, Paris. At Napoleon's request, Pius VII attended, but at the last moment the Emperor took the crown from the pope and set it on his own head. (Musée du Louvre, Paris)

against it. He defeated the Austrians at Ulm, occupied Vienna, and won his most brilliant victory over the combined Austrian and Russian forces at *Austerlitz (1805). The naval victory by Nelson at the battle of *Trafalgar (1805) led Napoleon to seek Britain's defeat by the introduction of the *Continental System, which aimed to stop all trade between Britain and France and its allies on the continent of Europe.

In 1806 the Holy Roman Empire was dissolved and Napoleon consolidated his domination of the continent. The difficulty of enforcing the Continental System, the ill-fated invasion of Russia (1812), and the set-backs of the *Peninsular War (1807-14), all contributed to Napoleon's decline and, following his defeat in the battle of *Leipzig and the proclamation by *Talleyrand of the deposition of the emperor, came his abdication in 1814. After a brief exile on Elba he returned, but defeat at the battle of *Waterloo (1815) ended his rule after only a *Hundred Days. He spent the rest of his life in exile on St Helena. In 1796 he married *Joséphine de Beauharnais, whose failure to give him a son led to their divorce. In 1810 Napoleon married the Austrian princess Marie-Louise. Their only child, Joseph-François-Charles, crowned as the Roi de Rome, died aged 21.

Napoleonic Wars, the campaigns carried out between *Napoleon I and the European powers, including Britain (1796-1815). The first great Italian campaign (1796)

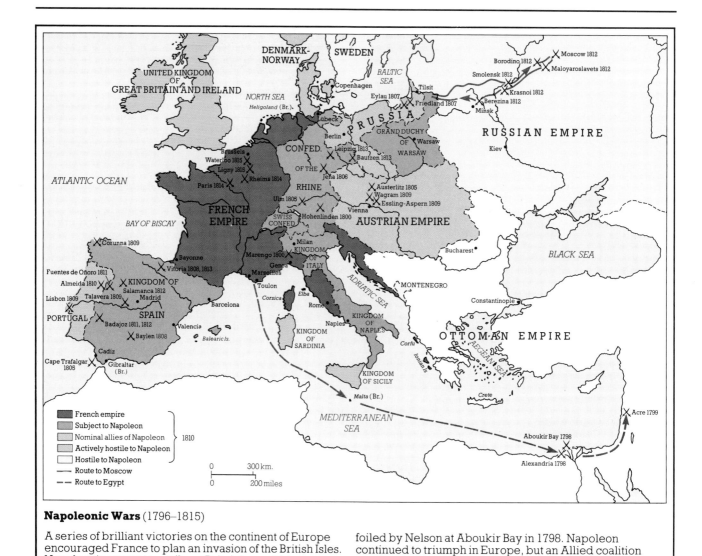

Napoleonic Wars (1796–1815)

A series of brilliant victories on the continent of Europe encouraged France to plan an invasion of the British Isles. Napoleon's attempt to strike at Britain's wealth by occupying Egypt and threatening the route to India was foiled by Nelson at Aboukir Bay in 1798. Napoleon continued to triumph in Europe, but an Allied coalition resulted in Napoleon's defeat at Leipzig (1813), and final surrender at Waterloo (1815).

under Napoleon secured a series of decisive victories for the French over the Austrians in northern Italy. In 1798 he led an expedition to Egypt, but the British fleet under Admiral *Nelson destroyed the French fleet in Aboukir Bay. In 1799 he led an army over the Alps to win the battle of Marengo (1800) over the Austrians. Britain, apprehensive of Napoleon's threat in the Mediterranean and in continental Europe, was by 1803 once more at war with France. Nelson destroyed the combined Spanish and French fleets at *Trafalgar (1805), and in the same year Napoleon swung his *grande armée* towards Austria, which, with Russia and Sweden, joined Britain in the Third Coalition. Napoleon's forces encircled the Austrians at Ulm, forcing them to surrender without a battle. Napoleon fought and defeated the emperors of Austria and Russia at the battle of *Austerlitz (1805) and forced Austria to sue for peace. In the following year Prussia joined the Third Coalition but, in a campaign that lasted twenty-three days, Napoleon broke the Prussian armies at *Jena and Auerstadt and accepted the surrender of Prussia. The Russian emperor *Alexander I concluded a treaty of friendship and alliance with Napoleon at *Tilsit in July 1807. In 1808 a revolt broke out in Spain, which by now was also under French rule. Napoleon sent a large force to quell it, but was confronted by the British army under Sir Arthur Wellesley, later Duke of *Wellington. Britain won a series of victories in the *Peninsular War which, though not conclusive, tied up 300,000 French soldiers when they were needed elsewhere. In 1812 Napoleon defeated the Russians at *Borodino and occupied Moscow, but instead of suing for peace, Alexander's forces withdrew further into the country. Napoleon's *grande armée* was forced to retreat from Moscow in the severest winter conditions, which cost the lives of nearly half a million men. After a crushing defeat at *Leipzig the following year, Napoleon abdicated and retired to Elba (1814). Next year he returned to France and was finally defeated by Wellington and Blücher at the battle of *Waterloo (1815).

Napoleon III (Charles-Louis Napoléon Bonaparte) (1808–73), Emperor of the French (1852–70). He was the third son of Louis Bonaparte (1778–1846), brother of *Napoleon I and King of Holland (1806–10), and of Hortense de Beauharnais. After the fall of Napoleon I he began a long period of exile in Switzerland, where he was associated with the *Carbonari. On the death of Napoleon I's only son, the Roi de Rome, in 1832, he became Bonapartist pretender to the French throne, and twice attempted to overthrow *Louis-Philippe. The first time (1836) he was deported to the USA. Four years later he embarked upon the disastrous 'Boulogne Conspiracy'. Imprisoned in the fortress of Ham, he escaped to London (1846) disguised as a mason by the name of 'Badinguet', which thereafter became his nickname. During the *Revolutions of 1848, he returned to France, and in December under the new constitution was elected President of the French Republic. In 1852, following a coup against Parliament, he had himself accepted as Emperor of the French. Napoleon took part in the *Crimean War and presided over the Congress of *Paris (1856). He at first supported the *Risorgimento, but concluded a peace treaty with Austria at Villafranca in 1859. His 'Liberal Empire' (1860–70) widened the powers of the legislative assembly and lifted restrictions on civil liberties. His attempts at extending French colonial interests to *Mexico (1861–7) ended in failure. Underestimating *Bismarck, he allowed the latter's belligerent *Ems Telegram to provoke him into fighting the *Franco-Prussian War, whose outcome brought ruin to the Second Empire. He was captured by the Prussians and deposed, spending the rest of his life in exile in England.

Nash, Sir Walter (1882–1968), New Zealand statesman. A life-long Christian Socialist, he joined the Labour Party, becoming the most important spokesman for its moderate wing. He (and Peter *Fraser) made the Labour Party into a national organization and formulated the

Carry **Nation**, described as the 'celebrated bar-room smasher', outside a US theatre. Public appearances by this formidable woman helped to publicize her anti-liquor crusade.

policies which led to the election of the first Labour Government in 1935. He played a major role in piloting through Parliament in 1938 the great system of child allowances, and 'free' medicine, which was the most extensive system of social security in the world at that time. He became leader of the Opposition in 1950. Nash led Labour to a narrow victory in the 1957 election, serving as Prime Minister until the party's defeat in 1960. During this period of financial stringency he introduced further important social reforms.

Nasser, Gamal Abdul (1918–70), Egyptian statesman. Together with three other officers in the Egyptian army, he founded the revolutionary Free Officers' Movement with the objective of expelling the British and the Egyptian royal family. In 1952, with eighty-nine Free Officers, he achieved an almost bloodless coup, forcing the abdication of King *Farouk. A republic was declared and a Revolutionary Command Council set up, with Major-General Muhammad Neguib as President. In 1954 he deposed Neguib, and became head of state. In 1956 he promulgated a one-party constitution. With massive Russian aid he launched a programme of domestic modernization. Failing to receive British and US support for a project to extend the Aswan High Dam, he nationalized the Suez Canal Company (1956), whose shares were mainly owned by British and French investors; his object was to use the canal dues to pay for the Aswan Dam project. Britain, France, and Israel invaded Egypt, but the *Suez War was halted, mainly by US intervention. His attempt to unite the Arab world in a United Arab Republic, a federation with Syria, failed. In 1967 Egypt was disastrously defeated by Israel in the *Six-Day War, but his reputation as a Pan-Arab leader and social reformer emerged untarnished. He died in office.

Natal, the eastern coastal province of the Republic of *South Africa. Natal had been settled, probably since the first millennium, by black farmers. The region was devastated during the *Nguni Difagane Wars (1819–38). During this period, European settlers arrived. In 1840 the Boers set up an independent Republic of Natal, but Britain annexed it in 1843. Many Boers then migrated to the *Orange Free State. In 1856 it became a self-governing colony. *Zululand was annexed to it in 1897. Natal took an active part in fighting the British in the Second *Boer War and joined the Union in 1910.

Nation, Carry (Amelia Moore) (1846–1911), US *temperance agitator. Born in Kansas, and married to an alcoholic, she became an evangelical Christian and a militant temperance advocate. During the 1890s she became convinced of her mission to destroy the illegal bars in nominally 'dry' states, such as Kansas. A tall, powerfully-built woman, she was often involved in fights and was arrested thirty times for violent behaviour. Her exploits helped to create a climate for the introduction of *prohibition. An advocate of *women's suffrage, she received little support from other reform organizations because of her unorthodox conduct.

National government, a term used to describe the British coalition governments (1931–5). In August 1931 a financial crisis led to a split within the Labour government, nine ministers resigning rather than accepting cuts in unemployment benefits. The Liberal leader Herbert Samuel suggested that the Prime Minister, *MacDonald, create a 'government of national salvation', by inviting Conservatives and Liberals to replace them, and the first National government was formed on 24 August. An emergency budget was introduced which increased taxes and proposed to reduce both benefits and public sector salaries. When naval ratings at Invergordon refused duty in protest, this so-called *Invergordon Mutiny caused further financial panic and sterling fell by 25 per cent. Britain abandoned the *gold standard and *free trade, adopting a policy of protection. The Labour Party split, supporters of the government being regarded as traitors. In October MacDonald won a general election and formed a second National government, but its balance was now strongly towards the Conservative Party. The governments of Stanley *Baldwin (1935-7) and Neville *Chamberlain (1937-40) retained the term National, but they were effectively Conservative governments.

national insurance (or social insurance), a state insurance scheme financed by compulsory contributions from employee and employer. It seeks to provide economic protection against various risks, including sickness and unemployment. Such schemes help to fund social provision within the *welfare state. Pioneered in Germany by *Bismarck, national insurance schemes were introduced in other European countries, including Britain (1911), and in New Zealand before World War I, for state assistance in sickness, accident, unemployment, and old age. In Britain as a result of the *Beveridge Report (1942), national insurance was extended to all adults in employment. As part of the *New Deal in the USA a federal national insurance scheme was introduced by the Social Security Act of 1935. The majority of US citizens, however, rely more on private insurance schemes for health, accident, pension, and other provision.

National Party, a South African political party. It was originally founded in 1913-15 by General J. B. M. *Hertzog after his secession from Botha's South African Party. In 1924 it became the Nationalist–Labour alliance under Hertzog, who joined *Smuts in the *United Party in 1934. In the same year D. F. *Malan founded the Purified Nationalist Party, which was reunited in 1939 with Hertzog, emerging as the Afrikaner-dominated party of *apartheid. It has held uninterrupted power since 1948. Recent attempts by President P. W. Botha to meet the twin threats of domestic unrest and international condemnation of apartheid with a programme of mild reforms have led to some defections to extreme right-wing groups.

National Party of Australia, an Australian political party which largely represents rural interests. Farmers' representatives were elected to colonial (later state) parliaments from the 1890s onwards. A number of farmers' candidates were elected to the federal parliament in 1919. The following year, they formed the Australian Country Party. It has governed federally in coalition with the Nationalist Party (1923-9), the United Australia Party (1934-9, 1940-1), and the *Liberal Party (1949-72, 1975-83). Several of its parliamentary leaders have been Prime Ministers, albeit briefly. They were Page (1939), Fadden (1941), and John McEwen (1967-8). It has also governed (under various names), mostly in

coalition with other parties, for periods in most states. The party's national name was changed to the National Country Party in the mid-1970s, and to the National Party of Australia in 1982.

Nations, battle of the *Leipzig, battle of.

NATO (North Atlantic Treaty Organization), defence alliance between Western powers. Founded in 1949, it was established primarily to counter the perceived military threat from Soviet power in Eastern Europe. Its original members were: Belgium, Canada, Denmark, France, Great Britain, Iceland, Italy, Luxemburg, the Netherlands, Norway, Portugal, and the USA. Greece, Turkey (both 1952), West Germany (1955), and Spain (1982) joined later. In 1966 France withdrew its forces from the NATO Military Committee, though remaining a nominal member of the Council.

Nauru, a small island in the south-west Pacific, inhabited mainly by Polynesians and discovered by the British in 1798. In 1899 a British company, the Pacific Islands Company in Sydney found that the island comprised the world's richest deposits of phosphate of lime. The company began mining the deposits in 1906. From 1888 to 1914 the island was part of Germany's Marshall Islands protectorate. Thereafter, apart from three years of Japanese occupation during World War II, Nauru was a trust territory of Britain, Australia, and New Zealand, before achieving independence and a limited membership of the Commonwealth in 1968.

navy, a fleet of ships and its crew organized for war at sea. By 1800 many countries had developed fleets of warships as well as continuing the practice of arming merchant ships in time of war. They were officered by professionals, but relied for their crews on men recruited by various forms of press-gangs. At the time of the *Napoleonic Wars, naval vessels were sailing ships, built of wood and armed with cannon that fired broadsides. They engaged at close quarters and ratings were armed with muskets and hand-grenades. Following the battle of *Trafalgar (1805), the British navy dominated the oceans of the world for a century. Change came slowly. Steam power replaced sail only gradually, while in 1859 the French navy pioneered the protection of the wooden hull of a ship with iron plates (*ironclads). With the development of the iron and steel industry in the late 19th century, rapid advances were made in ship design and the armament of ships. At the same time the submarine, armed with torpedoes, emerged as a fighting vessel. When Germany challenged the supremacy of the British navy, the latter responded with the huge steel *Dreadnought battleships (1906), equipped with guns with a range of over 32 km (20 miles). During World War I the German submarine (U-boat) fleet was only checked by the *convoy system to protect Allied merchant shipping, but the major British and German fleets only engaged in the inconclusive battle of *Jutland (1916). Between the wars aircraft were rapidly developed and naval warfare in World War II was increasingly fought by aircraft from aircraft carriers, particularly in the great naval battles of the *Pacific Campaign. Since World War II, the development of submarines armed with nuclear missiles has reduced the number of surface ships. Most countries retain fleets of small, fast vessels for coastal

Navy

a) HMS *Victory*

b) HMS *Alecto*

c) *La Gloire*

d) SMS *Rheinland*

e) US Essex-class carrier

f) US nuclear submarine

Sketch (a) is of HMS *Victory*, Nelson's flagship, launched in 1765. Steam-power was only slowly adopted by the world's navies, (b) showing a sloop (1845) powered by both sail and paddle. The first ironclad, *La Gloire* (c) powered by propeller and sail, was protected against exploding shells. Very rapid change from 1880 resulted in vast surface fleets, with Dreadnought-class battleships as in sketch (d), of the early 20th century. Aircraft became the principal means of naval battle in World War II. The submarine (sketch (f)) has nuclear-powered engines and nuclear-armed missiles of strategic capability.

patrol. The USA and the Soviet Union, however, have continued to compete in the size and armament of their navies. The *Falklands War (1982) revealed the extent to which there remained a place for a conventional navy but also showed how exposed surface ships now are to missile attack.

Naxalite Movement, an Indian revolutionary movement named after the village of Naxalbari in the Himalayan foothills in West Bengal, where it first began. The theoretician and founder of the movement, Charu Majumdar, a veteran communist, broke away from the *Communist Party of India (Marxist) and established the Communist Party of India (Marxist–Leninist). The CPI (M–L) first organized several armed risings of landless agricultural labourers, especially in eastern India. It developed into an urban guerrilla movement, especially in Calcutta. Its programme of terror was suppressed with considerable violence. The CPI (M–L) eventually split up into several factions, one of which has adopted a policy of participating in constitutional politics.

Nazi, a member of the Nationalsozialistische Deutsche Arbeiterpartei or National Socialist German Workers' Party. It was founded in 1919 as the German Workers' Party by a Munich locksmith, Anton Drexler, adopted its new name in 1920, and was taken over by *Hitler in

A **Nazi** rally at Nuremberg, Germany, in the 1930s. Massed rallies, orchestrated to arouse patriotic fervour, were used by the Nazis to gain popular support. Their use of the swastika symbol, their claims of German racial superiority, and their dramatic military demonstrations combined into powerful propaganda.

1921. The Nazis dominated Germany from 1933 to 1945. In so far as the party had a coherent programme it consisted of opposition to democracy. It promulgated theories of the purity of the Aryan race and consequent *anti-Semitism, allied to the old Prussian military tradition and an extreme sense of nationalism, inflamed by hatred of the humiliating terms inflicted on Germany in the *Versailles Peace Settlement. Nazi ideology drew on the racist theories of the comte de *Gobineau, on the national fervour of Heinrich von Treitschke, and on the superman theories of Friedrich Nietzsche. It was given dogmatic expression in Hitler's *Mein Kampf* (1925). The success of the National Socialists is explained by the widespread desperation of Germans over the failure of the *Weimar Republic governments to solve economic problems during the Great *Depression and by a growing fear of *Bolshevik power and influence. Through Hitler's oratory they offered Germany new hope. Only after Hitler had obtained power by constitutional means was the *Third Reich established. Rival parties were banned, terrorized, or duped, the institutions of state and the German army were won over. Thereafter they were all-powerful agents of Hitler's aim to control the minds of the German people and to launch them on a war of conquest. In the period leading up to *World War II Nazi ideology found many adherents in countries throughout the Western world. Nazi systems and dogmas were imposed on occupied Europe from 1938 to 1945, and over six million Jews, Russians, Poles, and others were incarcerated and exterminated in *concentration camps. The German Nazi Party was disbanded in 1945 and its revival officially forbidden by the Federal Republic of Germany.

Nazi–Soviet Pact (23 August 1939), a military agreement signed in Moscow between Germany and the Soviet Union. It renounced warfare between the two countries and pledged neutrality by either party if the other were attacked by a third party. Each signatory promised not to join any grouping of powers which was 'directly or indirectly aimed at the other party'. The pact also contained secret protocols whereby the dictators agreed to divide *Poland between them, and the Soviet Union was given a free hand to deal with the *Baltic states.

Ndebele, a Bantu-speaking people in southern Africa. *Nguni people from the Natal area, they were forced to flee north by *Shaka, King of the Zulu, under their leader *Mzilikazi (1823). After continued raids by the Zulu and also by *Boer settlers in the *Transvaal, they crossed the Limpopo River, extending their influence over both the Shona and Sotho tribes. Under their leader *Lobengula they were defeated (1893) by troops of the British South Africa Company. They accepted British control over what was at first called Matabeleland, before it was incorporated into Southern Rhodesia (*Zimbabwe).

Nehru, Jawaharlal (named Pandit, Hindi, 'teacher') (1889–1964), Indian statesman. The son of Motilal *Nehru, he became a leader of the Indian National *Congress, where he attached himself to Mohandas *Gandhi. He conducted campaigns of *civil disobedience which led to frequent imprisonment by the British. His conviction that the future of India lay in an industrialized society brought him into conflict with Gandhi's ideal of a society centred on self-sufficient villages. On his release

Jawaharlal **Nehru**, son of Motilal, was an active agitator against British rule in India before becoming the first Prime Minister of independent India.

from prison (1945) he participated in the negotiations that created the two independent states of India and Pakistan, becoming the first Prime Minister of the independent Republic of India in 1947. As Prime Minister (1947–64) and Minister of Foreign Affairs he had to contend with the first Indo-Pakistan war (1947–8), which ended in the partition of Kashmir, and the massive influx of Hindu refugees from Pakistan. His government also faced the challenges of integration (sometimes by force) of the *Princely States and of a communist government (1957–9) in Kerala, as well as the planning and implementation of a series of five-year economic plans to underpin the new state. In 1961 he annexed the Portuguese colony of *Goa. In foreign affairs he adopted a policy of non-alignment, but sought Western aid when China invaded India in 1962. His daughter, Indira *Gandhi, succeeded him.

Nehru, Motilal (1861–1931), Indian political leader. Together with C. R. Das (1870–1925), he organized the *Swaraj (Independence) Party in 1922. This set out to participate in the Indian legislative councils but aimed to oppose the British by wrecking the councils from within, as an alternative to Gandhi's *non-co-operation movement. In 1928 he chaired the All Parties' Committee which produced the Nehru Report, setting out a proposed new Indian constitution with dominion status for India.

Nelson, Horatio, Viscount (1758–1805), British admiral. He entered the Royal Navy at the age of 12 and became a captain at the age of 20. On the outbreak of war with France in 1793 he was given command of the battleship *Agamemnon* and served under Admiral Hood. He lost the sight of his right eye during a successful attack on Corsica in the following year. In 1797 he played a notable part in the defeat of the French and Spanish fleets at the battle of Cape St Vincent and was subsequently promoted rear admiral. Later the same year he lost his right arm while unsuccessfully attempting to capture Santa Cruz de Tenerife in the Canary Islands. In 1798, after pursuing the French fleet in the eastern Mediterranean, he achieved a resounding victory at the battle of the Nile. While stationed at Naples he began his life-long love affair with Lady Emma Hamilton, the wife of the British ambassador there. In 1801 Nelson was promoted vice-admiral and, ignoring a signal from his commander, Sir Hyde Parker, defeated the Danish fleet at the battle of *Copenhagen. Following this engagement he was created a viscount. In 1803, after the renewal of war with France, Nelson was given command of the Mediterranean and for two years blockaded the French fleet at Toulon. When it escaped he gave chase across the Atlantic and back, finally bringing the united French and Spanish fleets to battle at *Trafalgar in 1805. This decisive victory, in which Nelson was mortally wounded, saved Britain from the threat of invasion by Napoleon.

Nepal, a country in southern Asia. The country was conquered by the Gurkhas in the 18th century; their incursion into north-west India led to a border war (1814–16) and to territorial concessions to the British. Effective rule then passed to a family of hereditary prime ministers, the Ranas, who co-operated closely with the British. Gurkhas were recruited to service in the British and Indian armies. Growing internal dissatisfaction led in 1950 to a coup, which reaffirmed royal powers under the king, Tribhuvan (1951–5). His successor, King Mahendra (1955–72), experimented with a more democratic form of government. This was replaced once more with monarchic rule (1960), which continues under his son, King Birendra Bir Bikram (1972–).

Netherlands (often called Holland), a country in western Europe. From 1795 to 1814 the Netherlands came increasingly under the control of France. During those years Britain took over the colonies of Ceylon (now Sri Lanka) and Cape Colony in South Africa, important trading posts of the Dutch East India Company. At the settlement of the Congress of *Vienna the entire Low Countries formed the independent kingdom of the Netherlands (1815). Despite the secession in 1830 of *Belgium, the Netherlands flourished under the House of Orange, adopting in 1848 a constitution based on the British system. It remained neutral during World War I, suffered economic difficulties during the Great *Depression, and was occupied by the Germans during World War II, when many Jews were deported to *concentration camps. Until World War II it was the third largest colonial power, controlling the Dutch East Indies, various West Indian islands, and Guiana in South America. The Japanese invaded the East Indian islands in 1942 and installed *Sukarno in a puppet government for all *Indonesia. In 1945 he declared independence and four years of bitter war followed before the Netherlands transferred sovereignty. Guiana received self-government as *Surinam in 1954 and independence in 1975, but Curaçao and other Antilles islands remained linked to the Netherlands. Following the long reign of Queen *Wilhelmina (1890–1948) her daughter Juliana became queen. She retired in 1980 and her daughter succeeded her as Queen Beatrix.

Neutrality Acts (1935–9), US laws to prevent the involvement of the country in non-American wars. They grew out of the investigations of the *Nye Committee, and included the prohibition of loans or credits to belligerents and a mandatory embargo on direct or indirect shipments of arms or munitions. In a spirit of *isolationism, the USA declared that it would take no stand on issues of international morality by distinguishing between aggressor and victim nations. In World War II the Roosevelt administration fought for repeal of the Acts on the ground that they encouraged *Axis aggression and ultimately endangered US security. Gradually they were relaxed.

New Australians, immigrants to Australia immediately after World War II. Before 1939 the population of the continent consisted of *Aborigines and people of predominantly British descent. The war greatly stimulated industrial production, and a labour shortage was met by a programme of assisted immigration (£10 from London to Sydney). Of some 575,000 new arrivals between 1947 and 1952 over half were Polish, Austrian, Italian, Maltese, Dutch, Greek, and Yugoslav, and they introduced cultural variety into Australia.

New Caledonia, a group of islands in the South Pacific. The island of New Caledonia was inhabited for at least 3,000 years before being discovered in 1774 by Captain James Cook. The French annexed the island in 1853 and began using it as a penal colony in 1864. With the discovery of nickel in 1863 New Caledonia assumed economic importance for France. The island group was occupied by US troops in 1942–5; and in 1946 it was proclaimed a French Overseas Territory, since when there has been a growing movement for independence.

New Deal, US term applied to the programme of Franklin D. *Roosevelt (1933–8), in which he attempted to salvage the economy and raise the Great *Depression. The term was coined by Judge Samuel Rosenman, used by Roosevelt in his 1932 speech accepting the presidential nomination, and made popular by the cartoonist Rollin Kinby. New Deal legislation was proposed by progressive politicians, administrators, and Roosevelt's 'brains trust'. It was passed by overwhelming majorities in Congress. The emergency legislation of 1933 ended the bank crisis and restored public confidence; the relief measures of the so-called first New Deal of 1933–5, such as the establishment of the *Tennessee Valley Authority, stimulated productivity; and the *Works Project Administration reduced unemployment. The failure of central government agencies provoked the so-called second New Deal of 1935–8, devoted to recovery by measures such as the Revenue Act, the *Wagner Acts, the Emergency Relief Appropriation Act, and the *Social Security Act. Although the New Deal cannot be claimed to have pulled the USA out of the Depression, it was important for its revitalization of the nation's morale. It extended federal authority in all fields, and gave immediate attention to labour problems. It supported labourers, farmers, and small businessmen, and indirectly blacks, who were beneficiaries of legislation designed to equalize opportunity and to establish minimum standards for wages, hours, relief, and security.

New Democratic Party, a political party in Canada. It grew out of the Canadian Co-operative Commonwealth Federation (CCF), a political party of industrial workers and small farmers formed in 1932 during the Great *Depression. In 1956 a Canadian Labour Congress of Trade Unions was formed, and in 1961 this amalgamated with the CCF to form the New Democratic Party, a mildly socialist party, committed to a planned economy and extension of social benefits. Leaders of the NDP have been T. C. Douglas (1961–71), David Lewis (1971–5), and Ed Broadbent (1975–   ).

New Economic Policy, a policy introduced into the Soviet Union by *Lenin in 1921. It represented a shift from his former 'War Communism' policy, which had been adopted during the *Russian Civil War to supply the Red Army and the cities, but had alienated the peasants. The NEP permitted private enterprise in agriculture, trade, and industry; encouraged foreign capitalists; and virtually recognized the previously abolished rights of private property. It met with success which Lenin did not live to see, but was ended (1929) by *Stalin's policy of five-year plans.

New Guinea   *Papua New Guinea.

New Hebrides   *Vanuatu.

Ne Win (1911–), Burmese general and statesman. A member of the extreme nationalist Dobama Asiayone (We-Burmans' Association) from 1936, he served as chief of staff to *Aung San's Burma National Army from 1943, defecting with it to the Allied side in 1945. As commander-in-chief after Burmese independence in 1948, he led the campaign against insurgent hill tribes and communist guerrillas, and served as Prime Minister (1958–60) in a caretaker government. In 1962 he led a coup against the U Nu administration, abolished the parliamentary system, and proclaimed the Socialist Republic of the Union of Burma, going on to expel 300,000 foreigners in an attempt to regain Burmese control of the economy. Constantly harassed by guerrilla disturbances, he has succeeded in maintaining his power as military dictator, and, although he stepped down as President in 1981, his influence remains considerable.

Newman, John Henry (1801–90), British theologian. He was a leading figure in the Oxford Movement, a group of people who in the 1830s attempted to reform the Church of England by restoring the high-church traditions, and became a prominent convert (1845) to Roman Catholicism. The publication in 1841 of *Tract 90*, which argued that the Thirty-nine Articles of the Church of England could be reconciled with Roman Catholic doctrine, caused a major scandal. In 1846 he went to Rome, where he was ordained priest. A gifted writer, in 1864 he published *Apologia pro Vita Sua*, a justification of his spiritual evolution. He was created cardinal in 1879. His cause for beatification is being examined in Rome.

New Orleans, battle of (8 January 1815), a battle in the *War of 1812, fought outside the city of New Orleans.

A numerically superior British attempt led by Sir Edward Pakenham to seize New Orleans was brilliantly repelled by US forces commanded by Andrew *Jackson. The battle proved of little military significance, the treaty of *Ghent having formally ended the war two weeks earlier, but Jackson's triumph made him a national hero.

Newport Rising (1839), a political insurrection which took place in Newport, Wales, in 1839, following the dissolution of the National Convention of the *Chartists. John Frost, a former mayor of Newport, planned to capture the town and release a Chartist leader, Henry Vincent, from gaol. However, the authorities were forewarned and soldiers ambushed the insurgents, killing several of them. Frost and others were arrested and received death sentences, later commuted to transportation.

New South Wales, a state of the Commonwealth of *Australia. Captain James Cook took British possession of the eastern part of Australia (1770), naming it New South Wales. In 1788 Britain established a penal colony, Australia's first white settlement, at Sydney Cove in the Pacific inlet of Port Jackson. Conflicts between governors

Above: This portrait of Cardinal **Newman** by Emmeline Deane was painted in 1889, a year before Newman's death. (National Portrait Gallery, London)

Below: The American victory at the battle of **New Orleans**, 1815, inspired an outburst of patriotic feeling. Andrew Jackson is shown directing his troops, firing from defensive position on the Mississippi river bank. They were presented with easy targets as the British advanced in massed ranks. British naval support was of little help to their poorly managed land force, 2,000 of whom were killed. (The Mariners' Museum, Virginia)

and officers culminated in the *Rum Rebellion (1808). Settlement was confined to an area around Sydney until the crossing of the Blue Mountains in 1813. Increasing numbers of free settlers and *emancipists led to changes in the colony's nature, as did the *Bigge Inquiry (1819–21), which recommended liberal land grants to settlers and extensive use of convict labour to open up the country. After the Molesworth Report (1838), convict transportation to New South Wales ended (1840). Partial representative government was granted in 1842, responsible government in 1855. Originally New South Wales comprised all of known Australia. Van Diemen's Land (present-day *Tasmania) was separated from New South Wales in 1825, the Port Phillip District (present-day *Victoria) in 1851, the Moreton Bay district (present-day *Queensland) in 1859, and land for the Australian Capital Territory (the site of Canberra) was transferred to the Australian Commonwealth in 1911.

New Zealand, a country in the southern Pacific ocean. First peopled by the Polynesian *Maori from about AD 800, European contact began in 1642 with the exploration of the Dutch navigator Abel Tasman. Captain James Cook, in successive explorations from 1769, thoroughly charted the islands, and brought them within the British ambit. Commercial colonization began from New South Wales in Australia and from the *New Zealand Association (later Company) (1837) of E. G. *Wakefield. Humanitarian pressures contributed to the decision formally to annexe the islands as the colony of New Zealand in 1840 on the basis that the rule of law was necessary to regulate Maori–settler relations (Treaty of *Waitangi). In 1846 the British government conferred a limited constitution (rescinded in 1848) on New Zealand, divided into the provinces of New Munster and New Ulster, and in 1852 granted the islands representative government. Responsible self-government came in 1856. Settlement of the South Island prospered, assisted by the *gold rushes of the 1860s. In the North Island, following the rapid acquisition of Maori land by settlers and by the government, the population was drawn into the disastrous *Anglo-Maori wars, following which most Maori land was settled. Regulations of 1881 restricted the influx of Asians, who were resented as a threat to the ethnic purity of the New Zealand people. They were confirmed by the Immigration Restriction Act (1920), whose terms were gradually liberalized. The property qualification for voting was abolished and women were enfranchised in 1893. In 1931 New Zealand became an independent dominion, although it did not choose to ratify the Statute of *Westminster formally until 1947. In 1891–1911 (under the Liberal-Labour Party) and 1935–47 (under Labour) New Zealand won a world reputation for state socialist experiment, providing comprehensive welfare and education services. New Zealand actively supported the Allies in both World Wars. The country has since then enjoyed a remarkable stability and a high standard of living. After World War II the country concentrated its defence policy on the Pacific and Far East, participating in *ANZUS and sending a military force to Vietnam. Since then it has pursued an active policy of creating a nuclear-free zone around its shores.

New Zealand Company (originally the New Zealand Association), an organization formed in 1837 by E. G. *Wakefield to colonize New Zealand. Despite powerful figures such as Lord Durham, Wakefield's patron, it was denied a charter by the British government, largely because of fears that it would come into conflict with the *Maori. Nevertheless, in May 1834 the company began sending out agents and settlers, buying land from the Maori and founding the settlements of Wellington, Nelson, and New Plymouth. The establishment of the crown colony in 1840 led to a review of the company's grandiose land claims. This, and Maori resistance, prevented the settlements developing as planned and in 1843 a rash attempt at Wairau, near Nelson, to assert authority over the powerful Te Rauparaha resulted in the deaths of Captain Arthur Wakefield and twenty-two settlers. By 1846 the Company had secured recognition from the Colonial Office, a loan, and a settlement of its land claims, but it became commercially unviable and was dissolved in 1858.

Ney, Michel, duc d'Elchingen (1769–1815), Marshal of France. The most famous and popular of *Napoleon's generals, he served Napoleon in the brilliant campaigns of 1794 and 1795, commanded the army of the Rhine (1799), and conquered the Tyrol. His support was decisive in Napoleon's victory at Friedland (14 June 1807). In the retreat from Moscow (1812) he commanded the defence of the *grande armée* against the Russians, and was created (1813) prince of Moscow by Napoleon. After the battle of *Leipzig he urged Napoleon to abdicate (1814). He agreed to take the oath of allegiance to the restored monarchy, but, when sent to check Napoleon's advance (1815) during the *'Hundred Days', he joined him instead, fighting heroically at *Waterloo, after which he was tried for treason and shot.

Ngo Dinh Diem (1901–63), South Vietnamese statesman, President (1955–63). A Catholic aristocrat, he became a minister in *Bao Dai's government in the 1930s. He was exiled by the French in 1947 after forming the anti-French and anti-communist National Union Front. He returned to South Vietnam in 1954 with joint American and French support, and in the following year, after a dubious election, became President of an anti-communist government of South Vietnam. He had commenced military resistance against the *Vietcong by 1960 and had achieved some degree of success with both social and economic reform, but his harshly repressive regime, in which his brother, Ngo Dinh Nhu, earned particular notoriety as head of political police, aroused strong local resentment and he was killed in a military coup, launched with apparent US collusion, on 2 November 1963.

Nguni, related ethnic groups in southern Africa. In the 1820s, the Zulu in the Natal area, under their king *Shaka, developed a superior military force, made up of regiments (or impis) which attacked neighbouring peoples (Difagane Wars). Refugees from Shaka, copying the military discipline and strategy of their Zulu conquerors, established themselves in the *Ndebele state in Zimbabwe, the Gaza state in Mozambique, the Swazi state in Swaziland, and a group of Nguni states in Tanzania, Zambia, and Malawi. Nguni people came into conflict with European settlers: the British in the Cape Colony moved into the lands of the Xhosa, precipitating the *Xhosa (Kaffir) Wars; Boer settlers in Natal clashed with the Zulus, as later did the British. Urbanization in the

Ngo Dinh Diem inspects a 'strategic hamlet' in South Vietnam, 1963. These villages were intended to counteract Vietcong infiltration of the countryside. With Diem is a US military adviser. Three months later the South Vietnamese leader was killed in a coup.

20th century has been accompanied by the policy of *apartheid. The *Bantustans created in South Africa have little connection with original Nguni culture.

Nguyen Van Thieu (1923–), South Vietnamese statesman, President (1967–75). He participated in the post-war struggle against the French, but left the *Vietminh because of its communist policies and thereafter became a general in the army of South Vietnam. He participated in the overthrow of *Ngo Dinh Diem's government in 1963 and together with another military strongman, Nguyen Cao Ky, dominated the politics of his country. Elected President in 1967, and re-elected in 1971, he continued to press hard in the war against the *Vietcong and their North Vietnamese allies despite growing opposition to his dictatorial methods. In the early 1970s the war began to run against him and his US allies drastically reduced their support. In April 1975, with his army in ruins and enemy forces closing on Saigon, he finally resigned his office.

Nicaragua, the largest Central American country. The country was colonized by the Spaniards in the 16th century, and achieved its independence in 1821. Nicaragua was briefly annexed into the Mexican empire of Agustín de *Iturbide, and with the collapse of that experiment formed part of the United Province of *Central America. It became independent in 1838. In 1848 the British seized San Juan del Norte, known as the Mosquito Coast, after a tribe of American Indians. In

1855 a US adventurer, William Walker, seized control of the country and made himself President (1856–7). His ousting helped unite the country, which made peace with Britain and recognized a separate Mosquito kingdom. The 20th century opened with the country under the vigorous control of the dictator José Santos Zelaya, who extended Nicaraguan authority over the Mosquito kingdom. The USA, apprehensive of his financial dealings with Britain, supported the revolution which overthrew him in 1907. The US presence, including two occupations by the marines, dominated the country until 1933. In 1937 Nicaragua fell under the control of Anastasio *Somoza, who ruled until his assassination in 1956. He was succeeded by his son, General Anastasio Debayle Somoza (1956–72, 1974–9). In 1962 a guerrilla group, the Sandinistas, was formed. It gained increasing support from the landless peasantry and engaged in numerous clashes with the National Guard, ending in civil war (1976–9). Once established as a ruling party, the Sandinistas expropriated large estates for landless peasants. Their dispossessed and exiled owners then organized opposition to the regime, recruiting a Contra rebel army, funded and organized by the *CIA. Mines and forests were nationalized and relations with the USA deteriorated. In 1981 US aid ended and the regime was accused of receiving aid from Cuba and the Soviet Union. The *Reagan administration sought increasing support from the US Congress to give aid to the exiled Contra forces in Honduras and Miami, but was seriously embarrassed by exposure in 1986–7 of illegal diversion of money to the Contras from US sale of arms to Iran.

Nicholas I (1796–1855), Emperor of Russia (1825–55). The third son of Paul I, he succeeded his brother *Alexander I, having crushed a revolt by the *Decembrists, who favoured his elder brother Constantine. His rule was authoritarian and allowed for little social reform. Russia was ruled by the army bureaucracy and police, and intellectual opposition expressed itself in study circles and secret societies. These polarized into 'Slavophiles', who held that Russian civilization should be preserved through the Orthodox Church and the village community, and 'Westernizers' who wished to see western technology and liberal government introduced. Nicholas embarked on the *Russo-Turkish Wars and brutally suppressed the uprising (1830–1) in Poland.

Sandinista soldiers in **Nicaragua**, 1979. The initial civil war waged by the Sandinistas had ended, but a second-phase conflict, against the Contra rebels, was beginning.

Religious minorities, including Jews, were persecuted. In the *Revolutions of 1848 he helped Austria crush the nationalists in Hungary, and his later attempts to dominate Turkey led to the *Crimean War (1853–6). He was succeeded by his son *Alexander II.

Nicholas II (1868–1918), last Emperor of Russia (1894–1917). In 1894 he formalized the alliance with France, but his Far Eastern ambitions led to disaster in the *Russo-Japanese War (1904–5), an important cause of the *Revolution of 1905. He was forced to issue the October Manifesto promising a representative government and basic civil liberties. An elected Duma and an Upper Chamber were set up. Although Russia was prosperous under *Stolypin's premiership (1906–11) and he won popular support for the war against Germany (1914), he unwisely took personal command of the armies, leaving the government to the empress Alexandra and *Rasputin. Mismanagement of the war and government chaos led to his abdication in February 1917 and later imprisonment. On 16–17 July 1918 the *Bolsheviks, fearing the advance of counter-revolutionary forces, murdered him and his family at Ekaterinburg (now Sverdlovsk).

Niemöller, Martin (1892–1984) German Protestant churchman. A U-boat commander in World War I, he

Nicholas II, the last emperor of Russia, pictured with three of his daughters while under arrest at Tsarskoye Selo in 1917. Following his abdication, Nicholas hoped to be allowed to leave for exile in Britain, but instead he and the royal family were moved east to Siberia, to be murdered at Ekaterinburg in July 1918.

became a priest in 1924. In 1933 he founded the Pastors' Emergency League to help combat rising discrimination against Christians of Jewish background. His opposition to the Nazification of the Church led to his arrest and confinement in a *concentration camp (1938–45). A controversial pacifist, he became a president of the World Council of Churches (1961–8).

Niger, a landlocked country of West Africa lying mainly in the Sahara. The French first arrived in 1891, but the country was not fully colonized until 1914. A French colony (part of *French West Africa) from 1922, it became an autonomous republic within the *French Community in 1958 and fully independent in 1960, but there were special agreements with France, covering finance, defence, technical assistance, and cultural affairs. Since 1974 it has been governed by a Supreme Military Council, and all political associations have been banned.

Nigeria, a large West African country, consisting of a federation of twenty-one states, with the highest population (ninety-three million) of any African country. The island of Lagos and other coastal trading stations were important centres for the 18th-century slave trade of Lagos. In 1851 the British attacked and burnt the city and ten years later bought it from King Dosunmu, administering it first from Freetown and then from the Gold Coast, until in 1886 a separate protectorate (later colony) of Lagos was formed. Explorers worked their way inland, but until the discovery of quinine (1854) to provide protection against malaria, the region remained known as 'the white man's grave'. During the second half of the 19th century trading companies were established, forming the Royal Niger Company in 1886, which was then taken over by the British Colonial Office to become the Niger Coast protectorate in 1893. Following the conquest of the kingdom of *Benin, this became the protectorate of Southern Nigeria (1900). The protectorate of Northern Nigeria was proclaimed in 1900. In 1906 the colony of Lagos was absorbed into the southern protectorate and in 1914 the two protectorates were merged to form the largest British colony in Africa, which, under its governor Frederick *Lugard, was administered indirectly by retaining the powers of the chiefs and emirs of its 150 or more tribes. In Northern Nigeria Muslim chiefs of the *Fulani tribes maintained a conservative rule over the majority of the country's Hausa population. In the West, the *Yoruba dominated; the Ibo tribe was centred in the East.

Under the constitution of 1954 a federation of Nigeria was created, consisting of three regions: Northern, Eastern, and Western, together with the trust territory of Cameroons and the federal territory of Lagos. In 1960 the federation became an independent nation within the *Commonwealth of Nations, and in 1963 a republic. In 1967 the regions were replaced by twelve states, further divided in 1976 into nineteen states. Oil was discovered off Port Harcourt and a movement for Ibo independence began. In January 1966 a group of Ibo army majors murdered the federal Prime Minister, Sir Alhaji Abubakar Tafawa *Balewa, the Premiers of the Northern and Western regions, and many leading politicians. In July a group of northern officers retaliated and installed General *Gowon as Head of State. A massacre of several thousand Ibo living in the North followed. Attempts to work out constitutional provisions failed, and in May

1967 the military governor of the Eastern region, Colonel Ojukwe, announced his region's secession and the establishment of the republic of *Biafra. Civil war between the Hausa and Ibo peoples erupted, and Biafra collapsed in 1970. General Gowon was deposed in 1975. In 1979 the military government organized multi-party elections. Corruption and unrest precipitated more military takeovers, in 1983 and 1985. In spite of wealth from its oil revenues Nigeria has continued to face deep social and economic problems.

Nigerian Civil War   *Biafra.

Nightingale, Florence (1820-1910), British hospital reformer. After considerable opposition from her family she received some training as a nurse at Kaiserswerth in Prussia and in Paris in the early 1850s. Her success as superintendent of the Hospital for Invalid Gentlewomen in London led her to be invited in 1854 to take a team of nurses to Scutari in Turkey to look after British soldiers wounded in the *Crimean War. In the face of much official hostility, she succeeded in improving beyond recognition the state of the British military hospitals, whose insanitary conditions had been largely responsible for the high mortality rate among the wounded. In 1856

Florence **Nightingale** in 1857, a year after her return from the Crimea. Pencil drawing by Sir George Scharf. (National Portrait Gallery, London)

she returned home a national heroine, and in 1860 used a gift of money raised by a grateful public to found a training school for nurses at St Thomas's Hospital in London. Throughout the rest of her life, although an invalid, she took an active interest in nursing and hospital matters.

'Night of the Long Knives' (29-30 June 1934), the name coined by *Hitler for a weekend of murders throughout Germany. It followed a secret deal between himself and the *SS units. Precise details remain unknown, but the army is believed to have promised to support Hitler as head of state after *Hindenburg's death in return for destroying the older and more radical Nazi private army known as the SA (Sturmabteilung), or *brownshirts, led by Ernst Röhm. Hitler announced that seventy-seven people had been summarily executed for alleged conspiracy. Subsequent arrests by the SS all over Germany, usually followed by murder, numbered many hundreds including some non-party figures, and the former Chancellor Schleicher.

nihilism, the total rejection of authority as exercised by the church, the state, or the family. More specifically, the doctrine of a Russian extremist revolutionary party active in the late 19th and early 20th centuries. In their struggle against the conservative elements in Russian society, the nihilists justified violence, believing that by forcibly eliminating ignorance and oppression they would secure human freedom. The government of *Alexander II repressed the revolutionaries severely, and they sought vengeance by assassinating the emperor near his palace on 13 March 1881. After 1917 the small and diffuse cells of nihilists were themselves destroyed by better co-ordinated revolutionaries.

Nimeiri, Gaafar Muhammad al- (1930-), Sudanese statesman. He led campaigns against rebels in the southern Sudan in the 1950s and joined in leftist attempts to overthrow the civilian government. Following a coup, he became Prime Minister in 1969 and Chairman of the Revolutionary Command Council. Elected President (1971-85), in 1972 he ended the civil war in the southern Sudan, granting it local autonomy. He switched from socialist economic policies to capitalism, to make the Sudan a major food producer. A devout Muslim, he proposed a new Islamic Constitution in December 1984 which was to make Islamic law apply to everyone. This was opposed in southern Sudan, where the majority are non-Muslim, and a military coup overthrew him in 1985.

Nimitz, Chester William (1885-1966), US admiral. After various surface ship commands and shore appointments he took over command of the Pacific Fleet in 1941 after *Pearl Harbor. From his Hawaii headquarters, he deployed his forces to win the battle of Midway, and subsequently supervised the moves in the *Pacific Campaigns, leading to successful actions off Guadalcanal and in the *Leyte Gulf. To a large extent he was responsible for making the Pacific Fleet, weakened by Pearl Harbor, the instrument of Japan's defeat. After the war he was briefly chief of naval operations.

Nixon, Richard Milhous (1913-), US lawyer, and thirty-seventh President of the USA. Elected to the US House of Representatives (1947, 1949), he was prominent in the investigations that led to the indictment of Alger Hiss in the *McCarthy era. He was elected as Vice-President under *Eisenhower and as such (1953-60) earned a reputation for skilful diplomacy. He was narrowly defeated in the presidential election of 1960 by John F. *Kennedy and lost (1962) the election for governor of California. In 1968 he was chosen as Republican presidential candidate, when he narrowly defeated the Democrat Hubert Humphrey. In his first term he achieved many successes, especially in foreign affairs, and was re-elected. His second term was marred by the *Watergate Scandal (1973-4), and he became the first President to resign from office. He was granted a pardon by President Ford for any crimes he may have committed over Watergate. He returned to politics in 1981 as a Republican elder statesman.

Nixon presidency (1969-74), the term of office of Richard M. *Nixon as thirty-seventh President of the USA. His administration initiated a New Economic Policy (1971) to counteract inflation, which included an unprecedented attempt to control prices and wages in peace-time, as well as the reversal of many of the social policies of President *Johnson. In an attempt to achieve a balance of trade, the dollar was twice devalued in 1971 and 1973. The presidency is best remembered for its achievements in foreign affairs, for which the Secretary of State Henry *Kissinger was at least partly responsible. Having inherited the *Vietnam War, Nixon began by extending it, by invading Cambodia (1970) and Laos

Nixon presidency. Richard Nixon waves farewell as he boards the presidential aircraft for the last time, having resigned as President of the USA on 8 August 1974.

(1971), and by saturation bombing. From 1971 onwards, however, a policy of gradual withdrawal of US troops began, while negotiations were taking place, ending with the cease-fire accord of 1973. At the same time support was being given to the policy of *Ostpolitik with a presidential visit to the Soviet Union bringing about agreements on trade, joint scientific and space programmes, and nuclear arms limitation. Recognition was given to the communist regime of the People's Republic of China as the official government of China, and in February 1972 Nixon paid a state visit to China. Although Nixon was re-elected President in 1972, his second term was scarred by the *Watergate scandal.

Nkomo, Joshua Mqabuko Nyongolo (1917-), Zimbabwe politician. He was Secretary-General of the Rhodesian Railways African Employees Association and President of the African National Congress (1957-9), when it was banned in Rhodesia. In 1960 he founded the National Democratic Party. When this was banned he instituted the Zimbabwe African People's Union (ZAPU). He was twice detained (1962-4) and then imprisoned (1964-74). On release he travelled widely to promote the nationalist cause. His ZAPU, mainly supported by the Ndebele in south-western Zimbabwe allied uneasily with *Mugabe's ZANU as the Patriotic Front. He lost the 1980 election to Mugabe and became his foremost opponent.

Nkrumah, Kwame (Francis Nwia Kofi) (1909-72), African statesman, Prime Minister of Ghana (1952-60) and first President (1960-6). After studying in the USA and Britain he returned to Ghana in 1947 as General Secretary of the United Gold Coast Convention, an African nationalist party founded by J. B. Danquah. In 1949 Nkrumah founded the Convention People's Party, and led a series of strikes and boycotts for self-government. He became Prime Minister after a short imprisonment by the British for sedition, and led his country to independence (1957) as *Ghana, the first British African colony to achieve this. His style of government was autocratic, but in his first years in power he was immensely popular with his policy of Africanization. In 1964 he was declared President for life. Economic pressures led to political unrest and in 1966, while he was on a visit to China, a military coup deposed him. He took refuge in Guinea, where President Sekou *Touré made him 'Co-President'. An outstanding African nationalist and a firm believer in *Pan-Africanism he died in exile.

NKVD (initial Russian letters for 'People's Commissariat for Internal Affairs'). It was the Soviet secret police agency responsible from 1934 for internal security and the labour *prison camps, having absorbed the functions of the former *OGPU. Mainly concerned with political offenders, it was especially used for *Stalin's purges. Its leaders were Yagoda (1934-6), Yezhov (1936-8), and *Beria until 1946, when it was merged with the MVD (Ministry of Interior). After Beria's fall in 1953 the Soviet secret police was placed under the *KGB (Committee of State Security).

non-co-operation (in British India), a political campaign by the Indian National *Congress organized and led by M. K. *Gandhi (1920-2). Its aims were to force further concessions from the British government by

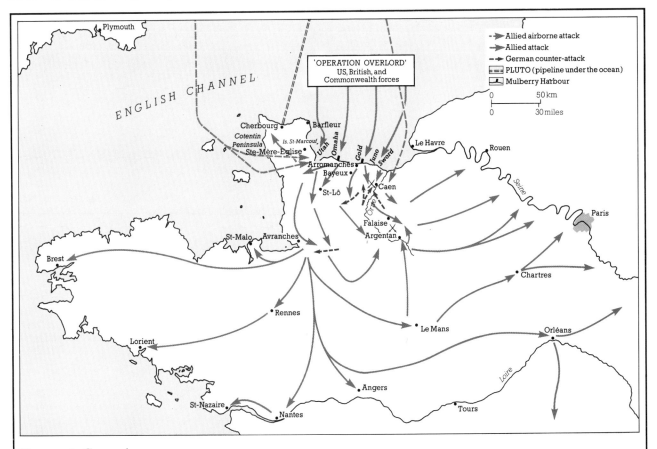

Normandy Campaign

The Allied plan 'Operation Overlord' succeeded in 1944 largely as a result of meticulous planning. U-boat bases in Brittany were captured in a lightening campaign by US troops, who then swung east. Other US forces, together with British and Commonwealth troops, defeated German defenders at the battle of the Falaise Gap. A swift drive through Normandy led to the capture of Rouen and Paris, and the advance into north-west Europe.

organizing the boycotting of the legislative councils, courts and schools, and other symbolic acts. The movement, inspired by Gandhi's *satyagraphia campaign, was intended to be non-violent but it degenerated into violence and was called off by Gandhi in February 1922 after the murder of a number of policemen by a mob at Chauri Chaura in the United Provinces. The movement failed to win enough support to paralyse government; its chief effect was to mobilize mass support for Congress and to consolidate Gandhi's position in the leadership of the national movement.

Normandy Campaign (June–August 1944), Allied counter-offensive in Europe in World War II following the *Normandy Landings. The US forces under General *Bradley cut off the Cotentin Peninsula (18 June), and accepted the surrender of Cherbourg. The British army attacked towards Caen, securing it after heavy fighting (9 July) before advancing on Falaise. US troops broke through the German defences to capture the vital communications centre of Saint-Lô, cutting off the German force under *Rommel. The Germans launched a counter-attack but were caught between the US and British armies in the 'Falaise Gap' and lost 60,000 men in fierce fighting. Field-Marshal Model, transferred from the Eastern Front, was unable to stem Patton's advance, which now swept across France to Paris, while Montgomery moved his British army up the English Channel.

Paris was liberated by General Leclerc on 26 August, and Brussels on 3 September. By 5 September more than two million troops, four million tonnes of supplies, and 450,000 vehicles had been landed, at the cost of some 224,000 Allied casualties.

Normandy Landings (June 1944), a series of landings on the beaches of Normandy, France, in World War II. Five beaches had been designated for the Allied invasion, code-name 'Operation Overlord', for which General *Eisenhower was the supreme commander. All the beaches, given code-names, had been carefully re-connoitred by commandos and at dawn on 6 June 1944 (D-Day) five separate groups landed between St Marcouf and the River Orne: at 'Utah', 'Omaha', 'Gold', 'Juno', and 'Sword'. British and Canadian troops fought across the eastern beaches, the Americans the western. Four beaches were taken easily, but 'Omaha' encountered fierce German resistance. Allied airforces destroyed most of the bridges over the Seine and the Loire, preventing the Germans from reinforcing their forward units. At the height of the fighting, *Rommel, who commanded Germany's western defences, was seriously wounded and was recalled. Meanwhile old ships had been towed across the Channel and sunk to provide more sheltered anchorages. On D-Day plus 14 two vast steel-and-concrete artificial harbours (code-name 'Mulberry') were towed across the English Channel. One was sunk by a freak storm, but the second was established at Arromanches, on beach 'Gold'. It provided the main harbour for the campaign. Meanwhile a series of twenty oil pipelines (code-name 'Pluto') was laid across the Channel to supply the thousands of vehicles now being landed. After months of detailed and meticulous preparation, the greatest amphibious landing in history was complete and the *Normandy Campaign launched.

North African Campaigns (June 1940–May 1943), a series of military campaigns in Africa in World War II. When Italy declared war in June 1940, General *Wavell in Cairo with 36,000 Commonwealth troops attacked first, the Italians giving up Sidi Barrani, Tobruk, and Benghazi between September 1940 and January 1941. In July 1940 the Italians had occupied parts of the Sudan and British Somaliland, but in January 1941 the British counter-attacked and on 6 April 1941 Ethiopia and all of Italian East Africa surrendered, thus opening the way for Allied supplies and reinforcements to reach the Army of the Nile. In March 1941 General *Rommel attacked, and the British withdrew, leaving *Tobruk besieged. Under General *Auchinleck, an offensive (Operation Crusader) was planned. At first successful, the campaign swung back and forth across the desert, both German and British tank casualties being high. Tobruk fell in June 1942 and the British took up a defensive position at El *Alamein in July. From there in October the reinforced 8th Army of 230,000 men and 1,230 tanks now under General *Montgomery launched their attack, and Rommel fell back to Tunisia. Meanwhile 'Operation Torch' was launched, an amphibious landing of US and British troops (8 November) under General *Eisenhower near Casablanca on the Atlantic and at Oran and Tunis in the Mediterranean, where it was hoped to link up with *Free French forces in West Africa. The *Vichy French troops of General *Darlan at first resisted, but after three days acquiesced. From November 1942 to May 1943 German armies, although reinforced, were being squeezed between the 8th Army advancing from the east and the Allied forces advancing from the west. On 7 May Tunis surrendered. Some 250,000 prisoners were taken, although Rommel skilfully succeeded in withdrawing the best troops of his Afrika Korps to Sicily.

North Atlantic Treaty Organization *NATO.

Northcliffe, Alfred Harmsworth, Viscount *Harmsworth, Alfred.

Northern Expedition (1926–8), military campaign in China waged by nationalist forces under the leadership of *Chiang Kai-shek to extend their power from their base in southern China to much of northern China by defeating local *warlord armies, initially with military assistance from the Soviet Union. Shanghai and Nanjing were captured in March 1927, and Beijing finally fell on 8 June 1928. A nationalist government was established in Nanjing from 1928 to 1932. The Northern Expedition was notable both for the final emergence of Chiang Kai-shek as the sole leader of the nationalist *Kuomintang, and for his purge of the communists, which resulted in a series of unsuccessful communist risings in August 1927 and the first ten-year phase of the nationalist–communist civil war.

Northern Ireland, the six north-eastern counties of Ireland, established as a self-governing province of the United Kingdom by the Government of Ireland Act (1920) as a result of pressure from its predominantly Protestant population. Discrimination by the Protestant majority against the largely working-class Catholics (about one-third of the population) over electoral reforms erupted in violence in the 1960s. The civil rights movement (1968) led to outbreaks of violence, and paramilitary groupings such as the *Irish Republican Army clashed with 'loyalist' militant organizations such as the Ulster Defence Association (UDA) and the Ulster Defence

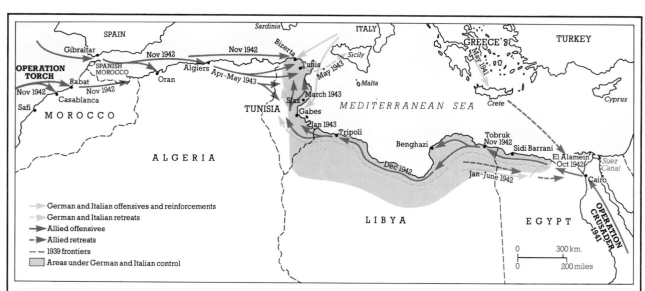

North African Campaigns (1941–3)

In January 1941 British and Commonwealth troops liberated Italian-occupied East Africa. By June 1942 German Panzer troops under General Rommel had reached the Egyptian border, where they were defeated at El Alamein. Rommel was steadily pushed west, while an Allied advance from Casablanca moved east through Algeria. In May 1943 German troops withdrew to Italy.

Force (UDF). In 1969 extra British military forces were sent to the province at the request of the *Stormont government, and have remained there ever since. The British government suspended (1972) the Northern Irish constitution and dissolved the Stormont government, imposing direct rule from London. A more representative Northern Ireland Assembly was elected (1973), but collapsed through extremist unionist opposition. Leaders such as the Revd Ian Paisley, together with the Ulster Workers Council, which organized a general strike in 1974, paralysed the province, forced the collapse of the non-sectarian Northern Ireland Executive, and foiled attempts at a new governmental framework for power-sharing between both sides. Since 1979 closer co-operation between the Republic of Ireland and Britain has developed, leading to the Anglo-Irish Accord (the Hillsborough Agreement) signed in 1985, giving the republic a consultative role in the government of Northern Ireland. Attempts to organize an agreed and permanent system of government have so far met with failure.

Northern Territory, a territory of north-central Australia. The Dutch ship *Arnhem* landed the first Europeans on its coast in 1623. The French explorer Nicolas Baudin (1801, 1803), and the British hydrographers Matthew Flinders (1803) and Phillip Parker King (1818) followed. Attempts to establish a British military settlement in the north failed. In 1824 the first party from *New South Wales arrived at Port Essington, and, in an attempt to forestall French colonization, took formal possession of the area. In 1862 the explorer John McDouall Stuart crossed Australia from Adelaide to the northern coast. The following year the Northern Territory was annexed to South Australia. Gold discoveries at Pine Creek in the 1870s attracted people to the Northern Territory. In 1911 the Australian federal government took over the administration of the Northern Territory. A Legislative Assembly was created in 1974. The Northern Territory was granted internal self-government in 1978.

North Korea, north-east Asian country. Consisting of the northern half of the Korean peninsula, above the 38th parallel, North Korea was formed from the zone occupied by the Soviet Union at the end of World War II, an independent Democratic People's Republic being proclaimed on 1 May 1948. Intent on reuniting Korea, North Korea launched a surprise attack on *South Korea in June 1950, suffering considerable damage and loss of life in the following three years of the indecisive *Korean War. Since the war, the ruling communist party of Kim Il Sung (President and General Secretary since 1948) has undertaken a programme of reconstruction, using the country's mineral and power resources to finance economic development. After many years of tension, relations with the westernized regime in South Korea have improved slightly in recent years. North Korea has generally succeeded in not becoming too closely identified with either Chinese or Russian interests.

North Vietnam   *Vietnam.

North-West Europe Campaign (September 1944–May 1945), a military campaign in World War II. Following the *Normandy Campaign, *Montgomery's forces captured Antwerp (4 September) and crossed the Albert Canal. The US 1st Army captured Namur and

Aachen, while the US 3rd Army moved east and reached the Moselle. Montgomery's attempt to seize the lower Rhine by dropping the 1st Airborne Division at *Arnhem ended in failure. In November the Germans consolidated and in December launched a counter-attack in the *Ardennes, the battle of the Bulge. In January 1945 Montgomery's forces pushed forward to the Rhine. In March a massive bombardment at Wesel preceded a successful crossing of the lower Rhine by Montgomery's troops. The US 7th Army pushed east towards Munich, French forces moved up the upper Rhine to Lake Constance, and the US 3rd Army advanced to Leipzig and across the Austrian border into Czechoslovakia. On 11 April Montgomery reached the River Elbe. Following the capture of Berlin by the Red Army and the suicide of Hitler and Eva Braun (30 April), Montgomery received the surrender of the German forces in north-west Europe on Lüneburg Heath on 4 May. Four days later (V-E Day), the war in Europe was declared at an end.

North-West Frontier, the mountainous area, inhabited by *Pathan and Baluchi peoples, forming that part of Pakistan which abuts Afghanistan. Under British rule from 1849 to 1947, a policy of extending British control into the tribal territories began with the acquisition of Quetta (1876) and of Pishin, Sibi, and Kurram (1879). This led to a number of tribal wars, including those against Hunza and Nagar (1891), Chitral (1895), and Tirah (1897–8). The most troubled area, however, was Waziristan, where resistance was led by the Mulla Powindah until 1913 and by the Fakir of Ipi during the 1930s. To control the frontier the British employed several forces, including the Punjab Irregular Force (founded 1849), various militias, and the Frontier Constabulary. The border with Afghanistan was defined by the Durand Line (1893). Administratively the area was constituted a

An Indian army expedition on the **North-West Frontier**; a watercolour titled *Looking back from the Kahanuk Pass* by Sir Edward Durand, mapper of the Durand Line which established the border between India and Afghanistan. The isolation and rugged landscape of the Frontier posed severe problems for the British. For the troops, mostly Indian infantry, such expeditions were hazardous tests of endurance. (India Office Library, London)

province in 1901 with a number of tribal agencies, and after 1947 the Pakistan government withdrew military forces from tribal territory.

Norway, a country in northern Europe. After centuries of Danish suzerainty, Norway was ceded to Sweden in 1814, the country having established its own parliament (Storting) from 1807 onwards. A literary revival and a new national consciousness brought demands for complete independence. Responsible government was granted in 1884, and universal male suffrage in 1898. Finally, union with Sweden was unilaterally declared dissolved in June 1905, and Prince Charles of Denmark elected as *Haakon VII. A Liberal Party government introduced women's suffrage and social reform, and maintained neutrality during *World War I. In *World War II, the Germans invaded, defeating Norwegian and Anglo-French forces at Narvik in 1940 and imposing a puppet government under Vidkun *Quisling. In 1945 the monarchy, and a Labour government, returned. Norway withdrew from the *European Economic Community (1972) after a national referendum. The exploitation of North Sea oil in the 1970s gave a great boost to the economy.

Nuclear Test-Ban Treaty (1963), an international agreement not to test nuclear weapons in the atmosphere, in outer space, or under water, signed by the USA, the Soviet Union, and Britain (but not France). The issue of *disarmament had been raised at the Geneva Conference (1955), and discussions on the ban of nuclear testing had begun in Geneva in 1958. In spite of the treaty the spread of nuclear weapons became a major preoccupation of the *NATO powers in the 1960s, and in 1968 a Non-Proliferation Treaty was signed. However, with the availability of uranium, other countries, such as China, Israel, and India, have become nuclear powers.

Nuffield, William Richard Morris, Viscount (1877–1963), British motorcar manufacturer and philanthropist. He set up his own bicycle business at Cowley, Oxford, when he was 16. In 1902 he designed a motorcycle and, by 1912, he was repairing motorcars. He bought a factory at Cowley, where the Morris Oxford, the first British car planned for middle-class family use, began production in 1913. *Mass production kept Morris cars cheap. Business acumen earned him a vast fortune, much of it given away. Nuffield College, Oxford, and the Nuffield Foundation, a trust to encourage educational developments, were established by him.

nullification, a US doctrine holding that a state had the right to nullify a federal law within its own territory. The Kentucky Resolutions of 1798–9 asserted the right of each state to judge whether acts of the federal government were constitutional. The nullification theory was fully developed in South Carolina, especially by John *Calhoun, in response to the high protective tariffs of 1828 and 1832. South Carolina's Ordinance of Nullification in 1833 prohibited the collection of tariff duties in the state, and asserted that the use of force by the federal government would justify secession. Although this ordinance was repealed after the passing of new federal legislation in the same year, the sentiments behind nullification remained latent in Southern politics, and were to emerge again in the secession crisis prior to the *American Civil War.

Nuremberg Trials (1945–6), an international tribunal for Nazi war criminals. The trials were complex and controversial, there being few precedents for using international law relating to the conduct of states to judge the activities of individuals. The charges were: conspiracy against peace, crimes against peace, violation of the laws and customs of war, crimes against humanity. As a result of the trials several Nazi organizations, such as the *Gestapo and the *SS, were declared to be criminal bodies. Individual judgments against the twenty-four war-time leaders included death sentences, imprisonment, and not guilty. Ten prisoners were executed, while *Goering and Ley committed suicide. Rudolf *Hess was sentenced to life-imprisonment.

Nyasaland *Malawi.

Nye Committee (1934–6), a US Senate committee, chaired by Gerald P. Nye of North Dakota, to investigate the dealings of the munitions industry and bankers and their reputed profits from promoting foreign wars. The findings revealed high profits and a studied hostility to disarmament, but no evidence to support the theory that President *Wilson had at any time been influenced by the financial 'stake' in his relations with Germany. However, so strong and widespread was the spirit of *isolationism that the Senate, in an effort to remain aloof from global problems, passed a series of *Neutrality Acts (1935–9).

Nyerere, Julius Kambarage (1922–), African statesman, the first Prime Minister of independent Tanganyika (1961), and first President of *Tanzania (1964–86). In 1954 he organized the Tanganyika African National Union (TANU). In 1956 the British administration nominated him as TANU representative in the Legislative Council. In 1957 he resigned, complaining of slow progress, but on Tanganyika's independence (1961) Nyerere became Prime Minister, surrendering his premiership a month later. In 1962 he was elected President of the Tanganyika Republic. In 1964 following a revolution in *Zanzibar, he effected union between it and Tanganyika as the Republic of Tanzania, bringing it (1967) into the East African Community, a customs union with Uganda and Kenya. In the *Arusha Declaration (1967) he outlined the socialist policies that were to be adopted in Tanzania. He has been a major force in the *Organization of African Unity and over the broad range of African politics, especially in relation to Uganda, Zimbabwe, and South Africa. He resigned the presidency in 1986.

OAS *Organization de l'Armée secrète; *Organization of American States.

Oastler, Richard (1789–1861), British social reformer. He began his agitation in 1830 with the support of John Wood, a Bradford manufacturer, who revealed some of the worst abuses of *child labour in factories. Oastler, a Tory radical, combined his attack on the factory system with a condemnation of the *Poor Law Amendment Act of 1834. He attained some of his objectives with the Ten Hours Act in 1847, which limited the hours worked to ten a day.

OAU *Organization of African Unity.

Obote, Milton (1924–), Ugandan statesman. Politically active throughout the 1950s, Obote served as Prime Minister of Uganda (1962–6), when he overthrew *Mutesa II, Kabaka of Buganda, and assumed full power as President. Himself overthrown by Idi *Amin in 1971, Obote returned from exile in Tanzania to resume the presidency in 1980. However, he failed either to restore the economy or stop corruption and tribal violence, and was once again overthrown in 1985, seeking refuge in Zambia.

Obregón, Alvaro (1880–1928), Mexican general and statesman. President of Mexico (1920–4), he was elected after the most violent decade of the *Mexican Revolution. He succeeded in bringing about some measure of agrarian, educational, and labour reform, although *peonage remained strong. His implementation of the revolutionary programme of 1917 brought him into bitter conflict with the Catholic Church. After turning the presidency over to his successor Plutarco Elías *Calles in 1924, he was elected to a second term in 1928 but was assassinated prior to taking office.

Obrenović, Serbian dynasty (1817–1903). It was founded by a cattle drover, Milos Obrenović I (1780–1860), who became a revolutionary fighting the Turks under *Kara George. He persuaded the Turks, after the murder (probably at his instigation) of Kara George, to accept his election in 1817 as Prince of Serbia. His tyrannical rule led to his abdication and the brief reign of Milan Obrenović II. Michael Obrenović III (1823–68) succeeded in 1840 but was forced into exile two years later. Milos was reinstated in 1858 but died in 1860. Michael resumed his rule, and proved an able and effective leader. He was assassinated in 1868, and his cousin, Milan Obrenović IV (1854–1901), began thirty years of unpopular rule. He announced himself king of Serbia in 1882 but was forced to abdicate in favour of his son Alexander (1876–1903). Alexander's murder in June 1903 ended the dynasty, and the Karageorgević dynasty again came to power.

OCAS *Organization of Central American States.

O'Connell, Daniel (1775–1847), Irish nationalist leader and social reformer. A lawyer by training, he first came to prominence with his condemnation of the *Act of Union (1801). A powerful orator and a skilled organizer, in 1823 he founded the Catholic Association to press for the removal of discrimination against Catholics in tithes, education, the electoral franchise, and the administration of justice. In 1828 he was elected Member of Parliament at the *Clare Election, although as a Catholic he was ineligible for membership of the House of Commons. To avoid the risk of civil disorder the British government passed the Roman Catholic Relief Act (1829), which granted *Catholic emancipation and enabled O'Connell to take his seat. In the 1840s his campaign for the repeal of the Act of Union by constitutional methods was unsuccessful and this lost him the support of many nationalists and of the radicals in the *Young Ireland movement.

O'Connor, Feargus Edward (1794–1855), Irish radical politician and Chartist leader. He was elected Member of Parliament for County Cork in 1832 as a supporter of Daniel *O'Connell but lost his seat in 1835. In 1837 he founded a radical newspaper the *Northern Star* in England, and it was largely through his tireless energy and his ability as an orator that *Chartism became a

Daniel **O'Connell** (*foreground*), in a detail from a painting by Joseph Haverty, at one of the 'monster meetings' by which he skilfully mobilized Irish Catholic opinion in favour of Catholic emancipation. (National Gallery of Ireland, Dublin)

mass movement. After a term of imprisonment for seditious libel, he was elected Member of Parliament for Nottingham in 1847.

October Revolution   *Russian Revolution (1917).

October War   *Yom Kippur War.

Oder–Neisse Line, the frontier, formed by these two rivers, between Poland and the German Democratic Republic. It had marked the frontier of medieval Poland and, as a result of an agreement at the *Potsdam Conference, nearly one-fifth of Germany's territory in 1938 was reallocated, mainly to Poland. Germans were expelled from the territories, which were resettled by Poles. The frontier was finally accepted by the Federal Republic of Germany as part of the *Ostpolitik.

OGPU (initial Russian letters for 'United State Political Administration'), a security police agency established in 1922 as GPU and renamed after the formation of the *Union of Soviet Socialist Republics (1923). It existed to suppress counter-revolution, to uncover political dissidents, and, after 1928, to enforce *collectivization of farming. It had its own army and a vast network of spies. It was absorbed into the *NKVD in 1934.

O'Higgins, Bernardo (1778–1842), South American revolutionary and ruler. He was the illegitimate son of Ambrosio O'Higgins (c.1720–1801), governor of *Chile, and became the most famous leader of the Chilean movement for independence. His first major military effort against royalist forces ended unsuccessfully at the battle of Rancagua (1814). He led the remnants of his army across the Andes and joined the Argentine independence leader José de *San Martín. The two commanders prepared their combined army for an invasion of Chile and defeated the Spanish troops at Chacabuco in February 1817. The following year, with the independence of the new republic secured, O'Higgins was chosen by the leading citizens of Santiago as supreme director of Chile. The financial, political, and social reforms that he attempted to introduce during his five-year term met with much opposition, and he was forced to resign (1823). He went into exile in Peru.

Okinawa, an island situated between Taiwan and Japan, captured from the Japanese in World War II by a US assault that lasted from April to June 1945. With its bases commanding the approaches to Japan, it was a key objective and was defended by the Japanese almost to the last man, with *kamikaze air attacks inflicting substantial damage on the US ships. After the war it was retained under US administration until 1972, when it was returned to Japan, following a vociferous campaign.

Oklahoma Indian Territory, area of early Indian reservation in western USA. Most of the area of modern Oklahoma came to the USA by the *Louisiana Purchase, which stimulated President *Jefferson to think that one answer to Indian–white relations would be to transfer Indians to these newly acquired lands. In 1817 some Cherokees made the journey west to join original inhabitants like the Kiowa, Shawnee, Comanche, and Pawnee. After the Indian Removal Act of 1830 increasing numbers of Indians were transferred west and in 1867

During the First **Opium War** of 1839–42, British naval superiority proved decisive against the Chinese. This painting by E. Duncan shows the armed steamer *Nemesis* (*right, background*) attacking and destroying eleven Chinese junks near Guangzhou (Canton). (National Maritime Museum, London)

the general Oklahoma Reservation was established. By the end of the 1880s many whites had been persuaded that the Indians were not using all the land and called for the opening of much of it to white settlement. In 1889 Congress authorized settlement, and in 1890 it organized the white-controlled areas into Oklahoma Territory, which became the 46th state in 1907.

Oman, a country on the Arabian Peninsula. Formerly known as Muscat and Oman, it was, under *Said ibn Sultan Sayyid, the most powerful state in Arabia in the early 19th century, controlling *Zanzibar and the coastal regions of Iran and Baluchistan. Tension frequently erupted between the sultan of Oman and the interior tribes. Oil, now the country's major product, began to be exported in 1967. In 1970 the present ruler, Sultan Qaboos bin Said (1940–), deposed his father Said bin Taimur in a palace coup. An uprising by left-wing guerrillas was defeated in 1975.

OPEC   *Organization of Petroleum Exporting Countries.

Opium Wars (1839–42; 1856–60), two wars between Britain and China. In the early 19th century British traders were illegally importing opium from India to China and trying to increase trade in general. In 1839 the Chinese government confiscated some 20,000 chests of opium from British warehouses in Guangzhou (Canton). In 1840 the British Foreign Secretary, Lord *Palmerston, sent a force of sixteen British warships which besieged Guangzhou and threatened Nanjing and communications with the capital. It ended with the Treaty of *Nanjing (1842). In 1856 Chinese officials boarded and searched a British flagged ship, the *Arrow*. The French joined the British in launching a military attack in 1857, at the end of which they demanded that the Chinese agree to the Treaty of Tianjin in 1858. This opened further ports to western trade and provided freedom of travel to European merchants and Christian missionaries inland. When the emperor refused to ratify the agreement, Beijing was occupied, after which, by the Beijing Convention (1860), the Tianjin Agreement was accepted. By 1900 the number of *treaty ports had risen to over fifty, with all European colonial powers, as well as the USA, being granted trading concessions.

Oppenheimer, Sir Ernest (1880–1957), South African financier. With the help of Herbert *Hoover, he linked US finance and South African mining enterprise in the Anglo-American Corporation of South Africa in 1917. He extended his interests to diamonds, and established the Diamond Producer's Association (1934). He helped to finance the Northern Rhodesian (Zambian) copperfield and sat in the South African Parliament from 1924 to 1938. A passionate believer in the British Commonwealth and in African advancement, he made many benefactions.

Orange Free State, a province of the Republic of *South Africa. In the early 19th century it was inhabited mainly by the Bantu-speaking Sotho people. Except for frontiersmen, it was not occupied before 1836. In 1848 the British annexed the region as the Orange River Sovereignty, with the result that the Boers crossed into *Transvaal. In 1852 the British government ordered the province to be relinquished, and by the Bloemfontein Convention (1854) recognized it formally, granting it independence as the Orange Free State. The discovery of diamonds in 1867 caused a rush of mainly British immigrants and helped to cause the First *Boer War. After the Peace of *Vereeniging the state, as the Orange River Colony, became a crown colony, becoming self-governing in 1907. In 1910 it became a founding province in the Union of South Africa under its earlier title.

Oregon Boundary dispute (1843–6), a dispute between the USA and Britain. Since 1818, Britain and the USA had agreed on joint occupation of the disputed territory. As settlers moved up the *Oregon Trail, pressure mounted for the area west of the Rocky Mountains running down into the valley of the Columbia River, known as Oregon country, to become part of the USA. The settlers wanted their northern boundary with the British to be to the north of Vancouver Island on the 54° 40′ parallel, and in 1844 President *Polk used the slogan '54 40 or fight' in his victorious campaign. Discussions took place with the British, who at first insisted on the Columbia River as boundary. A compromise agreement between the British Foreign Secretary Lord *Aberdeen and President Polk was reached in 1846. This was to accept a line well to the south of Polk's original demands, extending the 49th parallel boundary to the Pacific, but excluding Vancouver Island. In 1848, Congress created the Oregon Territory. This was later split up into the state of Oregon (1859), the state of Washington (1889), the state of Idaho (1890), and parts of Montana and Wyoming.

Oregon Trail, a wagon trail across the US Rocky Mountains. News of the pleasing climate and rich soils began to draw settlers from the east. Each year from 1842 to 1846 some 100 wagons gathered at Independence, on the Missouri River, to make the 3,200 km. (2,000 mile) trip to the Pacific. As numbers grew, hostilities arose with the British in Vancouver (the *Oregon Boundary Dispute, settled in 1846). When gold was found in Montana (1862) John Bozeman (1831–67) charted a branch trail to the goldfield, leaving at Fort Laramie and running up into the Big Horn Mountains. After this gold rush the Bozeman Trail became a *cattle trail in the 1880s.

Organization de l'Armée secrète (OAS) a French secret terrorist organization based in Algeria, formed in 1961. Its aim was the destruction of the French Fifth Republic in the interest of French colonial control of Algeria. It plotted an unsuccessful assassination attempt on President *de Gaulle in 1962. Its action had little effect on the French government, which by now was determined to grant independence to Algeria. Subsequent riots in Algiers were suppressed, and the OAS itself eliminated (1963) by the capture or exile of its leaders.

Organization for Economic Co-operation and Development (OECD), an association of Western states to assist the economy of member nations and to promote world trade. It was established in 1961 as a replacement for the Organization for European Economic Co-operation, which in turn had been created in 1948 by those countries receiving aid under the *Marshall Plan. Membership has risen from the original twenty full members to twenty-four.

Organization of African Unity (OAU), an association of African states. It was founded in 1963 for mutual co-operation and the elimination of colonialism. All African states except South Africa and Namibia have at one time belonged. The leaders of thirty-two African countries signed its charter at a conference in Addis Ababa in 1963. There is an annual assembly of heads of state and government, a council of ministers, a general secretariat, and a commission for mediation, conciliation, and arbitration.

Organization of American States (OAS), a regional international organization. Originally founded in 1890 on US initiative for mainly commercial purposes, the OAS adopted its present name and charter in 1948. The major objective of the thirty-two American states which comprise the OAS is to work with the United Nations to ensure the peaceful resolution of disputes among its members, to promote justice, to foster economic development, and to defend the sovereignty and territorial integrity of the signatory nations.

Organization of Central American States (OCAS) (1951–60), a regional grouping comprising Costa Rica, El Salvador, Guatemala, Honduras, and Nicaragua. Established in 1951, its purpose was to establish the *Central American Common Market. This goal was reached in 1960, but OCAS members co-operated on little else. The San Salvador Charter (1962) expanded the trade and fiscal provisions of the original treaty, envisaging permanent political, economic, and defence councils.

Organization of Petroleum Exporting Countries (OPEC), an international organization seeking to regulate the price of oil. The first moves to establish closer links between oil-producing countries were made by Venezuela, Iran, Iraq, Kuwait, and Saudi Arabia in 1949. In 1960, following a reduction in the oil price by the international oil companies, a conference was held in Baghdad of representatives from these countries, when it was decided to set up a permanent organization. This was formed in Caracas, Venezuela, next year. Other countries later joined: Qatar (1961), Indonesia (1962), Libya (1962), United Arab Emirates (1967), Algeria (1969), Nigeria (1971), Ecuador (1973), and Gabon (1975). OPEC's activities extend through all aspects of oil negotiations, including basic oil price, royalty rates, production quotas, and government profits. Following a crisis with the oil companies (1973) the price of crude oil was raised by some 200 per cent over three months. This steep increase was to have vast world repercussions, not only in making some Arab states extremely rich, but adversely affecting the cost of living both in developed and developing countries. The appearance of new, non-OPEC oil producers, such as Britain and Norway, somewhat reduced the influence of the organization on oil-pricing and production, but it continues to play a major part in influencing world prices and production.

Orlando, Vittorio Emanuele (1860-1952), Italian statesman and jurist. He had supported Italy's entry into World War I and after the *Caporetto disaster (1917) became Premier. At the *Versailles Peace Settlement he clashed with President *Wilson over what Wilson thought were excessive claims by Italy to former Austrian territory. In 1922 he at first supported *Mussolini, but after the murder of *Matteotti he resigned from Parliament in protest (1925) and fled the country. After the fall of Mussolini (1943) he became a leader of the Conservative Democratic Union.

Orsini, Felice (1819-58), Italian revolutionary. After being implicated in revolutionary plots, he was condemned in 1844 to life imprisonment. He was later pardoned by Pius IX but took part in Italy in the *Revolutions of 1848. In 1849 he joined *Mazzini in Rome and then went to Hungary, where he was arrested. After his escape in 1854 he formed a plot to assassinate *Napoleon III, seen as the principal obstacle to Italian independence. His attempt to blow up the imperial carriage failed (14 January 1858) and he was executed.

Orthodox Church (or Eastern Orthodox Church), the historic Christian churches of the Near East and Asia Minor. Each Church is independent, but a special honour is accorded to the Ecumenical Patriarch of Constantinople. There is a sizeable Orthodox population in the USA, but 85 per cent of the 150 million Orthodox in the world live under varying conditions of persecution and discrimination in communist states, the majority of them in the Soviet Union. The Orthodox Church is the overwhelmingly dominant church in Greece. In recent years there has been some rapprochement between the Orthodox Churches and the *Roman Catholic Church, following the lifting in 1965 of mutual excommunications imposed in 1054.

Osborne judgment, a British court decision in 1909. It stemmed from an action brought by a Liberal trade unionist, W. V. Osborne, against the trade union practice of using part of trade union subscriptions to pay salaries to Labour Members of Parliament. The courts found in favour of Osborne. Up to this time Members of Parliament received no Parliamentary salary. This was remedied in 1911. The Trade Union Act (1913) authorized unions to have a political fund but subscriptions to it were optional, members being able to 'contract out'.

Osceola (c.1800-39), American Seminole Indian chief. During the *Seminole War of 1835-42, Osceola, an extremely capable military leader, held off for three years US attempts to remove his tribe to the west. Frustrated by his successful resistance in the Florida Everglades, General Jessup offered Osceola safe conduct to a peace conference, but treacherously imprisoned him at Fort Moultrie near Charleston, where he died.

Ostpolitik (German, 'eastern policy'), a term used in the Federal Republic of *Germany (West Germany) to describe the opening of relations with the Eastern bloc. It was a reversal of West Germany's refusal to recognize the legitimacy of the German Democratic Republic (East Germany) as propounded in the Hallstein Doctrine. This asserted that West Germany would sever diplomatic relations with any country (except the Soviet Union) that recognized East German independence. The policy of Ostpolitik was pursued with particular vigour by Willy *Brandt, both as Foreign Minister and as Chancellor of the Federal Republic. A General Relations Treaty (1972) normalized relations between the two Germanys, while treaties between West Germany and both the Soviet Union and Poland gave formal recognition to the *Oder-Neisse frontier (1970-2).

Oswald, Lee Harvey (1939-63), alleged US assassin of President *Kennedy at Dallas, Texas, in 1963. He was arrested leaving the scene, but before he could be tried, he was himself killed by another civilian, Jack Ruby. Many theories have been aired that Oswald had accomplices, but the *Warren Commission concluded that he had acted on his own.

Ottawa Agreements (1932), a series of agreements on tariffs and trade between Britain and its *dominions. They were concluded at the Imperial Economic Conference, held at Ottawa, and constituted *tariff reforms by Britain and the dominions based on the system of imperial preferences to counter the impact of the Great *Depression. They provided for quotas of meat, wheat, dairy goods, and fruit from the dominions to enter Britain free of duty. In return, tariff benefits were granted by the dominions to imported British manufactured goods.

Lee Harvey **Oswald** was being moved from the basement of the Dallas County Jail when he was shot by nightclub owner Jack Ruby (*with back to camera*) on 24 November 1963. Oswald died soon after the shooting, and no one ever stood trial for the assassination of the President.

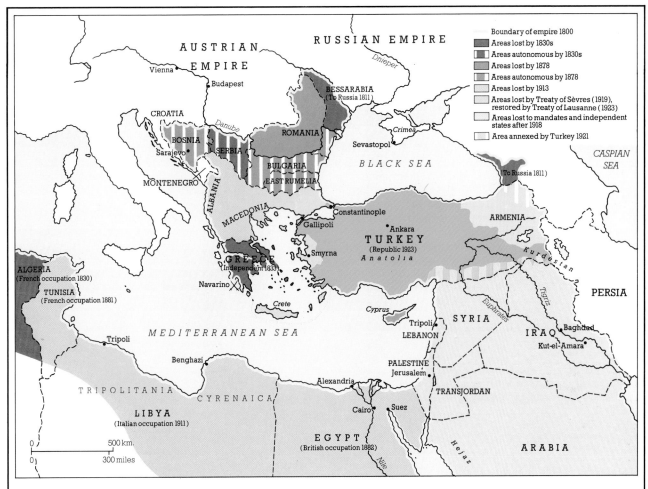

Ottoman empire

In the early 19th century the European lands of the Ottoman empire suffered from maladministration. The Austrian and Russian empires sought to extend their power by exploiting nationalism. Western influence helped produce administrative and judicial reforms (the Tanzimat) and from 1856 western capital was increasingly invested in Ottoman Turkey, producing nationalist reaction in the Young Turk movement. The African lands of the empire fell victim to European imperialism: Algeria and Tunisia being conquered by the French, Egypt by the British, and Libya by Italy. In the late 18th century in Arabia a movement had developed (the Wahhabist) for the purification and reform of Islam, aimed against the corrupt ways of the sultans of Constantinople, who as caliphs claimed allegiance from all Muslims. In 1923 the frontiers of the new Republic of Turkey were accepted in the Treaty of Lausanne, when the Ottoman Arabian provinces were re-created as nations and mandated to France and Britain. These gradually gained independence.

The economic gains were helpful but not massive. After World War II the benefits were steadily eroded, and, with the prospect of British entry into the *European Economic Community, the agreements became increasingly dispensable. Although seriously considered during the 1961-3 negotiations, they played little part in the 1971-2 terms of entry, apart from New Zealand dairy products.

Ottoman empire, an Islamic empire originally created by Turkish tribes from Anatolia. During the 18th century cultural links with Europe developed, while territorially Russia made steady advances into the Caucasus. In 1807 the Sultan Mustafa IV led a reactionary movement against western influence, but he was succeeded in 1808 by *Mahmud II (1808-39), who began a long process of reform, which included the destruction in 1827 of the military corps of the Janissaries and the creation of a new, more westernized army. This did not, however,

succeed in preventing Greece from becoming independent in 1833. The *Tanzimat reform movement accelerated under Mahmud's two sons Abdulmecid I (1839-61) and Abdulaziz (1861-76). Abdulhamid II (1876-1908) agreed in 1878 to a western-style constitution, the first in any Islamic country. The Parliament it created only met once (1878) and was not reconvened until the revolution of 1908. Through the 19th century *Russo-Turkish Wars steadily reduced the empire in Europe, and at the Congress of *Berlin in 1878 it abandoned all claims over *Romania, *Serbia, *Montenegro, *Bulgaria and *Cyprus, while from 1882 Egypt effectively passed into British control. In the later 19th century a movement for more liberal government produced the *Young Turks revolution in 1908 and the deposition of Abdulhamid II. During World War I Britain and France occupied much of what remained of the empire, encouraging Arab nationalism and creating, after the war, such successor states as *Jordan, *Syria, *Lebanon, and *Iraq, as well as promising (1917) a Jewish national home in *Palestine. The *Versailles Settlement attempted to reduce the empire to only part of Anatolia, together, reluctantly, with Istanbul. Turkish nationalist feelings rejected the proposals, forcibly expelling Greeks and *Armenians and adopting the present frontiers in 1923. By then the last sultan, Mehmed VI, had been overthrown and the caliphate abolished, and the new republic of Turkey proclaimed under Mustafa Kemal *Atatürk.

Outback, a colloquial Australian term to describe the inland grazing lands beyond the Great Dividing Range of New South Wales. The *Bigge Inquiry had recommended that *emancipists and poor settlers should be excluded from taking up land, and in 1829 a vain effort was made to forbid land settlement beyond the 'nineteen counties' of New South Wales, anybody going beyond this boundary being an illegal *squatter. However, as the demand for wool grew so did the numbers of such squatters, pressing further west, south-west, and north. In 1836 a £10 per year licence fee gave them rights over any tract of land over which they were already grazing sheep. Vast sheep stations of many thousands of acres developed, and the rich pastoralists became and remain a powerful force in Australian society.

Overland Stage, stage-coach mail-carrying service in the USA, established in 1850 as a government service to run from Independence, Missouri, south-west to Santa Fe and west to Salt Lake City. In 1858 the Southern (Butterfield) Overland Mail was established, running from St Louis to San Francisco. The latter service was expanded into Oregon and Montana before being sold to *Wells Fargo in 1866. By the 1860s thousands of miles of overland route were in use, and, although the expansion of the railways reduced long-distance business, the Overland Mail continued to serve areas not reached by rail for much of the rest of the century.

Owen, Robert (1771–1858), British social reformer, industrialist, and pioneer of the co-operative movement. In 1799 he acquired some textile mills at New Lanark in Scotland and proceeded to create a model community, radically improving the conditions in which his employees worked and lived. He provided amongst other things educational facilities for children, shorter working hours, and better housing. During the period of economic misery after 1815 he proposed the formation of 'villages of co-operation', in which the unemployed would be mutually self-supporting instead of relying on poor relief. He also played an important part in the passing of the 1819 Factory Act. From 1825 to 1829 he lived in the USA, where he attempted, unsuccessfully, to establish co-operative communities run on socialist lines. In 1833 he founded the Grand National Consolidated Trades Union, which collapsed in the following year.

Oxford Movement   *Newman, John Henry.

Pacific, War of the (1879–84), a conflict pitting Peru and Bolivia against Chile over a disputed region in the northern Atacama Desert that was rich in potassium nitrate (saltpetre). British-backed mining companies moved from Chile into the region, whereupon Bolivia and Peru agreed to resist by force. The combined military forces of Peru and Bolivia were no match for the Chileans. The Chilean navy commanded the Pacific coast and its army was better trained and better equipped. Bolivia withdrew, leaving Peru at the mercy of Chilean forces, who occupied and sacked Lima in 1881. In the Treaty of Ancón (1883) Chile took most of the disputed territory and left Peru a humiliated nation. Bolivia lost Antofagasta, its Pacific port, and was left a landlocked nation.

Pacific Campaigns (1941–5), naval and amphibious engagements in World War II. The war spread to the Pacific when Japanese aircraft attacked the US naval base of *Pearl Harbor in 1941. Their landforces quickly occupied Hong Kong, French Indochina, Malaya, *Singapore, and Burma. Other Japanese forces captured islands in the Pacific, while convoys sailed to occupy Borneo and the Dutch East Indies following the Japanese naval victory at the battle of the Java Sea (27 February– 1 March 1942). By April the Philippines were occupied, followed by northern New Guinea, and General *MacArthur withdrew to Australia, where he organized a counter-attack. The battle of the Coral Sea (5–8 May) between Japanese and US carriers was strategically a US victory. It prevented Japanese landings on southern New Guinea and ended their threat to Australia. It was followed (3–6 June) by the decisive battle of Midway Island, which, under Admiral *Nimitz, shifted the balance of naval power to the USA. In August 1942 US marines landed on Guadalcanal and Tulagi in the Solomon Islands, where fighting raged until February 1943. During 1943, the remaining Solomon Islands were recaptured, with Bougainville falling in November, followed by New Britain early in 1944. In June 1943 MacArthur had launched his campaign to re-occupy New Guinea, and through 1944 US forces gradually moved back towards the Philippines. On 19 June 1944 the Japanese lost some 300 planes in the battle of the Philippine Sea and in July the Mariana Islands were recaptured, from which US bombing raids on Tokyo were then organized. In October 1944 the battle of *Leyte Gulf marked the effective end of Japanese naval power, while on the mainland the *Burma Campaign had reopened land communication with China and begun the process of reoccupation of the short-lived Japanese empire. Manila fell in March 1945, and in April US forces reoccupied *Okinawa against fierce *kamikaze air raids, at the cost of very high casualties on both sides. Plans to invade Japan were ended by the decision to drop atomic bombs on *Hiroshima and *Nagasaki (6 and 9 August), which resulted in Japanese surrender.

Pacific scandal (1873), a Canadian political scandal. British Columbia had entered the dominion of *Canada in 1871 on the understanding that a trans-continental railway-line would be built within ten years. In 1872 a contract for such a railway was awarded to a syndicate headed by Sir Hugh Alan, a banker, shipowner, and financial contributor to the Conservative Party. Following a general election in 1872, the defeated Liberals accused the Prime Minister, John *Macdonald, of having given the contract as a reward. Macdonald resigned and the contract was cancelled. The Conservatives were heavily defeated in a new election but the *Canadian Pacific Railway was finally completed in 1885.

Páez, José Antonio (1790–1873), Venezuelan revolutionary and statesman. He was a leader of Venezuela's movement for independence, controlling (1810–19) a band of *Ilaneros* (plainsmen) in guerrilla warfare against the Spanish. He led the separatist movement against *Bolívar's Colombian republic and became Venezuela's

first President. During his first term (1831–5) he governed within the provisions of the Venezuelan constitution, but he became increasingly oligarchic in his subsequent terms of office (1839–46, 1861–3). He was exiled (1850–8), but returned in 1861 to become supreme dictator. In 1863 he again went into exile. He encouraged economic development, promotion of foreign immigration, and construction of schools.

Pahlavi, Muhammad Reza Shah *Muhammad Reza Shah Pahlavi.

Pai Marire (Maori, 'goodness and peace'), a Maori political and religious movement. In 1862 Horopapera Te Ua of south Taranaki began teaching his people to worship the new Pai Marire god whom he saw as the Old Testament Jehovah. His followers danced around *niu* (decorated poles) seeking the gift of prophecy and powers

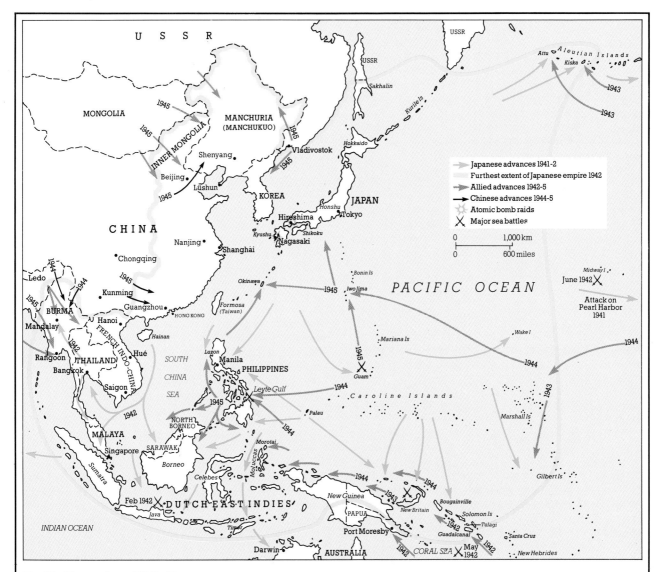

Pacific Campaigns (1941–5)

The rapid Japanese conquest of the countries of South-East Asia and of the Pacific islands was at first supported by anti-colonial nationalist movements. By April 1942 the Philippines were occupied and Australia threatened. The US Pacific Fleet had suffered badly at Pearl Harbor

(December 1941), but won a decisive victory at Midway Island (June 1942). From then on a series of naval battles, fought largely by aircraft-carriers, brought US forces steadily nearer to Japan, which was heavily bombed throughout 1945.

to heal. Renewed war from 1865 saw the movement take a violent turn, becoming known as the Hau-hau. The ritual now involved the exhibition of heads of white soldiers and missionaries, and cannibalism revived. The movement affected much of the North Island, but gradually subsided after 1872.

Pakistan, a country in southern Asia. Following the British withdrawal from the Indian sub-continent in 1947, Pakistan was created as a separate state, comprising the territory to the north-east and north-west of India in which the population was predominantly Muslim. The 'Partition' of the subcontinent of India led to unprecedented violence between Hindus and Muslims, costing the lives of more than a million people. Seven and a half million Muslim refugees fled to both parts of Pakistan from India, and ten million Hindus left Pakistan for India. Muhammad Ali *Jinnah became the new state's first governor-general. The country's liberal constitution was opposed by the orthodox Muslim sector, and in 1951 the Prime Minister, *Liaqat Ali Khan, was assassinated by an Afghan fanatic. In 1954 a state of emergency was declared and a new constitution adopted (1956). When attempts to adopt a multi-party system failed, Ayub Khan (1907-74) imposed martial law (1958). His decade of power produced economic growth, but also political resentment. The two wings of Pakistan were separated by a thousand miles of Indian territory. Allegations by the Bengalis in East Pakistan against West Pakistan's disproportionate share of the state's assets led to demands by the Awami League, led by *Mujibur Rahman, for regional autonomy. In the ensuing civil war (1971), the Bengali dissidents defeated a Pakistani army, with Indian help, and established the new state of *Bangladesh (1971). In 1970 the first ever general election brought to power Zulfikar Ali *Bhutto (1928-79), leader of the Pakistan People's Party, who introduced constitutional, social, and economic reforms. In 1977 he was deposed, and later executed. The regime of General *Zia ul-Haq (1977-) stands committed to an Islamic code of laws. Martial law was lifted in 1986, anticipating a slow but steady return to democracy.

Palacký, František (1798-1876), Czech nationalist and historian. A leading figure in the Czech cultural and national revival in the mid-19th century, he presided over the first *Pan-Slav Congress in Prague in 1848, advocating Czech autonomy within a federal Austria. After the suppression of the liberal and nationalist uprising of 1848, Palacký retired from active politics until 1861, when he became a deputy to the Austrian Reichstag. After the foundation of the *Austro-Hungarian empire in 1867, he advocated complete Czech independence. His influence on Czech political thought, and on later leaders such as Tomáš *Masaryk, was immense.

Palestine, a territory in the Middle East. It was part of the *Ottoman empire from 1516 to 1918, when Turkish and German forces were defeated by the British at Megiddo (19 September 1918). The name 'Palestine' was revived as an official political title for the land west of the Jordan, which became a British *mandate in 1923. Following the rebirth of *Zionism, Jewish immigration, encouraged by the *Balfour Declaration of 1917, became heavy, and Arab-Jewish tension culminated in a revolt in 1936. The Peel Commission (1937) recommended

Yasser Arafat, leader of the **Palestine Liberation Organization** since 1969, and co-founder of al-Fatah, the leading military component of the PLO.

partition into Jewish and Arab states, but neither group would accept this. Renewed pressure for Jewish immigration in 1945 inflamed the situation, with acts of anti-British violence. Britain ended the mandate in 1948 when the state of *Israel was established. In spite of the United Nations plan of 1947 for separate Arab and Jewish states, Palestine ceased to exist as a political entity after the Arab-Israeli War of 1948, being divided between Israel, Egypt (the Gaza strip), and Jordan (the West Bank of the River Jordan). The West Bank and the Gaza strip were occupied by Israel in 1967. The name continues to be used, however, to describe a geographical entity, particularly in the context of Arab aims for the resettlement of nearly three-quarters of a million people who left the area when the state of Israel was established.

Palestine Liberation Organization (PLO), a political and military body formed in 1964 to unite various Palestinian Arab groups in opposition to the Israeli presence in the former territory of *Palestine. From 1967 the organization was dominated by al-*Fatah, led by Yasser Arafat. The activities of its radical factions caused trouble with the host country, Jordan, and, following a brief civil war in 1970, it moved to Lebanon and Syria. In 1974 the organization was recognized by the Arab nations as the representative of all Palestinians. The Israeli invasion of Lebanon (1982) undermined its military power and organization, and it regrouped in Libya. Splinter groups of extremists, such as the 'Popular Front for the Liberation of Palestine' and the 'Black September' terrorists, have been responsible for kidnappings, hijackings, and killings both in and beyond the Middle East.

Palmer Raids (1919-21), action taken by the US attorney-general A. Mitchell Palmer against political radicals. After World War I, fears in the USA of a Bolshevik-style revolution, added to the wartime suspicion of aliens, resulted in harassment and the arrest of numerous political radicals believed to be plotting the overthrow of the US government. They began in the autumn of 1919, and were at a peak in January 1920.

Palmerston, Henry John Temple, 3rd Viscount (1784-1865), British statesman. An Irish peer, he entered the House of Commons as a Tory in 1807, serving as Secretary for War from 1809 to 1828. In 1830 he became Foreign Secretary in the new Whig government, holding this position almost continuously until 1841 and then again from 1846 to 1851. During this period he helped *Belgium to win its independence from the Netherlands, supported the constitutional monarchies of Spain and Portugal against absolutist pretenders, and opposed Russian expansion at the expense of the Ottoman empire in the Near East. However, Palmerston's conduct of foreign policy was to be increasingly characterized by an aggressive nationalism. He initiated the First *Opium War against China in 1840 in defence of British commercial interests and in 1850 intervened against Greece in the *Don Pacifico affair. In 1855 he succeeded Lord Aberdeen as Prime Minister and brought the Crimean War to a successful conclusion. He kept Britain neutral during the *American Civil War, but was unable to save Denmark from defeat by Prussia in 1864. Palmerston's high-handed methods often led him into conflict with his colleagues, but brought him much popular support.

Pan-African Congress, South African political movement. A militant off-shoot of the *African National Congress (ANC), it was formed in 1959 by Robert Sobukwe. He advocated forceful methods of political pressure, such as strikes and boycotts, and sponsored (1960) a demonstration at *Sharpeville, in which 67 black Africans were killed and 180 wounded by police. The South African government outlawed both the PAC and the ANC, and imprisoned Sobukwe and other leaders. Some PAC members went into exile, continuing their campaign under the Secretary of the Party, Potlako Leballo.

Pan-Africanism, a movement seeking unity within Africa. It became a positive force with the London Pan-African Conference of 1900. An international convention in the USA in 1920 was largely inspired by the Jamaican Marcus *Garvey. A series of further conferences followed. The invasion of Ethiopia by Italy in 1935 produced a strong reaction within Africa, stimulating anti-colonial nationalism, particularly in British West African colonies, as well as keeping the ideals of Pan-Africanism alive. The Pan-African Congress in Manchester in 1945 was dominated by Jomo *Kenyatta and Kwame *Nkrumah, and by the 'father of Pan-Africanism', the American W. E. B. *Du Bois. In 1958 a conference of independent African states was held in Accra, followed by two further conferences in Monrovia in 1959 and 1961. In 1963 in Addis Ababa thirty-two independent African nations founded the *Organization of African Unity, by which time Pan-Africanism had moved from being an ideal into practical politics.

Panama, the southernmost country of Central America, situated on the isthmus which connects North and South America. In 1821 it gained independence from Spain as a province of Gran Colombia. Despite many nationalist insurrections against Colombia in the 19th century, the area only became independent as the republic of Panama in 1903 as a protectorate of the USA. The latter had aided Panama's independence struggle in return for a Panamanian concession to build a canal across the isthmus and a lease of the zone around it to the USA. The volatile, élite-dominated politics which have characterized Panama during much of the 20th century have led to its occupation by US peace-keeping forces in 1908, 1912, and 1918. From 1968 to 1981, General Omar Torrijos controlled Panama, working to diversify the economy and to reduce US sovereignty over the Canal Zone, an object of long-standing national resentment. In 1977, Torrijos signed treaties with the USA providing for Panama's gradual takeover of the *Panama Canal.

Panama Canal, a canal about 81 km. (51 miles) long, across the isthmus of Panama, connecting the Atlantic and Pacific oceans. It was begun by a French company under Ferdinand de Lesseps in 1882, abandoned through

Excavating the **Panama Canal** (1904–14). To complete the project, US engineers used the most powerful technology available, such as this giant steam shovel. This photograph shows work at Pedro Miguel in the Canal Zone, as railway trains haul away the spoil excavated by the diggers.

bankruptcy, and completed in 1914 by the USA at a cost of nearly $400 million. In the Hay-Bunau-Varilla Treaty of 1903, Theodore *Roosevelt gained for the USA the concession from *Panama for a 16-km. (10-mile) wide Canal Zone under perpetual control of the US government. After World War II Panamanians became more hostile to US sovereignty over the Canal. After years of negotiations, in 1977 the US President, Jimmy *Carter, succeeded in obtaining congressional approval of the Panama Canal Treaties which provide for relinquishing total control of the Canal Zone to Panama by 1 January 2000, while assuring the Canal's perpetual neutrality.

Pan-Americanism, the movement towards economic, military, political, and social co-operation among the twenty-one republics of South, Central, and North America. The first Pan-American conference was held in 1889 in Washington, DC, to encourage inter-American trade as well as the peaceful resolution of conflicts in the region. The seventh conference (Montevideo 1933) was important because the USA, in harmony with Franklin D. *Roosevelt's 'Good Neighbor' policy, finally adopted the long-espoused Latin American principle of non-intervention, while the conference at Buenos Aires in 1936 adopted a treaty for the peaceful resolution of conflicts between American states. The Conference at *Chapultepec (1945) agreed on a united defence policy for the signatory nations. At the conference held in Bogotá in 1948, the *Organization of American States (OAS) was established, transforming the Pan-American system into a formal regional organization within the framework of the United Nations. The Alliance for Progress conference (1961), attended by representatives of all the American states except Cuba, pledged the members to support one another, by co-ordinating the economies of Latin America and by resisting the spread of communism. Some member states saw the OAS as a cover for extending US influence in Latin America and instead gave their support to the *Central American Common Market and the Latin American Free Trade Association, both founded in 1960.

Pancho Villa   *Villa, Francisco.

Pankhurst, Emmeline (1858–1928), British feminist and leader of the *suffragette campaign. She founded the Women's Social and Political Union in 1903 in Manchester. Moving to London, she limited suffragette tactics at first to attending processions, meetings, and heckling leading politicians. Then the suffragettes, under her direction, turned to more militant methods. Frequently imprisoned for causing disturbances, she responded by refusing to eat, drink, or sleep, until she was released, only to be re-arrested. Her daughter, Christabel, shared with her mother the planning of tactics. Another daughter, Sylvia, sought support for *women's suffrage among working-class women in the East End of London.

Pan-Slavism, the movement intended to bring about the political unity of all Slavs. It should be distinguished from Slavophilism, which was purely cultural and acted as a powerful stimulus towards the revival of Slavonic languages and literature, and from Austro-Slavism, which sought to improve the lot of Slavs within the *Austro-Hungarian empire. The aim of Pan-Slavism was to destroy the Austrian and *Ottoman empires in order to

Emmeline **Pankhurst** is carried away by a London policeman after being arrested outside Buckingham Palace in May 1914. With the outbreak of World War I the campaign for women's suffrage was postponed. Mrs Pankhurst turned her energies to recruiting for the army, and went to the USA and Russia on propaganda missions.

establish a federation of Slav peoples under the aegis of the Russian emperor. The ideology was developed in Russia, where it took on a militant and nationalistic form and helped provoke the *Russo-Turkish War (1877–8). Another manifestation was the Balkan League of 1912 by which Russia supported nationalist aspirations of the *Balkan States against Austrian ambitions. This led to the crisis that precipitated *World War I. The Bolshevik government of the newly established Soviet Socialist Republic (1917) renounced Pan-Slavism, but during and after World War II the concept was revived as a justification for dominance by the Soviet Union in Eastern Europe.

Papal States   *Vatican City.

Papen, Franz von (1879–1969), German politician. A member of the Catholic Centre Party, he had little popular following, and his appointment as Chancellor (1932) came as a surprise. To gain *Nazi support he lifted the ban on the *brownshirts, but *Hitler remained an opponent. Attempts to undermine Nazi strength failed and he resigned. He persuaded *Hindenburg to appoint Hitler (January 1933) as his Chancellor, but as Vice-Chancellor he could not restrain him. He became ambassador to Austria (1934), working for its annexation (*Anschluss) in 1938, and to Turkey (1939–44). He was tried as a war criminal (1945) but released.

Papineau's Rebellion (1837–8), a French-Canadian uprising of those seeking democratic reforms and protesting against the proposed union of *Upper and Lower

Canada. The Speaker of the Lower Canada Assembly, Louis-Joseph Papineau (1786–1871), led the reformist movement in French Canada (now *Quebec), and in 1837 he agitated for armed insurrection against the British, but fled to the USA before fighting broke out. Clashes between a few hundred of his supporters and regular troops occurred at Saint Denis in November. The rebellion broke out again in 1838, but was suppressed, and twelve supporters were executed. Papineau received an amnesty in 1844 and returned to Canada in 1845, by which time the establishment of parliamentary 'responsible' government had been achieved by moderate reformers such as Robert *Baldwin.

Papua New Guinea, the eastern half of the island of New Guinea and many off-shore islands in the south-west Pacific. Contact with Europe goes back to the 16th century, when the Portuguese Jorge de Meneses named the island Ilhas dos Papuas (Malay, 'frizzy-haired') and the Spaniard Ortiz Retes christened it New Guinea because he was reminded of the Guinea coast of Africa. In 1828 the Dutch annexed the western half of the island, followed, in 1884, by the German and British division of the eastern half. In 1904 the British transferred their territory, now called Papua, to Australia, and at the

Rebels exchange fire with regular troops (*left*) near the church of St Eustache in Quebec, 14 December 1837, during **Papineau's Rebellion**. The church and nearby buildings are shown on fire, as more troops arrive to help disperse the rebels. This lithograph, published in London in 1840, is from a sketch of the incident made by Lord Charles Beauclerk. (McCord Museum of Canadian History, McGill University, Montreal)

outbreak of World War I an Australian expeditionary force seized German New Guinea (Kaiser-Wilhelmsland). During World War II Australian troops fought off a determined Japanese invasion. Formal administrative union of the area as Papua New Guinea was achieved in 1968. Self-government was attained in 1973 and in 1975 Papua New Guinea became an independent nation within the Commonwealth of Nations. The Western part of the island forms part of Indonesia.

Paraguay, a land-locked country in south America. Part of the Spanish empire, it achieved its independence (1811) when local Paraguayan military leaders led a bloodless revolt against the Spanish governor. The dictator José Gaspar Rodriguez de *Francia ruled the new republic from 1813 to 1840, but the rest of the 19th century was dominated by corruption, coups, and chronic bankruptcy. Francisco Solano López led the country to disaster in the *Paraguayan War (1864–70). Political turmoil continued into the 20th century with the exception of the presidency of the liberal Edvard Schaerer (1912–17), which was marked by foreign investment and economic improvements. In the *Chaco War (1932–5), Paraguay won from Bolivia the long-contested territory believed to have oil reserves. In 1954 General Alfredo *Stroessner seized power. At the price of the repression of civil liberties his period in office has been one of peace and some material progress.

Paraguayan War (1864–70), also known as the War of the Triple Alliance, a conflict resulting from rivalries between Paraguay, Uruguay, Brazil, and Argentina. The Paraguayan President Francisco Solano *López, alarmed

by Brazilian intervention in Uruguay and harbouring desires for Paraguayan territorial expansion and access to the sea, initiated hostilities against Brazil in 1864 Despite traditional rivalry between Brazil and Argentina, the latter joined Brazil and its puppet government in Uruguay in the Triple Alliance pact (May 1865) against Paraguay. Paraguay's well-trained army of 600,000 men did not prove equal to the task, and López's death in March 1870 ended one of the most destructive wars in Latin American history. In addition to losing more than half of its population, Paraguay was also stripped of considerable territory as a result of the war.

Paris, Commune of (15 March–26 May 1871), a revolutionary government in Paris. It consisted of ninety-two members, who defied the provisional government of *Thiers and of the National Assembly. The Commune, which had no connection with communism, was an alliance between middle and working classes. Suspicious of royalist strength and opposing the armistice made with Prussia, the Communards wanted to continue the war and were determined that France should regain the principles of the First Republic. With the victorious German army encamped on the hills outside Paris, government troops were sent to remove all cannons from the city. They were bitterly resisted; Paris, demanding independence, broke into revolt. Thiers decided to suppress the revolt ruthlessly. For six weeks Paris was bombarded by government troops and its centre destroyed. Early in May its defences were breached and a week of bitter street fighting followed. Before surrendering, the Communards murdered their hostages, including the Archbishop of Paris. Over 20,000 people were massacred by the government forces, leaving France deeply divided.

During the **Commune of Paris** of 1871, the Communards burned public buildings and pulled down statues of rulers, including this one of Napoleon I, before the French army recaptured the city.

Paris, Congress of (1856), a conference held to negotiate the peace after the *Crimean War, attended by Britain, Austria, Russia, Turkey, and Sardinia. It marked a defeat for Russia, which conceded part of Bessarabia to Moldavia and Wallachia in the Balkans. The revival of the Straits Convention of 1841 meant that the Black Sea was again closed to all warships and neutralized while navigation of the Danube was to be free. The *Ottoman empire was placed under joint guarantee of the West European powers and the sultan agreed to recognize the rights of his Christians. The decline of the Ottoman empire, however, was not halted and Russia, determined to retrieve its Balkan supremacy, was to break the Black Sea clause in 1870.

Park, Mungo (1771–1806), British explorer and surgeon. In 1795 he was employed by the African Association to explore the course of the River Niger. Travelling east from the Gambia River, he then crossed to the Senegal Basin and the Upper Niger. His *Travels in the Interior Districts of Africa* (1799) was an immediate success. Although his second expedition to the Niger reached Bamako and Ségou, it was attacked at Boussa and all of Park's party died.

Parkes, Sir Henry (1815–96), Australian politician. He was active in several New South Wales political campaigns, including one against convict transportation, before entering Parliament in 1854, where he remained almost continuously until 1895. He was Premier of New South Wales five times. Parkes, the 'Father of Federation', publicly declared his support for *Australian federation, notably in his now-famous Tenterfield speech in 1889. He died before federation was achieved.

Parliament, British, the supreme legislature in *Britain and Northern Ireland comprising the sovereign, as head of the state and the two Chambers which sit in the Palace of Westminster—the House of *Lords and the House of *Commons. The Prime Minister and the cabinet (a

selected group of ministers from either House) are
responsible for formulating the policy of the government.
Acts of Parliament in draft form, known as Bills, each of
which have to be 'read' (debated) three times in each
House, are referred in the House of Commons (and
occasionally in the House of Lords) for detailed consideration to parliamentary standing or select committees.
The sovereign's powers of government are dependent on
the advice of ministers, who in turn are responsible
to Parliament. The monarch's prerogatives, exercised
through the cabinet or the Privy Council, include the
summoning and dissolution of Parliament. The Treaty of
Rome, which Britain accepted in 1972 when joining
the *European Community, provided for a gradual
development of Community institutons. The Single European Act (1986) laid down that the considerable powers
of those institutions take precedence over those of
member-states. The British parliamentary system was
adopted by many European countries and by most
countries of the *Commonwealth of Nations when they
gained dominion status or independence.

Parnell, Charles Stewart (1846–91), Irish nationalist.
He was the leader of the Irish Members of Parliament
agitating for *Home Rule. The son of a Protestant
Anglo-Irish landowner, he was elected to Parliament in

Sydney Prior Hall's painting of the Irish patriot Charles
Stewart **Parnell**, who was seen as both hero and martyr to
his cause. (National Gallery of Ireland, Dublin)

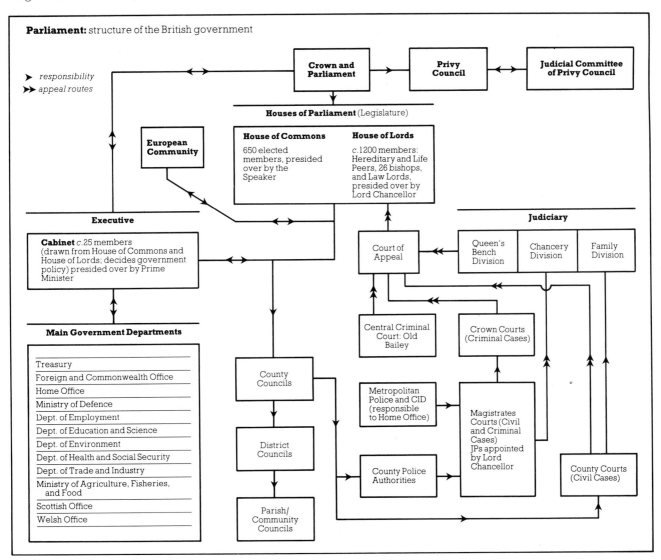

Parliament: structure of the British government

1875, where his obstructive tactics, notably the filibuster, drew attention to Irish grievances. He became President of the *Land League in 1879, successfully organizing the boycotting of unjust landlords. His advocacy of non-co-operation led to his imprisonment in Kilmainham gaol in 1881. He was released (1882) partly on the understanding that he would curb violence. The *Phoenix Park murders of two British politicians, following his release, damaged his reputation although a letter allegedly linking him to the murders was shown in court to be a forgery. In 1890 Parnell's involvement in a divorce case ended his career.

Pašić, Nikola (1845–1926), Serbian statesman and a founder of *Yugoslavia. Suspicious of Croats on both political and religious grounds, his ideal was a 'Greater Serbia', including much of Croatia and Dalmatia, with Serbs the master race. However, he signed the Corfu Pact (1917) which resulted in a union of Serbs, Croats, and Slovenes into a new kingdom which he represented at the *Versailles Peace Settlement (1919). He was twice Premier of the new kingdom (1921–4 and 1924–6), which in 1929 was to adopt the name Yugoslavia.

Passchendaele, battle of (31 July–10 November 1917), this, the third battle of Ypres, was fought on the *Western Front in World War I. The name of this Belgian village has become notorious for the worst horrors of *trench warfare and failure to achieve any strategic gain for over 300,000 British casualties. *Haig, the British commander-in-chief, without French help, remained convinced, despite the *Somme, that frontal assaults in superior numbers must succeed. Torrential rain and preliminary bombardment reduced Flanders to a sea of

The battlefield of **Passchendaele** came to symbolize the full horror of World War I. In Belgian countryside transformed into an alien landscape, Australian troops use a duckboard track to cross the morass created by shellfire at Château Wood, its trees reduced to skeletons.

mud, making advance impossible. Only on the final day did Canadians reach the ruined village of Passchendaele. Even this nominal gain was surrendered in the retreat before *Ludendorff's final offensive (April 1918).

passive resistance, non-violent opposition to a ruling authority or government. It frequently involves a refusal to co-operate with the authorities or a defiant breach of laws and regulations and has been a major weapon of many nationalist, resistance, and social movements in modern times. One of the most successful campaigns was that waged by *Gandhi against British rule in India, when widespread and large-scale civil disturbances and protests persuaded the British to make major concessions. Gandhi's example was an inspiration for the *Civil Rights movement in the USA from the 1950s, where passive resistance, large-scale demonstrations, and the deliberate breaking of *segregation laws brought considerable improvements for the black population. Similar methods have failed in less liberal regimes, notably South Africa and the Soviet bloc.

Patagonia, that portion of Argentina south of the 44th parallel, the southernmost region of South America. Argentine and Chilean rangers began settling in the territory, occupied by Indians, in the late 19th century; they fell into bitter boundary disputes which were resolved by an agreement in 1901. Europeans, notably of Basque, Welsh, and Scottish origin, immigrated in the early 20th century. The pastoral economy of the region was challenged in the period after World War II by the discovery of iron ore, petroleum, uranium, and natural gas deposits, which are now being exploited.

Patel, Sardar Vallabhbhai (1875–1950), Indian statesman. Deeply influenced by Mohandas *Gandhi, he became the principal organizer of many civil disobedience campaigns, suffering frequent imprisonment by the British. He was elected (1931) President of the Indian National *Congress. He played an important role in the negotiations that led to the partition of the sub-continent into India and Pakistan. As Deputy Prime Minister (1947–50) he initiated a purge of communists, and with the assistance of V. P. Menon (1894–1966) he integrated the *Princely States into the Indian Union.

Pathan, a Pashto-speaking people of Pakistan and Afghanistan, more especially the tribesmen of the mountainous regions along the *North-West Frontier of Pakistan. After 1849, when control of the region passed from the Sikhs to the British, a series of frontier uprisings took place. In 1893 an international frontier, the Durand Line, was established, with probably about two-thirds of Pathans in British India and one-third in Afghanistan. A Pathan rebellion against this frontier occurred in 1897–8, necessitating an extensive British military occupation, but the frontier was to be inherited by Pakistan in 1947 when the Pathans of the British North-West Frontier Province voted to join the newly independent Pakistan. An Afghan movement for a Pathan homeland, Pathanistan, has been a factor in the civil war in Afghanistan since 1979.

Pathet Lao, Laotian communist movement. In the independence struggle after World War II, Pathet Lao forces co-operated with the *Vietminh against French colonial power. After the *Geneva Agreement, it emerged as a major political and military force within Laos, seeking the alignment of their country with communist China and North Vietnam. Between the mid-1950s and mid-1970s the Pathet Lao and its political wing, the Neo Lao Haksat (Patriotic Party of Laos) under the leadership of Prince Souphanouvong, waged a constant political and military struggle for power with non-communist government forces, eventually emerging triumphant with the formation of the People's Democratic Republic of Laos in 1975.

Patiño, Simon Iture (1860–1947), Bolivian capitalist. He pioneered the development of Bolivia's tin resources and died one of the world's richest men. Although his modest origins impeded his entry into the Bolivian social élite, by the 1900s Patiño controlled over half of the tin production in Bolivia, and exerted considerable influence on his country's government. He is still criticized for not investing in the development of his own country.

Patton, George Smith (1885–1945), US general. In World War II he commanded a corps in North Africa and then the 7th Army in Sicily. He lost his command in 1944 after a publicized incident in which he hit a soldier suffering from battle fatigue, but later led the 3rd Army in the *Normandy Campaign. His tendency to make rapid military advance, at times with no regard for supporting units or allies, became evident in 1944 in his spectacular sweep through France, across the Rhine, and into Czechoslovakia. As military governor of Bavaria, he was criticized for his leniency to Nazis. He was killed in a road accident while commanding the US 15th Army.

Paulus, Friedrich von (1890–1957), German field-marshal. As deputy chief of staff he planned the German invasion of Russia (Operation 'Barbarossa') in World War II. In 1942 he failed to capture *Stalingrad, was cut off, and surrendered (February 1943). In captivity, he joined a Soviet-sponsored German organization, and publicly advocated the overthrow of the Nazi dictatorship. He lived in East Germany until his death.

Paul VI (Giovanni Battista Montini, 1897–1978), Pope (1963–78). He continued the work of his predecessor *John XXIII by reconvening the Second Vatican Council (1963–5) of the *Roman Catholic Church. Following the recommendations of the Council, important post-conciliar commissions (for example, for liturgical reform, Christian unity, greater lay participation, and the reform of the curia) were set up. A traditionalist by conviction, he was suspicious of any innovation which might undermine the authority of the Church, insisting on the necessity of priestly celibacy and condemning artificial methods of birth control.

Paz Estenssoro, Victor (1907–), Bolivian statesman. In 1941 he helped to form a left-wing political party, the Movimiento Nacionalista Revolucionario (MNR). In the same year he became Minister of Finance (1941–4), but then went into exile until 1951. In 1952, when the MNR came to power, he became President of Bolivia (1952–6). During this time the tin-mines were nationalized, adult suffrage introduced, and many large estates broken up and transferred to Indian peasants. He was re-elected President (1960–4), when he reached an understanding

The US destroyer *Shaw* explodes in flames after a direct hit on the ship's magazine. Silhouetted to the right are the big guns of another US warship, helpless against the surprise Japanese air attack on **Pearl Harbor**.

with international financiers for the re-organization of the tin industry. Elected a third time in 1964, he was overthrown by the army and went into exile until 1971, when he returned to Bolivia. Military government ended in Bolivia in 1982, and in 1985 he was re-elected President in succession to Dr Herman Zuazo.

Peabody, George (1795–1869), US financier and philanthropist. He established a prosperous trading business in the eastern USA before settling in London in 1837. He became one of the leading international bankers of his age, amassing a vast fortune, a substantial part of which he devoted to philanthropic ends, including the first educational foundation in the USA, the Peabody Education Fund, set up in 1867 to promote education in the South. He also gave large sums for slum clearance in Britain.

Pearl Harbor, a harbour on the island of Oahu in Hawaii. It is the site of a major US naval base where a surprise attack by Japanese carrier-borne aircraft (7 December 1941) delivered without a prior declaration of war, brought the USA into World War II. A total of 188 US aircraft were destroyed, and 8 battleships were sunk or damaged. The attack was a strategic failure because the crucial element of the US Pacific fleet, its aircraft carriers, were out of harbour on that day.

Pearson, Lester Bowles (1897–1972), Canadian diplomat and statesman. He served as delegate to the United Nations, where he became President of the General Assembly. Leader of the Canadian *Liberal Party in 1958, he was Prime Minister from 1963 to 1968. His administration saw the implementation of a medical care programme and the adoption of a new Canadian flag (the red maple leaf) signifying the growing Canadian sense of national identity. Pearson's administration, which had no overall majority, was troubled by the problems of Quebec's separatist aspirations and the US struggle in Vietnam. He retired in 1968.

Pedro I (1798–1835), the first Emperor of Brazil (1822–31). The son of John VI of Portugal, he fled from Napoleon to Brazil. Recognizing that Brazilian independence from Portugal was inevitable, Pedro I led the revolt himself (1822) and then governed the new Brazilian monarchy under the executive powers of the constitution of 1824. Republican-inspired uprisings and nationalist resentment over his Portuguese connections undermined Pedro's rule; he abdicated (1831) in favour of his son *Pedro II.

Pedro II (1812–91), Emperor of Brazil (1831–89). The son of *Pedro I, he succeeded under a regency when his father abdicated (1831). The central government was unable to quell the uprisings at Balaiada (1838–41) and elsewhere until the General Assembly declared Pedro to be of age and confirmed (1840) his emperorship. Within eighteen months he had established order throughout the country. A popular, moderate leader, he was dedicated to the economic progress of Brazil. After a long rule, with only occasional revolts and foreign conflicts, the emperor eventually alienated his military officers by his refusal to grant them privileges, and the planters by his gradual abolition of slavery (completed in 1888). The army and the Republican Party engineered his overthrow (1889) and he spent the rest of his life in exile in Europe.

Peel, Sir Robert (1788–1850), British statesman. He entered Parliament as a Tory in 1809, holding office as Secretary for Ireland (1812–18). As Home Secretary (1822–7 and 1828–30) he made sweeping reforms in the penal code, removing about 100 offences from the list of crimes punishable by death, founding the Metropolitan Police, and helping to secure the passage of the act granting *Catholic emancipation. He opposed the Whig government's *Reform Bill of 1832 but in the *Tamworth Manifesto (1834) accepted the principle of moderate reform within the framework of existing institutions which was later to characterize Disraeli's new *Conservative Party. He was briefly Prime Minister in 1834–5, and

An 1829 cartoon attacks Robert **Peel** for his support of Catholic emancipation. As Home Secretary he is shown (*left*) waiting his turn behind Wellington, the Prime Minister, to do homage to the pope by kissing the papal foot. The unpopularity of the issue cost Peel his parliamentary seat for Oxford in the 1830 election.

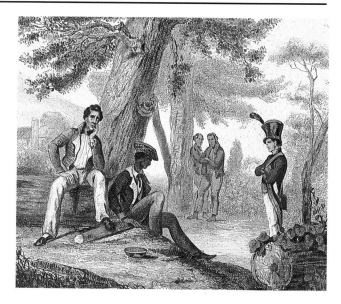

A 19th-century illustration of a group of convicts under guard in an Australian **penal settlement** in Van Dieman's Land (later renamed Tasmania). Transportation offered British authorities a seemingly expedient means of removing unwanted elements from society, at little cost.

during his second term of office from 1841 to 1846 he advanced the cause of *free trade by reducing import duties. He introduced the Bank Charter Act of 1844 which monopolized the issue of paper money and tried to link notes issued to the gold reserve. He is, however, best remembered for his decision in 1846, when faced with the miseries of the *Irish Famine, to repeal the *Corn Laws. This action split the Tories, and Peel was forced to resign.

penal settlements, Australian, settlements in 19th-century Australia for convicts who had committed further crimes within the colonies. Convicts at Newcastle, New South Wales, (1804–24) worked as coalminers, cedar-cutters, and lime-burners. Port Macquarie (1821–30) and Moreton Bay (1824–39) also were used as penal settlements. Norfolk Island, re-settled in 1825 as a penal settlement, became notorious. It held an average of 1,500 to 2,000 convicts, considered to be of the worst type. Punishment was harsh and a number of mutinies occurred. The last convicts left Norfolk Island in 1856. Port Arthur, in Van Diemen's Land, begun in 1830, was finally closed in 1877.

Pendleton Act (1883), an Act to reform the US civil service. Its purpose was to curb the discredited *spoils system and establish the principle of merit in public service. The assassination of *Garfield by a disappointed office seeker in 1881 stimulated agitation for the reform. The Act, named after its sponsor, Senator George H. Pendleton of Ohio, called on the President to appoint a standing Civil Service Commission of three members (not more than two from any one party), who were required to organize competitive examinations for prospective federal employees in Washington and elsewhere.

Peng Pai (1896–1929), founder member of the *Chinese Communist Party who pioneered peasant organizations. Peng ran the Haifeng Peasants' Association (1921–4) and set up a Peasants' Bureau which *Mao Zedong later directed and which successfully organized thousands of peasants in Guangdong province. Peng was executed by the nationalist authorities in 1929.

Peninsular War (1807–14), one of the *Napoleonic Wars, fought in Spain and Portugal. War was caused by *Napoleon's invasion of Portugal (1807) in order to compel it to accept the *Continental System. In 1808 the conflict spread to Spain, whose king was forced to abdicate, Napoleon's brother Joseph Bonaparte being placed on the throne. In June the Spanish revolted and forced the French to surrender at Baylen, whereupon Joseph fled from Madrid. In August Wellesley (later the Duke of *Wellington) landed in Portugal and routed a French force at Vimeiro and expelled the French from Portugal. In November Napoleon personally went to Spain, winning a series of battles, including Burgos, and restoring Joseph to the throne. British hopes of pushing the French out of Spain were destroyed in January 1809, after *Moore's retreat to Corunna. Despite his victory at Talavera, Wellesley withdrew to Lisbon. Here he built a strong defensive line which he centred at Torres Vedras.

In 1810 Napoleon sent Massena to reinforce *Soult and drive the British into the sea. Massena attempted to lay siege to Torres Vedras, but after four months his army, starved and demoralized, was forced to retreat. Soult, jealous of Massena's command, was slow in coming to his support, but managed to capture Badajoz. Wellington, who had pursued Massena and defeated him at Almeida, withdrew from invading Spain and turned to face Soult. During 1812 Wellington recaptured Badajoz and after defeating Massena's replacement, Marmont, at Salamanca, entered Madrid. The following year he defeated Joseph at the decisive battle of Vitoria. He went on to defeat Soult at Orthez and Toulouse (1814), having driven the French out of Spain.

penny post   *Hill, Sir Rowland.

Pentagon Papers, an official study of US defence policy commissioned (1967) to examine US involvement in south-east Asia. Leaked by a former government employee, they revealed miscalculations, deceptions, and unauthorized military offensives. Their publication provoked demands for more open government.

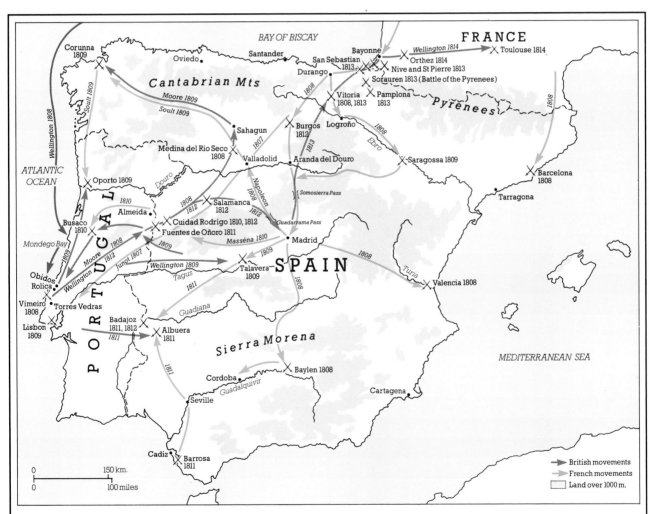

Peninsular War (1807–14)

In 1807, in order to impose the Continental System, Napoleon occupied Portugal. In 1808 he installed his brother Joseph on the Spanish throne. This sparked off a long guerrilla struggle. Napoleon personally intervened in 1809, but after his withdrawal the struggle continued. When French troops were withdrawn to fight in Russia, the British commander, Wellington, entered Madrid. In 1813 he won the decisive battle of Vitoria.

Pentrich Rising (1817), a quasi-political insurrection which took place in Derbyshire, England. Led by Jeremiah Brandreth, a framework knitter, a group of about 200 men from Pentrich and other nearby villages, armed only with primitive weapons, began to march on Nottingham in a protest against the government. Brandreth had been tricked by a government spy into believing that they were taking part in a nation-wide insurrection. They were dispersed by a troop of cavalry and Brandreth went into hiding. He was betrayed and later executed, together with two of his associates.

peon (South-American Spanish, a day-labourer), an unskilled labourer in rural areas of Latin America. Usually the term is associated with agricultural labour or peasants, and it is often linked to the more specific institution of debt peonage, a means of holding agricultural labourers on an *hacienda. By advancing wages to cover such expenses as religious ceremonies or clothing, landowners encouraged the accumulation of debts by

A 1947 photograph of Eva **Perón** reading the newspaper *Democrasia*, which she owned. Her political ability, allied to her natural glamour, made her a formidable figure. Through her re-allocation of financial resources, many hospitals, schools, and other welfare institutions were established.

workers who would then be obliged to remain in service until the debt was paid off. The degree to which debt actually signified bondage depended upon market conditions and the availability of labour.

'People's Budget', a controversial British budget in 1909, introduced by *Lloyd George, Chancellor of the Exchequer, to raise revenue for naval defence and social reform, particularly the funding of Old Age Pensions. Increased death duties and the imposition of taxes on land provoked the Conservative-dominated House of *Lords to reject this Liberal budget. By constitutional convention the Lords should have automatically approved all financial bills passed by the Commons. Their rejection led to the Parliament Act (1911), a statute that curtailed the power of the House of Lords and asserted the supremacy of the Commons on finance.

People's Party (USA) *Populist Party.

Perceval, Spencer (1762–1812), British statesman. He entered Parliament in 1796 as a supporter of William Pitt in the war against France. He was appointed solicitor-general in 1801 and attorney-general in the following year, holding this post until 1806. In 1807 he became Chancellor of the Exchequer in the Duke of Portland's administration, succeeding him as Prime Minister (1809–12). A moderately competent man and a strong evangelical, Perceval was shot dead in the lobby of the House of Commons by a deranged bankrupt with a grievance against the government.

Perkins, Frances (1882–1965), US politician and reformer. The first US woman cabinet member, she was made state Industrial Commissioner (1929–33) by Franklin D. *Roosevelt when he was governor of New York. In the *New Deal, against bitter business and political opposition, she became his Secretary of Labor (1933–45), playing an active role in minimum wage and maximum hours legislation and helping draft the Fair Labor Standards Act. Her influence was particularly seen in the second New Deal after 1935, where she gained increased responsibilities and much respect for her skilful administration.

Perón, Eva (Duarte de) (1919–52), Argentine political leader. A minor actress, 'Evita', as she was called, became active in politics and organized the mass demonstration of workers (1945) that secured Juan *Perón's release from prison. After her marriage to him (1945) she in effect ran the Ministries of Labour and of Health. By 1947 Eva Perón owned or controlled almost every radio station in Argentina, and had closed or banned over 100 newspapers and magazines. A gifted orator, she was a militant champion of women's rights. Her bid for the vice-presidency (1951) was blocked by the army, who opposed her. Her death from cancer at the age of 33 contributed to the decline of Perón's regime.

Perón, Juan (Domingo) (1895–1974), Argentine statesman. President of Argentina (1946–55, 1973–4), he was first elected with the support of labour and the military. An army officer who had favoured the fascist governments of Germany and Italy, he fashioned (1946) a revolutionary movement (*peronismo*), calling for a rapid economic build-up leading to self-sufficiency based on

the expansion and organization of the urban working class at the expense of agriculture. While trying at first with the support of his second wife, Eva *Perón to implement his programme, Perón became increasingly dictatorial. In September 1955, he was deposed by the armed forces. Despite powerful military opposition, Perón was recalled from exile to be re-elected in 1973. He died in office in July 1974, and was succeeded (1974-6) by his third wife, María Estela (Isabel) Martínez de Perón (1931-), who in turn was deposed (1975), and replaced by a military triumvirate.

Perry, Matthew Calbraith (1794-1858), US naval officer and pioneer of Western contact with Japan. Perry served under his older brother Oliver Perry during the *War of 1812. Given the rank of commodore, he headed an expedition to Japan entering the fortified port of Uraga in 1853, and Edo (modern Tokyo) Bay in 1854. His display of Western technology, both military and civil, forced the *shogunate to open two Japanese ports to US trade in the Treaty of *Kanagawa. Perry's mission initiated the process which, within half a century, would transform Japan from an isolated feudal country to a highly industrialized world power.

Pershing, John J(oseph) ('Black Jack') (1860-1948), US general. He served in the *Spanish-American War and later in the Philippines. In 1913 he became commander of the 8th Cavalry Brigade, and led the US expedition against Mexico in 1916. In May 1917 he was appointed commander of the American Expeditionary Forces in France, and his talent for organization was largely responsible for the moulding of hastily trained US troops into well-integrated combat troops. In 1919 he became general of the armies of the USA, and from 1921 to 1924 was army chief of staff.

Persia *Iran.

Peru, a country in South America. Sporadic Inca revolts followed the establishment of Spanish rule in the 16th century, the last in 1814. In 1821 José de *San Martín captured Lima, proclaiming an independent republic and issuing a constitution (1823). In 1824 José de *Sucre won the battle of Ayacucho, and Spanish troops were withdrawn. Political quarrels in the new republic led to an invitation to Simón *Bolívar to accept the powers of a dictator. He aimed unsuccessfully to bring Peru into his Gran Colombia. A long period of civil war followed, the situation stabilizing under President Ramón *Castilla (1844-62), who ended slavery, established an education system, and sponsored the guano industry (natural nitrates), which was to bring thirty years of prosperity. The loss of nitrate revenue and the cost of the War of the *Pacific (1879-84) with Chile led to national bankruptcy in 1889. Civilian politics had emerged in the 1870s with two parties, the Democrats and the Civilians, alternating in office. The latter, led by Augusto *Leguia, held power (1908-30), introducing much progressive legislation and settling the *Tacna-Arica Dispute. After World War I a more radical group, the Alianza Popular Revolucionacia Americana (APRA), led by *Haya de la Torre, sought to obtain greater participation in politics by the Indians. President Manuel Prado, elected in 1939, aligned Peru with US policies in World War II. Terry Belaúnde gained office in 1963. In 1968 a left-wing military junta seized power, seeking to nationalize US-controlled industries. A more moderate junta succeeded in 1975, and in 1979 elections were again held. In 1980 Belaúnde was re-elected President, when a new constitution was established. In the face of severe economic problems Belaúnde succeeded in re-democratizing the country, and in 1985 President Alan Garcia was elected. Confronted by massive rescheduling requirements for Peru's foreign debts, his regime imposed an austerity programme and was engaged in a guerrilla war against a strong ultra-left Maoist group, Sendero Luminosa ('Shining Path').

Above: The French general Henri-Philippe **Pétain** (*right*) in conversation with the Premier, Georges Clemençeau, at the front in 1917. During the battle of Verdun he had inspired in his troops a heroism that saved France from defeat. His grasp of military tactics was to help the Allies to final victory in World War I.

Below: The **Peterloo massacre** in St Peter's Fields, Manchester, in 1819 was seized upon by critics of the government. Dramatic illustrations, such as this, published in London in January 1820, showed innocent men, women, and children being trampled down by the 15th Hussars. The incident helped to fuel demand for a less authoritarian attitude towards popular demonstrations.

Pétain, Henri-Philippe (1856–1951), French general and head of state. Acclaimed a military hero for halting the German advance at *Verdun (1916), he replaced Nivelle as French commander-in-chief (1917). He later entered politics, becoming Minister of War (1934). In 1940 he succeeded *Reynaud as Premier, and concluded an armistice with Nazi Germany, which provided that the French forces be disarmed and that three-fifths of France be surrendered to German control. The French National Assembly established its seat at *Vichy and conferred on him the power to establish an authoritarian government. He designated *Laval as his Vice-Premier and Foreign Minister, but later dismissed him for too closely collaborating with Germany. German forces entered unoccupied France, and Pétain was forced to reinstate Laval. Thereafter his equivocal dealings with Allies and Germans can only be excused on grounds of failing powers. Arrested (1945) and tried as a collaborator, the death sentence was commuted to life imprisonment by General *de Gaulle.

Peterloo massacre (16 August 1819), the name given by government critics to a violent confrontation in Manchester, England, between civilians and government forces. A large but peaceable crowd of some 60,000 people had gathered in St Peter's Fields to hear the radical politician Henry 'Orator' *Hunt address them. After he had begun speaking the local magistrates sent in constables to arrest him. In the mistaken belief that the crowd was preventing the arrest, the magistrates ordered a body of cavalry to go to the assistance of the constables. In the ensuing riot 11 civilians were killed and over 500 injured. The immediate response of *Liverpool's government was the passing of the repressive *Six Acts, but the incident provoked widespread criticism, cartoonists comparing the 'massacre' to that of Waterloo.

Philippines, south-east Asian country. A Spanish colony since the 16th century, a nationalist uprising against Spain broke out in Manila in 1896, led by José *Rizal. In 1898, during the *Spanish–American War, General Emilio *Aguinaldo, acting with the support of the USA, declared the country's independence. After Spain's defeat, however, the nationalists found themselves opposed by the Americans, and after a brief war (1899–1901), the islands passed under US control. Internal self-government was granted in 1935, and, after the Japanese occupation during World War II, the Philippines became an independent republic in 1946 under the Presidency of Manuel Roxas, with the USA continuing to maintain military bases. Successive administrations proved incapable of dealing with severe economic problems and regional unrest. In 1972, using the pretext of civil unrest, in particular the communist guerrilla insurgency conducted by the New People's Army in Luzon, and violent campaigns of Muslim separatists, the Moro National Liberation Front, in the southern Philippines, President *Marcos declared martial law, assuming dictatorial powers. While the Marcos regime achieved some degree of success in dealing with both economic problems and guerrilla activities, the return to democratic government was never satisfactorily achieved. After the murder of the opposition leader, Benigno Aquino Jr, in 1983, resistance to the Marcos regime coalesced behind his widow Corazon Aquino and the United Nationalist Democratic Organization. US support for the Marcos government waned and in 1986, after a disputed election and a popularly backed military revolt, Marcos fled, and Corazon Aquino became President in his place, restoring the country to a fragile democracy.

Philippines Campaign (1944–5), the US campaigns that recaptured the Philippines in World War II. In the battle of the Philippine Sea, fought in June 1944 by aircraft carriers while US forces were securing required bases in the Marianas, the Japanese naval air service suffered crippling losses. A further Japanese naval defeat was incurred at *Leyte Gulf in October, in a vain attempt to prevent US forces landing in the Philippines. In July 1945 *MacArthur announced that the territory was liberated, although detached groups of Japanese, in accordance with their instructions to fight to the last man, were still at large after the war ended.

Phoenix Park murders (1882), the assassination in Phoenix Park, Dublin, of the British Chief Secretary for Ireland, Lord Frederick Cavendish, and his Under-Secretary, T. H. Burke, by members of the 'Invincibles', a terrorist splinter group of the *Fenians. In the subsequent climate of revulsion against terrorism, *Parnell was able to gain ascendancy over the Irish National League and strengthen the more moderate *Home Rule Party.

Pibul Songgram (1897–1964), Thai statesman. A career soldier, he took part in the bloodless coup which ended absolute rule by the Chakri dynasty in 1932. Emerging as the leader of militarist nationalist forces, he became head of state in 1938 and in January 1942 brought Thailand into World War II on the Japanese side. Overthrown in 1944, he returned to power in 1948 and controlled Thailand dictatorially until 1957, taking an anti-communist line, until overthrown after the corruption of his regime had aroused widespread resentment.

Piedmont, a region of Italy centred around Turin, ruled by the dukes of Savoy, who in 1720 became also the kings of Sardinia. Following the Congress of *Vienna in 1815 Victor Emanuel I returned to Turin from Sardinia. From 1831 under King Charles-Albert, Piedmont became the centre of the movement for Italian unification. Parts of Savoy were forfeited to France in 1860, but in 1861 *Victor Emanuel II was proclaimed king of a united Italy.

Pierce, Franklin (1804–69), fourteenth President of the USA (1853–7). He gained the Democratic presidential nomination in 1852 as a compromise candidate at a time of bitter party divisions. In office, Pierce supported an expansionist foreign policy and was generally under the influence of the southern wing of the Democratic Party. His support for the *Kansas–Nebraska Act lost him the support of northern Democrats and any chance of renomination in 1856.

Pilsudski, Joseph Klemens (1867–1935), Polish general and statesman. Early revolutionary activity against Tsarist Russia had led to his imprisonment. In World War I he raised three Polish legions to fight Russia, but German refusal to guarantee the ultimate independence of Poland led him to withdraw his support of Germany. After the war Poland was declared independent with Pilsudski as Chief of State (1918–22) and Chief of the Army Staff (1918–27). He successfully commanded the Poles in the war against the *Bolsheviks (1919–20). In 1926, after a military revolt, he assumed the office of Minister of Defence, establishing a virtual dictatorship, and tried to guarantee Poland's independence by signing non-aggression pacts with Germany and the Soviet Union in 1934. He died in office.

Pinckney, Charles Cotesworth (1746–1825), US statesman. He was appointed minister to France in 1796 and a year later participated in the unsuccessful diplomatic peace mission to avert a naval war with France, known as the XYZ Affair. Pinckney ran with John Adams for the *Federalist Party and was the unsuccessful Federalist candidate for President in 1804 and 1808. His brother Thomas was also a diplomat and was responsible for negotiating the Treaty of San Lorenzo (popularly known as Pinckney's Treaty) with Spain in 1795, winning US trading rights at New Orleans, and gaining Spanish acceptance of US frontier claims east of the Mississippi and in Florida.

Pindaris, groups of Indian mounted marauders in central India during the early 19th century, often in the service of the Maratha princes. After 1807 they extended their depredations to British India, and under Lord Hastings (governor-general 1813–23) a major campaign (*Maratha Wars) was conducted (1817–18) to wipe out the Pindaris. The campaign developed into a struggle involving some of the Maratha princes and contributed to the establishment of British paramountcy.

Pinkerton, Allan (1819–84), US detective, the first in the Chicago police force (1850). Later that year he opened a private detective agency. During the *American Civil War, he organized an intelligence network behind the lines of the *Confederacy, which was one of the forerunners of the US secret service. His inflated estimation of enemy strength influenced General McClellan. The Pinkerton detective agency itself continued to flourish, despite gaining a reputation as a force of strikebreakers in the late 19th century.

Pinochet, Augusto (1915–), Chilean statesman. Bitterly opposed to the left-wing policies of President *Allende, he master-minded a military coup against him. When Allende died (September 1973) in the revolt, Pinochet became President of the Council of Chile (a junta of military officers) and imposed a harsh military rule for three years, during which some 130,000 people were arrested, many tortured, and thousands never seen again. Proclaimed President of Chile in 1974, he held plebiscites in 1978 and 1980 to confirm this office. Under a new constitution of 1980 he was proclaimed President again for a seven-year term.

Pioneer Column, a force formed by Cecil *Rhodes in 1890 in Rhodesia to occupy Mashonaland (*Zimbabwe). With the British explorer F. C. Selous as guide, 180 men under the command of Major F. Johnson crossed the River Maklautsi in Bechuanaland (Botswana) and raised the British flag on a small hill, naming the place Salisbury (now Harare) after the then Prime Minister. On 1 October 1890 the force was disbanded, each member being awarded a farm, together with other privileges.

Pitt, William (Pitt the Younger) (1759–1806), British Prime Minister, second son of Pitt the Elder. He became Chancellor of the Exchequer at the age of 23, when he began his long rivalry with *Fox. On the overthrow of the Fox–North coalition in 1783, Pitt became Prime Minister, a post he was to hold until 1801 and again from 1804 to 1806. During the years of peace until 1793 Pitt was nurturing Britain's economic recovery from the American War of Independence, and he showed his brilliance in his handling of finance. From 1793 he symbolized Britain's resistance to the French Revolution and Napoleon, but although he raised three European coalitions and gave subsidies to Britain's allies, victory eluded him. When George III refused to consider *Catholic emancipation in exchange for the *Act of Union, he resigned (1801). Pitt suppressed British radicalism as readily as he suppressed Irish revolution, and although he called himself an independent Whig, his followers, after his early death, were to form the nucleus of the emerging Tory party.

Pius VII (Luigi Barnaba Chiaramonti, 1740–1823), Pope (1800–23). His predecessor, Pope Pius VI, had seen the Papal States occupied by the French and died a prisoner in France (1799). Pius VII restored papal fortunes by signing a *concordat with Napoleon I in 1801, which re-established the Roman Catholic faith as the national religion of France. However, his attempts to increase papal influence by refusing to support the *Continental System against Britain led to the annexation of the Papal States in 1808–9. He, too, was imprisoned by Napoleon in France, but returned in triumph after Napoleon's downfall. He renounced his earlier liberalism, condemned revolution and secret societies, and re-established the Society of Jesus (Jesuits) in 1814.

Pius IX (Giovanni Maria Mastai-Ferretti, 1792–1878), Pope (1846–78). Elected as a moderate progressive, he relaxed press censorship, freed political prisoners, set up a council of ministers which included laymen, and opened negotiations for an Italian customs union. After the *Revolutions of 1848 his Prime Minister was murdered and Pius himself fled in disguise. On the establishment of a Roman republic Pius appealed to the French to come to his aid, and, with their help, he re-entered Rome, which retained a French garrison until 1870. In that year Italian forces occupied Rome, which was then incorporated into the kingdom of Italy. Despite government assurances, Pius saw himself as a prisoner in a secular state, and never set foot outside the Vatican again. In 1854 he defined the doctrine of the Immaculate Conception of the Virgin Mary (the Mother of Jesus Christ) and encouraged the Marian cult. He presided over the First Vatican Council (1869–70) which proclaimed the infallibility of the Pope when speaking *ex cathedra*.

Pius XII (Eugenio Maria Giuseppe Giovanni Pacelli, 1876–1958), Pope (1939–58). As Secretary of State for *Pius XI he negotiated a *concordat (1933) with Nazi Germany, whose repeated violations by *Hitler led to the encyclical, *Mit brennender Sorge* (1937), branding Nazism as fundamentally anti-Christian. As pope he remained politically impartial between the Allied and the *Axis Powers in World War II. He supervised a programme for the relief of war victims through the

The end of the battle of the River **Plate**. The German battleship *Graf Spee* is left a burning hulk after being scuttled by her crew off Montevideo, 1939.

Pontifical Aid Commission and made the *Vatican City an asylum for refugees. While unwilling to speak out in public against Nazi atrocities, he gave assistance to numbers of individual Jews. Politically he inveighed against communism, threatening (1949, 1950) its supporters with excommunication, and concluded accords advantageous to the Roman Catholic Church with Portugal (1950) and Spain (1955).

PKI (after 1924, Partai Kommunis Indonesia), Indonesian Communist Party. Formed in 1920, the PKI was active in trade-union activities, but Dutch repression and the abortive communist uprisings in Banten and West Sumatra in 1926–7 led to its eclipse until 1945. During the 1950s, the PKI became one of the largest communist parties outside China and the Soviet Union, winning 20 per cent of the vote in the 1955 general elections. Rivalry with the army and communist-inspired actions against Muslim landlords in Java, provoked the military to move against it after the coup attempt of 1965. Up to one million PKI members were killed and the party all but wiped out.

Place, Francis (1771–1854), British radical reformer. In 1799 he opened a tailoring shop in London which became a meeting place for leading radicals. In 1814 he began his campaign for the repeal of the anti-trade union *Combination Acts, which were eventually abolished in 1824. Place later took a prominent part in the agitation leading to the 1832 *Reform Act and also helped draft the People's Charter of 1838 (*Chartism).

Plaid Cymru, a political party devoted to the cause of Welsh nationalism. Founded in 1925 as Plaid Genedlaethol Cymru (Welsh Nationalist Party), it seeks to ensure independent recognition for *Wales in matters relating to its culture, language, and economy. It became active in the 1960s and 1970s, but its hope that Wales would be able to have a separate representative assembly was rejected by a referendum in Wales in 1979. Plaid Cymru has not succeeded in wooing the majority of Welsh electors, particularly in the towns, from their support of the major British parties.

Plate, battle of the River (13 December 1939), a naval action between British and German forces in the South Atlantic. It was the first major naval surface engagement of World War II, in which the German battleship *Graf Spee*, which had sunk many cargo ships, was damaged by three British cruisers and forced into the harbour of Montevideo, from which she emerged only to be scuttled by her crew on Hitler's orders.

Platt, Thomas Collier (1833–1910), US politician. He was a prominent member of the 'Stalwart' faction supporting the *spoils system. Platt was elected to Congress, serving in the US Senate in 1881, though he resigned shortly afterwards in protest at President *Garfield's appointment of reform Republicans to federal posts. Again in the Senate (1897–1909), he was largely responsible for the election (1898) of Theodore *Roosevelt as governor of New York. Threatened by the latter's drive against corrupt political practices, Platt attempted to suppress his initiative by securing him the vice-presidency (1901). Thereafter Platt's power declined.

Plekhanov, Georgi Valentinovich (1857–1918), Russian revolutionary and Marxist theorist. He became a leader of the *populist organization, Land and Liberty, in 1877, but when this turned increasingly to terrorist methods, he formed an anti-terrorist splinter group to continue mass agitation. Exiled in Geneva, he became one of the founders of the League for the Liberation of Labour, the first Russian Marxist revolutionary organization (1883), which merged (1898) with the Russian Social Democratic Workers' Party. In the 1903 split with *Lenin he supported the Mensheviks but always tried to re-unite the party. He returned to Russia (1917), but failed to prevent the *Bolsheviks from seizing power.

PLO *Palestine Liberation Organization.

'Plug' strikes (1842), the name given to an outbreak of strikes accompanied by rioting in the north of England. The agitation began as a protest by Staffordshire miners against cuts in wages, but by August of that year had spread to industries in other parts of the country. The men involved went from factory to factory, removing the boiler plugs from steam engines so that these could not be operated by anyone else. There were also attempts to use the strike as a weapon to compel the government to accept the *Chartist petition. Hundreds of strikers were arrested and the ringleaders transported to Australia.

pocket borough, a British Parliamentary borough effectively under the control of a single wealthy individual or family. Before the 1832 *Reform Act there was no uniform basis for the parliamentary franchise in towns, and the right to vote tended to be limited to a small number of people. This made it easier for a local magnate to ensure the election of any candidate he chose to put forward. The borough was thus said to be 'in his pocket'. Such pressure on voters was not effectively ended until the introduction of secret ballots in 1872.

pogrom (Russian, 'riot' or 'devastation'), a mob attack approved or condoned by authority, frequently against religious, racial, or national minorities—most often against Jews. The first occurred in the Ukraine following the assassination of *Alexander II (1881). After that, there were many pogroms throughout Russia, and Russian Jews began to emigrate to the USA and western Europe, giving their support to Theodor *Herzl's *Zionist campaign. After the unsuccessful revolution of 1905, *anti-Semitic persecutions increased in number and force. Conducted on a large scale in Germany and eastern Europe after Hitler came to power, they led ultimately to the *holocaust.

Poincaré, Raymond (1860–1934), French statesman. As President (1913–20) he strove to keep France united during *World War I and in 1919 supported stringent *reparations against Germany. When Germany defaulted, as Premier he ordered French troops to occupy the *Ruhr (1923) until Germany paid. He could not sustain this policy and he resigned (1924). Premier again (1926), he lessened an acute economic crisis by introducing a deflationary policy, balancing the budget, and securing the franc (1928) at one-fifth of its former value.

Point Four Program, a US aid project, so called because it developed from the fourth point of a programme set forth in President *Truman's 1949 inaugural address, in which he undertook to make 'the benefits of America's scientific and industrial progress available for the improvement and growth of under-developed areas'. In the following years Congress annually provided technical assistance for the long-term development of industries, agriculture, health and education in developing countries. The project also encouraged the flow of private investment capital to the poor nations.

Poland, a country in eastern Europe. Poland was at its largest in the 16th century, but was eradicated in the partitions by Prussia, Russia, and Austria in 1772, 1793, and 1795. Following the treaties of *Tilsit in 1807 Napoleon created the Grand Duchy of Warsaw, under the King of Saxony, introducing the *Code Napoléon, but retaining serfdom and the feudal nobility. The duchy collapsed after the battle of *Leipzig and at the Congress of *Vienna, when Poland was represented by Count *Czartoryski, parts of the duchy reverted to Prussia and Austria, but the bulk became the kingdom of Poland, with its own administration but with the Russian emperor Alexander I as king. Revolutions took place in 1830, 1846–9, and 1863. Serfdom was ended in 1864, but policies of repression followed in both Russian and Prussian Poland. This did not, however, prevent the development of political parties demanding democratic government. After World War I in 1918 full independence was granted and Poland became a republic. War against Bolshevik Russia (1920–1) was followed by the dictatorship of Marshal *Pilsudski. Poland was to have access to the port of Danzig (Gdańsk) via a *Polish Corridor. The status of Danzig and the existence of this corridor provided an excuse for the Nazi invasion in 1939, which precipitated World War II. As a result of the *Nazi–Soviet Pact, Poland lost territory to both countries. After 1945 two million Germans left East Prussia (now in Poland) for the Federal Republic of Germany, and Poles, mainly from those Polish territories annexed by the Soviet Union, were re-settled in their place. Following the *Warsaw Rising a provisional Polish government was established under Red Army protection, which co-operated with *Stalin to bring the country within the Soviet bloc. Political opposition was neutralized, and in 1952 a Soviet-style constitution was adopted. In 1956 Polish workers went on strike to protest against food shortages and other restrictions. Under Wladyslaw Gomulka (1956–70) rigid control by the government was maintained, leading to further strikes (1970), which were again suppressed by military force. The election of a Polish pope, Karol Wojtyla, as John Paul II in 1978, strengthened the influence of the *Roman Catholic Church in the country. Strikes, organized by the illegal free trade union *Solidarity erupted at Gdańsk shipyard (1980). Martial law was declared (1981) under General Wojciech Jaruzelski. Despite an official end to martial law, military tribunals continue to operate. In 1987 the government, beset by severe international debt problems, put forward plans for limited decentralization.

police, body of civilian officers with specific supervisory duties to maintain public order. By the late 18th century large cities, such as Paris and London, had increasing problems of lawlessness and crime. In France, there had developed a centralized system of police spying, which developed into a national police force in the 19th century. In London the Magistrates' Court at Bow Street pioneered from c.1750 a system of unarmed uniformed 'runners' employed to apprehend criminals. These were developed by Sir Robert *Peel in 1829 by the formation of the London Metropolitan Police Force, soon copied in most other major British cities. All countries have developed police forces, and in most the role of the military is now sharply distinguished from that of the police. The majority of West European countries and all East European countries followed the French pattern of a national police force, and this was adopted by most developing countries. However, in Britain, Denmark, the USA, Canada, and elsewhere a series of local forces has survived, in spite of strong pressures for more centralization. Non-uniformed 'plain-clothes' detective and intelligence agencies also developed in all countries, these usually being controlled nationally, as with the *FBI in the USA and, sometimes, as with the *KGB in the Soviet Union, having specific political functions. Police states, totalitarian in practice, control the activities of their citizens by means of a national police force.

Polish Corridor, the belt of territory separating East Prussia from the rest of Germany and granted to *Poland by the *Versailles Peace Settlement (1919) to ensure access to the Baltic Sea. It contained the lower course of the River Vistula, except for Danzig (Gdańsk), and three other towns. Historically, the territory had belonged to Polish Pomerania in the 18th century, but it had been colonized by a German minority. In 1939 Hitler's forces annexed the Polish Corridor, Danzig, Posen, and districts along the Silesian frontier, and placed the rest of Poland under a German governor. As a result of this Britain and France declared war on Germany, and World War II began. After the war the territory reverted to Poland.

Politburo, the highest policy-making committee of the USSR and of some other communist countries. The Soviet Politburo was founded, together with the Ogburo (Organizational Bureau), in 1917 by the leading *Bolsheviks to provide continuous leadership during the *Russian Revolution. After the revolution both bureaux were re-formed to control all aspects of Soviet life. They were disbanded in 1952 and the Politburo was renamed the Praesidium. In 1966 its name reverted to Politburo, or bureau of party leadership, to distinguish it from the Praesidium of the Supreme Soviet, elected by universal suffrage.

Top-hatted Peelers of the London **police** force, instituted by Sir Robert Peel in 1829, take to the streets. Their arrival creates alarm among London's street boys.

Pol Pot (*second in line*), whose name is now synonymous with the genocide of the Kampuchean people, seen here with Khmer Rouge followers in Kampuchea, June 1979. After his overthrow, the former Premier continued a guerrilla war against the new Vietnamese-backed regime.

Polk, James Knox (1795–1849), eleventh President of the USA (1845–9). A Jacksonian Democrat, Polk won the presidential election of 1844 on a *Manifest Destiny ticket. His term of office resulted in major territorial additions to the USA. Apart from the successful prosecution of the *Mexican–American War, and the acquisition of California and the south-west, he witnessed the re-establishment of an independent treasury and the settlement of the *Oregon Boundary Dispute (1846).

Pol Pot (1928–), Kampuchean leader. Trained as a Buddhist monk and educated at a French university, he joined the anti-French resistance under *Ho Chi Minh and rose to a high position within the Cambodian communist movement, supported by the People's Republic of China. After the *Khmer Rouge had overthrown the Lon Nol regime, he succeeded *Sihanouk as Prime Minister in 1976 and presided over the 'reconstruction' of the country in which as many as two million Cambodians may have been killed. Overthrown in 1979, he led the Khmer Rouge until ill-health forced his semi-retirement in 1985.

Pompidou, Georges Jean Raymond (1911–74), French statesman. He served in the *resistance movement in World War II and, from 1944, became an aide and adviser to *de Gaulle. While the latter was President, Pompidou held the post of Prime Minister (1962–8) and played an important part in setting up the *Évian Agreements. The strikes and riots of 1968 prompted de Gaulle's resignation (1969) and Pompidou was elected President. In a swift and decisive policy change he devalued the franc, introduced a price freeze, and lifted France's veto on Britain's membership of the *European Economic Community.

Pony Express (1860–1), horse-borne mail delivery system in the 19th-century US west. It was founded in 1860 by the Missouri freight company of Russell, Majors, and Waddell to prove that there was a viable alternative to the southern route into California for the year-round transportation of overland mail. It operated between St Joseph, Missouri, and Sacramento, California, through Cheyenne, Salt Lake City, and Carson City. It used a relay of fresh ponies and riders, and took two weeks to cover the full distance of nearly 3,200 km. (2,000 miles). High costs made the operation unprofitable, and the coming of the telegraph made it unnecessary, but the Pony Express is still remembered as one of the most picturesque episodes in the opening up of the US frontier.

Poor Laws, legislation in Britain concerned with relief of the poor. By the late 18th century the Elizabethan Poor Law system, administered through parish overseers, was increasingly inadequate when people began travelling many miles in search of employment. In some rural areas, in addition to direct relief to those without work, there developed the so-called Speenhamland system (named after the district in Berkshire where the system originated in 1795), whereby the wages of low-paid workers were supplemented by a parish allowance. Such a system was clearly unsatisfactory, since it tended both to depress wages and to subsidize landlords and employers. In 1834 a Poor Law Amendment Act tried to end the giving of assistance outside the workhouse; it established the principle that all citizens should have the right to relief from destitution through accommodation. But conditions in the workhouse, often with families separated by sex and age, were to be so severe as to discourage all but the most needy from its care. The workhouses were run by locally elected Boards of Guardians, who raised money through a poor-rate. The system proved inadequate in the growing cities, where the Guardians sometimes resorted to relief without the guarantee of accommodation. The Poor Law was gradually dismantled by social legislation of the 20th century, particularly that of the Liberal governments (1906–14) by important Acts in 1927, 1929 (when Boards of Guardians were abolished), 1930, 1934 (when Unemployment Assistance Boards were created), by *social security legislation following the *Beveridge Report (1942), and by the establishment of the *welfare state.

Popular Front, a political coalition of left-wing parties in defence of democratic forms of government believed threatened by right-wing fascist attacks. Such coalitions were made possible by the strategy adopted by the *Comintern in 1934. In France such an alliance gained power after elections in 1936, under the leadership of Leon *Blum, who implemented a programme of radical social reforms. In Spain the Popular Front governments of Azaña, Caballero, and Negrin were in office from 1936 to 1939, and fought the *Spanish Civil War against *Franco and the Nationalists. A Popular Front government ruled in Chile (1938–47).

population, world, the number of people alive in the world at a given moment in time. Between the collapse of the Roman empire in the 5th century AD and the late 18th century world population appears to have remained fairly stable, with high death rates and occasional plagues. Nineteenth-century Europe saw an unprecedented rise in population, largely because of improved public health and control over infectious *disease. In 1803 Thomas *Malthus propounded the theory that the rapidly growing population would soon increase beyond the capacity of the world to feed it, and that controls on population were therefore necessary to prevent catastrophe. Japan instituted the first modern official birth-control policy after World War II, setting an example which other countries followed. Population growth has slowed down in the more developed countries, while rising steeply in the rest of the world. Since 1950 there has been a population explosion in Asia, Africa, and Latin America. In Kenya, which has the highest birthrate in the world, more than 50 per cent of the population is now under 15. By the late 1980s the total world population was 5,000 million. Although the steep increase is expected to decline, world population is likely to reach 6,000 million by AD 2000, posing serious political and economic problems.

population migrations and emigrations, large movements of peoples across and between continents, sometimes under duress. The major 19th-century movement was the great Atlantic migration from Europe. This took some 5 million people to Canada, mostly from Britain; some 15 million to South America, mostly from the Mediterranean countries; and some 36 million to the USA (1800–1917), which received succeeding waves from Ireland (c.4.4 million), and the continent of Europe. In

the first half of the 19th century the white settlers in turn forced the migration of American Indian tribes to reservations. In total the movement across the Atlantic involved the greatest number of people ever to migrate, the climax being in the decade 1901–10, when c.9 million arrived, many via *Ellis Island in the USA. Australian and New Zealand emigration was mostly from Britain until the *New Australians after World War II. The empty lands of the Russian empire east of the Urals in the 19th and 20th centuries steadily attracted settlers (some 20 million), of whom some 2 million were political exiles in *prison camps, as convict labour. Chinese migration south accelerated in the 19th century (perhaps 12 million), often with incentives from the European colonial powers, to Vietnam, Thailand, Malaysia, Singapore, and Indonesia. The European colonial powers also moved 'indentured labour' between colonies, where there were labour shortages; for example, Tamils to Ceylon (now Sri Lanka) and other Indians to South Africa and the West Indies. Most 20th-century migrations have been forced, either by political coercion or famine. *Stalin's policy of forcible resettlement of ethnic groupings, such as the Crimean Tartars to Central Asia, left a legacy of national unrest within the Soviet Union. Some 17.5 million refugees resulted from the partition (1947)

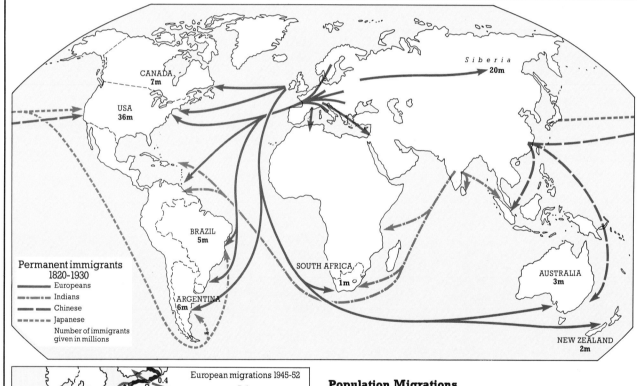

Permanent immigrants 1820-1930
- Europeans
- Indians
- Chinese
- Japanese

Number of immigrants given in millions

European migrations 1945-52
- Poles
- Germans
- Baltic peoples
- Russians
- Czechs
- Italians and Turks
- 'Iron Curtain'

Figures are given in millions

Population Migrations

The last two centuries have seen population shifts on an unprecedented scale. While 19th-century migrants were mostly impelled to leave their countries by the promise of seemingly unlimited land to be cultivated, grazed, or mined in the Americas, Australia and, to a lesser degree, Africa, their 20th-century counterparts were to a large degree victims of wars and persecutions: c.17 million were refugees within the sub-continent of India, c.30 million lost their homelands in Europe, and c.2 million Jews entered the newly created state of Israel after 1948.

of India and *Pakistan; over 10 million Germans were moved from eastern Europe by Soviet authorities after World War II; the creation of *Israel (1948) resulted in some 700,000 Arab refugees. African countries have witnessed many refugees, especially those from famine, while the process of *urbanization has drawn millions of people from rural hinterlands all over the world into the ever-expanding cities.

Populist Movement (Russia), a group of agrarian socialists in Russia devoted to radical reform and government by small economic units resembling village communes. It was active in the latter half of the 19th century, first under the name of Land and Liberty, when members such as *Herzen, *Bakunin, Lavrov, *Plekhanov, and Chernyshevsky plotted the overthrow of the Tzarist regime. Dissatisfied with the government reforms, the radicals were soon persecuted by the police. In 1879 the most radical wing of the Populist group re-formed under the name of the People's Will Movement, and began to adopt terrorist tactics which culminated, two years later, in the assassination of *Alexander II.

Populist Party (USA), a US agrarian organization. It began in 1889 as a grouping of southern and western interests seeking to remedy the lot of debtor farmers. It drew on the *Granger Movement, the Farmers' Alliances, the *Greenbacks, and other protest groups who met in Cincinnati to create the People's Party of the USA. Its members called for a flexible currency system under government control, a graduated income tax, and political reforms including direct election of US Senators. In 1892 its candidate for President, James B. Weaver, won over a million popular and 22 electoral votes. From this point the movement declined, largely because its objectives seemed more likely to be realized by other parties.

Portal, Charles Frederick Algernon, Viscount Portal of Hungerford (1893–1971), Marshal of the Royal Air Force. In July 1915 he joined the Royal Flying Corps, and by 1937 he was an air vice-marshal and Director of Organization at the Air Ministry. In April 1940 he was placed in charge of Bomber Command. The aircraft available had many technical deficiencies, especially in navigation, but by carrying the *bombing offensive into Germany, they disrupted munitions factories, power plants, and railway junctions. While introducing technical improvements, he also pressed for a policy of 'area bombing' to replace that of specific targets. After the war he became Controller of Atomic Energy in Britain (1945–51).

Portales, Diego (1793–1837), Chilean statesman. He entered the interim conservative government in 1830 and held several ministerial offices. He was responsible for reforms in the army, treasury, civil service, commerce, and industry. He assisted in the drafting of the new constitution of 1833, a conservative document, giving great powers to the President and limiting the suffrage to the larger property owners. He returned to office in 1835, but was increasingly autocratic. He persuaded the President to declare war against Peru in June 1837 over the non-payment of debt. This was unpopular in the army, which mutinied and assassinated him.

Portland, 3rd Duke of　*Bentinck.

Portsmouth, Treaty of (1905), treaty ending the *Russo-Japanese War (1904–5). Although the Russians had been decisively defeated on land and at sea in 1904, it was the intervention of the US President Theodore *Roosevelt which finally brought a successful end to the Russo-Japanese War. The treaty, signed at Portsmouth, New Hampshire, allowed for the mutual evacuation of Manchuria but granted Japan railway rights in southern Manchuria, Russian acknowledgement of Japanese supremacy in Korea, and the ceding to Japan of the Liaodong Peninsula (including Port Arthur, now Lüshun) and the southern half of Sakhalin. Russian eastward expansion was thus halted and Japanese hegemony in north-east Asia confirmed.

Portugal, a European country in the western part of the Iberian peninsula. During the *Napoleonic Wars the Prince Regent John (King John VI from 1816), together with the Braganza royal family, fled to Brazil. Here he met demands for political and economic freedom, Brazil emerging peacefully as an independent empire in 1822. Through most of the rest of the 19th century there was considerable political instability until 1910, when a republic was established. In 1926 there was a military coup which was followed in 1932 by the establishment of Antonio de Oliveira *Salazar as Prime Minister, Minister of Finance, and virtual dictator (1932–68), strongly supported by the Roman Catholic Church. Portugal supported the Allies in World War I and in World War II remained theoretically neutral while allowing the Allies naval and air bases. After the war Goa, Diu, and Damao were lost to India, but Macao in South China was retained. Salazar's autocratic policies were continued by Marcello Caetano until a military coup in 1974. Increasingly bitter guerrilla warfare had developed in Portuguese Africa, especially in *Angola and *Mozambique. These gained independence in 1975, although both then experienced civil war, while the tiny state of *Guinea Bissau was created in 1974. After two years of political instability at home, a more stable democracy began to emerge following the election of Antonio Eanes as President in 1976. Moderate coalition governments both left and right of centre have alternated, all struggling with severe economic problems. President Mario Soares was elected in 1986, having been Prime Minister since 1983.

Potatau Te Wherowhero (d. 1860), Maori leader. Widely respected for his chiefly status, learning, and warrior prowess, Potatau was chosen in 1858 by Waikato and central North Island tribes as their first king in the hope of uniting all the Maori tribes. Potatau supported the *Kingitanga's moves to resist land selling and to secure recognition from the government of *rangatiratanga* (chieftainly authority). Up to his death he tried to avoid involving the Kingitanga directly in the *Taranaki war which broke out in 1860. Potatau's lineal descendants still head the Kingitanga.

Potemkin, the battleship whose crew mutinied in the *Russian Revolution of 1905 when in the Black Sea. The crew killed some officers, and took control of the ship. As other ships in the squadron were close to mutiny it was thought unwise to open fire on her. Unhindered, she bombarded Odessa before seeking asylum in Romania. The incident persuaded the emperor to agree to the election of a Duma, but unrest continued until the issue of the October Manifesto and a new Russian constitution.

Potsdam Conference (17 July–2 August 1945), the last of the World War II summit conferences. Held in the former Hohenzollern palace at Potsdam, outside Berlin, it was attended by *Churchill (replaced by *Attlee during its course), *Stalin, and *Truman. It implicitly acknowledged Soviet predominance in eastern Europe by, among other things, accepting Polish and Soviet administration of certain German territories, and agreeing to the transfer of the German population of these territories and other parts of eastern Europe (over ten million people) to Germany. It established a Council of Foreign Ministers to handle peace treaties, made plans to introduce representative and elective principles of government in Germany, discussed reparations, decided to outlaw the Nazi Party, de-monopolize much of German industry, and decentralize its economy. The final agreement, vaguely worded and tentative, was consistently breached in the aftermath of German surrender, as the communist and capitalist countries polarized into their respective blocs. The Potsdam Declaration (26 July 1945) demanded from Japan the choice between unconditional surrender or total destruction.

Potter, Beatrice *Webb, Sidney.

Prasad, Rajendra (1884–1963), Indian nationalist politician. A lawyer by profession, he began working with Mohandas *Gandhi in 1917. He was imprisoned by the British (1942–5) for supporting the *Congress opposition to the British war effort in World War II. He represented the conservative wing within Congress, of which he was President on four occasions between 1932 and 1947. Prasad became President of India (1950–62) when the republic was proclaimed.

Prempeh I (d. 1931), African leader, Asantehene (Chief) of the *Asante. He was elected in 1888, but deposed by the British in 1896. In 1924 he was allowed to return to Kumasi, and in 1926 was installed as Kumasihene, a simple divisional chief. On his death in 1931 his nephew, Prempeh II, was elected Kumasihene, and then Asantehene in 1935, when the Golden Stool was returned and the traditional Asante Confederacy was restored.

Pretorius, Andries (Wilhelmus Jacobus) (1798–1853), Boer leader and general. After several frontier campaigns against the Xhosa he took part in the *Great Trek, and became commandant-general of the Boers after Piet Retief's murder (1838). He defeated the Zulu at the *Blood River (1838). In 1847 he organized protests against the British annexation of the land between the Orange and Vaal, but in 1848 he was defeated at Boomplaats by Sir Harry Smith, when the Orange River Sovereignty was established. In 1852 he was instrumental in negotiating the Sand River Convention, which recognized the land beyond the Vaal as the South African Republic (*Transvaal).

Pretorius, Martinus Wessel (1819–1901), Boer statesman. He was President of the South African Republic (*Transvaal) (1857–71), having followed his father, Andries *Pretorius, in the *Great Trek. After fighting the *Zulu, he became one of the four Transvaal commandant-generals, and was elected President. He was also elected President of the Orange Free State (1859–63). His claim to diamond fields (1867) on the Vaal River brought him into conflict with British interests. Following the annexation of the Transvaal by Britain in 1877 he was imprisoned. With the outbreak of the First *Boer War (1880–1) he proclaimed with *Kruger and Joubert a new Boer republic (January 1881). After the victory of Majuba Hill, he was a signatory of the Treaty of Pretoria, which re-established the independent states of Transvaal and Orange Free State.

primary elections, elections in the USA for the selection of candidates for public office, most significantly for the Presidency. In the later 19th century, each state selected delegates to send to a national party convention that nominated the party's candidate for President. These delegates were chosen by a closed 'caucus system', that is, senior party members in the state chose the delegates and instructed them how to vote. Beginning with Wisconsin in 1903, however, primary elections steadily replaced the 'caucus primaries'. They are held by the state and the results are legally binding. There are both 'open' and 'closed' presidential primaries. In the former, any adult voter in a state may take part, regardless of his or her own party preference. In the latter, only those who are registered members of the party may vote.

Primo de Rivera, Miguel (1870–1930), Spanish general and statesman, dictator of Spain (1923–30). Believing that widespread disorder was a product of corrupt and inefficient parliamentary government, he led a coup (1923). His authoritarian regime attempted to reunite the nation around the motto 'Country, Religion, Monarchy'. Apart from successfully ending the Moroccan war (1927) he achieved little. His reliance on the landlord class prevented urgently needed agricultural reforms. Although he survived three attempts to remove him (1926), increasing lack of army support forced his resignation (1930).

Primrose League, an organization founded by Sir Drummond Wolf and Lord Randolph *Churchill in 1883, devoted to the cause of Tory democracy. The League used the emblem of *Disraeli's favourite flower to focus on his concept of Conservatism. This involved defence of traditional features of British life, but also a wish to broaden support for Conservatism by showing its capacity to improve living and working conditions for the masses.

Princely States, more than 500 Indian kingdoms and principalities during the *British Raj period (1858–1947). Although their rulers preserved some autonomy they were bound by treaty to the British. The states, although scattered, made up two-fifths of India's territory. Their princes were Hindu and Muslim (a few also Sikh and Buddhist), some, like *Hyderabad and *Kashmir, ruling majorities of other faiths. Most ruled autocratically, but a few, like Mysore, were regarded as progressive. Many princes had been forced to accept indirect British rule during the era of *East India Company expansion and paramountcy between 1757 and 1857. Mutual rivalries, historical, religious, and social, prevented co-ordinated resistance to British predominance. After 1857, when control of India passed to the crown, their collaboration was deliberately sought by confirmation of their internal autonomy. After 1877, when Queen Victoria was proclaimed Empress of India, they participated in imperial *durbars*. On British withdrawal, in 1947, they came under pressure to join either India or Pakistan. In Kashmir, Hyderabad, and Junagadh crises occurred, but most acceded peacefully, hoping some of their privileges, particularly financial, would be upheld. Many of the smaller states were grouped together into unions, for example the United States of Rajasthan. Legislation of 1970 abolished the special privileges of their ruling families.

prison camps, Soviet, punitive institutions for forced labour in the Soviet Union. The tradition of exiling political protesters and reformers to Siberia was well established in 19th-century Russia. By a decree of 1919 *Lenin maintained such punishment, operating through his police agency, the *Cheka. During *Stalin's rule millions were arrested by the *MVD, including peasants who resisted *collectivization, Christians, Jews, intellectuals, and political protesters. The prisoners were passed to GULAG (acronym for the Main Administration of Corrective Labour Camps), established in 1930 and responsible for administering the forced labour system. The camps were mostly situated in the east of the Soviet Union, and were referred to metaphorically as the 'Gulag archipelago'. Estimates of numbers confined to Gulag camps in the years of Stalin vary, ranging between six and fifteen million. After the worst years of the purges in the 1930s thousands continued to be sent to the camps. Among the noted authors whose writings are a testament to their experiences in the camps have been Evgenija Ginzburg, *Journey into the Whirlwind* (1975, 1985), and

Federal agents examine bottles of illicit alcohol from a 3,000-bag consignment in New York harbour during the **Prohibition era**. The manufacture, importation, and sale of bootleg alcohol provided a powerful incentive to the growth of organized crime in the USA.

Aleksandr Solzhenitsyn, *One Day in the Life of Ivan Denisovich* (1962). After the arrest and execution of *Beria (1953) and the de-Stalinization policy of Khrushchev, there was a steady decline in the worst excesses of the camps, which were formally replaced in 1955 by Corrective Labour Colonies. Many distinguished Soviet citizens have been either 'exiled' to Siberia or placed in 'psychiatric hospitals', or in other ways restrained. In 1987 in the wake of a new policy of *glasnost* or openness, *Gorbachev ordered the release of some intellectual dissidents.

Progressive Movement (US) (1890–1914), a US movement that sought to provide the basic political, social, and economic reforms necessary for the developing industrial economy. In both the Republican and Democratic parties Progressives were distinguished by a commitment to popular government, free trade, and control of competition-stifling trusts. To secure these ends they advocated direct *primaries for the nomination of candidates, the popular election of Senators (secured in 1913), and *anti-trust legislation. Social reforms were also demanded, for example legislation improving conditions of employment, and *prohibition attracted much support. Under Progressive pressure, government extended its activity at municipal, state, and federal levels in the pursuit of equality, efficiency, and social harmony. In different ways, progressive policies were adapted by the Republican Theodore *Roosevelt and the Democrat Woodrow *Wilson.

Progressive Parties (US) (1912, 1924, and 1948), the name of three US political organizations. In 1912 the first Progressive (the 'Bull Moose') Party, led by the former President Theodore *Roosevelt, polled more votes in the presidential elections than the Republican candidate, President *Taft. By splitting the Republican vote, it allowed the Democrats to win, on a platform equally progressive. In 1924 a revitalized Progressive Party, based on Wisconsin and other farm states, challenged the conservative outlook of both the Republican and Democratic parties. Although securing five million votes, the party carried only Wisconsin. In 1948 Henry A. *Wallace, formerly Democratic Secretary of State for agriculture, campaigned for a more conciliatory policy towards the Soviet Union. His Progressive Party, however, appeared too sympathetic to communism, and failed to challenge either main party.

Prohibition era (US) (1920–33), the period of national prohibition of alcohol in the USA. A culmination of the *Temperance Movement, it began when the Eighteenth Amendment to the Constitution went into effect by the passing of the *Volstead Act (1919). Despite the securing of some 300,000 court convictions between 1920 and 1930, drinking continued. Speakeasies (illegal bars) and bootlegging (illegal distilling of alcohol) flourished. The success of gangsters like Al *Capone, who controlled the supply of illegal alchohol, led to corruption of police and city government. After the Wickersham Commission in 1931 reported that the prohibiton laws were unenforceable and encouraged public disrespect for law in general, the Eighteenth Amendment was repealed by the Twenty-First Amendment. A number of states and counties retained full or partial prohibition, but by 1966 no state-wide prohibition laws existed.

Proudhon, Pierre Joseph (1809-65), French social theorist. In 1840 he published the pamphlet *What is Property?* This began with the famous words '*La propriété, c'est le vol*' ('property is theft'), and went on to maintain that property was the denial of justice, liberty, and equality, since it enabled some men to exploit the labours of others. For this he was tried by the assize court of Besançon but acquitted. In 1848 he was elected to the National Assembly and took to journalism; he was condemned to three years' imprisonment (1849-52) for attacking the President of the republic. His writings considerably influenced later supporters of *anarchism and federalism, but *Marx criticized him as a sentimentalist.

Provisional IRA *Irish Republican Army.

Prussia, a former German kingdom in north-eastern Europe. *Frederick William III (1770-1840) tried to maintain neutrality in the *Napoleonic Wars, but was unsuccessful. Prussia was forced to yield territories, which became part of the *Confederation of the Rhine; but the Treaty of *Tilsit saved the kingdom from extinction. In 1813 it joined the *Quadruple Alliance and in 1815 at the Congress of *Vienna gained part of Saxony and important lands on the Rhine. For the rest of his reign Frederick William followed *Metternich's policy of repression. The sponsorship of the customs union or *Zollverein was important for the economic development of Germany. His son, *Frederick William IV (1840-61), was a romantic, whose concessions to the Prussian nobility and Catholic Church alienated the liberal middle classes. When the *Revolution of 1848 began he promised a constitution, and a Prussian assembly met in Berlin, but in April 1849 he refused the crown of a united Germany without Austria. In 1858 he was declared insane and his brother became regent. From now on Prussia vied with Austria for the leadership in the restored *German Confederation. This struggle was successfully orchestrated by Prussia's chief minister, Otto von *Bismarck. Following the *Franco-Prussian War (1870), the *German Second empire was founded in 1871, when *William I was proclaimed emperor. After the defeat of Germany in 1918 a reduced Prussia became a German state in the *Weimar Republic. Prussia was eventually abolished in 1947, most being divided between Poland, the Soviet Union, and the new German Democratic Republic.

Puerto Rico, an island in the Caribbean. It was maintained by Spain as a garrison protecting trade routes until the loss of Mexico removed its strategic importance. In 1887 the Autonomist Party was founded to protect home rule under Spanish sovereignty. In 1898, during the *Spanish-American War, the island came under US military rule and was ceded to the USA at the end of the war. In 1917 an Act of the US Congress (Jones Act) declared Puerto Rican inhabitants to be US citizens. Since the 1940s, with a decline in the sugar industry, there have been successful efforts at industrialization and diversification of the economy. In 1946, Jesus T. Pinero was appointed governor. In 1952 the Commonwealth of Puerto Rico was proclaimed, and ratified by a plebiscite. The party which has dominated politics since the 1940s, the Popular Democratic Party (PPD) is demanding greater autonomy within the existing status, but this position has been increasingly challenged by the demands of the New Progressive Party (PNP) for statehood and political representation, and by the Puerto Rican Independence Party (PIP) who demand the creation of an independent republic.

Pullman strike (1894), a US labour dispute that began when the Pullman Palace Car Company of Chicago laid off men and cut the wages of others, blaming the economic depression, and refused to discuss grievances with its employees. The cause of the workers was taken up by the powerful American Railway Union, led by Eugene V. *Debs. The strike threatened to paralyse the entire railway network unless Pullman went to arbitration. President *Cleveland's sympathies were with the company, and the federal circuit court at Chicago issued an injunction declaring the strike illegal. Rioting and bloodshed ensued, and Debs was gaoled (1895). The injunction remained open to misuse until amended by the Norris-La Guardia Anti-Injunction Act of 1932.

Punjab, formerly a north-western province of British India, now partitioned between India and Pakistan. In 1799 the Sikhs under *Ranjit Singh established a kingdom there. After their defeat in the *Sikh Wars the province remained under British rule until 1947. As the British withdrawal in 1947 approached, latent tensions between Punjab's Muslim, Hindu, and Sikh communities surfaced. Muslims constituted more than half the province's population, and the *Muslim League's capture of Punjabi votes in the 1945-6 election strengthened its demand for a separate Muslim homeland. Growing communal violence pushed Britain and the *Congress Party into a reluctant acceptance of partition. Violence, during which hundreds of thousands died, increased as Hindus and Sikhs crossed to India and Muslims to Pakistan. Subsequently two wars (1965 and 1971) between India and Pakistan have involved fighting on the Punjabi frontier. Since 1967 Punjabis have been dominant in Pakistan's political life, and a new national capital was built at Islamabad in 1966. The Indian state of Punjab was divided on a linguistic basis to provide a predominantly Punjabi-speaking state of Punjab (with a Sikh majority) and a Hindi-speaking state of Maryana, with Chandigarh as a joint capital. Militant *Sikhs continue to campaign for an independent Sikh state in the Punjab.

Puyi (or P'u-i) (1906-67), last *Qing emperor of China (1908-12). Proclaimed Xuantung Emperor at the age of two by the empress dowager *Cixi (his great aunt), he reigned until the *Chinese Revolution forced his abdication in 1912. He continued to live in the imperial palace with extensive privileges, serving as a focus for monarchist movements and even experiencing a 12-day restoration in 1917, before being forced to flee by a local *warlord to the Japanese concession of Tianjin in 1924. After the Japanese seizure of Manchuria, Puyi was placed at the head of the puppet state of *Manchukuo. Deposed and captured by Soviet forces in 1945, he was later handed over to the communist Chinese, and, after a period of imprisonment, was allowed to live out his life as a private citizen.

Q

Muammar al-**Qaddafi**, who holds the ceremonial title, 'Leader of the Revolution', in Libya.

Qaddafi, Muammar al- (1942–), Libyan statesman. He served in the Libyan army and overthrew Idris I in a military coup in 1969. He became Chairman of the Revolutionary Command Council (RCC) and, in 1970, Prime Minister. He then nationalized the majority of foreign petroleum assets, closed British and US military bases and seized Italian and Jewish properties. He has used his nation's vast oil wealth to support the *Palestine Liberation Organization and other revolutionary causes. He several times intervened in Chad, Sudan, and Uganda by military force. In retaliation for alleged acts of terrorism against US nationals President *Reagan authorized the bombing of the Libyan capital in 1986. In 1979, while remaining actual head of state with the title of chairman, Qaddafi re-organized the Libyan constitution in a series of committees whose intention was to involve all citizens.

Qajar, Turkic tribe in north-east Iran from which came the Qajar dynasty which ruled Persia (Iran) from 1794 to 1925. The dynasty was established by Agha Muhammad (1742–97), a eunuch who made Tehran his capital and was crowned Shah in 1796. He was succeeded by his nephew Fath Ali Shah (1797–1834), during whose reign Iran was forced to cede the Trans-Caucasian lands to Russia. The constitutional revolution of 1906 established a parliament. Muhammad Ali (1907–9) was deposed for attacking the constitution and after a lengthy regency, Muhammad Ali's son, Ahmad Shah (1914–25), became the last Qajar ruler, being deposed by an army officer, *Reza Shah Pahlavi, in 1925.

Qatar, a country on the west coast of the Persian Gulf. Historically linked with *Bahrain, it was under Bahraini suzerainty for much of the 19th century. In 1872 it came under Ottoman suzerainty, but the Ottomans renounced their rights in 1913. In 1916 Qatar made an agreement with Britain which created a *de facto* British protectorate. Oil was discovered in 1939 and exploited from 1949. The agreement with Britain was terminated in 1968 and Qatar became fully independent in 1971.

Qing (or Ch'ing) dynasty (1644–1912), the last dynasty of the Chinese empire. It began with the Manchu invasion of Ming China in 1644 and reached its peak during the reign of Qianlong (1736–96). Faced with major internal revolts, most notably the *Taiping Rebellion (1850–64) and a succession of Muslim uprisings in the far west, the Qing proved unable to contend simultaneously with increasing intrusions from western powers interested in the economic exploitation of China. During the long dominance of the conservative empress dowager *Cixi, young, ineffectual emperors failed to inject sufficient force into modernization schemes such as the *Self-Strengthening Movement and the *Hundred Days Reform to prevent increasing foreign intervention. Humiliating defeat in the *Sino–Japanese War (1894–5) and the *Boxer Rising (1900) weakened Qing power, and after the *Chinese Revolution of 1911, the last Qing emperor *Puyi was forced to abdicate in 1912.

Quadruple Alliance, an alliance formed in 1813 by Britain, Prussia, Austria, and Russia which committed them to the defeat of Napoleon. The resulting battle of *Leipzig (1813), was decisive and the war ended. In November 1815 *Castlereagh at the Congress of *Vienna arranged for the Alliance to be maintained to form a permanent league to safeguard 'the general tranquillity of Europe'. However, it became less a means of containing France (the latter joined in 1818 to form the Quintuple Alliance), and more a means of containing revolution. A series of Congresses was held, but Britain became steadily less involved, and, after the death of Castlereagh (1822), the alliance had little effective influence.

Quebec, a Canadian province. A French colony until 1763, some territory was lost in the American War of Independence (1775–83), and in 1791 the province was divided into two, *Upper and Lower Canada, the latter having a majority of French-speaking Canadians. The *War of 1812 confirmed' the frontiers with the USA. In 1822 British merchants in Montreal sought a reunion of the two provinces in order to gain an overall English-speaking majority. In 1837 the unsuccessful rebellion led by Louis Joseph *Papineau was a protest as much against the increasing dominance of British-Canadians as against any lack of political representation. In the face of French-Canadian opposition, the Act of Union of 1840 joined the two provinces into the new Province of Canada, with a majority of British-Canadians. Lower Canada became Canada-East. The *British North America Act (1867) once again divided the Province of Canada, creating a separate Province of Quebec within the Confederation of Canada. The Province was expanded by the acquisition of New Quebec in the north in 1912. Particularly since the early 1960s French-Canadian separatism has remained a factor in Canadian politics.

Quebec Liberation Front, a French Canadian separatist movement. Set up in the early 1960s, it launched a terrorist and bombing campaign to secure the separation of *Quebec Province from Canada. The Front de Libération du Québec (FLQ) was greatly encouraged when *de Gaulle used the separatist slogan *Vive le Québec Libre* (Long Live Free Quebec) while visiting Canada in 1967. But its terrorist activities proved unpopular; much more support was given to the constitutional Parti Québeçois, which won a majority of the seats in the Quebec legislative assembly in 1976.

Queensland, a state of the Commonwealth of *Australia. In its early years of white settlement, when it formed part of New South Wales, it was known as the Moreton Bay district and served as a penal colony from 1824 to 1839. It was separated from New South Wales, re-named Queensland, and granted responsible government in 1859. Gold discoveries in the 1860s and 1870s attracted people to Queensland. The pastoral industry also developed in those years. In 1901 it became a state in the Commonwealth of *Australia.

Quezon, Manuel Luis (1878–1944), Filipino statesman. He followed *Aguinaldo in the Philippine wars against Spain and the USA (1896–1901). Later he served in the Philippines Assembly and became resident commissioner for the Philippines in Washington (1909–16). His successful conduct in this post made him a national hero and he was elevated to the office of President of the Philippine Senate. In 1935 he became first President of the newly constituted Philippine Commonwealth and ruled his country dictatorially until forced into exile by the Japanese invasion in 1942. He headed a government in exile in the USA until his death, and was succeeded by his Vice-President, Sergio Osmena.

Quisling, Vidkun Abraham Lauritz Jonsson (1887–1945), Norwegian fascist leader. An army officer, he founded the fascist Nasjonal Samling (National Unity) Party, and in 1940 helped Hitler to prepare the conquest of Norway. He became head of a new pro-German government and was made Premier in 1942. He remained in power until 1945, when he was arrested and executed. By this time 'Quisling' had become a derogatory term to describe politicians who supported invaders of their countries.

Rabeh, az-Zubayr (d. 1900), Sudanese military leader. Born a slave, he first appeared in the Bahr al-Ghazal region as a lieutenant of the slave-trader Zubayr Pasha. In 1879, when General *Gordon defeated Zubayr's son in the course of stamping out the slave trade, Rabeh found himself without a master. Rallying 400 soldiers, he gradually assembled an efficient army. In 1893 he sacked Bagirmi, and then Kukawa, capital of Borno. Now master of the Chad basin, he attacked Sokoto, but was repulsed. He was killed in 1900, when he was defeated by a French army, and his territories divided between Britain, France, and Germany.

Race Relations Act (1976), British Act of Parliament. It repealed the Acts of 1965 and 1968, strengthened the law on racial discrimination, and extended the 1968 ban on discrimination to housing, employment, insurance, and credit facilities. The Act also established (1977) a permanent Race Relations Commission to eliminate discrimination and to promote equality of opportunity and good relations between different racial groups within Britain. Large numbers of West Indians had emigrated to Britain in the 1950s, meeting labour shortages, and in the early 1960s immigrants from India and Pakistan also increased dramatically, rising from an average of 7,000 per annum to 50,000 before the first Immigration Act restricted entry in 1962. General economic problems, particularly increased unemployment, have tended to produce racial inequality, discrimination, and prejudice, the problems which the Race Relations Act was intended to ameliorate. In spite of continuing efforts by the Commission, racial tensions have continued to flare up in a number of inner-city areas, for example Brixton and Tottenham in London (1981, 1985), Handsworth in Birmingham (1985), Toxteth in Liverpool (1981), and St Paul's in Bristol (1980).

Radek, Karl (1885–c.1939), international communist leader. He joined the Polish Social Democratic Party and participated in the *Russian Revolution of 1905 in Warsaw. He crossed Germany with *Lenin after the outbreak of the *Russian Revolution (1917) and took part in the *Brest-Litovsk peace negotiations. He helped the communist uprising in Germany (1918–19) and returned in 1923 as an agent of the *Comintern to organize another communist rising, whose failure contributed to his declining influence. Expelled from the party (1927) for alleged support of *Trotsky, he was readmitted after recanting. He assisted *Bukharin in drafting *Stalin's new constitution (1936). Accused of treason at the second show trial (1937), he was sentenced to prison where, it is presumed, he died.

Radetzky, Josef, Count of Radetz (1766–1858), Austrian field-marshal. He fought in the *Revolutionary Wars, and in 1805 was promoted to a command in Italy. After Austria's defeat at *Wagram, he was appointed to assist *Schwarzenberg with army reorganization. In 1813 he joined Schwarzenberg in the field and his proposed tactics for the battle of *Leipzig were decisive in the defeat of Napoleon. At the beginning of the Italian *Risorgimento he was placed in command of the Austrian forces in Lombardy (1831–57). He constructed a quadrilateral of fortresses: Peschiera, Mantua, Verona, and Legnano. Here in 1849 he successfully withstood the forces of the Risorgimento under Charles Albert of Piedmont. He continued until the age of 91 to rule the Lombardo-Venetian territories of the Austrian *Habsburgs.

Radhakrishnan, Sir Sarvepalli (1888–1975), Indian scholar and statesman. A professor of philosophy at Mysore and Calcutta, and professor of eastern religion and ethics at Oxford, he wrote extensively on Hindu religious and philosophical thought. He also served as Indian ambassador to the Soviet Union (1949–52) and as Vice-President (1952–62) and President (1962–7) of India, succeeding Rajendra *Prasad, who was the first President of independent India. As a scholar without political affiliations he occupied a rare and detached position in Indian political life, stressing the need for India to establish a classless and casteless society.

Raeder, Erich (1876–1960), German admiral. He was Admiral Hipper's chief of staff in World War I, and from 1928 was commander-in-chief of the German navy, secretly rebuilding it in violation of the *Versailles Peace Settlement. In the 1930s he elaborated his 'Z-Plan' for building a fleet capable of challenging Britain, but World War II began before this was achieved. He resigned and was replaced by Doenitz in January 1943, after Hitler became outraged by the apparently poor performance of the surface fleet against Allied convoys. His part in unrestricted U-boat warfare led to his post-war imprisonment after trial at *Nuremberg.

Raffles, Sir Thomas Stamford (1781–1826), British colonial administrator. He joined the English East India Company in 1795 and in 1805 was appointed to Pinang. After participating in the capture of Java (1811) he served as its lieutenant-governor (1811–16), instituting wide-ranging and only partially successful administrative and social reforms. As lieutenant-governor of the Sumatran port of Bengkulu (1818–24) he recognized the commercial potential of *Singapore. In 1819 he took advantage of a disputed succession to the sultanate of Johore, within whose territory Singapore was, to found a British settlement without permission from his superiors. Ill-health forced him to return home in 1824, but Singapore went on to become one of the most important trading centres in Asia.

Raglan, FitzRoy James Henry Somerset, 1st Baron (1788–1855), British soldier. Joining the army in 1804, he served as aide-de-camp to Arthur Wellesley (Duke of *Wellington) during the *Peninsular War, and lost an arm at the battle of Waterloo. Appointed to lead the British expeditionary force in the *Crimean War, he won a victory at Inkerman (5 November 1854) with French assistance, but was criticized for his general conduct of the campaign.

Rahman, Tungku Abdul (1903–), Malaysian statesman. He entered the Kedah state civil service in 1931, and in 1952 succeeded Dato Onn bin Jafaar as leader of

A portrait of Thomas Stamford **Raffles** by G. F. Joseph, 1817. Two years later Raffles established the port of Singapore in the belief that it would become a major trading centre. (National Portrait Gallery, London)

the *United Malays National Organization (UMNO). He played a central role in organizing UMNO's alliance with the moderate Malayan Chinese Association (founded in 1949 by Tan Cheng Lock), which provided the political base for the achievement of independence. After becoming the leader of the Federal Legislative Council in 1955, he became Malaya's first Prime Minister (1957–63), and in 1963 he successfully presided over the formation of the Federation of *Malaysia, which he led as Prime Minister (1963–70). He remained in office until the political crisis caused by the riots of 1969 between the Malays and the Chinese forced him to stand down.

railway (US, railroad), a form of transport, rapidly developed in the 19th century for moving people and goods on a large scale. Railways used horse- or man-drawn trucks for moving coal and minerals in Europe from the 16th century onwards; one of the first of these being found in Alsace in 1550. The first steam locomotive to run on rails was made by Richard Trevithick in Wales in 1804. The Stockton and Darlington Railway (1825) was the first to carry both freight and passengers. In 1830 it was followed by the Liverpool and Manchester Railway, which, with the introduction of George Stephenson's 'Rocket' locomotive, heralded the beginning of the railway era. By 1880 some 70 per cent of railway revenue in Britain came from passenger traffic. British engineers built railroads all over Europe, America, India, Africa, and Australasia during the second half of the 19th century. In the USA railroad companies were the main agents of *westward expansion. In Europe cheap and easy travel helped to break down provincial differences, while in Switzerland and the Mediterranean the holiday industry steadily developed. Railways were important for

Railways: a lithograph by T. M. Baynes of the opening of the Canterbury to Whitstable branch-line on 3 May 1830, which boasted the first railway tunnel in the world. Most railways were initially planned for freight, rather than passengers, as this was expected to be their principal source of revenue.

both sides in the American Civil War, for moving troops and supplies. Strategically they were equally important on all fronts in World War I. Since then in many countries they have been developed as nationalized industries, but their importance has declined with competition from road transport.

Rajagopalachariar, Chakravarti (1878–1972), Indian nationalist politician. He joined the Indian National *Congress after World War I, became a close associate of Mohandas *Gandhi, and was imprisoned on several occasions for *non-co-operation with the British. Himself a Hindu, he was tolerant of the right of Indian Muslims to demand special minority safeguards and of the creation of the separate state of Pakistan. He served as governor-general of India (1948–50), and chief minister of the Madras government (1952–4). In 1959 he was one of the founders of the conservative Swatantra Party.

Rakosi, Matyas (1892–1971), Hungarian politician. He played an important role in the Hungarian communist revolution led by Béla *Kun in 1919. After four years in Moscow (1920–4), Rakosi returned to Hungary but was later arrested, to be released only in 1940. In 1944 he became First Secretary of the Hungarian Communist Party and during this time established a ruthless Stalinist regime. He was Prime Minister (1952–3). Opposition to his Stalinist policies led to his resignation as Party Secretary and return to the Soviet Union in 1956. The brutality of his secret police contributed to the *Hungarian Revolution of 1956.

Randolph, Asa Philip (1889–1979), US black labour leader. He was prominent in many of the struggles for *civil rights. His threat of a march on Washington in

1941 contributed to the end of race restrictions on employment in the defence industries, and his activities in 1948 helped to persuade President *Truman to end segregation in the armed forces. In 1957, as leader of the Brotherhood of Sleeping-Car Porters, he became a vice-president of the *American Federation of Labor. In 1963 he helped to organize the march on Washington for Jobs and Freedom, one of the largest civil rights demonstrations ever held in the USA.

Ranjit Singh (1780–1839), ruler of a Sikh kingdom in the Punjab (1799–1839). His state was based on Lahore, which he secured in 1799. In 1801 he took the title 'Maharaja of the Punjab' and in 1802 extended his rule

Ranjit Singh, nicknamed 'Lion of the Punjab': an illustration of the Sikh leader by Godfrey Thomas, 1837. (India Office Library, London)

over the Sikh holy city of Amritsar. By agreement with Britain (Amritsar 1809), the eastern boundary of his state remained on the Sutlej River, but with his large army, trained by French officers, he expanded westwards. Between 1813 and 1821 he extended his rule westwards and also took control of Kashmir. In 1834 he annexed Peshawar. At the end of the *Sikh Wars following his death most of his territory was taken by Britain.

Rapallo, Treaties of (1920, 1922). The first settled differences between Italy and the kingdom of Serbs, Croats, and Slovenes (Yugoslavia). Italy obtained the Istrian peninsula while Dalmatia went to Yugoslavia. Fiume (Rijeka) became a free city. The second and more important treaty recorded an agreement between Germany and the Soviet Union. The two countries agreed to abandon any financial claims which each might bring against the other following *World War I. Secretly, in defiance of the *Versailles Peace Settlement, German soldiers were to be permitted to train in the Soviet Union.

Rashid Ali al-Ghailani (1892–1965), Iraqi statesman. Member of a well-known religious family, he served as Prime Minister on four occasions. In April 1941 he seized power by a military coup which also deposed Abd al-Ilah, regent for the child-king, Faisal II. Believing the new government to be pro-Axis, Britain intervened to expel Rashid Ali in May 1941, and restored the regent.

Rasputin, Grigori Yefimovich (1871–1916), Russian religious fanatic. A Siberian peasant and mystic with healing and hypnotic powers, he earned his nickname, meaning 'debauchee', from his immoral life. He came to St Petersburg (1903), meeting the royal Romanov family. His beneficial treatment of the haemophilic crown prince won him a disastrous hold over the empress. His influence increased when *Nicholas II left the court to command the army (1915). He and the empress virtually ruled Russia, and were responsible in a large measure for the emperor's failure to respond to the rising tide of discontent which eventually resulted in the *Russian Revolution. Rasputin was murdered by a group of nobles led by Prince Yusupov.

Ratana, Tahupotiki Wiremu (1870–1939), Maori political and religious leader. In the tradition of Maori prophets such as *Te Kooti and *Te Whiti, Ratana emerged as a teacher and healer in 1918. He attacked the traditional fears of *tapu* ('power of the spirit world') and sorcery, and sought to introduce a heterodox belief in the Christian Trinity. Rejected by the formal Churches, he established his church with its liturgy, temples, and clergy. In 1936 nearly one-fifth of Maori were adherents. Politically he sought advancement of Maori rights through recognition of the Treaty of *Waitangi. Ratana candidates contested the Maori seats in the New Zealand Parliament. In 1931 the Labour Party accepted Ratana candidates as its own. Both the religious and political aspects of this movement survived Ratana's death in 1939. Since 1943 the Ratana/Labour candidates, with rare intervals, have held all four Maori seats and have influenced Labour legislation on Maori affairs.

Rathenau, Walther (1867–1922), German industrialist and statesman. He was responsible for directing Germany's war economy (1916–18) and later became

Rasputin, self-proclaimed 'holy man' at the Russian court who was to have a disastrous influence on the Empress Alexandra.

Minister of Reconstruction (1921) and Foreign Minister (1922) in the *Weimar Republic. He believed that Germany must fulfil its obligations under the *Versailles Peace Settlement, including payment of *reparations. Convinced of Germany's ability to gain ascendancy in Europe he negotiated the treaty of *Rapallo (1922) with Russia, establishing military and trade links. He was assassinated by *anti-Semitic nationalists in 1922.

Reagan presidency (1981–), the terms of office of Ronald W. Reagan (1911–) as fortieth President of the USA. A Hollywood actor, he became Republican governor of California (1966–74), and won a landslide victory in the 1980 presidential election on a programme of reduced taxation and increased defence expenditure against world communism. He cut federal social and

President **Reagan** at his desk in the Oval Office at the White House with Secretary of State, George Schultz.

welfare programmes, reduced taxes, and increased defence spending by heavy government borrowing. He campaigned against alleged Soviet involvement in Latin America, especially *Nicaragua and *Grenada. Relations with China steadily improved during the Presidency, with a large increase in trade. There was domestic legislation to strengthen Civil Rights, but increasing disagreements over budget policies. This, together with balance-of-payments deficits, precipitated a serious stock-market collapse in October 1987. Congress steadily obliged the President to reduce proposed defence expenditure in order to give more balanced budgets. Reagan stood for a second term in 1984, overwhelmingly defeating the Democrat, Walter Mondale. Intransigence on the *Strategic Defense Initiative blocked advance on nuclear arms control in 1986, at the end of which the so-called 'Iran-gate' scandal broke. This revealed that in spite of strong anti-terrorist talk, the administration had begun secret negotiations for arms sales to Iran, with profits going illegally to Contra forces in Nicaragua. In the 1986 mid-term elections the Democrats gained control of the Senate. Talks on nuclear *arms control began in Geneva in 1985, continued at Reykjavik in 1986, and Washington in 1987, when an Intermediate Nuclear Forces (INF) Treaty was signed with the Soviet Union, eliminating all ground-based intermediate-range nuclear missiles.

Rebecca riots (1839, 1842–3), a series of agrarian riots in south-west Wales. They were a protest against the toll-gates introduced by turnpike trusts. Bands of rioters disguised in women's clothes attacked and broke the gates. Each band was led by a 'Rebecca' after the Old Testament story of Rebecca, 'be thou the mother of millions, and let thy seed possess the gate of those which hate them'. In 1843 rioting and the breaking of toll-gates was followed by a series of massed meetings in the autumn. Troops and a contingent of the Metropolitan Police were sent from London, while a commission to investigate grievances took evidence. In 1844 an Act to 'consolidate and amend the Laws relating to Turnpike Trusts in Wales' ended the protest.

Reconstruction Acts (1867–8), legislation passed by the US Congress dealing with the reorganization of the South in the aftermath of the *American Civil War. The

Welsh rioters, disguised in women's clothes, use axes to break down a toll-gate during the **Rebecca riots** in their protests against turnpikes.

question of the treatment of the defeated *Confederacy raised conflicting priorities between reconciliation with white Southerners and justice for the freed slaves. In 1866 an impasse developed between President Andrew *Johnson, and the Republican majorities in Congress. In 1867 Congress passed, over the President's veto, a Reconstruction Act which divided the South into military districts, and required the calling of a new constitutional convention in each state, elected by universal manhood suffrage. The new state governments were to provide for black suffrage and to ratify the *Fourteenth Amendment as conditions for readmittance to the Union. Further Reconstruction Acts were passed in the following twelve months to counter Southern attempts to delay or circumvent the implementation of the first measure.

Red Army, Soviet army formed by *Trotsky as Commissar for War (1918–25) to save the *Bolshevik revolution during the *Russian Civil War. His energy and oratory restored discipline and expertise to the new recruits, most of whom were workers and peasants. For trained officers, Trotsky had to rely on former officers of the Imperial Army. To ensure the reliability of officers and to undertake propaganda among the troops, political commissars were attached to units, often resulting in dual commands. A major offensive against *Poland failed in 1920 when *Pilsudski successfully organized national resistance. After the Treaty of *Rapallo (1922) close co-operation with Germany led to greater efficiency. Progress was checked by *Stalin's purge of army leaders (1937–8) to remove possible opposition, resulting in a lack of leadership in the *Finnish–Russian War. After *Hitler's invasion of the Soviet Union (1941) the Red Army became the largest in the world—reaching five million by 1945. Precise figures remain unknown, but Red Army casualties in World War II have been estimated as high as seven million men. The name fell into disuse shortly after World War II and was replaced by that of Soviet Armed Forces.

Red Brigades, an *anarchist grouping of Italian urban guerrillas. They were especially active in the period c.1977–81, when they achieved notable publicity, and the security forces seemed powerless against them, though some arrests were subsequently made. The Red Brigades were responsible for the kidnapping and murder of the Italian statesman Aldo *Moro (1978) and the bomb attack on civilians at Bologna railway station (1980), in which eighty-five people were killed.

Red Cross, international agency concerned with the alleviation of human suffering. Its founder, the Swiss philanthropist Henri Dunant (1828–1910), horrified by the suffering he saw at the battle of *Solferino, proposed the formation of voluntary aid societies for the relief of war victims. In 1863 the International Committee of the Red Cross was established and in the following year twelve governments signed the *Geneva Convention. This drew up the terms for the care of soldiers and was extended to include victims of naval warfare (1906), prisoners of war (1929) and, twenty years later, civilians. Its conventions have now been ratified by almost 150 nations. The Red Cross flag, a symbol of neutrality is, in honour of Dunant's nationality, a red cross on a white background—the Swiss flag with colours reversed. In Muslim countries the cross is replaced by a red crescent.

Red Feds, members of the National Federation of Labour, an association of militant unions formed in 1909 in New Zealand. Never winning the support of a majority of the country's unionists, it briefly became an important force because of the presence of strong unions, such as the miners, in its ranks. In 1912 confrontation between government and Red Feds occurred at the Waihi goldmine, when a strike resulted in violence and one death before the miners accepted defeat. The United Federation of Labour was formed, which was soon involved in a bitter strike, centred on the Wellington docks. Again the government intervened. Thousands of mounted police were recruited and there were violent clashes. This strike was also broken. The Red Feds and their confrontation methods were permanently discredited. Since then the New Zealand trade unions have worked amicably through a process of arbitration, with a series of Acts being passed to amend the basic Industrial and Conciliation Arbitration Act of 1894.

Red Guards, militant young supporters of *Mao Zedong during the Chinese *Cultural Revolution (1966–9). Taking their name from the army units organized by Mao in 1927, the Red Guards, numbering several million, provided the popular, paramilitary vanguard of the Cultural Revolution. They attacked supposed reactionaries, the Communist Party establishment, China's cultural heritage, and all vestiges of Western influence, maintaining the momentum of the movement through mass demonstrations, a constant poster war, and violent attacks on people and property. Fighting between opposing Red Guard groups led to thousands of deaths. After the Cultural Revolution, many were sent into the countryside for forced 're-education'.

Redl, Alfred (1864–1913), Austrian spy. A colonel from a modest background in the Austro-Hungarian army, he was a specialist in counter-intelligence and security. By introducing modern criminological methods his department in Vienna became one of the best equipped intelligence centres in Europe. But he was a secret homosexual needing considerable sums of money to pay blackmail for his affairs. This he obtained by becoming himself a spy, selling military secrets to Russia. When eventually detected, he was ordered to commit suicide.

Redmond, John Edward (1856–1918), Irish politician. Less intransigent than Charles *Parnell, he sought to achieve the same objective of *Home Rule, but by co-operation with British politicians. By 1900, he had united under his leadership the Irish Nationalist Party in the British House of Commons. In return for Redmond's support over the Parliament Act (1911), the Liberal Prime Minister, *Asquith, introduced the third Home Rule Bill in 1912. However, the outbreak of World War I, in which Redmond supported Britain, postponed implementation of the Bill.

Red River Settlement, an early 19th-century agricultural colony in the Red River (now Manitoba) area of central Canada granted by the Hudson's Bay Company to Thomas Douglas, 5th Earl of Selkirk (1771–1820). Selkirk endeavoured to settle the dispossessed of Scotland and Northern Ireland there. But his first group of settlers succumbed to North West Company pressure to abandon the area soon after their arrival in 1812. The colony was re-established in 1816, but twenty-two settlers were killed in a massacre at Seven Oaks, led by North West Company men. Other attacks followed. Selkirk himself went bankrupt, but the publicity attracted by the affair led to the forced merger (1821) of the North West Company and Hudson's Bay Company and cleared the way for more successful settlement in the area.

Red Shirts, a nationalist organization in British India in the *North-West Frontier province. Formed in 1929 by Abdul Ghaffar Khan, a follower of *Gandhi, it was correctly entitled Khudai Khidmatgar (Servants of God). It provided the main support for Ghaffar Khan's control of the province until 1946, during which time it was deployed in support of *Congress policies. Opposed to partition in 1947 Ghaffar Khan and the Red Shirts campaigned for a separate state of Pakhtunistan. The new government of Pakistan banned the Red Shirts and imprisoned Ghaffar Khan for thirty years.

Reeves, William Pember (1857–1932), New Zealand statesman and journalist. As Minister of Labour in the Liberal government of *Ballance and *Seddon, Reeves was responsible for a sweeping code of humane labour and factory laws, then for the introduction of compulsory arbitration in industrial disputes. In 1896 Reeves went to Britain as agent-general and then high commissioner (1905–8). He became an associate of the *Fabians and director of the London School of Economics (1908–19).

Reform Acts, a series of legislative measures which extended the franchise in 19th- and 20th-century Britain. In the 1780s William *Pitt the Younger had proposed to remove some of the worst abuses in the electoral system, but abandoned his attempts in the face of considerable opposition. In the 1820s there was a renewed call for change and the Whigs, who were returned to power in 1830, were pledged to reform. However, the government succeeded in getting its measures through Parliament only after a fierce struggle. The Reform Act of 1832 eliminated many anomalies, such as *rotten boroughs, and enfranchised the new industrial towns, which had hitherto been unrepresented. The lowering of the property qualification gave voting rights to the middle classes, but left effective power in the hands of the landed aristocracy. The bulk of the population (all women and five-sixths of the men) were still without a vote, and agitation continued. The Reform Act of 1867 doubled the size of the electorate and gave many urban working-class men the vote. However, agricultural labourers and domestic servants had to wait a further seventeen years to be enfranchised: the Reform Act of 1884 increased the electorate to about five million. The Representation of the People Act (1918) gave the vote to all men over the age of 21 and conceded some of the demands of the *suffragettes by enfranchising women over 30, but on a property qualification. Universal adult suffrage for everyone over 21 was finally achieved in 1928, when women between the ages of 21 and 30 secured the right to vote and the property qualification was abolished. In 1969 the voting age was lowered to 18.

Regency, the, the period in Britain from 1811 to 1820 when the Prince of Wales, later *George IV, acted as regent for his father, *George III. At first the prince's powers were restricted by an Act of Parliament in case

George Cruikshank's savage 1816 satire on the **Regency**. The artist shows the decline of the Prince Regent, later, George IV from 'Gent', as a colonel of the 10th Hussars; to 'No Gent', carousing the night away; and finally to 'Regent', a gout-ridden libertine. (British Museum, London)

his father should recover, but the king's disability proved to be permanent. Among the major events of the Regency were the *War of 1812 involving Britain and the USA, the successful conclusion of the *Napoleonic Wars, and the Congress of *Vienna (1814–15). In Britain the post-war period was marked by a slump in the economy, which caused much social unrest. The Tory government used severe measures to quell popular discontent, which culminated in the *Peterloo Massacre of 1819. The

The **Reichstag** building, the seat of the German parliament, burning on the night of 27–8 February 1935. The fire, and its political aftermath, gave Hitler the opportunity to suspend the constitution and strengthen his own grip on the government.

Regency period was aesthetically characterized by a type of architecture and domestic furniture notable for restraint and simplicity and strongly influenced by classical styles. An important architect was John Nash, who designed Regent Street and Regent's Park in London, naming them in honour of his patron, the prince.

Reichstag (German, 'imperial parliament'), the legislature of the *German Second empire and of the *Weimar Republic. Its origins reach back to the *Diet of the Holy Roman Empire. It was revived by *Bismarck (1867) to form the representative assembly of the constituent states of the North German Confederation and, from 1871, of the German Second empire. Its role was confined to legislation, being forbidden to interfere in federal government affairs and having limited control over public spending. Under the *Weimar Republic it enjoyed greater power as the government was made responsible to it. On the night of 27 February 1933 the Reichstag building was burnt. *Goering and *Goebbels allegedly planned to set fire to the building, subsequently claiming it as a communist plot. The arsonist was a half-crazed Dutch communist, van der Lubbe. The subsequent trial was an embarrassment as the accused German and Bulgarian communist leaders were acquitted of complicity and only van der Lubbe was executed. But the fire had served its political purpose. On 28 February a decree suspended all civil liberties and installed a state of emergency, which lasted until 1945. Elections to the Reichstag were held on 5 March 1933, but by the Enabling Act of 23 March 1933 the Reichstag effectively voted itself out of existence.

Reid, Sir George Houston (1845–1918), Australian statesman. He was Premier of New South Wales from 1894 until 1899. His ambivalent attitude towards *Australian federation resulted in his being dubbed 'Yes-No' Reid. He led the Free Traders in the first federal parliament. He was Prime Minister, leading a coalition of Free Traders and Protectionists (1904–5). After the defeat of his government, Reid led the Opposition until 1908. He was the first Australian High Commissioner in London (1910–16), after which he was elected to the British House of Commons.

Reith, John Charles Walsham, 1st Baron (1889–1971), first director-general of the British Broadcasting Corporation (1927–38). His strongly Calvinistic temperament moulded the early years of broadcasting, with an emphasis on programmes which were educational in the widest sense: classical music, book reviews, news, and drama. His aim was that the BBC should earn respect for its impartiality and sense of responsibility, the more so because of its monopoly of radio broadcasting at that time in Britain. In 1936 he inaugurated British television. In 1940 he was elected National Member of Parliament for Southampton and was appointed Minister of Works by Winston *Churchill. During 1943–4 Reith worked at the Admiralty, planning the movement of supplies, war materials, and transport for the invasion of Europe.

reparations, compensation payments for damage done in war by a defeated enemy. They were a condition of the armistice for World War I, and part of the *Versailles Peace Settlement. France, who had paid reparations to Germany in 1871, secretly hoped to bankrupt Germany.

British civilians had sustained little damage and so *Lloyd George claimed only the cost of war pensions. The US Senate did not ratify the Versailles Treaty, and waived all claims on reparations. A sum of £6,500,000,000 was demanded from Germany, a figure which the British economist *Keynes argued was beyond German capacity to pay without ruining the interdependent economies of Europe. Hungary, Austria, and Bulgaria were also to pay huge sums. Turkey, being more or less bankrupt, agreed to an Allied Finance Commission. To enforce its claims France occupied the *Ruhr (1923), precipitating an inflationary crisis in Germany. With Britain and the USA unhappy about reparations, various plans were devised to ease the situation. The *Dawes Plan (1924) permitted payment by instalments when possible; the *Young Plan (1929) reduced the amount demanded; the Lausanne Pact (1932) substituted a bond issue for the reparation debt, and German repayments were never resumed. After World War II reparations took the form of Allied occupation of Germany and Japan. Britain, France, and the USA ended reparation collections in 1952. Stalin systematically plundered the East German zone by the removal of assets and industrial equipment. In Japan the USA administered the removal of capital goods, and the Soviet Union seized Japanese assets in Manchuria. Since 1953 West Germany has paid $37 billion (£20·7 billion) as reparations to Israel for damages suffered by Jews under Hitler's regime.

Representatives, US House of, lower house of the US Congress. The powers and composition of the House of Representatives are set out in Article I of the Constitution, and the House first met in 1789. Its members are apportioned among the states according to their population. A Congressman is elected for a two-year term, and must be 25 or older, hold US citizenship for at least seven years, and be an inhabitant of his or her state and electoral district. The presiding officer of the House, the Speaker, is elected by members and is third in line for executive power after the President and Vice-President. The House and the *Senate have an equal voice in legislation, though all revenue bills must originate with the former. The size of the House has increased with the country's population, until fixed (in 1929), at 435. Seats are apportioned among the states every ten years, after the federal census.

Republican Party, a major political party in the USA. The term republican was first used in the USA by *Jefferson's Democratic-Republican Party, founded in 1796 and the antecedent of the *Democratic Party. The present Republican Party was formed in 1854, being precipitated by the *Kansas–Nebraska Act and by the agitation of the *Free Soil Party; it brought together groups opposed to slavery but supporting a protective trade tariff. The party won its first presidential election with Abraham *Lincoln in 1860 and from then until 1932 lost only four such contests, two each to *Cleveland and Woodrow *Wilson. Its early success was based on the support of the agricultural and industrial workers of the north and west, and its conservative financial policies, tied to tariffs and the fostering of economic growth, remained pre-eminent. The opulence of the *'Gilded Age', contrasted with increasing poverty among immigrants and the urban proletariat, led to the *Progressive Movement and a split in the party when Theodore

*Roosevelt formed his *Progressive Party. After World War I the policy of *isolationism brought the party back to power, but from 1932 onwards Republicans lost five successive presidential elections, only returning to power through the massive popularity of President *Eisenhower in 1952. Under the recent Republican Presidents *Nixon, *Ford, and *Reagan, it has become strongly associated with military spending and a forceful assertion of US presence world-wide, especially in Central America. Strongly backed by corporate business, it has nevertheless failed to maintain a grip on Congress, which has sometimes had a Democratic majority even when the President has been Republican.

Repudiation Movement, an attempt to set aside land purchase contracts in New Zealand. Ownership of *Maori land had been steadily proceeding through purchases from individual Maori, not tribal communities, ignoring the guarantees of the Treaty of *Waitangi (1840). In 1873 Henare Matua, a Maori chief, appealed to the Hawkes Bay Native Lands Alienation Commission to repudiate land purchase contracts drawn up in the Hawkes Bay area. He received some support from settler-politicians anxious to embarrass the large landed interests. The movement met with little success in over-turning contracts, but it did contribute to the growing separatist movement among Maoris, the Kotahitanga.

resistance movements, underground movements that fought against Nazi Germany and Japan during World War II. Their activities involved publishing underground newspapers, helping Jews and prisoners-of-war to escape, conveying intelligence by secret radios, as well as committing acts of sabotage. In Germany itself resistance to the Nazi regime was active from 1934 onwards, at first expressed by both Protestant and Catholic Churches, but also from 1939 onwards by groups such as the Roman Catholic student group Weisse Rose, and the communist Rote Kapelle, which carried out sabotage and espionage for Russia until betrayed in 1942. Admiral Wilhelm Canaris, head of German Counter-Intelligence (Abwehr), was a key resistance figure until betrayed and hanged after the *July Plot. In occupied Europe there were often deep divisions between communist and non-communist organizations, notably in France, where the *Maquis was active, as well as in Belgium, Yugoslavia, and Greece. Communist parties had at first remained passive, but following the German invasion of the Soviet Union (June 1941), they formed or joined underground groups. Dutch, Danish, and Norwegian resistance remained unified and worked closely with London, where in 1940 the British Special Operations Executive (SOE) was set up to co-ordinate all subversive activity, both in Europe and the Far East, and to supply arms and equipment by secret air-dropping. In eastern Europe the long German lines of communication were continually harassed by partisans, and the Polish resistance was almost certainly the largest and most elaborate in Europe. Eastern European resistance later turned against the Red Army as it advanced west (1944–5), the Polish *Warsaw Risings being a tragic example of the tensions between communist and non-communist forces. In the Far East clandestine operations were carried out through British and American intelligence organizations. Much of their effort was devoted to intelligence gathering, psychological warfare, and prisoner-of-war recovery, while the actual sabotaging of selected installations and communication lines was conducted by native-born, nationalist, and often communist-inspired guerrillas. Their leaders, such as *Ho Chi Minh in Vietnam, went on to form the core of the post-war independence movements against the colonial powers.

Reuter, Paul Julius, Baron (original name, Israel Beer Josaphat, 1816–99), German-born founder of one of the first news agencies. He established a pigeon-post service between Aachen and Brussels in 1849 to relay commercial information. In 1851 he settled in London, where he opened a telegraph office near the Stock Exchange. Linked by telegraph with correspondents in other countries, he was able to supply the daily news-papers with information about share prices, eye-witness reports of foreign wars such as the Crimean, and other international news. By the 1870s his agency had become a world-wide organization.

Reuther, Walter Philip (1907–70), US labour leader. A foreman in a Detroit automobile plant, he was dismissed (1932) for his union activity. He helped to organize the United Automobile Workers (UAW) and served as its president (1946–70). He pioneered negotiations for guaranteed employment, wage increases tied to pro-ductivity, and welfare provisions for his members. An anti-communist, he was president of the Congress of Industrial Organizations (1952–5), and fought strenu-ously to rid the unions of racketeers. He also helped to organize non-union workers through a short-lived Al-liance for Labor Action, formed from the UAW and the *Teamsters in 1969.

revivalism (US), a recurrent Protestant movement of religious evangelization in the USA, initiated by the 'Great Awakening' in the 18th century. A 'Second Awakening' (1797–1805), most spectacular in the West, began with the frontier camp meetings (for prayer and exhortation) and the preachings of James McGready. Revivalism in the early 19th century became the accepted method of worship among Congregationalists, Pres-byterians, Baptists, and Methodists. In the early 20th century came the professional evangelists, preaching fundamentalism, with large organizations and soph-isticated publicity. Revivalists such as Billy Graham have continued to attract mass audiences to their meetings, missions, and television shows.

Revolutionary Wars (1792–1802), a series of wars in Europe following the French Revolution. In 1791 Louis XVI attempted unsuccessfully to escape from France to Germany, to win support from Austria and Prussia. In April 1792 France declared war on Austria, which then ruled Belgium (the Austrian Netherlands). A series of French defeats followed until, on 20 September, an invading Prussian army was defeated at Valmy. In February 1793 war was declared against Britain, Spain, and the United Provinces of the Netherlands. For a year a Reign of Terror operated in France, but, at the same time, under the skill of *Carnot, armies had been steadily raised and trained. At first the aim was to consolidate the frontiers of France along the 'natural frontiers' of the Rhine and the Alps, but from 1795, these armies were to conquer Europe. A number of brilliant young officers emerged, for example Bernadotte (later *Charles XIV of Sweden), Barthélemy Joubert (killed in battle 1799), and above all *Napoleon Bonaparte. All the Netherlands were conquered, Belgium being annexed, and the Re-public of Batavia created from the United Provinces; French armies advanced across the Rhine and into South Germany. Switzerland was made into the Helvetic Republic (1798). In 1796–7 Napoleon took an army into Italy, defeated the Austrians at Arcola and occupied Venice, creating the Cisalpine and Ligurian Republics. In 1798 he led an expedition to Egypt, but the British fleet under *Nelson destroyed his fleet at Aboukir Bay, and Napoleon returned to Paris. Meanwhile Austrian and Russian troops had re-occupied Italy and in 1799 Napoleon again marched across the Alps to win a crushing victory over the Austrians at Marengo. At the same time General Moreau won a second great victory at Hohenlinden. The peace treaties of Lunéville (1801 with Austria) and Amiens (1802 with Britain) were then negotiated, ending the Revolutionary Wars.

Revolutions of 1848, a series of revolutions in western and central Europe. They sprang from a shared back-ground of autocratic government and, economic unrest, as well as from the failure of conservative governments to grant representation to the middle classes, and the awakened nationalism of minorities in central and eastern Europe. Revolution erupted first in France, where sup-porters of universal suffrage and a socialist minority under Louis *Blanc caused the overthrow of the *July monarchy and established the Second Republic. In most German states there were popular demonstrations and uprisings, and a movement for an elected national parliament to draft a constitution for a united Germany. Rioting in Austria caused the flight of both *Metternich and the emperor, and the formation of a constituent assembly and the emancipation of the peasantry. A movement for Hungarian independence, headed by *Kos-suth, led to a short-lived republican government from Budapest for all Hungarian lands; but Magyar refusal to consider independence for its own minorities resulted in an insurrection by Croat, Serb, and Transylvanian forces and in Hungary's defeat by Austrian and Russian forces. In the Italian states there was a series of abortive

The Second Republic in France came into being in February 1848 with the overthrow of Louis-Philippe. The caps and tricolours of 1789 reappeared in Paris during the **Revolutions of 1848** as revolutionaries hailed the new republic. Here they are depicted symbolically burning the throne of the July monarchy.

revolutions which led to the temporary expulsion of the Austrians and the flight of Pope *Pius IX from Rome, but the united, democratic republic dreamt of by *Mazzini did not come about. A *Pan-Slav Congress in Prague inspired Czech nationalist demonstrations to demand autonomy within a federal Austria. By 1849 counterrevolutionary forces had restored order, but the concept of absolute monarchy and the feudal rights of a land-owning aristocracy had been tacitly abandoned.

Reynaud, Paul (1878–1966), French politician. He was Finance Minister (1938–40), and Prime Minister in the emergency of 1940, but, having appointed *Pétain and *Weygand, he was unable to carry on the war when these two proved defeatist. He resigned in mid-June 1940. After the war he was Finance Minister (1948) and Vice-Premier (1953) in the Fourth Republic. He assisted in the formation of the Fifth Republic, but later quarrelled with *de Gaulle.

Reza Shah Pahlavi (Reza Khan) (1878–1944), Shah of Iran (1925–41). An officer of the Persian Cossack Brigade, he achieved power through an army coup (1921) and established a military dictatorship. He was successively Minister of War and Prime Minister before becoming Shah. He followed a policy of rapid modernization, (constructing a national army, a modernized administrative system, new legal and educational systems), and economic development, notably through the Trans-Iranian Railway (1927–38). He crushed tribal and other opposition to his policies. In World War II his refusal to expel German nationals led to the invasion and occupation of Iran by Soviet and British forces. He was forced to abdicate in favour of his son, *Muhammad Reza Shah, and died in exile in South Africa.

Rhee, Syngman (1871–1965), Korean statesman. He was an early supporter of Korean independence from Japan, and after a spell of imprisonment (1897–1904) for nationalist activities, he went to the USA and became President of a 'government-in-exile' formed by a small group of his supporters. After World War II he returned to become leader of *South Korea during the US occupation, and in 1948 he became the first President of the Republic of Korea, advocating the unification of Korea both before and after the *Korean War (1950–3). Rhee was re-elected in 1952 and 1956, but opposition to his corrupt and autocratic government grew more intense as economic conditions deteriorated and a third re-election in 1960 caused accusations of rigging and serious rioting, which forced Rhee into exile.

Rhineland, a former province of Prussia. The success of the French revolutionary armies brought the left bank of the Rhine to France in 1794, but this was ceded by the Congress of *Vienna to Prussia as a bulwark against French expansion. With the formation of the *German Second empire in 1871 the nearby French provinces of *Alsace and Lorraine were annexed, both being rich in iron and coal. In 1918 these were restored to France and the Rhineland 'demilitarized' but allowed to remain within the *Weimar Republic. In 1936 *Hitler's troops 're-militarized' the area, but met with no effective resistance from France or its allies. The scene of heavy fighting in 1944, it was recaptured by US troops in early 1945 and now forms part of West Germany.

Cecil **Rhodes** was seen by his contemporaries as a colossus of colonial enterprise and ambition. This 1892 *Punch* drawing shows him astride the African continent from Cape Town to Cairo, determined to keep the influence of other imperial powers—Belgium, Germany, Holland, France, and Portugal—subordinate to Britain.

Rhodes, Cecil (John) (1853–1902), British businessman and colonial administrator. He was sent to *Natal in 1870 after a protracted illness. In 1871 diamonds were found at Kimberley, and he and his brother were successful in prospecting, which made him financially independent at the age of 19. By the age of 35 he controlled the largest diamond mining and trading companies in South Africa. He fought determinedly for British interests against the *Afrikaner exclusivism of President *Kruger and against German expansion in southern Africa. He was instrumental in acquiring Bechuanaland (*Botswana), and then, through his British South Africa Company, the vast territories of Matabeleland and Mashonaland (*Zimbabwe), which were eventually renamed Rhodesia. He aimed to 'make Africa British from Cape to Cairo' with Afrikaner consent. He became Prime Minister of the Cape (1890–5). Although he was acquitted of responsibility, the abortive *Jameson Raid had his financial support and he resigned, devoting the rest of his life to the development of Rhodesia. In his will he left enough money to provide some 200 annual scholarships for Commonwealth, US, and German students to study at Oxford University.

Rhodesia, the former name of a large area of southern Africa. Rhodesia was developed for its mining potential by Cecil *Rhodes and the British South Africa Company from the last decade of the 19th century. It was administered by the company until Southern Rhodesia became a self-governing British colony in 1923 and Northern Rhodesia a British protectorate in 1924. From 1953 to 1963 the two Rhodesias were united with Nyasaland to form the *Central African Federation. After Northern Rhodesia became the independent state of *Zambia in 1964, the name Rhodesia was used by the former colony of Southern Rhodesia until the proclamation of the Republic of *Zimbabwe in 1980.

Rhodesia, Northern   *Zambia.

Rhodesia, Southern   *Zimbabwe.

Rhodesia and Nyasaland, Federation of   *Central African Federation.

Ribbentrop, Joachim von (1893–1946), German Nazi statesman. He joined the Nazi Party in 1932 and became a close associate of Hitler. In 1936–8 he was ambassador in London. As Foreign Minister (1938–45), Ribbentrop conducted negotiations with states destined to become Hitler's victims. The *Nazi–Soviet Pact was regarded as his masterpiece, opening the way for the attack on *Poland and the *Baltic states. He was responsible for the Tripartite Pact (1940) between Germany, Italy, and Japan. After trial at *Nuremberg he was executed.

Ricardo, David (1772–1823), British economist. His fame rests on his *Principles of Political Economy and Taxation* (1817), in which he made a systematic attempt to establish how wealth was distributed, and put forward his theory that the value of a commodity rests on the amount of labour required for its production—a theory later adopted by Karl *Marx. In 1819 he entered Parliament, where he supported *free trade, a return to the *gold standard, and the repeal of the *Corn Laws.

Riel Rebellions (1869, 1885), two uprisings of the métis or half-Indian population of Manitoba against the Canadian government, led by the French-Indian Louis Riel (1844–85). Expansion westwards led to the uprising in 1869 in which a provisional government under Riel was set up. Riel aroused outrage in the east by executing an Ontario settler, but the arrival of British and Canadian troops coincided with negotiations in Ottawa leading to the area's inclusion within the confederation, with all the local rights demanded by the métis. Riel escaped but increasing resentment of eastern domination and economic dislocation produced a second insurrection in 1885. Riel, who returned from the USA to lead it, was supported by several Indian tribes, but alienated most of the white population and was defeated by the Canadian militia. He was then tried and executed for treason.

Ripon, George Frederick Samuel Robinson, 1st Marquess of (1827–1909), British statesman. He entered Parliament as a Liberal in 1853, supporting a scheme to provide working men with opportunities for education. He served as Secretary for War (1863–6), and Secretary for India (1866–8). As President of the Council (1868–73) he was responsible for the 1870 Education Bill which his deputy, W. E. *Forster carried through the House of Commons. In 1873 he became a Roman Catholic and resigned from public office. *Gladstone

appointed him viceroy of India (1880–4). There he introduced a system of local self-government and ended restrictions on the freedom of the vernacular press. His Ilbert Bill (1883) gave qualified Indians jurisdiction over Europeans and established trial by a jury, of which half should be Europeans. On his return to Britain he again held ministerial office in Liberal governments.

Risorgimento (c.1831–61) (Italian, 'resurrection' or 'rebirth'). A period of political unrest in *Italy, during which the united kingdom of Italy emerged. Much of Italy had experienced liberal reforms and an end to feudal and ecclesiastical privilege during the *Napoleonic Wars. The restoration of repressive regimes led to uprisings in Naples and *Piedmont (1821), and in Bologna (1831), then part of the Papal States. Following the French *July Revolution of 1830, Italian nationalists began to support *Mazzini and the *Young Italy movement. In this they were encouraged by the liberal Charles Albert, who succeeded to the throne of Sardinia, and became ruler of Piedmont in 1831. In 1847 Count *Cavour started a newspaper, *Il Risorgimento*; this had a considerable influence on Charles Albert, who in 1848 tried to drive the Austrians out of Lombardy and Venetia. He was defeated at Custozza (1848) and Novara (1849) and abdicated. He was succeeded by his son *Victor Emanuel II. During

Risorgimento

The vision of a united and liberated Italy inspired the 'Young Italy' movement in 1831 and triumphed in 1861, when Victor Emanuel II was proclaimed King of Italy.

the *Revolutions of 1848 republicans held power briefly in Rome, Florence, Turin, and Venice and hoped to create a republic of Italy, but were also defeated. Under the guidance of Cavour, Prime Minister of Piedmont from 1852, the French emperor *Napoleon III was encouraged to ally with Piedmont, in return for promises of Nice and a part of the Alpine region of Savoy, and Austria was defeated in the battles of *Magenta and *Solferino in 1859. Austria evacuated Lombardy and much of central Italy. *Garibaldi liberated Sicily, marched north and almost reached Rome. Plebiscites were held and resulted in a vote to accept Victor Emanuel II as first King of Italy (1861).

Rizal, José (1861–96), Filipino nationalist. While training as a doctor in Spain he wrote two novels attacking Spanish repression in the *Philippines, which marked him out as one of the leading spokemen of the nationalist movement and of the publicity campaign known as the Propaganda Movement. A reformist rather than a revolutionary, Rizal fell foul of the Spanish authorities when he formed a reform society, the *Liga Filipina*, in 1892, and was exiled to Mindanao. Although not involved in the nationalist uprising of 1896, he was executed by the Spanish authorities for supposed complicity in the rebellion.

'robber baron', a ruthlessly aggressive businessman. The title was first applied in the USA in the *'Gilded Age', and was conferred on railway operators by aggrieved Kansas farmers in 1880. The earlier generation of 'robber barons' were financiers who made fortunes in the American Civil War and, thereafter, exploited the stock market, railways, and public utilities, a notable example being Jay Gould (1836–92), who made a fortune of $25 million out of railway and bullion speculation. Cornelius *Vanderbilt made an even larger fortune out of railways ($100 million) which his son, William Henry, further extended. Other financial giants who obtained monopolies over industry or market were J. P. *Morgan, Andrew *Carnegie, and J. D. *Rockefeller.

Robertson, Sir William Robert (1860–1933), British field-marshal. After service in India and South Africa he became commandant of the Staff College, Camberley, Surrey, in 1910. He was chief of the Imperial General Staff from 1915 until criticisms by *Lloyd George of British strategy led to his resignation in 1918. He subsequently commanded British troops on the Rhine (1919–20).

Rockefeller, John D(avison) (1839–1937), US industrialist and philanthropist. He became a partner in a produce business in 1858. His firm entered the oil-refining business in 1862 and became the nucleus of the Standard Oil Company, incorporated in Ohio in 1870. By 1879, due to debatable practices, it controlled 90 to 95 per cent of the nation's oil-refining capacity. The US government fought to limit this monopoly, dissolved eventually in 1911. A pious Baptist, Rockefeller had turned to philanthropy by the 1890s. He gave away $600 million in his life-time, including $183 million to the Rockefeller Foundation. His son, **John D(avison), Jr** (1874–1960), bought for the United Nations the site for its headquarters in New York. His son, **Nelson Aldrich** (1908–79), was attracted by public life and was elected to four consecutive terms as governor of New York (1959–73), also serving as Vice-President (1974–7).

Rokossovsky, Konstantin Konstantinovich (1896–1968), Polish-born Soviet field-marshal. He enlisted first in the Tzarist army, then joined the *Red Army in 1919. Arrested during *Stalin's purges, he was released from prison camp to become one of the outstanding generals of World War II, taking part in the battles of Moscow, *Stalingrad, *Kursk, and others. His Red Army troops stood by (August–September 1944) on the outskirts of Warsaw, without helping in the *Warsaw Rising against the German occupying forces. After the war he was transferred to the Polish army, and became Deputy Premier and Minister of Defence under President Bierut. Rokossovsky led the army in a bloody suppression (June 1956) of Polish workers in Poznań, who were demonstrating for 'bread and freedom'. On 20 October Polish and Soviet troops exchanged fire; Rokossovsky's troops were recalled to Moscow, and *Gomulka's new nationalist government was able to claim some independence from interference by the Soviet Union.

Roman Catholic Church, the largest branch of the Christian Church in the world, with about 750 million members, under the jurisdiction of the pope (the Bishop of Rome). Under Pius VII the papacy signed a *concordat (1801) with Napoleon, restoring Catholicism in France, but Pius's refusal to support the *Continental System against Britain led Napoleon to occupy Rome and the *Vatican. During the period of religious decline in the 19th century, the Church and the papacy reacted defensively to the modern world, responding to challenges to its teaching and authority by stressing strict obedience and uniformity of belief. In Prussia, the *Kulturkampf, reflecting the antagonism between the Church and the state, failed because of the passive resistance of the clergy and the Catholic population. The First Vatican Council (1869–70) under *Pius IX, declared the infallibility of the pope. At the same time Catholic *missionaries, notably the Society of Jesus (Jesuits) carried the faith beyond Europe, into Latin America, Asia and Africa. During World War II Pope *Pius XII, faithful to the *Lateran Treaties, retained the strict neutrality of the Church. A second Vatican Council (1962–5) was summoned by *John XXIII and reconvened by his successor *Paul VI; it set out to modernize the Church's teaching, discipline, and organization. Under John Paul II (1978–) (Karel Wojtyla, 1920–), a Pole, the global mission of the papacy has been emphasized. Contacts between the Catholic Church, the *Eastern Orthodox and *Protestant Churches, and other faiths have been established. In 1984 the Church concluded with the Italian government a revision of the Lateran Treaties, formalizing the separation of the Church and state. By the year 2000 the majority of Roman Catholics will be living in Latin America, where 'liberation theology', the identification of the priesthood with the poor, has become a source of controversy within the Church.

Romania, a country in south-east Europe. Part of the *Ottoman empire since the 15th century, Turkish domination began to be challenged by both Russia and Austria. In 1812 Russia gained control of north-east Moldavia (Bessarabia), and during the next forty years Romanian nationalism precipitated many insurrections.

Following the *Crimean War, Walachia and Moldavia proclaimed themselves independent principalities and in 1861 united to form Romania, electing a local prince, Alexander Cuza, as ruler. On his deposition (1866) Prince Carol Hohenzollen-Sigmaringen was elected. At the Congress of *Berlin independence was recognized, and Prince Carol crowned king as *Carol I (1881–1914). His pro-German policy led in 1883 to its joining the Triple Alliance of 1882 (Germany, Austria, and Italy). In World War I Romania remained neutral until, in 1916, it joined the Allies and was rewarded at the *Versailles Peace Settlement with the doubling of its territories, mostly from Hungary. Carol I was succeeded by Ferdinand I (1914–27) and then by *Carol II (1930–40), who imposed a fascist regime. He was forced to cede much territory to the *Axis powers in 1940. Romanian forces co-operated with the German armies in their offensives (1941–2), but after *Stalingrad the Red Army advanced and Romania lost territory to the USSR and Bulgaria. During the next twenty years it became a Soviet satellite. However, it retained a degree of independence, which increased when Nicolae Ceausescu became President (1967–). In the last twenty years the country has benefited from progressive industrialization assisted by its own oil deposits. Despite this, stringent economic measures had to be enforced in 1987. In 1971 it was the only Soviet-bloc country to join the *International Monetary Fund.

Rome, Treaties of (1957), two international agreements signed in Rome by Belgium, France, Italy, Luxemburg, the Netherlands, and the Federal Republic of Germany. They established the *European Economic Community and Euratom (the European Atomic Energy Community). The treaties included provisions for the free movement of labour and capital between member countries, the abolition of customs barriers and cartels, and the fostering of common agricultural and trading policies. New members of the European Community are required to adhere to the terms of these treaties.

Rommel, Erwin (1891–1944), German field-marshal. He entered the army in 1910 and rose through the ranks. Having attracted Nazi Party attention as commander of the forces assuring the security of Hitler's headquarters, he was soon entrusted with field commands, and in 1940 led a Panzer division in a brilliant assault through the Ardennes to the Channel. In 1941 he commanded the Afrika Korps, an élite tank formation which bore the brunt of the battle in Libya, earning for himself the name 'the Desert Fox'. In 1942 he advanced to El *Alamein, but British resistance and lack of supplies impeded him, and he eventually had to retreat from *North Africa. In 1944 he was entrusted with the defence of the Channel coast in northern France against a possible Allied invasion. Wounded in the *Normandy Campaign, he was recalled to Germany. The Gestapo believed he was connected with the *July Plot against Hitler. No accusations were made publicly against him, but he was forced to commit suicide by taking poison.

Roosevelt, Franklin D(elano) (1882–1945), thirty-second President of the USA (1933–45). He studied law before entering politics, and was Assistant Navy Secretary under President *Wilson (1913–20). In 1921 he was stricken with polio, and henceforth operated from a wheelchair. A reforming Democrat and governor of New

Field-marshal Erwin **Rommel**, photographed while commanding the German Afrika Korps, 1941–2.

York from 1928, he began his long presidency in 1933, having beaten the Republican incumbent, *Hoover. His *New Deal programme tackled with confidence the crisis of the *Great Depression. In the 1936 Presidential election he won a crushing victory, thanks partly to his more intimate relationship with voters through such innovations as his regular radio 'fireside chats'. In his second term (1936–40), inherent weaknesses of his New

President Franklin D. **Roosevelt** signs the declaration of war, 8 December 1941, committing the USA to fight against the Axis powers in World War II.

Deal became more obvious, and hostility towards him from business and other sections of the community grew. However, he carefully steered his country away from policies favoured by the *isolationists. After the fall of France in 1940 he was able to make the USA a powerful supporter of Britain's war effort while remaining, until *Pearl Harbor, a non-belligerent. Measures such as the *Destroyer-Bases Deal and *Lend-Lease were typical instruments of this policy. Elected for a third term in the 1940 election and for a fourth term in 1944, he died before the war ended. He wife **Eleanor** (1884–1962) was actively involved in humanitarian projects. As a delegate to the United Nations, she was chairman of the UN Commission on Human Rights, and played a major role in the drafting and adoption of the Universal Declaration of Human Rights (1948).

Roosevelt, Theodore (1858–1919), twenty-sixth President of the USA (1901–9). He served as a Republican member on the New York State Legislature (1881–4), a member of the Civil Service Commission (1889–95), and Assistant Secretary of the Navy (1897–8), before helping to form the *Rough Riders regiment to fight in the *Spanish–American War (1898). A popular and successful governor of New York (1899–1900), his reform policies threatened to disrupt corrupt political practices there, and Republicans under T. C. *Platt attempted to suppress his initiative, nominating him as Vice-President. However, the assassination of *McKinley (1901) brought him to the Presidency, and his period of office was notable for the concessions he made to the *Progressive Movement. He arbitrated in the coal strike (1902), instituted an anti-monopoly move against the Northern Securities Company (1902), and established the Department of Commerce and Labor (1903). He secured US control of the construction of the *Panama Canal. By the Roosevelt corollary (1904) to the *Monroe Doctrine, he claimed for the USA the right to collect bad debts from Latin America. When *Taft, his successor, failed to maintain Roosevelt's 'Square Deal', he formed a splinter group, the *Progressive Party, popularly named the Bull Moose Party. Yet despite his enormous personal popularity he failed to regain the Presidency in 1912 on the Progressive Party ticket. Instead, by dividing the Republicans, he enabled Woodrow *Wilson to capture the Presidency for the Democrats.

Root, Elihu (1845–1937), US statesman and diplomat. A Republican, he became Secretary of War (1899–1904) and made drastic reforms in the organization of the army. He formulated the *Platt Amendment (1902) giving the USA greater control over Cuba. He was Secretary of State under Theodore *Roosevelt and reorganized the consular service. He negotiated the Root–Takahira Agreement (1908) with Japan, upholding the Open Door Policy of US commercial interests in China. From 1909 to 1915 he was US Senator for New York. He opposed US neutrality in 1914 and supported the Allied cause and, with reservations, the League of Nations.

Rosas, Juan Manuel de (1793–1877), Argentine dictator (1835–52). Reacting to the failure of the liberals, who dominated Argentina after independence, Rosas brutally repressed his political enemies and suppressed civil liberties. Often depicted as a *caudillo, he was a consummate politician who contributed to the es-

tablishment of national unity in Argentina and who stood up to foreign powers like Britain and France when they imposed two blockades (1838–40, 1845–50) as a result of disputes over Paraguay and Uruguay. In February 1852 Rosas was overthrown by another caudillo, Justo José de Urquiza, and he fled to England.

Rosebery, Archibald Philip Primrose, 5th Earl of (1847–1929), British statesman. He served in Gladstone's cabinet as Foreign Secretary (1892–4) and was briefly Prime Minister (1894–5) at the wish of Queen Victoria rather than of the Liberal Party. His support for British imperialism alienated many Liberal supporters, but his concept of regular meetings of colonial Premiers led to the establishment of Imperial Conferences. He was a strong critic of the *entente cordiale (1904).

Rosenberg case (1953), a US espionage case, in which Julius Rosenberg and his wife Ethel were convicted of obtaining information concerning atomic weapons in 1944–5 and passing it on to Soviet agents. They became the first American civilians to be sentenced to death for espionage by a US court. The only seriously incriminating evidence had come from a confessed spy, and the lack of clemency shown to them was an example of the intense anti-communist feeling that gripped the USA in the 1950s.

Rothschild, a Jewish banking family. Its members exerted considerable influence on both economic and political affairs during the 19th and early 20th centuries. As a banking house it was founded in the 18th century by Mayer Amschel Rothschild of Frankfurt, who became financial adviser to the Landgrave of Hesse-Kassel. He and his five sons prospered in the years of the *Revolutionary and *Napoleonic Wars, coming to London in 1804. They loaned money for the raising of mercenary armies and negotiated means of bypassing Napoleon's *Continental System. After 1815 Rothschild houses were opened as a banking group in all the great cities of Europe. Of Mayer's sons, Anselm (1773–1855) became a member of the Prussian privy council, Salomon (1774–1855) became financial adviser to Metternich in Vienna, and Nathan (1777–1836) established a branch of the bank in London. His son, Lionel (1808–79), became the first Jew to sit in the British House of Commons (1858). It was he who was able to lend to the British government £4 million in 1875 to buy the *Suez Canal shares. His son Nathan (1840–1915) was the first British Jewish Peer and became regarded as the unofficial head of both French and British Jews. It was to his son Lionel Walter (1868–1937), the second Baron and distinguished scientist and scholar, that the *Balfour Declaration was addressed in 1917.

rotten borough, a British Parliamentary borough whose population had virtually disappeared by 1832. At that time there were more than fifty such boroughs with two Members of Parliament. Among the most notorious were Old Sarum with a handful of electors and Dunwich, mostly submerged under the North Sea. They were abolished by the *Reform Act of 1832.

Rough Riders, popular name for the 1st Regiment of US Cavalry Volunteers. They were largely recruited by Colonel Leonard Wood and Lieutenant-Colonel Theo-

Theodore Roosevelt at the head of a small unit of **Rough Riders**, here flamboyantly depicted by a magazine artist of the day, W. G. Read. In fact, the Riders' most celebrated battle was fought on foot, at San Juan Hill. (Library of Congress, Washington)

dore *Roosevelt for service in the *Spanish–American War of 1898. Comprising rangers, cowboys, Indians, and college students, their most notable exploit was the successful charge up San Juan Hill in Cuba, on foot (their horses having been left behind in Florida).

Round Table Conferences, meetings held in London in 1930–2 between Britain and Indian representatives to discuss Indian constitutional developments. The procedure was suggested by the viceroy, Lord Irwin, in 1929. *Congress boycotted the first session (November 1930–January 1931), but, following the Gandhi–Irwin Pact (March 1931), *Gandhi attended the second session (September–December 1931). With the renewal of the *non-co-operation campaign, Gandhi was imprisoned and Congress took no part in the final session (November–December 1932). The constitutional discussions formed the basis of the 1935 Government of India Act, with its plan for a federal organization involving the Indian *Princely States.

Rowlatt Act (1919), repressive legislation in British India, following the report of a committee under Mr Justice Rowlatt. The report had recommended the continuation of special wartime powers for use against revolutionary conspiracy and terrorist activity. The Act aroused opposition among Indian nationalists, and this was channelled by Mohandas *Gandhi into a nationwide *satyagraha, known as the Rowlatt agitation, which ended with the *Amritsar massacre.

Rowntree, Joseph (1801–59), British businessman and philanthropist. A Quaker, he founded the family cocoa and chocolate manufacturing firm in York. He was keenly interested in civic affairs and in the development of educational opportunities. His son **Joseph** (1836–1925), chairman of the firm (1897–1923), was distinguished for his philanthropy and for his care and the welfare of his employees. He created a charitable social service and village trusts, consulting his employees about conditions and providing for housing, unemployment insurance, and pensions. An eight-hour day was introduced in 1896, a pension scheme in 1906, and a works doctor was appointed in 1904. 'Social helpers' were recruited to deal with the problems of women workers, and works councils were set up in 1919. His son **Seebohm** (1871–1954) was chairman of the firm (1923–41) and achieved national prominence for his surveys of poverty in York (1897–8 and 1936) and for his studies of management.

Roy, Ram Mohan (1772–1833), Indian religious and social reformer. He devoted his life to reforming Indian society on the basis of a selective appeal to ancient Hindu tradition. He founded the Atmiya Sabha (Friendly Association) to serve as a platform for his liberal ideas. He evolved a monotheistic form of worship, adapting the ethical and humanitarian aspects of Christianity. He attacked idolatry and popular practices, including the burning of widows (suttee) and polygamy, discrimination against women, and the caste system. He also helped to found the Hindu College in Calcutta (1817) and several secondary schools in which English educational methods were employed. In 1828 he founded the *Brahmo Samaj (Society of God), whose influence on Indian intellectual, social, and religious life has been profound.

Ruanda-Urundi   *Burundi; *Rwanda.

Ruapekapeka (Maori, 'the bats' nest'), a strongly fortified village (*pa*) in New Zealand defended by the Maori chief Kawiti, an ally of Hone Heke, who, in 1844–5, challenged white sovereignty. The *pa* was attacked by about 1,000 British troops, with artillery, and 500 Maori allies. Kawiti and his people abandoned it, with little loss, after about 24 hours' fighting.

Ruhr, German river which flows south through North Rhine-Westpahlia into the Rhine. In 1802 the Ruhr valley was occupied by Prussia and five years later by Napoleon. Its restoration to Prussia in 1815 marked the beginning of its industrial development, which was helped by the *Zollverein (customs union), the development of a railway network, and the establishment of the Krupps armament works in Essen. In 1851 new laws, which cut the coal tax, encouraged investment, and within fifteen years its coal output was second only to that of Britain. It was the industrial heart of *Bismarck's united Germany after the creation of the *German Second empire (1871). After World War I France feared that the Ruhr valley might again become an armaments centre. In 1923 the Ruhr was occupied by French and Belgian troops when Germany defaulted on *reparations payments. The loss of resources, production, and confidence resulted in soaring inflation in Germany that year. Two years later the French accepted the *Dawes Plan and withdrew. After 1933 war industries were re-established in the Ruhr as Germany re-armed. During World War II it was an important target for bombing, and after the war its recovery was monitored by an international control commission. Control passed to the European Coal and Steel Community in 1952 and to the Federal Republic of Germany in 1954.

Rum Rebellion (1808), a revolt in Australia, when colonists and officers of the New South Wales Corps (later known as the Rum Corps because of its involvement in the rum trade) overthrew Governor William *Bligh. It was fuelled by Bligh's drastic methods of limiting the rum traders' powers and his attempts to end the domination of the officer clique, while an immediate cause

Gun barrels being made at a Krupp factory in the **Ruhr** before World War I. In the years leading up to both world wars, the Ruhr's armaments factories played a key role in Germany's military build-up.

was the arrest of the sheep-breeder John *Macarthur in his role as liquor merchant and distiller. The officers induced the commander, Major George Johnston, to arrest Bligh as unfit for office. When Governor *Macquarie took office in 1810, the Corps was recalled, George Johnston was court-martialled in England and cashiered in 1811. Bligh, although exonerated, was removed from office.

Rundstedt, (Karl Rudolf) Gerd von (1875–1953), German field-marshal. He was called from retirement in 1939 to command army corps in the Polish and French campaigns of World War II. In 1941 he commanded Army Group South in the invasion of the Soviet Union but was dismissed after he had withdrawn from Rostov against Hitler's orders, in order to improve his chances of resisting a Soviet counter-offensive. From 1942 to 1945 he commanded the forces occupying France and launched the battle of the Bulge in the *Ardennes Campaign in December 1944. Relieved of his command in March 1945, he was captured by US troops in May. He was released in 1949.

Russell, John, 1st Earl (1792–1878), British statesman. As Lord John Russell he entered the House of Commons as a Whig in 1813 and first came into prominence with his opposition to the suspension of the Habeas Corpus Act in 1817. A firm advocate of *Catholic emancipation, and of the removal of all religious disabilities, he urged the repeal of the Corporation Act and the Test Act (1828) under which no Catholic or Protestant Non-conformist could hold public office. He was largely responsible for drafting the first *Reform Bill. As Home Secretary (1835–9), he brought in the Municipal Reform Bill of 1835 and reduced the number of crimes punishable by death. For

two years (1839–41) he was Secretary for War and the Colonies, becoming Prime Minister (1846–52) after the fall of Sir Robert *Peel. Russell's government was overshadowed by the dominant personality of his Foreign Secretary, Lord *Palmerston, whom he eventually dismissed in 1851. He later served as Foreign Secretary in Palmerston's government from 1859, succeeding him as Prime Minister (1865–6). However, he resigned in the following year when his proposal for a new Reform Bill split his party.

Russia, a country in eastern Europe and northern Asia. At the beginning of the 19th century Russia was drawn in to the *Napoleonic Wars; its peace treaty at *Tilsit enabled it to acquire *Finland from Sweden, while the early *Russo-Turkish Wars led to territorial acquisitions in Bessarabia and the Caucasus. By the Treaty of *Vienna Russia and Austria were confirmed as the leading powers on the continent of Europe. Attempts at liberal reform by the *Decembrists were ruthlessly suppressed, and Russia helped Austria to quell Hungarian nationalist aspirations in the *Revolutions of 1848. Rivalry of interests, especially in south-east Europe, between Russia and the Western powers led to the *Crimean War. Serfdom was abolished in 1861, and attempts at changes in local government, the judicial system, and education were partially successful, though they fell short of the demands made by the *Populists and other radical reform groups. Russian expansionism, curtailed by the Congress of *Berlin, led to its abandonment of the *Three Emperors' League and, later, to a Triple Entente with Britain and France (1907). Defeat in the unpopular *Russo-Japanese War led to the *Russian Revolution of 1905. A *Duma (Parliament) was established, and its Prime Minister, *Stolypin, attempted a partial agrarian reform. The beginning of the 20th century saw a rapid growth in Russian industry, mainly financed by foreign capital. It was there, among the urban concentration of industrial workers, that the leftist Social Democratic Party won support, although split since 1903 into *Bolsheviks and

Mensheviks. Support for *Balkan nationalism led Russia into *World War I. The hardship which the war brought on the people was increased by the inefficient government of *Nicholas II. A series of revolts culminating in the *Russian Revolution of 1917 led to the overthrow of the Romanov dynasty and to the *Russian Civil War, after which the *Union of Soviet Socialist Republics was established.

Russian Civil War (1918–21), a conflict fought in Russia between the anti-communist White Army supported by some Western powers, and the *Red Army of the *Soviets in the aftermath of the *Russian Revolution of 1917. It is sometimes referred to as the War of Allied Intervention. Counter-revolutionary forces began organized resistance to the *Bolsheviks in December 1917, and clashed with an army hastily brought together by *Trotsky. In northern Russia a force made up of French, British, German, and US units landed at Murmansk and occupied Archangel (1918–20). Nationalist revolts in the *Baltic States led to the secession of Lithuania, Estonia, Latvia, and Finland, while a Polish army, with French support, successfully advanced the Polish frontier to the Russian Ukraine, gaining an area not re-occupied by the Soviet Union until World War II. In Siberia, where US and Japanese forces landed, Admiral *Kolchak acted as Minister of War in the anti-communist 'All Russian Government' and, with the aid of a Czech legion made up of released prisoners-of-war, gained control over sectors of the Trans-Siberian Railway. He, however, was betrayed by the Czechs and murdered, the leadership passing to General *Denikin, who sought to establish (1918–20) a 'United Russia' purged of the *Bolsheviks. In the Ukraine Denikin mounted a major

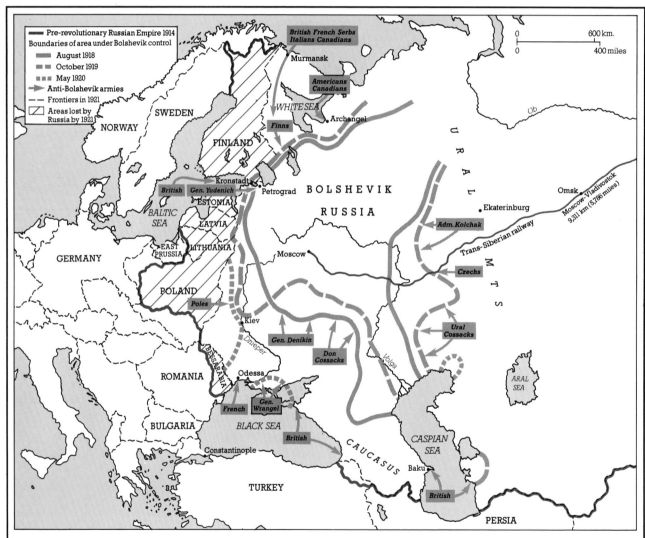

Russian Civil War

Following the Bolshevik Revolution, groups of anti-Bolshevist troops (the Whites) emerged to fight Trotsky's Red Army. In Siberia Admiral Kolchak, in control of the Trans-Siberian Railway, was supplied by a US, British, and Japanese force in Vladivostok. In the Ukraine and the Caucasus Cossacks resisted unsuccessfully, while Generals Denikin and Wrangel, supplied by the British and the French, mounted an offensive. In the north a British fleet threatened Petrograd, and an Allied force supported General Yudenich, who operated from Estonia. A Red Army counter-offensive carried the revolution beyond the Urals, into the Caucasus, and to the Polish frontier, where only sustained resistance preserved that country's independence. Wrangel retreated to the Crimea, then to Turkey. By 1921 the frontier of the new Soviet Union had stabilized.

Revolutionary troops in Petrograd (Leningrad) during the first days of the **Russian Revolution** in March (February, old style) 1917. Soldiers and sailors deserted from the imperial armed forces to join the rioters, while the country slipped into political, economic, and military chaos.

offensive in 1919, only to be driven back to the Caucasus, where he held out until March 1920. In the Crimea the war continued under General *Wrangel until November 1920. A famine in that year caused further risings by the peasants against the communists, while a mutiny of sailors at Kronstadt (1921) was suppressed by the Red Army with heavy loss of life. To win the war, *Lenin imposed his ruthless policy of 'war communism'. Lack of co-operation between counter-revolutionary forces contributed to their final collapse and to the establishment of the *Union of Soviet Socialist Republics.

Russian Revolution (1905), a conflict in Russia between the government of *Nicholas II and industrial workers, peasants, and armed forces. Heavy taxation had brought mounting distress to the poor, and Russia's defeat in the *Russo-Japanese War aggravated discontent. A peaceful demonstration in St Petersburg (now Leningrad) was met with gunfire from the imperial troops. Agitation continued to mount, mutiny broke out on the battleship *Potemkin*, and a *soviet or council of workers' delegates was formed in St Petersburg. The emperor yielded to demands for reform, and granted Russia a constitution which included a legislative *Duma. The *Social Democrats continued to fight for a total overthrow of the system, and were met with harsh reprisals by the government. Soon, democratic freedoms were curtailed and the government once more became reactionary.

Russian Revolution (1917), the overthrow of the government of *Nicholas II in Russia and its replacement by *Bolshevik rule under the leadership of *Lenin. It was completed in two stages—a liberal (Menshevik) revolution in March (February, old style), which overthrew the imperial government, and a socialist (Bolshevik) revolution in November (October, old style). A long period of repression and unrest, compounded with the reluctance of the Russian people to continue to fight in World War I, led to a series of violent confrontations whose aim was the overthrow of the existing government. The revolutionaries were divided between the liberal intelligentsia, who sought the establishment of a democratic, Western-style republic, and the socialists, who were prepared to use extreme violence to establish a *Marxist proletarian state in Russia. In the March Revolution strikes and riots in Petrograd (now Leningrad), supported by imperial troops, led to the abdication of the emperor and thus to the end after more than 300 years of Romanov rule. A committee of the *Duma (Parliament) appointed the liberal Provisional Government under Prince Lvov, who later handed over to the Socialist revolutionary *Kerensky. He faced rising opposition from the Petrograd Soviet of Workers' and Soldiers' Deputies. The October Revolution was carried through in a nearly bloodless coup by the Bolsheviks under the leadership of Lenin. Workers' Councils (*soviets) took control in the major cities, and a cease-fire was arranged with the Germans. A Soviet constitution was proclaimed in July 1918 and Lenin transferred the government from Petrograd to Moscow. The *Russian Civil War continued for nearly three more years, ending in the supremacy of the Bolsheviks and in the establishment of the *Union of Soviet Socialist Republics.

Russo-Japanese War (1904–5), an important conflict between these two countries over Manchuria and Korea. The Japanese launched a surprise attack on Russian warships at anchor in their naval base in Port Arthur

The battle of Tsushima (May 1905) was a triumphant naval victory for Japan in the **Russo-Japanese War**. This print shows a Japanese destroyer in action at high speed, while in the background Russian ships suffer heavily from the guns of their adversaries (*right*). (Victoria and Albert Museum, London)

(now Lüshun), Manchuria, without declaring war, after Russia had reneged on its agreement to withdraw its troops from Manchuria. Port Arthur fell to the Japanese, as did Mukden, the capital of Manchuria. The Russian Baltic fleet sailed 28,000 km. (18,000 miles) from its base to the East China Sea, only to be sunk in the Tsushima Straits by the Japanese fleet led by Admiral Togo. The victory was important to Japan since it had for the first time defeated a Western power both on land and at sea. The war was ended by the Treaty of *Portsmouth. For Russia, it was a humiliating defeat, which contributed to the *Russian Revolution of 1905.

Russo-Turkish Wars (1806-12, 1828-9, 1853-6, 1877-8), a series of wars between Russia and the *Ottoman empire, fought in the Balkans, the Crimea, and the Caucasus for political domination of those territories. The wars enabled the Slavonic nations of *Romania, *Serbia, and *Bulgaria to emerge and stimulated nationalist aspirations throughout the area to develop. In 1806-12 a vigorous campaign under Marshal *Kutuzov in the Balkans compelled the Turks to make peace, recognizing the autonomy of Serbia and ceding Bessarabia to Russia. The war of 1828-9 was an aspect of the *Greek War of Independence. Russian ships fought at the battle of Navarino. One Russian army invaded Wallachia and Moldavia and, advancing through the Balkans, threatened Constantinople; a second army crossed the Caucasus to reach the Upper Euphrates. The Treaty of Adrianople (1829), which ended the war, gave Wallachia and Moldavia effective independence and granted Russia control over a part of *Armenia. Russia was opposed in the *Crimean War of 1853-6 by Britain and France as well as Austria and Turkey, and, at the Treaty of *Paris, ceded territories. In 1876 the Turks quelled an uprising in *Bulgaria, causing a European outcry against the 'Bulgarian atrocities'. Russian forces invaded in 1877, allegedly to protect Bulgarian Christians; they again threatened Constantinople. The Treaty of San Stefano (March 1878) (*Three Emperors' League) which ended the war, provoked criticism from Britain and Germany and was modified by the Congress of *Berlin (June 1878), as it was alleged to have given too much influence to Russia in the Balkans.

Rwanda, a country in central Africa. It obtained its present boundaries in the late 19th century under pastoral Tutsi kings who ruled over the agriculturalist Hutu. In 1890 Germany claimed it as part of German East Africa, but never exercised effective control. Belgian forces took it in 1916, and administered it under a League of Nations *mandate. Following civil war (1959) between the Tutsi and Hutu, Rwanda was declared a republic in 1961, and became independent in 1962. The now dominant Hutu forced large numbers of Tutsi into exile, but since the accession to power of President Juvénal Habyarimana in 1973 domestic stability has gradually improved.

ryotwari system (from the Arabic *ra'iya*, 'subject' or 'peasant'), a term introduced in Madras in India to denote direct settlement of land taxation between the government and the cultivators, without the intervention of a landlord. In the Madras Presidency the ryotwari system was devised by Governor Thomas Munro in 1820. Elsewhere in India taxation systems were often via the landlord, a system which often worked against the interests of the ryot or peasant. Tenancy legislation was later passed to protect the interests of the peasants, adopting the ryotwari system.

S

SA *Brownshirts.

Saar (French, Sarre), river of France and south-west Germany. In 1792 the Saar valley was occupied by the French, but after the defeat of Napoleon in 1815 most of the area was ceded to Prussia. Its progress as a major industrial area really began after the unification of Germany in 1871 and the acquisition of *Alsace-Lorraine, with its coal and iron deposits. A rapid economic development took place, which was interrupted after World War I, when, as part of the *Versailles Settlement, the area was placed under the administration of the League of Nations and its mines awarded to France. In 1935 a plebiscite restored it to Germany, but it was again occupied by French troops in 1945. In 1955 a referendum voted for restoration to Germany, and in 1959 the Saarland became the tenth state of the Federal Republic of Germany.

Sabah, state of Malaysia. Controlled in the 19th century by the sultan of Brunei, the area now known as Sabah was leased in 1881 to the British North Borneo Company and in 1888 was taken under British protection as North Borneo. Occupied by the Japanese during World War II, it was united with Labuan as a crown colony in 1946. In 1963 it became a state within the Federation of *Malaysia. It was subject to Indonesian attempts at subversion during the *Konfrontasi (1963-6) and to unsuccessful Filipino claims of sovereignty.

Sacco–Vanzetti Case, US legal case (1920-7). Two Italian immigrants, Nicola Sacco and Bertolomeo Vanzetti, were found guilty of murder amid allegations that their conviction resulted from prejudice against them as immigrants, *anarchists, and evaders of military service. There were anti-US demonstrations in Rome, Lisbon, and Montevideo, and one in Paris, where a bomb killed twenty people. For six years efforts were made to obtain a retrial without success, although the judge was officially criticized for his conduct during the trial; the two men were electrocuted in August 1927. The affair helped to mobilize opinion against the prevailing *isolationism and conservatism of post-war America. Later evidence pointed to the crime having been committed by members of a gang led by Joe Morrelli.

Sadat, (Muhammad) Anwar (1918-81), Egyptian statesman. An original member of *Nasser's Free Officers association committed to Egyptian nationalism, he was imprisoned by the British for being a German agent during World War II, and again (1946-9) for terrorist acts. He took part in the coup (1952) that deposed King *Farouk and brought Nasser to power. He succeeded Nasser as President of Egypt (1970-81). By 1972 he had dismissed the Soviet military mission to Egypt and, in 1974, following the *Yom Kippur War, he recovered the Suez Canal Zone from Israel. In an effort to hasten a Middle East settlement he went to Israel in 1977. This marked the first recognition of Israel by an Arab state

The murder of Egypt's President **Sadat** on 6 October 1981. Sadat was reviewing a military parade commemorating the Arab–Israeli War of 1973 when four assassins shot him. For his efforts to negotiate peace in the Middle East he, together with Menachem Begin, Prime Minister of Israel, was awarded the Nobel Peace Prize of 1978, but opposition within the Arab world was to cost him his life.

and brought strong condemnation from most of the Arab world. He met the Israeli Prime Minister, Menachim *Begin, again at *Camp David, Maryland, USA (1978), under the chairmanship of President Carter, and a peace treaty between Israel and Egypt was finally signed at Washington in 1979. Mounting Egyptian disillusionment led to his assassination in 1981.

Sadowa, battle of (3 July 1866), fought near the Bohemian town of Sadowa (Königgratz), between the Prussian army under *Moltke and Benedek's Austrian army. The Prussians were able to overcome Benedek by their superior mobility and weapons. This battle decided the *Seven Weeks' War and marked the end of Austrian influence in Germany. Prussian domination in north Germany was confirmed.

Said ibn Sultan Sayyid (1791-1856), ruler of Oman and Zanzibar (1806-56). In 1806 he became ruler (Sayyid) of *Oman, with his capital at Muscat on the Persian Gulf. In 1822, assisted by the British, he sent an expedition to Mombasa, whose rulers, the Mazrui family, owed him nominal allegiance, but who were seeking independence. He himself visited Mombasa in 1827 and in the next decade brought many East African ports under his control. In 1837 he ended Mazrui rule in Mombasa and signed commercial agreements with

Britain, France, and the USA. He first visited *Zanzibar in 1828, buying property and introducing clove production. In 1840 he took control of Zanzibar. Said sent trading caravans deep into Africa, seeking ivory and slaves, and Zanzibar became the commercial capital of the East African coast. Although an ally of the British, he was under constant pressure from them to end his trade in slaves, and he signed an agreement to do this in 1845. When he died, he divided the Asian and African parts of his empire between his two sons.

Saigo Takamori (1828–77), Japanese soldier and statesman. A member of a lowly but prestigious *samurai family, he played a central role in the overthrow of the *shogunate and the establishment of the *Meiji imperial state. Showered with the highest honours, he initially retired from public life, but in 1871 was persuaded to return to the government as commander of the Imperial Guard. Fearing for the decline of the samurai way of life in the face of the introduction of conscription, Saigo promoted a war of redemption against Korea, to be triggered off by his own murder at Korean hands, but retired in 1873 when this plan was vetoed. Subsequently his private school at Kagoshima became a centre for samurai dissatisfaction, and in 1877 he was forced into rebellion by the actions of his followers. Defeated by government forces under *Yamagata, he had himself killed by one of his own men.

St Laurent, Louis Stephen (1882–1973), French-Canadian lawyer and statesman. As Minister of Justice (1941–6) he upheld limited military conscription in 1944 in the face of widespread French-Canadian opposition. In 1946 he became Secretary of State for External Affairs. He succeeded Mackenzie *King as Prime Minister of Canada (1948–57). He played a significant part in setting up the *NATO alliance, and did much to raise the international reputation of Canada. Significant constitutional changes were also made during his administration, with the word 'Dominion' being dropped and with Newfoundland becoming the tenth Province in 1949. After overwhelming victories in 1949 and 1953, he was defeated in the election of 1957. As only the second French-Canadian to become Prime Minister, St Laurent gave notable service in the promotion of good relations between English- and French-speaking Canadians.

Saint-Simon, Claude Henri de Rouvroy, comte de (1760–1825), French social philosopher and founder of French socialism. He fought in the American War of Independence and was imprisoned in France during the Terror. Between 1814 and 1825 he published his doctrines on the organization of society, in which he envisaged government not by landowners, lawyers, and priests but by industrialists, scientists, and poets, representing the three faculties of action, thought, and feeling. After his death his theories, summed up in his major work, *Nouveau Christianisme* (1825), exerted a far-reaching influence, and a generation of Saint-Simonian technocrats emerged committed to the industrial development of France.

Salazar, Antonio de Oliveira (1889–1970), Portuguese statesman. He was Prime Minister and, in effect, dictator of Portugal (1932–68). A successful Finance Minister (1928–32), he was invited to become Premier (1932–68). He introduced a new constitution in 1933, creating the New State (*Estado Novo*) along fascist lines, using his authority to achieve social and economic reforms. During the *Spanish Civil War and *World War II Salazar was Minister for Foreign Affairs and maintained a policy of neutrality. His policy of defending Portugal's African colonies in the face of mounting nationalism embittered his military leaders, who were forced to wage difficult battles in Africa. He was succeeded by *Caetano.

Salisbury, Robert Arthur Talbot Gascoigne-Cecil, 3rd Marquess of (1830–1903), British statesman. He became Prime Minister of a 'caretaker' Conservative government in 1885 when the Liberal leader, William *Gladstone, resigned. Having won an election he remained as Prime Minister (1886–92) and again headed governments (1895–1900 and 1900–2). In foreign affairs he refused to allow Britain to become embroiled in the alliance-making provoked by Franco-German rivalries, and supported the policies which resulted in the second *Boer War (1899–1902). Domestically his main achievement was administrative, for example in providing a new organizational framework for local government between 1888 and 1899.

Sālote Tupou III   *Tonga.

SALT   *Strategic Arms Limitation Talks.

A portrait of the Marquess of **Salisbury** by George Richmond. It shows the future Conservative Prime Minister in his robes as the Chancellor of the University of Oxford, a post he held from 1869 to 1903. (Hatfield House)

In one of the most spectacular campaigns in Gandhi's non-violent campaign against the British Raj, he led the 320-km. (200-mile) **Salt March** in 1930, to extract illegally salt from the sea. Its immediate result was the imprisonment of more than 60,000 people.

Salt March (12 March–6 April 1930), a march by Indian nationalists led by Mohandas *Gandhi. The private manufacture of salt violated the salt tax system imposed by the British, and in a new campaign of civil disobedience Gandhi led his followers from his ashram at Sabarmati to make salt from the sea at Dandi, a distance of 320 km. (200 miles). The government remained inactive until the protesters marched on a government salt depot. Gandhi was arrested on 5 May, but his followers continued the movement of *civil disobedience.

Salvation Army, international Christian evangelical and charitable organization, founded in London in 1865 by William *Booth as the Christian Revival Association. Preaching in the slums of London, Booth used unconventional methods to win people to Christianity, insisting on militant teetotalism and revivalism. In 1878 his mission took on the name Salvation Army, run on military lines, and with a fundamentalist approach to religion. The Army placed emphasis on welfare work and devoted much time to helping the destitute. It expanded rapidly not only in Britain but also overseas; in the USA Ballington Booth, a son of the founder, set up (1896) a splinter group, the Volunteers of America.

Samoa, a chain of volcanic islands in the southern Pacific. The indigenous population is Polynesian. The first European to visit the islands was the Dutch navigator Jacob Roggeveen, in 1722. International rivalry over the islands was settled in 1899, when Western Samoa was annexed to Germany and Eastern Samoa to the USA. During World War I New Zealand seized Western Samoa, guiding it through serious nationalist disturbances in the 1920s to eventual independence in 1962. American Samoa remains an unincorporated territory of the USA.

Samori Touré (1830–1900), African military leader. A Muslim, he began to amass a personal following in the mid-1850s, establishing a military base on the Upper Niger. By 1870 his authority was acknowledged throughout the Kanaka region of the River Milo, in what is now eastern Guinea. By 1880 he ruled a vast Dyula empire,

The **Salvation Army** parading through city streets. Drawing a crowd, through music and song, was a most effective way of spreading their religious message, and the 19th-century Salvationists established a musical tradition that survives to the present.

from the Upper Volta in the east to the Fouta Djallon in the west, over which he attempted to create a single Islamic administrative system. His imperial ambitions clashed with those of the French and there were sporadic battles between 1882 and 1886. His attempts to impose Islam on all his people resulted in a revolt in 1888. A French invasion in 1891–2 forced him to move eastwards to the interior of the Ivory Coast, where he established himself in Bondoukon (1891–8). French forces, however, captured him and he was exiled to Gabon.

samurai, a member of the feudal warrior class of Japan. The samurai were influential in Japanese society under the *shogunate, but lost their privileged position when the *Meiji restoration finally abolished feudalism in 1871. Rebellions by discontented former samurai during the 1870s were put down by the newly formed national army.

Sandinista Liberation Front   *Nicaragua.

Sandino, Augusto César (1893–1934), Nicaraguan revolutionary general. A guerrilla leader, he tenaciously resisted US intervention in Nicaragua from 1926 to 1933. His anti-imperialist stance attracted wide support in Latin America. After US marines withdrew, Sandino became leader of a co-operative farming scheme. Seen as a liberalizing influence, he was assassinated by Anastasio *Somoza's National Guard. The Sandinista Liberation Front, which defeated the Somoza dynasty in 1979, considers itself the spiritual heir of Sandino.

San Francisco Conference (1951), conference held to agree a formal peace treaty between Japan and the nations against which she had fought in World War II. When the treaty came into force in April 1952, the period of occupation (*Japan, occupation of) was formally ended and Japanese sovereignty restored. Japan recognized the independence of Korea and renounced its rights to Taiwan, the Pescadores, the Kuriles, southern Sakhalin, and the Pacific islands mandated to it before the war by the League of Nations. The country was allowed the right of self-defence with the proviso that the USA would maintain its own forces in Japan until the Japanese were able to shoulder their own defensive responsibilities. The Soviet Union did not sign the treaty, but diplomatic relations were restored in 1956, while peace treaties with Asian nations conquered by the Japanese in the war were signed through the 1950s as individual problems with reparations were resolved.

San Jacinto, battle of (21 April 1836), the last important battle of *Texas's brief struggle to establish an independent republic. Sam *Houston, with 800 Texans, defeated a Mexican force of 1,400 at the San Jacinto River and captured the Mexican leader *Santa Anna. The armistice terms dictated by Houston established *de facto* independence for Texas, and Houston himself was installed as President.

San Martín, José de (1778–1850), South American revolutionary. Born in Argentina, he trained as a military officer in Spain. On his return to Argentina in 1812, he joined the forces in rebellion against Spain and was sent to take command of the nationalist army in Mendoza. In an outstanding military feat he led his soldiers across the Andes and assured Chilean independence in the

battle of Maipú (April 1818). After landing his troops on the Peruvian coast in 1820, he took Lima unopposed on 28 July 1821. He became Protector of Peru (1821–2). In late July 1822, San Martín met with Simón *Bolívar at Guayaquil and decided to withdraw, leaving the liberation of the rest of Peru to Bolívar. In 1824 he sailed for Europe, where he died in 1850.

Santa Anna, Antonio López de (1794–1876), Mexican military adventurer and statesman. He entered the Spanish colonial army and served as one of the Creole supporters of the Spanish government until 1821, when *Iturbide made him governor of Vera Cruz. At first a supporter of the Federal Party, he subsequently overthrew (1822) Iturbide and himself became (1833) President of Mexico. His policies led to the uprising at *Alamo, to his defeat and capture in the battle of *San Jacinto (1836), and to the secession of *Texas. He was released, and returned to Vera Cruz, where he defended the city against the French (1836–9). In his next presidential tenure during the early 1840s, he discarded the liberal constitution of 1824 and ruled as a dictator. Subsequently, despite defeat in the *Mexican–American War and the loss of half of Mexico's territory to the USA, Santa Anna was recalled to the presidency in 1853 by Mexican conservatives. In 1855 the liberal revolution of Ayutla deposed him.

Sanusi, popular name for the *Sanusiyyah*, a Muslim brotherhood. It was founded in Mecca in 1837 as a Sufi religious order by an Algerian, Sidi Muhammad al-Sanusi al-Idrisi, but in c.1843 he retired to the desert in Cyrenaica. The movement spread in *Libya under the son and grandson of al-Sanusi; by 1884 there were 100 *zawiyas* or daughter houses, scattered through North Africa and further afield. It became important politically in both world wars, becoming more militant and attacking the British occupation of Egypt in World War II, and opposing the Italians in Libya. When Libya became independent in 1951, the leader of the order at that time, Idris I, became the country's first king.

Sarajevo   *Francis Ferdinand.

Sarawak, state of Malaysia in western Borneo. Ceded by the sultan of Brunei to Sir James *Brooke in 1841 after the latter had put down a revolt, Sarawak became an independent state under Brooke (the 'White Raja'). Although it became a British protectorate in 1888, it remained under the effective control of the Brooke family until the Japanese occupation of 1942–5. The Brooke family ceded it to Britain in 1946, and it became a crown colony. After a guerrilla war in 1962–3 during the run up to independence, it joined the Federation of *Malaysia in 1963. Some fighting occurred there during the *Konfrontasi with Indonesia (1963–6).

Sarekat Islam, Indonesian Islamic political organization. Formed in 1911 as an association of Javanese *batik* traders to protect themselves against Chinese competition, it had developed, by the time of its first party congress in 1913, into a mass organization dedicated to self-government through constitutional means. Its leader H. Q. S. Cokroaminoto (1882–1934), was viewed by many as a latter-day Messiah, but the organization was weakened from within by the political challenge posed by the emergent *PKI in the early 1920s, and gradually faded away as more radical nationalist parties, most prominently *Sukarno's PNI, were formed.

Sato Eisaku (1901–75), Japanese statesman. As a supporter of *Yoshida Shigeru, he advocated co-operation with the USA in the immediate post-war period. Forced from the cabinet over allegations of corruption in 1954, he returned four years later and between 1964 and 1972 served as Prime Minister. He overcame a period of student violence, oversaw the extension of the revised United States Security Treaty (1970), negotiated with the USA for the return of *Okinawa and the other Ryukyu islands, and normalized relations with South Korea. After leaving office he received a Nobel Peace Prize for his efforts to make Japan a nuclear-free zone.

satyagraha (Hindu, 'holding to the truth'), a campaign of civil disobedience employing *passive resistance, developed by Mohandas *Gandhi in South Africa, and widely used in India as a weapon against British rule. Commonly, campaigns of civil disobedience degenerated into violence, but the method had some success against a liberal government normally reluctant to use force. The technique continued to be employed in India and elsewhere after 1947, for example in Goa in 1955, when the satyagrahis were fired on and defeated.

Saud, the name of the ruling family of Saudi Arabia. Originally established at Dariyya in Wadi Hanifa, Nejd, in the 15th century, its fortunes grew after 1745 when Muhammad ibn Saud allied himself with the Islamic revivalist 'Abd al-Wahhab (*Wahhabism), who later became the spiritual guide of the family. The first wave of Saudi expansion ended with defeat by Egypt in 1818, but Saudi fortunes revived under Abd al-Aziz ibn Saud (c.1880–1953), who captured Riyadh (1902), al-Hasa (1913), Asir (1920–6), Hail (1921), and the *Hejaz (1924–5), thus assembling the territories which formed the kingdom of *Saudi Arabia in 1932. Abd al-Aziz was succeeded by his sons, Saud IV ibn Saud (1953–64), *Faisal II (1964–75), Khalid (1975–82), and Fahd (1982–), as rulers of the richest oil state in the world.

Saudi Arabia, a state in south-west Asia occupying most of the Arabian peninsula. It was formed from territories assembled by the *Saud family, who were followers of *Wahhabism, and proclaimed as the kingdom of Saudi Arabia in 1932. The early years of the kingdom were difficult, when revenues fell as a result of the declining Muslim pilgrim trade to Mecca and Medina. An oil concession was awarded to the US firm Standard of California in 1933 and oil was exported in 1938. In 1944 the oil company was re-formed as the Arabian American Oil Company (ARAMCO), and Saudi Arabia was recognized as having the world's largest reserves of oil. Since the death of Abd al-Aziz ibn Saud (1953) efforts have been made to modernize the administration by the passing of a series of new codes of conduct to conform both with Islamic tradition and 20th-century developments. The Saudi Arabian Minister for Petroleum and Natural Resources, Sheikh Ahmad Yemani, ably led the *OPEC in controlling oil prices in the 1970s. King Fahd succeeded to the throne after the death (1982) of his half-brother, Khalid. Since then the political stability of the country has been threatened by Islamic revivalists.

Savage, Michael Joseph (1872–1940), New Zealand statesman. Settling in New Zealand in 1907, he joined the Labour Party on its foundation, entering Parliament in 1919, and becoming deputy-leader in 1923. He took over as leader in 1933 on the death of Harry Holland, and became Prime Minister in 1935 after Labour's landslide victory. Savage is best remembered for his insistent advocacy of the Social Security Act and was one of the most popular of the country's political leaders.

Saxony, former German duchy and kingdom. In 1806 Frederick Augustus III allied with Napoleon, joining the *Confederation of the Rhine and being elevated to the throne. Although losing territory to Prussia at the Congress of *Vienna, Saxony remained the fifth largest kingdom in the *German Confederation. Under moderately liberal rulers its constitution of 1831 survived until 1918, while industrialization gained from membership of the *Zollverein. The kingdom survived the *Revolutions of 1848, declaring for Austria in the war of 1866. This resulted in an indemnity of 10 million thalers to Prussia, and an obligation to join the North *German Confederation. Incorporated into the *German Second empire in 1871, its industrial workers provided strong support for the Social Democratic Party, which earned it the title 'red Saxony'. After 1918 it became a state of the *Weimar Republic and then of the *Third Reich. In 1945 it became a part of the German Democratic Republic.

scalawag, a white supporter of the Republican *Reconstruction programme in the American South in the early years after the *American Civil War. Like the *carpetbaggers with whom they associated, the scalawags were a diverse group including some profiteers, but also businessmen, genuine reformers, many former Southern Whigs, and poor yeoman farmers who supported the Republican regime out of opposition to the old ruling planter class.

Scapa Flow, a stretch of sea in the Orkney Islands, Scotland. In May 1919 the terms of the *Versailles Peace Settlement were submitted to the Germans, who protested vigorously. As an act of defiance, orders were given under Admiral von Reuter to scuttle and sink the entire German High Seas Fleet, then interned at Scapa Flow. In October 1939 the defences of Scapa Flow were penetrated when a German U-boat sank HMS *Royal Oak*.

Schacht, Hjalmar (1877–1970), German financier. As Commissioner of Currency (1923) his rigorous monetary policy stabilized the mark after its collapse in that year. He took part in *reparations negotiations but rejected the *Young Plan (1929). Under *Hitler he became Minister of Economics (1934–7), responsible for Nazi programmes on unemployment and rearmament. Rivalry with *Goering caused his resignation. In 1944 he was imprisoned in a concentration camp for his alleged conspiracy in the *July plot to assassinate Hitler. At the *Nuremberg Trials (1946) he was acquitted.

Scharnhorst, Gerhard Johann David von (1755–1813), Prussian general and military reformer. He served in the Hanoverian army before entering the Prussian army in 1801. Following *Napoleon's defeats of Prussia (1806–7) Scharnhorst began his reform of the Prussian army, converting it from a mercenary to a conscripted force. He abolished capital punishment and promoted non-aristocrats to the officer corps. He resigned in 1812 when Prussia was forced into an alliance with Napoleon against Russia. France's defeat before Moscow enabled Prussia to join the anti-French coalition (1813), and Scharnhorst returned as chief of staff to *Blücher.

Scheer, Reinhard (1863–1928), German admiral. After winning fame as a submarine expert, he commanded the High Seas Fleet (1916–18). His hopes of dividing and defeating the British Grand Fleet at *Jutland (1916) failed, but his brilliant manoeuvring saved his own fleet. In October 1918 the German fleet at Kiel mutinied under him, refusing to put out to sea. The mutiny spread rapidly to north-west Germany, and by November Germany had accepted an end to World War I.

Schirach, Baldur von (1907–74), German Nazi youth leader. An enthusiastic Nazi while still a student, from 1933 to 1945 he led the *Hitler Youth. In 1940 he was appointed governor of Vienna, where he took part in plans to ship Vienna's Jews to *concentration camps. He was found guilty at the *Nuremberg trials and sentenced to twenty years' imprisonment.

Schleswig–Holstein, a state of north Germany within the Federal Republic of Germany. Both Schleswig and Holstein were originally duchies owing allegiance to the Danish crown. At the Congress of *Vienna Holstein was incorporated into the *German Confederation. In 1848 Denmark incorporated Schleswig, but the German-speaking population gained support from the German Parliament at Frankfurt and from Prussian troops, which invaded Denmark. Britain, Russia, and France intervened to oblige Prussia to agree to an armistice, and under the London Protocol of 1852 Denmark retained its rights in the duchies. However, Denmark in 1864 again incorporated Schleswig. Prussian and Austrian troops invaded and defeated the Danish army. In 1866, following war with Austria, Prussia annexed both duchies. After World War I there were plebiscites and much of north Schleswig passed to Denmark as the province of South Jutland. Between the wars the existence of a German minority in the province created considerable tension. After World War II over three million refugees from East Germany crowded into Schleswig-Holstein and the area was reorganized to become a West German state.

Schlieffen, Alfred, Graf von (1833–1913), German field-marshal and strategist. He developed the Schlieffen Plan, which formed the basis for the German attack in 1914. According to the plan, Germany could fight on two fronts by descending through Belgium and neutralizing France in a swift campaign, and then attacking Russia. The plan failed due to French resistance and German lack of military manœuvrability. It was abandoned when Germany's leaders decided to withdraw forces from the *Western Front to stem Russian advances into East Prussia. In 1940 Hitler successfully employed the principles of the Schlieffen Plan in his *Blitzkrieg or 'lightning war' in the west.

Schmidt, Helmut (1918–), German statesman. A member of the Social Democratic Party, he was elected to the Bundestag (Parliament of the Federal Republic of Germany) in 1953. He was Minister of Defence (1969–72) and of Finance (1972–4). Elected federal Chancellor in 1974, following the resignation of Willy *Brandt, he served for a second period (1978–82), during which he increasingly lost the support of the left wing of his party and of the *Green Party. He sought to continue the Brandt policy of Ostpolitik or dialogue with the *German Democratic Republic and the Soviet Union.

school systems, systems for the provision of public primary and secondary education. Early systems were established by religious orders, but before 1800 the concept of education for all had only emerged in a few Calvinist countries such as Scotland, the Netherlands, and the New England colonies of America. From the late 18th century, however, state systems developed. France (1791) and Prussia (1807) were the first two countries to establish secular state primary and secondary schools, Napoleon imposing the French system on most of Western Europe (in France the system reverted to Church control after 1814). From the early 19th century each state in the USA was obliged by the federal government to provide a secular school system. In England church schools and *Sunday schools were the main providers of primary education and mass education was pioneered by the monitorial system of Joseph *Lancaster. There were not sufficient schools in towns, however, and the Elementary Education Act (1870) required local authorities to provide schools where needed. Secondary education was provided largely by fee-paying grammar schools and 'public' (in fact, private) schools. The Education Act of 1902 established a state system for both primary and secondary education, later supplemented by the Butler Act (1944). Systems vary in terms of administration, curriculum, and teaching methods, and they broadly divide between the centralized Prussian and French systems and the more decentralized US and British systems. These four systems have acted as models: the Prussian, for example in mid-19th-century Russia and Turkey; the French, for example in much of western

School systems: As mass primary education became the norm, basic hygiene as well as the 'three Rs' appeared on the curriculum. In this classroom, c.1898, girls are being taught how to clean their teeth. The blackboard admonition reads, 'Spare the brush and spoil the teeth'.

Europe, Egypt, the Middle East, French-speaking Africa, and much of South-east Asia; the US system throughout much of South America, China, and Japan, where a national system of universal education was established in 1871; and the British in India and throughout the *Commonwealth of Nations. Curricula have been affected by political ideology, as in some communist countries, or by religious belief. In the late 20th century primary education has become almost universal throughout the world while secondary and technical education for all are high priorities for developing countries.

Schuman Plan (9 May 1950), a proposal drafted by Jean *Monnet and put forward by the French Foreign Minister Robert Schuman. It aimed initially to pool the coal and steel industries of France and the Federal Republic of Germany under a common authority which other European nations might join. The Plan became effective in 1952 with the formation of the European Coal and Steel Community, to which Italy, Belgium, Holland, and Luxembourg as well as France and West Germany belonged. Britain declined to join. Its success ultimately led to the formation of the *European Economic Community.

Schuschnigg, Kurt von (1897–1977), Austrian statesman. He became Chancellor following the murder of *Dollfuss (1934). He considered his main task to be the prevention of German absorption of Austria. Although an Austro-German Agreement (July 1936) guaranteed Austrian independence, Hitler accused him of breaking it. In February 1938 Hitler obliged him to accept Nazis in his cabinet. His attempt to hold a plebiscite on Austrian independence was prevented and he was forced to resign. On 12 March German troops invaded Austria without resistance in the *Anschluss.

Schwarzenberg, Felix, Prince of (1800–52), Austrian statesman. A career diplomat, Schwarzenberg joined the army of Field-Marshal Joseph *Radetsky on the outbreak of the *Revolutions of 1848. He persuaded the ageing Ferdinand I to abdicate in favour of his nephew, *Francis I. Opposed to granting autonomy to Austria's many states, Schwarzenberg drew up a constitution (1849) which transformed the Habsburg empire into a unitary, centralized, and absolutist state with strengthened imperial powers. The *Hungarian nationalist uprising was crushed (1849) with Russian aid, and Habsburg supremacy was restored in northern Italy. He secured the revival of a strengthened *German Confederation which maintained a precarious balance of power with *Prussia.

Schwarzenberg, Karl Philipp, Prince of (1771–1820), Austrian field-marshal. He entered the imperial cavalry in 1787. His courage at the battles of Hohenlinden (1800) and Ulm (1805) saved many Austrian lives, and he was then appointed vice-president of the supreme imperial war cabinet in Vienna, where he was responsible for raising a popular militia to defend Austrian homelands. As general of cavalry he fought at the unsuccessful battle of *Wagram (1809), after which Austria made peace with Napoleon at the Treaty of Schönbrunn. Schwarzenberg negotiated (1810) the marriage between Napoleon and Marie-Louise, daughter of the Austrian emperor, who in 1811 agreed to assist Napoleon in his forthcoming campaign against Russia. After Napoleon's failure to capture Moscow (1812) Schwarzenberg skilfully withdrew his troops back to Austria. Next year, when Austria joined Russia, Prussia, and Sweden to fight Napoleon (August 1813), he was appointed commander-in-chief of the Austrian Army and was the senior commander at the battle of *Leipzig in October. He attended the Congress of *Vienna and then retired.

Schweitzer, Albert (1875–1965), German medical missionary, musician, and theologian. In 1913 he went as a Lutheran missionary doctor to Lambaréné in Gabon, where he set up a hospital. He organized his hospital as an African village, in which he played the role of chief. His success in this won him world-wide fame.

Scopes case (July 1925), a US trial. A biology teacher in Dayton, Tennessee, John T. Scopes was charged with violating state law by teaching Darwin's theory of evolution. The state legislature had enacted (1925) that it was unlawful to teach any doctrine denying the literal truth of the account of the creation as presented in the Authorized (King James) Version of the Bible. The judge ruled out any discussion of constitutional legality, and since Scopes clearly had taught Darwin's theory of evolution he was convicted and fined $100. On appeal to the state Supreme Court the constitutionality of the state's law was upheld, but Scopes was acquitted on the technicality that he had been 'fined excessively'. The law was repealed in 1967.

Scotland, the northern part of Great Britain. In the late 18th century Scotland's industrial revolution began in the Lowlands, and the 19th century brought explosive growth in heavy industries. The increasing market for meat and wool led to the *Highland Clearances. In the 20th century Scotland's heavy industries have declined, and new industries, such as microelectronics and North Sea oil, have slowed but not halted the country's relative decline. Partly as a result, the future of the union with England has been subject to periodic bouts of questioning. The *Scottish Nationalists narrowly won a referendum (1979) for a greater degree of independence from Britain and a proposal to introduce devolution (limited home rule), but failed to carry it in the House of Commons.

Scott, Winfield (1786–1866), US general. He joined the army in 1808 and distinguished himself in the *War of 1812. After serving in the *Black Hawk War, acting as presidential emissary during the South Carolina *Nullification Crisis, and supervising the removal of the *Cherokee Indians to the south-west, he was appointed general-in-chief of the US Army (1841–61). His successful conduct of the march on Mexico City during the *Mexican–American War of 1846–8 made him a national hero. An opponent of secession of the Southern states, Scott was still in post at the outbreak of the *American Civil War, but retired some six months later.

Scottish Nationalist Party, a Scottish political party, formed in 1934 from a merger of the National Party of Scotland and the Scottish Party. The party gained its first parliamentary seat in 1945 at a by-election in Motherwell. In the October 1974 general election eleven of its candidates won parliamentary seats. In 1979 a referendum in Scotland on a Scottish representative assembly failed to elicit the required majority, and in the 1979 general election all but two of the SNP candidates were defeated. Three were elected in 1987.

Scottsboro case (1931), US litigation, in which nine black youths were falsely accused by two white girls of multiple rape on a train near Scottsboro, Alabama. They were found guilty and sentenced to death or long-term imprisonment. The sensational case highlighted race relations in Alabama and across the USA. The intervention of the Supreme Court and a series of retrials returned a verdict of not proven and all the Scottsboro boys were released in the years 1937–50.

'Scramble for Africa', a term used loosely to describe the period between the French occupation of Tunis (1881)

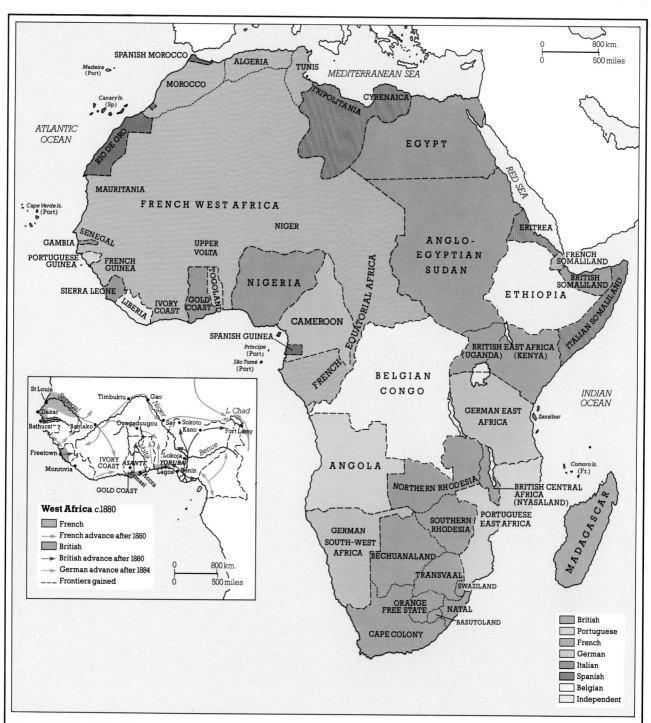

'Scramble for Africa'

Attempts in the 19th century to suppress the slave trade led to increasing colonial involvement. This was stimulated by missionary activity, the discovery of raw materials—minerals, oils, timber, ivory—and a need for markets. In 1884 a conference in Berlin of those European powers interested in Africa produced broad agreement on administrative frontier lines, irrespective of cultural, linguistic, or tribal affinities of the people involved.

and the end of the Second *Boer War (1902). During it almost all Africa was partitioned between Belgium, Britain, France, Germany, Spain, and Portugal. Later Italy ousted Turkey from *Tripolitania, having already taken possession of *Eritrea; while France and Spain partitioned Morocco. The USA maintained an economic foothold in Liberia. Conferences in Berlin (1884) and Brussels (1890) settled the outlines of the new states. In all the events of the period the people concerned, the inhabitants of the new colonies, were seldom consulted. After World War II resistance movements, until then largely muffled or suppressed, erupted in nationalist movements demanding independence.

Scullin, James Henry (1876–1953), Australian statesman. He was a goldminer, shopkeeper, and organizer for the Australian Workers' Union before becoming a Labor Member of the House of Representatives (1910–13). He was re-elected in 1922, led the Opposition (1928–9), and in 1929 became Prime Minister. In the *Depression he faced deepening divisions within his own party, and deflationary measures brought electoral defeat. From 1932 he led the Opposition until his resignation as leader of the Labor Party in 1935.

SEATO   *South-East Asia Treaty Organization.

Second Front, the term used in World War II to describe the return of Allied forces to fight in continental Europe. The First Front (a term not used) was that fought by the Soviet Union and it was the Soviet

government which pressed for an early opening of the Second Front as a means of relieving heavy German pressure in the east. The hope that it would be opened in 1942 was ended by Churchill's insistence that there was insufficient shipping. The disaster of the *Dieppe Raid (August 1942) confirmed this, although the Soviet and the US governments continued to criticize British hesitancy through 1943. When the *Normandy Landings eventually opened the Second Front in June 1944 it was clear that the operation was an immense enterprise that could easily have failed if undertaken too hastily.

Second Reich   *German Second empire.

Security Council, United Nations, principal council of the *United Nations. It is charged with the responsibility of keeping world peace and is composed of five permanent members, Britain, the United States, the Soviet Union, the People's Republic of China, and France, and ten members elected to two-year terms by the *General Assembly. The Security Council may investigate any international dispute, and its recommendations, which might involve a peaceful settlement, the imposition of trade sanctions, or a request to UN members to provide military forces, are to be accepted by all member countries. In deciding upon a course of action, the Security Council requires the votes of nine members, but each of the five permanent members can veto a resolution by voting against it. This veto has been a controversial issue, for in many instances it has prevented UN action.

Sedan, battle of (1 September 1870), a battle fought on the River Meuse, near the Belgian frontier, between French and Prussian forces during the *Franco-Prussian War. The Prussians, discovering that *MacMahon's army had set out to relieve Metz, diverted two armies marching on Paris and encircled the army of *Napoleon III at

Rocket-firing aircraft of the Royal Air Force bombard German armour at the Falaise Gap in Normandy, 1944, in a scene painted by Frank Wootton. The Allied invasion of France, opening up the long-awaited **Second Front**, forced Germany to fight two major land campaigns at the same time. (Imperial War Museum, London)

French troops fight a losing battle at the border fortress of **Sedan**. Encircled by the Prussian army, the French emperor surrendered and the next morning he and 83,000 French soldiers became prisoners-of-war.

Sedan. The French, under heavy shellfire, surrendered unconditionally. Napoleon III was taken prisoner, together with a large army. In World War II the Germans breached the *Maginot Line when they crossed the River Meuse at Sedan (1940).

Seddon, Richard John (1845–1906), New Zealand statesman. Arriving in New Zealand in 1866, he became the miners' advocate and was elected (1881) as parliamentary member for Kumara. He was Minister of Public Works in *Ballance's first Liberal government and Premier from Ballance's death in 1893. Seddon oversaw the introduction of a range of radical legislation including low-interest credit for farmers, women's suffrage, *Reeves's Industrial Conciliation and Arbitration Act, old age pensions, free places in secondary schools, and a State Fire Insurance Office.

Seeckt, Hans von (1866–1936), German general. He had gained his experience of warfare in eastern Europe and the Balkans in World War I and skilfully rebuilt the German army during the *Weimar Republic (1919–33). Although this was limited by the *Versailles Peace Settlement to 100,000 men, he trained his soldiers as an efficient nucleus for a much larger army. The secret agreement concluded after the Treaty of *Rapallo (1922) permitting German troops to train in the Soviet Union enabled him to circumvent the peace treaty. His work enabled Hitler to expand the army rapidly.

Seku Ahmadu Lobbo (c.1775–1845), West African religious leader. A student of *Uthman dan Fodio, Ahmadu participated in Uthman's *jihad* (or holy war) before settling in the province of Macina (in Mali), where he founded an independent Muslim community. Expelled from Macina by the pagan king of Segu, he established a new capital at Hamdullahi and in 1818 proclaimed a *jihad*, capturing Macina, and extending his authority around it. He established a strictly theocratic Muslim Fulani state which survived until 1859, when it was absorbed in the Tukulor empire of *Umar ibn Said Tal.

selectors, small farmers in Australia during the second half of the 19th century. By the 1850s, *squatters had acquired much of the best agricultural land. Increasing demands were made, especially by those who had come during the gold rushes of the 1850s, for remaining land to be made available for small farms, at low cost. Selection before survey (hence the name 'selectors') was introduced in all of the colonies between 1858 and 1872. Factors causing the failure of many selectors included the unsuitability of much Australian land for agriculture, lack of capital, and opposition from squatters.

Self-help *Smiles, Samuel.

Self-Strengthening Movement, Chinese military and political reform movement of the second half of the 19th century. Initiated in the early 1860s by Feng Guifen and supported by *Zeng Guofan, *Zuo Zongtang, *Li Hongzhang, and Prince Gong, the Self-Strengthening Movement attempted to adapt western institutions and military innovations to Chinese needs. Prominent among the innovations introduced were the Zongli Yamen (1861), an imperial office established to manage relations with foreign countries, the Jiangnan Arsenal (1865), the Nanjing Arsenal (1867), the Beiyang fleet (1888), (China's first modern navy), and various government-sponsored modern industries. Such reforms, however, were superficial and failed to solve deep-seated institutional problems, as was made clear by China's humiliation in the *Sino–Japanese War of 1894–5.

Seminole Wars (1816–18, 1835–42), two Indian wars in the US south-east. Natives of Florida, the Seminole retaliated against US military forces sent into their area in search of escaped slaves. Andrew *Jackson's subsequent punitive expedition forced the Seminole south into the Everglades. In 1819 Spain ceded east Florida to the USA, and in 1832 the Seminole were forced to sign a treaty involving their removal to the Indian Territory west of the Mississippi (*Trail of Tears). A substantial part of the tribe under *Osceola refused to move and held out in the Everglades until Osceola was treacherously captured and most of his followers exterminated. General William T. Worth then ordered (1841) that the Seminoles' crops be burned and their villages destroyed. Starved into surrender, the Indians signed a peace treaty (1842) and accepted their deportation westwards.

Senanayake, Don Stephen (1884–1952), Sinhalese statesman. The chief architect of the independence of *Sri Lanka, he entered politics in 1915. He became Vice-President of the State Council in 1936, leader of the constitutional movement and, as head of the United National Party, the country's first Prime Minister (1947–52). He was succeeded as Prime Minister by his son, **Dudley Shelton Senanayake** (1911–73), who also became leader of the United National Party. Dudley resigned in 1954 in the face of growing pressure from the socialist Sri Lanka Freedom Party associated with Solomon *Bandaranaike, but led his party again in the elections of 1965 and formed a new government which endured until 1970, pursuing a policy of communal reconciliation.

Senate, US, upper house of the US Congress. The powers and composition of the Senate are set out in Article I of the US Constitution, and the Senate first met in 1789. Senators, two from each state, have six-year terms and were chosen by the state legislatures until 1913, when the Seventeenth Amendment provided for their direct election. The terms of one-third of the Senators expire every two years. A Senator must be at least 30 years old, not less than nine years a US citizen, and a resident of the state he or she represents. The Vice-President presides over the Senate, voting only in the case of a tie. The Senate must ratify all treaties, confirm important presidential appointments, and take an equal part with the House of *Representatives in legislation.

Senegal, a country in West Africa with the *Gambia as an enclave. Founded by France in the 17th century, the colony was disputed by Britain in the Napoleonic Wars. The interior was occupied by the French governor L. L. Faidherbe (1854–61); in 1871 the colony sent its first Deputy to the French Assembly. It became part of French West Africa in 1895, and in 1958 it was made an autonomous republic within the *French Community. It became part of the Federation of *Mali (1959–60). Under the leadership of Léopold Sédar *Senghor it became independent in 1960. In 1982 it federated with the *Gambia as Senegambia. The confederation shares certain joint institutions and the integration of defence and security, but each country remains a sovereign and independent state.

Senghor, Léopold Sédar (1906–), African statesman and poet, President of *Senegal (1959–80). In 1946 he was elected to the French National Assembly as a Socialist Deputy, and when Senegal became independent in 1959 he was elected President. Together with the writers Aimé Césaire and Léon Damas he formulated the concept of '*négritude*', which he defined as 'the sum total of cultural values of the Negro-African world'. In 1960 he sought unsuccessfully to achieve federation among what had been French West African colonies and in 1975 Senegal joined the *West African Economic Community.

Serbia, a constituent republic of Yugoslavia, formed from the former kingdom of Serbia. In 1804 a massed rebellion against the *Ottoman Turks under *Kara George was followed by short-lived independence, which, after a second insurrection and Russian interference resulted in autonomy (1817) under the suzerainty of the

sultan. This was confirmed by the Treaty of *Adrianople (1829) and in 1830 Milos *Obrenovich became hereditary Prince of Serbia, although his autocratic ways led to his abdication in 1839. In 1877 the Serbs allied themselves with the *Pan-Slav movement. In 1878, at the Congress of *Berlin, Serbia gained sovereign nationhood under Prince Milan Obrenovich, who had ruled since 1872. In 1882 he was proclaimed king. A period of political unrest followed, including war with Bulgaria (1885–6), culminating in the assassination of Alexander Obrenovich in 1903. His successor Peter Karageorgević allowed liberalization and parliamentary government. Austrian fears of Serb expansion into neighbouring *Bosnia-Hercegovina led it to annex the latter in 1908 and attempt to control Serbia. These policies led to the assassination in 1914 of the Austrian archduke *Francis Ferdinand by a Serbian nationalist, precipitating World War I. In 1918 Serbia absorbed Bosnia and Hercegovina and joined with Croatia and Slovenia; in 1929 it took the name of *Yugoslavia.

Sevastopol, Russian port and naval base on the south west coast of the Crimean peninsula. By the outbreak of the *Crimean War its strong fortifications had been completed and it was able to sustain an eleven-month siege after the battle of *Balaklava. It was almost completely destroyed after the Russian withdrawal in September 1855 and its fortifications ordered not to be rebuilt at the Congress of *Paris.

Seven Weeks War   *Austro-Prussian War.

Sèvres, Treaty of (1920), a treaty, part of the *Versailles Peace Settlement, signed between the Allies and Turkey, effectively marking the end of the *Ottoman empire. Adrianople and most of the hinterland to Constantinople (now Istanbul) passed to Greece; the Bos-

The siege of **Sevastopol**, as seen from the Anglo-French side. Heavy mortars (*foreground*) and cannon pound the strong Russian fortifications. Sevastopol withstood the Allied assault for eleven months.

Lord **Shaftesbury** (*centre*) goes underground to see for himself the appalling conditions in which children worked in the coal mines. During a visit to the 'Black Country' in central England in 1840 he saw young boys hauling trucks along shafts too small for men or pit ponies.

porus was internationalized and demilitarized; a short-lived independent *Armenia was created; Syria became a French *mandate; and Britain accepted the mandate for Iraq, Palestine, and Transjordan. The treaty was rejected by Mustafa Kemal *Atatürk, who secured a redefinition of Turkey's borders by the Treaty of *Lausanne (*Versailles Peace Settlement).

Seward, William Henry (1801–72), US statesman. He served as Whig governor of New York (1839–42) and then as Senator (1849–61). A convinced opponent of slavery, he joined the newly formed *Republican Party in 1855 and served as Secretary of State under *Lincoln during the *American Civil War. Wounded in a separate attack at the time of Lincoln's assassination, Seward recovered and stayed in office during the Presidency of Andrew *Johnson, generally supporting him against the radical Republicans. Seward believed in the need for the USA to expand its influence in the Pacific and was responsible for the US purchase of *Alaska from Russia (1867). He advocated friendly relations with China and pressed for the annexation of Hawaii and other islands to act as coaling stations for a US Pacific fleet.

Seychelles, a country of ninety-two islands in the Indian Ocean. They were captured from the French by Britain during the Napoleonic Wars and were administered from *Mauritius before becoming a separate crown colony in 1903. The islands gained universal suffrage in 1970, becoming an independent republic in 1975. In 1977 there was a coup, the Prime Minister, France Albert René proclaiming himself President.

Seyss-Inquart, Arthur (1892–1946), Austrian Nazi leader. As Interior Minister in Vienna, he organized the *Anschluss with Germany in 1938, and was made governor of Austria by Hitler. He later became the Nazi commissioner in the occupied Netherlands, where he was responsible for thousands of executions and deportations to *concentration camps. He was sentenced to death at the *Nuremberg Trials.

Shaba *Congo Crisis.

Shaftesbury, Anthony Ashley Cooper, 7th Earl of (1801–85), British politician and reformer. A Tory in politics, his first parliamentary campaign was over the abolition of 'suttee', the Hindu widow's practice of committing suicide on her husband's funeral pyre. He persuaded Parliament to pass an Act reforming the treatment of lunatics, and agitated for the abolition of slavery. He was largely responsible for the Ten Hours Factory Act of 1847, which shortened the working day in textile mills to ten hours. He supported charity schools (the 'Ragged Schools') for children in slums, and championed the abolition of boy chimney sweeps. An active reformer in urban housing for the poor, he pleaded for parks and playgrounds, and for the reduction of working hours. He opposed the development of trade-unionism and the *Reform Bills of 1832 and 1867, fearing that they might provoke class warfare.

Shahs of Iran *Muhammad Reza Shah Pahlavi; *Reza Shah Pahlavi.

Shaka (or Tshaka, or Chaka) (1787–1828), Zulu chief (1818–28). He was conscripted into *Dingiswayo's army c.1809. Rising rapidly, he became Chief of the Zulu in 1816. He re-organized his army of 40,000 Zulu warriors into regiments (impi), arming them with a stabbing spear, issuing them with distinctive dress, and training them to go barefoot for mobility. Save for veterans, marriage was forbidden, and training was rigorous. In 1818 Shaka profited by Dingiswayo's death to extend his dominions. He subjugated all of what is now Natal. Women's regiments were organized, and the whole nation placed on a war footing against the Boers. In 1828 his half-brother *Dingaan assassinated him.

Shamil, (c.1798–1871), leader of Muslim resistance to the Russian occupation of the Caucasus from 1834 to 1859. He became Imam of a branch of the Sufi Naqshbandi order known as Muridism which recommended strict adherence to Islamic law and preached *jihad* (holy war) against Russia. After the Crimean War Russia employed some 200,000 troops in the Caucasus to encircle and subdue Shamil and his followers. He was captured (1859), and imprisoned, but allowed to go on a pilgrimage to Mecca (1870), where he died.

sharecropping system (US), a system of farm tenancy in the USA. The system developed in the Southern states after the abolition of slavery, involving both black and white tenant farmers who lacked the resources to provide their own equipment or stock. In return for the labour of the farmer and his family, a half share of the crop was provided. Generally, this was diminished in value by the need to obtain credit from the landlord for family needs. As late as 1940, some 750,000 sharecroppers remained, but after World War II their numbers declined as a result of farm mechanization and a reduction in land devoted to cotton cultivation.

Sharpeville massacre (21 March 1960), an incident in the South African township of Sharpeville. The police opened fire on a demonstration against *apartheid laws, killing sixty-seven Africans, and wounding 180. There was widespread international condemnation, and a state of emergency was declared in South Africa. 1,700 persons were detained, and the political parties, the *African

The mass funeral after the **Sharpeville massacre** of 1960, in which South African security forces killed sixty-seven Africans. Their burial in a communal grave was a rallying point for the black community.

National Congress and *Pan-African Congress were banned. Three weeks later a white farmer attempted to assassinate the Prime Minister, Verwoerd, and, as pressure from the Commonwealth against the apartheid policies mounted, South Africa became a republic and withdrew from the Commonwealth (1961).

Shearers' strikes (1891, 1894), major strikes in Queensland and New South Wales, Australia. Sheep shearers were fighting for the principles of unionism and the 'closed shop' (an establishment in which only trade-union members are employed). Sheep farmers were fighting for 'freedom of contract' (the right to employ anyone). The strikes were marked by violence and bitterness on both sides. Non-union labour was used. Union leaders, including some from the Barcaldine shearers' camp of 1891, were arrested on charges such as conspiracy, seditious language, and riot. Some were gaoled. The unions were defeated.

Shere Ali (1825–79), amir (ruler) of Afghanistan (1863–79). He succeeded his father, *Dost Muhammad. During the early part of his reign Afghanistan experienced civil war and his authority was not confirmed until 1868, when he was given British assistance. Shere Ali introduced a number of reforms in Afghanistan including the establishment of a regular, European-style army. In 1878 he admitted a Russian mission to Kabul but refused to accept a British mission, resulting in the Second *Anglo-Afghan War. Shere Ali fled to northern Afghanistan seeking Russian support, and died.

Sheridan, Philip Henry (1831–88), US general. He emerged as the outstanding cavalry leader on the Union (Northern) side in the *American Civil War, distinguishing himself in Tennessee and in the *Chattanooga campaign (November 1863) before being appointed in April 1864 to command the cavalry of the Army of the Potomac. His campaign in the Shenandoah Valley (September–October 1864) laid waste one of the south's most important supply regions, while his victory at Five

Forks on 1 April 1865 effectively forced Robert E. *Lee to abandon Petersburg and Richmond. After the war Sheridan commanded the 5th military district in the South, and in 1884 he succeeded *Sherman as commander-in-chief of the US Army.

Sherman, William Tecumseh (1820–91), US general. He served on the Union (Northern) side in the *American Civil War, commanding a brigade at the first battle of Bull Run and a division at Shiloh, before participating in the *Vicksburg and *Chattanooga campaigns as General *Grant's most trusted subordinate. He was appointed commander in the western theatre in March 1864 and conducted a successful campaign against Atlanta, which fell to him in September. Determined to carry the war into the heart of the Confederacy, he marched his army through Georgia to the sea and then northward through the Carolinas, taking the surrender of the forces of the *Confederacy there (1865). His devastation of the territory through which he marched gravely damaged the Southern war effort and earned him a reputation as a proponent of total war. He became commander-in-chief of the US Army (1869–83).

Shiite *Islam.

A portrait of General **Sherman**, 1866, by G. P. A. Healy. In a series of campaigns his Union forces crushed the South in the American Civil War. His speech in Ohio on 11 August 1880 warned, 'There is many a boy here today who looks on war as all glory, but, boys, it is all hell.' (Smithsonian Institution, Washington)

Shimonoseki, Treaty of (17 April 1895), treaty between China and Japan, ending the *Sino-Japanese War (1894–95). With her navy destroyed and Beijing in danger of capture, China was forced to grant the independence of Korea, pay a large indemnity, grant favourable trade terms, and cede Taiwan, the Pescadores Islands, and the Liaodong peninsula (including the naval base at Port Arthur, now Lüshun). International pressure forced the return of Port Arthur and the abandonment of the claim to the Liaodong peninsula shortly afterwards, but Japanese domination over north China had been established.

shogunate, system of Japanese government, abolished in 1868, in which the shoguns exercised civil and military power in the name of the emperors, who became figure-heads. Japan had been effectively ruled by the Tokugawa shogunate since the beginning of the 17th century, but from the 1840s it was progressively undermined by political pressures unleashed by increasing foreign incursions into Japanese territory. Resistance to the shogunate's conservative policies coalesced around advocates of a return to full imperial rule, and between 1866 and 1869 the Tokugawa armies were gradually defeated by an alliance of provincial forces from Choshu, Satsuma, and Tosa acting for the *Meiji emperor, who formally resumed imperial rule in January 1868.

Siam *Thailand.

Sicilies, Kingdom of the Two, a southern Italian kingdom, consisting of Sicily and mainland southern Italy centred on Naples. In 1799 a French Revolutionary army occupied Naples, declaring it the Parthenopean Republic, and King Ferdinand sought refuge in Palermo in Sicily, under British protection. He returned to Naples in June, escorted by Admiral *Nelson, ordering a mass execution of 100 Italian patriots accused of collaborating with the French. In 1806 he had again to seek British protection in Sicily, when Napoleon established first his brother Joseph and then General Joachim *Murat as kings of Naples. The nine years of French rule of the mainland were enlightened, and the power of the aristocracy was broken, but no constitution was granted and control was maintained by 40,000 men. In Sicily, under British pressure, Ferdinand issued a constitution in 1812 creating a Parliament along British lines, but was less successful in breaking the power of the feudal lords. In 1815 Ferdinand was restored again in Naples, now as Ferdinand I, King of the Two Sicilies, but his repressive regime was unpopular, as was that of his son Ferdinand II. The kingdom became a centre for nationalist societies seeking Italian unification. A revolt broke out in Naples (1820) and the first 1848 revolution in Italy took place in Palermo. *Garibaldi and his 'Thousand' landed at Marsala in Sicily in 1860 and defeated Ferdinand at a battle near his palace of Caserta. The kingdom voted by plebiscite to join the rest of the kingdom of Italy in October 1861.

Sidmouth, 1st Viscount *Addington, Henry.

Siegfried Line, a fortified defensive line in France. It was erected from Lens to Rheims in World War I by the Germans, after their failure to capture Verdun. Sometimes known as the Hindenburg Line, it proved

The **Siegfried Line** was of little help to German defenders attempting to stem the Allied tide of advance in 1944–5. Here US infantrymen pass 'dragon's teeth' anti-tank defences as they go through the line at Habscheid, Germany, in February 1945.

useful in 1917, enabling a front to be maintained with depleted forces. In World War II Hitler applied the term to the fortifications along Germany's western frontier. Some of it was briefly used by German troops retreating into Germany in 1944–5.

Sierra Leone, a small West African country. In 1772 Britain declared that any escaped slave who came to Britain would automatically become free. British philanthropists organized their transport to Cape Sierra Leone, where in 1788 Freetown was established, becoming the first British crown colony in Africa in 1806. After 1815 British warships who captured slave ships brought freed captives there. During the 19th century the hinterland of Sierra Leone was gradually explored and in 1896 it became a British protectorate, which remained separate from the colony of Freetown until 1951. The country gained its independence under Prime Minister Sir Milton Margai (1895–1964) in 1961, but after his death electoral difficulties produced two military coups before some stability was restored by the establishment of a one-party state under Dr Siaka Stevens. Food shortages, corruption, and tribal tensions produced serious violence in the early 1980s, and in 1985 Stevens retired in favour of Major-General Joseph Momoh, who, as head of state, retained a civilian cabinet.

Sihanouk, Norodom (1922–), Cambodian statesman and King of Cambodia (1941–55). He exploited the complicated political situation immediately after the *French Indo-Chinese War to win full independence for Cambodia (now *Kampuchea) in 1953. He abdicated in 1955 to form a political union with himself as Prime Minister and became Head of State in 1960. After attempting to remain neutral in the *Vietnam War, he became convinced that communist forces would win and began to lend covert assistance, earning US enmity, which contributed to his overthrow in Lon Nol's military coup in 1970. He supported the *Khmer Rouge from exile in China and returned as nominal head of state

The charge of the 16th Queen's Own Lancers against Sikh infantry at the battle of Aliwal, January 1846, during the **Sikh Wars**. This battle ended in one of the two decisive British victories in the first Sikh War. (National Army Museum, London)

following their victory in 1975, but was removed from office in the following year. In exile he has sought since 1979 to overthrow the Vietnamese-backed Heng Samrin regime in collaboration with Son Sann's nationalist forces.

Sikh Wars (1845-9), two conflicts between the Sikhs of Lahore and the English *East India Company India. The First Sikh War (1845-6) took place when Sikh troops crossed the Sutlej River into British India. After the drawn battles of Mudki and Firuzshah, the British defeated the Sikhs at Aliwal and Sobraon. By the Treaty of Lahore (1846) Britain obtained the cession of the Jullundar Doab, took Kashmir for Gulab Singh, and established control of the Lahore government through a Resident. Sikh discontent led to the Second Sikh War (1848-9); the bloody battle of Chillianwallah was followed by the decisive British victory at Gujerat over an army of 60,000 Sikhs. The governor-general, Lord Dalhousie, annexed the Punjab in 1849. The battles of the Sikh wars were the toughest which the British fought in India, the Sikh forces created by *Ranjit Singh being well trained in the European mode of war and determined in battle.

Sikkim, a small state in India in the eastern Himalayas. Until 1975 it was a protectorate state, ruled feudally by *chogyals* (kings) of the Namgyal dynasty. In the past Sikkim suffered continual invasions from Himalayan neighbours, especially Bhutan and *Nepal. Its strategic interest to Britain resulted in the Anglo-Sikkimese Treaty (1861), which made it a protectorate of British India. In spite of criticism of his feudal rule, the *chogyal* hoped to retain internal autonomy when Britian left India, but a referendum in 1975 demanded transfer to the Indian Union, which then followed.

Sikorski, Vladislav (1881-1943), Polish general and statesman. He commanded divisions against the *Bolsheviks (1919-20) and during 1922-3 headed a non-parliamentary coalition government in Poland. In 1939 he fled to France and organized a Polish army in exile

that fought with the Allies in World War II. As head of the exiled Polish government in London he succeeded in maintaining tolerable relations with Moscow until news of the *Katyn massacre broke. During his ascendancy Polish prisoners-of-war in the Soviet Union were recruited to form the 'Polish Army in Russia' under General Wladyslaw Anders to fight with the Allies. He was killed in an air crash.

Silesia, a region of eastern Europe. By 1800 Prussia had won control from Poland and Austria of both Upper and Lower Silesia, Austria retaining the province of Austrian Silesia. This was confirmed at the Congress of *Vienna. After the defeat of the Central Powers in 1918 and a series of plebiscites, Upper Silesia (the coal- and steel-producing area) went to Poland, and most of Austrian Silesia to Czechoslovakia; Germany was left with Lower Silesia. During the years between the wars there was heavy French financial investment in Polish Silesia, but also civil unrest as the interests of Germany, Poland, and Czechoslovakia were disputed. In 1939 Upper Silesia was occupied by the German *Third Reich and with its defeat in 1945 the *Potsdam Conference decreed that the whole area should pass to Poland. German nationals were repatriated, mainly to the Federal Republic of *Germany.

Singapore, south-east Asian island state. It was acquired for the English *East India Company by Sir Stamford *Raffles in 1819 from the sultan of Johore, and rapidly developed into an important trading port. In 1867 it was removed from British Indian administration to form part of the new colony of the *Straits Settlements, its commercial development, dependent on Chinese immigrants, proceeding alongside its growth as a major naval base. In 1942 it fell to Japanese forces under General Yamashita (*Singapore, fall of) and remained in Japanese hands until the end of World War II. The island became a separate colony in 1946 and enjoyed internal self-government from 1959 under the leadership of *Lee Kuan Yew. It joined the Federation of *Malaysia in 1963, but Malay fears that its predominantly Chinese population would discriminate in favour of the non-Malays led to its expulsion in 1965, since when it has been ruled as an independent republic by Lee Kuan Yew and his People's Action Party. A member of the *Commonwealth of Nations, and the *Association of South-East Asian Nations, it maintains close ties with Malaysia and Brunei.

Singapore, fall of (8–15 February 1942), one of the greatest Japanese victories in World War II. Although *Singapore had strong coastal defences, no fortifications had been built against land attack from the Malay peninsula, apart from providing for the causeway across the 1.6 km. (1 mile) Strait of Johore to be blown up. After swiftly overrunning Malaya, Japanese forces under General *Yamashita massed opposite the island at the beginning of February 1942. During the night of 7/8 February armoured landing-craft crossed the Strait of Johore, surprising the garrison of Australian troops opposite. Many Japanese troops followed by swimming across the water. The causeway was blown up and the defenders retreated. The garrison of British, Indian, and Australian troops numbered some 80,000 under General A. E. Percival. However, incessant air-attack by the Japanese destroyed oil tanks and supplies and reduced morale. Having repaired the causeway, further Japanese were sent in and Percival continued to retreat south-east towards the residential area. On 15 February attempts were made to evacuate key personnel by boat, but few survived, and Percival surrendered. The defeat was a significant milestone in the ending of British imperial interests in south-east Asia.

Sinn Fein (Gaelic, 'we ourselves'), an Irish political party dedicated to the creation of a united Irish republic. Originally founded by Arthur *Griffith in 1905 as a cultural revival movement, it became politically active and supported the *Easter Rising in 1916. Having won a large majority of seats in Ireland in the 1918 general election, Sinn Fein Members of Parliament, instead of going to London, met in Dublin and proclaimed Irish independence in 1919. An independent parliament (Dáil Éireann) was set up, though many of its MPs were in prison or on the run. Guerrilla warfare against British troops and police followed. The setting up of the *Irish Free State (December 1921) and the partition of Ireland were bitterly resented by Sinn Fein, and the Party abstained from the Dáil and the Northern Ireland parliament for many years. Sinn Fein today is the political wing of the Provisional *Irish Republican Army and has the support of the uncompromising Irish nationalists.

Sino–French War (1884–5), conflict between France and China over *Vietnam. China had assisted Vietnam in partial resistance to French expansion since the 1870s, first with irregular forces of the Black Flag Army and after 1883 with regular forces. In 1884, after both governments had rejected the compromise Li-Fournier

agreement, war broke out. The Chinese were unable to resist the French navy, which attacked Taiwan and destroyed Fuzhou (Foochow) dockyard in south-east China, and the *Qing dynasty's reputation was weakened. In the treaty signed in 1885, France won control of Vietnam.

Sino-Japanese War (1894–5), war fought between China and Japan. After Korea was opened to Japanese trade in 1876, it rapidly became an arena for rivalry between the expanding Japanese state and neighbouring China, of which Korea had been a vassal state since the 17th century. A rebellion in 1894 provided a pretext for both sides to send troops to Korea, but the Chinese were rapidly overwhelmed by superior Japanese troops, organization, and equipment. After the Beiyang fleet, one of the most important projects of the *Self-Strengthening Movement, was defeated at the battle of the Yellow Sea and Port Arthur (now Lüshun) captured, the Chinese found their capital Beijing menaced by advancing Japanese forces. They were forced to sign the Treaty of *Shimonoseki, granting Korean independence and making a series of commercial and territorial concessions which opened the way for a Japanese confrontation with Russia, the other expansionist power in north-east Asia.

Sino–Japanese War (1937–45), conflict on the Chinese mainland between nationalist and communist Chinese forces and Japan. China had been the target of Japanese expansionism since the late 19th century, and after the *Mukden Incident of 1931 full-scale war was only a matter of time. Hostilities broke out, without any formal declaration of war by either side, after a clash near the Marco Polo bridge just west of Beijing in 1937. The Japanese overran northern China, penetrating up the Yangtze and along the railway lines, capturing Shanghai, Nanjing, Guangzhou, and Hankou by the end of 1938. In the 'Rape of Nanjing', over 100,000 civilians were

Japanese troops in action during the **Sino-Japanese War** early in 1937. Though Chinese troops fought well, the Japanese enjoyed a crucial advantage in equipment that secured their initial victory, and enabled them to occupy large areas of eastern China in 1937–8.

massacred by Japanese troops. The invaders were opposed by both the *Kuomintang army of the nationalist leader *Chiang Kai-shek and the communist 8th Route Army, the former being supplied after 1941 by Britain and the USA. By the time the conflict had been absorbed into World War II, the Sino–Japanese War had reached a state of near stalemate, Japanese military and aerial superiority being insufficient to overcome tenacious Chinese resistance and the problems posed by massive distances and poor communications. The Chinese kept over a million Japanese troops tied down for the entire war, inflicting a heavy defeat upon them at Jiangxi in 1942 and successfully repelling a final series of offensives in 1944 and 1945. The Japanese finally surrendered to Chiang Kai-shek on 9 September 1945, leaving him to contest the control of China with *Mao Zedong's communist forces.

Sino–Soviet border dispute (March 1969), brief conflict between China and the Soviet Union over possession of an island in the Ussuri River. The exact position of the border between north-east China and the Soviet Union had long been a subject of dispute. The disagreement turned into a military confrontation because of the ideological dispute between China and the Soviet Union after 1960 and the militant nationalism which was part of the *Cultural Revolution. In March 1969 two battles were fought for possession of the small island of Zhen Bao (also known as Damansky). The Chinese ultimately retained control of the island, and talks in September 1969 brought the crisis to an end.

Sitting Bull (c.1834–90), Dakota Sioux Indian chief. He began his career as a warrior against the Crow Indians, and opposed white incursions on to the Great Plains in the 1860s and 1870s. It was his resistance to enforced settlement on a reservation that led to General *Custer's expedition of 1876 and the battle of the *Little Big Horn. Despite Custer's defeat, Sitting Bull had to flee to Canada, where he remained until 1881, attracted back by an amnesty. In 1885 he appeared in Buffalo Bill's 'Wild West' Show, but he continued to lead the Indians in their refusal to sell their lands to the white settlers, and advocated the Ghost Dance religion. This preached the coming of an Indian messiah who would restore the country to the Indians. In an uprising that followed, Sitting Bull was killed while 'resisting arrest'.

Six Acts (1819), legislation in Britain aimed at checking what was regarded as dangerous radicalism, in an immediate response to public anger over the *Peterloo Massacre. It dealt with procedures for bringing cases to trial, the prohibition of meetings 'for military exercises', the issue of warrants to search for arms, powers to seize seditious and/or blasphemous literature, the extension of a stamp-duty on newspapers and periodicals, and the regulation and control of all public meetings. The last three were particularly resented and regarded as a threat to freedom. The Acts proved counter-productive by provoking much opposition; three years later the government of Lord *Liverpool began to move towards more liberal policies.

Six-Day War (5–10 June 1967), Arab–Israeli war, known to the Arabs as the June War. The immediate causes of the war were the Egyptian request to the UN

Sitting Bull posing in costume, probably during his engagement with Buffalo Bill's 'Wild West' show in 1885, in which he gained international fame. Sitting Bull remained the leader of his people, who were among the last to rebel against white domination. He returned from the show to active resistance, as chief of the Sioux nation.

During the **Six-Day War**, Israel occupied Jordanian territory on the West Bank and on 9 June 1967 Israeli forces captured Bethlehem. Here an Israeli soldier escorts a Jordanian army prisoner through the town.

Emergency Force in Sinai to withdraw from the Israeli frontier, the increase of Egyptian forces in Sinai, and the closure of the Straits of Tiran (the Gulf of Aqaba) to Israeli shipping. An Egyptian, Syrian, and Jordanian military alliance was formed. The war was initiated by General Dayan as Israel's Minister of Defence, with a pre-emptive air strike which was followed by the occupation of Sinai, Old Jerusalem, the West Bank, and the Golan Heights (9–10 June). The Arab-Israeli conflict erupted again in the *Yom Kippur War of 1973.

Slaves, US Proclamation for the Emancipation of (January 1863), the executive order abolishing slavery in the 'rebel' (Confederate) states of the USA. The Proclamation, issued by President *Lincoln as commander-in-chief of the US Armed Forces, was partly a measure designed to win international support for the Union cause. It was of doubtful constitutional validity. Lincoln had issued a preliminary proclamation on 22 September 1862 advising that all slaves would be legally free as from 1 January 1863. This was now confirmed, to be enforced by military authority without compensation. After the war the US Congress passed (1865) the Thirteenth Amendment to the Constitution, abolishing slavery throughout the USA and thus confirming the constitutionality of the President's action.

slave trade, abolition of, the ending of trade in slaves. The slave trade reached its peak in the 18th century on the West African coast, where merchants from Europe worked in co-operation with native chieftains and slave raiders who were willing to exchange slaves for Western commodities. Denmark made participation in the Atlantic slave-trade illegal in 1792 and the USA did so in 1794. In Britain a group of humanitarian Christians, including Thomas *Clarkson and William *Wilberforce (members of the so-called Clapham Sect), argued that if the Atlantic slave-trade were abolished, with its appalling cruelties, plantation owners would treat their slaves more humanely, as being more valuable. They succeeded in getting Parliament to pass a Bill abolishing the British trade in 1807 and at the Treaties of Ghent (1814) and

US **slaves** parade in jubilation, displaying copies of President Lincoln's Emancipation Proclamation ending slavery, as proof that they are free. For many blacks, emancipation was to bring homelessness and a long trek to Northern states to look for work.

Vienna (1815) Britain agreed to use the Royal Navy to try to suppress the trade, most European countries now supporting the abolition of slavery. However, for the next forty years illegal smuggling, mainly from Africa, continued. Even after slavery was abolished in the British West Indies (1834) trade in slaves between the southern US slave states and Cuba, Costa Rica, Brazil, and elsewhere continued. Within the US southern states the breeding, transport, and sale of slaves became highly profitable, there being some four million slaves on the plantations by the time of the Emancipation Proclamation (1863). Only after such countries as Cuba (1886) and Brazil (1888) had ended slavery did trade in the Caribbean end. In the Muslim world the Arab slave trade from Africa operated from Morocco and Zanzibar, and stretched throughout the Ottoman empire into Persia and India. In 1873 the Sultan of Zanzibar was finally persuaded by Britain and Germany to close his markets, while the Moroccan trade gradually dwindled with French and Spanish occupation. However, as late as 1935 there was still evidence of trade in slaves in Africa.

Slim, William Joseph, 1st Viscount (1891–1970), British field-marshal. He commanded an Indian division in the 1941 conquest of the *Vichy French territory of Syria. In early 1942 he joined the *Burma campaign, and in 1943 took command of the 14th Army. After the victory at Kohima he pushed down the Irrawaddy River to recapture Rangoon and most of Burma. After the war he became Chief of the Imperial General Staff (1948–52) and governor-general of Australia (1953–60).

Smiles, Samuel (1812–1904), Scottish author and journalist. In 1845 he began delivering lectures to a group of young working-class men in Leeds, who had set up their own evening school for mutual improvement. The lectures were so popular that Smiles had them published as *Self-help* in 1859. The message of the book was that for poor people who wanted to improve themselves the remedy lay in their own hands. If a person worked hard, practised thrift, and tackled problems with determination, he could do almost anything he wanted. The success of *Self-help* was largely due to its celebration of those virtues which some in the Victorian age held dear—the gospel of work and an optimistic belief in material progress.

Smith, Alfred E(manuel) ('Al') (1873–1944), US politician. He rose through his association with *Tammany Hall and his identification with Irish-American, Roman Catholic, and new immigrant interests in the Democratic Party, to become an able and incorruptible four-term Democratic governor of New York state (1918–20, 1922–8). However, his Catholicism, opposition to *prohibition, and Tammany connections prevented him gaining the Democratic nomination for President in 1924, and then from winning the Presidency when he was at last nominated in 1928. Nevertheless, his cultivation of new urban interests helped to develop the Democratic coalition that paved the way for Franklin D. *Roosevelt's victories of 1932 and 1936, victories Smith deeply resented.

Smith, Jedediah Strong (1798–1831), US explorer. Born in New York state, he became one of the most famous of the *Mountain Men who opened up the

American north-west. In 1824 he led the third *Ashley expedition into Wyoming and in subsequent years made the first west-east crossing of the Sierra by a white man and, on a separate expedition, journeyed up the entire Pacific coast and returned through the length of Idaho. Having opened more territory and broken more trails than any other explorer of the north-west, he was killed by Comanche Indians on the Santa Fe Trail.

Smith, Joseph (1805-44), US founder of the *Mormon Church. In 1823 he claimed to have experienced visions and in 1827 a revelation of the existence of mystical religious writings, which he published as the *Book of Mormon* (1829). Smith organized the first Mormon community at Fayette, New York, but persecution as well as disagreements within the community drove him west, first to Ohio, then to Missouri, and finally in 1840 to Nauvoo, Illinois. Opposition increased following Smith's sanctioning of polygamy, and in June 1844, following a general breakdown in order and an ill-considered declaration of martial law, Smith was lynched by a mob in Carthage. Leadership of the Mormons then passed to Brigham *Young, who led them to Utah.

Smuts, Jan Christian (1870-1950), South African statesman, soldier, and scholar. In 1898 he became state attorney in Johannesburg and a member of the *Kruger government. In 1899 he contributed to a propaganda pamphlet, *A Century of Wrong*, explaining the Boer case against Britain, and rose to prominence in the Second *Boer War as a guerrilla leader of exceptional talent. He was a leading negotiator at the Treaty of *Vereeniging, believing that the future lay in co-operation with Britain. He held a succession of cabinet posts under President *Botha, but in 1914 rejoined the army and served in South Africa's campaign against German East Africa. In 1917 he joined the Imperial War Cabinet in London, and helped to establish the Royal Air Force. He was an advocate of the *League of Nations at the Versailles Peace Conference, returning to South Africa in 1919 to become Prime Minister (1919-24). He led the Opposition until 1933; and was Deputy Prime Minister, 1933-9, and Prime Minister, 1939-48. Among his many achievements was the drafting of the *United Nations Covenant. His struggle against extreme nationalism found expression in his philosophic study, *Holism and Evolution* (1926).

Snowden, Philip, 1st Viscount (1864-1937), British politician. Permanently crippled in a bicycle accident, he became a socialist and worked for the *Independent Labour Party as a journalist. He was elected to Parliament for Blackburn in 1906. Opposing British intervention in World War I, he always advocated self-government for India. He became Chancellor of the Exchequer in 1924 and 1929-31, and again in 1931-2 in the *National government. He did not support the *General Strike of 1926, and his cautious approach in welfare spending alienated many Labour supporters. His budget in 1931, reducing unemployment benefits because of the alarming international financial crisis, further antagonized them. The abandonment of *free trade at the *Ottawa conference caused his resignation.

social credit, a theory advanced by the social economist Clifford *Douglas, to eliminate the concentration of economic power. It became popular in Canada and New Zealand, particularly among hard-pressed farmers and small businessmen at the time of the Great *Depression. In Canada a Social Credit Party, led by William Aberhart, won an overwhelming victory in Alberta in 1935 and remained in power until 1971 without, however, implementing many of Douglas's ideas. In 1952 it won an election in British Colombia, but never gained more than a handful of federal seats in Ottawa. A New Zealand Social Credit Party was formed in 1953 and has held from one to three seats in the New Zealand Parliament.

social Darwinism, a 19th-century theory of social and cultural evolution. Even before Charles Darwin published the *Origin of Species* (1859), the writer Herbert *Spencer had been inspired by current ideas of evolution to write the *Principles of Psychology* (1855), where he first applied the concept of evolution to the development of society. The theory, based on the belief that natural selection favoured the most competitive or aggressive individual, was often used to support political conservatism. It justified inequality among individuals and races, and discouraged attempts to reform society as an interference with the natural processes of selection of the fittest. Social Darwinism lost credibility during the 20th century.

Social Democrat, a member of a political party using that term. This was first adopted by Wilhelm *Liebknecht and August Bebel in Germany, when they founded the German Social Democratic Labour Party (1869), based on the tenets of Karl *Marx, but advocating evolutionary reform by democratic and constitutional means. In 1875 it was fused with the German Workers' Association, founded (1863) by Ferdinand Lasalle, to form at Gotha the Social Democratic Party of Germany, which was then subjected to anti-socialist legislation by Bismarck. Other parties followed, for example in Denmark (1878), Britain (1883; *Hyndman's Social Democratic Federation), Norway (1887), Austria (1889), the USA (1897), under Eugene *Debs, later fusing with the Socialist Labor Party to form a Socialist Party, and Russia (1898), where a split came in 1903 into *Bolshevik and Menshevik. In other countries, for example France, Italy, and Spain, the term Socialist Party was more commonly adopted. The German SDP was the largest party in the Weimar Republic, governing the country until 1933, when it was banned. It was reformed in West Germany after World War II, with a new constitution (1959), ending all Marxist connections. It entered a coalition with the Christian Democrats in 1966, and headed a coalition with the Free Democrats between 1969 and 1982. In East Germany a revived SDP was fused with the Communist Party. In Sweden the SDP, socialist and constitutional in outlook, has been the dominant party since the 1930s, although it was out of office from 1976 to 1982. In Britain four prominent members of the Labour Party resigned in 1981 to form a short-lived, moderate Social Democratic Party.

socialism, a political and economic theory of social organization. It advocates that the community as a whole should own and control the means of production, distribution, and exchange to ensure a more equitable division of a nation's wealth. The word first appeared in France and Britain in the early 19th century in the writings of *Saint-Simon and Fourier, and in Robert *Owen's experiments at his New Lanark works in co-operative control of industry. Its main intention was to replace competition and exploitation by association and harmony. Socialism as a political ideal was revolutionized by Karl *Marx in the mid-19th century, who tried to demonstrate scientifically how *capitalist profit was derived from the exploitation of the worker, and argued that a socialist society could be achieved only by a mass movement of the workers themselves. Both the methods by which this transformation was to be achieved and the manner in which the new society was to be run have remained the subject of considerable disagreement and have produced a wide variety of socialist parties, ranging from moderate reformers to ultra left-wing *communists dedicated to upheaval by violent revolution.

Socialist League, a British political organization, set up in 1884 by the designer William Morris, to re-create society on *socialist principles. The League published pamphlets, but its main activity took the form of processions and demonstrations, at times resulting in clashes with police and troops. Its membership, consisting partly of moderates seeking working-class progress through parliamentary methods, but partly, also, of revolutionary socialists and *anarchists, was too miscellaneous to endure. By 1890 most moderates had joined the *Fabian Society, leaving only an extremist and ineffective minority.

social security, state assistance to those lacking adequate means or welfare. *Bismarck introduced into Germany the first comprehensive scheme in Europe (1881-9), providing for the payment of insurance benefits by the state in the event of accident, sickness, and old age. Similar schemes were introduced in a number of European countries and Australia and New Zealand later in the 19th century and in Britain by *Lloyd George in 1908-11. The coalition government of Winston Churchill appointed a committee (1941-2) headed by William *Beveridge, to review social insurance schemes. Its recommendations, generally known as the Beveridge Report (1942), proposed a comprehensive national insurance scheme that formed the basis of the post-war *welfare state with a *National Insurance Act, 1946. In the USA (where the term 'social security' originated), the huge increase in unemployment and social distress during the Great *Depression prompted F. D. *Roosevelt to introduce the *New Deal programme and in particular the Social Security Act (based on a payroll tax) in 1935, which provided for old age, widows or widowers, and disability. Social security in the USA was extended in the 1960s but reduced in the 1980s.

Society Islands   *Tahiti.

Sokoto   *Fulani empire of Sokoto.

Solferino, battle of (24 June 1859), fought in Lombardy between the armies of France and Piedmont, and Austria. Piedmont, under the leadership of *Cavour, persuaded France to give military support against Austria. The French and Piedmont-Sardinians had defeated the Austrians at *Magenta, and their armies, commanded by *Napoleon III, captured the elevated position at Solferino and successfully defended it after a fierce counter-attack by the Austrians. The latter, led by *Francis Joseph, began to retreat. A meeting between the two emperors

Polish supporters of **Solidarity** demonstrate. The rapid spread of the movement among Poles induced the Soviet Union to place severe pressures on the military leadership in Warsaw to outlaw it.

took place shortly afterwards at *Villafranca, after which hostilities ceased.

Solidarity (Polish, *Solidarnosc*), an independent trade-union movement in Poland. It emerged out of a wave of strikes at Gdańsk in 1980 when demands included the right to a trade union independent of Communist Party control. Solidarity's leader is Lech Walesa (1943-). Membership rose rapidly, as Poles began to demand political as well as economic concessions. In 1981, following further unrest aggravated by bad harvests and poor distribution of food, General Jaruzelski was appointed Prime Minister. He proclaimed martial law and arrested the Solidarity leaders. Solidarity was outlawed in 1982, but continues as an underground movement.

Solomon Islands, a double chain of six large and many smaller islands in the south-west Pacific. Occupied for at least 3,000 years, European missionaries and settlers arrived throughout the 18th and 19th centuries, and in 1885 the German New Guinea Company established control of the north Solomons. Britain declared a protectorate over the southern islands in 1893. During World War II the Solomons witnessed fierce battles between Japanese and Allied forces. The Solomon Islands became an independent member of the *Commonwealth of Nations on 7 July 1978.

Somalia, a country in north-east Africa. The area of the Horn of Africa was divided between British and Italian spheres of influence in the late 19th century. The modern Somali Republic is a result of the unification of the former British Somaliland Protectorate and the Italian Trusteeship Territory of Somalia. Since independence,

Somalia has been involved in border disputes with Kenya and Ethiopia. In 1969 President Shermarke was assassinated in a left-wing coup and the Marxist Somali Revolutionary Socialist Party took power, renaming the country the Somali Democratic Republic. Since the mid-1970s Somalian affairs have been dominated by intermittent war with Ethiopia over the Ogaden Desert and a related change in military reliance from the Soviet Union to the USA.

Somme, battle of the (July–November 1916), fought between British and German forces in northern France in World War I. The battle was planned by *Joffre and *Haig. Before it began the Germans attacked *Verdun, the defence of which nearly destroyed the French army. To relieve pressure on Verdun the brunt of the Somme offensive fell on the British. A preliminary eight days' bombardment poured 52,000 tonnes of ammunition on the German positions. On 1 July the British advanced from their trenches, fighting almost shoulder to shoulder, a perfect target for German machine gunners. Although the Germans retreated a few kilometres, they fell back on the Hindenburg Line defences (a barrier of concrete pillboxes armed with machine guns), while for the loss of some 600,000 men the Allies had gained a sea of mud.

Somoza, a family dynasty which dominated *Nicaragua from the 1930s until 1979. **Anastasio Garcia Somoza** (1896–1956), as chief of the National Guard, engineered a successful coup against the liberal regime and took over the presidency in 1936, exercising dictatorial control until his assassination in 1956. Somoza family rule continued under his sons **Luis** and **Anastasio (Tachito) Somoza Debayle** (1956–63, 1967–79, respectively). The Somozas used the National Guard to eliminate political opposition (although Luis was more conciliatory) while they accumulated vast amounts of Nicaragua's agrarian and industrial resources. Military and economic assistance from the USA helped maintain the Somozas in power until 1979, when economic problems and world outcry against human rights abuses undermined Tachito's control and the *Sandinista National Front defeated the National Guard, and took power.

Sonderbund (1845–7; German, 'separate league'), a league formed by seven Swiss Roman Catholic cantons. Its aim was to safeguard Roman Catholic interests and preserve the federal status of the cantons against the movement by the Radical Party to establish a more centralized government in Switzerland. The Radicals had closed (1841) all monasteries in the Aargau, and a posse had invaded (1844) the canton of Lucerne. The Radical majority in the federal Diet declared the Sonderbund dissolved in 1847, and sent an army against the separatists. In an almost bloodless campaign, the Sonderbund capitulated, and a new federal constitution in 1848 ended the virtual sovereignty of the cantons. The Jesuits were banned and no new religious orders were allowed to be established in the country.

Soult, Nicholas Jean de Dieu (1769–1851), Marshal of France and statesman. He served with distinction in the *Revolutionary Wars and was severely wounded in 1799. In 1805 he was victor at Landsberg and delivered the decisive blow to the Austrian and Russian armies during the battle of *Austerlitz. He was given command

in the *Peninsular War, but was finally beaten by *Wellington at Orthez in 1814. The following year he fought at *Waterloo. After the restoration of the monarchy *Louis XVIII made him Minister of War. He rejoined Napoleon in the *Hundred Days. Exiled after the second restoration, he was reinstated by *Charles X. Under *Louis Philippe he served in several ministerial offices, including that of Premier.

South Africa, a country in southern Africa. Formed as a self-governing *dominion of the British crown in 1910, the Union of South Africa comprised the former British colonies of the *Cape and *Natal, and the Boer republics of the *Transvaal and *Orange Free State recently defeated in the *Boer Wars. Politically dominated by its small white minority, South Africa supported Britain in the two World Wars, its troops fighting on a number of fronts. After 1948 the right-wing Afrikaner-dominated National Party formed a government. It instituted a strict system of *apartheid, intensifying discrimination against the disenfranchised non-white majority. South Africa became a republic (1960) and left the Commonwealth (1961). Although its economic strength allowed it to dominate the southern half of the continent, the rise of black nationalism both at home and in the surrounding countries (including the former mandated territory of *Namibia) produced increasing violence and emphasized South Africa's isolation in the diplomatic world. In 1985 the regime of P. W. Botha began to make some attempts to ease tension by interpreting apartheid in a more liberal fashion. This failed, however, to satisfy either the increasingly militant non-white population or the extremist right-wing groups within the small white élite. In 1986 a state of emergency was proclaimed and several thousands imprisoned without trial. The domestic and international sides of the problem remain inseparable, with South African troops fighting against *SWAPO guerrillas in Namibia and Angola, and support by surrounding states for the forces of the outlawed *African National Congress producing a series of cross-border incidents.

South America, the southern half of the American land mass. Between 1816 and 1825 most of Spanish South America achieved independence under the leadership of Simón *Bolívar and José de *San Martín, and subsequently broke into nine separate countries: Venezuela, Colombia, Ecuador, Peru, Bolivia, Chile, Argentina, Uruguay, and Paraguay. Brazil gained independence from Portugal in 1822. British Guiana gained independence from Britain in 1966 as Guyana. Surinam, the Dutch colony, became independent in 1975, and French Guiana continues under French rule. The continent remained politically independent during the 19th century, partly as a result of the *Monroe Doctrine, which prevented European expansion. At the same time it received some fifteen million immigrants from Europe, and was continuously receptive to both cultural and ideological influences from the USA and Europe. Economic investment was considerable, particularly by Britain, in primary production such as minerals and beef, leading to a dependence on such trade. The continent remains predominantly Roman Catholic; in the 19th and early 20th centuries the Church occupied a central political and social position as a conservative force. More recently its stand has been challenged by priests of the 'liberation theology' movement, which has sought to involve the Church actively in the politics of poverty and deprivation. Rapid *urbanization has overtaken the supply of housing and employment. In an effort to stimulate trade and production, economic groupings such as the *Central American Common Market, the Latin-American Free Trade Association (1960), and the Latin-American Integration Association (1981) have been established. Extensive development projects and the rapid increase in oil prices in the 1970s have burdened many South American countries with debts that their economies, heavily dependent on the world commodity market, find almost impossible to service.

South Australia, a state of the Commonwealth of *Australia. Sealers began to use Kangaroo Island in the

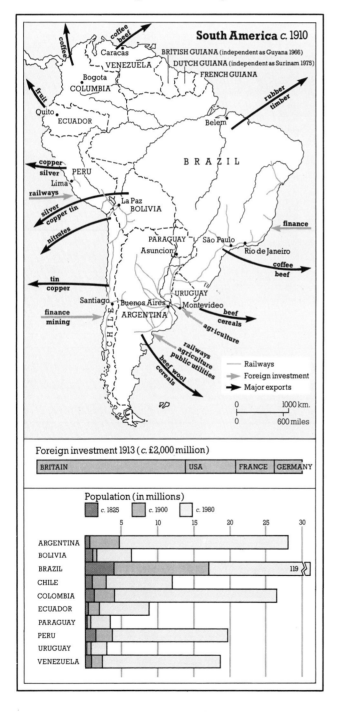

Foreign investment 1913 (c. £2,000 million)

BRITAIN		USA	FRANCE	GERMANY

Population (in millions)

early 1800s. Settlement of the British province of South Australia began in 1836. The Colonial Office and a Colonization Commission shared authority for it. The province went bankrupt, and, after its debts were met by Britain, it became a colony in 1842. In 1851 it became the first British colony to dissolve the connection between church and state. South Australia, which never received convicts, was granted responsible government in 1856. It administered the *Northern Territory (1863-1911).

South-East Asian decolonization, the process by which European and US Asian colonies gained independence. The *Philippines were granted self-government in 1934 by the USA, and after 1945 the process of decolonization accelerated in the European colonies. It was more or less complete by 1975, apart from a few Pacific islands.

South-East Asia Treaty Organization (SEATO), a defence alliance established under the South-East Asia Collective Defence Treaty, signed at Manila in 1954, as part of a US policy of *containment of communism. The signatories were Australia, Britain, France, New Zealand, Pakistan, the Philippines, Thailand, and the USA. The treaty area covered south-east Asia and part of the south-west Pacific. Pakistan and France withdrew from the organization in 1973 and 1974 respectively. The Organization was dissolved in 1977.

South Korea, north-east Asian country. Consisting of the southern half of the Korean peninsula, beneath the 38th parallel, South Korea was formed from the zone occupied by US forces after World War II, an independent republic being proclaimed on 15 August 1948. Badly damaged by the *Korean War (1950-3), the South Korean economy was initially restricted 'by its lack of industrial and power resources and by a severe post-war refugee problem. Unemployment and inflation damaged the reputation of the government of President *Rhee, and its increasing brutality and corruption finally led to its overthrow in 1960. After a second civilian government had failed to restore the situation, the army, led by General Park Chung Hee, seized power in 1961. Park, who assumed the powers of a civilian president (1953-79) organized an extremely successful reconstruction campaign which saw South Korea emerge as a strong industrial power, but his repressive policies soon engendered serious unrest. He was assassinated by the head of the South Korean Central Intelligence Agency in 1979. His successor, General Chun Doo Hwan, continued his policies until forced to partially liberalize the political system after widespread student unrest in 1987.

South Vietnam   *Vietnam.

South-West Africa   *Namibia.

South West Africa People's Organization *SWAPO).

South Yemen, a state on the south-west Arabian peninsula, formed in 1967 from the former British-controlled territory of *Aden and the Aden Protectorates. Civil War between royalist and republican forces in the area, which British forces attempted to control, followed World War II. It ended with the British withdrawal and

South Yemen's declaration of independence in 1967, when the forces of the National Liberation Front (NLF) under Qahtan al-Snaabi took control. In 1970 the name of the state was changed to the People's Democratic Republic of Yemen. From 1967 to 1976 it was involved in helping the Dhofar rebels of *Oman. It remains preoccupied with the question of union with the *Yemen Arab Republic.

Souvanna Phouma, Prince (1901–84), Laotian statesman. He was a member of the provisional government (1945–6) opposed to French recolonization and was first elected Premier in 1951–4. He formed a brief coalition (1962–3) with the *Pathet Lao, led by his half-brother, Prince Souphanouvong, and after the return of civil war, he continued as Premier. He tried to maintain a neutral policy during the Vietnam war, but this proved impossible. In 1973 he signed a ceasefire agreement with the Pathet Lao and remained Premier until the People's Democratic Republic of Laos was declared in 1975.

Soviet (Russian, 'council'), an elected governing council in the Soviet Union. The Soviets gained their revolutionary connotation in 1905 when the St Petersburg (now Leningrad) Soviet of Workers' Deputies was formed to co-ordinate strikes and other anti-government activities in factories. Each factory sent its delegates, and for a time other cities were dominated by Soviets. Both *Bolsheviks and Mensheviks realized the potential importance of Soviets and duly appointed delegates. In 1917 a Soviet modelled on that of 1905, but now including deserting soldiers, was formed in Petrograd (previously St Petersburg), sufficiently powerful to dictate industrial action and to control the use of armed force. It did not at first try to overthrow *Kerensky's Provisional Government but grew increasingly powerful as rep-

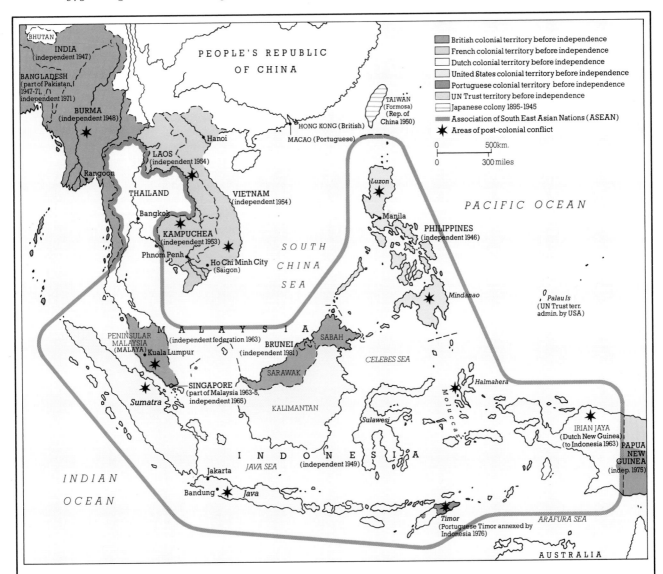

South-East Asian decolonization

Following the defeat of Japan in 1945, British policy was to advance rapid independence throughout its Asian empire: within two years it had been granted to India, Pakistan, and Burma, and to Malaya in 1957. The French and Dutch reluctantly agreed to independence only after two bloody wars, in the Dutch East Indies (Indonesia) in 1949, and in French Indo-China (Laos, Cambodia, and Vietnam) between 1949 and 1954. Various groups of Pacific islands have also won independence, although the USA retains a stong naval presence.

resenting opposition to continuing Russian participation in World War I against Germany. Consisting of between 2,000 and 3,000 members, power was exercised by the executive committee. Soviets were established in the provinces and in June 1917 the first All Russia Congress of Soviets met. The Bolsheviks gradually dominated policy, leading to their seizure of power in the *Russian Revolution (1917). During the *Russian Civil War village Soviets controlling local affairs and agriculture were common. The national Soviet is called the Supreme Soviet, comprising delegates from all the Soviet republics.

Soviet Union   *Union of Soviet Socialist Republics.

Soweto (amalgamation of several townships), a predominantly black township, south-west of Johannesburg in South Africa. In January 1976 black schoolchildren demonstrated against legislation proposing to make Afrikaans the compulsory language of instruction, and police broke up the demonstration, using guns and tear gas. It triggered off a wave of violence. By the end of 1976 some 500 blacks and coloureds had been killed by the police, many of them children. The plans for compulsory teaching in Afrikaans were dropped. Since then the anniversary of the demonstration has led to further riots and violence on both sides.

Spaak, Paul-Henri (1899–1972), Belgian statesman. He practised law before entering Parliament as a socialist in 1932. He was Prime Minister (1938–9, 1947–9). Spaak was the first President of the *United Nations General Assembly, and of the consultative assembly of the *Council of Europe (1949–51); he was also Secretary-General of *NATO (1957–61). Spaak was a firm supporter of a united Europe; his proposals for an economic association based on free trade and movement of labour, as well as joint social and financial policies, formed the basis of the *European Economic Community.

The first child to die in the **Soweto** riots of 1976, Hector Petersen, aged 13, is carried by grieving relatives from the scene of the violence.

space exploration, exploration of space in the 20th century. In 1903 the Russian physicist Konstantin Tsiolkovsky was developing ideas for space rockets fuelled by liquefied gas and by 1926 Robert Goddard in the USA had successfully designed the first liquid fuelled rocket. There followed considerable German research into rockets, culminating in the launch of the V-2 rocket in 1944. In 1957 the Soviet Union surprised the USA by putting the first artificial satellite, *Sputnik I*, in orbit; this was followed by the US *Explorer I* in 1958. Yuri Gagarin was the first man in space in 1961, followed by John Glenn in 1962. In 1961 President *Kennedy proposed the *Apollo* project to achieve a manned lunar landing by 1970, and in 1969 Neil Armstrong and Edwin ('Buzz') Aldrin landed on the moon. The Soviet Union concentrated on unmanned flights, *Luna IX* achieving a soft landing on the moon in 1966. In the early 1970s space stations were launched by both the USA and the Soviet Union, and in 1975 an *Apollo* capsule linked up with a Soviet *Soyuz* capsule. Unmanned flights have been made to Venus and Mars, while the US probe, *Voyager 2*, launched in 1977, should reach Neptune in 1990. In 1981 the USA launched a space shuttle, the first reusable space craft, but its commercial and scientific programme was interrupted by the explosion of the shuttle, *Challenger*, on lift-off in 1986. In 1986 the giant orbiting laboratory, *Mir*, was launched, with astronauts being ferried to the stations by *Soyuz* spacecraft, followed in 1987 by the placing in space of the powerful *Energiya* station. In 1987 Romanenko set a space endurance record of 326 days in orbit. Space technology has resulted in numerous applications, and telecommunication satellites have greatly improved global *communications; while meteorological satellites provide advance weather information, and reconnaissance satellites register the earth's resources and military information.

Spain, a country in south-west Europe, occupying the greater part of the Iberian Peninsula. In the early 19th century Spain suffered as a result of the *Napoleonic Wars, when contact with its empire in South America and the Pacific was lost. The *Spanish–South American Wars of Independence led to the emergence of Argentina, Bolivia, Peru, Venezuela, and Mexico. Spain subsequently remained peripheral and undeveloped in a Europe which was fast becoming industrialized. From 1814 the absolutist monarchy was involved in a struggle with the forces of liberalism, and from 1873 to 1875 there was a brief republican interlude. In 1898 the *Spanish–American War resulted in the loss of Puerto Rico, the Philippines, and Guam, while Cuba, which had been more or less in revolt since 1868, became a US protectorate in 1903. In 1923 General Miguel *Primo de Rivera established a virtual dictatorship, which was followed by another republican interlude (1931–9), scarred by the savage *Spanish Civil War (1936–9). Nationalist victory resulted in the dictatorship of General Francisco *Franco (1939–75). His gradual liberalization of government during the late 1960s was continued by his successor Juan Carlos I, who has established a liberal, democratic constitutional monarchy. Separatist agitation, often violent, by Eta, an organization seeking independence for the Basque provinces, continued throughout the period. Of its remaining colonies Spain granted independence to Spanish Sahara in 1976, which was divided between Morocco and Mauritania.

Space exploration

1903	Russian physicist Konstantin Tsiolkovsky advocates use of liquid-fuelled rockets for space exploration.
1923	Hermann Oberth publishes *The Rocket into Interplanetary Space* in Germany.
1926	US rocket pioneer Robert Goddard tests first liquid-fuelled rocket.
1942	First launch of Germany's V2 rocket missile, designed by Wernher von Braun.
1957	Soviet Union launches first artificial sattelite, *Sputnik I*, with a payload of 83 kg. (183 lb.). *Sputnik II* carries the first space traveller, the dog Laika.

Sputnik II with the dog Laika, the first space traveller.

1958	*Explorer I* is first US satellite to reach orbit.
1959	Soviet *Luna I* craft is first to fly beyond the pull of earth's gravity. *Luna II* is first spacecraft to hit the moon.
1960	*TIROS I* (USA) is the first weather sattelite.
1961	*Vostok I* (Soviet Union) carries the world's first astronaut, Yuri Gagarin, who makes one orbit of the earth. Gherman Titov in *Vostok II* makes 17 orbits.
1962	John Glenn, in *Mercury-Atlas 6*, is first US astronaut to orbit the earth, with three orbits.
1963	*Vostok VI* (Soviet Union) carries the first female astronaut, Valentina Terechkova.

Valentina Terechkova, the first woman in space.

1965	Alexei Leonov (Soviet Union) performs first 'spacewalk'.
1965	*Mariner I* (USA) sends back television pictures of Mars. *Venera 3* (Soviet Union) makes first soft landing on Venus. France becomes third nation to launch a satellite using its own launcher.
1966	*Luna IX* (Soviet Union) makes first soft landing on moon.
1968	The US *Apollo 8* spacecraft carries three astronauts on a flight around the moon.
1969	The climax of the US Apollo programme: the moon landing by *Apollo 11* astronauts Neil Armstrong and Edwin ('Buzz') Aldrin. Their stay lasts 21½ hours.

Apollo 11 astronaut Edwin ('Buzz') Aldrin, the second human to step on the moon on the first manned moon landing.

1970	*Lunokhod I* (Soviet Union) lands lunar rover on the moon. Japan and China become the fourth and fifth satellite-launching nations.
1971	*Mars 1* and *Mars 2* probes (Soviet Union) land scientific instruments on Mars. Britain becomes the sixth satellite-launching nation.
1972	*Apollo 17* is the sixth and last *Apollo* manned moon mission by the USA. Launch of *Landsat 1*, the first earth resources technology satellite (ERTS).
1973	First of three *Skylab* missions (USA); three astronauts set a US duration record of 84 days in space.

Skylab in earth orbit.

1975	*Viking 1* and *2* spacecraft (USA) land instruments on Mars, but find no conclusive evidence of life on the planet. India is seventh nation to launch its own satellite. In joint US and Soviet flight, *Apollo* and *Soyuz* spacecraft dock in earth orbit.
1977	Launch of *Voyager 2* (USA), which flies by Jupiter (1979), Saturn (1981), and Uranus (1986).
1981	First flight of US space shuttle, the first reusable spacecraft.

The launch of the first space shuttle, *Columbia*, from Kennedy Space Center, Florida, USA, on 4 April 1981.

1983	First flight of *Spacelab*, a scientific laboratory carried into orbit by the US shuttle. First operational flight of Europe's *Ariane* multi-stage rocket, intended to compete with the shuttle as a satellite launcher.
1986	Soviets launch new orbital space station *Mir*. European *Giotto* satellite passes close to the nucleus of Halley's comet. US shuttle *Challenger* explodes immediately after launch killing seven astronauts and halting the shuttle programme.
1987	Soviet astronaut Yuri Romanenko spends a record 326 days in space.

Spanish–American War (1898), a conflict between Spain and the USA. It had its roots in the struggle for independence of *Cuba, and in US economic and imperialist ambitions. Sympathetic to Cuban rebels whose second war of independence against Spain had begun in 1895, the USA used the mysterious blowing up of its battleship, the *Maine, in Havana harbour as a pretext for declaring war. The Spanish navy suffered serious defeats in Cuba and the Philippines, and a US expeditionary force (which included the future President Theodore Roosevelt and his *Rough Riders) defeated Spanish ground forces in Cuba and in Puerto Rico. Spain surrendered at the end of 1898, Puerto Rico being ceded to the USA, and Cuba placed under US protection. The Pacific island of Guam was also ceded while the Philippines were bought by the USA for $20 million. The war signalled the emergence of the USA as an important world power as well as the dominant power in the Caribbean.

Spanish Civil War (1936-9), a bitter military struggle between left- and right-wing elements in Spain. After the fall of *Primo de Rivera in 1930 and the eclipse of the Spanish monarchy in 1931, Spain was split. On the one hand were the privileged and politically powerful groups like the monarchists and *Falange Party, on the other were the Republicans, the Catalan and Basque separatists, socialists, communists, and anarchists. The elections of February 1936 gave power to a left-wing *Popular Front government and strikes, riots, and military plots followed. In July 1936 the generals José Sanjurjo and Francisco *Franco in Spanish Morocco led an unsuccessful coup against the republic, and civil war, marked by atrocities on both sides, began. In 1937 Franco's Nationalist troops overran the Basque region which, in hope of ultimate independence, supported the Republicans. Nationalists also held the important town of Teruel against Republican attacks, which enabled Franco, with German and Italian assistance, to divide the Republican forces by conquering territory between Barcelona and Valencia (1938). The Republicans, weakened by internal intrigues between rival factions and by the withdrawal of Soviet support, attempted a desperate counter-attack. It failed, and Barcelona fell to Franco (January 1939), quickly followed by Madrid. Franco became the head of the Spanish state and the Falange was made the sole legal party. The civil war inspired international support on both sides: the Soviet Union sent advisers and military supplies to the Republicans, while soldiers from Italy fought with Franco. Germany supplied some 10,000 men to the Nationalists, mostly in the aviation and tank services. Bombing of civilians by German pilots and the destruction of the Basque town of Guernica (1937) became the symbol of fascist ruthlessness and inspired one of Picasso's most famous paintings. As members of the *International Brigades, left-wing and communist volunteers from many countries fought for the Republican cause. The war cost about 700,000 lives in battle, 30,000 executed or assassinated, and 15,000 killed in air raids.

Spanish-South American Wars of Independence (1810-25). The roots of the wars of independence are to be found in the attempts made by Spain after 1765 to re-establish imperial control over its American colonies. This was resented by the Creoles (colonial descendants of Spanish settlers) whose local political authority, growing economic prosperity, and increasing sense of national identity were threatened. The precipitant for the armed conflict was the series of international wars in Europe after 1796 which culminated in the *Peninsular War (1807-14). Creoles in Spanish America achieved de facto economic independence, and with the abdication of *Ferdinand VII (1808), political independence. In 1811 the first declarations of independence were made. Initially the movements were hampered by a counter-revolutionary drive by Spanish royalists. In 1816 Simón *Bolívar returned to Venezuela from exile and united with José Antonio *Páez and the Ilaneros (plainsmen) of the interior. With the assistance of British mercenaries Bolívar crossed the Andes and won the battle of Boyaca, and proclaimed the United States of *Colombia (1819). The victories of Carabobo (1821) and Pichincha (1822) brought Venezuela and Ecuador into the Colombian Federation. Bolívar then linked up with the independence movement in the south under the leadership of *San Martín, who had crossed the Andes from the United Provinces of La Plata (Argentina) and won the battles of Chabuco (1817) and Maipo (1818) and liberated Chile. Both movements now closed in on the bastion of the Spanish empire, Peru. The battles of Junín and Ayacucho (1824) were the final victories in the liberation of the continent. Bolivia's Federation of Gran Colombia survived until 1829 when it began to disintegrate. The Creole élites throughout South America were divided over the constitutional foundations of the new nations, every one of which had to grapple with the question of federalism: how tightly regional autonomy should be subordinate to the central state. These constitutional problems could not hide the fact that the social and economic foundations in the post-independence period demonstrated a remarkable degree of continuity with those of the late Bourbon period. A conservative revolution had been achieved.

Spartakist Movement, a group of German radical socialists. Led by Karl *Liebknecht and Rosa *Luxemburg, it was formed in 1916 in order to overthrow the German imperial government and replace it with a communist regime. The name was used as a pseudonym by Liebknecht in his publications denouncing international warfare as a capitalist conspiracy and calling on the modern 'wage slave' to revolt like the Roman gladiator Spartacus. In November 1918 the Spartakists became the German Communist Party and attempted to seize power in Berlin. In January 1919, Gustav Noske, as leader of the armed forces, ordered the suppression of all radical uprisings throughout Germany. Within days, a second rebellion in Berlin was brutally crushed and the two leaders murdered without trial. There was a further Spartakist rising in the Ruhr in 1920.

Speer, Albert (1905-81), German Nazi leader. He became the official architect for the Nazi Party, designing the grandiose stadium at Nuremberg (1934). An efficient organizer, he became (1942) Minister for Armaments and was mainly responsible for the planning of Germany's war economy, marshalling conscripted and slave labour in his Organization Todt to build strategic roads and defence lines. He was imprisoned after the war.

Speke, John Hanning (1827-64), British explorer. He reached Lake Victoria (1858) on *Burton's expedition,

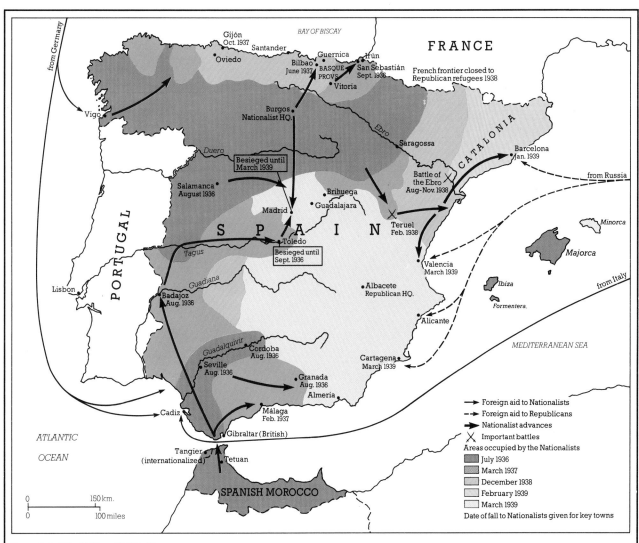

Areas occupied by the Nationalists:
→ Foreign aid to Nationalists
--→ Foreign aid to Republicans
▶ Nationalist advances
✕ Important battles
Areas occupied by the Nationalists
■ July 1936
■ March 1937
▨ December 1938
▨ February 1939
□ March 1939
Date of fall to Nationalists given for key towns

Spanish Civil War (1936–9)

The early thrust of Nationalist forces under General Franco from Spanish Morocco was westward towards Barcelona, held by the Republicans, and north against the Basques. The latter suffered air-raids including one on the civilian population of Guernica which was to inspire the painting (*below*) by Picasso. Internationally, the Civil War was seen as an ideological struggle; its brutalities caused deep bitterness in Spain, which lasted for a generation.

identifying it as the source of the White Nile. He returned in 1860 with James Grant to confirm the discovery. Working their way up the western shore of Lake Victoria they came to its northern tip, where Speke found the Ripon Falls. Speke and Grant then followed the Nile down to Khartoum, which they reached in 1863.

Spence, William Guthrie (1846–1926), Australian trade unionist and politician. Spence helped to found several unions, including the Amalgamated Shearers' Union of Australasia (1886) and the Australian Workers' Union (1894). He was President of the former (1886–93), and the latter (1898–1917), then Australia's largest union. He played a prominent role in the Maritime Strike of 1890 and the Queensland *Shearers' Strike of 1891. He was elected to the Federal House of Representatives for Darling (1901–19).

Spencer, Herbert (1820–1903), British philosopher and sociologist. He received little formal education, and began his career as a railway engineer before turning to writing and the study of philosophy, publishing his first book, *Social Statistics*, in 1851. He welcomed Charles Darwin's *Origin of Species* (1859) and coined the phrase 'survival of the fittest'. He sought to trace the principles of evolution in all branches of knowledge in a projected ten-volume work *A System of Synthetic Philosophy*, of which volume 3, *The Principles of Sociology* (1896), was his most influential. He attacked all forms of state interference, which he believed would lead to the loss of individual freedom. His optimistic belief in human progress through evolution won him a large following.

Speransky, Mikhail Mikhailovich, Count (1772–1839), Russian statesman and chief adviser to *Alexander I. After the defeat of Russia by Napoleon and the Treaty of *Tilsit, he drew up, at the emperor's request, a constitution that proposed popular participation in legislation; this was only partially implemented. He increased the burden of taxation on the nobility and sought to educate the bureaucracy, and established promotion on the basis of merit. He incurred the enmity of both the aristocracy and the bureaucrats, and was charged (1812) with treason and secret dealings with the French, and sent into exile. Reinstated four years later, he rejoined the council of state (1821) and spent his final years codifying Russian law.

spoils system (or patronage system), a term used in US politics to describe the practice whereby a victorious political party rewards its supporters with public appointments. The term was coined by Senator William Marcy of New York in 1832 in connection with appointments made by President Andrew *Jackson, who replaced 20 per cent of federal office-holders by his political supporters during his two terms. A President or state governor has considerable patronage at his disposal. After the American Civil War attempts were made to reduce patronage in the Civil Service, for example by the *Pendleton Act (1883), which created the Civil Service Commission. In addition to the awarding of public office, the term 'spoils system' also refers to the award of contracts, especially defence contracts, to a state in return for the support of its representatives for presidential policies in Congress, and the granting of public contracts to party contributors on favourable terms.

sports, spectator, sporting events played before massed spectators. In the early 19th century most sporting events were local and rural, one of the most popular being horse-racing. They were often violent, such as boxing, and cruel, such as cock-fighting. One of the few events for massed spectators was the bull-fight in southern France and Spain. In the second half of the 19th century sporting stadiums began to be built in cities, followed by many more in the 20th century. One of the most popular spectator sports became association football or 'soccer'. In its modern form soccer developed at some English 'public' (i.e. private) schools. Its rules were first codified at Cambridge in 1843 and the Football Association was founded in 1863, with inter-city contests soon following. In 1904 the Fédération Internationale de Football Association (FIFA) was founded in France, and international soccer became the world's most popular sport both to watch and play. Its rival in English schools was rugby football, which began at Rugby School in 1823 and spread in varying forms all over the world, and from which are descended American, Canadian, and Australian football. Cricket has been played in England since the 13th century (the Marylebone Cricket Club (MCC), which is the world governing body of the sport, was founded in 1787). The game gained in popularity within the British empire, for example in Australia where the first Test (international) Match was played in 1877. Tennis, originating from a 12th-century French handball game, was first played in its modern form in Wales in 1873. In 1877 the All England Croquet Club at Wimbledon sponsored the first World Tennis Championship. Athletics contests based on those of ancient Greece were revived by the Olympic Games movement in 1896 for track and field events. The Games gradually extended to all sports—attracting ever higher skills and larger crowds—although they have been marred by political tensions and boycotts since the massacre in 1972 of Israeli participants in the Munich Olympics. Violence between rival European supporters, particularly of football teams, has increased in recent years.

squatter, a term with several meanings. In the USA, from the late 18th century, a squatter was a settler having no normal or legal title to the land he occupied, particularly in a district not yet surveyed. In New South Wales, Australia, in the early 19th century, the first squatters, often ex-convicts, occupied land without authority and stole stock. By the 1830s, its meaning had begun to change, with many pastoralists settling beyond the official 1829 limits of settlement. They were mostly involved in the wool industry, and in 1836 were granted grazing rights for an annual licence fee. The squatters demanded security of tenure and pre-emptive rights, which they gained in 1847, securing the land most suitable for agricultural and pastoral purposes. Squatters became a very powerful group, socially, economically, and politically. Squatters and *selectors struggled bitterly over land during the second half of the 19th century. Squatters continued to be known by that name even after they acquired their land freehold. Eventually, the term was applied to all large pastoralists in Australia.

Sri Lanka (formerly Ceylon), an island state in the Indian Ocean off the south-east tip of India. Its early history was shaped by Indian influences and its modern identity by three phases of European colonization. It was successively dominated by the Portuguese, Dutch, and British, and finally annexed (1815) by the British, who called it Ceylon. The colonial era saw the introduction of new plantation products, notably tea, coffee, and rubber, considerable conversion to Christianity, and the emergence of a small, westernized middle class, which through the early 20th century was pressing for self-government. The Donoughmore Commission (1928) recommended a new constitution, established in 1931, but racial tensions prevented its full implementation. The island, although now granted an element of self-government, remained a crown colony until 1948, when it was granted independence as a dominion within the *Commonwealth of Nations. A government was established by the United National Party under Don *Senanayake, who was succeeded (1952) by his son, Dudley Senanayake. The Socialist Sri Lanka Freedom Party was in power from 1956 to 1965, and Solomon *Bandaranaike (1899-1959) was its dominant force until his death in 1959. His widow, Sirimavo Bandaranaike (1916-), succeeded him as Prime Minister (1960-5, 1970-7). A new constitution in 1972 established the island as the Republic of Sri Lanka. Tensions have re-emerged between the majority Sinhalese, traditionally Buddhist, and the minority Tamil, chiefly Hindu, who had come from southern India and live in northern Sri Lanka. A ceasefire was arranged by the Indian government in 1987 between Tamil guerrilla groups and the Sri Lankan government.

SS (abbr. for *Schutzstaffel*, German, 'protective echelon'), the élite corps of the German Nazi Party. Founded (1925) by *Hitler as a personal bodyguard, the SS was schooled in absolute loyalty and obedience, and in total ruthlessness towards opponents. From 1929 until the dissolution of the *Third Reich the SS was headed by Heinrich *Himmler, who divided it mainly into two groups: the Allgemeine SS (General SS), and the Waffen-SS (Armed SS). Initially subordinated to the SA (*brownshirts) the SS assisted Hitler in the *Night of the Long Knives massacre (1934) which eliminated its rivals. By 1936 Himmler, with the help of Reinhard *Heydrich, had gained control of the national police force. Subdivisions of the SS included the *Gestapo and the Sicherheitsdienst, in charge of foreign and domestic intelligence work. The Waffen-SS served as an élite combat troup alongside but independent of the armed forces. It also administered the *concentration camps.

Stalin, Josef Vissarionovich (b. Dzhugashvili) (1879-1953), Soviet dictator. The son of a shoemaker, he was born in Georgia, where he attended a training school for priests, from which he was expelled for holding revolutionary views. An early member of the *Bolshevik Party, he was twice exiled to Siberia. He escaped after the start of the *Russian Revolution and he rose rapidly to become *Lenin's right-hand man. After Lenin's death he won a long struggle with *Trotsky for the leadership, and went on to become sole dictator. Features of his rule were: rapid industrialization of the Soviet Union under the five-year plans (which was eventually to turn the Soviet Union into the world's second industrial and military power); the violent *collectivization of agriculture that led to famine and the virtual extermination of many peasants; a purge technique that not only removed, through show trials and executions, those of his party colleagues who did not agree with him but also

This official propaganda painting of **Stalin** by J. Yefanov, 1936, shows the Soviet leader presenting flowers to a factory manager's wife while Molotov and other communist leaders applaud. The despotic nature of Stalin's rule and the large-scale liquidation of his foes, real or imaginary, were concealed from the Soviet people.

placed millions of other citizens in *prison camps. In 1939 Stalin signed the *Nazi-Soviet Pact with *Hitler, and, on the latter's invasion (1941) of the Soviet Union, Stalin entered *World War II on Britain's side, signing the Anglo-Soviet Treaty in 1942. He met with *Roosevelt and *Churchill at the conferences of *Teheran (1943), *Yalta (1945), and *Potsdam (1945). By skilful diplomacy he ensured a new Soviet sphere of influence in eastern Europe, with communist domination in all its neighbouring states. Suspicious of any communist movement outside his control, he broke (1948) with *Tito over party policy in *Yugoslavia. Increasingly the victim of his own paranoia, he ordered the arbitrary execution of many of his colleagues. After his death the 20th All-Party Congress (1956) under *Khrushchev attacked the cult of Stalin, accusing him of terror and tyranny. The term Stalinism has come to mean a brand of communism that is both national and repressive.

Stalingrad, battle of (1942-3), a long and bitter battle in World War II in which the German advance into the

Soviet troops advance through the ruined city of **Stalingrad** during the battle that was crucial to the outcome of World War II. House-to-house fighting had reduced the city to rubble before the German army surrendered.

Soviet Union was turned back. During 1942 the German 6th Army under General von *Paulus occupied Kursk, Kharkov, all the Crimea, and the Maikop oilfields, reaching the key city of Stalingrad (now Volgograd) on the Volga. Soviet resistance continued, with grim and prolonged house-to-house fighting, while sufficient Soviet reserves were being assembled. The Germans were prevented from crossing the Volga and in November Stalin launched a winter offensive under Marshalls *Zhukov, *Koniev, Petrov, and Malinovsky. By January 1943 the Germans were surrounded and von Paulus surrendered, losing some 330,000 troops killed or captured. The Russians now advanced to recapture *Kursk, and this German defeat marked the beginning of the end of German success on the *Eastern Front.

Stalwarts, US politicians, a faction of conservative Republicans. Led by Roscoe *Conkling during the Presidency (1877–81) of Rutherford *Hayes, they supported the *spoils system, but opposed both the final ending of *reconstruction in the South, symbolized by the withdrawal of Federal troops in 1877, and any reform of the Civil Service. In 1880 they sought a third term for President *Grant, but failed. They dubbed their opponents, the anti-Grant wing of the Republican Party, as 'Half-breeds'. These were led in Congress by James *Blaine and they succeeded in getting their candidate *Garfield elected (1880). The nickname 'Stalwart' was dropped after President Garfield was shot (July 1881) by a Stalwart who was a disappointed office-seeker. A direct result of this tragedy was the *Pendleton Civil Service Act in 1883, which sought to make entry into the service dependent on merit rather than on reward.

Stanley, (Sir) Henry Morton (b. John Rowlands) (1841–1904), British/American explorer and journalist. An illegitimate orphan, he was brought up in a Welsh workhouse and ran away to the USA in 1859. He was adopted by a New Orleans merchant, Henry Stanley, whose name he took. He served as a soldier on the Confederate side in the *American Civil War and as a seaman on US ships before becoming a journalist on the

Steam power

1698	Thomas Savery (UK) patents a steam-operated pump.
1712	Thomas Newcomen (UK) builds steam engine for pumping at Dudley, Worcestershire.
1765	James Watt (UK) improves the steam engine.
1769	Nicholas-Joseph Cugnot (France) builds a steam-driven road vehicle.
1781	Watt adapts his steam engine for factory use.
1786	Edmund Cartwright (UK) establishes factory with his power looms driven by a steam engine.
1801	Richard Trevithick (UK) builds a steam-driven road vehicle.
1802	William Symington (UK) experiments with a steam-driven tug, the *Charlotte Dundas*, on the Forth and Clyde Canal.
	The two-cylinder compound steam engine, developed by Arthur Woolf (UK), represents the most significant advance since Watt's improvements.
1804	Trevithick demonstrates his steam-driven railway locomotive in Wales.
1805	Oliver Evans (USA) designs a steam-driven dredger, the *Orukter Amphibolos*, able to travel both on land and in water.
1806	Evans's factory manufactures steam engines for industrial use. They power screw-presses in tobacco, cotton, and paper mills.
	William Deverell (UK) patents the steam hammer.
1807	The world's first regular passenger steamship service (USA) begins on the Hudson river. The vessel is Robert Fulton's *Clermont*.
1812	Henry Bell (UK) begins a paddle steamer service on the River Clyde, with the *Comet*.
1814	Fulton's *Demologos* is the first warship to use steam power.
1825	The world's first steam railway (UK), the Stockton and Darlington, opens using steam locomotives for freight haulage.
1829	Introduction of steam pumps by London's firefighters.
	George Stephenson's steam locomotive *Rocket* (UK) wins the Rainhill competition between steam locomotive and stationary steam engine.

George Stephenson's steam locomotive *Rocket*.

1830	First steam passenger line opens, Manchester to Liverpool.
	Baltimore and Ohio Railroad opens, the first in the USA.
1838	First crossing of the Atlantic Ocean by a ship (the *Sirius*) using steam alone, from London to New York via Cork.

1839	James Nasmyth (UK) designs an improved steam hammer, originally to forge the drive shaft of I. K. Brunel's steamship *Great Britain*, launched 1843.
1850s	Steam tractors begin to be used on farms, for haulage and as power sources for other linked machines such as loaders, ploughs, and threshing machines.
1852	Henri Giffard (France) develops a steam-driven airship.
1853	East Indian Railway Company opens its first line, between Bombay and Thana.
1858	Brunel's *Great Eastern*, the largest ship of its day, uniquely combines steam-driven screws and paddles with sails.

A Currier and Ives lithograph (1866) of a Mississippi steamboat race.

1859	W. J. Rankine's *Manual of the Steam Engine* (UK) is first attempt to systematize the theory of steam power.
1869	The USA is linked by steam railway with the meeting of the Union Pacific and Central Pacific railroads.
1880	Thomas Edison builds an electricity generating station, powered by steam engines, in New York.
1884	Charles Parsons (UK) designs the first steam turbine to generate electricity.
1885	The transcontinental Canadian Pacific Railway is completed.
1891	Russia begins constructing the Trans-Siberian Railway linking Moscow and Vladivostock.
1897	Parsons's *Turbinia* of 2100 horsepower sets marine speed record of 34½ knots (63 km./hour).
	The Stanley steamer (USA) is one of a line of successful if short-lived steam automobiles.
1900s	Electricity generated by steam turbines begins to replace steam engines as a source of power.
1906	HMS *Dreadnought* is world's first battleship powered by steam turbines.
1931	At Larderello, Italy, geothermal power (using underground steam heat) is used to make electricity in the world's first geothermal generating station.
1938	The British locomotive *Mallard* sets the all-time speed record for a steam locomotive: 203 km./hour (126 m.p.h.).
1968	British Rail withdraws its last steam locomotive from passenger service.

New York Herald, which then sent him to find *Livingstone. The meeting took place on Lake Tanganyika (10 November 1871). Stanley later explored Uganda and the Congo and helped to found, under the auspices of the King of the Belgians, the *Congo Free State.

states' rights, a US political doctrine. It upholds the rights of individual states against the power of the federal government. The framers of the US Constitution produced a federal system in which the delineation of power between the federal government and the states was open to interpretation, and from the very beginning divergent views on this issue have influenced US politics. In the early years of the USA Hamilton and the *Federalist Party saw the Constitution as a sanction for strong central (federal) government, while *Jefferson and his followers believed that all powers not specifically granted to the federal government should be reserved to the states. It was behind the *Nullification Crisis of 1828-33 and it provided the constitutional basis of the Southern case in the dispute prior to the *American Civil War. In recent years the doctrine of states' rights has been central to controversies over *civil rights and welfare expenditure.

Stauffenberg, Claus Graf von   *July 1944 Plot.

steam power, the use of steam to power machinery. In 1781 James Watt adapted his improved steam engine to drive factory machinery, thus providing a reliable source of industrial power. Before this many factories depended on water power and were sited in the countryside near swiftly flowing streams, where transport was difficult, and production dependent upon the weather. However, steam engines were expensive and only large businesses could afford to install them. Factories were sited near coal mines and large towns grew up to house the factory workers. The use of steam engines in the textile industry and in other manufacturing processes led to a growth in the size of factories, while their application to railways and steamships led to both faster and cheaper travel and transport of goods. The steam hammer (1808) enabled much larger pieces of metal to be worked while such developments as the steam-driven threshing machine reduced farmers' reliance on wind- and water-mills. The direct use of steam engines began to decline in the early 20th century with the development of petrol and diesel engines and the use of steam-driven turbines to generate electricity, an energy source that can be applied more cleanly and easily in industry.

steamship   *Transport Revolution.

Stein, Heinrich Friedrich Karl, Baron von (1757-1831), Prussian statesman and reformer. After various diplomatic and administrative appointments he became Minister of Commerce (1804-7). He was dismissed by *Frederick William III for attempting to increase the responsibilities of the ministers of state. After the Prussian defeat at *Jena, Stein was recalled to begin his enlightened reforms. He persuaded the king to abolish the serf system, to end the restrictions against the sale to non-nobles of land owned by nobles, and to end the monopoly of the sons of the nobility in the Prussian officers corp. He wanted the king to authorize a national insurrection against the French and mobilize patriotic energies by the grant of a 'free constitution', but this alarmed Napoleon,

who persuaded the king to dismiss him again (1808). His pleas for a united Germany were ignored at the Congress of *Vienna.

Stern Gang, British name for a *Zionist terrorist group ('Lohamei Herut Israel Lehi', Fighters for the Freedom of Israel). It campaigned actively (1940-8) in Palestine for the creation of a Jewish state. Founded by Abraham Stern (1907-42), the Stern Gang numbered no more than a few hundred. They operated in small groups and concentrated on the assassination of government officials. Their victims included Lord Moyne, the British Minister for the Middle East (1944), and Count *Bernadotte, the United Nations mediator in Palestine (1948).

Stevens, Thaddeus (1792-1868), US statesman. He served in the Pennsylvania state legislature as a supporter of the *Anti-Masonic Party before his election to Congress as a Whig in 1849. He left the House in 1853 as a result of his strong *abolitionist views and helped to organize the *Republican Party before returning to Congress as one of its representatives (1859-68). His most influential period came after 1865 when he was one of the main champions of the radical reconstruction programme, and one of the chief architects of the *Fourteenth Amendment and the *Reconstruction Acts. He chaired the committee that prepared impeachment charges against Andrew *Johnson.

Stevenson, Adlai E(wing) (1900-65), US statesman. He served in various government posts and in 1948 he was elected governor of Illinois with the largest majority in the state's history. Chosen as the Democratic candidate for the Presidency in the elections of 1952 and 1956, on both occasions he was badly beaten by Dwight D. *Eisenhower. A liberal reformer and internationalist, his Presidential campaigns were marked by brilliant and witty speeches. President *Kennedy appointed him US ambassador to the United Nations (1961-5) with cabinet rank.

Stilwell, Joseph Warren (1883-1946), US general. Popularly known as 'Vinegar Joe' on account of his tactlessness, he served in China between the wars. In World War II, he commanded US and Chinese forces in south China and Burma, co-operating with the British in the *Burma campaign. Technically, his authority was by virtue of his appointment as chief-of-staff in this region by *Chiang Kai-shek. Difference of opinion with Chiang led to his recall, and he later commanded the US 10th Army at Okinawa.

Stimson, Henry Lewis (1867-1950), US statesman. He was Secretary of War for *Taft (1911-13) and served in *World War I. While governor-general of the Philippines (1927-9) he pursued an enlightened policy of conciliation. He entered *Hoover's cabinet as Secretary of State (1929-33), in which post he promulgated the Stimson Doctrine (or Doctrine of Non-Recognition) in response to the Japanese invasion of Manchuria (1931): a refusal to grant diplomatic recognition to actions which threatened the territorial integrity of China or violated the *Kellogg-Briand Pact. As Secretary of War (1940-5) he served under Franklin D. *Roosevelt and *Truman, and in 1945 he made the recommendation to drop the atomic bomb.

Stock Market Crash: panic continues on Wall Street on 24 October 1929 as the pace of selling shares nears its peak while anxious investors throng the sidewalks of New York's financial centre.

Stock Market Crash (USA) (1929), also known as the 'Great Crash', the 'Wall Street Crash', the 'Great Panic'. During the first half of 1929 an unprecedented boom took place on the New York Stock Exchange. Prices, however, began to fall from late September and selling began. In less than a month there was a 40 per cent drop in stock value, and this fall continued over the next three years. Its causes were numerous. Although the post-war US economy seemed to be booming, it was on a narrow base and there were fundamental flaws. The older basic industries, such as mining and textiles, were weak; agriculture was depressed; unemployment at four million was unacceptably high; international loans were often poorly secured. A new rich class enjoyed a flamboyant life-style, which too many people tried to copy by means of credit and stock-market speculation, with an unsound banking system. Once the business cycle faltered, a panic set in. The effects of the crash were hugely to accelerate a downward spiral: real estate values collapsed, factories closed, and banks began to call in loans, precipitating the world-wide Great *Depression.

Stolypin, Piort Arkadevich (1862–1911), Russian statesman. The last effective statesman of the Russian empire, he was Premier (1906–11). He was hated for his ruthless punishment of activists in the *Russian Revolution of 1905, for his disregard of the *Dumas, and for his treatment of Jews. His constructive work lay in his agricultural reforms. Believing that a contented peasantry would check revolution, he allowed peasants to have their land in one holding instead of strips which were periodically re-allocated within the peasant commune. Those taking advantage of this became prosperous, but were not powerful enough to stem the revolutionary tide. He was assassinated in a Kiev theatre.

Stopes, Marie Charlotte Carmichael (1880–1958), British scientist and writer on parenthood and birth control. She was appointed lecturer in palaeobotany at Manchester University in 1904 and then taught in London. It was, however, her books, particularly *Married Love* (1918) and *Wise Parenthood* (1918), with her clear views on birth control, which made her famous. With her second husband, H. Verdon-Roe, she founded a clinic for birth control in London in 1921. Her activities roused opposition but also steadily increasing support among the medical profession and the general public.

Stormont, suburb of Belfast and former seat of the parliament of *Northern Ireland. Created by the Government of Ireland Act (1920) as a subordinate body to Westminster, the Stormont Parliament was dominated by the *Unionist Party until, following the breakdown in law and order in the late 1960s it was suspended in 1972. Direct rule from Westminster was imposed, to be administered by civil servants of the Northern Ireland Office based in Stormont Castle.

Straits Settlements, former British crown colony comprising territories bordering on the strategic Malacca Strait in South-East Asia. The three English East India colonies of Penang, Malacca, and *Singapore were combined in 1826 as the Straits Settlements. After 1858, they passed to British Indian control, and in 1867 became a crown colony, to which Labuan was added in 1912. The colony was dismantled in 1946, Singapore becoming a separate colony and Penang, Malacca, and Labuan joining the Malayan Union.

Strategic Arms Limitation Talks (SALT), agreements between the USA and the Soviet Union, aimed at limiting the production and deployment of nuclear weapons. A first round of meetings (1969–72) produced the SALT I Agreement, which prevented the construction of comprehensive anti-ballistic missile (ABM) systems and placed limits on the construction of strategic (i.e. intercontinental) ballistic missiles (ICBM) for an initial period of five years. A SALT II Treaty, agreed in 1979, sought to set limits on the numbers and testing of new types of intercontinental missiles, but it was not ratified by the US Senate. *Arms-control talks resumed in 1982.

Strategic Defense Initiative (SDI, Star Wars), a proposed US defence system against potential nuclear attack. Based partly in space, it is intended to protect the USA from intercontinental ballistic missiles (ICBMs) by intercepting and destroying them before they reach their targets. Critics of the programme argue that in order to be effective, it would need to be technically infallible, that its costs would be excessive, and that it would escalate the arms race between the superpowers.

Streicher, Julius (1885–1946), German Nazi leader and propagandist. Originally a school-teacher, he expounded his anti-Semitic views in his periodical *Der Stürmer*. He was Party leader (*Gauleiter*) in Franconia (1933–40), and continued to function as a propagandist. He was sentenced to death at the *Nuremberg Trials.

Stresa Conference (April 1935), a conference between Britain, France, and Italy on Lake Maggiore. It proposed measures to counter *Hitler's open rearmament of Germany in defiance of the *Versailles Peace Settlement. Together these countries formed the 'Stresa Front' against German aggression, but their decisions were never implemented. In June Britain negotiated unilaterally a naval agreement with Germany. In November 1936 *Mussolini proclaimed his alliance with Hitler in the Rome–Berlin *Axis.

Stresemann, Gustav (1878–1929), German statesman. He was Foreign Minister (1923–9) in the *Weimar Republic, which he supported despite monarchist sympathies. He ended passive resistance to the French and Belgian occupation of the *Ruhr. He readily accepted both the *Dawes and *Young Plans on *reparations. Personal friendship with *Briand and Austen *Chamberlain enabled him to play a leading part at *Locarno (1925) and to negotiate the admission of Germany to the *League of Nations (1926). In 1928 he signed the *Kellogg–Briand Pact. Even so, he was adamant about wanting revision of Germany's eastern frontier, and advocated that Danzig (Gdańsk), the *Polish Corridor, and Upper Silesia should be returned by Poland.

Stroessner, Alfredo (1912–), Paraguayan military leader and President (1954–). Son of a German immigrant, he fought in the *Chaco War (1932–5). He rose from the ranks to become commander-in-chief of the armed forces (1951–4) and was responsible for the overthrow of President Frederico Chavez (1949–54). Basically supportive of the large landowners and international commercial interests, as President he has used foreign aid to develop schools, hospitals, highways, and hydro-electric power. His regime remains strongly backed by the army and is essentially totalitarian in that, while allowing for some political dissent, it has been accused of using harsh methods against political opponents.

Stuart, James Ewell Brown (1833–64), US general in the army of the *Confederacy. He resigned from the US Army to join the Confederacy at the outbreak of the *American Civil War. In command of a cavalry brigade after the first battle of Bull Run (July 1861), his daring raids behind Union (Northern) lines made him the South's pre-eminent cavalry leader. 'Jeb' Stuart served in all the major campaigns of the Army of North Virginia until he was mortally wounded at Yellow Tavern on 11 May 1864.

student revolts, social or political protests by student groups. Students have played an important part in almost every major revolution of the 19th and 20th centuries. In the early 19th century, the German universities produced student movements (*Burschenschäfter*) supporting German nationalism and opposing the rule of *Metternich. In Tsarist Russia students who agitated for liberal reforms were imprisoned, exiled, or executed. In the period between the two World Wars the universities in Germany and Japan had movements supporting mainly right-wing causes and revolutions. After World War II universities in the developing countries often fostered strong nationalist and Marxist movements, while in the 1960s left-wing movements were predominant in many universities and colleges in Europe, the USA, and Japan. The protests at the University of California's Berkeley campus (1964) and the nationwide strike at approximately 200 US campuses (1970) challenged US policy in Vietnam. In Paris French students and workers joined in the movement (1968) to challenge the *de Gaulle regime, while in Japan students acted militantly against the westernization of Japanese society and the US military presence in the country. Students in Hungary,

The most publicized **student revolt** of modern times was that in France in May 1968. French students, whose initial demonstration calling for the reform of higher education escalated into a crisis of national proportions, provided inspiration for students elsewhere in Europe and North America.

Poland, and Czechoslovakia have at times been in the forefront of protests against their countries' authoritarian regimes, while demonstrations by South Korean university students (1987) led to constitutional amendments and the release of political prisoners.

Sucre, Antonio José de (1795-1830), South American revolutionary. As *Bolívar's chief-of-staff, Sucre was entrusted with the liberation of Guayaquil and Quito in 1822, and his victory at the battle of Ayacucho on 9 December 1824 represented the defeat of the bulk of Spanish forces remaining in Peru. In 1825, Sucre drove the last Spanish forces from Bolivia and was chosen one year later as that nation's first elected President. He introduced legislation to implement fiscal and educational reforms, but soon became the target of factional divisions. He resigned the Presidency after an invasion (1828) by Peruvian troops, and retired to Ecuador. While working to preserve the union of Gran Colombia (Venezuela, Colombia, and Ecuador), he was assassinated.

Sudan, a country in north-east Africa. In 1800 northern Sudan consisted of the Muslim empire of the Funji, where an Islamic revival was occurring. The Funji were conquered by *Mehemet Ali from Egypt (1820-3). In 1874 Khedive Ismail, viceroy of Egypt, offered the post of governor of the Egyptian Sudan to Charles *Gordon. His anti-slave administration was not popular. Ismail's government in Egypt collapsed in 1879, Gordon retired in 1880, and in 1881 Muhammad Ahmad declared himself *Mahdi and led an Islamic rebellion in the Sudan. Britain occupied Egypt in 1882 and invaded the Sudan where Gordon was killed (1885). The Mahdists resisted Anglo-Egyptian forces until Kitchener defeated them at Omdurman in 1898. Following the *Fashoda incident, an Anglo-Egyptian condominium was created for the whole Sudan (1899) under a British governor. A

constitution was granted in 1948 but in 1951 King Farouk of Egypt proclaimed himself King of Sudan. After his fall, Egypt agreed to Sudan's right to independence; self-government was granted in 1953 and full independence in 1956. North–South political and religious tension undermined stability until General *Nimeiri achieved power in 1969 and negotiated an end to the civil war in the south in 1972. Peace broke down again in the early 1980s with the collapse of the economy, widespread starvation, and a renewal of separatist guerrilla activity in the south. Nimeiri was overthrown by the army in April 1985, and in 1986 a civilian coalition government was formed.

Sudetenland, the north-western frontier region of *Czechoslovakia. The region had attracted German settlers for centuries, but their claim to self-determination (1918) was denied and the land awarded to Czechoslovakia. The inhabitants had some cause for complaint against the Czech government, but this was whipped up by Konrad Henlein, the *Nazi leader. His demands for incorporation with Germany gave leverage to Hitler at *Munich (1938), and he annexed the region into the *Third Reich. In 1945 Czechoslovakia regained the territory and by the *Potsdam Agreement was authorized to expel most of the German-speaking inhabitants.

Suez Canal, a shipping canal 171 km. (106 miles) long connecting the Mediterranean (at Port Said) with the Red Sea. It was constructed under the supervision of the Frenchman Ferdinand de Lesseps (1805-94) and opened in 1869. A substantial shareholding (172,602 shares) was purchased in 1875 by Britain from the Khedive Ismail for £4 million. The operation of the Canal was regulated by the 1888 Constantinople Convention, which provided for the free transit of all vessels in peace and war. From 1882 to 1953 Britain held military control of the Canal, but evacuated the base in 1955. In 1956 Egypt nationalized the Suez Canal Company, precipitating the *Suez War, which temporarily closed the Canal. The Canal was again closed from 1967 to 1975.

Suez War (1956), a military conflict involving British, French, Israeli, and Egyptian forces. It arose from the nationalization of the *Suez Canal Company by Egypt in 1956. When attempts to establish an international authority to operate the Canal failed, Britain and France entered into a military agreement with Israel. The latter, concerned at the increasing number of *fedayeen* or guerrilla raids, was ready to attack Egypt. On 29 October Israel launched a surprise attack into Sinai, and Britain and France issued an ultimatum demanding that both Israel and Egypt should withdraw from the Canal. This was rejected by President *Nasser. British and French planes attacked Egyptian bases, and troops were landed at Port Said. Under pressure from the USA, with the collapse of the value of sterling, and mounting criticism of most other nations, the Anglo-French operations were halted and their forces evacuated. A UN peace-keeping force was sent to the area. The US Secretary of State, J. F. Dulles, formulated the short-lived *Eisenhower Doctrine (1957), offering US economic and military aid to Middle East governments whose independence was threatened. Israeli forces were withdrawn in March 1957 after agreement to install a UN Emergency Force in Sinai and to open the Straits of Tiran to Israeli shipping.

The aftermath of the **Suez War** led to a diminished role in the Middle East for Britain and France. The Canal, blocked from October 1956 to April 1957 by sunken and damaged shipping, is being cleared by a British task force.

suffragette, a member of a British militant feminist movement that campaigned for the right of adult British women to vote in general elections. The Women's Social and Political Union, which was founded by Emmeline *Pankhurst in 1903, gained rapid support, using as its weapons attacks on property, demonstrations, and refusal to pay taxes. There was strong opposition to giving women the vote at national level, partly from calculations of the electoral consequences of enfranchising women. Frustration over the defeat of Parliamentary bills to extend the vote led the suffragettes to adopt militant methods to press their cause; Parliamentary debates were interrupted, imprisoned suffragettes went on strike, and one suffragette, flinging herself in front of the king's horse in the 1913 Derby horse-race, was killed. These tactics were abandoned when Britain declared war on Germany in 1914 and the WSPU directed its efforts to support the war effort. In 1918, subject to educational and property qualifications, British women over 30 were given the vote (the age restriction was partly to avoid an excess of women in the electorate because of the deaths of men in the war). In 1928 women over 21 gained the vote.

Suharto (1921–), Indonesian statesman and general. Having played a prominent role in the *Indonesian Revolution, he became chief-of-staff of the army in 1965. He crushed a communist coup attempt by the *PKI in 1965 and in 1966 *Sukarno, implicated in the coup, was forced to give him wide powers. He united student and military opponents of the Sukarno regime and became acting President in 1967, assuming full powers as President in 1968. He ended the *Konfrontasi with Malaysia and revitalized the Indonesian economy, as well as restoring the country to the Western capitalist fold. Increasingly dictatorial, he faces domestic opposition, most notably from fundamentalist Muslim organizations.

Sukarno, Achmad (1901–70), Indonesian statesman and founder of Indonesia's independence. A radical nationalist, he emerged as leader of the PNI (Indonesian Nationalist Party) in 1926. He spent much of the 1930s either in prison or exile. During the Japanese occupation, he consolidated his position as the leading nationalist, and claimed the title of President of Indonesia in 1945. He led his country during the *Indonesian Revolution (1945–9), remaining President after the legal transfer of power from Holland in 1949. He became a spokesman for the non-aligned movement and hosted the *Bandung Conference in 1955, but after that his dictatorial tendencies aroused increasing resistance. Economic difficulties, and the *Konfrontasi with Malaysia, further undermined his position in the mid-1960s. Seeking increasing support from the communists, he finally lost power to the army after the abortive left-wing officer coup of 1965. Officially stripped of his power in 1967, he was succeeded as President by General *Suharto.

Sumner, Charles (1811–74), US statesman. Elected to the Senate in 1851, he served in it for the rest of his life, his powers of oratory making him one of the leading political reformers of his day. He joined the new *Republican Party and emerged as a leading member of the anti-slavery campaign. In 1856 he was assaulted in the Senate Chamber and brutally caned by Congressman P. S. Brooks, whose father, a South Carolina Senator, he had attacked in an anti-slavery speech. He did not resume

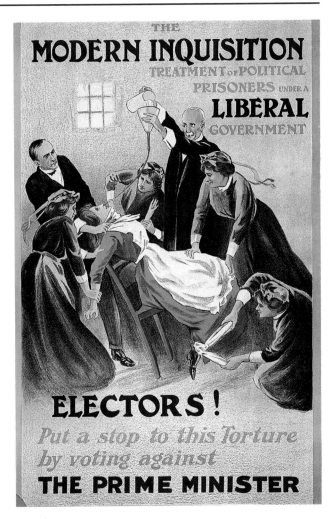

Suffragettes who resorted to hunger strike were imprisoned and forcibly fed. Such treatment aroused much criticism and inspired posters calling on voters to show their disapproval of the Liberal government.

his seat in the Senate for three years. After 1861 he served as chairman of the Foreign Relations Committee, and pressed hard for emancipation of the slaves. In the years after the *American Civil War, he supported the radical *reconstruction programme. He constantly attacked the administration of President *Grant for corruption and inefficiency.

Sunday school, a school for general and religious instruction of children on Sundays. Probably the first to be so called was established in Gloucester in 1780 by a journalist, Robert Raikes, to teach poor children (who were often at work for the rest of the week) reading and writing and knowledge of the scriptures. Raikes's ideas quickly spread, greatly assisted by the founding of the London Sunday School Union in 1803. By the early 1800s several hundred thousand children were enrolled in Sunday schools, which were to play an important part in elementary education before William *Forster's Education Act of 1870. In the USA the Sunday school movement developed rapidly in the 1790s, and the American Sunday School Union, formed in 1817, set out to establish schools all over the country.

Sunni *Islam.

Sun Yat-sen, pictured in November 1924 on board ship, with his wife, Soong Ching-ling (1892–1981), who herself became a notable political figure in China, both before and after the Communist Revolution.

Sun Yat-sen (or Sun Yixian) (1866–1925), Chinese revolutionary. In 1895 he organized an unsuccessful rising against the *Qing dynasty and fled the country. Briefly imprisoned (1896) in the Chinese legation in London, his release was negotiated by the British government. In 1905, in Tokyo, he formed a revolutionary society, the Tongmenghui (United League), which became the nucleus of the *Kuomintang. When the *Chinese Revolution broke out (1911), Sun returned to China and was declared Provisional President (1912) of the republic. He shortly resigned in favour of *Yuan Shikai. When Yuan suppressed the Kuomintang (1913), Sun, with *warlord support, set up a secessionist government in Guangzhou (Canton). In 1923 he agreed to accept Russian help in re-organizing the Kuomintang, thus inaugurating a period of uneasy co-operation with the *Chinese Communist Party. He died in Beijing, trying to negotiate for a unified Chinese government. Sun's Three Principles of the People, nationalism, democracy, and 'people's livelihood', are the basic ideology of *Taiwan, and he is regarded as the founder of modern China by both nationalists and communists.

Supreme Court (Federal), the highest body in the US judicial system. Established by Article III of the Constitution as a third branch of government, independent of the legislative and executive branches, the Supreme Court has become the main interpreter of the Constitution. Members are appointed by the President, with advice and consent of the Senate. Between 1789 and 1869, the number of Supreme Court justices varied between five and ten, but since 1869 it has remained at nine. Early in its history, the Supreme Court established its right to judge whether laws passed by Congress or by the state legislatures conform to the provisions of the Constitution, but it can do so only when specific cases arising under the laws are referred to it. The decisions of the Court have played a central role in the development of the US political system, not only as regards the fluctuating balance of power between the executive and legislative branches, and between the states and the federal government, but also concerning the evolution of social, economic, and legal policies.

Surinam, a country on the north-east coast of South America, known until 1948 as Dutch Guiana. The territory alternated between British and Dutch control until the Netherlands received it in a treaty settlement of 1815. In the 17th century African slaves had begun to be imported. By the late 19th century plantation labour was recruited from India and Java. The ethnic diversity of Surinam resulted in increasing racial and political strife after World War II. In 1954 Surinam became an equal partner in the Kingdom of the Netherlands, and full independence was granted in 1975. After several years of party strife the military took over in 1980. In February 1986 civilian rule was restored.

SWAPO, South West Africa People's Organization. Nationalist feeling in South West Africa (*Namibia) began to grow in the early 1960s as the South African government began to try to extend formal authority in the region, and SWAPO was formed in 1964–6 out of a combination of existing nationalist groups. Driven from the country, SWAPO, under the presidency of Sam Nujoma, began a guerrilla campaign, operating largely from neighbouring Angola. Efforts by the United Nations at mediation have persistently failed to find an agreeable formula for Namibian independence and the guerrilla war continues, with the situation further complicated by South African raids on SWAPO bases in Angola.

Swaraj Party (Sanskrit, 'self-ruling'), a nationalist party in India. In 1922 it was formed by C. R. Das, Motilal *Nehru, and Vallabhbhai *Patel to win control of the legislative councils for *Congress and to paralyse their working by a policy of obstruction. In 1923 the Swarajists won a majority in the central legislature and in several states, but the policy of obstruction was generally unsuccessful and the party was weakened by the resignation of members who objected to Congress's failure to safeguard Hindu interests. The Party came to an end in 1929, although it was revived during the 1937 elections.

swastika (from the Sanskrit, *svastika*, 'conducive to well-being'), an emblem in the form of an even-length cross, with the arms bent at right angles, clockwise or anti-clockwise. A symbol of prosperity and good fortune, it was used in ancient Mesopotamia, in early Christian and Byzantine art, in South and Central America, and among the Hindus and Buddhists of India. In 1910 the German poet Guido von List proposed the swastika (German, *Hakenkreuz* 'hooked cross') as a symbol for all *anti-Semitic organizations in the mistaken belief that it was Teutonic in origin. The *Nazi Party adopted it in 1919 and incorporated it (1935) into the national flag of the *Third Reich.

Swaziland, a landlocked country in southern Africa. It takes its name from the Swazis, who occupied it from the mid-18th century. A South African protectorate from 1894, it came under British rule as a High Commission Territory in 1902 after the Second *Boer War, retaining its monarchy. In 1968 it became a fully independent kingdom under Sobhuza II, King of the Swazi (1921–82). Revisions of the constitution in 1973 in response to requests from its Parliament, and again in 1978, have given the monarchy wide powers. The country's economy relies on co-operation with South Africa.

Sweden, a northern European country occupying the eastern part of the Scandinavian peninsula. During the *Napoleonic Wars Sweden joined the Third Coalition against France (1805), but France defeated Russia, and the latter in turn took Finland from Sweden as compensation (1809). In that year the pro-French party in the Swedish estates overthrew the existing monarch, Gustav IV, and elected the aged and childless Charles XIII (1809–18). In 1810 they invited Jean-Baptiste Bernadotte to become crown prince. He subsequently ruled as *Charles XIV (1818–44), and his descendants have remained monarchs of Sweden ever since. From 1814 to 1905 Norway was united with Sweden. Pursuing a policy of non-alignment, Sweden kept out of both World Wars, and by the 1950s it had developed into one of the world's wealthiest and most socially progressive states.

'Swing, Captain', a name used to sign threatening letters to landowners and farmers during an outbreak of agricultural disturbances in Britain (1830–1). In the years after 1815 many farm labourers were either unemployed or receiving very low wages. Their conditions were made worse by the bad harvests of 1829 and 1830 and in the latter year there was a spontaneous rising of poor labourers across southern and eastern England. Hay ricks were burnt and the newly introduced threshing machines destroyed. The riots were ruthlessly suppressed.

Switzerland, a country in central Europe consisting of a Federation of twenty-three cantons. In 1798 French Revolutionary armies entered the country and established the Helvetic Republic. But at the Congress of *Vienna (1815) Swiss control was restored and the European powers guaranteed the confederation's neutrality. In 1847 a separate Roman Catholic league within the federation, the *Sonderbund, was formed after radicals took power in one of the cantons. After a brief civil war, peace and stability were restored by the new, democratic, federal constitution of 1848. During World War I the country maintained its neutrality despite the contradictory affections of the French and German sections of its population. In World War II the Swiss again preserved their armed neutrality, and have continued since then to enjoy a high level of economic prosperity. In 1979 the twenty-two cantons of the confederation were joined by the new Canton of Jura. Women were not allowed to vote on a federal basis until 1971, and suffrage remains restricted in some cantons.

Sykes–Picot Agreement (1916), a secret Anglo-French agreement on the partition of the Ottoman empire after World War I. It was negotiated by Sir Mark Sykes (1879–1919) and François Georges-Picot and provided for French control of coastal Syria, Lebanon, Cilicia, and Mosul, and for British control of Baghdad and Basra and northern Palestine. The rest of Palestine was to be under international administration; independent Arab states were to be created in the remaining Arab territories. The agreement was the consequence of the British and French desire to compensate themselves for Russian gains under the secret Constantinople Agreement of 1915 between Russia, Britain, and France, in which the Dardanelles and the Bosporus were to be incorporated into the Tsarist empire in return for British and French spheres of influence in the Middle East.

syndicalism, a movement among industrial workers to abolish, by direct action, *capitalism and the capitalist state and to replace them with groups of workers called 'units of production'. Syndicalists were influenced by *Proudhon and Sorel, whose seminal *Refléxions sur la violence* was published in 1908. They differed from *socialists in believing essentially in *anarchism. Advocating direct action, especially a general strike, they were active in the early years of the 20th century, mainly in France, Italy, Spain, Russia, and in the USA. The growing complexity of industrial organization and the attraction of *communism reduced their influence after 1918.

Syria, a country in the Middle East. It became a province of the *Ottoman empire in 1516, and after the Turkish defeat in World War I Syria was mandated to France. Controlled by *Vichy France at the outbreak of World War II, the country was invaded and occupied by British and *Free French forces, and declared its independence in 1941. Political stability proved elusive, with three army-led coups in 1949 and others in 1951 and 1954. An abortive union with Egypt in the *United Arab Republic provided no solution and was terminated by a further army coup. A leading political grouping, the Ba'ath Socialist Party, remained split by personal and ideological rivalries, though one successful and two abortive coups in 1963 did see a swing to policies of nationalization. Further coups in 1966 and 1970 saw the eventual emergence of General Hafiz al-Assad as the leader of a new regime, capable not only of crushing internal opposition but also of asserting significant influence over neighbouring war-torn Lebanon. But, aspiring to a role of regional dominance, Syria suffered major reverses in the 1967 *Six-Day War and the *Yom Kippur War of 1973 against Israel. It became deeply involved in the civil war in Lebanon (1975 onwards). Syria has remained generally antagonistic towards Iraq and has developed close ties with the Soviet Union and its allies.

T

Tacna–Arica Conflict (1883–1929), a territorial dispute between Peru and Chile. The provinces of Tacna and Arica belonged to Peru at the time of its independence from Spain, but after the War of the *Pacific (1879–83), Chile appropriated Arica and Tacna. In 1929 negotiations between Peru and Chile produced a settlement which returned Tacna to Peru with an indemnity of $6 million and left Arica under Chilean control.

Taff Vale case (1901), a British court action that established the principle that trade unions could be sued for damages. Following a strike by railwaymen employed by the Taff Vale Railway Company, the company sued the Amalgamated Society of Railway Servants for loss of revenue. The House of Lords, on appeal, awarded the company damages and costs. The resentment felt by workers at this contravention of the Trade Union Act of 1871, which had, they thought, established the immunity of union funds, was an important factor in the increased support given to the *Labour Party. The Trade Disputes Act, passed in 1906, effectively reversed the decision by exempting trade unions from this type of action; this Act was amended in 1984.

Taft, William Howard (1857–1930), US jurist and twenty-seventh President of the USA (1909–13). He was appointed (1890) solicitor-general by President Benjamin *Harrison and later served as President Theodore *Roosevelt's Secretary of War (1904–8). It was Roosevelt who ensured that Taft gained the Republican nomination in 1908. His Presidency is remembered for its *dollar diplomacy in the field of foreign affairs, and tariff laws which were attacked by the *Progressive Movement as too sympathetic to big business. Roosevelt and Taft drifted apart and when he ran again in 1912 Taft had to share the Republican vote with Roosevelt running as a Progressive. As a result the Democrat, Woodrow *Wilson, was elected. Taft was appointed Chief Justice of the Supreme Court (1921–30), during which time he kept the Court on a conservative course.

Taft–Hartley Act (1947), an act of the US Congress to curb the power of trade unions. It banned the closed-shop in trade unions and the secondary boycott, allowed employers to sue unions for breach of contract and for damages inflicted on them by strikes, empowered the President to order a sixty-day 'cooling-off period' before strike action, and required union leaders to take oaths stating that they were not communists. Despite protests from the unions, it has remained relatively unchanged.

Tahiti, an island in the southern Pacific. It is part of the overseas territory of French Polynesia. At the beginning of

A Chinese representation of the attack on Nanjing (Nanking) in 1864 which crushed the **Taiping Rebellion**. Non-combatants try to escape by water while the battle rages. The rebels fought fiercely to avoid defeat and thousands preferred death to capture.

the 19th century Protestant missionaries settled on the island, converting the Tahitian chief Pomare II to Christianity in 1815. The French government established a protectorate in 1842 and annexed Tahiti as a colony in 1880. In 1940 the overwhelming majority of Tahitians backed the *Free French government of General de Gaulle, who made French Polynesia an overseas territory of France in 1946. At that time many Polynesians, led by Pouvanaa a Oopa, demanded independence from France. In 1958 Pouvanaa was sentenced to imprisonment and exile. Since 1977 French Polynesia has had considerable powers of self-government.

Taiping Rebellion (1850–64), revolt against the Chinese *Qing dynasty. Led and inspired by Hong Xiuquan (1813–64), who claimed to be the younger brother of Jesus Christ, the Taiping Rebellion began in Guangxi province. It developed into the most serious challenge to the Qing, bringing most of the central and lower Yangtze region under rebel control, and costing twenty million lives. The rebels captured Nanjing in 1853 and established their capital there before launching an unsuccessful attack on Beijing. Qing resistance depended on such provincial forces as the Hunan Army of *Zeng Guofan and the Ever-Victorious Army, formed by the American F. T. Ward and later commanded by the British soldier Charles *Gordon. Taiping resistance was crushed with the capture of Nanjing in 1864, Hong Xiuquan having died in the siege, but the Qing regime never really recovered from the long civil war. Taiping ideology was a mixture of Christianity and radical, egalitarian social policies which later revolutionaries, including the Communist Party, drew upon.

Taiwan, island in the China Sea, previously known by its Portuguese name of Formosa. A part of the Chinese *Qing empire since the 17th century, Taiwan was occupied by Japan as a result of the Treaty of *Shimonoseki in 1895 and remained under Japanese control until the end of World War II. The island was occupied by the Chinese forces of *Chiang Kai-shek in September 1945, but Taiwanese resentment at the administration of Chiang's governor Chen Yi produced a revolt which had to be put down by force of arms. When the *Chinese Civil War began to turn against the Kuomintang in 1948, arrangements were made to transfer Chiang's government to Taiwan, a move completed in the following year, and by 1950 almost two million refugees from the mainland had also arrived on the island. Supported militarily by the USA, Taiwan maintained its independence from communist China, as the Republic of China, and, until expelled in 1971, sat as the sole representative of China in the United Nations. Chiang Kai-shek remained its President until his death in 1975, and was succeeded by his son, Chiang Ching-kuo. Since 1950 it has undergone dramatic industrialization, becoming a major industrial nation.

Talleyrand-Périgord, Charles Maurice de (1754–1838), French diplomat and statesman. He entered the priesthood in 1778; two years later he became agent-general of the clergy of France and in 1789 was installed as bishop of Autun. He was one of a minority of French clergy who tried to reform the Church to serve the nation. In 1791 he left the Church, began a diplomatic career in London and, six years later, was appointed

foreign minister. Involved in the coup that brought *Napoleon to power, he became the latter's trusted adviser and foreign minister (1799–1807), and took part in most of the peace negotiations during the *Napoleonic Wars. He resigned his ministerial office in 1807 and engaged in secret activities with the Allies to have Napoleon deposed. When the Allies entered Paris in 1814, Talleyrand persuaded the Senate to depose him and, as head of the new government, recalled *Louis XVIII to the throne. As the king's Foreign Minister, he represented France at the Congress of *Vienna. Towards 1830, aware of the growing unpopularity of the government of *Charles X, he entered into diplomatic relations with *Louis Philippe. After the 1830 *July Revolution, as French ambassador to London, he did much to shape the future course of events in Europe. He was a signatory to the Quadruple Alliance (1834) between Britain, France, Spain, and Portugal, aimed at the support of constitutional monarchy in the last two countries against pretenders supported by Austria.

Tammany Hall, headquarters of a political organization in New York City. Founded in 1789, it was named after a late 17th-century Indian chief, and based its rites and ceremonies on pseudo-Indian forms. It acquired, under the control of Aaron *Burr, a political importance that endured until the 1950s. The Society's head, the Grand Sachem, was invariably an Irish Democrat of great influence in New York City politics. Under 'Boss' *Tweed, corruption reached a high point between 1867 and 1872. Thereafter, the word Tammany became synonymous with machine politics, graft, corruption, and other abuses in city politics.

Tamworth Manifesto (1834), election address of Sir Robert *Peel to his constituents at Tamworth, Staffordshire. Peel pledged himself to accept the Whig government's *Reform Act of 1832. He declared his adherence to a policy of moderate reform, while stressing the need to preserve what was most valuable from Britain's past. This concept of change, where necessary, within existing institutions marked the shift from the old, repressive Toryism to a new, more enlightened Conservatism.

Tanaka Kakuei (1918–), Japanese statesman. First elected to the House of Representatives in 1947, his career was briefly interrupted by a bribery scandal soon after, but from 1957 he served successively as Minister of Communications, Minister of Finance (in three different cabinets) and Minister of International Trade, and in 1972 he became Japan's youngest post-war Prime Minister (1972–4). He was forced to resign as a result of a bribery scandal in December 1974 and in 1976 had to face accusations of responsibility for the Lockheed scandal, relating to the corrupt sale of Lockheed aircraft to Japan. Despite lengthy legal proceedings against him he remained a powerful force within the ruling *Liberal Democratic Party until disabled by a severe stroke in 1985, his intra-party faction disintegrating in July 1987.

Tanganyika   *Tanzania.

Tan–Zam Railway, a railway line in Central Africa. Its construction was formally agreed between *Zambia, *Tanzania, and China in August 1970. Its purpose

Tanaka Kakuei (*left*), with Nakasone Yasuhiro (*centre*) and Deng Xiaoping (*right*). The new style of leadership introduced by Tanaka stimulated the conservative, business-oriented elements in Japan.

Pasha (1800–58) a programme of reform was steadily developed. The army was reorganized, on the Prussian model. The slave trade was abolished. Abuses in the taxation system were to be eliminated; provincial representative assemblies were created; new codes of commercial, land, and criminal law, based on French models, were introduced, with new state courts, separate from the Islamic religious courts. In his last years sultan Abdulaziz lost interest in the Tanzimat and, from 1871, became increasingly autocratic.

Taranaki Wars, a major part of the *Anglo-Maori wars in New Zealand. In 1859, Governor Browne accepted an offer of land on the Waitara River from Teira (a Maori right-holder), despite the veto of the senior chief, Wiremu Kingi. When the survey was resisted, Browne sent troops to Waitara. Many Maori supported Kingi, believing the purchase to be a breach of the Treaty of *Waitangi. Fighting was inconclusive in 1860–1 and, after a two-year truce, resumed in 1863. Maori resistance on the coast was overcome and much land confiscated. Maori resistance in the interior, increasingly led by the *Pai Marire, continued through the 1860s.

tariff reform, a British fiscal policy designed to end the nation's adherence to *free trade by the use of protective duties on imported goods. Joseph *Chamberlain believed that the use of tariffs would strengthen Britain's revenue and its trading position; it would also strengthen links within the British empire by making possible a policy of imperial preference (the application of lower rates of duty between its member countries). Chamberlain's campaign (1897–1906) failed, dividing the Conservatives

Chinese technicians assist in construction of the **Tan–Zam Railway**, instructing Africans in railway engineering. Chinese aid, financial and technical, assured completion of the railway line from Tanzania to Zambia in 1975.

was to free Zambia from dependence on neighbouring countries for railway transport of its vital copper and other exports. China provided an interest-free loan of £169 million, with technical aid. It spanned 1,860 km. (1,162 miles), and was completed in October 1975, a year ahead of schedule.

Tanzania, a country in East Africa. It consists of the former republic of Tanganyika and the island of *Zanzibar. A German colony from the late 19th century, Tanganyika became a British *mandate after World War I, and a trust territory, administered by Britain, after World War II. It became independent in 1961, followed by Zanzibar in 1963. The two countries united in 1964 to form the United Republic of Tanzania under its first President, Julius *Nyerere. In the *Arusha Declaration of 1967 Nyerere stated his policy of equality and independence for Tanzania. In 1975 the *Tan–Zam railway line was completed. Recurring economic problems have caused some political problems, as has Tanzania's involvement in the troubled affairs of its neighbour, Uganda.

Tanzimat reforms (1839–71), a series of reforms in the *Ottoman empire. They were promulgated under sultans Abdulmecid I (1839–61) and Abdulaziz (1861–76) in response to western pressure. Under Mustafa Resid

A coloured lithograph of Hobart Town, **Tasmania**, printed in 1824, when the island was still called Van Diemen's Land. At the time there were around 2,000 Tasmanian Aborigines, but clashes with the settlers led to their extinction by 1876.

and rejected by the Liberals. Tariff reform was rejected again in 1923 when Stanley *Baldwin and the Conservatives failed to secure an overall majority in an election primarily on that issue. However, the shock caused by the international financial crisis of 1929–31, and the intensification of nationalist political and economic rivalries, made Britain's free trade policy even more of an anachronism. The adoption of protectionism by the MacDonald *National government from 1931 signalled the ultimate success of the tariff reform policy.

Tasmania, a state of the Commonwealth of *Australia. British convict settlements were established in 1803 and 1804. Van Diemen's Land, as it was then called, remained a part of New South Wales until 1825, when it became a separate colony. Sizeable numbers of free settlers began to arrive from the 1820s onwards. The indigenous inhabitants embarked on the Black War (1825) against the settlers, and the latters' efforts to contain them on Flinders Island (1835) failed. Convicts were transported to Van Diemen's Land until 1853, but Port Arthur, the penal settlement, was not closed until 1877. Partial self-government was granted in 1850, extended in 1855 to a cabinet system responsible to an elected Parliament. Its new name, Tasmania, was proclaimed in 1855.

Tata, an Indian Parsi commercial and industrial family. It is one of the two (with the *Birlas) most important merchant families in modern India. The family began in Far East trade, but diversified their operations creating, under Jahangir Ratanji Dadabhai Tata (1904–), an airline which later became Air India. The Tatas also spent extensively on scientific and other work, founding the Tata Institute of Fundamental Research in Bombay.

Tawhiao, son of the Maori king *Potatau, and head (1860–94) of the *Kingitanga movement. He led his people in the Second *Anglo-Maori War. Following a truce in 1868 an uneasy peace existed in the 'King Country'. When the government sought to build a railway concessions were made to the Maoris. Tawhiao returned to Lower Waikato in 1883 and continued to press claims for the return of confiscated land and for recognition of Maori *rangatiratanga* (chieftainship).

Taylor, Zachary (1784–1850), twelfth President of the USA (1849–50). One of the leading US soldiers of the mid-19th century, he fought in the *Black Hawk and *Seminole wars. His victories in the *Mexican–American War, particularly at Buena Vista (23 February 1847), made him a national hero and carried him to the White House as leader of the Whig Party. In the crisis preceding the *Compromise of 1850, he took a firm stand against appeasement of the South. He died after only sixteen months in office.

Teamsters, International Brotherhood of, a trade union in the USA. Formed in 1903, its members are workers in the transport industry. In the late 1950s its president, David Beck, was indicted for having links with criminals, and his successor, James R. Hoffa, was found guilty of attempting to influence a federal jury while on trial in 1964 for misusing union funds. The revelations of corruption did immense damage to the reputation of unions, but the Teamsters sought to recover by co-operating with more responsible union leaders like Walter *Reuther of the Union of Auto Workers.

Teapot Dome scandal (1922–4), US fraud perpetrated by the 'Ohio gang' surrounding President *Harding. It involved the siphoning of oil, intended for the US navy, from the oil reserves at Teapot Dome, Wyoming, to the Mammoth Oil Company. A second diversion allowed oil from Elk Hills, California, to be siphoned to the Pan-American Petroleum and Transportation Company. Harding died before the full extent of the involvement of the Secretary of the Interior, Albert B. Fall, was exposed by Senator Thomas J. Walsh of Montana in the years 1922–4. Fall was found guilty of accepting a $100,000 bribe and imprisoned (1929–32).

Technological Revolution, a term loosely applied to a variety of technological changes since World War II that have, for example, affected industry, commerce, agriculture, and medicine. The steady advance of computer technology has transformed the economies of the industrialized countries by increasing industrial productivity and commercial efficiency. This has resulted in higher standards of living for those in employment, but also increased unemployment as a result of *automation and the decline of traditional industries. Such unemployment tended to be concentrated in the 19th-century industrial conurbations of Europe and the USA, while newer industries, centred around the computer industry, developed elsewhere as, for example, in the so-called 'sun-belt' of the southern USA. Developments in agriculture, biology (including genetic engineering), and medicine have resulted in both improved food supplies, especially in developing countries, and greater life-expectancy. The Technological Revolution has produced considerable debate over such issues as genetic engineering and transplant surgery, nuclear power, atmospheric pollution, and excessive consumption of the world's natural resources.

Tecumseh (c.1768–1813), American Shawnee Indian chief in the Ohio Valley. He emerged as the most formidable opponent of the white *westward expansion, believing that Indian land was a common inheritance, which could not be ceded piecemeal by individual tribes. Together with his half-brother, the Prophet

Tenskwatana, he formed a confederacy of tribes to negotiate a peaceful settlement with the whites. This confederacy was defeated at the battle of *Tippecanoe in 1811, and Tecumseh then sided with the British in the *War of 1812, being killed at the battle of the *Thames in 1813. This marked the end of Indian resistance in the Ohio Valley. Tenskwatana retired to Canada with a British pension, but returned in 1826 and accompanied the Shawnee Indians when they were moved, first to Missouri, and then to Kansas, where he died (c.1837).

Teheran Conference (28 November–1 December 1943), a meeting between *Churchill, *Roosevelt, and *Stalin in the Iranian capital. Here Stalin, invited for the first time to an inter-Allied conference, was told of the impending opening of a *Second Front to coincide with a Soviet offensive against Germany. The three leaders discussed the establishment of the *United Nations after the war, and Stalin pressed for a future Soviet sphere of influence in the *Baltic States and Eastern Europe, while guaranteeing the independence of Iran.

Te Kooti Rikirangi Te Turuki (c.1830–93), Maori spiritual leader. A member of the Aitanga-a-Mahaki tribe, he was accused in 1865 of complicity with the *Pai Marire and its militant off-shoot, the Hau-hau, and was deported to the Chatham Islands. In exile he evolved a variation of Pai Marire ritual and belief, the Ringatu. In 1868 his group escaped, seized a government ship, and returned. Challenged by the military, he attacked the settlement at Poverty Bay, killing some Europeans and many more collaborating Maori. The subsequent pursuit and skirmishing lasted until 1872 when Te Kooti found sanctuary in the 'King Country' until pardoned. The Ringatu church survives today.

Tel-el-Kebir, battle of (12–13 Steptember 1882), a battle between British and Egyptian forces, 83 km. (52 miles) east of Cairo. British troops under Sir Garnet Wolseley defeated an Egyptian army led by *Arabi Pasha. Cairo fell on the following day, thus confirming the British conquest of Egypt.

Temperance Movements, campaigns to restrict the consumption of alcohol. Early temperance associations appeared first in New England in the USA, where, by 1833, there were some 6,000 local temperance societies. In 1829 the Ulster Temperance Society was formed and the temperance movement then spread to Scotland, Wales, England, Norway, and Sweden. In 1874 the Women's Christian Temperance Union was founded in Cleveland, Ohio. It quickly spread across the continent, criticizing the masculine 'saloon world', with its hard drinking and prostitution, and coming to the belief that only *women's suffrage could end the social degradation of women. In 1883 the Union became a world organization, the first international women's movement, its influence spreading beyond the USA to Australia and New Zealand. A *Prohibition Party had been founded in 1869 in the USA, quickly becoming an effective force in American politics, where the Anti-Saloon League was founded in 1893. In 1920 Prohibition was enforced nationally by Congress. In Britain a strong campaigner for temperance in the 1890s was the *Salvation Army, and temperance legislation was introduced in 1904, 1906, and 1916 in the *Defence of the Realm Act, to reduce

A propaganda broadsheet of the 1880s, produced in support of the British **Temperance Movement**, records the grim consequences, of addiction to drink.

the number of public houses, exclude people under 18, and impose 'drinking hours'. These acts were confirmed in 1953 and 1961. Many countries have some form of legislation concerned with alcohol consumption, and it is illegal in most Muslim countries.

Temple, William (1881–1944), British churchman and educationalist. He became a priest in 1909 and was Archbishop of Canterbury (1942–4). The achievement of greater equality in the educational system, springing, in part, from his early activities with the Workers' Educational Association, was an important objective for him. He worked with R. A. (later Lord) *Butler on his Education Bill which became law in 1944. He also sought to secure a greater sense of common purpose between the different religious denominations and his work led to the foundation of the *World Council of Churches.

Templer, Sir Gerald (1898–1979), British field-marshal. He commanded the 6th Armoured Division in World War II and after the war served as vice-chief of

Archbishop William **Temple** arrives to preach at St Mary Woolnoth, London, in 1942. He exercised greater influence on public life in Britain, and on the attempts made by governments to solve the social and educational problems of the day, than any other Anglican bishop in the 20th century.

the Imperial General Staff before being appointed high commissioner and commander-in-chief in Malaya (1952-4) at the peak of the *Malayan Emergency. Through a combination of military efficiency, adaptability to local circumstances, and the fostering of good relations with village populations, Templer turned the tide of war decisively against the communist guerrillas.

Tennessee Valley Authority (TVA), an independent federal government agency in the USA. Created by Congress (1933) as part of the *New Deal proposals to offset unemployment by a programme of public works, it set out to provide for the development of the whole Tennessee River basin. It took over a project (begun in 1916) for extracting nitrate at Muscle Shoals, Alabama. In addition the TVA was authorized to construct new dams and improve existing ones, to control floods and generate cheap hydro-electric power, to check erosion, and to provide afforestation across seven states.

Te Puea Herangi (1883–1952), sometimes called 'Princess' Te Puea, a niece of the Maori king, *Tawhiao. Deeply educated in *Maori language and culture, she took direction of the *Kingitanga movement. She built Turangawaewae, south of Auckland, as a leading *marae* (social, political, cultural, and spiritual centre) and secured recognition of the Kingitanga from the New Zealand government. Her leadership strengthened Maori values and institutions while fostering a controlled accommodation of European influences, including commercial farming and education.

terrorism, the use or threat of violence for political purposes. In the 19th and 20th centuries acts by terrorists (also described as guerrillas or freedom fighters) have included indiscriminate bombing, kidnapping, hijacking, and assassination to produce fear amongst opponents and the general public. In 19th-century Russia the *anarchists and *nihilists used bombings and assassinations against the Tsarist government, while in the USA organizations such as the *Ku Klux Klan used terrorism and lynchings to intimidate the black population after the American Civil War. In the 20th century dictators such as Mussolini and Hitler have come to power through the use of terror tactics, while the period after World War II witnessed the growth of nationalist or liberation groups which used terrorism as part of their struggle against an occupying power: for example in *Cyprus, *Palestine, *Ireland, and many countries in Africa, Asia, and the Middle East. A further development was the appearance of terrorist groups struggling against their countries' social and political structure, including the *Baader-Meinhof gang and the Red Army Faction in Germany, the *Red Brigade and its offshoots in Italy, Action Directe in France, Eta in Spain, and the Tupamaros in Uruguay. These and other groups presented an international problem as they began to co-ordinate attacks across frontiers, and many countries developed anti-terrorist units.

Tet Offensive (29 January–25 February 1968), offensive launched in the *Vietnam War by Vietcong and regular North Vietnamese army units against US and South Vietnamese forces. In a surprise attack timed to coincide with the first day of the Tet (Vietnamese Lunar New Year) holiday, North Vietnamese forces under General Giap took the war from the countryside to the cities of South Vietnam. After initial successes, the attackers were repulsed with heavy losses on both sides, but the offensive seriously damaged South Vietnamese morale and shook US confidence in their ability to win the war and brought them to the conference table in Paris in 1969. This led to the Paris Peace Accords of 1973 and the withdrawal of US forces from Indochina.

Te Whiti, Maori religious leader (c.1820–1907). In 1877, when government officials began land surveys in South Taranaki without first creating reserves as guaranteed by the Treaty of *Waitangi, Te Whiti organized non-violent resistance. He prophesized success through continued non-violent resistance and 2,000 people flocked to his settlement at Parihaka. In 1881 government forces arrested Te Whiti and dispersed the settlement. After a year in custody Te Whiti rebuilt and modernized Parihaka, where his teachings were promoted for many decades.

Texas, republic of (1836–45), a short-lived independent republic in the south-west of the USA. Texas had only been lightly colonized by the Spanish, and in 1821 the Mexican government granted Stephen Austin the right to bring US settlers into the region. Pressure began to build up for independence from Mexican control, and a revolt broke out in 1835–6. After defeat at the *Alamo, Texan forces under *Houston captured the Mexican general *Santa Anna at *San Jacinto. An independent republic of Texas was proclaimed, which was recognized by the USA as the 'Lone-Star' state. The republic lasted for almost a decade before it was admitted to the Union as the twenty-eighth state, an event which helped to precipitate the *Mexican–American War in the following year (1846).

Texas Rangers, a para-military US police force. The Texas Rangers were first organized in the 1830s to protect US settlers in Texas against Mexican Indians. After the formation of the republic of *Texas (1836), they were built up by *Houston as a mounted border patrol of some 1,600 picked men. They became renowned for their exploits against marauders and rustlers in the heyday of the great *cattle drives after the *American Civil War.

Thailand, south-east Asian country, known until 1939 as Siam. After the collapse of the Thai state of Ayuthia in the mid-18th century, the Chakri dynasty came to power in 1782 and began to revitalize Thai power in the face of Burmese rivalry and increasing European pressure. The Chakri succeeded in maintaining their country's independence through a policy of conciliation, ceding their vassal state in Laos and Cambodia to France in the late 19th and early 20th centuries. In the reigns of *Mongkut (1851–68) and *Chulalongkorn (1868–1910) Thailand achieved substantial modernization in both the administrative and economic spheres. The middle class produced by the modernization process became intolerant of absolute royal rule, and an economic crisis in 1932 produced a bloodless coup which left the Chakri dynasty on the throne but transferred power to a constitutional government. Although technically allied to Japan during World War II, Thailand retained western friendship because of prolonged guerrilla resistance to Japanese forces. Until the early 1970s the country was largely ruled by the army, Marshal *Pibul Songgram maintaining near personal rule from 1946 to 1957. Severe rioting resulted in a partial move to civilian government in 1973 and the introduction of a democratic constitution in 1974, but the threat of communist aggression, particularly on its borders with *Kampuchea, has allowed the continuation of pronounced military influence. A brief military uprising was forcibly suppressed by the government in 1981, and a delicate balance between civilian and military power has survived.

Thames, battle of the (5 October 1813), a military engagement in the *War of 1812, fought in present-day south-western Ontario, Canada. Following their abandonment of Detroit, British and Indian forces under General Proctor and *Tecumseh were overtaken and decisively defeated near Chatham on the Thames River by a US force under General *Harrison. The US victory, together with Tecumseh's death in the battle, destroyed the Indian confederacy and the British and Indian alliance, and secured the US north-west frontier.

Thatcher ministries, British Conservative governments (1979-) with Margaret Thatcher (1925-) as Prime Minister. Replacing Edward *Heath as Leader of the Conservative Party in 1975, she became, after the general election of 1979, the first woman Prime Minister in European history. During her first term of office (1979- 83) the main thrust of government policy lay in tackling

Prime Minister Margaret Thatcher at a press conference in 1982 during the first of the **Thatcher ministries**, at the height of the Falkland War.

inflation and industrial and public sector inefficiency in Britain. A severe monetary policy was adopted while government control and intervention in industry was reduced. This was the signal for a general reduction in overmanning, and resulted in many bankruptcies and a reduction in manufacturing, made worse by an over-valued pound and high interest rates. A corollary to this policy was the determination to curb public spending, which led to increasing friction between central and local governments, and the curbing of trade-union power. Unemployment rose to levels not seen since the Great Depression, but the *Falklands War produced a mood of national pride that helped to secure a landslide victory for the Conservatives in the 1983 election. During the second term (1983-7) Nigel Lawson as Chancellor of the Exchequer adopted a less rigid economic policy. Inflation was brought under control, helped by lower commodity prices, while a lower pound benefited manufacturing industry. Major trade-union legislation was challenged by the National Union of Mineworkers (1984-5), which conducted an unsuccessful and often violent strike. Lower taxation and a wide-ranging programme of public-asset sales, including council housing, helped to extend house ownership and share ownership as well as to reduce public borrowing. In spite of *Irish Republican Army activity, including an attempt to blow up the cabinet at the 1984 Conservative Party Conference, the Hillsborough Agreement was signed (1985) with the Republic of *Ireland. In 1987 the Conservatives were returned for a third term with a majority of 101, making Margaret Thatcher the first party leader to face three consecutive new parliaments as Prime Minister.

Thiers, Louis Adolphe (1797-1877), French statesman and historian. His fame came with the *Franco-Prussian War when his diplomatic skill helped in negotiations with *Bismarck, which resulted in the Treaty of *Frankfurt. He ordered the ruthless destruction of the Commune of *Paris (1871). He was elected President of the Third Republic in 1871, for, although a monarchist, he believed that national unity demanded a republic. In 1873 he was overthrown by right-wing deputies.

Third Reich (1933-45), the period covering the *Nazi regime in Germany. Adolf *Hitler accepted the Chancellorship of Germany in January 1933, after a period of political and economic chaos and assumed the Presidency and sole executive power on the death of *Hindenburg in 1934. He almost immediately engineered the dissolution of the *Reichstag after a fire, blamed on the communists, led to the Enabling Act, which gave the government dictatorial powers. The Third Reich proved to be one of the most radical reversals of democracy in European history. Germany became a national rather than a federal state, non-Aryans and opponents of Naziism were removed from the administration, and the judicial system became subservient to the Nazi regime with secret trials that encompassed a wide definition of treason and meted out summary executions. *Concentration camps were set up to detain political prisoners. All other political parties were liquidated and the National Socialists declared the only party. *Anti-Semitism was formalized by the Nuremberg Laws. Both Protestant and Catholic Churches were attacked. The *Hitler Youth movement was formed to indoctrinate the young. Most industrial workers were won over by the rapid end to unemployment through

The Third Reich

Hitler's control over Europe was at its peak in 1941, but the invasion of the Soviet Union proved fatal. From December 1942 Nazi Germany was steadily weakened, losing control of vital oil supplies to the east, and deprived of its industries by Allied saturation bombing.

rearmament and other public spending. Much of industry was brought under government control, while the small farmer found himself tied more securely to the land. A four-year plan of 1936 set out to attain self-sufficiency in the event of war. Hitler reintroduced compulsory military service in 1935, following the return of the *Saar Basin by the League of Nations. Having withdrawn from the disarmament conference, Hitler broke the *Locarno Treaties by re-occupying the Rhineland; he annexed Austria (*Anschluss) and the *Sudetenland in Czechoslovakia, and sought to break up any system of alliance within Eastern Europe. During the spring and summer of 1939 he made a political and military alliance with *Mussolini's Italy, and brought the old dispute over the *Polish Corridor to a head, arranging a *Nazi–Soviet pact with the Soviet Union. On 1 September 1939 he invaded Poland without a declaration of war, and Britain and France declared *World War II. German military occupation of most of continental Europe followed rapidly, until by 1941 Nazi-controlled territory stretched from the Arctic Circle and the English Channel to North Africa and Russia. Britain remained its sole

adversary from June 1940 to June 1941, when Hitler invaded the Soviet Union. Total mobilization was introduced early in 1942, and armaments production was increased despite heavy air attacks on industrial and civilian targets. Under Himmler, the *SS assumed supreme power. After 1943, the German armies fought a rearguard action, and by May 1945 the Third Reich lay in ruins.

Thistlewood, Arthur *Cato Street Conspiracy.

Thorez, Maurice (1900-64), French politician. He helped to form the French Communist Party, of which he became General Secretary in 1930. His party supported but never joined the Popular Front Coalition government (1936) which enacted important social and labour reforms. Conscripted in World War II, he deserted and went to Moscow. Sentenced *in absentia*, he was pardoned (1944) and re-elected as a Deputy. He led the Communist Party in the elections of 1945 and 1946, and became Vice-Premier (1946-7). By the late 1950s his authority was reduced by his Stalinist associations and his support of the Soviet invasion of Hungary.

Three Emperors' League (German, Dreikaiserbund), an alliance between Prussia, Austria, and Russia. In 1872, following the creation of the new *German Second empire, Bismarck persuaded the emperors of Austria and Russia into an unofficial alliance. It was strained by the Treaty of San Stefano (1878) which, in conclusion of the *Russo-Turkish Wars, assigned Russia the eastern part of *Armenia and created a large state of *Bulgaria. In June 1881, following the assassination of *Alexander II, the league was revived as a more formal alliance. Renewed in 1884, it finally expired three years later, as tension between Russia and Austro-Hungary mounted in the *Balkan States.

'Three Fs', the nickname given to the second Irish Land Act (1881) introduced by the *Gladstone government. The 'Three Fs' were fair rents (to be settled by a tribunal), freedom to sell improvements (involving compensation to tenants vacating a property for improvements they had made), and fixity of tenure (so long as rents were paid). The Act failed to win the support of Charles *Parnell and his followers.

thug (Hindi, *thag*, 'swindler'), a devotee of the Hindu goddess Kali, who was worshipped through ritual murder and sacrifice of travellers. The thuggee centre was in remote central India, where victims were strangled. Eradication of the brotherhoods was difficult because of the secrecy of the cult. It was largely suppressed in the 1830s by the detective skills of William Sleeman, appointed to the task by Lord William *Bentinck. Indians welcomed the intervention, and there has been no revival, but the term passed into the English language.

Tianjin, Treaty of (1858) *Opium Wars.

Tibet (or Xizang), an autonomous region of the People's Republic of China. Tibet came nominally under Chinese control in the 18th century, but by the late 19th century had become virtually independent under the Buddhist leadership of the Dalai Lama. Fears of Russian influence led to a British invasion in 1904, and the negotiation of

Thug devotees attack a traveller on horseback as he rides through a mountain pass in central India. Ritual assassination by strangulation was the mark of this secret society, as recorded in a drawing made in 1838 by Fanny Eden, herself travelling between Kanpur and Mussoorie. (British Library, London)

The Dalai Lama (wearing spectacles) escapes to India from **Tibet** with his followers in 1959, following the Chinese communist invasion. For most Tibetans, the common form of travel was still on foot or with pack horses, using coracle boats to cross rivers. After the Chinese take-over a network of tarmacked roads was constructed.

an Anglo–Tibetan trade treaty. Tibet became autonomous under British control when the Chinese empire collapsed in 1911, and remained so until Chinese troops returned in 1950, completely occupying the country a year later. After a rebellion in 1959 the Dalai Lama (1935–) and thousands of his subjects fled. Tibet was administered as a Chinese province until 1965, when it was reconstituted as an autonomous region within the People's Republic. A further revolt was suppressed in 1987.

ticket-of-leave, in Australia, a certificate which could be granted to a convict during the period of *convict transportation. It allowed a convict to be excused from compulsory labour, to choose his or her own employer, and to work for wages. There were some restrictions and the ticket-of-leave could be withdrawn. It usually was granted for good conduct.

Tilak, Bel Gangadhar (1856-1920), Indian scholar and politician. Known as Lokamanya ('revered by the people'), he owned and edited *Kesari* ('Lion'), a weekly Marathi nationalist paper. He was imprisoned for sedition by the British (1897). Released in 1899, he continued to advocate radical policies within *Congress, playing an important role in the radical/moderate split of 1907. In 1914 with Annie *Besant he formed the Indian Home Rule League, but subsequently advocated more moderate, co-operative policies with Muslims in the Lucknow Pact (1916), which recognized separate electorates for Muslim minorities.

Tillett, Benjamin (1860-1943), British trade unionist and politician. His powers as a speaker made him influential in the *London Dockers' Strike (1889), when he succeeded in gaining an assurance for a minimum wage of sixpence an hour and eightpence for overtime. He concentrated on union activities, organizing another dock strike in 1911. A critic of the weakness of the *Labour Party in its early days, he was a Labour Member of Parliament (1917-24, 1929-31).

Tilley, Sir Samuel Leonard (1818–96), New Brunswick and Canadian statesman. He became a Liberal member of the New Brunswick House of Assembly in 1850 and Premier in 1861. A Father of Confederation, he was an important advocate at the Westminster Conference, which drafted the provisions of the *British North America Act (1867). In the same year he was elected to the new dominion Parliament and served in John A. *Macdonald's cabinet until 1873. On Macdonald's return to office in 1878 he was appointed Minister of Finance. He formulated the protective tariff plan known as the National Policy for Canada.

Tilsit, Treaties of (7 and 9 July 1807), agreements between Russia and France, and Prussia and France. Near this East Prussian town (now Sovetsk in the Soviet Union) Napoleon I, having won the battle of Friedland, agreed to meet *Alexander I on a raft on the River Niemen. Their negotiations, joined by *Frederick William III of Prussia, led to the two treaties. Prussia lost over a third of its possessions, had to pay heavy indemnities to France, and was forced to support a large French army on its soil. The Polish lands annexed by Prussia under the partitions were turned into a French puppet state, the Grand Duchy of Warsaw. Russia recognized the *Confederation of the Rhine and was forced to join the *Continental System. Prussia rescinded its treaty in 1813 when it deserted the French, following the latter's invasion of Russia and joined the Russian emperor in a campaign against Napoleon in Germany.

Timor, south-east Asian island. The eastern part of Timor was first colonized by the Portuguese in the 16th century and the western part by the Dutch in the 17th, and the island remained divided between the two powers until after World War II (except during the Japanese occupation of 1942–5). Dutch Timor became part of Indonesia in 1949, but Portuguese Timor remained an overseas province of Portugal until 1975, when the leftist Fretilin movement proclaimed independence after a brief civil war. Shortly afterwards Indonesian troops invaded, occupying it after heavy fighting, and, despite continuing guerrilla resistance, the eastern part of Timor was formally integrated into Indonesia in 1976.

Tippecanoe, battle of (7 November 1811), a skirmish between US forces and Shawnee Indians, fought near the Wabash River 240 km. (150 miles) north of Vincennes. Governor *Harrison of the Indiana Territory engineered a conflict with the British-supported Indian confederacy of the Shawnee chiefs *Tecumseh and his brother the Prophet Tenskwatana in order to end Indian resistance to westward US expansion. In the resulting skirmish, Harrison sustained considerable losses but drove away the Indians. Tippecanoe was hailed as a major victory, but the British and Indian threat to the north-west frontier was not destroyed until Tecumseh fell in the battle of the *Thames two years later.

Tirpitz, Alfred von (1849–1930), German grand-admiral. As Secretary of State for the Navy (1897–1916) his first Navy Bill in 1898 began the expansion of the German navy and led to the naval race with Britain. In 1907 he began a large programme of *Dreadnought-class battleship construction for the High Seas Fleet. During World War I he made full use of submarines, but, following the sinking of the *Lusitania* (1915), unrestricted submarine warfare was temporarily abandoned. He resigned in 1916 and the policy was resumed in 1917, resulting in US entry into the war.

Tito (b. Josip Broz) (1892–1980), Yugoslav statesman of Croatian origin. He was Prime Minister (1945–53) and President (1953–80). During World War I he served with the Austro-Hungarian infantry and was taken prisoner in Russia. He escaped and fought for the *Russian Revolution. After returning to Yugoslavia he became involved in the Communist Party and was imprisoned for six years. After the German invasion of Yugoslavia (1941), Tito organized partisan guerrilla forces into a National Liberation Front. Tito emerged as the leader of the new federal government. He rejected *Stalin's attempt to control the communist-governed states of eastern Europe. As a result Yugoslavia was expelled from the *Cominform and Tito became a leading exponent of non-alignment in the *Cold War. Normal relations were resumed with the Soviet Union in 1955, though Tito retained his independence, experimenting with different communist styles of economic organization, including worker-participation in the management of factories. On Tito's death he was replaced by a collective government with rotational leadership.

Tobruk, siege of (1941–2), German siege of British and Commonwealth troops in Tobruk in North Africa in World War II. When General *Wavell's army captured Tobruk in January 1941 some 25,000 Italian troops were taken prisoner. The Afrika Korps of General *Rommel

Marshal **Tito**, photographed in August 1942 during World War II when he led the Yugoslav communist partisans.

then arrived (April 1941), and the British withdrew east, leaving a largely Australian garrison to defend Tobruk, which was subjected to an eight-month siege and bombardment. In November 1941, after being reinforced by sea, the garrison broke out, capturing Rezegh and linking up with the 8th Army troops of General *Auchinleck. But the Germans counter-attacked, and in June 1942, after heavy defeats, the British again withdrew leaving a garrison of two divisions, mostly South African and Australian, in Tobruk, which was then subjected to massed attack by German and Italian troops. On 20 June it capitulated, the garrison of 23,000 men surrendering, with vast quantities of stores. It was a major Allied defeat, but Tobruk was recaptured on 13 November 1942 by the troops of General *Montgomery.

Tocqueville, Alexis, comte de (1805-59), French statesman and political analyst. Sent to the USA in 1831 to study its penal system, de Tocqueville carried out a systematic survey of US political and social institutions, publishing the results in *De la démocratie en Amerique* (1835 and 1840). The book immediately found a large readership, and became one of the most influential

The Japanese military leader General **Tojo** photographed during his trial for war crimes Tokyo, 1948. Under his direction, a series of swift victories (1941–2) were made against the colonial powers in south-east Asia.

political writings of the 19th century. It remains probably the greatest of all European commentaries on American politics and society.

Togliatti, Palmiro (1893-1964), Italian politician. He was Secretary of the Italian Communist Party (1926-64). After the fascist take-over he lived mainly in Moscow (1926-44), and became chief of the *Comintern in Spain during the *Spanish Civil War. After World War II he made the Italian Communist Party the largest in Western Europe. Togliatti was undogmatic in his communism: he recognized Roman Catholicism as the state religion of Italy and propounded the doctrine of 'polycentrism', asserting national differences in the implementation of communism.

Togo, a country in West Africa. Annexed by Germany in 1884 as a colony, Togoland was *mandated between France and Britain after World War I. The western British section joined *Ghana on the latter's independence in 1957, and became known as the Volta region. The remainder of the area became a UN mandate under French administration after World War II and achieved independence, as Togo, in 1960. After two civilian regimes were overthrown in 1963 and 1967, Togo achieved stability under President Eyadéma and in 1979 staged its first election in sixteen years.

Tojo Hideki (1884-1948), Japanese general and statesman. He participated in the war against China in the 1930s, was leader of the militarist party from 1931 onwards, and became War Minister in 1940. He urged closer collaboration with Germany and Italy, and persuaded *Vichy France to sanction Japanese occupation of strategic bases in Indo-China (July 1941). He succeeded *Konoe as Prime Minister (1941-4), and he gave the order to attack *Pearl Harbor, precipitating the USA into World War II. In 1942 he strengthened his position in Tokyo, gradually taking increased powers, as War Minister, and creating a virtual military dictatorship. He resigned in 1944 after the loss of the Marianas to the USA. He was convicted at the *Tokyo Trials and hanged as a war criminal in 1948.

Tokugawa   *shogunate.

Tokyo Trials, war crimes trials of Japanese leaders after World War II. Between May 1946 and November 1948, twenty-seven Japanese leaders appeared before an international tribunal charged with crimes ranging from murder and atrocities to responsibility for causing the war. Seven, including the former Prime Minister *Tojo, were sentenced to death and sixteen to life imprisonment (two others receiving shorter terms), but General *MacArthur refused to allow the trial of the Emperor *Hirohito for fear of undermining the post-war Japanese state.

Tolpuddle Martyrs, the name given to six English farmworkers, who were charged in 1834 with taking illegal oaths, while establishing a local trade union branch of the Friendly Society of Agricultural Labourers in the Dorset village of Tolpuddle with the aim of obtaining an increase in their wages (then seven shillings a week). The six were found guilty and condemned to seven years' transportation to Australia. The severity of the sentence provoked a storm of protest and mass demonstrations

A mass meeting of trade-union supporters gathers in Copenhagen Fields, Islington, April 1834, to send a petition to the king on behalf of the **Tolpuddle Martyrs**. The petition, asking for remission of the transportation sentence passed on the Dorset labourers, is escorted into the city of London.

were held in London. After two years, in the face of continuing public hostility, the government was obliged to pardon the men.

Tonga, an archipelago in the southern Pacific consisting of more than 150 islands. Named the Friendly Islands by Captain James Cook, who visited them in 1773, the country was soon receiving missionaries. King George Tupou I (1845-93) unified the nation and gave it a constitution. In 1900 his son signed a treaty, making the islands a self-governing British protectorate. During World War II Queen Sālote Tupou III (1900-65) placed the island's resources at the disposal of the Allies; in 1968 British controls were reduced, and in 1970 Tonga became independent within the *Commonwealth of Nations.

Tonkin Gulf Resolution (1964), a resolution by the US Congress, giving the President authority to take all necessary measures to repel any attack against the forces of the USA. It was in response to an alleged attack by North Vietnam patrol boats against the US destroyer *Maddox* in the Gulf of Tonkin. The US involvement in the fighting in the *Vietnam War followed. Subsequent investigation revealed that the intelligence information on which it was based was inaccurate and, following the

war, the War Powers Act was passed in 1973. This restricts the time a President can commit US troops without Congressional approval to sixty days.

Tonypandy, a mining town in Mid-Glamorgan, Wales, which witnessed in 1910, a violent dispute over pay rates for miners. Miners interfered with pit machinery and there was looting and disorder in the town. The local police requested government help and the Home Secretary, Winston Churchill, sent 300 extra police from London and placed military detachments on stand-by. In a subsequent incident in Llanelli a year later troops mobilized by Churchill opened fire on strikers, killing four. Trade union hostility to Churchill was intense.

Touré, (Ahmed) Sékou (1922-84), African statesman, President of *Guinea (1958-84). In 1946, together with other African leaders, including *Houphouët-Boigny, he was a founder of the Rassemblement Démocratique Africain. He became Secretary-General of the CGT (Confédération Générale de Travail) for Africa in 1948. In 1955 he was elected Mayor of Conakry and took his seat in the French National Assembly in 1956. In 1957 he became Vice-President in the Guinea cabinet. He was elected President when Guinea became independent in 1958, and broke all links with the *French Community. A convinced Marxist, he received aid for Guinea from the Soviet bloc. Following his death in 1984 the armed forces staged a coup.

Toussaint-l'Ouverture, François Dominique (c.1743-1803), Haitian patriot. The self-educated son of

A lithograph of **Toussaint-l'Ouverture**, the slave who assumed governor-generalship of Haiti. His extraordinary military ability, coupled with republican idealism, inspired admiration among European liberals, and awe and adulation among the slaves he had freed.

slave parents, he organized the successful struggle against French planters in Haiti (1791), and was given the name l'Ouverture (the opening) in 1793 after a series of fast-moving campaigns that secured the emancipation of slaves. He briefly joined the Spanish and British when they invaded the island in 1793, but, with the help of generals *Dessalines and Christophe, secured their withdrawal and established Haiti as a black-governed French protectorate. He suppressed a mulatto insurrection and, in 1800, made himself governor-general of Haiti for life. In 1801 he conquered Santo Domingo, reorganized the government of the island, and instituted civic improvements. Napoleon sent (1802) a military force under General Leclerc to reassert French control. Toussaint was betrayed, arrested, and transported to France, where he died in prison.

Townsend, Francis Everett (1867–1960), US physician and reformer. He is mainly remembered for his Old Age Revolving Pension scheme (the Townsend Plan), meant to help the elderly and assist the USA out of the Great *Depression. The plan called for payments of $200 a month to all aged 60 or more. The funds were to be provided by a federal tax on commercial transactions. The popularity of this and other programmes (he secured at least ten million signatures to his petitions) may have persuaded Franklin D. *Roosevelt to adopt more far-reaching social policies.

Toynbee Hall, a settlement supported by universities, in the east end of London, aimed at improving the lives of the urban deprived. Founded in 1884 by Samuel and Henrietta Barnett, it was named after the young Oxford philosopher Arnold Toynbee. The settlement has attracted reformers, among them William *Beveridge and J. M. *Keynes. Its programme of educational and social activities influenced such far-reaching reforms as the National Health Insurance Act (1911) and the Old Age Pension Plan (1908), as well as a changed attitude to young offenders and penal reform, and the establishment of Labour Exchanges (1908). The practical and theoretical reforms initiated there have resulted in similar developments world-wide.

Trades Union Congress (TUC), an organization of British trade unions. It was founded in 1868 with the purpose of holding national conferences on trade union activities. In 1871 it set up a Parliamentary Committee to advance the interests of unions with Members of Parliament. From 1889 onwards, it began to be more politically militant and in 1900 helped to found the Labour Representation Committee, known from 1906 as the *Labour Party, with whom it has had links ever since. The General Council, elected by trade union members, replaced the Parliamentary Committee in 1920. The Congress can urge support from other unions, when

The virtues of industry are extolled amid scenes displaying the benefits which **trade-union** membership confers in times of disaster. The Brotherhood of Locomotive Firemen of North America also honours George Stephenson, pioneer of the locomotive engine, in this early banner. (Library of Congress, Washington)

a union cannot reach a satisfactory settlement with an employer in an industrial dispute, but it has no powers of direction. Since the *General Strike relations between the Congress and government (of whatever party) have been both cautious and conciliatory. It was closely involved in British industrial planning and management during World War II and under successive Labour and Conservative governments until 1979. Conspicuous failure has meant that since then it has tended to be on the defensive, particularly against legislation designed to weaken trade union power in industrial disputes.

trade union, an organized association of workers in a particular trade or profession. In the USA they are referred to as labor unions. In Britain in the late 18th century groups and clubs of working-men in skilled trades developed, to regulate admission of apprentices and sometimes to bargain for better working-conditions. During the wars with France (1793–1815) *Combination Acts suppressed any such activity, but on their repeal in 1824 limited trade union activity became possible in certain crafts. By 1861 a number of trade unions of skilled workers existed in Britain, forming the *Trades Union Congress in 1868, gaining some legal status in 1871, and the right to picket peacefully in 1875. A parallel development had proceeded in the USA, small local unions appearing in the 1820s. A national organization, the *Knights of Labor, flourished (1869–86), having as a main aim the abolition of child labour. It was succeeded by the *American Federation of Labor (AFL) (1886), an organization of skilled workers. With the development of *mass-production methods in the industrialized countries large numbers of semi-skilled and unskilled workers were recruited, and from the 1880s attempts were made to organize these into unions. These attempts were more successful in Britain and in Europe than in the USA, where cheap immigrant labour was for long available. Unions emerged in Australia and New Zealand and in other British dominions in the 19th century, first among skilled workers and later among semi-skilled and unskilled. As industrialization has proceeded in other countries so have trade unions developed, although in South Africa trade union activity among black workers was illegal until 1980. In the Soviet Union and Eastern Europe 90 per cent of industrial workers belong to government-controlled unions which concern themselves with training, economic planning, and the administration of social insurance. The Polish independent union *Solidarity was outlawed in 1982.

Trafalgar, battle of (21 October 1805), a naval engagement between the combined French and Spanish fleets, and the British, fought off Cape Trafalgar near the Spanish port of Cadiz. After failing to lure the British fleet away from Europe to enable Napoleon to transport his army to England, Admiral Villeneuve returned to Cadiz, and the English Channel fleet, commanded by Collingwood blockaded the port. On 29 September *Nelson arrived in his flagship, *Victory*, to take command. Twenty days later Villeneuve, ordered by Napoleon to leave Cadiz and threatened by the loss of his command,

The death of Nelson on board his flagship *Victory* at the battle of **Trafalgar**, 1805. A storm of grapeshot from the *Victory* had foiled the attempt by the French ship *Redoubtable* to board the British decks, but Nelson was wounded by a shot from the *Redoubtable*, and died three hours later. (National Maritime Museum, London)

finally put to sea but hoped to avoid a battle. Nelson, who had kept his main fleet out of sight, divided his fleet of twenty-seven ships and signalled at the beginning of the battle that 'England expects every man to do his duty'. The British lost no ships but took twenty from the French and Spanish. Nelson was mortally wounded by a shot from the French ship *Redoubtable* but British naval supremacy was secured for the remainder of the 19th century.

Trail of Tears, the route of enforced westward exile for American Indians. As more settlers moved into Georgia and to the states of Alabama, Mississippi, Louisiana, and Florida in the 1830s, it was US policy forcibly to expel the eastern Indian tribes from their lands and move them to *Oklahoma territory west of the Mississippi River. The tribes concerned were the *Cherokee, Creek, Choctaw, Chickasaw, and *Seminole (known as the Five Civilized Tribes). Bad weather, neglect, and limited supplies of food caused much suffering and death before the move was completed and the Trail of Tears closed in 1838.

Transjordan *Jordan.

transport revolution, the change in methods of moving goods and people from place to place. At the beginning of the 19th century wind, water, and horse power were relied on for transport. In Britain, as a result of the *Industrial Revolution, industrialists required improved roads and inland waterways to transport their goods. Turnpike trusts were created, an arbitrary system of road maintenance, paid for by fees collected from travellers at tollgates. Civil engineers, such as Thomas Telford and John Macadam, greatly improved the building of roads and bridges, and with the abolition of the turnpikes, revenue for road maintenance was derived from taxation. The first steam locomotive to run on rails was made by Richard Trevithick in 1804 and the first *railway to carry goods and passengers was the Stockton and Darlington railway, opened in 1825. By 1851 there were rail networks in seventeen other countries, including France (1832), Germany (1835), and the 650 km. (404 mile) link between St Petersburg and Moscow (1851).

Steamships were at first only used in river estuaries and for coastal transport; the first crossing of the Atlantic by a ship using steam alone was in 1838. Steam packet companies were formed by merchants such as Samuel *Cunard to carry passengers and cargo across the Atlantic, and from Europe to the colonial empires.

Later in the 19th century bicycles were developed to give convenient personal transport. In 1876 the safety bicycle was invented and in 1889 pneumatic tyres were introduced. In 1884 a German gunsmith, Gottlieb Daimler, invented an engine that burnt petrol and soon afterwards Karl Benz made one of the earliest petrol-driven motor cars. The mass production of motor cars was pioneered by Henry *Ford with the construction (1909) of his Model T car. Motor transport created further demands for improved roads and, together with the tramways and underground train system, greatly increased personal mobility and accelerated the growth of suburbia.

Modern air flight was pioneered by two American brothers, Wilbur and Orville Wright, who built and flew the first manned, power-driven flying machine in 1903. Other records followed fast: the Frenchman Louis Blériot flew the English Channel (1909), the Englishmen John

Alcock and Arthur Brown flew from Newfoundland to Ireland (1919) and the commercial exploitation of air transport soon followed. After World War II air routes penetrated to all parts of the world.

Trans-Siberian Railway, a railway which opened up Siberia and advanced Russian interest in East Asia. It was begun with the aid of French loans in 1891 and was virtually completed in 1904. The suspicion it aroused in Japan was one factor leading to the *Russo-Japanese War. From Moscow, it runs east around Lake Baikal to Vladivostok on the Sea of Japan, a distance of 9,311 km (5,786 miles). The express now takes six days to reach Vladivostok.

Transvaal, a province in the north-east of the Republic of *South Africa. Inhabited by *Ndebele Africans, the first white settlement was led by Andries Potgieter in 1842. In 1848, after Britain annexed the *Orange Free State, more *Boers crossed the River Vaal under Andries *Pretorius. The Ndebele leader Mzilikazi (*Zimbabwe) emigrated northwards, and the Boers were granted self-government by the Sand River Convention (1852). In 1877 the Transvaal was annexed by Britain, in an attempt to impose federation, and the First *Boer War followed. Internal self-government as a republic was regained by the Treaty of Pretoria (1881) under the presidency of *Kruger. After the discovery of gold on the Witwatersrand in 1886, *Rhodes and others tried unsuccessfully to unite Transvaal with the Cape. The Boers' denial of political rights and imposition of taxation on foreign workers contributed to the outbreak of the Second Boer War. In 1900 the Transvaal was annexed by Britain, and in 1906 self-government was granted. Under Louis *Botha as Prime Minister, it became a founding province of the Union of South Africa (1910).

Transylvania, a region in *Romania. Under Austrian domination throughout the first half of the 19th century, Transylvania became an integral part of Hungary at the establishment of the *Austro-Hungarian empire (1867). In 1918 the Romanians of Transylvania proclaimed their adhesion to Romania. This was confirmed by the Treaty of *Trianon (1920). Hungary annexed about two-fifths of the land during World War II, but was made to return it again to Romania in 1947. The redistribution of land and the policy of enforced cultural assimilation in turn by Romanians and Hungarians have remained causes of friction between the two countries.

treaty ports, Asian ports, especially Chinese and Japanese, opened to foreign trade and habitation as a result of a series of *Unequal Treaties in the 19th century. In China, the first five treaty ports were opened as a result of the Treaty of *Nanjing (1842), eleven more as a result of the Treaty of Tianjin (1858) and the Conventions of Beijing (1860), and approximately thirty-five more opened before the *Chinese Revolution (1911), some on the Yangtze River. Foreigners living in their own concessions in treaty ports had the protection of their home governments and were not required to pay Chinese taxes or to be subject to Chinese laws. This was strongly resented by the nationalist government, and all privileges were surrendered by 1943. After the Treaty of Kanagawa (1858), Japan established five treaty ports, but foreign powers were obliged to surrender their privileges in 1899.

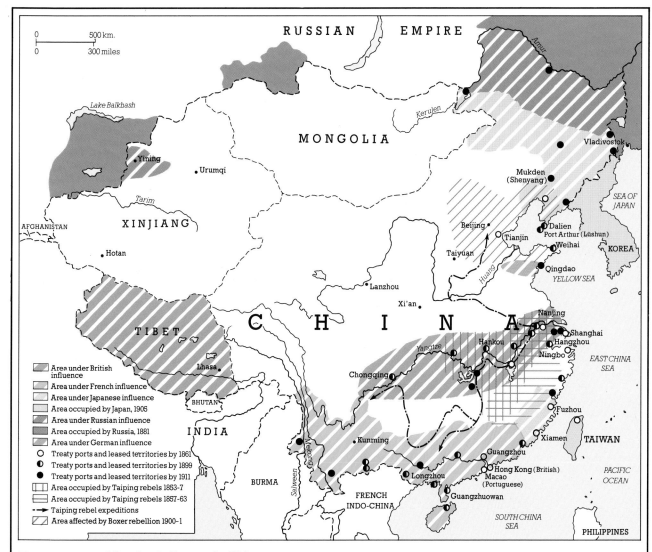

Treaty ports and foreign influence in China

Following the first Opium War (1839–42), imperial China at the Treaty of Nanjing (1842) ceded Hong Kong to Britain and agreed that Guangzhou (Canton), Xiamen (Amoy), Fuzhou (Foochow), Ningbo (Ningpo), and Shanghai be opened to western trade, with attendant privileges. These allowed an independent judicial, legal, and taxation system to be developed in each of the ports. The number of treaty ports steadily increased, until there were over fifty by 1911. In 1898 further territories were leased to European powers, including Guangzhuowan to France, Qingdao to Germany, and Port Arthur (Lüshun) to Russia, Japan taking over the latter in 1905. In 1899 the USA proclaimed an 'Open Door' policy for its own China trading operations. After World War II the remaining European colonies, with the exception of Hong Kong and Macao, reverted to China.

Trenchard, Hugh Montague, 1st Viscount (1873–1956), creator of the British Royal Air Force. In 1913 he joined the Royal Flying Corps (*airforce), a branch of the army. In August 1915 he became RFC commander in France, where he organized fighter battles against the superior German Fokker monoplane. In addition he began to develop the use of bombers aimed at military targets in Germany and occupied France. In April 1918 he won his fight for the RFC to become independent of the army as the Royal Air Force. As the first Chief of Air Staff he built up the Royal Air Force, continually resisting inter-service rivalry from the army and navy. In 1927 he was created the first Air Marshal. As Commissioner of Police (1932–5), he reorganized the Metropolitan Police, establishing a Police College and Forensic Laboratories at Hendon.

trench warfare, a form of fighting conducted from long, narrow ditches in which troops stood and were sheltered from the enemy's fire. At the beginning of World War I the prevailing belief that victory came from mass infantry charges dominated military thinking in spite of the introduction of rapid-firing small arms and artillery. After the first battle of the *Marne thousands of miles of parallel trenches were dug along the Western Front, linked by intricate systems of communication trenches and protected by barbed wire. With such trenches stretching from the North Sea to Switzerland, a stalemate existed and to break it various new weapons were introduced: hand-grenades, poison gas, trench mortars, and artillery barrages. Consequently casualties hitherto undreamed of followed every mass infantry attack. Not until 1918, with an improved version of the tank

A graphic illustration of the rigours of **trench warfare**, at the battle of the Somme, July 1916. A British rifleman on the alert for enemy movement keeps his head below the parapet while his exhausted companions snatch some sleep.

(invented in 1915), was it possible to advance across the trenches. World War II by contrast was a war of movement with no comparable trench fighting. Slit trenches, manned by two or three machine gunners, replaced them. In the *Korean War and in *Vietnam fortified bunkers were used.

Trent affair (November–December 1861), an incident between the USA and Britain during the *American Civil War. In November 1861 the US warship *San Jacinto* stopped the British mail packet *Trent* at sea and forcibly removed two Confederate (Southern) diplomats, and their secretaries, bound for Europe. News of the incident produced widespread demands in Britain for war against the Union (the North), but the crisis was averted partly through the intervention of Prince *Albert, and by the decision of US Secretary of State William *Seward to release the two diplomats on the grounds that the captain had erred in not bringing the *Trent* and its 'personal contraband' to port.

Triad Societies, Chinese secret societies, originally formed in the late 17th century to overthrow the Manchu *Qing dynasty and restore its Chinese Ming predecessor. The name was given to various related organizations such as the Three Dot Society, the Three Harmonies Society, and the Society of Heaven and Earth. The societies shared a similar ritual and acted both as fraternal and criminal organizations. They grew in strength during the *Taiping Rebellion, and thereafter played an erratic and violent role in China. Some Triad branches assisted *Sun Yat-sen, while others exerted strong political influence in cities like Shanghai.

Trianon, Treaty of   *Versailles Peace Settlement.

Trieste, Italian city at the northern end of the Adriatic Sea. As the sole port of the *Austrian empire it flourished, but became a target of *Irredentism, and after World War I was annexed (1919) by Italy. During World War II it was occupied by German troops. In a decision disputed by Italy it was awarded to Yugoslavia in 1945.

As a compromise, the city and a part of the coastal zone of Istria were made (1947) a 'free territory' of Trieste under the protection of the United Nations. The deadlock between the Italian and Yugoslav claims was resolved after negotiations in London in 1954, when the territory was divided between the two countries, Italy receiving the city of Trieste.

Trinidad   *West Indies; *Williams, Eric.

Triple Alliance (1882), an alliance between Germany, Austria, and Italy. This was a secret alliance signed in May 1882 at the instigation of *Bismarck. The three powers agreed to support each other if attacked by either France or Russia. It was renewed at five-yearly intervals, but Italy reneged in 1914 by not coming to the support of the Central Powers.

Triple Alliance, War of the   *Paraguayan War.

Tripolitan War (1800–15), a conflict between the USA and the Karamanli dynasty of Tripoli, which in 1796 had obtained from the USA the annual payment of $83,000 for the protection of its commerce from piracy. In 1801 the Bey of Tripoli demanded an increase; the USA declined, and sent a naval force to blockade the port of Tripoli. In 1803 the Tripolitanians captured the US ship *Philadelphia*. US forces then captured the port of Derna and the Bey agreed to peace, which was concluded in 1805. The Bey received $60,000 as ransom for the *Philadelphia* and renounced all rights to levy tributes on American ships. In 1815, following breaches of the agreement, a US squadron under Captain Stephen Decatur again visited North Africa and compelled the Bey of Algiers to renounce payments for immunity.

Trotsky, Leon (b. Lev Bronstein) (1879–1940), Russian communist revolutionary and military leader. After the split in the Social Democratic Party (1903) he sided with the Mensheviks and in the *Russian Revolution of 1905 was leader of the St Petersburg *Soviet. In 1917 he joined the *Bolsheviks, becoming the principal organizer of the successful October Revolution. With Lenin he now faced two dangers—war with Germany and internal civil war. In the first Soviet government he was Commissar for Foreign Affairs and negotiated the Peace of *Brest-Litovsk (1918) by which, on Lenin's insistence, Russia withdrew from World War I. As Commissar for War (1918–24) his great achievement was the formation of the *Red Army; his direction of the *Russian Civil War saved the Bolshevik revolution. On the death of Lenin (1924) he was the obvious successor, but he lacked Lenin's prestige within the party compared to its General Secretary, *Stalin. He was an internationalist, dedicated to world revolution, and strongly disagreed with Stalin's more cautious policy of 'Socialism in one country'. Steadily losing influence, he was expelled from the Party in 1927 and exiled. Shortly after founding the Fourth *International (1937) he was murdered in Mexico. The term Trotskyism has come to be used indiscriminately to describe all forms of left-wing communism.

Trucial States, term applied to seven Arab emirates on the Persian Gulf from the early 1820s until 1971, when they were established as the *United Arab Emirates. The name was derived from the annual 'truce' obtained by the British in the 1820s, by which the local rulers undertook to abstain from maritime warfare. Other treaties with Britain extended the ban to the arms and slave trades, and the Exclusion Agreements of 1892 provided for British control of the external affairs of the states.

Truck Acts, measures passed by the British Parliament in the 19th century regarding the method of payment of wages. Certain employers paid their workmen in goods or in tokens which could be exchanged only at shops owned by the employers—the so-called truck system. The Truck Act of 1831 listed many trades in which payment of wages must be made in coins. It was amended by an Act of 1887, which extended its provisions to cover virtually all manual workers. In 1896 a further Act regulated the amounts that could be deducted from wages for bad workmanship. The Payment of Wages Act of 1960 repealed certain sections of the Truck Acts to permit payment of wages by cheque.

Trudeau, Pierre Elliott (1919–), French-Canadian statesman, Prime Minister of Canada (1968–79, 1980–4). As Minister of Justice and Attorney General (1967) he opposed any separation of *Quebec from the rest of Canada. Elected leader of the *Liberal Party and succeeding Lester *Pearson as Prime Minister in 1968, he led his government to victory. In his first period as Prime Minister he sought to secure economic growth by increased government expenditure, but government deficits increased, while inflation and unemployment rose throughout the 1970s. His Bilingual Languages Act (1968) gave French and English equal status throughout Canada and helped to improve relations between English- and French-speaking Canadians. He improved relations with France, but made little real progress in his efforts to make Canada more independent of the USA. By 1979 Canada was experiencing serious economic problems, and Trudeau lost the election of that year. The Progressive *Conservatives briefly took office with a minority government. In his second period (1980–4) he continued with his opposition to separatism in Quebec, his policies

Canadian Premier Pierre **Trudeau** speaking at a Liberal Party meeting. A popular and colourful personality, his terms of office were marked by final constitutional independence from the British Parliament and the formation of a new Canadian constitution.

being supported by a referendum there rejecting sovereignty for the province. He also secured the complete national sovereignty of Canada in 1982, with the British Parliament accepting the 'patriation' of the British North America Act to Canada, thus abolishing formal links with Britain. He retired in 1984.

Trujillo (Molina), Rafael (Léonidas) (1891–1961), dictator of the Dominican Republic (1930–61). As commander of the army, he seized power in 1930 and his regime dominated all aspects of Dominican life, including the economy, employing authoritarian measures to accomplish some material progress, and using terrorist methods to repress opposition. In 1937, fearing Haitian infiltration, Dominican troops crossed the border and massacred between 10,000 and 15,000 Haitians. After alienating all of Latin America in an attempt to assassinate the liberal reformer Rómulo *Betancourt of Venezuela, Trujillo himself was assassinated in 1961.

Truman, Harry S. (1884–1972), thirty-third President of the USA (1945–53). From 1935 to 1944 he was a Democratic senator, and then became Vice-President. On Franklin D. *Roosevelt's death in 1945, he automatically succeeded as President. At home, he largely continued Roosevelt's *New Deal policies, but he was immediately faced with new problems in foreign affairs. He authorized the use of the atom bomb against Japan. His abrupt termination of *Lend–Lease in 1945 was damaging to East–West relations and the *Truman Doctrine was adopted in response to a perceived threat of Soviet expansion during the *Cold War period. He defeated *Dewey in the 1948 presidential election. His programme, later labelled the 'Fair Deal', called for guaranteed full employment, an increased minimum wage and extended social security benefits, racial equality, price and rent control, and public health insurance. Although Congress allowed little of this to pass into law, he did manage to achieve his 1949 Housing Act, providing for low-cost housing. By his executive authority he was able to end racial segregation in the armed forces and in schools financed by the federal government. He took the USA into its first peacetime military pact, *NATO, tried to give technical aid to less-developed nations with his *Point Four scheme, and in the *Korean War ensured that western intervention would, formally, be under *United Nations rather than US auspices. In 1951 he dismissed General *MacArthur from his Far Eastern command for publicly advocating a war with communist China. He did not run for re-election in 1953, although he remained active in politics long after his retirement.

Truman Doctrine (1947), a principle of US foreign policy aimed at containing communism. It was enunciated by President *Truman in a message to Congress at a time when Greece and Turkey were in danger of a communist take-over. Truman pledged that the USA would 'support free peoples who are resisting attempted subjugation by armed minorities or by outside pressures'. Congress voted large sums to provide military and economic aid to countries whose stability was threatened by communism. Seen by communists as an open declaration of the *Cold War, it confirmed the awakening of the USA to a new, global responsibility.

Tseng Kuo-fan   *Zeng Guofan.

Tshombe, Moise (Kapenda) (1920–69), African leader in the Belgian Congo. He founded the Conakat political party, which advocated an independent but loosely federal Congo. He took part in talks that led to Congolese independence in 1960, but then declared the province of Katanga independent of the rest of the country (*Congo Crisis). He maintained his position as self-styled President of Katanga (1960–3) with the help of white mercenaries and the support of the Belgian mining company, Union Minière. Briefly Prime Minister of the Congo Republic (1964–5), he was accused of the murder of *Lumumba, and of corruptly rigging the elections of 1965, and fled the country when General *Mobutu seized power. In 1967 he was kidnapped and taken to Algeria, where he died in prison.

Tso Tsung-tang   *Zuo Zongtang.

Tubman, Harriet (c.1821–1913), black American abolitionist and social reformer. An escaped Maryland slave, she became one of the most effective 'conductors' on the *Underground Railroad. During the *American Civil War she served as a nurse and a Union (Northern) spy behind *Confederacy lines. After the conflict she worked in the cause of black education in North Carolina.

Tubman, William Vacanarat Shadrach (1895–1971), Liberian statesman, President (1944–71). A member of an Americo-Liberian family, he was elected to the Liberian Senate in 1924 and became President in 1944. He encouraged economic development to remove Liberia's financial dependence on the USA and successfully integrated the inhabitants of the country's interior into an administration which had hitherto extended little beyond the coastline.

TUC   *Trades Union Congress.

Tunisia, a country in North Africa. Part of the Ottoman empire from the 16th century, the Bey of Tunis became increasingly independent. Corsairs operated from Tunis, leading to the *Tripolitan War with the USA. During the 19th century, the Bey's control weakened and, in 1881, France declared it a protectorate. The rise of nationalist activity led to fighting between the nationalists and the colonial government in the 1950s. Habib *Bourguiba, the nationalist leader, was imprisoned, but was released (1955) when the country achieved independence. The Bey of Tunis abdicated (1956) and the country became a republic led by Bourguiba and the neo-Destour Party. In the 1970s the government's refusal to allow the formation of other political parties caused serious unrest, while subsequent attempts at liberalization were interrupted by fresh outbreaks of rioting in 1984–5.

Tupamaro, a member of the Movimento de Liberación Nacional (National Liberation Movement) in Uruguay. An urban guerrilla organization, it was founded in Montevideo in 1963 and led by Raúl Sendic. It sought the violent overthrow of the Uruguayan government and the establishment of a socialist state. Its robberies, bombings, kidnappings and assassinations of officials continued until the early 1970s, when the movement was severely weakened by police and military repression. The Tupamaros derived their name from the 18th-century Inca revolutionary against Spanish rule, Tupac Amarú.

Tupper, Sir Charles (1821–1915), Canadian statesman. He became Conservative Member of the Nova Scotia Assembly in 1855, and Premier in 1863. A Father of the Confederation, he entered the dominion Parliament in 1867 and served under John A. *MacDonald (1870–3 and 1878–84). He became High Commissioner to Britain, holding the post, except for the years 1887–8, until 1896. Returning to Canadian politics, he was Prime Minister for a little over two months. He led the opposition until 1900, when he retired.

Turkey, a country in south-west Asia. Modern Turkey evolved from the *Ottoman empire, which was finally dissolved at the end of World War I. By the *Sèvres Treaty at the Versailles Peace Conference parts of the east coast of the Aegean around the city of Izmir (Smyrna) were to go to Greece, and the Anatolian peninsula was to be partitioned, with a separate state of *Armenia created on the Black Sea. The settlement triggered off fierce national resistance, led by Mustafa Kemal. A Greek army marched inland from Izmir, but was defeated. The city was captured, Armenia occupied, and the new Treaty of Lausanne negotiated. This recognized the present frontiers, obliging some one and a half million Greeks and some half-million Armenians to leave the country (July 1923). In October 1923 the new Republic of Turkey was proclaimed, with Kemal as first President. His dramatic modernizing reforms won him the title of *Atatürk, 'Father of the Turks'. The one-party rule of his Republican People's Party continued under his lieutenant Ismel Inonu until 1950, when in the republic's first open elections, the free-enterprise opposition Democratic Party entered a decade of power, ending with an army coup. Civilian rule was resumed in 1961, but there was a further period of military rule (1971–3). Atatürk's neutralist policy had been abandoned in 1952 when Turkey joined NATO. Relations with allies, however, were strained by the invasion of *Cyprus (1974). A US trade embargo resulting from this was only lifted in 1978. Tension between left-wing and right-wing factions, hostility to Westernization by the minority Shiites, who seek to enforce Islamic puritanism, and fighting between Turks, Kurds, and Armenians, continue to trouble the country. Since 1971 there have been a succession of military and civilian governments.

Turkistan, an area of central Asia north of the Himalayas, divided between China (*Xinjiang province), Afghanistan, and the Soviet Union. Early in its history it developed a series of Persian trading centres or khanates, for example, Tashkent and Samarkand, along caravan routes. Semi-autonomous khanates survived until the 19th century when Russia, China, and Afghanistan began to impose centralized government, Tashkent and the other khanates being conquered by Russia in 1867. Following the *Russian Civil War, four Central Asian Soviet Union Republics were formed, Turkmenia and Uzbekistan (1925), Tadzhikistan (1929), and Kirghizstan (1936). In 1931 the Turkistan–Siberian railway was completed. In recent years Islamic revivalism has become an important factor within the region.

Turner, Nat (1800–31), black American leader of the Virginia slave revolt of 1831. Believing himself a divine instrument to guide his people out of bondage, Turner led about sixty slaves into revolt in Southampton County,

The beginning of Nat **Turner**'s revolt: the black slave leader kills the family of his master, Joseph Travis, in August 1831.

Virginia, killing fifty-seven whites before he, his followers, and a number of innocent slaves were killed. Turner's rebellion exacerbated Southern fears of insurrection and led to a tightening of police measures against slaves.

Tuscany, a region in Italy, centred on Florence. Occupied by French troops in the *Revolutionary Wars, it was ruled by Napoleon's sister Elisa (1809–14). The grand-duke Ferdinand III was restored in 1814 and confirmed many of the reforms introduced by the French. In the *Revolutions of 1848 Ferdinand's son Leopold II granted a liberal constitution, but revolutionary agitators proclaimed a republic, Leopold fled, and was restored by an Austrian army (1849), after which his regime became more oppressive. He was obliged to flee again in 1859 when Tuscan liberals declared for unification with *Piedmont. A plebiscite in 1860 confirmed Tuscany's annexation into the new Kingdom of Italy.

Tuvalu, a group of nine atolls in the southern Pacific, formerly called the Ellice Islands after a 19th-century British politician, Edward Ellice. In the 19th century whalers, traders, missionaries, and 'blackbirders' (*Kanaka catchers) for the *Queensland sugar plantations began to take an interest in these atolls, which the British were to include in the Gilbert and Ellice Islands Protectorate in 1892. In 1974 the Ellice Islanders, who are of Polynesian descent, voted to separate from the Micronesian Gilbertese. They achieved independence in 1978, establishing a constitutional monarchy. The USA claims sovereignty over four of the islands.

Tweed Ring, corrupt group in New York City, USA. It revolved around William Marcy Tweed (1823–78), the New York city political 'boss' and state Senator who had built his power through the influence of *Tammany Hall. The ring, renowned for corrupt and dishonest dealing and for fraudulent city contracts and extortion, was exposed in the *New York Times* in 1871. Tweed was arrested and convicted but fled to Spain, from which he was extradited. He died in prison in 1876. The operations of the Tweed Ring cost New York City some $100 million.

Twentieth Congress (February 1956), the Congress of the Communist Party of the Soviet Union, noted for *Khrushchev's denunciation of *Stalin. After the first and open session of the Congress, Khrushchev, as First Secretary, made three significant doctrinal points: that peaceful co-existence between East and West was possible; that war between them was not inevitable; and that there were 'different roads to socialism', not only the Soviet route. More dramatic was the speech he delivered in the secret session, when he denounced the Stalinist cult of personality and *Stalin's acts of terror. The speech was carefully constructed to emphasize Stalin's treatment of the Party rather than of the country at large. A fervour of de-Stalinization and demands for liberalization swept through Eastern Europe as well as the Soviet Union. Khrushchev's 'secret' speech was an important contributory factor in prompting the uprisings in *Poland and *Hungary in 1956 and in the Sino-Soviet quarrel from 1960.

Twenty-One Demands, Japanese attempt to impose domination on China in January 1915. Taking advantage of its favourable international position after entering World War I on the Allied side and capturing the German base of Qingdao on the Chinese mainland, Japan attempted to impose virtual protectorate status on China, then diplomatically isolated and torn by civil war. Although one group of demands dealing with the appointment of Japanese advisers throughout the Chinese government was not enforced, threat of war left China no choice but to concede the others, including extension of Japanese leases in Manchuria, takeover of former German concessions in Jiaozhou, substantial interests in Chinese mining concerns, and an embargo on future coastal territorial concessions to any other foreign power. The Twenty-One Demands greatly extended Japanese power in China, but provoked serious resentment within China and aroused US fears of Japanese expansionism.

Tyler, John (1790-1862), tenth President of the USA (1841-5). Tyler entered politics as a Democrat and served in Congress (1817-21), as governor of Virginia (1825-7), and in the Senate (1827-36) before leaving the Party over the financial policies of Andrew *Jackson and the *Nullification Crisis. He ran successfully as *Whig candidate for Vice-President in the election of 1840, succeeding *Harrison on the latter's death a month after taking office. His disagreement with the Whig Party leader, Henry *Clay, resulted in the resignation of almost his entire cabinet in 1841. His subsequent alliance with *Calhoun and the Southern Democrats on the issue of *states' rights aggravated the geographical polarization of politics between the North and South. At the end of his Presidency, he secured the annexation of *Texas. In 1861 he chaired the Washington Peace Convention, an unsuccessful attempt to find a compromise to avert the *American Civil War.

Tyrol (German, Bundesland), Austrian Alpine province. In 1803 the province was enlarged by the addition of the Italian-speaking province of Trentino, but two years later, after the Treaty of Pressburg, Austria was forced to cede the Tyrol to France's ally, Bavaria. In 1809, when Austria renewed the war on France, Andreas Hofer led a successful revolt against Bavaria and the French. Hofer was made governor of the Tyrol by the Austrians. But Austria was defeated at *Wagram and forced, by the Treaty of Schönbrunn (1809), to cede the Tyrol to Napoleon once more. Hofer's resistance led to his betrayal to the French, and his execution, on Napoleon's orders, at Mantua. In 1810 Napoleon annexed most of South Tyrol to Italy, but the area was restored (1815) to Austria by the Congress of *Vienna. Tyrol became a centre of *Irredentist claims; after World War I the whole of South Tyrol was handed over to Italy, and an agreement between Hitler and Mussolini (1938) provided for extensive forced migration of the German-speaking population to Germany. An international meeting in 1946 gave South Tyrol the status of an autonomous region, and decided not to change its frontier. Serious tension between Austria and Italy continued. In 1971 a treaty concerning the Trentino–Alto-Adige region was ratified, stipulating that disputes would be referred to the *International Court of Justice at The Hague.

Tz'u hsi *Cixi.

Uganda, a landlocked country in East Africa. During the 19th century the kingdom of Buganda on Lake Victoria became the dominant power in the area under its kabaka (king) *Mutesa I. Although a Muslim he welcomed the explorer John *Speke in 1856 and the British-American explorer and journalist H. M. *Stanley in 1875, hoping for protection against Arab slave and ivory traders. Following Mutesa's death (1884) tensions developed between Christians and Muslims, and also between British and German interests. In 1890 there was an Anglo-German agreement that the area be administered by the British, and the newly formed British East Africa Company placed Buganda and the western states Ankole and Toro under its protection. In 1896 the British government took over the protectorate. After World War II nationalist agitation for independence developed, with *Mutesa II being deported (1953-5) for allegedly refusing to co-operate with the government of the protectorate. In 1962 full internal self-government was granted. Uganda was to be a federation of the kingdoms of Ankole, Buganda, Bunyoro, Busoga, and Toro. In September the Prime Minister, Milton *Obote, renounced this constitution and declared Uganda a republic, with an elected president. Mutesa II was elected first President, but in 1965 he was deposed by Milton Obote, who became President himself, only to be deposed in turn by General Idi *Amin (1971). Amin's rule was tyrannical, and in 1980, after the invasion by Tanzanian forces, he fled the country. Obote returned in 1981, but his failure to restore order led to a coup in 1985, the resulting military regime lasting only six months before being overthrown by the National Resistance Army of Yoweri Musevani, who became President in 1986.

U-2 Incident   *Eisenhower Presidency.

Uitlanders (Cape Dutch, 'outlander'), term for non-Boer immigrants into the *Transvaal, who came after the discovery of gold (1886). To the Boers the immigrants, with foreign capital and different life-styles, constituted a cultural and economic threat. The Transvaal government denied the Uitlanders citizenship, taxed them heavily, and excluded them from the government.

Ukraine, a region of eastern Europe. Ruled by Russia after the late 18th-century partition of Poland, Ukrainian nationalism, despite repression, remained strong. In 1918 independence was proclaimed, but by 1922 the area had been re-conquered by Soviet forces. Stalin imposed *collectivization on the region, which was devastated during the German occupation of 1941-4. Territorial gains from Romania, Poland, and Czechoslovakia completed the reunion of all Ukrainian lands into one republic by 1945, the Crimea being added in 1954.

Ulbricht, Walter (1893-1973), German statesman. He helped to found the German Communist Party in 1919, and became a communist member of the *Reichstag (1928-33), fleeing to the Soviet Union to escape Nazi persecution. After World War II he became a leading member of the communist-dominated Socialist Unity Party in the Soviet zone of Germany, subsequently the *German Democratic Republic. He was Party Secretary (1950–60) and Chairman of the Council of State (1960–71). Ulbricht's Stalinist regime was stern and its unpopularity was revealed by a serious uprising in 1953.

Ulster *Northern Ireland.

Ulster Unionist Party, a political party in *Northern Ireland. In 1886 Lord Harlington and Joseph *Chamberlain formed the Liberal Unionists, allying with the Conservatives and pledging to maintain the Union of Ireland with the rest of the United Kingdom. In 1920, with the division of Ireland, the majority party in Northern Ireland was the Unionist wing of the Conservative Party, now calling itself the Ulster Unionists, under Sir James Craig, who was Prime Minister (1921–40). The Party, supported by a Protestant electorate, continued to rule under his successors, until the imposition of direct rule from Westminster in 1972. The policy for handling the increased violence between Nationalists and Unionists after the civil rights campaign of 1968 led to divisions in the Party, and in 1969 it split into the Official Unionist Party and the Protestant Unionist Party. The latter, led by the Revd Ian Paisley, was renamed in 1972 the Democratic Unionist Party, with policies more extreme than those of the Official Unionists. Following the Hillsborough Agreement with the Republic of *Ireland (1985) neither Unionist Party has close links with the Conservative Party.

Ulster Volunteers, an Irish para-military organization, formed in 1913 to exclude Ulster from the *Home Rule Bill. Its supporters pledged themselves 'to use all means' to resist this. They were given every encouragement by Sir Edward *Carson and several prominent English Conservatives. The Volunteers were drilled and armed: thousands of rifles were smuggled into Ireland for their use. A clash between these Volunteers and the nationalist Irish Volunteers (formed in Dublin in 1913) became probable but was averted by the start of World War I.

Umar ibn Said Tal (or al-Hajj Umar or Umar Tal) (c.1797–1864), Muslim ruler of a state in Mali (1848–64). Born among the Tukolor people, he established the Tukolor empire. He set out on the pilgrimage to Mecca c.1820, where he was designated caliph for black Africa. He married the daughter of the Sultan of Sokoto and acquired a reputation as a scholar and mystic. He established himself in the Senegal River area and in 1854 proclaimed a *jihad* (holy war) against all pagans. He created a vast Tukolor empire, but failed to win many converts. He became a harsh ruler. In 1863 he captured Timbuktu but soon lost it to a combined force of Fulani and Tauregs. He was killed in 1864. The Tukolor empire survived until 1897, ruled by his son Ahmadu Seku.

UNCTAD *United Nations Conference on Trade and Development.

Underground Railroad, a secret network in the USA for aiding the escape of slaves from the South in the years before the *American Civil War. While the Railroad helped only a small number of slaves a year (perhaps 1,000 per annum after 1850), it served as a valuable symbol for the abolitionist cause and was viewed in the South as a far greater menace than its actual size merited.

unemployment assistance, payments made to unemployed persons as part of a system of social insurance. Such a system was first introduced in Germany by *Bismarck: for health (1883), unemployment (1884), and old age (1889), financed in different proportions by employer, employee, and the state. Similar schemes were introduced in Denmark, Sweden, Austria and, in 1911, in Britain, whose first National Insurance Act provided for unemployment assistance for a limited time, while the person concerned was finding another job. In the USA a Social Security Act of 1935 sought to co-ordinate social insurance schemes created by the different US states, while the British scheme was enlarged following the *Beveridge Report (1942). By the mid-20th century all industrialized nations had some form of social insurance scheme providing assistance for short-term unemployment. In some countries, such as Britain, long-term unemployment is met by a scheme of supplementary benefit (the dole), first introduced in 1934 and assessed on an individual's financial position. Many developing countries cannot afford a system of state social insurance.

Unequal Treaty, term used to describe treaties made between China and various Western powers in the 19th century. The *Qing dynasty was generally unable to resist foreign pressure for commercial and territorial concessions, and in such agreements as the Treaty of *Nanjing (1842) was forced to agree to Western demands. By the late 19th century, some European powers were expanding into 'spheres of influence' and the new imperial power of Japan was pressing for major new territorial concessions from the crumbling Chinese state.

Unilateral Declaration of Independence *Zimbabwe.

Union of Soviet Socialist Republics (USSR), a country occupying the northern half of Asia and part of eastern Europe, made up of fifteen constituent republics. The overthrow of *Nicholas II in the *Russian Revolution of 1917 led, after the *Russian Civil War, to the triumph of the *Bolsheviks under *Lenin. At a congress of the first four republics in 1922 the new nation was named the USSR. It was to base its government on the national ownership of land and of the means of production, with legislative power in the hands of the Supreme *soviet. Under *Stalin, *collectivization of agriculture was carried out, and a series of political purges took place; a total estimate of seven to nine million people died as a result. The Soviet Union signed a *Nazi–Soviet Pact (1939), and shared with the *Third Reich in the annexation of *Poland. The *Baltic States were annexed (1939), and Finland was invaded in the *Finnish–Russian War. After Germany's invasion of the Soviet Union in 1941 the latter fought on the side of the Allies in World War II. The Soviet Union declared war on *Japan (1945) and took part in the *Teheran, *Yalta, and *Potsdam conferences. It joined the *United Nations and, during the *Cold War, formed the *Warsaw Pact as a defensive alliance. In foreign affairs, the economic and energy supplies of the Eastern bloc *Comecon countries have remained closely tied to the Soviet Union. Soviet troops were sent to *Hungary and *Poland (1956), and to *Czechoslovakia (1968), to reinforce those countries' governments against liberalization programmes. Ideological differences have aggravated relations with *China since the late 1950s. In the developing world, the Soviet Union has given aid to pro-Soviet governments and political movements. *Afghanistan was invaded (1979) by Soviet troops, and a pro-Soviet government under Soviet military protection installed. A pervasive element of Soviet society has remained the high degree of police surveillance and state control of private citizens' lives. The number of political dissidents in *prison camps has been reduced, but, although the Soviet Union is a signatory to the *Helsinki Accord, agitation for greater human rights continues. The appointment of Mikhail *Gorbachev as Secretary-General in 1985 heralded a new style of Soviet leadership, committed to the modernization of Soviet technology, to partial de-collectivization, liberalization, a drive against corruption, and to international *arms control.

Union Pacific Railroad, a US railway forming the eastern part of the link between the Missouri River and the Pacific. The Union Pacific was chartered in 1862, with construction beginning from Omaha westward in 1865. In 1869 the Union Pacific joined the *Central Pacific west of Ogden, Utah, thus completing the first transcontinental rail connection. The Union Pacific went on to acquire large holdings in eastern railways.

Unitarianism, an undogmatic sect based on freedom, reason, tolerance, and a belief in the goodness of human nature. Modern Unitarianism derives from 16th-century Protestant Christian thinkers who rejected the doctrine of the Trinity and stressed the unity of God. Holding Unitarian views was technically a legal offence in Britain until 1813. The movement has never attracted many adherents in that country. In the USA, however, the influence of Unitarianism has been stronger, especially in New England and above all in Harvard University. Five Presidents of the USA have been Unitarians. In 1961 the Unitarian Universalist Association was founded in the USA by the union of Unitarianism and Universalism, the latter, founded in 1778, having members of diverse religious opinions.

United Arab Emirates, the federation of Arab Gulf States formed (1971) by the former *Trucial States of Abu Dhabi, Dubai, Sharjah, Ajman, Umm al-Qaiwain, and Fujairah. Ras al-Khaimah joined in 1972. The emirates came together as an independent state when they ended their individual special treaty relationships with the British government, and signed a Treaty of Friendship with Britain in 1971. The great wealth of the UAE is derived from the oil of Abu Dhabi, first discovered in 1958.

United Arab Republic, the union of *Syria and *Egypt (1958), which was dissolved in 1961 following an army

The final link between the **Union Pacific Railroad** and the Central Pacific was made on 10 May 1869 at Promontory Point in Utah. The UP chief engineer, Grenville M. Dodge (right), shakes hands with his CP counterpart, Samuel S. Montague, as the first US trans-continental railway is completed.

coup in Syria. The United Arab Republic was open to other Arab states to join, but only Yemen entered a loose association (1958), which lasted until 1966. Egypt retained the name United Arab Republic until 1971, when it adopted the name Arab Republic of Egypt.

United Democratic Front, a South African non-racial political organization. As part of its campaign to defuse the country's political crisis, the South African government proposed to give the Coloured and Indian communities a limited role in government. Opponents of such a compromise launched the United Democratic Front in 1983, and by 1985 the UDF had become a significant opposition group, an affiliated membership of about 2.5 million. Its activities were banned in 1985.

United Malays National Organization (UMNO), Malaysian political party. Formed by Dato Onn bin Jaafar, then Prime Minister of Johore, in 1946 in response to British attempts to form the Union of Malaya, UMNO's aim was to fight for national independence and protect the interests of the indigenous population. Since independence in 1957 UMNO has been the dominant party in Malaysia, forming the cornerstone of successive electoral alliances, notably the Alliance Party of the 1960s and its successor, the National Front.

United Nations, an international organization of countries with its headquarters in New York. It was established in 1945 in succession to the *League of Nations to work for world peace, security, and co-operation. The Allied Powers in World War II called a conference at San Francisco in 1945 to draw up a document for such an organization, and this document, known as the Charter of the United Nations, was signed by fifty nations in the summer of 1945, and the UN came into existence on 24 October, 1945. Since that date, more than 100 other nations have joined, the chief exceptions being Switzerland and North and South Korea. To carry out its functions, the United Nations has various organs and institutions, including a *General Assembly in which each member state has one vote; a *Security Council composed of five permanent members and ten members elected for a two-year term, and with powers to execute and carry out UN policies; a Secretariat headed by the Secretary-General to administer the organization; an Economic and Social Committee to co-ordinate and establish commissions on specific issues; an *International Court of Justice, based at The Hague, to deal with legal disputes; an *International Monetary Fund to promote monetary co-operation and expansion; and a number of specialized agencies to deal with social, educational, health, and other matters. With a shift in membership in the 1950s, the General Assembly aligned itself into new voting blocks, including the NATO nations, the Arab nations, and the (numerically largest) Afro-Asian nations. The office of Secretary-General, which reached its peak under Dag *Hammarskjöld, has declined in power. Resolutions passed by the General Assembly have little effect on world politics, due largely to a decreased support of the UN by the world powers, who have dealt with each other outside the UN framework.

United Nations Conference on Trade and Development (UNCTAD), a permanent agency of the *United Nations, with its headquarters in Geneva. It

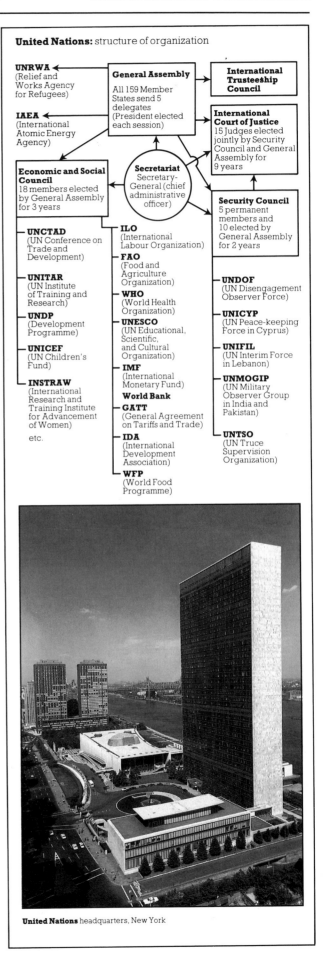

United Nations headquarters, New York

was established in 1964 to promote international trade and economic growth. The Conference, which meets every four years, called for discrimination in favour of the developing countries, since their industrial products are often subject to quotas and tariffs. In 1968 it proposed that developed countries should give 1 per cent of their gross national product in aid to developing countries, but the gap between rich and poor countries has continued to widen (*Brandt Report), aggravated by a steady decline in the price of many basic world commodities which the developing countries produce.

United Party, a South African political party. Officially the United South African National Party, it was established in 1934 as a coalition between the followers of *Hertzog's *National Party and *Smuts's South African Party. Although it had *Afrikaner and English backing, it was soon weakened by the defection of *Malan's 'purified' National Party. In 1939 it split when Hertzog attempted to declare South Africa neutral when war broke out. Until the mid-1970s it was the principal opposition. With the rise of the Progressive Party (later known as the Progressive Federal Party) support for the United Party fell and it was dissolved in 1977.

United Provinces of Central America   *Central America.

United States of America, a North American country, consisting of fifty states. In 1775 thirteen British North American colonies revolted, officially adopting the name United States of America one year later in the Declaration of Independence, and successfully defending their new status in the War of Independence ending in 1783. A structure of government was set out in the Constitution of 1787, which established a federal system, dividing power between central government and the constituent states, with an executive President, a legislature made up of two houses, the *Senate and the House of *Representatives, and an independent judiciary headed by the *Supreme Court. Territorial expansion followed with the *Louisiana Purchase of 1803, the acquisition of Florida in 1810–19, and of *Texas, California, and the south-west following the *Mexican–American War of 1846–8. The western lands of the Louisiana Purchase and those seized from Mexico were at first territories of the USA, administered by officers of the federal government and defended by detachments of the federal army. When the population reached some 60,000 an area of territory negotiated to be admitted to the Union as a new state. The mid-19th century was dominated by a political crisis over slavery and *states' rights, leading to the secession of the Southern states and their reconquest in the *American Civil War of 1861–5. The final decades of the century saw the *westward expansion of European

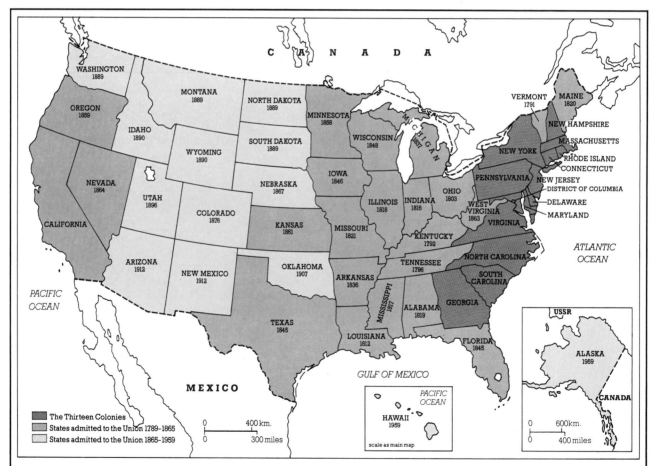

United States of America

An Ordinance of 1787 laid down that when a US territory area reached a population of 60,000 free inhabitants it could petition Congress for admission to the Union as a state. Vermont and Kentucky were early admissions, and many western states followed in the course of the 19th century.

settlement, the purchase of Alaska (1867), and the acquisition of Spanish overseas territories after the *Spanish–American War of 1898. In the 20th century the USA has participated in the two World Wars and has gradually emerged from *isolationism to become a world power, a process accelerated by the *Cold War division of the world into two spheres of influence dominated by two super-powers, the USA and the Soviet Union.

university, centre of higher education with responsibilities for teaching and conducting research. In Europe, by 1800 the great medieval universities such as Bologna, Paris, and Oxford had been augmented considerably, particularly in Italy and Germany, where individual cities or princes wanted the prestige of a foundation. In both Europe and America traditional courses in grammar, logic, and rhetoric, Greek, Latin, and mathematics still dominated, and graduates went into either the Church or government service. It is estimated that some 160 universities existed in 1800. Higher education expanded at a remarkable rate during the 19th and 20th centuries, and by 1980 some 1,320 universities existed, including some 400 in the USA, together with several thousand colleges and institutes of higher education. In the early years of the 19th century many German universities pioneered new curricula with a greater emphasis on science, which became the model for the numerous 19th-century foundations, particularly in the USA. After World War II there was accelerated university expansion, with some 500 new foundations between 1945 and 1980, particularly in developing countries. The Open University system, using radio and television for advanced distance-learning, was launched in Britain in 1971 and has since then been successfully introduced in other countries.

Upper and Lower Canada, two British North American colonies or provinces (1791–1841). Following the American War of Independence (1775–83) many loyalists to the British crown came north into the British colony of *Quebec. Pressure developed among the settlers in the west for separate status, which was granted by the Constitutional Act of 1791. Quebec was to be divided along the Ottawa River: the eastern area, with its predominantly French population to be known as Lower Canada (now Quebec); the western part to be known as Upper Canada (now Ontario), adopting English common law and freehold land tenure. Government in both provinces would remain in the hands of a governor appointed by the British crown, advised by an appointed executive council and a legislature consisting of an appointed upper house and a lower assembly of elected representatives, who in fact would wield little power. In both provinces movements for reform developed in the 1830s and, on the accession of Queen Victoria in 1837, two abortive rebellions took place led by Louis Joseph *Papineau in Lower Canada and William Lyon *Mackenzie in Upper Canada. The former protested against a proposal that the two provinces be reunited, and demanded more responsible government. In the wake of the *Durham Report (1838), an Act of Union was passed (1840) by the British Parliament, and the two provinces united to form United Canada, with a legislature in which Canada West and Canada East would enjoy equal representation. The objective of reformers for cabinet government on the British pattern, directly responsible

to the legislature, was not, however, to be achieved until 1848, when it was conceded by the new governor-general, Lord *Elgin.

Upper Volta   *Burkina Faso.

urbanization, the process of social change from a rural to an urban economy. Although large cities existed before 1800, the vast majority of the world's population at that time still lived in small, often self-sufficient village communities. *Industrialization and *population growth in Europe in the 19th century resulted in radical change: sometimes the excess peasant population moved to towns to seek paid work, being often obliged to live in unhygienic slums and dying of infectious *diseases; alternatively, where there was no industrial growth, *population migration to the New World occurred. In the 20th century *agricultural improvements have resulted in further rural depopulation, while the *transport revolution has ensured that many villages near large cities in Western Europe or the USA have become little more than dormitories for urban workers, who commute daily to the city. Urbanization has become a feature of many developing countries during the 20th century. Often, as in South American cities, very high levels of population growth have resulted in high unemployment and political instability. It is estimated that over 50 per cent of the world's population now lives in an urban environment.

Uruguay, a country in South America. During the colonial period, it was known as the Banda Oriental and became a part of the Spanish vice-royalty of Rio de la Plata. In 1814 the leaders of the Banda Oriental, notably *Artigas, broke with the military junta in *Argentina and led a struggle for Uruguayan independence until occupied by Brazil in 1820. In 1825 an independent republic of Uruguay was declared, which was recognized by the treaty between Argentina and Brazil, signed at Rio de Janeiro in 1828. Under a republican constitution, the liberals (*Colorados*, redshirts) and the clerical conservatives (*Blancos*, whites) struggled violently throughout the 19th century for political control. In 1872 the *Colorados* began a period of eighty-six years in office. During the first three decades of the 20th century, José *Batlle y Ordóñez, while in and out of the presidency, helped mould Uruguay into South America's first welfare state. Numerous measures for promoting governmental social services and a state-dominated economy were enacted. In 1958 the elections were won by the *Blancos*. Economic and political unrest plagued the nation throughout the 1960s and saw the emergence of the Marxist terrorist group, the *Tupamaros. The military took over in the 1970s, and a return to civilian rule took place in 1985. It has emerged as one of the most prosperous and literate nations in the continent, though falling world commodity prices and high inflation have caused renewed problems.

USSR   *Union of Soviet Socialist Republics.

U Thant (1909–74), Burmese statesman and third Secretary-General of the United Nations (1961–71). He entered the Burmese diplomatic service in 1948 and served at the United Nations from 1957 until he succeeded Dag *Hammarskjöld as Secretary-General in 1961. He filled the post with great distinction, his achievements including assistance in the resolution of the *Cuban

Missiles Crisis, the formation of a UN peace-keeping force in *Cyprus, the negotiation of an armistice to end the Arab–Israeli *Six-Day War of 1967, and the admission of communist China to full UN and Security Council membership in 1972.

Uthman dan Fodio (or Usuman dan Fodio) (1754–1817), a West African religious and political leader. A Muslim Fulani, he began teaching c.1775 among the Hausa, and established the Emirates of Northern Nigeria (1804–8) after waging a *jihad* (holy war). He conceived the latter as a primary duty, not only against infidels, but against any departure, public or private, from the original and austere ideals of Islam. Under his rule as caliph, and that of his son, Muhammad Bello (d. 1837), Muslim culture flourished in the *Fulani empire. He retired from public life c.1815.

utilitarianism, a 19th-century ideology. It was first outlined by Jeremy *Bentham in his *Introduction to the Principles of Morals and Legislation*, published in 1789. His theory was that all human activity should be directed so as to bring about the greatest happiness of the greatest number. Bentham devoted much of his time to attacking the abuses in the legal system, but both he and his followers were to have a profound influence on all aspects of political and social reform in Britain during the 19th century. Acting as a pressure group on both Conservative and Liberal governments, they often gave a lead to public opinion. J. S. *Mill's essay, *Utilitarianism* (1863), gave perhaps the clearest expression to the doctrine.

Van Buren, Martin (1782–1862), eighth President of the USA (1837–41). He became leader in 1820 of the *Democratic Party in New York. He served in the Senate (1821–8) and then as governor of New York, resigning his post in 1829 to become Secretary of State, only to leave in 1831. Van Buren ran successfully as *Jackson's Vice-President in the election of 1832, and was himself elected President in 1836. However, he lost support when the Democrats split over financial policies in 1837, and he accomplished little in the legislative field. He lost the election of 1840 to *Harrison, failed to secure the nomination in 1844, and ran unsuccessfully as nominee of the breakaway *Free Soil Party in 1848.

Vanderbilt, Cornelius (1794–1877), US railroad magnate. Starting by ferrying passengers and freight around New York City, he later gaining a virtual monopoly over the ferry lines along the coast. When the *gold rush created a demand, he connected New York and California by running his own shipping line to San Francisco via Nicaragua, constructing his own roads for the overland part of the journey. Turning to railways with the *American Civil War, he quickly came to dominate the network in and out of New York and as far west as Chicago. Amassing a vast fortune, he made an endowment of $1 million to found Vanderbilt University.

Vandernberg, Arthur Hendrick (1884–1951), US politician. He entered the US Senate in 1928 as a Republican. He supported much of Franklin D. *Roosevelt's domestic legislation, but he was an avowed *isolationist. In 1945, however, he responded to changing world conditions by working for a bipartisan foreign policy and supporting US membership of the *United Nations. Thereafter he was instrumental in securing the Senate's approval of the *Marshall Plan and the *North Atlantic Treaty Organization.

Van Diemen's Land   *Tasmania.

Vanuatu, a double chain of eighty islands in the south-west Pacific, formerly the New Hebrides. During the 19th century, thousands of *Kanakas were taken from Vanuatu to work on the *Queensland, Australia, sugar plantations. The population was decimated and took many years to recover. The islands were placed under an Anglo-French naval commission in 1887. In World War II they served as a major Allied base. They became an independent republic and member of the Commonwealth of Nations in 1980.

Vargas, Getúlio (1883–1954) Brazilian statesman. He governed *Brazil first as head of a provisional government (1930–4), then as constitutional President elected by Congress (1934–7), then as dictator (1937–45), and finally as constitutional President elected by universal suffrage (1950–4). The fraudulence of the 1930 elections, political corruption, and the growing impact of the Great *Depression on Brazil's vulnerable agricultural economy combined to bring an end to the first or Old Republic (1889-1930) through a co-ordinated military and civilian coup. Vargas assumed power, formally incorporating emerging sectional interest groups into national political organizations. Claiming a desire to rise above the factional strife, in 1937 Vargas announced a state of emergency, a ban on all political organizations, the dissolution of Congress, and the promulgation of a new Constitution which would create a nationalist, corporate, unified 'New State' (Estado Novo), backed by the military. His economic strategy concentrated on the diversification of agricultural production, improvements in transport and communication, the promotion of technical education, the implementation of a new labour code, the national ownership of mineral resources and key industries, and the promotion of industrial expansion. World War II offered a favourable climate for economic growth, and growing commercial and diplomatic co-operation with the USA led to Brazilian participation in the war (1942). The defeat of the Axis brought renewed pressure on Vargas to relax the authoritarianism of the Estado Novo. His reluctance prompted a military coup in 1945, but Vargas' national popularity and his courting of the left resulted in his return to power by popular vote in 1950. Thwarted by a growing economic crisis after 1952, and accused of political corruption, Vargas committed suicide in 1954.

Vatican City, an independent papal state in Rome, the seat of the Roman Catholic Church. Following the *Risorgimento, the former Papal States became incorporated into a unified Italy in 1870 while, by the Law of Guarantees (1871), the Vatican was granted extraterritoriality. The temporal power of the pope was suspended until the Lateran Treaty of 1929, signed between Pope Pius XI and *Mussolini, which recognized the full and independent sovereignty of the Holy See in the City of the Vatican. It covers an area of 44 hectares (109 acres), and has its own police force, diplomatic service, postal service, coinage, and radio station.

Venezuela, a country in South America. Colonized by the Spanish, by the mid-18th century wealthy Creoles (Spaniards born in the colony) were protesting against trade restrictions imposed by Madrid. It was in its capital Caracas that the Colombian Independence Movement began (1806), resulting in the creation by Simón *Bolívar of Gran Colombia. When this collapsed (1829), Venezuela proclaimed itself a republic under its first President, General José Antonio *Páez (1830-43), who, while preserving the great estates, provided a strong administration, allowed a free press, and kept the army under control. The period that followed (1843-70) was politically chaotic and violent. Under President Guzmán *Blanco (1870-88) moves were made towards democracy, with the first election in 1881, and there was growth in economic activity. Despotic government returned under the *caudillos Cipriano Castro (1899-1908) and Juan Vicente *Gómez (1909-35). Oil was discovered before World War I, and by 1920 Venezuela was the world's leading exporter of oil. Military juntas continued to dominate until Rómulo Betancourt completed a full term as a civilian President (1959-64), to be peacefully succeeded by Dr Raul Leoni (1964-9). Since then, democratic politics have continued to operate, with two parties, Accion Democratica and Christian Democrat,

alternating in power, even though extremists of left and right have harassed them with terrorism. A post-war oil boom brought considerable prosperity, but rising population and inflation have caused many of the problems faced by President Dr Jaime Lusinchi (1983-).

Venizélos, Eleuthérios (1864-1936), Greek statesman. Active in the anti-Turkish movement of 1895-1905, he became Premier (1910), modernizing Greek political institutions and joining the Balkan League against Turkey. In 1914 his wish to join the Allies was thwarted by the pro-German King Constantine, who later abdicated, thus enabling Greek troops to fight Germany. At the *Versailles Peace Settlement he negotiated promises of considerable territorial gains—the Dodecanese, western Thrace, Adrianople, and Smyrna in Asia Minor. In the event, following the challenge of *Atatürk's army only western Thrace was gained, and he resigned. Greece now alternated between monarchy and republic and his periods as Premier alternated with periods in exile.

Vereeniging, Treaty of (31 May 1902), the peace treaty that ended the Second *Boer War. It provided for the acceptance by Boers of British sovereignty; the use of Afrikaans in schools and law courts; a civil administration leading to self-government; a repatriation commission, and compensation of £3 million for the destruction inflicted during the war on Boer farms.

Versailles, 1871 Treaty of   *Frankfurt, Treaty of.

Versailles Peace Settlement (1919-23), sometimes referred to as the Paris Peace Settlement, a collection of peace treaties between the Central Powers and the Allied Powers ending World War I. The main treaty was that of **Versailles** (June 1919) between the Allied Powers (except for the USA, which refused to ratify the treaty) and Germany, whose representatives were required to sign it without negotiation. Germany had concluded an armistice in 1918 based on the *Fourteen Points of President *Wilson. By a new 'war-guilt' clause in the treaty Germany was required to accept responsibility for provoking the war. Various German-speaking territories were to be surrendered, including *Alsace-Lorraine to France. In the east, *Poland was resurrected, and given parts of Upper Silesia and the *Polish Corridor to the Baltic Sea, while Gdańsk (Danzig) was declared a free city. Parts of East Silesia went to Czechoslovakia; Moresnet, Eupen, and Malmedy to Belgium; and the *Saar valley was placed under international control for fifteen years, as was the Rhineland, which, together with Heligoland, was to be demilitarized. Overseas colonies in Africa and the Far East were to be *mandated to Britain, France, Belgium, South Africa, Japan, and Australia. Germany was henceforth to keep an army of not more than 100,000 men and to have no submarines or military aircraft, *Reparations were fixed in 1921 at £6,500 million, a sum which was to prove impossible to pay. Many aspects of the treaty were criticized as excessive, and its unpopularity in Germany created a political and economic climate that enabled *Hitler to come to power. The treaty established the *League of Nations and the *International Labour Organization.

A second treaty, that of **St Germain-en-Laye** (September 1919), was between the Allied powers and the new republic of *Austria. The *Habsburgs had been deposed

The most important treaty of the **Versailles Peace Settlement** was signed on 28 June 1919 in the Hall of Mirrors. It embodied the results of the long and acrimonious negotiations of the Paris Peace Conference of 1919. Among the Allied leaders in this painting by Sir William Orpen are (*centre, left to right*): Woodrow Wilson of the USA, Georges Clemenceau of France, and Lloyd George of Great Britain. Alone, and with his back to the viewer, is the German representative Johannes Bell. (Imperial War Museum, London)

and the imperial armed forces disbanded. Austria recognized the independence of Czechoslovakia, Yugoslavia, Poland, and Hungary. Eastern Galicia, the Trentino, South Tyrol, Trieste, and Istria were ceded by Austria. There was to be no union (Anschluss) with Germany. Austria, like Germany, was to pay reparations for thirty years. A third treaty, that of **Trianon** (June 1920), was with the new republic of *Hungary, whereby some three-quarters of its old territories (i.e., all non-Magyar lands) were lost to Czechoslovakia, Romania, and Yugoslavia, and the principle of reparations again accepted. A fourth treaty, that of **Neuilly** (November 1919), was with *Bulgaria, whereby some territory was lost to Yugoslavia and Greece, but some also gained from Turkey; a figure of £100 million reparations was agreed, but never paid. These four treaties were ratified in Paris during 1920. A fifth treaty, that of *Sèvres (August 1920), between the Allies and the old *Ottoman empire was never implemented as it was followed by the final disintegration of the empire and the creation by Mustafa Kemal *Attatürk of the new republic of Turkey. The treaty was replaced by the Treaty of **Lausanne** (July 1923), whereby Palestine, Transjordan, and Iraq were to be mandated to Britain, and Syria to France, together with much of Arabia. Italy was accepted as possessing the Dodecanese Islands, while Turkey regained Smyrna from Greece. The Dardanelles Straits were to be demilitarized and Turkey would pay no reparations.

Verwoerd, Hendrik Frensch (1901–66), South African statesman. As Minister of Native Affairs (1950–8) he was responsible for establishing the policy of *apart-

heid. He became Nationalist Party leader and Prime Minister (1958–66). During his government, in the aftermath of the *Sharpeville massacre, South Africa became a republic and left the Commonwealth. Harsh measures were taken to silence black opposition, including the banning of the *African National Congress. He was assassinated in Parliament.

Vesey, Telemaque ('Denmark') (*c*.1767–1822), leader of a projected American slave revolt. A slave who had purchased his freedom, he managed in 1822 to inspire a group of slaves with the idea of seizing the arsenals of Charleston, South Carolina, as a prelude to a mass escape to the West Indies, but the plan was betrayed and Vesey and thirty-six of his followers were arrested and hanged. The incident led directly to the tightening of the restrictive 'black codes' as a means of controlling the slave population.

Vichy government (1940–5), the French government established after the Franco-German armistice in World War II. The Germans having occupied Paris, it was set up under Marshal *Pétain in the spa town of Vichy by the French National Assembly (1940) to administer unoccupied France and the colonies. Having dissolved the Third Republic, it issued a new constitution establishing an autocratic state. The Vichy government was never recognized by the Allies. It was dominated first by *Laval, as Pétain's deputy (1940), then by *Darlan (1941–2) in collaboration with Hitler, and once more (1942–4) by Laval as Pétain's successor after German forces moved in to the unoccupied portions of France. After the Allied liberation of France (1944), the Vichy government established itself under Pétain at Sigmaringen in Germany, where it collapsed when Germany surrendered in 1945.

Vicksburg Campaign (November 1862–July 1863), a military campaign in the *American Civil War. By the autumn of 1862 the stronghold of the *Confederacy forces at Vicksburg in western Mississippi was the last remaining obstacle to Union (Northern) control of the Mississippi River. In late 1862, advances by Generals *Grant and *Sherman failed to capture the city. In May 1863, Grant started a siege of the city. After six weeks of resistance, Vicksburg surrendered on 4 July. With its capture, the Confederacy was effectively split in half. The Union success at Vicksburg and *Gettysburg in July 1863 marked a major turning-point in the Civil War.

Victor Emanuel II (1820–78), King of Sardinia (Piedmont) (1849–61), and first King of Italy (1860–78). He succeeded to the throne of Sardinia after the abdication of his father, Charles Albert. He fought in the *Revolutions of 1848 against Austrian rule and, on his accession to the throne, appointed *Cavour as Premier (1852). The central figure of the *Risorgimento, he sought the support of Britain and France in his bid for the reunification of Italy by entering the *Crimean War as their ally. Proclaimed King of Italy in 1861, he supported Prussia in the *Austro-Prussian War (1866). His troops seized the Papal States (1870), and Rome was made the capital of Italy in 1871.

Victor Emanuel III (1869–1947), King of Italy (1900–46). Succeeding Humbert I, he retained good relations with France and Britain although a member of the *Triple

Victor Emanuel II, in a portrait that symbolizes his role as a unifying figure in the Italian nationalist movement. His judicious wartime alliances strengthened the growth of the embryonic Italian state and secured Emanuel the throne of the new kingdom. (Museo del Risorgimento, Milan)

Alliance, and maintained neutrality in World War I, until joining the Allies in 1915. With the breakdown of parliamentary government after World War I, he refused to suppress a fascist uprising and asked *Mussolini to form a government (1922), fearing the alternative to be civil war and communism. He was created Emperor of Ethiopia (1936) and King of Albania (1939). He dismissed Mussolini (1943), replacing him with *Badoglio, and concluding an armistice with the Allies soon after. He declared war on Germany in October 1943. In 1946 he abdicated, dying in exile in Egypt a year later.

Victoria (1819–1901), Queen of Great Britain and Ireland and of dependencies overseas (1837–1901) and (from 1876) Empress of India. The last of the House of Hanover, she was the only child of George III's fourth son, Edward, Duke of Kent. She came to the throne in 1837 on the death of her uncle, *William IV. She was guided in the performance of her duties as a monarch by the Prime Minister, Lord *Melbourne. Her marriage to Prince *Albert of Saxe-Coburg-Gotha in 1840 was to prove a happy one; his early death in 1861 was a blow from which she never fully recovered, and her withdrawal from public life during the early years of her widowhood did not enhance her popularity. Benjamin *Disraeli persuaded her to take her place once more in the life of the nation, but it was largely at her own instigation that she was declared Empress of India by the Royal Titles Act of 1876. By the 1880s she had won the respect and admiration of her subjects at large. The Golden and Diamond Jubilees were great imperial occasions. Her death in 1901 marked the end of an era to which she had given her name, the Victorian Age, during which Britain had become the world's leading industrial power at the centre of the *British empire.

Victoria, a state of the Commonwealth of *Australia. In its early years of white settlement, it formed part of New South Wales. Convict settlements there were short-lived (1803–4, 1826–8). The first permanent settlements did not begin until 1834 and 1835. Soon after, *squatters settled. It was separated from New South Wales, and named Victoria, in 1851. *Gold rushes, notably to Ballarat and Bendigo, began that year. The *Eureka Rebellion occurred in 1854. Restrictions on Chinese immigration, imposed in 1855, marked the beginning of the *White Australia Policy. Attempts were made, between 1860 and 1890, to 'unlock the lands' for *selectors. Victoria became a state of the newly created Commonwealth of *Australia in 1901. It has been particularly influenced by *New Australian immigration immediately after World War II.

Vienna, Congress of (1814–15), an international peace conference that settled the affairs of Europe after the defeat of *Napoleon. It continued to meet through the *Hundred Days of Napoleon's return to France (March–

The six decades that spanned the reign of Queen **Victoria** (1) saw the proliferation not only of Britain's possessions overseas, but also of the queen's descendants on the thrones of Europe. Of her nine children, the eldest, the Princess Royal (2) became the Empress of Germany and mother of Emperor William II (3). Victoria's eldest son, the Prince of Wales (4) was to succeed her on the British throne as Edward VII, while of her grandchildren, Princess Marie (5) married Prince Ferdinand (6), later the King of Romania, and Princess Alice of Hesse (7) married the Russian Emperor, Nicholas II (8). This photograph was taken in 1894 near the town of Coburg in eastern Germany, at the ducal country seat of Rosenau where Prince Albert, consort of Queen Victoria, was born.

Legend (Territorial changes determined by the Congress of Vienna):
- To Austria
- To Baden, Bavaria or Württemberg
- To Hanover
- To Prussia
- To Russia
- To Sardinia
- To Sweden
- To United Kingdom
- German Confederation

0 ___ 300 km.
0 ___ 200 miles

Congress of Vienna

Convened in 1814, it was attended by representatives from all the major European powers, but was dominated by the Austrian Chancellor Metternich. Its guiding principle was the notion of legitimacy, that is, restoration and strengthening of hereditary rulers. This resulted in three decades of despotic rule and the suppression of liberal and nationalist sentiments on the continent of Europe, particularly in Italy.

June 1815). The dominant powers were Austria, represented by *Metternich, Britain, represented by *Castlereagh, Prussia, represented by *Frederick William III, and Russia, represented by *Alexander I. *Talleyrand represented Louis XVIII of France. The Congress agreed to the absorption by the new kingdom of the Netherlands of what had been the Austrian Netherlands (now Belgium), but otherwise the Habsburgs regained control of all their domains, including Lombardy, Venetia, Tuscany, Parma, and *Tyrol. Prussia gained parts of Saxony as well as regaining much of Westphalia and the Rhineland. Denmark, which had allied itself with France, lost Norway to Sweden. In Italy the pope was restored to the *Vatican and the Papal States, and the Bourbons were re-established in the Kingdom of the Two *Sicilies. The *German Confederation was established, and Napoleon's Grand Duchy of Warsaw was to be replaced by a restored Kingdom of Poland, but as part of the Russian empire with the Russian emperor also king of Poland. The Congress restored political stability to Europe, but often at the cost of nationalist and liberal sentiments.

Vietcong, communist guerrilla organization operating in South Vietnam (1960–75). Opposition to the Saigon-based regime of *Ngo Dinh Diem had already produced widespread guerrilla activity in South Vietnam when communist interests founded the National Front for the Liberation of South Vietnam (known to its opponents as the Vietcong) in 1960. As US military support for the Saigon government broadened into the full-scale *Vietnam War so Vietcong forces were supplied with arms and supported by North Vietnamese forces brought to the south via the Ho Chi Minh Trail which passed through neighbouring Laos and Cambodia. They maintained intensive guerrilla operations, and occasionally fought large set-piece battles. They finally undermined both US support for the war and the morale of the South Vietnamese army and opened the way for communist triumph and the reunification of Vietnam in 1975.

Vietminh, Vietnamese communist guerrilla movement. Founded in 1941 in south China by *Ho Chi Minh and other exiled Vietnamese members of the Indo-Chinese Communist Party with the aim of expelling both the French and the Japanese from Vietnam, the Vietminh began operations, with assistance from the USA, against the Japanese in 1943–5 under the military leadership of Vo Nguyen Giap. After the end of World War II, it resisted the returning French, building up its strength and organization through incessant guerrilla operations and

finally winning a decisive set-piece engagement at *Dienbienphu in 1954. This forced the French to end the war
and grant independence to Vietnam, partitioned into two
states, North and South.

Vietnam, south-east Asian country. In 1802 the two
states of Annam and Tonkin were reunited by the Annamese general Nguyen Anh, who became emperor Gïa-
Long. Gia-Long was given French assistance and French
influence increased in the 19th century. By 1883 Vietnam
was part of *French Indo-China, although a weak monarchy was allowed to remain. In World War II the Japanese occupied it but allowed *Vichy France to
administer it until March 1945. In September 1945 *Ho
Chi Minh declared its independence, but this was followed
by French reoccupation and the *French Indo-Chinese
War. The *Geneva Conference (1954), convened to seek
a solution to the Indochina conflict, partitioned Vietnam
along the 17th parallel, leaving a communist Democratic
Republic with its capital at Hanoi in the north, and, after
the deposition of the former emperor *Bao Dai in 1955, a
non-communist republic with its capital at Saigon in the
south. Ho Chi Minh, the North Vietnamese leader, remained committed to a united communist country, and
by the time the South Vietnamese president *Ngo Dinh
Diem was overthrown by the military in 1963, communist
insurgents of the *Vietcong were already active in the
south. Communist attempts to take advantage of the political confusion in the south were accelerated by the in-
fusion of massive US military assistance, and in the late
1960s and early 1970s, the *Vietnam War raged throughout the area, with the heavy use of US airpower failing to
crush growing communist strength. Domestic pressures
helped accelerate a US withdrawal and after abortive
peace negotiations, the North Vietnamese and their Vietcong allies finally took Saigon in April 1975, a united
Socialist Republic of Vietnam being proclaimed in the
following year. Despite the severe damage done to the
economy, Vietnam adopted an aggressively pro-Soviet
foreign policy, dominating Laos, invading Kampuchea
to overthrow the *Khmer Rouge regime (1975-9), and
suffering heavily in a brief border war with China (1979).
Attempts to reorder society in the south of the country
also produced a flood of refugees, damaging Vietnam's
international standing and increasing its dependence on
the Soviet Union.

Vietnam War (1964-75), name generally given to that
part of the civil war in Vietnam after the commencement
of large-scale US military involvement in 1965. Guerrilla
activity in South Vietnam had become widespread by
1961, in which year President *Ngo Dinh Diem proclaimed a state of emergency. Continued communist activity against a country perceived in the USA as a bastion
against the spread of communism in south-east Asia led
to increasing US concern, and after a supposed North
Vietnamese attack on US warships in the Gulf of Tonkin
in 1964, President Johnson was given congressional approval (*Tonkin Gulf Resolution) to take military action.
By the summer of 1965 a US army of 125,000 men was
serving in the country, and by 1967 the figure had risen
to 400,000, while US aircraft carried out an intensive
bombing campaign against North Vietnam. Contingents
from South Korea, Australia, New Zealand, and Thailand
fought with the US troops. Although communist forces
were held temporarily in check, the war provoked massive

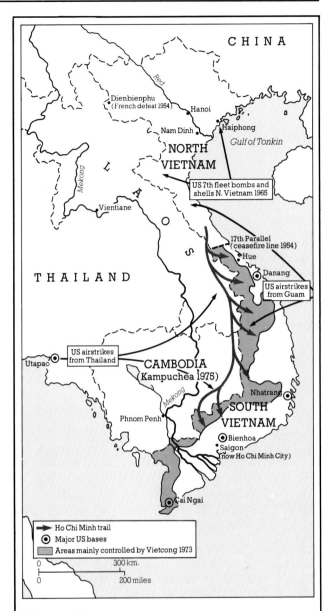

Vietnam War

By 1963 the communist leader of North Vietnam, Ho Chi
Minh, had established a network of Vietcong
(communist) insurgents in South Vietnam. In the USA
strategists argued that all south-east Asia was at risk
unless South Vietnam was buttressed against
communism, and by 1967 some half million US troops
were fighting there, with Thailand providing air-bases
for US raids against Vietcong strongholds. Two years
after the US withdrawal (1973) Saigon was captured and
renamed after Ho Chi Minh.

resentment within the USA, and after the *Tet Offensive
of February 1968 had shaken official belief in the possibility of victory, the bombing campaign was halted and
attempts to find a formula for peace talks started. US
policy now began to emphasize the 'Vietnamization' of
the war, and as increasing efforts were made to arm and
train the South Vietnamese army, so US troops were gradually withdrawn, although they were still periodically
caught up in heavy fighting in the early 1970s and the
bombing campaign was briefly resumed on several occasions. US troops were finally withdrawn after the Paris

Peace Accords of January 1973, but no lasting settlement between North and South proved possible, and in early 1975 North Vietnamese forces finally triumphed, capturing Saigon to end the war on 30 April 1975. The war did enormous damage to the socio-economic fabric of the Indochinese states, devastating Vietnam and destabilizing neighbouring Kampuchea (Cambodia) and Laos.

vigilante, a member of a self-appointed body for the maintenance of law and order in the US West. With the slow development of official policing, vigilance committees, organized and manned by local citizens, frequently took the law into their own hands, meting out rough justice and sometimes resorting to lynch law. They disappeared in the USA with the growth of official institutions in the last decades of the 19th century.

Villa, Francisco ('Pancho', b. Doroteo Arango) (1877–1923), Mexican revolutionary leader. He pursued an early career as a bandit and a merchant before taking up the cause of Francisco *Madero in the *Mexican Revolution of 1910. Along with Pascual Orozco, Villa provided the military leadership that was responsible for Madero's defeat of Porfirio *Díaz. After Madero's assassination in 1913, Villa first joined and then broke with the constitutionalist opposition to the usurper Victoriano Huerta. During 1914–15, Villa held sway in Chihuahua with his cavalry, *los dorados*, expropriating the holdings of large landowners and using their revenues to equip the revolutionary army in Mexico. By 1916 he had been defeated by Carramaza and other revolutionary factions. With *Zapata, he moved north, and for reasons still controversial, ordered an attack (1916) on the US town of Columbus, New Mexico, killing seventeen Americans and provoking retaliation from a punitive expedition under General *Pershing. He escaped and lived in retirement until he was assassinated.

Villafranca di Verona, Treaty of (1859), an agreement between France and Austria. After the battles of *Magenta and *Solferino, *Napoleon III and *Francis Joseph II met at Villafranca, where the Austrians agreed to an armistice. Austria handed Lombardy over to France, who later passed it to Sardinia (Piedmont) but retained Venetia. The rulers of the central Italian duchies were restored. Piedmont acquiesced and *Cavour resigned.

Vimy Ridge, battle of (9 April 1917), an Allied attack on a German position in World War I, near Arras in France. One of the key points on the *Western Front, it had long resisted Allied attacks. Canadian troops under General Byng and commanded by General Horne launched an offensive. In fifteen minutes, despite heavy casualties, most of Vimy Ridge was captured and 4,000 prisoners taken. The Allied offensive was unsuccessful elsewhere and by 5 May had ground to a halt.

Virginia Campaigns (July 1861–5), a series of engagements and campaigns in the *American Civil War. The first engagement of the Civil War was fought on 21 July 1861 at the first battle of Bull Run. In a confused mêlée the *Confederacy was saved from defeat by the brigade of 'Stonewall' *Jackson. In the Peninsula Campaign of April–June 1862 Union (Northern) forces under General McClellan attempted to advance up the peninsula between the James and York rivers to capture Rich-

mond, but in the Seven Days battle (26 June–2 July) he was forced to withdraw by the Southern commander, General Robert E. *Lee. At the same time in the Shenandoah Valley a brilliant campaign by 'Stonewall' Jackson pinned down Union forces. The second battle of Bull Run followed (29–31 August), when Lee forced the Union army to retreat to Washington. The way was open for an invasion of the North, but it ended in defeat at the battle of *Antietam (17 September). Following the Confederate army's escape back to Virginia, the new Union commander, General Ambrose Burnside, launched an assault on Lee's positions above Fredericksburg (13 December 1862). Burnside withdrew, the reverse severely shaking the Union war effort. In the spring of 1863 a reinforced Union army under General Joseph Hooker, resumed the offensive in the battle of Chancellorsville (2–4 May 1863). Lee withstood the assault, but suffered heavy casualties, including the death of Jackson. Lee now invaded Pennsylvania, but suffered the major defeat of *Gettysburg, after which he was on the defensive for the rest of the war. A series of engagements was fought in May and June 1864 in the 'wilderness' region of Virginia, when General *Grant was defeated three times before retreating across the River James, to renew his attacks in the Petersburg Campaign. This last campaign was launched in June 1864 and continued into 1865. Three assaults on Richmond by Grant were repelled by Lee, after which Union forces besieged the Confederate capital through the winter. Lee was thus prevented from sending reinforcements south to repel *Sherman's advance through Georgia and on 1 April he was defeated at Five Forks and forced to abandon both Richmond and Petersburg. All but surrounded, he surrendered at *Appomattox on 9 April, bringing the campaign, and the war, to an end.

Virgin Islands, a group of Caribbean islands at the eastern extremity of the Greater Antilles. Discovered by Columbus in 1493, effective settlement, primarily by British and Danish planters, did not occur until the 17th century. Descendants of African slaves imported for the sugar plantation economy account for the majority of the population of the islands today. In 1917 Denmark sold its possessions to the USA, interested in their strategic value. The British Virgin Island group at the northern end of the Leeward Islands is smaller, with Tortola the largest island. From 1872 to 1956 this group was part of the British colony of Leeward Islands, and since 1956 has been administered separately by British governors or administrators who have gradually extended self-government.

Virginius incident, (1873), the capture of an arms-running ship fraudulently flying the US flag during the Cuban rebellion against Spain (1868–78). Seized off the coast of Jamaica by a Spanish gunboat, the *Virginius* was taken to Santiago de Cuba. Subsequently her commander, Captain Fry, a US citizen, and fifty-two others, claiming to be Americans, were executed as *filibusters (persons engaged in unauthorized warfare against a foreign state). Despite the angry reaction of the USA the dispute was resolved by compromise and Spain paid the USA an indemnity of $80,000.

Vitoria, battle of (21 June 1813), fought between the French and the British near the Basque city of Vitoria, during the *Peninsular War. *Wellington decisively defeated the French under Joseph Bonaparte and Jourdan. News of this victory inspired Austria, Russia, and Prussia to renew their plans to attack France and they declared war on *Napoleon on 13 August.

Vittorio Veneto, battle of (October 1918), the scene of a decisive victory in World War I by the Italians over the Austrians. A town in north-east Italy, it is named after *Victor Emanuel II, in whose reign Venetia was regained from Austria after the *Six Weeks War (1866). Italian forces under General Diaz avenged the *Caporetto disaster (1917) by routing the Austro-Hungarian army, which resulted in an Austrian request for an armistice.

Vivekananda, Swami (b. Narendranath Datta) (1863–1902), Hindu monk who preached the Hindu philosophy of Vedanta in Europe and the USA and established the Ramakrishna Mission, named after his guru (spiritual teacher) Ramakrishna Paramahamsa (1836–86), a mystic who preached that all religions contained the same fundamental truth. A mystic, he renounced the world and, as a wandering *sadhu* (religious man), discovered the poverty of the Indian masses. He went to the USA to raise funds to establish an institution for educating the masses. After his speech at the Chicago Parliament of Religions (1893), he became a celebrity in the West. His patriotic speeches and writings inspired youthful Indian nationalists, especially the early revolutionaries.

Vogel, Sir Julius (1835–99), New Zealand statesman. He came to Otago with the gold rushes, and soon dominated provincial politics. He entered national politics in 1863 and, as Colonial Treasurer from 1869 and Premier (1872–4), was responsible for a bold and successful policy of borrowing to promote immigration, road and railway building, and land development. He was responsible for the establishment of the Government Life Insurance Office and the Public Trust, thus launching a tradition of state involvement for which New Zealand is noted. He served as Treasurer again (1884–7).

Volstead Act (1919), US federal *prohibition act, enforcing the Eighteenth Amendment, banning the manufacture, distribution, and sale of alcohol. The Act was devised by the Anti-Saloon League counsel Wayne Wheeler but named after Congressman Andrew Volstead of Minnesota. It proscribed beer and wine as well as distilled spirits, to the surprise of those moderate prohibitionists who wanted the prohibition of spirits only. It became void by the passage of the Twenty-first Amendment in 1933.

Voortrekkers *Great Trek.

Wade, Benjamin Franklin (1800–78), US statesman. He served in the Senate from 1851 to 1869, first as a Whig, and then as a Republican, generally on the radical wing of the Party. He sponsored the Wade–Davis Bill (1864), an abortive blueprint for the *reconstruction of the South which was vetoed by President *Lincoln. Wade played an active role in the attempted impeachment of Andrew *Johnson, and as acting president of the Senate would have succeeded him in the White House had the move succeeded; but its failure and Wade's own defeat in the next election marked the end of his political career.

Wafd (Wafd al-Misri, Arabic, 'Egyptian Delegation'), Egyptian nationalist party. Under the leadership of Zaghlul Pasha it demanded freedom from British rule. When Egypt won nominal independence in 1922, the Wafd organized a political party, demanding full autonomy and control of the *Sudan and the *Suez Canal. After 1924 there were frequent Wafdist governments, in conflict with the monarchy. In 1930 the constitution was suspended and Egypt became a royal dictatorship until the Wafdists succeeded in restoring the constitution in 1935. In 1950 the Wafd formed a one-party cabinet and the struggle between King *Farouk and his government intensified. The monarchy fell in 1952 and the new Revolutionary Command Council under Colonel Gamal *Nasser dissolved all political parties.

Wagner–Connery Act (1935), US labour Act. Its official title was the National Labor Relations Act, and it was introduced by Senator Robert Wagner of New York. It aimed to outlaw employer-dominated trade unions and to provide for enforcement of the right of free collective bargaining. It established a National Labor Relations Board, with powers to supervise and conduct elections in which workers would select the union to represent them. The Board survived various attempts to secure a Supreme Court ruling of its unconstitutionality, and it served to increase trade union membership.

Wagram, battle of (5–6 July 1809), a battle in the *Napoleonic Wars fought between the combined French and Italian forces led by *Napoleon, and the Austrians under the Archduke Charles, at the village of Wagram, near Vienna. Napoleon, determined to offset earlier setbacks, ordered a massive attack on the well-chosen Austrian position. The first day's fighting was inconclusive, and the following day Napoleon renewed his bloody assault and the Austrian army finally began to retreat. The French, who had been close to defeat, claimed the victory, but their losses outnumbered those of the Austrians. It was followed by the Treaty of Schönbrunn, in which Austria lost territory and agreed to join the *Continental System against Britain. Napoleon married the Austrian Princess Marie-Louise in 1810.

Wahhabism, the doctrine of an Islamic reform movement. Founded by Muhammad ibn 'Abd al-Wahhab (1703–92) in Nejd, Saudi Arabia, it is based on the Sunni teachings of Ibn Hanbal (780-855), involving puritanism, monotheism and rejection of popular cults. Under the leadership of the *Saud family, the Wahhabis raided into the Hejaz, Iraq, and Syria, capturing Mecca in 1806. They were crushed by Ottoman forces in a series of campaigns (1812-18), but the *Saud family gradually consolidated its power within the peninsular.

Waitangi, Treaty of (6 February 1840), a treaty between *Maori chiefs and the British government signed at Waitangi, New Zealand. Some 500 local Maori chiefs of the North Island signed the document drawn up by the governor, recognizing Queen Victoria's sovereignty over New Zealand in return for recognition of the Maori's *rangatiratanga* (chieftainship) and land rights, and their rights as British subjects. The treaty cleared the way for a formal declaration of sovereignty on 21 May 1840. Subsequent encroachment on their lands led to the *Anglo-Maori Wars of 1860-72 in which Maori independence was overcome. Since its recognition by New Zealand statutes the treaty has since 1975 assumed new importance as a basis of relations between Maori and non-Maori New Zealanders.

Wakefield, Edward Gibbon (1796-1862), British colonial reformer and writer. In 1829 he published his *Letter from Sydney* using information he had obtained while serving a sentence in Newgate gaol in London. Concerned that Australian settlements were failing because land could be acquired so easily, he proposed a 'sufficient price' for land, which would finance the regulated emigration of labourers and oblige them to work to buy their own land. This would give a balanced colonial society and provide some relief to unemployment in Britain. His ideas were taken up and implemented from 1831, with some 70,000 migrants travelling to Australia in the next ten years. In 1837 Wakefield founded the *New Zealand Association (later Company). He was largely responsible for the succession of systematic settlements in New Zealand. He wrestled for years for self-government for the colonists, emigrating to New Zealand in 1853 and involving himself in colonial politics until ill-health intervened.

Wales, a country in the western part of Britain. Since the political union of Wales with England (1536), the separate history of Wales has been mainly religious and cultural. The strong hold of Nonconformity, especially of the Baptists and Methodists, made the formal position of the Anglican Church there the dominant question of Welsh politics in the later 19th century, leading to the *disestablishment of the Church from 1920. The social unrest of rural Wales, voiced in the *Rebecca riots, resulted in significant emigration. The *Industrial Revolution brought prosperity to South Wales, shown, for example, in the boom in coal-mining, which by the late 19th century had transformed Wales into the world's chief coal exporting region. The depletion of the coal seams brought about the closure of most of the coalfields by the 1980s. The introduction of a more diversified industry has alleviated some unemployment problems. Political, cultural, and linguistic nationalism survive, and have manifested themselves in the *Plaid Cymru Party, the National Eisteddfod, and Welsh-language campaigns. A Welsh referendum in 1979 voted overwhelmingly against partial devolution from the United Kingdom.

Wallace, George Corley (1919-), US politician and state governor. He became an Alabama state congressman (1947-53) and a District Judge (1953-8). When first elected governor of Alabama (1963-6) he resisted the *desegregation of state schools and universities. He stood in the Presidential campaign of 1968 as leader of the newly established *Dixiecrat American Independent Party. His main support was in the South, and he polled over ten million votes. In 1972 he sought the Democratic Party's presidential nomination, but his campaign ended when an assassination attempt left him paralysed. He stood again unsuccessfully in 1976, by which time he was becoming reconciled to the issue of *civil rights. For the 1982 election as governor he publicly recanted his opposition to desegregation, polled a substantial number of black votes, and was re-elected.

Wallace, Henry Agard (1888-1965), US agricultural reformer and statesman. Having developed successful varieties of corn grown throughout the USA, in 1932 he helped swing the state of Iowa to the Democratic Party. In 1933 he became Franklin D. *Roosevelt's Secretary of Agriculture, in 1940 his Vice-President, and in 1944 Secretary of Commerce in Roosevelt's last administration, continuing under *Truman. A visionary liberal, Wallace soon fell out with Truman's *Cold War policy, and resigned in 1946. Wallace then moved considerably to the left, exposing himself to charges of 'fellow travelling' with the communists. In 1948 he formed his own 'Progressive Party' and ran against Truman for the Presidency. He won only 1.2 million votes and carried no state. He then retired to continue his agricultural research.

Ward, Sir Joseph George (1856-1930), New Zealand statesman. A minister of the first and successive Liberal cabinets, he was noted for successful loan raising, and for provision of low-interest credit to farmers. As Prime Minister (1906-12) and in coalition with *Massey (1915-19) he supported waning concepts of empire unity in defence and foreign affairs. He won office as Prime Minister again in 1928 as head of the United Party, partly on his reputation as a 'financial wizard', but was unable to solve the crises brought on by the Great *Depression.

warlord, Chinese regional military ruler of the first half of the 20th century. Following the death of *Yuan Shikai in 1916, China was divided among many local rulers who derived their power from control of personal armies. In origin, the warlords were mostly former soldiers of the imperial and republican armies, bandits, or local officials. They depended on revenue from towns and agricultural areas in their own spheres of influence to feed the well-equipped troops with which they sought to establish their primacy over local rivals. The most successful generally controlled easily defended areas, and the largest of the many wars between rival cliques of warlords witnessed the mobilization of hundreds of thousands of soldiers. *Chiang Kai-shek's Nanjing government (1928-37) re-established central authority over most warlord areas, but military rulers persisted in the far west of China into the 1940s.

War of 1812 (1812-15), a war between Britain and the USA. US frustration at the trade restrictions imposed by

Britain in retaliation for Napoleon's *Continental System, together with a desire to remove British and Canadian obstacles to US westward expansion, led the US Congress to declare war on Britain (June 1812). The USA–British North American (Canadian) border was the main theatre of war. In July, the US General, William Hull, advanced into Upper Canada, but in early August withdrew to Detroit, which was soon after captured by Major-General Isaac Brock. In October 1812 a second invading US force crossed the Niagara River and stormed Queenston Heights, but it too was driven back by a British force, under Brock, who was killed. In October 1813 another US army under General William Harrison won the battle of the *Thames, in southwestern Ontario, at which the Indian leader *Tecumseh was killed. Later that month a US army, intending to take Montreal, was defeated at Châteauguay; in November US troops were defeated by a mnuch smaller British force at Crysler's Farm on the St Lawrence. In July 1814, at the battle of Lundy's Lane, a US force under General Jacob Brown briefly fought at night a British force under General Drummond and then withdrew, after which no more attempts were made to invade Canada. On Lake *Erie in September 1813 a US force captured a British squadron of six ships, while the following year (September 1814), in a similar victory on Lake Champlain, a British squadron of sixteen ships was forced to surrender. At sea US warships won a series of single-ship engagements, but they were unable to disrupt the British naval blockade, which by 1814 was doing considerable harm to the US economy. In June 1814 a British expeditionary force landed in Chesapeake Bay, Virginia, marching north and burning the new city of Washington. War-weariness now brought the two sides to the conference table and in December 1814 the Treaty of *Ghent was signed, restoring all conquered territories to their original owners.

Warren, Earl (1891–1974), US judge and public official. He had been attorney-general and governor of California when in 1948 he became Republican Vice-Presidential candidate. Eisenhower appointed him Chief Justice of the US Supreme Court (1953–69). In 1954 he wrote the Supreme Court ruling that racial segregation in public schools was unconstitutional. In 1964 he was appointed chairman of the Commission to investigate the assassination of President *Kennedy. By establishing the sole responsibility of Lee Harvey *Oswald, the Commission went far to allay the nation's fears of either communist or extreme right-wing conspiracies.

Warsaw Pact, formally Warsaw Treaty of Friendship, Co-operation and Mutual Assistance, a military alliance between Soviet-bloc powers. It was signed in 1955 by Albania, Bulgaria, Czechoslovakia, the German Democratic Republic, Hungary, Poland, Romania, and the Soviet Union after the Paris agreement between the Western powers admitting the Federal Republic of Germany to *NATO. Albania formally withdrew in 1968. The pact provides for a unified military command, the maintenance of Soviet Army units on member states, and mutual assistance, the latter provision being used by the Soviet Union to launch a multi-national invasion of *Czechoslovakia against the *Dubček regime in 1968.

Warsaw Rising (August–October 1944), Polish insurrection in Warsaw in World War II, in which Poles tried to expel the German Army before Soviet forces occupied the city. As the Red Army advanced, Soviet contacts in Warsaw encouraged the underground Home Army, supported by the exiled Polish government in London, to stage an uprising. Polish *resistance troops led by General Tadeusz Komorowski gained control of the city against a weak German garrison. Heavy German air-raids lasting sixty-three days preceded a strong German counter-attack. The Soviet Army under *Rokossovsky reached a suburb of the city but failed to give help to the insurgents, or allow the western Allies to use Soviet air bases to airlift supplies to the hard-pressed Poles. Supplies ran out and on 2 October the Poles surrendered. The Germans then systematically deported Warsaw's population and destroyed the city itself. The main body of Poles that supported the Polish government in exile was thus destroyed, and an organized alternative to Soviet political domination of the country was eliminated. As the Red Army resumed its advance into Poland the Soviet-sponsored Polish Committee of National Liberation was able to impose a Communist Provisional Government on Poland (1 January 1945) without resistance.

Washakie (c.1804–1900), Shoshone American Indian chief. He became the leader of the eastern band of his tribe in the 1840s. He chose to ally with whites in the Indian wars of the 1870s, and was accorded a commission in the US army and a tomb in Fort Washakie.

Washington, Booker T(aliaferro) (1856–1915), US educator. The son of a Negro slave and a white father,

Booker T. **Washington**, US educator and leader of the movement to secure better schooling for blacks in the USA.

he was the undeclared leader of those blacks who favoured 'gradualism' as the route to integration and *civil rights. He created economic opportunities for blacks from his base at Tuskegee Industrial Institute, Alabama, by training them as farmers, mechanics, and domestics. In a widely publicized speech of 1895, known as the Atlanta Compromise, he abandoned the racial equality for blacks as a priority in favour of material progress. He was criticized for his emphasis on vocational skills to the detriment of academic development and civil rights by militant intellectuals such as W. E. B. *Du Bois. Yet his insistence on industrial education encouraged white businessmen to subsidize black institutions, to provide him with funds for the National Negro Business League of 1900, and to finance court cases against segregation.

Washington Conference, conference held in the USA between November 1921 and February 1922 to discuss political stability in the Far East and naval disarmament. Summoned on US initiative, the conference was attended by Belgium, Britain, China, France, Holland, Italy, Japan, Portugal, and the USA and resulted in a series of treaties including a Nine-Power Treaty guaranteeing China's independence and territorial integrity, a Japanese undertaking to return the region around Qingdao to Chinese possession, and an Anglo-French–Japanese-US agreement to guarantee each other's existing Pacific territories. Naval discussions resulted in a ten-year moratorium on capital-ship construction. The Washington Conference successfully placed restraints on both the naval arms race and Japanese expansionism, but by the 1930s both problems broke out afresh.

Watergate scandal, a major US political scandal. In 1972 five men were arrested for breaking into the headquarters of the Democratic Party's National Committee in the Watergate building, Washington, DC, in order to wire-tap its meetings. It was soon discovered that their actions formed part of a campaign to help President *Nixon win re-election in 1972. At first the White House denied all knowledge of the incident, but after intensive investigations, initially led by journalists on the *Washington Post*, it became apparent that several of the President's staff had been involved in illegal activities and an attempt to cover up the whole operation. Several White House officials and aides were prosecuted and convicted on criminal charges. Attention then focused on President Nixon, and as extracts from tapes of White House conversations were released, it became clear that he too had been involved. In August 1974 Nixon resigned to avoid impeachment, but he was pardoned for any federal offences he might have committed by the new President, Gerald R. Ford (*Ford Presidency).

Waterloo, battle of (18 June 1815), decisive battle between French and British and Prussian forces near the Belgian village of Waterloo. It was fought during the *Hundred Days of *Napoleon between his hastily recruited army of 72,000 men and *Wellington's Allied army of 68,000 men (with British, Dutch, Belgian, and German units) before the Prussians (45,000 men) arrived. There had been a violent storm in the night, and Napoleon postponed his attack until midday to allow the ground to dry. By 2 p.m. a first contingent of Prussians arrived, and attacked Napoleon on the right. At 6 p.m. Marshal *Ney ordered a co-ordinated attack and

captured La Haye Sainte, a farmhouse in the centre of the Allied line. The French artillery then began attacking the Allies from the centre. At 7 p.m. Napoleon launched his famous Garde Impériale in a bid to break Wellington's now weakened infantry. At this point, however, *Blücher appeared with the main Prussian forces, taking Napoleon in the flank, and Wellington ordered a general advance. The French were routed, with the exception of the Garde, who resisted to the end. In Wellington's words, the outcome of the battle was 'the nearest run thing you ever saw in your life'. On 22 June, Napoleon signed his second and final abdication.

Wavell, Archibald Percival, 1st Earl (1883–1950), British field-marshal and viceroy of India. In World War I he served in France (where he lost an eye) and in 1937–9 he commanded the British forces in Palestine. In 1939 he became commander-in-chief in the Middle East and won the victory of Sidi Barrani (1940) over the Italians. In 1941, forced to divert some of his forces to Greece, and facing new German formations, he had to retreat in *North Africa and was dismissed by Churchill. He then served in India, first as commander-in-chief from 1941, then as viceroy (1943–7), where he made it his main task to prepare India for independence.

Webb, Sidney James, Baron Passfield (1859–1947), social reformer and historian. Initially a civil servant he, together with his wife Beatrice (b. Potter), became a leading member of the *Fabian Society. As members of the Royal Commission on the Poor Law (1905–9) they were the moving spirit behind its minority report that poverty should be dealt with by setting up government organizations to concentrate on specific causes of poverty. This later became the basic approach by British governments to the problem. Two British institutions, the London School of Economics and Political Science (1895) and the journal, the *New Statesman*, owe their foundation to Sidney Webb. His writings, many of them joint studies with his wife, on trade unionism and local government, exerted considerable influence on political theory and

Fierce fighting took place around the farmhouse of La Haye Sainte during the battle of **Waterloo**, 1815. A French infantry attack was repulsed by Wellington's cavalry, and British infantry then advanced. This illustration shows British troops moving forward past the farmhouse, while cavalry engage at close quarters in the foreground.

social reform. He helped to found the Labour Party, served as a Labour Member of Parliament (1922-9), and was Dominion and Colonial Secretary in the Labour government (1929-31).

Webster, Daniel (1782-1852), US orator, lawyer, and statesman. A native of New Hampshire, he served as a Congressman (1813-17) and Senator (1827-41, 1845-50), and secured his reputation by winning a series of important cases before the Supreme Court. His Second Reply to Senator Hayne of South Carolina in 1830 was a staunch defence of the Union. Through the 1830s and 1840s Webster was one of the leaders of the Whig Party. He failed to fulfil his presidential ambitions in successive elections from 1836 to 1852, but served as Secretary of State (1841-3) under Harrison and Tyler, during which time he was responsible for the *Webster–Ashburton Treaty, and again as Secretary of State under Fillmore (1850-2).

Webster–Ashburton Treaty (1842), an agreement between Britain and the USA settling the present Maine-New Brunswick border. Negotiated by the US Secretary of State Daniel *Webster and the British minister Lord Ashburton, the treaty settled the disputed boundaries in the north-east, awarding the USA more than 18,000 sq. km. (7,000 sq. miles) of the 31,000 sq. km. (12,000 sq. miles) disputed area, and opening the St John River to free navigation. The treaty also fixed the Canadian–US boundary in the Great Lakes region, and served as a precedent for the successful settlement of other 19th-century Anglo-US border disputes.

Weimar Republic (1919-33), a term used to describe the republic of Germany formed after the end of World War I. On 9 November 1918 a republic was proclaimed in Berlin under the moderate socialist Friedrich Ebert. An elected National Assembly met in January 1919 in the city of Weimar and agreed on a constitution. Ebert was elected first President (1919-25), succeeded by *Hindenburg (1925-34). The new republic had almost at once to face the *Versailles Peace Settlement, involving the loss of continental territory and of all overseas colonies and the likelihood of a vast reparations debt, the terms being so unpopular as to provoke a brief right-wing revolt, the *Kapp putsch. Unable to meet reparation costs, the mark collapsed, whereupon France and Belgium occupied the Ruhr in 1923, while in *Bavaria right-wing extremists (including *Hitler and *Ludendorff) unsuccessfully tried to restore the monarchy. Gustav *Streseman succeeded in restoring confidence and in persuading the USA to act as mediator. The *Dawes Plan adjusted reparation payments, and France withdrew from the Ruhr. It was followed in 1929 by the *Young Plan. Discontented financial and industrial groups in the German National Party allied with Hitler's *Nazi Party to form a powerful opposition. As unemployment developed, support for this alliance grew, perceived as the only alternative to communism. In the presidential elections of 1932 Hitler gained some thirteen million votes, exploiting anti-communist fears and anti-Semitic prejudice, although Hindenburg was himself re-elected. In 1933 he was persuaded to accept Hitler as Chancellor. Shortly after the *Reichstag fire, he declared a state of emergency (28 February 1933), and on Hindenburg's death in 1934, made himself President and proclaimed the *Third Reich.

The swearing-in of Russian-born Chaim **Weizmann** as Israel's first President. He played a crucial role in persuading the US government to recognize the new state and grant it an initial loan of $100,000,000.

Weizmann, Chaim (Azriel) (1874-1952), Zionist leader and scientist. Born in Poland, he became a British subject in 1910. During World War I his scientific work brought him to the notice of *Lloyd George. Weizmann exploited his contacts to help to obtain the *Balfour Declaration in 1917. At the *Versailles Peace Conference Weizmann was the chief spokesman for *Zionism and thereafter, as President of the World Zionist Organization, he was the principal negotiator with the British and other governments. He played a major role in shaping the *Palestine mandate and in frustrating the British attempt to restrict Jewish immigration and land purchase in the 1930s. When the state of Israel came into being (1948), Weizmann became its first President (1948-52).

Welensky, Sir Roy (Roland) (1907-84), Rhodesian statesman. He entered politics in 1938, and founded the Federal Party in 1953, dedicated to 'racial partnership'. He was an advocate of the *Central African Federation, which was created largely as a result of his negotiations. He was Prime Minister of the Federation from 1956 to 1963. When the Federation was dissolved (1963) Welensky lost the support of the white Rhodesians, who gave their allegiance to the Rhodesian Front of Ian Smith.

The Duke of **Wellington**, a watercolour portrait by T. Heaphy. The victor of Waterloo held that the British army must always be recruited from 'the scum of the earth', and that corporal punishment was indispensable for maintaining discipline. (National Portrait Gallery, London)

welfare state, a term believed to have been coined by Archbishop *Temple in 1941, loosely used to describe a country with a comprehensive system of social welfare funded both by taxation and schemes of *national insurance. The emergence of the strong secular state in 19th-century Europe was characterized by the development of state involvement in an increasing number of areas of social activity, for example education, public health, and housing. Public education systems were first introduced in France and Prussia early in the 19th century, while the need for housing and public health measures accelerated as *urbanization increased in Europe. A scheme of social insurance against unemployment, sickness, and old age was pioneered in Germany under *Bismarck, and other European states soon followed. In Britain a similar scheme, together with other social welfare measures, began to be introduced under the Liberal governments (1906-14). Between the Wars significant developments towards its establishment took place in New Zealand under the New Zealand *Labour Party, while F. D. Roosevelt's *New Deal in the USA created a series of federal social welfare agencies. In 1942 a report by William *Beveridge proposed that the British system of national social insurance be extended to provide for the entire population 'from the cradle to the grave'. His proposals were implemented by the *Attlee Ministries after World War II which added other reforms such as the National Health Service. In the Soviet Union and East European states state welfare provision became an official part of the fabric of society. In the USA and elsewhere in the Western World, the concept of social welfare support remains selective, to be given only to 'those in need'. In Britain, the heavy public expenditure required to distribute social benefits irrespective of means was increasingly challenged from the mid-1970s, and selective health charges were introduced.

Wellington, Arthur Wellesley, 1st Duke of (1769-1852), British soldier and statesman. He joined the army in 1787. He saw action in Flanders in 1794-5 before being posted to India in 1796, where he was to distinguish himself both as a soldier and as an administrator, winning a notable victory over the *Marathas at Assaye in 1803. On his return home he was knighted. In 1808 he was sent to Portugal to lead an army against the French. Throughout the *Peninsular War he adopted defensive tactics, which aroused much criticism in Britain, but eventually brought about the expulsion of the French from Spain after the victories of Salamanca (1812) and Vitoria (1813). Created Duke of Wellington (1814) in acknowledgement of his services, he attended the Congress of *Vienna and subsequently commanded the forces which defeated Napoleon at the battle of *Waterloo in June 1815. In 1818 he embarked upon a political career, entering Lord *Liverpool's government as Master-General of Ordnance, and he represented Britain at the Congress of Verona in 1823. As Prime Minister (1828-30) he reluctantly agreed to *Catholic emancipation, but his refusal to contemplate parliamentary reform of any kind made him extremely unpopular and he resigned. He served briefly as acting Prime Minister in 1834 and then as Foreign Secretary (1834-5) under Sir Robert *Peel. In 1841 he again joined Peel's government, this time as a minister without portfolio. Wellington retired from public life in 1846 but in 1848 as commander-in-chief of the army he organized a military force to protect London against possible *Chartist violence.

Wells, Fargo and Company, US transport organization, founded by Henry Wells, William C. Fargo, and associates in 1852 to operate between New York and California. Wells and Fargo established a monopoly west of the Mississippi within a decade, succeeding the *Pony Express as the agency for transporting bullion to eastern markets, and for twenty years dominated the postal service in the West. In 1918 it merged with a number of other concerns to become the American Railway Express Company.

Welsh Nationalist Party   *Plaid Cymru.

Western Australia, a state of the Commonwealth of *Australia. In 1826, Governor Darling, acting on British instructions, sent soldiers and convicts to King George Sound, but they were withdrawn in 1831. In 1829, Britain founded the new free colony of Western Australia. The colony's progress was slow, inadequate labour being a major problem. The colonists eventually asked for convicts. Approximately 10,000 (all male) were transported to Western Australia between 1850 and 1868. The colony was granted parliamentary government in 1890. Gold discoveries, notably at Kalgoorlie, increased the population and boosted the economy. Western Australia became a state of the Australian Commonwealth in 1901. Western Australians voted in favour of secession at a referendum held in 1933, but Britain rejected the petition the following year. Western Australia experienced a significant mining boom in the 1960s.

Western European Union, a West European defence organization consisting of Belgium, France, West Germany, Britain, Italy, Luxemburg, and the Netherlands. Founded in 1954 after France had refused to ratify the treaty providing for a European Defence Community, the primary function of the WEU was to supervise the rearmament of the German Federal Republic and its accession to *NATO. The Union formally ended the occupation of West Germany and Italy by the Allies. The social and cultural activities initially envisaged by its founders were transferred to the *Council of Europe in 1960, leaving the Union with the task of improving defence co-operation among the countries of Western

Europe. In 1987 it actively concerned itself with a European nuclear defence policy and *arms control.

Western Front, line of fighting in World War I stretching from the Vosges mountains through Amiens in France on to Ostend in Belgium. Fighting in World War I began in August 1914 when German forces, adopting the *Schlieffen Plan, were checked in the first battle of the *Marne. The subsequent German attempt to reach the Channel ports was defeated in the first battle of Ypres (12 October–11 November). Thereafter both sides settled down to *trench warfare, the distinctive feature of fighting on this front. The year 1915 saw inconclusive battles with heavy casualties: Neuve Chapelle (March), the second battle of Ypres (April/May), when poison gas was used for the first time, and Loos (September). In 1916 Germany's heavy attack on Verdun nearly destroyed the French army but failed to secure a breakthrough. To relieve pressure on the French, the British bore the brunt of the *Somme offensive (July), gaining little ground for appalling casualties. Early in 1917 the Germans withdrew to a new set of prepared trenches, the *Siegfried (or Hindenburg) Line, and in

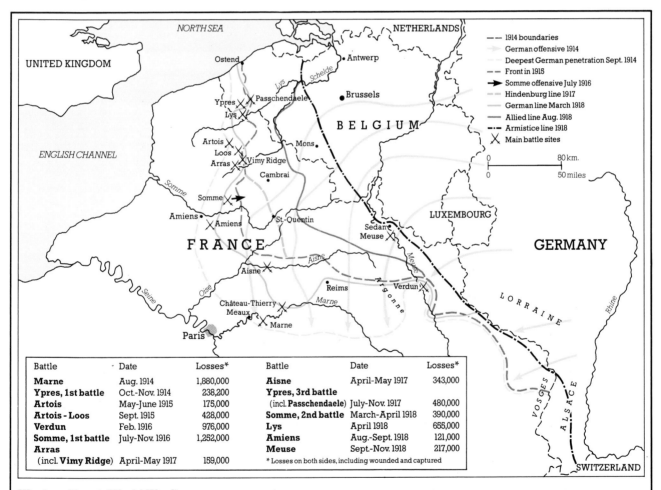

Battle	Date	Losses*	Battle	Date	Losses*
Marne	Aug. 1914	1,880,000	**Aisne**	April-May 1917	343,000
Ypres, 1st battle	Oct.-Nov. 1914	238,200	**Ypres, 3rd battle**		
Artois	May-June 1915	175,000	(incl. **Passchendaele**)	July-Nov. 1917	480,000
Artois - Loos	Sept. 1915	428,000	**Somme, 2nd battle**	March-April 1918	390,000
Verdun	Feb. 1916	976,000	**Lys**	April 1918	655,000
Somme, 1st battle	July-Nov. 1916	1,252,000	**Amiens**	Aug.-Sept. 1918	121,000
Arras			**Meuse**	Sept.-Nov. 1918	217,000
(incl. **Vimy Ridge**)	April-May 1917	159,000	* Losses on both sides, including wounded and captured		

Western Front (World War I)

The Schlieffen Plan, by which the German army planned to encircle and capture Paris early in the war, was checked by the Allies in the first battle of the Marne. An almost stationary situation of trench warfare then ensued. Stalemate continued into 1918, when a massive German offensive was again checked on the Marne. By now US troops and equipment were reinforcing the Allies, who finally advanced into Belgium in September, an armistice following on 11 November. Each year hundreds of thousands of men died for mere acres of mud, as the Western Front moved backwards and forwards.

1917 the Canadians captured *Vimy Ridge. In November the British launched yet another major offensive, the battle of *Passchendaele or third battle of Ypres, at the cost of 300,000 men lost. The entry of the USA into the war (1917) meant that the Allies could now draw on its considerable resources. US troops commanded by General Pershing landed in France in June 1917. In March 1918 *Ludendorff's final offensive began, with his troops again reaching the Marne before being stemmed by the Americans at Château-Thierry. *Foch, now Allied commander-in-chief, began the counter-offensive with the third battle of the Marne (July). British troops broke the Siegfried Line near St Quentin, while the Americans attacked through the Argonne region. By October Germany's resources were exhausted, and on 11 November Germany signed the armistice that marked the end of World War I.

West Indian independence, a movement towards independence among British West Indian colonies. Pressure towards greater participation in the government of British West Indian colonies developed in the 19th century. A Negro rising in Jamaica had been ruthlessly suppressed in 1865, but following a Jamaican deputation to London in 1884, elected legislatures, on a limited franchise to advise governors, were steadily introduced throughout the islands. After World War I there were further moves towards more representative government, for example, in Trinidad in 1923. By 1940 the British government, aware of the strategic importance of the area, established a Commission for Development and Welfare in the West Indies, substantial financial aid being given and the principle of self-government being accepted. This policy did not, however, prevent large-scale unrest after World War II. Britain believed that individual islands could never be viable as independent states, hence the concept of federation and the attempt to form the West Indies Federation (1958-62). When this failed, *Jamaica and Trinidad/Tobago were granted full independence in 1962, and Barbados in 1966. New attempts to create a Federation of the East Caribbean also failed (1967), when many smaller islands temporarily became 'associated states of the United Kingdom'. Since then six new member states of the *Commonwealth of Nations have emerged: *Grenada 1974; Dominica 1978; St Vincent 1979; St Lucia 1979; Antigua/Barbuda 1981; St Kitts-Nevis 1983. The island of Anguilla was first associated with St Kitts-Nevis, but unsuccessfully declared its independence in 1967, being obliged to resume the status of an independent territory of the United Kingdom in 1969.

West Indies, three groups of islands in the West Atlantic and the Caribbean Sea. They consist of: the Bahamas, the Greater Antilles, and the Lesser Antilles. The Bahamas are a string of some 500 islands running south-west off Florida, USA. Strategically important to the British navy in the 19th century, they have close links with the USA and, like Bermuda, reject West Indian cultural affiliation. They became the independent Commonwealth of the Bahamas in 1964. The Greater Antilles is a chain of mostly large islands running roughly east-west and consisting of *Cuba, *Jamaica, Hispaniola (shared by *Haiti and the *Dominican Republic), *Puerto Rico, and a few off-shore and small islands, notably the Cayman Islands (a British colony and tax-haven), Curaçao and

Aruba, part of the Netherlands Antilles, which retained the status of self-governing colonies. The Lesser Antilles are grouped into the Leeward and Windward Islands, running roughly north–south, and consisting of numerous small islands colonized in the 17th century by the Spanish, British, French, Dutch, and Danish. They include the US *Virgin Islands, about fifty small islands bought by the USA in 1917 from Denmark for strategic reasons and since developed for tourism; the British Virgin Islands, a smaller group, that retain colonial status with increasing self-government; Montserrat, Martinique, and Guadeloupe, overseas dependencies of France; Trinidad, Tobago, and a string of now independent Commonwealth states which were colonized by the British. Some of these were federally linked into the Leeward Islands Federation (1871-1956), and rather more into the brief Federation of the West Indies (1958-62) before *West Indian independence was granted. By the 17th century indigenous peoples of the islands such as the Carib and Arawak had largely disappeared through disease, and they were replaced by African slaves. Slavery was abolished in the British colonies (1834), in the French colonies (1848), and in the Dutch colonies (1863) and *free trade introduced; increasing economic hardship developed, and racial tension grew between white minorities and the poor, largely illiterate black majority. The development of sugar-beet in Europe reduced the demand for West Indian sugar-cane and resulted in widespread unrest in the early 20th century. Emigration, the growth of air transport and of the tourist industry, improved education, and attempts to diversify economies, for example oil in Trinidad, have all more recently helped to ameliorate the situation in the 1960s and early 1970s. However, the combined effects of oil price rises, sharp falls in commodity prices, the world recession, and US protectionism have been responsible in the 1980s for economic decline in most islands and a serious debt problem in several countries, notably Jamaica.

Westminster, Statute of (1931), legislation on the status of British *dominions. At the 1926 and 1930 Imperial Conferences pressure was exerted by the dominions of Canada, New Zealand, the Commonwealth of Australia, the Union of South Africa, Eire, and Newfoundland for full autonomy within the British *Commonwealth. The result was the Statute of Westminster, accepted by each dominion Parliament, which recognized the right of each dominion to control its own domestic and foreign affairs, to establish a diplomatic corps, and to be represented at the League of Nations. It still left unresolved certain legal and constitutional questions—not least the status of the British crown. The Consequential Provisions Act (1949) allowed republics such as India to remain members of the Commonwealth.

westward expansion, American, the growth of the USA from the original white settlements on the east coast to span the entire continent. Between the *War of 1812 and the *American Civil War, the Mississippi Valley and the Great Lakes region were settled, and westward expansion came to the fore as one of the dominant themes of the 19th century. The successful *Mexican-American War (1846-8) and the *gold rush in California (1848) played a major role in the opening up of the *'Wild West' and the Pacific coast, while in the aftermath of the American Civil War the new *Union Pacific Railroad, the spread of the railway network, the pacification of the Indian population, and ever-increasing immigration from Europe began to fill in the area of the Great Plains. By the 1890s the frontier as such had disappeared, although the movement of population within the continent remained generally westwards through the 20th century.

Weygand, Maxime (1867–1965), French general. He was *Foch's chief of staff in World War I, and in 1920 was sent by the French government to aid the Poles in their ultimately successful defence against the advancing Soviet *Red Army. In the military crisis of May 1940 Weygand was recalled to assume command of the French armies attempting to stem the German *Blitzkrieg attack. Advising capitulation, he later commanded the *Vichy forces in North Africa, was dismissed at the request of the Germans, arrested by the Gestapo, and then freed by the Allies. He was tried and acquitted under the *de Gaulle regime on a charge of collaboration with the Germans.

Whig Party, a US political party of the second quarter of the 19th century. The Whig Party was formed in the mid-1830s by those who opposed what was perceived as the executive tyranny of President Andrew *Jackson. Dominated by Henry *Clay and Daniel *Webster, the Party elected *Harrison to the White House in 1840 and

*Taylor in 1848, but disunity on *free-soil and slavery issues weakened it severely, and it broke up.

Whitbread, Samuel (1758–1815), British politician. A notable champion of reform and of the liberties of the individual, he began his parliamentary career in 1790 as a Whig. He allied himself with Charles James *Fox and became the dominant figure in the opposition to the government of William *Pitt the Younger. Whitbread was to prove himself a fervent advocate of the abolition of slavery and the extension of religious and civil rights.

White Australia policy, a restrictive immigration policy pursued in Australia. In the mid-19th century there was a shortage of labour and *squatters brought in Chinese and *Kanaka labour. By the 1880s developing trade unions were calling for a policy to protect the 'white working man'. By 1890 all states had legislation to preserve the purity of white Australia, Alfred *Deakin being one of its strongest advocates. The new Commonwealth government legislated to exclude non-Europeans (Immigration Restriction Act, 1901). The main device used was to be a dictation test in any European language (any 'prescribed' language from 1905), the language being chosen to ensure failure. This policy of exclusion continued until the 1950s. Since then, the policy has been reformed.

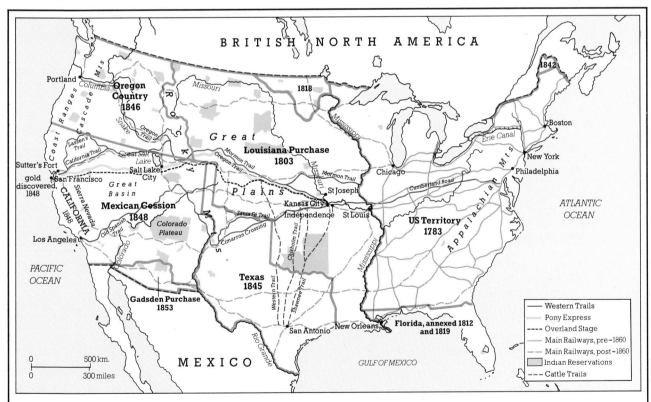

Westward expansion (USA)

The Louisiana Purchase (1803) and the Lewis and Clark Expedition (1804–6) stimulated interest in the West, and after the 'War of 1812' rivalry with colonists from British North America (later Canada) ended. The Erie Canal offered water-transport as an alternative to the wagon-train. In the 1840s Oregon country was attracting thousands of migrants each year; Texas was annexed and a vast area of Mexico seized after the Mexican–American War. This included California, to which goldminers flocked from all parts of the world. By now railroad development was pushing west, linking the east and west coasts after the Civil War in 1865. Cattle on the Great Plains were herded north-east for the expanding meat industry, and wheat farming steadily developed at the expense of reservation land allocated to Indians.

White Russians, those who fought against the Soviet *Red Army in the *Russian Civil War (1918–21). The name was derived from the royalist opponents of the French Revolution, known as Whites, because they adopted the white flag of the French Bourbon dynasty. The White Army, though smaller than the Red, was better equipped and had an abundance of Tsarist officers, some of whom offered to serve as ordinary soldiers. Its two main bases were in the south, where the army was successively led by Kornilov, *Denikin, and *Wrangel, and in Siberia where *Kolchak was nominally head of a provisional government at Omsk. The White Russians were ultimately defeated by their own internal quarrels, by their refusal to grant land reforms in the areas under their control, by Red control of railways, and by *Trotsky's organizing genius.

Whitlam crisis (1975), an Australian political and constitutional crisis. The House of Representatives, which had a Labor majority, and the Senate, which had a narrow non-Labor majority, reached a deadlock following the Loans Affair, which involved attempts by several Labor ministers to raise overseas loans without consulting the cabinet. The Prime Minister, Gough Whitlam, did not agree to a dissolution, which the Senate demanded before it would pass the Appropriation Bills. The governor-general, Sir John Kerr, withdrew Whitlam's commission as Prime Minister, and commissioned Malcolm Fraser, the Leader of the Opposition, instead. Fraser advised Kerr to dissolve both houses, and the Appropriation Acts were passed. Fraser's Liberal–National Country Party coalition secured majorities in both houses at the election.

WHO *World Health Organization.

Wilberforce, William (1759–1833), British philanthropist who played a leading part in the anti-slavery movement. Entering the House of Commons in 1780, he became a staunch supporter of William *Pitt the Younger. An effective parliamentary speaker, he lost no opportunity to denounce the horrors of the slave trade, which was eventually abolished in 1807. He also supported the campaign to outlaw slavery completely within the British empire, which culminated in the passing of the Emancipation Act of 1833, a month after his death.

'Wild West', a term used to describe frontier society in 19th-century USA. Around the masculine, saloon-bar world of the *gold rushes and the *cowboy cattle-drives of Texas, California, and the largely unsettled western territories there early developed a mythology. Bandits such as Billy the Kid and Jesse James were romanticized, as were General *Custer and his 'last stand'. An early perpetrator of the myth was Edward Z. C. Judson, who, under the pseudonym Ned Buntline, wrote penny (dime) novels romanticizing the exploits of his friend W. F. Cody as 'Buffalo Bill'. The latter in turn organized 'Wild West Shows' from 1883 onwards, which included the appearance of the Indian Chief *Sitting Bull and which travelled as far afield as Europe. There is no evidence that the West was much less law-abiding than the rest of the USA. None the less the 'Wild West' was no purposeless myth; it suggested an arena in which individuals struggled to make order out of chaos and to progress through individual effort and moral worth. The North American continent had had a succession of 'Wests', as its frontiers receded, and that known as the 'Wild West' was the last. It disappeared after 1890, with the end of Indian hostilities, the decline of the long-distance cattle drives, the building of the railways, and the steady growth of population.

Wilhelmina (1880–1962), Queen of the Netherlands (1890–1948). The daughter of William III, she probably had a large share in maintaining the neutrality of the Netherlands during World War I. When her country was invaded by Germany (1940), she and her ministers maintained a government-in-exile in London. Through frequent radio talks she became a symbol of resistance to the Dutch people. She returned in 1945, but abdicated (1948) in favour of her daughter Juliana.

Wilkins, Roy (1901–81), US social reformer and civil rights leader. The grandson of a slave, he worked for black newspapers before he joined the National Association for the Advancement of Colored People (NAACP) in 1931. He was an active leader of the *civil rights movement to improve the status of the black population. He held high executive office in the NAACP from 1955 to 1977, but came under increasing criticism from militant blacks towards the end of his life, for his commitment to non-violence.

Wilkinson, James (1757–1825), US general and adventurer. He distinguished himself in the early days of the War of Independence. Moving to Kentucky, he became a principal figure in the confused politics of the developing south-west. He re-entered the army in 1791 and went on to serve as governor of the Louisiana Territory (1805–6). While there he became involved with Aaron *Burr's conspiracy, but betrayed the latter and acted as prosecution witness in the subsequent treason

Emperor **William II** (centre) inspects German troops in trenches at the front shortly after the outbreak of hostilities in World War I, 1914.

trial. The failure of his campaign to capture Montreal in the early stages of the *War of 1812 led to his removal from command and eventual retirement.

Willard, Emma (Hart) (1787–1870), US pioneer of women's education. She opened a seminary at Middlebury, Vermont, in 1814 to teach subjects not then available to women (*feminism). Her appeal to the New York legislature, *Plan for Improving Female Education* (1819), led her to be invited by Governor Clinton to settle in that state, and in 1821 she opened the Troy Female Seminary, offering a college education which became a model for similar establishments in Europe and the USA.

William I (1797–1888), King of Prussia (1861–88) and German Emperor (1871–88). He devoted himself to the welfare of the Prussian army, assuming personal command in suppressing the *Revolution of 1848 in Baden. When he succeeded to the Prussian throne in 1861 he proclaimed a new 'era of liberalism', but this did not last for long. In 1862 he invited Otto von *Bismarck to become his Minister-President and from then on relied increasingly on Bismarck's policies, giving his approval to the growing influence of Prussia. During the *Franco-Prussian War he took command of troops, receiving the surrender of Napoleon III at *Sedan (September 1870). In January 1871 he was invited by the princes of Germany, at Bismarck's instigation, to become their emperor, thus creating the *German Second empire. Two unsuccessful assassination attempts strengthened his popularity, but also offered a pretext to clamp down on socialists.

William II (1859–1941), King of Prussia and Emperor of Germany (1888–1918). A grandson of Queen Victoria and of William I of Prussia, in 1890 he forced *Bismarck's resignation and embarked on a personal 'new course' policy regarded abroad as war-mongering. He supported *Tirpitz in building a navy to rival that of Britain. His congratulatory telegram to the Boer leader, *Kruger, on the failure of the *Jameson Raid (1896) offended public opinion in Britain. He made friendly overtures to Turkey and dangerously provoked France in the *Morocco crises of 1905 and 1911. His support of Austro-Hungary against *Serbia (1914) led to World War I, though his personal responsibility for the war is less than was once thought. He played little direct part in the war, and in 1918 was forced to abdicate.

William IV (1765–1837), King of Great Britain and Ireland and dependencies overseas, King of Hanover (1830–7). The third son of George III, his reign marked a decline in the political influence of the crown. He joined the navy as an able seaman in 1779, subsequently becoming a close friend of Horatio *Nelson. In 1790 he set up house with an actress, Dorothea Jordan, who was to bear him ten children. In order to secure the succession to the throne, in 1818 he married Adelaide of Saxe-Meiningen and had two daughters, who both died in infancy. Becoming king on the death of his brother, *George IV, in 1830, he overcame his natural conservatism sufficiently to help ensure the passage of the *Reform Act of 1832.

Williams, Eric (Eustace) (1911–81), West Indian statesman. In 1955 he founded the People's National Movement (PNM), which remained unsuccessful during the West Indies Federation (1958–62), but won a landslide victory in the national elections of 1961. In 1962 he led his country to independence, becoming the first Prime Minister (1961–81) of the colony and then of the republic of Trinidad and Tobago. An 'empirical' socialist, Williams attracted foreign capital through tax incentives and, by skilful use of foreign aid, made Trinidad and Tobago the wealthiest Commonwealth nation in the *West Indies. He faced increasing militant opposition to his government before his death in 1981.

Wilson, (Thomas) Woodrow (1856–1924), twenty-eighth President of the USA (1913–21). He entered an academic career in 1883 and was appointed president of Princeton University in 1902. He was responsible for major changes in the educational and social organization of Princeton. In 1910 he resigned to run as governor of New Jersey, and was elected. Wilson became a successful reform governor and earned a reputation that helped give him the Democratic nomination for the Presidency in 1912. Once in office, Wilson determined to effect a programme known as the 'New Freedom', designed to stimulate competition, promote equal opportunity, and

This campaign poster supporting US President Woodrow **Wilson** reflects the division of opinion in the USA over his war policy. Isolationists were hostile to Wilson's shift in 1917 from neutrality to participation in World War I, and to his later attempts to bring the USA into the League of Nations.

STAND BY THE PRESIDENT

check corruption. Faced with the outbreak of World War I in 1914, he at first concentrated on preserving US neutrality. Gradually, however, he came to a view that the USA should enter the war on the side of the Allies. German policy of unrestricted submarine warfare from January 1917 led to the declaration of war in April. From then on he worked to realize his vision, proposed in the *Fourteen Points, of a peaceful post-war world. Wilson's Presbyterian background and respect for legal traditions made him favour an international peacekeeping forum, but he fell foul of American *isolationism, which saw the proposed *League of Nations as a tool of British and French diplomacy. The isolationists in the Senate defeated Wilson on the issue of US participation in the League, while his exertions in negotiating the *Versailles Peace Settlement and trying to win its acceptance by the Senate brought on a severe stroke. He never fully recovered, and for the last year of his presidency Mrs Wilson, a lady of powerful personality, largely directed such business as could not be avoided or postponed.

Wilson ministries, British Labour governments (1964–70, 1974–6) with (James) Harold Wilson, Baron Wilson of Rievaulx (1916–) as Prime Minister. Wilson's policies were noted for their non-doctrinal content—pragmatic according to his supporters, unprincipled according to his detractors. Throughout their periods in office the Wilson cabinets were plagued by economic problems: in the first period these mainly took the forms of balance-of-payments deficits and sterling crises, the latter leading to a devaluation of the pound in 1967. Experiments to create a prices and incomes policy collapsed, and the White Paper *In Place of Strife* (1969), which sought to curb trade unions, to end unofficial strikes, and to establish a permanent Industrial Relations Commission, was withdrawn. In February 1968 the government passed an Immigration Act to prevent mass entry into Britain of Asians from Kenya and Uganda who held

Wilson ministries: Harold Wilson (*left*) and Denis Healey (*right*), campaigning in the 1970s. Wilson was adroit at holding together a Labour Party in which mixed-economy supporters strained against declared Marxists at a time when the country was deeply divided on British membership of the EEC.

British passports, and to restrict entry by all Commonwealth citizens. A *Race Relations Bill was introduced two months later. During the 1974–6 period the major problem was inflation, which the government unsuccessfully tried to cure by abating wage demands by a 'social contract' with the trade unions. However, important regional development and social reforms, particularly in education, were achieved in the 1960s: the introduction of comprehensive education, the expansion of higher education, changes in the law on sexual relations, divorce, and abortion, the end of the death penalty, and the reduction of the age of adulthood to 18. In 1968 the decision was made to withdraw forces from east of Suez. Overseas, the government failed to solve the problem of Rhodesian UDI in 1965 (*Zimbabwe). In 1975 the last Wilson administration reluctantly confirmed British membership of the *European Economic Community after a referendum. A perceived defect in all Wilson's governments was the array of private advisers and the highly personalized style of running public affairs.

'Wind of Change', a phrase in a speech by the British Prime Minister, Harold *Macmillan. He used it in his address to both Houses of the South African Parliament on 3 February 1960, to draw attention to the growth of national consciousness which was sweeping like a 'wind of change' across the African continent, and warned that *South Africa should take account of it.

Windsor, House of, the official designation of the British royal family since 1917. Anti-German feeling during World War I was sufficiently strong for George V to feel that it would be an appropriate gesture to remove all references to the German titles of Saxe-Coburg, derived originally from the marriage of Queen Victoria to Prince Albert of Saxe-Coburg-Gotha. 'Windsor' was adopted because Windsor Castle, Berkshire, has long been the main home of British monarchs.

Wingate, Orde Charles (1903–44), British major-general. A brilliant exponent of guerrilla warfare, in the 1930s he helped to establish and train Jewish irregular forces operating against Arabs in Palestine, and in 1941 he organized Sudanese and Abyssinian irregulars to fight the Italian occupiers and restore Emperor Haile Selassie to the throne. He created and led the *chindits*, a Burmese guerrilla group that operated behind Japanese lines. He died in an air crash in 1944 at the outset of his second, and greatly enlarged, *chindit* operation.

Witte, Sergei Yulyevich, Count (1849–1915), Russian statesman. As Finance Minister (1892–1903) and Premier (1905–6), he believed that if Russia was to become the equal of western industrial nations both government investment and foreign capital were essential. New railways linked the Donetz coalmines with St Petersburg and Moscow; the Trans-Siberian railway was built; steel production began; sufficient petroleum was produced to satisfy Russia's need and for export. Thus on the eve of political revolution Russia underwent a remarkable industrial revolution. Although his ideal was economic modernization combined with authoritarian rule, during the *Revolution of 1905 he urged *Nicholas II to issue the October Manifesto granting Russia a constitution, and to summon the *Duma. Nicholas disliked him and dismissed him.

Garnet **Wolseley**, in white-plumed hat, with mounted British officers at Alexandria, Egypt, in 1882 by Orlando Norie. Wolseley had just led his army to victory against Arabi Pasha. (Private Collection)

Wolseley, Garnet Joseph, 1st Viscount (1833–1913), British field-marshal. He served in Burma, and in the Crimea, the Indian Mutiny, China, and Canada. He furthered the army reforms of Edward *Cardwell at the War Office (1871–2) and in 1873 commanded the *Asante Expedition. His most famous achievement was his brilliant defeat of *Arabi Pasha in 1882. In 1884 he was too late to relieve *Gordon in Khartoum. An advocate of army reform, he successfully worked for the abolition of the purchase of commissions, shorter periods of enlistment, and the creation of an army reserve.

women's liberation, a radical movement demanding the improvement of women's status in society. Its analysis of the social relationship between the sexes is deeper than earlier demands (*women's suffrage, *feminism) for political and educational rights. Women's liberation was especially vocal and active as a movement in the USA during the 1960s and 1970s; in 1966 the National Organization for Women (NOW) was formed in the USA and has remained active since. Its demands were taken up in other industrialized countries, notably Britain and Australia. Women's Lib. (as the term is frequently abbreviated) argues that men, naturally sexist, dominate and exploit women. Practical demands have been focused on the right to equal opportunities and equal pay. In Britain the Sex Discrimination Act and the creation of the Equal Opportunities Commission in 1975 gave legal effect to some demands, although many employment practices and financial rewards remain tilted in favour of men. Women's liberation movements in Islamic countries suffered a setback with the revival of Islamic fundamentalism in the 1970s, which re-established the traditional segregation and restriction of women.

women's suffrage, the right of women to take part in political life and to vote in an election. The feminist movement (*feminism) of the 19th century demanded among other things the right of women to vote in a political election. This was first attained at a national level in New Zealand (1893). The state of Wyoming, USA, introduced women's suffrage in 1869; in 1893 the National American Woman Suffrage Association (NAWSA) was formed, which organized state-by-state activity, and in 1920 all women over 21 were given the vote in the USA. The first European nation to grant female suffrage was Finland in 1906, with Norway following in 1913, and Germany in 1919. In Britain, as a result of agitation by the suffragists and the *suffragettes, the vote was granted in 1918 to those over 30 and in 1928 to the 'flappers' (women over 21). In 1918 the Irish politician Constance Markievicz became the first woman to be elected to the British House of Commons though, as a member of *Sinn Fein, she never took her seat, becoming instead the world's first woman Minister of Labour in *de Valera's Dáil Éireann. The Roman Catholic Church was reluctant to support women's suffrage and in many Catholic countries it was not gained until after World War II. In France it was granted in 1944, in Belgium in 1948, while in Switzerland not until 1971. Women's suffrage was gained in the Soviet Union after the Revolution (1917) and in the new east European republics after World War I. Following World War II it came to the rest of eastern Europe. In Third World countries it was usually obtained with independence, and in most Muslim countries women have been granted suffrage. One result of the suffrage has been the emergence in the 20th century of some outstanding women politicians, for example Golda *Meir (1898–1978), Indira *Gandhi (1917–84), and Margaret *Thatcher (1925–), although the proportion of women taking an active part in politics remains low.

Works Project Administration (WPA), a US federal relief measure for the unemployed. An agency of the *New Deal, it was established by the Emergency Relief Appropriation Act of 1935. The initiators were Harold Lekes and Harry *Hopkins; they wanted the (estimated) 3.5 million unemployed but able-bodied to be given work and not a dole. Thus, through their wages, they would have money to spend and thereby help business to revive. The WPA employed about two million at one time, and by 1941 eight million (20 per cent of the labour force) were engaged in public works. It built playgrounds, airport landing fields, school buildings and hospitals, and ran a campaign against adult illiteracy. It was also, as even its friends had to admit, a useful source of employment for Democratic Party workers.

World Bank (International Bank for Reconstruction and Development), an organization linked to the *United Nations. It was proposed at the *Bretton Woods Conference in 1944, and constituted in 1945. It has over 130 members and is designed to finance enterprises that advance the economic interests of member nations. The Bank receives its funds from member countries (the USA being the largest contributor) and from borrowing in the world money markets. Initially it gave loans for reconstruction, but by 1949 it was concentrating on loans

for economic development, particularly in the Third World. Many early projects were very expensive and even though loan interest-rates are usually below the market rate, heavy debts were incurred. Sensitive to the problems of the North/South divide highlighted in the *Brandt Report, the Bank established a Special Fund (1977) to help least developed countries with debt-service relief. At the same time it has moved towards supporting projects which involve simpler 'intermediate technology', and, since 1970, has concentrated on agricultural and rural development, education, health, and public hygiene, as well as helping to plan strategies for industrialization.

World Council of Churches, an inter-denominational organization of Christian Churches, created in 1948. Apart from the *Roman Catholic Church and the *Unitarians, the Council includes all the major and many minor denominations and nearly all the Eastern Orthodox Churches. Since 1961 the Roman Catholic Church has sent accredited observers. Most of the work of the Council

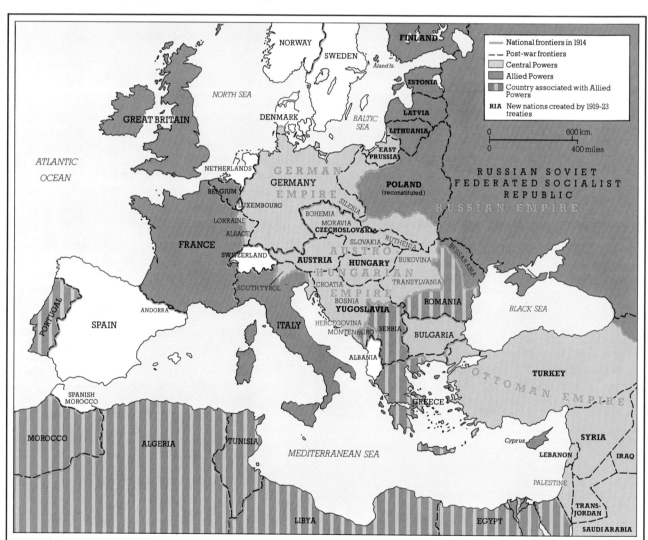

World War I

In 1914 the empires of the Central Powers—German, Austro-Hungarian, and Ottoman— extended through Central Europe into the Middle East. The Versailles Settlement (1919–23) sought to draw frontiers of the successor states along ethnic lines, and these largely survived, although Czechoslovakia's frontiers were to be challenged by Hitler. Following the Treaty of Brest-Litovsk and the Russian Civil War, the Baltic states of Latvia, Estonia, and Lithuania, together with a reconstituted Poland, became independent.

is advisory, but it has a number of administrative units; the largest of these is the division of Inter-Church Aid, Refugee, and World Service.

world fairs, displays, emphasizing the industrial, scientific, and technological achievements of the participating nations. They seek to promote trade and publicize cultural progress. Early examples were the *Great Exhibition at the Crystal Palace in London in 1851 and the Paris Exhibition of 1861. The first New York Fair was held in 1853-4 and the Philadelphia Centennial Exhibition in 1876. Fairs of particular note in the 20th century include those in New York in 1939-40 and 1964-5, the Brussels World Fair in 1958, the Seattle Century 21 Exposition of 1962, and Expo 67 held in Montreal.

World Health Organization, a specialized agency of the *United Nations, founded in 1948 with the aim of promoting 'the highest possible level of health' of all peoples. It undertakes the establishing of health services in the developing countries, the organizing of campaigns against epidemic diseases, the development of international quarantine and sanitation rules, the funding of international research programmes, and the training of medical specialists. A notable success has been its eradication of smallpox throughout the world.

World War I (1914-18). It was fought between the Allied Powers—Britain, France, Russia, Japan, and Serbia—who were joined in the course of the war by Italy (1915), Portugal and Romania (1916), the USA and Greece (1917)—against the Central Powers: Germany, Austro-Hungary, *Ottoman Turkey, and Bulgaria (from 1915). The war's two principal causes were fear of Germany's colonial ambitions and European tensions arising from shifting diplomatic divisions and nationalist agitation, especially in the *Balkan States. It was fought in six main theatres of war. On the *Western Front fighting was characterized by *trench warfare, both sides believing that superiority in numbers would ultimately prevail despite the greater power of mechanized defence. Aerial warfare, still in its early stages, involved mainly military aircraft in air-to-air combat. On the Eastern Front the initial Russian advance was defeated at Tannenberg (1914). With Turkey also attacking Russia, the Dardanelles expedition (1915) was planned in order to provide relief, but it failed. Temporary Russian success against Austro-Hungary was followed (1917) by military disaster and the *Russian Revolution. The *Mesopotamian Campaign was prompted by the need to protect oil installations and to conquer outlying parts of the Ottoman empire. A British in 1917 advance against the Turks in Palestine, aided by an Arab revolt, succeeded. In north-east Italy a long and disastrous campaign after Italy had joined the Allies was waged against Austro-Hungary, with success only coming late in 1918. Campaigns against Germany's colonial possessions in Africa and the Pacific were less demanding. At sea there was only one major encounter, the inconclusive battle of *Jutland (1916). A conservative estimate of casualties of the war gives 10 million killed and 20 million wounded. An armistice was signed and peace terms agreed in the *Versailles Peace Settlement.

World War II (1939-45), a war fought between the *Axis Powers and the Allies, including Britain, the Soviet Union, and the USA. Having secretly rearmed Germany, *Hitler occupied (1936) the Rhineland, in contravention of the *Versailles Peace Settlement. In the same year the Italian *fascist dictator, Benito *Mussolini, joined Hitler in a Berlin-Rome axis, and in 1937 Italy pledged support for the *Anti-Comintern Pact between Germany and Japan. In the 1938 *Anschluss, Germany annexed Austria into the *Third Reich, and invaded Czechoslovak *Sudetenland. Hitler, having secured the *Munich Pact with *Chamberlain in 1938, signed the *Nazi-Soviet Pact with *Stalin in August 1939. Germany then felt free to invade the *Polish Corridor and divide Poland between itself and the Soviet Union. Britain, which until 1939 had followed a policy of *appeasement, now declared war (3 September) on Germany, and in 1940 Winston *Churchill became head of a coalition government. The Soviet Union occupied the *Baltic States and attacked *Finland. Denmark, parts of Norway, Belgium, the Netherlands, and three-fifths of France fell to Germany in rapid succession, while the rest of France was established as a neutral state with its government at *Vichy. A massive *bombing offensive was launched against Britain, but the planned invasion of the country was postponed indefinitely after the battle of *Britain. Pro-Nazi governments in Hungary, Romania, Bulgaria and Slovakia now joined the Axis Powers, and Greece and Yugoslavia were overrun in March-April 1941. Hitler, breaking his pact with Stalin, invaded the Soviet Union, where his forces reached the outskirts of Moscow. Without declaring war, Japan attacked the US fleet at *Pearl Harbor in December 1941, provoking the USA to enter into the war on the side of Britain. In 1942 the first Allied counter-offensive began against *Rommel in *North Africa, and in 1943 Allied troops began an invasion of the Italian mainland, resulting in the overthrow of Mussolini's government a month later. On the *Eastern Front the decisive battles around *Kursk and *Stalingrad broke the German hold. The Allied invasion of western Europe was launched with the *Normandy landing in June 1944 and Germany surrendered, after Hitler's suicide in Berlin, in May 1945. The *Pacific Campaigns had eliminated the Japanese navy, and the heavy strategic bombing of Japan by the USA, culminating in the atomic bombing of Hiroshima and Nagasaki on 6 and 9 August 1945, induced Japan's surrender a month later.

   The dead in World War II have been estimated at 15 million military, of which up to 2 million were Soviet prisoners-of-war. An estimated 35 million civilians died, with between 4 and 5 million Jews perishing in *concentration camps, and an estimated 2 million more in mass murders in Eastern Europe. Refugees from the Soviet Union and Eastern Europe numbered many millions. The long-term results of the war in Europe were the division of Germany, the restoration to the Soviet Union of lands lost in 1919-21, together with the creation of communist buffer-states along the Soviet frontier. Britain had accumulated a $20 billion debt, while in the Far East nationalist resistance forces were to ensure the decolonization of south-east Asian countries. The USA and the Soviet Union emerged from the war as the two largest global powers. Their war-time alliance collapsed within three years and each embarked on a programme of rearmament with nuclear capability, as the *Cold War developed.

The burial of the dead at **Wounded Knee** after the one-sided battle between Indians and US troops on 29 December 1890. US soldiers pose beside the mass grave to which the Indian dead are being consigned.

Wounded Knee, battle of (1890), the last major battle between the US army and the Sioux Indians. The site is a creek on Pine Ridge reservation, North Dakota, where, after the killing of *Sitting Bull, the 7th US Cavalry surrounded a band of Sioux. These were followers of the Indian Ghost Dance religion, evolved around 1888 among the Paiute by Wovoka ('Jack Wilson'), who preached the coming of an Indian messiah who would restore the country to the Indians and reunite the living with the dead. In 1890 a Ghost Dance uprising in North Dakota culminated at Wounded Knee, when US troops massacred some 200 Teton Sioux. The massacre was recalled in 1973 when members of the American Indian Movement occupied the site. They were surrounded by a force of federal marshals; two Indians were killed, and one marshal seriously wounded. They agreed to evacuate the area in exchange for negotiation on Indian grievances.

Wrangel, Piotr Nikolayevich (1878-1928), Russian general. He became prominent as the leader of the counter-revolutionary armies in the *Russian Civil War (1918-21). Serving under *Kolchak and then under *Denikin he became commander-in-chief of all White armies after Denikin's withdrawal (March 1920) to the Caucasus. In the Crimea Wrangel maintained a base as head of a provisional government. In November 1920 the *Red Army broke through his defences and he fled to Turkey. His defeat ended White Russian resistance.

Wuchang Uprising (10 October 1911), revolt in the city of Wuchang which began the *Chinese Revolution of 1911. An accidental explosion forced republican revolutionaries to begin a planned uprising earlier than intended, but on the next day army units won over to the rebel cause seized the city. The Qing government failed to respond swiftly to the uprising, and further provincial risings followed, leading to the formation of a Provisional Republican Government on 1 January 1912.

Xhosa Wars (1779-1879), wars between the Xhosa (*Nguni) people and Dutch and British colonists along the east coast of Cape Colony, between the Great Fish and Great Kei rivers. From 1811 the policy of clearing the land of Xhosa people to make way for Europeans began and following a year of fighting (1818-19) some 4000 British colonists were installed along the great Fish river. As they pushed the frontier east, however, the colonists met greater resistance, cattle raids resulting in retaliation—the war of 1834-35 yielding 60,000 head of cattle to the colonists. It was followed by the longer struggle of 1846-53. The war of 1877-9, which yielded 15,000 cattle and 20,000 sheep was vainly fought by tribesmen returning from the diamond fields in a last bid to regain their land. Afterwards all Xhosa territory was incorporated as European farmland within Cape Colony (*Cape Province).

Xi'an incident (December 1936), the kidnapping of the Chinese leader *Chiang Kai-shek while visiting disaffected Manchurian troops at Xi'an. Chiang was captured by conspirators headed by Zhang Xueliang, who attempted to force him to give up his campaign against the communists and lead a national war against the Japanese, who had occupied Manchuria in 1931. After Chiang had refused to accede to their demands, the communists, headed by *Zhou Enlai, also became involved in the negotiations, and eventually Chiang was released, having promised to take a more active role against the Japanese, and to allow local autonomy to the communists. Zhang Xueliang was imprisoned by Chiang, but the incident led to limited co-operation between the communists and the *Kuomintang against the Japanese.

Xinjiang (or Sinkiang), area in the far west of China, also known as Chinese *Turkistan. Under *Qing military administration since the mid-18th century, Xinjiang was always a rebellious area, partly because of an Islamic revival among the Uighur tribesmen and partly because of the corruption of local Qing officials. Local Islamic leaders launched unsuccessful *jihads* (holy wars) against the Qing in 1815, 1820-8, and 1857. In 1865 Yakub Beg invaded Xinjiang from Kokand, receiving British support to stem growing Russian influence in the area. Chinese forces defeated Yakub Beg in 1877, reclaiming nearly all of Xinjiang, which became established as a province in 1884. Russian influence in the area remained strong and only after 1950 did it become more closely integrated into China with the settlement of many Chinese in the province.

Yalta Conference (4–11 February 1945), a meeting between the Allied leaders, *Stalin, *Churchill, and *Roosevelt, at Yalta in the Soviet Union. Here the final stages of World War II were discussed, as well as the subsequent division of Germany. Stalin obtained agreement that the Ukraine and Outer Mongolia should be admitted as full members to the *United Nations, whose founding conference was to be convened in San Francisco two months later. Stalin also gave a secret

undertaking to enter the war against Japan after the unconditional surrender of Germany, and was promised the Kurile Islands and an occupation zone in Korea. The meeting between the Allied heads of state was followed five months later by the *Potsdam Conference.

Yamagata Aritomo (1838–1922), Japanese soldier and statesman. A member of a samurai family, he was an early opponent of the westernization of Japan, but, having experienced western military supremacy, he became a strong advocate of the modernization of the recently created *Meiji state. Serving in a succession of senior posts, he was the prime architect of the modern Japanese army, shaping a mass conscript army organized on the principle of unswerving loyalty to the emperor. He served as the first Prime Minister (1889–91) after the

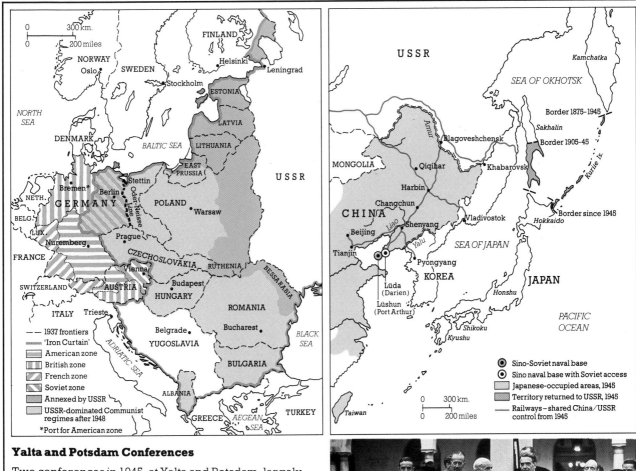

Yalta and Potsdam Conferences

Two conferences in 1945, at Yalta and Potsdam, largely determined the pattern of post-war power blocs. At Yalta (February), Stalin, Roosevelt, and Churchill (right) confirmed that Austria and Germany were to be divided into four zones of occupation, as was the Saarland. The Soviet Union agreed to enter the war against Japan, in return for territories lost during the Russo-Japanese War of 1905. The Potsdam Conference (July–August) confirmed that the Soviet Union was to keep the Polish territory which it had conquered in 1939, and Poland was to receive part of eastern Germany in compensation. Stalin's deft manipulation of agreements reached at the conferences were later to leave the Soviet Union in control of a broad area of 'buffer zones' stretching from the Baltic to the Adriatic, and to its emergence as a major World Power.

General **Yamashita** and his staff accepting the surrender of the British after the fall of Singapore in 1942.

introduction of the parliamentary system and held the post again (1898–1900). Serving also as chief of the general staff during the *Russo-Japanese War he exercised great influence and power, largely behind the scenes, in the years leading up to World War I. In 1921 he was publicly censured for meddling in the Crown Prince's marriage, dying in disgrace a year later.

Yamashita Tomoyuki('The Tiger of Malaya') (1888–1946), Japanese general. In World War II he led his forces in a lightning series of successes, capturing Malaya (1941–2), *Singapore (1942), and Burma. In 1944 he assumed control of the *Philippines Campaign, capturing Bataan and Corregidor. He surrendered to the Allies under General *MacArthur in 1945. After the war he was tried before a military commission for atrocities committed by his soldiers, and hanged.

Yaoundé Convention, a trade agreement (1969) reached in Yaoundé, capital of *Cameroon, between the *European Economic Community and eighteen African states, giving the developing countries aid and trade preferences. The *Lomé Convention superseded it.

Yemen Arab Republic, a country in the south of the Arabian peninsula. It was created in 1962 by a republican revolution led by Abdullah al-Sallal, army chief of staff to the imam Muhammad al-Badr. Subsequently, there began a protracted civil war in which the imam, with the support of several Zaidi tribes and of Saudi Arabia, fought Sallal and the republicans, who were supported by Egypt. Following Egypt's defeat in the *Six-Day War (1967) their support was withdrawn, Sallal resigned, and a new republican government under the moderate leadership of Qadi Iryani was formed. Political instability continued, owing to the divisions between townsmen and tribesmen, and left and right; it was exacerbated by the problem of relations with *South Yemen.

Yom Kippur War (1973), the Israeli name for the Arab-Israeli war called by the Arabs the October War. The war began on 6 October, the Feast of Yom Kippur, Israel's most important holiday, when Egyptian forces crossed the Suez Canal and breached the Israeli Bar Lev Line. Syrian troops threw back Israeli forces on the Golan Heights, occupied by the latter since the *Six-Day War. The war lasted three weeks, in which time Israel pushed Syrian forces back into Syria and crossed the Canal, encircling an Egyptian army. In the aftermath, disengagement agreements were signed by Israel with Syria in 1974 and with Egypt in 1974 and 1975. The Israeli withdrawal from Sinai was completed in 1982 after the 1978 Israeli–Egyptian peace treaty.

Yoruba empire of Oyo, a loose confederation of Yoruba kingdoms in West Africa. By the early 19th century the Yoruba empire of Oyo was beginning to disintegrate, a process accelerated by the decline of the slave trade and the rise of the *Fulani empire in the north. The Fulani destroyed the old city of Oyo, creating the Muslim emirate of Ilorin. Alafin Atiba, the new ruler of the Oyo empire (1836–59), built a capital, Ago Oja, and allied his empire with Ibadan; but on his death civil war developed between the smaller Yoruba kingdoms. At the same time the influence of the spiritual leader of the Yoruba people, the Oni of Ife, began to decline with the arrival of Christian missionaries. In 1888 a treaty was made with the then Alafin of Oyo, whereby all the Yoruba kingdoms were brought under British protection. In 1900 the empire was incorporated into the protectorate of Southern Nigeria.

Yoshida Shigeru (1878–1967), Japanese statesman. A liberal-conservative politician whose appointment as Foreign Minister had been blocked in 1936 by militarist interests, he was imprisoned for advocating surrender in the closing stages of World War II. He emerged after the war as the leader of the Liberal party. As Prime Minister (1946–7, 1949–54), Yoshida was a major architect of Japan's political rehabilitation and socio-economic recovery, working closely with *MacArthur and espousing pro-Western policies. His popularity peaked with the ending of Allied occupation in 1952 and thereafter declined rapidly.

Young, Brigham (1801–77), US *Mormon leader. Young joined the Mormons in 1832 and by 1835 had become one of the Council of Twelve, making a successful missionary journey to Britain to attract converts to the USA. After Joseph *Smith's death in Illinois in 1844, Young became the dominant figure of Mormonism, leading the migration west to Salt Lake City, ruling over the new community with autocratic firmness, and turning a desert waste into a flourishing and expanding city.

Young England, a British political movement of young Tory aristocrats in the early 1840s. It aimed at ending the political dominance of the middle classes by an alliance between the aristocracy and the working classes, which would carry out all necessary social reforms. The rather vague, romantic ideas of its members were given some substance by Benjamin *Disraeli, who defined its principles in his novel *Coningsby* (1844). The movement broke up in 1845 over the issue of *free trade and the disputed grant to Maynooth College, the principal institution in Ireland for training Roman Catholic clergy.

Young Ireland, an Irish nationalist movement of the 1840s. Led by young Protestants, including Smith O'Brien

(1803-64) and John Mitchel (1815-75), who, inspired by Mazzini's *Young Italy, set up their own newspaper, the *Nation*. It called for a revival of Ireland's cultural heritage. At first the members of Young Ireland were associated with Daniel *O'Connell in his campaign to repeal the *Act of Union, but later they turned to more radical solutions. In 1848 they attempted a rebellion, which was easily suppressed, O'Brien and Mitchel being sentenced to transportation.

Young Italy, an Italian patriotic society. Formed in 1831 by Giuseppe *Mazzini and forty other Italian exiles in Marseilles, it set out to replace earlier secret societies such as the *Carbonari as a prime force in the *Risorgimento. It was supposed to mobilize the 'people' against the Italian princes and nobles, and create a united Italian republic, but it appealed mainly to the middle classes; the dream of violent revolution ended with an unsuccessful uprising in Calabria in 1844. Its significance lay in the kindling of national consciousness and thus contributed towards Italian unification.

Young Plan, programme for the settlement of German *reparations payments after World War I. The plan was embodied in the recommendations of a committee which met in Paris (February 1929) under the chairmanship of a US financier, Owen D. Young, to revise the *Dawes Plan (1924). The total sum due from Germany was reduced by 75 per cent to 121 billion Reichsmark, to be paid in fifty-nine annual instalments. Foreign controls on Germany's economy were lifted. The first instalment was paid in 1930, but further payments lapsed until *Hitler repudiated all reparations debts in 1933.

Young Turks, European name applied to a number of late 19th- and early 20th-century reformers in the *Ottoman empire who carried out the Revolution of 1908. They should be distinguished from the Young Ottomans, a group of Ottoman reformers active from 1865 to 1876. After the 1908 Revolution the Young Turks organized themselves in political parties under the restored constitution. The most prominent party was the Committee of Union and Progress which seized power in 1913 and under the triumvirate of *Enver, Talat, and Jamal Pasha ruled the Ottoman empire until 1918, supporting the Central Powers in World War I.

Ypsilanti, Alexander (1792-1828), Greek nationalist leader. He served as a general in the Russian army, and was elected leader of the Philike Hetairia, a secret organization that sought Greek independence from the *Ottoman empire. He raised (1821) a revolt in Moldavia, proclaiming the independence of Greece, but lacking the support of Russia or Romania, was defeated by the Turks and imprisoned in Austria. Together with the successful Greek rebellion in the Peloponnese, his uprising marked the beginning of the *Greek War of Independence.

Yuan Shikai (or Yuan Shih-k'ai) (1859-1916), Chinese soldier and statesman. He established his military reputation in Korea and returned to China to undertake a programme of army reform. He supported the empress dowager *Cixi in her suppression of the *Hundred Days Reform. Dismissed from office after her death (1908), he retired to his old power base in northern China. He was recalled by the court when the *Chinese Revolution

began, but he temporarily sided with the republicans, and advised the emperor to abdicate. In 1912 he became President of the republic. Initially successful in restoring central control, his suppression of *Sun Yat-sen's *Kuomintang, dissolution of Parliament, and his submission to Japan's *Twenty-One Demands provoked a second revolution in the Yangtze region. He had himself proclaimed emperor in 1916, but died shortly afterwards, leaving China divided between rival *warlords.

Yugoslavia, a country in south-west Europe. At the end of World War I it was formed as the new Kingdom of the Serbs, Croats, and Slovenes from the former Slavic provinces of the *Austro-Hungarian empire: (Slovenia, Croatia, Bosnia and Hercegovina), together with Serbia and Montenegro, and with Macedonian lands ceded from Bulgaria. The monarch of Serbia, Peter I, was to rule the new kingdom and was succeeded by his son *Alexander I. At first the Serbian Premier Nikola *Pasic (1921-6) held the rival nations together, but after his death political turmoil caused the new king to establish a royal dictatorship, renaming the country Yugoslavia (January 1929). Moves towards democracy ended with his assassination (1934). During World War II Yugoslavia was overrun by German forces (1941), aided by Bulgarian, Hungarian, and Italian armies. The king fled to London and dismemberment of the country followed, with thousands of Serbs being massacred. A guerrilla war began, waged by two groups, supporters of the Chetnik *Mihailovic and *Tito's Communist partisans. Mutual suspicion ruined joint operations and led to fighting between them. Subasic, Premier of the exiled government, worked for a while with Tito, but resigned in November 1945, allowing Tito, supported by the Soviet Union, to proclaim the Socialist Federal Republic of Yugoslavia. Since 1945, as a communist federal state, it has refused to accept Soviet domination and in June 1948 was expelled by Stalin from 'the family of fraternal Communist Parties'. It became a leader of the non-aligned nations and the champion of 'positive neutrality'. Improved relations with the West followed and, after Stalin's death, diplomatic and economic ties with the Soviet Union were renewed (1955). Tito became President under a new constitution in 1953 and remained so until his death (1980). Since then the office of Head of State has been transferred to an eight-man collective Presidency, with the post of President rotating annually among its members.

Z

Zaharoff, Sir Basil (b. Zacharias Basileios) (1850–1936), international financier and munitions manufacturer. Originating from Anatolia, Turkey, he was known as the 'mystery man of Europe' and was accused of fomenting warfare and of secret political intrigue. He built up profitable connections with British, German, and Swedish armament firms and amassed vast wealth. His sympathies in World War I lay with the Allies, who rewarded him with civil decorations for supplying them with intelligence information.

zaibatsu, Japanese business conglomerates. The zaibatsu (literally 'financial clique') were large business concerns, with ownership concentrated in the hands of a single family, which grew up in the industrialization of late 19th-century Japan. They had their origins in the activities of the seisho ('political merchants'), who made their fortunes by exploiting business links with the newly restored Meiji government. After the government ceased to play a direct role in economic activity, zaibatsu like Mitsui and Mitsubishi expanded to fill the gap through the ownership of interrelated mining, transport, industrial, commercial, and financial concerns, dominating the business sector in a fashion which had no near equivalent elsewhere in the industrialized world. Despite efforts to break up their power in the aftermath of World War II, they continued in a modified form to provide the characteristic pattern for large Japanese industrial organizations into the 1980s. They are now more usually known in Japan as keiretsu.

Zaïre, Central African country. The pre-colonial 19th-century history of Zaïre was dominated by the Arab slave trade. *Livingstone was the first European explorer of the country. In 1871 *Stanley undertook to sail down the River Zaïre (Congo). His reports prompted King *Leopold II of Belgium to found the International Association of the Congo (later termed the Congo Free State). Stanley began to open up its resources. Maladministration by Leopold's agents obliged him to hand the state over to the Belgian Parliament (1908), but in the next fifty years little was done, except by Catholic mission schools, to prepare the country for self-government. The outbreak of unrest in 1959 led to the hasty granting of independence in the following year, but the regime of Patrice *Lumumba was undermined by civil war, and disorder in the newly named Congo Republic remained endemic until the coup of General *Mobutu in 1965. In 1967 the Union Minière, the largest copper-mining company, was nationalized and Mobutu achieved some measure of economic recovery. In 1971 the name of the country was changed to Zaïre. Falling copper prices and centralized policies undermined foreign business confidence, and two revolts followed in 1977 and 1978 in the province of Shaba (formerly Katanga), only put down with French military assistance.

Zambia, a landlocked country in Central Africa. It was settled by *Nguni people in flight from Zululand in 1835, but was also subject throughout much of the 19th century to Arab slave-traders. Agents from Cecil *Rhodes entered the country (known at this time as Barotseland) in 1890. Rhodes's British South Africa Company had been granted responsibility for it in its charter of 1889 and it began to open up the rich deposits of Broken Hill from 1902. The country was named Northern Rhodesia in 1911. It became a British protectorate in 1924 and between 1953 and 1963 was federated with Southern Rhodesia and Nyasaland, before becoming the independent republic of Zambia under President Kenneth *Kaunda in 1964. Dependent on its large copper-mining industry, Zambia has experienced persistent economic difficulties due to its lack of a coastline and port facilities and to low copper prices. It was also closely involved in the guerrilla war in Southern Rhodesia (*Zimbabwe).

zamindar (under the Mogul empire in India, a tax collector or landlord), the basis of a system of land-settlement developed in India under British rule. It fell into two distinct groups. In Lower Bengal, the government of Lord Cornwallis fixed the land revenue payable by the zamindars in perpetuity in 1793 in the hope of stabilizing the revenue, providing an incentive for improvement, and creating a class of loyal landlords. The effect was to create a privileged group of large and wealthy landlords. In the North Western Provinces, on the other hand, the government made a settlement during the 1830s, with much smaller landlords for thirty-year periods, in the hope of creating a class of small yeoman farmers and retaining a larger revenue for government.

Zanzibar, an island off the East African coast. Following its development by *Said ibn Sultan, his son Majid became ruler of Zanzibar strongly guided by the British consul Sir John Kirk (1866–87). German trading interests were developing in these years, but Britain and Germany divided Zanzibar's mainland territories between them and, by the Treaty of Zanzibar (1890), Germany conceded British autonomy in exchange for control of the North Sea island of Heligoland. Zanzibar became a British protectorate. In December 1963 it became an independent member of the Commonwealth, but in January 1964 the last sultan was deposed and a republic proclaimed. Union with Tanganyika, to form the United Republic of *Tanzania, followed in April. Zanzibar retained its own administration and a certain degree of autonomy, and, after the assassination of Sheikh Karume in 1972, Aboud Jumbe and the ruling Afro-Shirazi Party ruthlessly put down all forms of political oppposition until growing resentment forced Jumbe's resignation in 1984. A new constitution provided for more representative government in 1985.

Zapata, Emiliano (1879–1919), leader in the *Mexican Revolution. A mestizo peasant, Zapata forcefully occupied the land that had been appropriated by the *haciendas and distributed it among the peasants. He joined *Madero's revolution aimed at toppling the *Díaz regime. On the latter's overthrow, he sought to have the land returned to the ejidos (the former Indian communal system of ownership). When he saw that Madero was not prepared to embark upon major programmes of agrarian reform, he declared against him. For eight years he led his peasant guerrilla armies against the haciendas and successive heads of state before falling victim to an assassination plot at Chinameca. His creed, *zapatismo*, became one with *agrarismo*, calling for the return of the land to the Indians, and with the fundamental tenet of *indianismo*, the cultural, nationalist movement of the Mexican Indian.

Zeebrugge raid (23 April 1918), raid on a German U-boat base in Belgium in World War I. During the night of 22–23 April 1918 a daring raid by Admiral Keyes' force attacked it, sinking three blockships in the channel and nearly but not completely closing it. More effective but less dramatic was the line of deep mines which he laid across the Straits of Dover.

Zeng Guofan (or Tseng Kuo-fan) (1811–72), Chinese soldier and statesman. An imperial official and scholar, critical of the emperor's behaviour and the government's financial policies, in 1852 he reluctantly agreed to organize imperial resistance to the *Taiping Rebellion, raising the Hunan Army, and played a key role in wearing down resistance. With the crucial help of purchased modern European weapons, foreign military containment of the rebels along the eastern coast, and the capture of Nanjing in 1864, he finally broke their power. He became governor-general of Liang-Jiang, which gave him considerable powers in east-central China. In the 1860s and 1870s he supported the *Self-Strengthening Movement and developed the Jiangnan Arsenal for the manufacture of modern arms and the study of Western technical literature and Western languages.

Zhou Enlai (or Chou En-lai) (1898–1976), Chinese revolutionary and statesman, Premier (1949–76). Politically active as a student, he studied in France (1920–4) and became a communist. On his return he became deputy political director of the *Kuomintang Whampoa Military Academy. He organized an uprising in Shanghai in 1927, which was violently suppressed by *Chiang Kai-shek. Escaping to Jiangxi, he took part in the *Long March and became *Mao Zedong's chief adviser on urban revolutionary activity and his leading diplomatic

Emiliano **Zapata** (*seated second from right*) with fellow-revolutionaries at the presidential palace in Mexico City, December 1914. Beside Zapata is his comrade-in-arms, Pancho Villa.

Zhou Enlai, Premier of the People's Republic of China for twenty-seven years; he became a well-regarded figure in international affairs, enhancing Communist China's leadership among the African and Asian countries, and remaining a moderating force within his country.

envoy, representing the communists in the *Xi'an Incident and in US attempts to mediate in the civil war in 1946. He became Premier on the establishment of the People's Republic and served as Foreign Minister until 1958. He played a major role in the *Geneva Conference (1954) and the *Bandung Conference (1955), and was the main architect of Sino–US détente in the early 1970s. During the *Cultural Revolution he actively restrained extremists and helped restore order. He was one of the earliest proponents of the *Four Modernizations policy which later became associated with *Deng Xiaoping.

Zhu De (or Chu Teh) (1886–1976), Chinese revolutionary and soldier. He served as an officer in the imperial army and in the republican force which succeeded it. He became a communist in 1925, presenting his inherited wealth to the party. With *Mao Zedong he organized and trained early units of the People's Liberation Army (1931) in Jiangxi, and served as its commander-in-chief until 1954. He was a leader of the *Long March of 1934–5 and commanded the 8th Route Army against the Japanese between 1939 and 1945 before overseeing the communist victory in the *Chinese Civil War. He became marshal in 1955 and remained influential until purged during the *Cultural Revolution. He was restored to favour in 1967 and lived out his life in honoured retirement.

Zhukov, Georgi Konstantinovich (1896–1974), Soviet field-marshal. Of peasant origin, he joined the *Bolsheviks and fought in the *Russian Revolution (1917), as well as in the *Russian Civil War (1918–21). In 1939 he led the successful defence against Japanese incursions in the Far East. Much of the planning of the Soviet Union's World War II campaigns was done by him. He defeated the Germans at *Stalingrad (1943) and lifted the siege of *Leningrad. He led the final assault on Germany (1945), captured Berlin, and became commander of the Soviet zone in occupied Germany. He was demoted by Stalin to command the Odessa military district (1947), but after the latter's death rose to become Defence Minister (1955). He supported *Khrushchev against his political enemies in 1957 but was dismissed the same year, only to be reinstated after Khrushchev was deposed (1964).

Zia ul-Haq, General Mohammed (1924–), Pakistani army officer and statesman. In 1977 he led a military coup deposing Zulfilkar Ali *Bhutto, who was later tried and hanged. In 1978 he was proclaimed President of Pakistan, outlawing all political parties and industrial strikes, and enforcing press censorship. A zealous Muslim, he introduced a full Islamic code of laws and an Islamic welfare system, while insisting on greater Islamic control over the school system. Since the Soviet invasion of *Afghanistan in December 1979 millions of refugees have flooded Pakistan and Zia has received increasing amounts of economic and military aid, especially from the USA. He has been under some pressure to allow greater participation in government and in 1982 he formed a Federal Advisory Council of 350 nominated members.

Zimbabwe, a south-east African country. The Shona people of the country had been settled for centuries when, in the early 19th century, the Ndebele, under their leader *Mzilikazi (c.1795–1868), invaded the country from the south. He created a kingdom of Matabeleland, which for the next fifty years was to be in a state of permanent tension with the Shona to the north, in what came to be called Mashonaland. When Mzilikazi died he had obtained a peace treaty with the new *Transvaal Republic, and he was succeeded by his son *Lobengula. In 1889 the British South Africa Company of Cecil *Rhodes was founded, and in 1890 his Pioneer Column marched into Mashonaland. Following the *Jameson Raid and the Matabele War of 1893, Mashonaland and Matabeleland were united. Rebellion erupted in 1896-7, but it was ruthlessly suppressed. Rapid economic development followed, the country becoming the crown colony of Southern Rhodesia in 1911 and a self-governing colony in 1923. After the victory of the right-wing Rhodesia Front in 1962, the colony sought independence but refused British demands for black political participation in government and, under Prime Minister Ian Smith, issued the Unilateral Declaration of Independence (UDI) in 1965, renouncing colonial status and declaring Rhodesian independence. Subsequent British-sponsored attempts at negotiating a political compromise failed and nationalist forces waged an increasingly successful guerrilla campaign. Military pressure finally forced Smith to concede the principle of black majority rule, but the regime of the moderate Bishop Muzorewa could not come to an accommodation with the guerrilla leaders of the Patriotic Front, Robert *Mugabe and Joshua *Nkomo. Following the Lancaster House Conference (1979) Robert Mugabe was elected Prime Minister, and Rhodesia became the republic of Zimbabwe in 1980.

Zimmermann note (19 January 1917), a German secret telegram, containing a coded message from the German Foreign Secretary, Alfred Zimmermann, to the German minister in Mexico City. This instructed the minister to propose an alliance with Mexico if war broke out between Germany and the USA. Mexico was to be offered the territories lost in 1848 to the USA. The British intercepted the message and gave a copy to the US ambassador. The US State Department released the text on 1 March 1917, even as US–German relationships were deteriorating fast over submarine warfare. With the possibility of a German-supported attack by Mexico, the *isolationists lost ground and on 6 April 1917 Congress entered *World War I against Germany.

Zinoviev was a collaborator of Lenin and a leading Soviet communist. The 1924 'Zinoviev letter' became a *cause célèbre* in British politics. Zinoviev fell foul of Stalin and was executed during the great purges of 1936. (Portrait photograph by Moisei Nappelbaum.)

Zinoviev, Grigori Yevseyevich (1883–1936), Soviet communist leader. Despite originally opposing the *Russian Revolution, he became chairman of the *Comintern (1919–26). In 1924 he gained international notoriety from a letter, published in the British Conservative newspapers four days before the general election. Apparently signed by him and sent by the *Comintern to the British Communist Party, it urged revolutionary activity within the army and in Ireland, and may have swung the middle-class vote in favour of the Conservatives. Labour leaders believed that the letter was a forgery. On *Lenin's death he, with *Stalin and *Kamenev, formed a triumvirate, but Stalin intrigued against him. He lost power and was executed after Stalin's first show trial.

Zionism, a movement advocating the return of Jews to *Palestine and specifically a political movement founded in 1897 under the leadership of Theodore *Herzl, that sought and has achieved the re-establishment of a Jewish state in Palestine. After the Russian *pogroms of 1881, Leo Pinsker wrote a pamphlet, *Auto-Emanzipation*, appealing for the establishment of a Jewish colony in Palestine. With the rise of nationalist feeling in Europe, Zionism assumed a political character, notably through Herzl's *Der Judenstaat* (1896) and the establishment of the World Zionist Organization in 1897. A minority wished to accept the British offer of Uganda as an immediate refuge for Jews. The issue of the *Balfour Declaration in 1917 and the grant of a *mandate for Palestine to Britain gave a crucial impetus to the movement. During the mandate period (1920–48) the World Zionist Organization under Chaim *Weizmann played a major part in the development of the Jewish community in Palestine by facilitating immigration, by investment (especially in land), and through the Jewish Agency. Zionist activities in the USA were influential in winning the support of Congress and the Presidency in 1946-8 for the creation of the state of Israel.

The defence of Rorke's Drift, Natal, during the **Zulu War**. On 22 January 1879 the small British station, defended by some 80 men, mostly of the 24th Regiment, was attacked by a well-equipped, well-disciplined army of about 4,000 Zulu, fresh from victory at Isandhlwana. The defenders repulsed successive charges, and the Zulu withdrew. (National Army Museum, London)

Zog (b. Ahmed Bey Zogu) (1895–1961), King of Albania. An Albanian politician who supported Austria in World War I, he later served as Premier (1922–4), President (1925–8), and King (1928–39) after the Albanian throne had been refused by the English cricketer, C. B. Fry. The new constitution placed power in his hands, and he became the champion of the modernization of the country, instituting language reforms, educational development, and religious independence. He relied on Italy for financial help, and the Treaty of Tirana (1926) provided him with Italian loans in return for Albanian concessions. By 1939 *Mussolini controlled Albania's finances and its army and an Italian invasion ended Albania's independence, forcing him into exile.

Zollverein (German, 'customs union'), a customs union to abolish trade and economic barriers between the states of Germany. The Prussian Zollverein was founded in 1833 by merging the North German Zollverein with smaller customs unions, thus considerably increasing Prussian influence. After the Austro-Prussian War (1866) the newly formed North German Confederation entered the Zollverein, and by 1888 the union, which excluded Austria, had largely achieved the economic unification of Germany.

Zululand, former Zulu state in South Africa. Zulu and related *Nguni groups are thought to have occupied the region from about the 15th century. Under their chief *Shaka in the early 19th century they reduced rival tribes to vassalage and occupied their territories before meeting Boer settlers migrating north. In 1879 Britain initiated the *Zulu War, annexing the whole area in 1887. It became a crown colony until 1897 when it was incorporated into Natal. There were rebellions in 1888 and 1906, two-thirds of Zululand being confiscated, the inhabitants being confined to native reserves. These were developed by the Bantu Self-government Act of 1959 into the *Bantustan of KwaZulu.

Zulu War (1879), a war fought between Britain and Zululand. Until he occupied the *Transvaal in 1877, the policy of the Natal Secretary for Native Affairs, Theophilus Shepstone, had been to protect the Zulu empire of *Cetshwayo against Afrikaner aggression. After the annexation, he reversed this policy to placate the Afrikaner population, and a scheme was prepared to seize *Zululand. Frontier incidents provided opportunities, and the British High Commissioner ordered the disbandment of the Zulu army within thirty days. *Cetshwayo did not comply, and war began on 11 January 1879. On 22 January the British suffered disaster at *Isandhlwana, but with reinforcements the Zulu capital, Ulundi, was burnt, Cetshwayo was captured (28 August), and the war ceased on 1 September.

Zuo Zongtang (or Tso Tsung-t'ang) (1812–85), Chinese soldier and statesman. He rose to military prominence, assisting *Zeng Guofan in suppressing the *Taiping Rebellion, and was appointed governor-general of Zhejiang province (1862). As a powerful provincial leader and experienced soldier, he supported the *Self-Strengthening Movement. In 1868 he was appointed governor-general of Shaanxi and Gansu provinces and suppressed Nian and Muslim rebels. In 1877 he recaptured *Xinjiang from Yakub Beg, making possible its incorporation as a Chinese province.

Acknowledgements

Photographs

Abbreviations: t = top; b = bottom; c = centre; l = left; r = right.

The illustrations on pages 141, 217, 218, and 365b are reproduced by Gracious Permission of Her Majesty the Queen.

The illustration on page 230 is reproduced by kind permission of the Chief Royal Engineer.

The illustration on page 300 is reproduced by courtesy of The Marquess of Salisbury.

The illustration on page 122 is from the collection of Mr and Mrs Manuel A. Villafana, photograph courtesy of Kennedy Galleries, Inc., New York.

Associated Press, 223b, 236.

Dr Barnardo's, 31.

BBC Hulton Picture Library, 81lb, 81rt, 146, 260, 264b, 277, 97, 301t, 316, 116.

Bettmann Archive, 7, 82, 89.

Bibliothèque Nationale, 288, 308.

Bildarchiv Preussischer Kulturbesitz, 134.

Bridgeman Art Library, 199.

Britain/Israel Public Affairs Committee, 34t, 59b.

British Library/Bridgeman Art Library, 25.

British Museum, 47b, 286t.

Brown Brothers, 224.

Ann S. K. Brown Military Collection, 219.

Camera Press, 38t, 92, 145, 153t, 153b, 180, 181, 192t, 207, 227, 252, 338b.

Communist Party Library, 70b, 99b, 164, 268b.

Co-operative Union Library, 86.

Corcoran Gallery of Art, 171.

Crown Copyright, 328b.

© DACS 1988, 325.

E.T. Archive, 14t, 113, 140, 205, 264t, 336, 347, 385.

Mary Evans Picture Library, 46, 56b, 87b, 100t, 101t, 118t, 167b, 187b, 187t, 229t, 284.

Gernsheim Collection, Photography Department, Harry Ransom Humanities Research Center, University of Texas at Austin, 35.

Burt Glinn/John Hillelson Agency, 197.

The Trustees of the Imperial War Museum, London, 45, 50, 71, 105, 112, 120, 129, 155b, 177t, 194b, 262, 270, 292b, 307, 312, 314, 352, 364, 386.

The Trustees of the Imperial War Museum, London, photo: Bridgeman Art Library, 185.

India Office Library, 83b, 91l, 174t, 247, 282b, 344b.

International Defence and Aid Fund for Southern Africa, 5, 16, 209, 311t, 322.

David King Collection, 44, 54, 72t, 78, 90, 107t, 193, 210, 212, 242, 283t, 297, 327t, 390.

Fried. Krupp GmbH, 295.

Library of Congress, 15, 101b, 110, 126, 139, 148, 196r, 223t, 267, 294, 328r, 348b, 355.

London Borough of Richmond on Thames, 56t.

London Express News and Feature Services, 9.

Manchester City Art Gallery, 63.

Mansell Collection, 12, 52, 66t, 66b, 67l, 70t, 72b, 123t, 128, 196l, 201, 202, 257, 272, 304, 310, 340t, 348t, 378.

Marconi Co. Ltd., 81lc.

Mariners Museum, Newport News, Virginia, 239b.

MAS, 118b, 168.

McCord Museum of Canadian History, McGill University, Montreal, 259.

Metropolitan Museum of Art, 53.

Mississippi Department of Archives and History, 93.

Montana Historical Society, 137b.

Musée d'Histoire Contemporaine – BDIC (Université de Paris), 174b.

Musées Nationaux, 109.

Museum of the American Indian, Heye Foundation, 315t.

Museum of London, 258, 333.

National Archives of Canada, 204.

National Army Museum, 13, 28, 162, 206, 391.

National Army Museum, photo: Bridgeman Art Library, 313.

National Gallery of Ireland, 249, 261.

National Maritime Museum, 41, 87t, 189, 250, 349.

National Portrait Gallery, 57b, 239t, 243, 281, 374.

National Portrait Gallery, Smithsonian Institution, transfer from the National Museum of American Art: gift of P. Tecumseh Sherman, 1935, 311b.

Peter Newark's Western Americana, 39b, 40b, 61, 91r, 98, 135, 192b, 233, 330, 379, 384, 389b.

Novosti, 327b.

Oakland Museum, 225.

Österreichische Nationalbibliothek, Porträt-Sammlung und Bild-Archiv, 123b.

Pacemaker Press International, 167t.

Photographie Giraudon, 177b, 231.

Pierpont Morgan Library, 176.

Popperfoto, 1, 2, 21, 22, 30t, 30b, 34b, 37, 38b, 39t, 47t, 57t, 69, 75, 83t, 95, 99t, 103, 111t, 124, 131t, 131b, 143, 149, 152, 155t, 157, 159, 175t, 175b, 182b, 183t, 183r, 203t, 203b, 213, 214t, 226b, 228, 237, 241t, 244, 266, 292t, 315b, 332, 340b, 346, 371, 373.

Post Office, 151.

Press Association, 107b.

The Queen's Dragoon Guards, Carver Barracks, Essex, 200.

Rainbird Publishing Group Ltd., 17, 20, 119, 194t, 215, 230, 289, 300, 309, 339, 372.

Rex Features, 26, 65, 67r, 106, 111b, 115, 130, 138, 142, 165, 179, 211, 216, 221, 226t, 273, 279, 283b, 331, 338t, 342, 380, 389t.

Royal College of Surgeons of Edinburgh, 100lc.

Salvation Army Archives, 301b.

Scala, 214b, 365t.

Science Museum, photo: Bridgeman Art Library, 282t.

Science Photo Library, 81rb; SPL/NASA, 323rt, 323rc, 323rb; SPL/Novosti, 323lt, 323lb.

Scott Polar Research Institute, 14b.

Sheffield City Museum, 163.

Frank Spooner, 36, 59t, 96t, 182t, 191, 241b, 256, 299, 318, 353.

Stanley Collection, 229b.

Syndication International, 132.

TASS, 104.

Tibet Society in Britain, 344t.

Trades Union Congress Library, 198.

Ullstein Bilderdienst, 137t, 286b.

US Navy, 263.

United Nations, 359.

United Press International, 121.

US Information Agency, 96b.

Utah State Historical Society, 358.

Victoria and Albert Museum, 40t, 298, 381.

Roger-Viollet, 173, 222, 268t.

Weidenfeld and Nicolson, 156.

World Health Organization/J. Wickett, 100r.

Yugoslav Military Museum, 345.

Xinhua News Agency, 74, 76, 334.

The publishers have made every attempt to contact the owners of the photographs appearing in this book. In the few instances where they have been unsuccessful they invite the copyright holders to contact them direct.

Picture researchers: Sheila Corr and Catherine Blackie.

Maps and Illustrations

Creative Cartography Ltd/Terry Allen and Nick Skelton: 4, 11, 24, 42, 43, 50, 60, 67, 73, 134, 154, 161, 169, 186, 231, 245, 246, 253, 255, 265, 274, 290, 296, 306, 320, 3?, 325, 343, 351, 360, 366, 367, 375, 377, ?, 385.

Vanna Haggerty: 6, 19, 235.

Oxford University Press: 84, 261, 359.

36

AFRICA, NEAR AND MIDDLE EAST

1900 British protectorates of S and N Nigeria established; First Pan-African Conference
1903 Britain occupies Sokoto (Nigeria)
1906 Revolution in Iran
1908 Congo Free State (Zaïre) under Belgian government; Young Turk Revolution
1909 Oil drilling in Iran

1910 Union of South Africa formed
1911 Italy conquers Libya
1912 African National Congress formed
1914 Britain proclaims Egypt proctectorate
1915–16 Gallipoli Campaign
1916–18 Mesopotamia Campaign
1917 Balfour Declaration
1918 British capture Syria
1919 Arab Revolt in Egypt

1920 Mustafa Kemal leads Turkish Revolution
1921 Reza Khan Shah of Persia
1923 Ottoman empire and Caliphate formally ended; Palestine, Transjordan, and Iraq mandated to Britain; Syria mandated to France
1924 N Rhodesia becomes British protectorate
1926 France establishes Republic of Lebanon

1932 Kingdom of Saudi Arabia formed
1935 Italy invades Ethiopia; Prempeh made Asantehene in Kumasi
1936 Arab-Jewish revolt in Palestine

1940 Italians expelled from Somalia, Eritrea, Ethiopia
1942 Battle of El Alamein
1943 Casablanca and Teheran conferences

AMERICAS

1901 US President McKinley assassinated; Theodore Roosevelt succeeds
1901–10 Nine million immigrants to USA
1902 Republic of Cuba proclaimed
1903 Panama Canal Zone to USA; first powered flight by Wright brothers
1907 USA supports revolution in Nicaragua

1910–40 Mexican Revolution
1912 Wilson elected US President
1913 Model T Ford car produced
1914 Panama Canal opens
1917 USA enters World War I
1919 Atlantic flown by Alcock and Brown

1920 US Senate rejects Versailles Settlement; Marcus Garvey founds Universal Negro Improvement Association
1920–33 Prohibition Era
1921–2 Washington Conference
1928–30 War of Cristeros in Mexican Revolution
1929 Stock Market Crash on Wall Street; Great Depression begins

1930 Revolution in Brazil; Vargas becomes President
1932 F. D. Roosevelt elected US President
1933 Roosevelt's New Deal
1937 Vargas proclaims *Estado Novo*

1940 Batista President in Cuba
1941 Atlantic Charter; USA enters World War II
1942 Brazil enters World War II

ASIA, AUSTRALASIA

1900 Boxer Rising
1901 Commonwealth of Australia created; USA takes control of Philippines
1902 Anglo-Japanese Alliance
1904–5 Russo-Japanese War
1905 Muslim League formed

1910 Japan annexes Korea
1911 Revolution in China
1912 Kuomintang formed; Amundsen reaches South Pole (Antarctica)
1915 Japan imposes Twenty-one Demands
1919 Amritsar massacre

1920 Gandhi dominates Indian Congress
1921 Chinese Communist Party formed
1926 Chiang Kai-shek takes leadership of Kuomintang
1927–37 Chinese Civil War

1930 Salt March
1930–2 Round Table Conferences
1931 Jiangxi Soviet established; New Zealand independent dominion; Japan occupies Manchuria
1934–5 Long March
1935 Philippines granted self-government
1936 Japan signs Anti-Comintern Pact
1937 Chinese Civil War truce
1937–45 Sino-Japanese War

1941 Japan attacks Pearl Harbor and overruns SE Asia
1942 Singapore falls; USA wins naval battles in Pacific
1944 Battle of Leyte Gulf

EUROPE

1901 Queen Victoria dies
1903 Bolshevik/Menshevik split
1904 Franco-British *entente*
1905 Revolution in Russia
1906 Russian Duma; British Labour Party founded
1907 Anglo-Russian *entente*
1909 'People's Budget' (UK)

1912–13 Balkan Wars
1914–18 World War I
1917 Russian Revolutions (February and October)
1918 Spartakist Revolt
1919 Poland, Hungary, Czechoslovakia, Finland, Estonia, Latvia, and Lithuania independent
1918–21 Russian Civil War
1919–23 Versailles Peace Settlement

1920 League of Nations formed; Kapp putsch
1921 Irish Free State formed
1922 USSR formed; Mussolini takes power in Italy
1923 French occupy Ruhr; Munich 'beer-hall' putsch
1924 Stalin succeeds Lenin
1925 Locarno Treaties
1926 British General Strike; first television transmissio
1929 Yugoslavia proclaime
1929–33 Great Depression

1931 Spain becomes a republic
1933 Hitler appointed Chancellor
1934 Third Reich proclaimed; Stalin purg begin
1936 Anti-Comintern Pa
1936–9 Spanish Civil Wa
1938 German Anschluss Austria; Munich crisis; Czechoslovakia cedes Sudetenland
1939 Nazi/Soviet Pact invaded by German
1939–45 World War I

1940 Germany over Western Europe; enters war
1941 Germany inva
1942 'Final solutio Holocaust begins Beveridge Repor
1943 Italy surrend
1944 Normandy la